Birkhäuser

Static & Dynamic Game Theory: Foundations & Applications

Static & Dynamic Game Theory: Foundations & Applications aims to publish top quality state-of-the-art textbooks and research monographs at the graduate and post-graduate levels in game theory and its applications in a variety of fields, including biology, communications, computer and computational science, ecology, economics, environmental science, engineering, management science, networks, operations research, political science, security, and the social sciences. The scope of the series includes both cooperative and non-cooperative games; static, repeated, and dynamic games, defined in discrete time as well as in continuous time (so-called differential games); deterministic as well as stochastic games (such as Markov games); and complete or incomplete information games. The series will also include titles that study learning, evolution, and asymptotics in high player population regimes such as mean-field analysis.

The series will appeal to a broad interdisciplinary readership of mathematically-oriented engineers, physical and social scientists, computer and computational scientists, and economists, as well as to mathematicians at the graduate and post-graduate levels. Students and researchers in these fields will benefit from the comprehensive treatment of topical areas, groundbreaking theoretical exploration and development of computational tools, and applications covering the entire interdisciplinary spectrum where multi-agent decision-making is at the forefront. Theoreticians and practitioners in industry, government, and academia will find the titles in the series accessible across the varied terrain of game-theoretical research.

Tamer Başar • Boualem Djehiche •
Hamidou Tembine

Mean-Field-Type Game Theory II

Applications

Birkhäuser

Tamer Başar
Dept. Electrical & Computer Engineering
Coordinated Science Lab.
University of Illinois at Urbana-Champai
Urbana, IL, USA

Boualem Djehiche
Department of Mathematics
KTH Royal Institute of Technology
Stockholm, Sweden

Hamidou Tembine
Electrical Eng and Computer Science
Quebec University
Trois-Rivieres, QC, Canada

ISSN 2363-8516 ISSN 2363-8524 (electronic)
Static & Dynamic Game Theory: Foundations & Applications
ISBN 978-3-032-07030-2 ISBN 978-3-032-07031-9 (eBook)
https://doi.org/10.1007/978-3-032-07031-9

This book is published under the imprint Birkhäuser, www.birkhauser-science.com by the registered company Springer Nature Switzerland AG
The registered company address is: Gewerbestrasse 11, 6330 Cham, Switzerland

To my wife Tangül (T.B.),
To my grand children Adam, Selma, Bilal (B.D.),
To Yandai and Pama Tembine, Florence and Jean-Pierre and Marie-Claire Tixier and Anna Netzhammer for their unconditional support (H.T.).

Preface

The rapid advancements in the efficiency of computers and communications and the evolution of reliable software for numerical and data-driven computation during the past three decades have led to an astonishing growth in the theory, methods, algorithms, and strategic learning for game theoretic solutions. This body of knowledge has, in turn, motivated widespread applications of mean-field-type methods in many disciplines, e.g., geoscience, engineering, social science, ecology, and business, and led to problem solutions that were considered intractable not too long ago. Many processes can be described by action, state-space, and state-action mean-field-type systems. These include processes in computer science, engineering, economics, physics, chemistry, biology, and many other fields. If we can derive a data-driven model for a process, then we can use mathematical tools to control the process and obtain information about the process. This is why action, state-space, and their distribution systems are so important to computer scientists, economists, engineers, and industrial professionals. If we know the state of a system at the present time, and we know all of the present and future inputs, then we can deduce the values of all future outputs of the system for an associated action, control, or decision. State-space and mean-field-type models can be linear or nonlinear. Although most real processes are nonlinear, the mathematical tools that are available for estimation and control are much more accessible and better understood for linear systems. That is why nonlinear systems are often approximated as linear systems plus second-order perturbations. That way we can use the tools that have been developed for linear systems to derive estimation, adaptive, iterative, and control algorithms. In presence of multiple interacting components and multiple decision-makers in the system, the state dynamics become interdependent since the decision-makers can make their response via their payoffs or their states. By modeling interdependent decision-makers such as users, transmitters/receivers, (macro/femto/pico/small) base stations, radio devices, nodes, designers, operators, mean-field-type game theory allows us to model scenarios such as risk-awareness, quantile-based, variance-based, cooperative, centralized, collaborative, co-opetitive, coalitional mean-field-type decisions, decentralized decision-making, without full picture of the system conditions. The challenges in applying mean-field-type game theory to computer science, networks, communications, and Internet have attracted a lot of attention during the last two decades. Most of the mean-field-type game-theoretic models can abstract away important assumptions and mask critical unanswered questions. In absence of symmetry, and homogeneity, and within unknown dynamic environment, the predictions of the outcomes are less clear.

The purpose of this second volume is to provide mean-field-type game theory applications in existing and emerging technologies, networks, physical, social, chemical, and biological interactions. The book attempts to provide a balanced view of several important elements of mean-field-type games in real-world applications. It is our hope that this book will illuminate both the promise of mean-field-type games as a tool for analyzing network evolution and the potential pitfalls and difficulties likely to be encountered when mean-field-type game theory is applied by practicing engineers, undergraduate and graduate students, and researchers. We have organized the book in such a way that all the relevant topics can be covered at some minimum essential level of detail. This volume is organized in 17 chapters and 5 parts. Each part addresses a specific domain or problem, showcasing the versatility and wide-ranging applicability of mean-field-type games in various fields, from engineering and computer science to ecology, biochemistry, sociology, and network economics.

Part I: Engineering

The first part of Volume 2 of the book presents some engineering applications.

Chapter 1: Evacuation of Multi-Level Buildings In Chap. 1, we introduce engineering aspect of mean-field-type games, specifically focusing on the challenging task of evacuating multi-level buildings. It explores how localized congestion measures propagate toward the exits of the building and how mean-field game theory can be applied to optimize evacuation strategies for enhanced safety and efficiency.

Chapter 2: Signal Quality-Based CSMA Networks This chapter introduces CSMA (Carrier Sense Multiple Access) networks using mean-field-type Markov games, emphasizing signal quality as a key factor. It discusses how mean-field games can be used to optimize network performance, especially in scenarios where signal quality is crucial.

Chapter 3: Power Savings, Mobility, and Data Queues Dynamic power allocation is a critical aspect of wireless communication systems. The chapter explores the use of mean-field games to optimize power allocation strategies dynamically, leading to more efficient and adaptive wireless networks in extremely large regimes. We then focus on the Interference Field in Wireless Networks, which is a common challenge in such networks. We investigate how mean-field games can be applied to analyze and mitigate interference in wireless communication systems, ensuring smoother and more reliable connectivity. Following that, we examine mobility and power saving in wireless networks focusing on the intersection of mobility and power efficiency in wireless networks. It explores how mean-field games can be used to strike a balance between user mobility and power consumption, a crucial consideration in modern wireless systems. We then extend the study to Anti-jamming Mean-Field Games. Cybersecurity is paramount in wireless communication. We discuss the application of mean-field games to anti-jamming strategies, highlighting their role in enhancing the security of wireless networks. After that, we examine Battery-state Dependent Power Management. Efficient battery management is essential for many devices in wireless networks. We discuss the use of mean-field games to optimize power management strategies based on the state of batteries, enhancing device longevity and performance.

Chapter 4: User and Base Station Association in Next Generation Networks The evolution of networks toward the next generation presents challenges in user and base station association. This chapter investigates how mean-field games can optimize these associations for improved network performance.

Part II: Computer Science

Chapter 5: Malware Propagation Malware poses a significant threat in today's digital world. This chapter explores how mean-field games can be utilized to model and understand the propagation of malware in networks, aiding in the development of effective countermeasures.

Chapter 6: Rent-Seeking Games in Cloud Networking This chapter explores the concept of rent-seeking games in the context of cloud networking, shedding light on how mean-field games can be applied to optimize resource allocation and utilization in cloud environments.

Chapter 7: Auctions Auctions are fundamental in various economic and technological domains. This chapter presents the application of mean-field games in auction theory, offering insights into optimizing auction mechanisms.

Chapter 8: Network of Neural Networks This chapter explores the application of mean-field-type games to neural network interactions and their implications in the field of machine intelligence.

Part III: Evolutionary Biology and Biochemistry

Chapter 9: BioChemical Reaction Networks This chapter discusses the application of mean-field game dynamics in chemical reaction networks, transitioning from individual chemical reactions to mean-field modeling, exploring the concept of limit cycles, and introducing biochemical reaction mechanisms.

Chapter 10: COVID-19: Data-Driven Modeling The COronaVIrus Disease 2019 (COVID-19) pandemic has highlighted the importance of data-driven modeling. This chapter discusses the application of data-driven mean-field-type games in modeling the spread of infectious diseases, with a focus on COVID-19, shedding light on effective mitigation strategies.

Part IV: Sociology, Psychology, Ecology, and Environment

Chapter 11: Pollution Control and Fishing Strategies in Smarter Cities Environmental concerns are paramount in today's world. This chapter explores how mean-field games can be used to model pollution control and fishing strategies, offering innovative approaches to sustainability in smarter cities.

Chapter 12: Mean-Field Games for Marriage A unique and intriguing application, this chapter explores the use of mean-field games in the context of marriage. It describes the dynamics of marital relationships and how mean-field games can be used to analyze and optimize them.

Part V: Network Economics, Finance, and Insurance

Chapter 13: Economics of Networks Economics plays a crucial role in network design and operation. This chapter discusses the application of mean-field games in understanding and optimizing the economics of networks.

Chapter 14: Electricity Price Dynamics in the Smart Grid The smart grid is a key element of modern energy systems. This chapter explores how mean-field-type games can be used to model and optimize electricity price dynamics in smart grids, enhancing efficiency and sustainability.

Chapter 15: Blockchain-Based Distributed Power Networks Blockchain technology has transformative potential in various industries, including energy. This chapter presents the application of mean-field-type games in optimizing blockchain-based distributed power networks, revolutionizing the energy sector.

Chapter 16: Blockchain Token Economics Blockchain tokens have become central to many digital ecosystems. This chapter explores how mean-field-type games can be applied to analyze and optimize token economics, offering valuable insights for blockchain-based businesses.

Chapter 17: Hub-Based Airline Networks The airline industry relies on complex network structures. This chapter discusses how mean-field games can be used to optimize hub-based airline networks, potentially leading to more efficient air travel systems.

Part VI: Wrap-Up

Chapter 18: Conclusions and New Directions In this chapter, we present some summary and new directions.

Urbana, IL, USA — Tamer Başar
Stockholm, Sweden — Boualem Djehiche
Trois-Rivieres, QC, Canada — Hamidou Tembine
Paris, June 2025

Acknowledgments

We would like to express our sincere gratitude to everyone who contributed to the completion of the two volumes of this book. Some of the research described in the book has emerged from collaborations with others, and the authors are indebted to all of their co-authors on collaborative work cited in the bibliography sections. The authors would like to thank Eitan Altman, Rachid El Azouzi, Yezekael Hayel, Alonso Silva, Antonia Masucci, Xin Luo, Ahmed Hanif, Jian Gao, Yida Xu, Julian Barreiro-Gomez, Massa N'Dong, Salah Choutri, Tyrone E. Duncan, Bozenna Pasik-Duncan, Dario Bauso, Alain Bensoussan, and four anonymous reviewers of this book for their interesting questions and comments on mean-field-type game applications.

Over the last ten years, we received partial support from different grants and research projects:

- The Verg Foundation, Stockholm
- AFOSR: NonCooperative Dynamic Games with Large Population of Players
- Tommo-So Dogon: Audio-Centric Machine Intelligence for Tommo-So Dogon Using Mean-Field-Type Game Theory
- TIMADIE: MFTG for Risk-Aware Textless Audio-to-Audio Machine Intelligence
- GRABALTON: Risk-Aware Graph-Chain for Livestock & Poultry
- GUINAGATON: Graph-Chain for Agriculture: From Farm Sources to Consumers' Table
- TIMADIE: Risk-Aware Blockchain and Last Mile Integration
- MFTG4MI: Mean-Field-Type Game Theory for Machine Intelligence
- MCF: Data-Driven Mean-Field-Type Game Theory Modelling of SARS-CoV-2
- AFOSR: MFTG: Mean-Field-Type Games: Foundations
- SITE: Center on Stability, Instability and Turbulence
- MFTG4Evacuation: Mean-Field-Type Games for Multi-Level Building Evacuation
- MFTG4Water: Mean-Field-Type Games for Water Distribution Networks
- MFTG: Mean-Field-Type Games for Intelligent Transportation Systems
- MELANGE: Mean-Field-Type Games: Learning, Explicit Solutions and Implementation
- 5G: Learning and Incentives in Large Scale Millimeter Wave Wireless communications
- GOLD: Game Theory, Optimization, Learning and Dynamics
- Center for Uncertainty Quantification in Computational Science & Engineering
- MEAN-MACCS: MEAN-field games for Multi-Agent Allocation and Control in Communication Systems
- MORE: Advanced Mathematical Tools for Complex Network Engineering
- LEG: Learning, Evolution and Games
- RMG: Random Matrix Games in Network Engineering, Power Grid and Transportation Science

- COWINET: Evolutionary Coalitional Games in Wireless Networks
- TF: Evolutionary Games for Audio-Literacy

Artwork

The scientific graphs in the book are generated using the MATLAB software by Mathworks Inc. and the Mean-Field-Type Package for Simulation of Large Population games.

Competing Interests The authors have no competing interests to declare that are relevant to the content of this manuscript.

Contents

Acronyms

DPP	Dynamic Programming Principle
ESS	Evolutionarily Stable Strategy
MFTG	Mean-Field-Type Game
MFG	Mean-Field Game
MP	Maximum (Minimum) Principle
ODE	Ordinary Differential Equation
PDE	Partial Differential Equation
PIDE	Partial-Integro Differential Equation
SMP	Stochastic Maximum (Minimum) Principle

Part I
Engineering

Chapter 1
Evacuation of Multilevel Buildings

The evacuation of a multilevel building with various stairs, doors, and pedestrians aiming to reach exits quickly is a strategic game, because the time it takes for each pedestrian to exit depends on the overall congestion and efficiency of the evacuation, ultimately impacting the final exit time of the last person in the building. An individual's exit time during an evacuation depends on other pedestrians' choices, especially the ones who are behind in a directional neighborhood as they collectively affect the localized congestion, bottleneck formation, and the overall efficiency of exit routes. When multiple pedestrians are trying to exit a multilevel building simultaneously, the following dependencies come into play. If many pedestrians choose the same path or exit route simultaneously, congestion can occur, slowing down the movement of individuals. Those who end up in congested areas will experience longer exit times. If certain exit routes or stairs are narrower than others, pedestrians using these routes may experience delays as they wait for others to pass through. Those who choose wider paths may exit more quickly. Pedestrians often make decisions based on the behavior of others they see or know of these behaviors. If they see a group of people heading toward a specific exit, they may follow suit, potentially creating a queue at that exit, which can affect the exit time for everyone in that queue. The availability of different exit routes can impact individual exit times. If one route becomes blocked or congested, pedestrians may need to backtrack and choose an alternative route, resulting in longer exit times. If there's a lack of coordination, it can lead to delays. Control panel and pedestrians may share information with each other about which exits are accessible or less crowded, influencing the choices of others and potentially affecting their exit times. Clearly, the choices made by other pedestrians can significantly impact an individual's exit time during an evacuation by influencing congestion, route availability, and overall coordination, making it an interdependent process [460, 464].

This chapter presents a class of mean-field-type games that captures some of the key dynamic features of crowd and pedestrian flows in multilevel building evacuations. It considers a simple congestion-based model with both microscopic

T. Başar et al., *Mean-Field-Type Game Theory II*, Static & Dynamic Game Theory: Foundations & Applications, https://doi.org/10.1007/978-3-032-07031-9_1

and macroscopic route choices by strategic agents. To achieve this, we use an interaction model with a local congestion measure based on the location of the agent in the building. Including the local congestion measure and its evolution along the path causes a sort of dispersion of the flow: The agents will try to avoid high-density areas in order to reduce their overall walking costs and queuing costs at the stairs and exits. Each agent state is represented by a center of a box that follows a first-order dynamical system in an Euclidean space. Each agent will move to one of the "closest" exits that is safer and with less congested path. Firstly, we formulate the problem and derive optimality equations using maximum principle and dynamic programming with boundary conditions. Then, well-posedness conditions and existence results are examined. Numerical simulations are carried out to illustrate mean-field equilibria of a safer evacuation process. The methodology is shown to be flexible enough to include movement noises and stochastic structural components of the building.

The chapter is organized as follows. The next section presents the main model. Section 1.3 provides some analysis and structural results. Section 1.4 focuses on the implementation of the mean-field game scheme. Section 1.5 discusses and provides different interpretations of the mean-field game evacuation scheme. It also presents small noise and robustness components.

1.1 Building Evacuation Interactions

Building evacuation strategies is critical elements in building fire safety. Simulation and empirical research have been focusing on elevators, exit stairs, or sky bridges for tall buildings based on the short path to the exits. However, there is a need to apply this research to evacuation systems which may include combinations of these three exit strategies as well as the congestion created by the crowd which may slow down the evacuation process. Recently, mean-field games for the evacuation scenario of a single-level building have been proposed to capture pedestrian patterns, crowd, and congestion effect. With the recent high penetration of multilevel buildings in emerging smarter cities, towers, and multilayer malls, the single-level building game may not capture these scenarios. There are at least 16 completed buildings with 100 floors or more worldwide. In addition, more than 20 buildings with at least 100 floors are under construction. Designers and engineers have to care more about the behavior in situations of emergency, namely, evacuation and incorporate people's behavior and experience. Office blocks, sports stadiums, shopping malls, and underground stations are all examples of areas where there is a need to ensure uncomplicated transitions in normal operation and rapid evacuation in the event of an emergency. Optimization of people flow through design, modeling, and simulation can provide a more agreeable environment and more effective safety design in large and crowded places.

Driven these considerations, we introduce a mean-field game within a multilevel building. A number of questions arise:

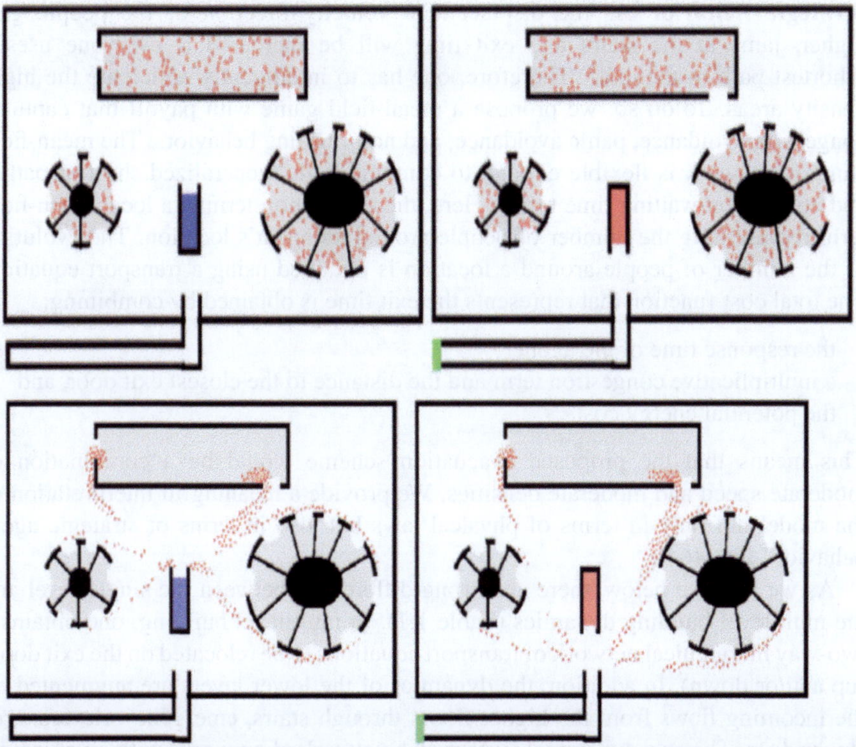

Fig. 1.1 Illustration of the evacuation flow in a two-level building. There are two levels. There is a stair connecting level two to level one. From level one, people can exit (the final exit door is represented in green)

- What would happen if a multilevel building has to be evacuated?
- Which escape routes would be mostly used?
- What is the global evacuation time for a given architecture and constraints in a multilevel building?

To investigate some of these aspects, we develop a simple and tractable theoretical framework for behavioral dynamics of people in multilevel building evacuation scenarios.

Network Effect We analyze the traffic at the stairs, sky bridges, and the main building exits and look at the bottlenecks. There is a network effect between the floors in the sense that the number of people in lower level floors are influenced by the people coming from upper level ones. For each block, the number of people at a given lower floor may increase with time up to a certain point, when most of the people have exited the building. Since the floors are interconnected through stairs and elevators, the system should be analyzed as whole network, and the flows per floor are correlated depending on the route choice as illustrated in Fig. 1.1.

Strategic Behavior As the displacement velocity/direction of the people gets higher, jams at the stairs and exit time will be increased if everyone uses a "shortest path to the exit." Therefore, one has to intelligently anticipate the high-density areas. To do so, we propose a mean-field game with payoff that captures congestion avoidance, panic avoidance, and non-pushing behaviors. The mean-field game framework is flexible enough to combine both "generalized shortest paths" and congestion/waiting time terms. Here, the congestion term is a local mean-field term representing the number of people around an agent's location. The evolution of the number of people around a location is obtained using a transport equation. The total cost function that represents the exit time is obtained by combining:

- the response time of the agent,
- a multiplicative congestion term and the distance to the closest exit door, and
- the potential energy cost.

This means that the proposed evacuation scheme would be a combination of moderate speed and moderate densities. We provide a meaningful interpretation of the model not only in terms of physical laws but also in terms of strategic agent behavior [460, 464].

As we will see below, there are strong differences between the single-level and the multilevel building dynamics (Table 1.1). In multilevel building, one obtains a two-way hierarchical network of transport equations to be relocated on the exit doors (up and/or down). In addition, the dynamics of the lower layers are augmented by the incoming flows from the higher floors through stairs, emergency elevators, or sky bridges. This is a dynamical system on a network of networks with an objective that combines waiting time cost and shortest path:

- In the single-level building case, the remaining number of people on the ground floor would be monotonically decreasing. It cannot increase even for a small period of time, since there is no incoming flux. However, we observe that the remaining number of people on ground floor of a multilevel building may increase for a certain period of time, because the incoming flux from other floors may be higher than the outgoing flux. This monotonicity property is not necessarily preserved when we detail it per floor.

Table 1.1 Differences between single and multiple level building evacuations

Number of Levels	Properties
Single	✓ The remaining number of people on the ground floor
	✓ Would be monotonically decreasing
	✓ No oscillation
	✓ No incoming flux (from other floors)
Multiple	✓ The remaining number of people per floor is not monotone
	✓ Oscillation may occur due to the presence of incoming flux
	✓ extra risk factors due to verticality

- In the single-level building case, the system of the ground floor (there would be only one floor in that case) does not oscillate as it is the case for buildings with two or more floors. This oscillation comes from the balance equation which characterizes the expected changes in a multilevel building. The variation is given by the difference between the incoming flux and the outgoing flux from the network.

Evacuating a multilevel building, particularly a skyscraper with 163 floors like the Burj Khalifa, presents complexities far exceeding those of a single-floor structure with interconnected compartments. Unlike a single-level scenario where horizontal movement to exits is the primary concern, vertical evacuation introduces a cascade of variables. The Burj Khalifa's design, featuring designated refuge floors like the 25th, 50th, 75th, and 100th accessible via sky bridges, highlights this significant difference. Instead of immediate ground-level egress, occupants might be directed to these temporary safe zones, potentially requiring subsequent vertical movement. This introduces decision-making under duress: Should one proceed immediately downward or await further instructions at a refuge floor? Moreover, the sheer height introduces psychological factors; the perceived risk of jumping from the 148th floor is drastically different from simply exiting through a ground-level door, leading to panic, less-rational behavior, and potentially hindering orderly evacuation. The vertical dimension also necessitates visibility and communication systems to guide occupants across multiple levels, managing congestion and preventing bottlenecks, unlike the relatively simpler flow control in a single-floor scenario with a better visibility. This interplay of vertical movement, refuge strategies, and heightened psychological stress makes multilevel evacuations significantly more challenging and riskier, demanding sophisticated modeling that accounts for these unique human behavior risk factors. While a building, in a purely topological sense, can be rotated or transformed without fundamentally altering its connectivity or structural relationships, making "vertical" and "horizontal" somewhat interchangeable concepts in abstract modeling, this analogy utterly fails when considering human evacuation. People are not mere topological points; they are more complex beings with physical limitations and psychological responses. Rotating a person, or equating a ground-level exit via broken glass with a 163rd-floor jump, disregards the fundamental reality of gravity, human physiology, and fear. The topologically invariant structures of a building do not translate to invariant human risk. Breaking a ground-floor window, while dangerous, offers a tangible path to safety. Conversely, a jump from the 163rd floor, regardless of any abstract topological equivalence, is a near-certain death sentence. These human behaviors should be measured properly using, for example, the vertical localized congestion term and included in the mathematical abstraction, but a visceral reaction to perceived danger, and the sheer height of a skyscraper introduces a terror that fundamentally alters evacuation dynamics, rendering any simple topological transformation irrelevant.

Measure of Congestion as a Key Feature The innovative component of the proposed framework is the local congestion feature. We capture the congestion around a location as the number of people surrounding that location. People adjust their behavior depending on their location and on the local congestion level. Note however that the local congestion term is not the probability density (if any) at that point. It is an integral of the measure around that point. It is important to note that the probability density/distribution at the location of the corresponding agent may not capture the congestion level surrounding the agent because of the spatial occupation of her body. For this reason, we count the number of people around each agent. This mean-field game model captures the microscopic multi-agent system with a spatiotemporal counting process without requiring any asymptotics in the range of the neighborhood. $B(x(t), \varepsilon)$ is the interaction neighborhood of the agent located at $x(t)$ at time t. It conveys the idea that the generic agent experiences interactions with some selected field agents, namely, those inside $B(x(t), \varepsilon)$.

Why do we not use a global congestion measure for all agents?

- A global mean-field term such as the average of all agent positions $\bar{m}(t) = \int y\rho(t, dy)$ can be introduced in order to build a simplified model. The average position of all agents in a multilevel building is not necessarily a good congestion measure, because it does not account for the non-convexity of movable areas, and it can be skewed by positions that are not actually reachable, such as walls or spaces between floors, making it an impractical basis for penalizing pedestrians based on distance to the average position in the context of evacuation time.
- Penalizing an agent who is closer to one of the final exit doors with respect to the barycenter $\bar{m}$ does not reflect what is observed in practice.
- The barycenter $\bar{m}$ is not necessarily in an allowed area of the building. For example, if some users are in the second floor and some others in the first floor, the barycenter may end up in the wall in between, where there is no user. Due to non-convexity of the domain, this is not a movable area.
- Running cost function

$$\langle Q_i(t)(x_i(t) - \bar{m}(t)), x_i(t) - \bar{m}(t)\rangle + \langle R_i(t)u_i(t), u_i(t)\rangle$$

  does not necessarily capture the behavior, because a user who has already exited from the building will be penalized for being outside. This is reason: The linear-quadratic model may not capture the evacuation scenario. The interdependence between the users should be carefully designed than just a global barycenter of their positions.

Why not consider the probability density (if any) at the agent position?

The local congestion term does not represent the probability density, if any, at that specific point; rather, it quantifies the presence of other agents around that point using an integral measure (a counting measure). It is crucial to understand that the probability density at an agent's location may not accurately reflect the congestion level surrounding the agent due to their physical presence, meaning that ε cannot be negligible in practical applications involving real individuals, ε being larger than the size of the person.

Pedestrians cannot be considered as existing at a fluid-like scale primarily, because they are not atoms. Each pedestrian is a distinct entity with a physical body, and they cannot occupy the same space simultaneously, unlike particles in a fluid. Pedestrians interact with their surroundings in a manner that depends on their size and shape. They must navigate around obstacles, adjust their paths, and consider personal space, which is not a characteristic of fluid particles. Pedestrians are subject to spatial constraints and cannot pass through each other or solid objects without taking physical paths. This contrasts with fluid particles that can flow through any available space. Pedestrians move at different speeds, and their movement is influenced by factors such as age, physical/health condition, or urgency, making their behavior highly heterogeneous compared to the flow of fluid particles [460, 464].

Based on the above, we propose a localized congestion measure for each agent present in the multilevel building.

The notation used in this chapter are listed in Table 1.2.

Table 1.2 Notation

Symbol	Meaning
I	Number of the people
K	Number of floors
D	Accessible domain of the building
m	Dimension of the domain
F	Cumulative distribution function of the positions
ρ	Probability density function
G	Number of people around a given position
$\mathbb{1}_{\{A\}}$	Indicator function over A
f_x	Derivative of f with respect to x

1.2 Deterministic Mean-Field Game Setup

The mean-field game modeling has several key features: the microscopic (atomic) mean-field game, the mesoscopic mean-field game, and the macroscopic (nonatomic) mean-field game. The mesoscopic scale is an intermediate one between the microscopic and the macroscopic scales. The difference between the micro and macro regimes is at the two extreme observation scales. On the microscopic mean-field game, the number of agents is finite, and the motion of each single agent is analyzed. In this case, the state variable is the position, and the control action is direction/velocity. The system consisting of states and actions of the agents forms a finite population dynamics, whose evolution in time is generally described by a system of differential equations, each for any pedestrian coordinate, in which behavioral route choices and movement are introduced to model the response of each individual to the local environment. This agent-based approach and the resulting strategic behavior seem having some attractive features. It is able to capture each agent's reaction (visibility, proximity of other agents, and exit door locations). Moreover, the possibility in the simulation to define different parameters for each agent allows the onset of complex heterogeneous motions. Also, the computation of entire path for each agent may quickly become expensive, in particular, in the areas of high congestion or rapidly changing environment. A possible way to overcome these difficulties is to consider the limiting regime (in population size) of macroscopic mean-field game where the dependent variables are written in condensed or compact forms. The macroscopic mean-field game approach still captures a generic agent's position and behavior, but the basic aggregate quantity is the probability finding other agents in a certain region of the position and action. The evolution of such a distribution function leads to a system of integrodifferential equations. Below we specify this setup.

Evacuation Domain as a Network

We consider a building with multiple floors and resolutions (several sub-buildings with connecting blocks) represented by a compact domain D in the m-dimensional Euclidean space $\mathbb{R}^m$, with $m \geq 1$. We assume that D is feasible; that is, there is a route from the movable areas to at least one of the exits. A typical value for m is 3 in the experimental setup below.

- There are $K \geq 2$ floors in the building.
- The domain at floor k is denoted by D_k.
- For $1 < k < K$,, the floor k is connected to the higher floor $k + 1$ using the intermediary domain I_k^+ but also the lower floor $k - 1$ using I_k^-. The sets I_k can be elevator zones or stairs.

1.2.1 Microscopic Setup

We consider finite number of agents distributed in a multilevel building with stairs, exit doors, and possibly sky bridges. The number of agents initially in the building is $I \geq 2$. The set of agents is denoted by $\mathcal{I}$. The length of the horizon of the game is T which needs to be minimized. At time $t \in [0, T]$,, each agent i knows her current location in the building, denoted by $x_i(t)$.

Individual State Dynamics
The state/location x_i of agent i changes depending on her control action u_i. The dynamics of her state follows a first-order differential equation with domain constraint:

$$\begin{cases} \dot{x}_i(t) = u_i(t), \ 0 < t < T, \\ x_i(0) \in D, \ 1 \leq i \leq I, \quad I \geq 2. \end{cases} \tag{1.1}$$

A stochastic extension of the individual dynamics is provided in Sect. 1.5.2.

Congestion Measure

$$G_{I,\varepsilon}(x_i(t)) = \frac{1}{vol(B(x_i(t), \varepsilon))} \frac{\sum_{j \neq i} \mathbb{1}_{\{d(x_j(t), x_i(t)) \leq \varepsilon\}}}{I - 1},$$

represents the (normalized) number of agents around the position x_i except i within a distance less than $\varepsilon > 0$, $vol(B)$ is the m-dimensional volume of the ball $B(x_i(t), \varepsilon)$,, i.e., a region in the m-dimensional Euclidean space consisting of all points within a distance ε from the point $x_i(t)$. By translation invariance of the Lebesgue measure, the volume of the ball $vol(B(x_i(t), \varepsilon))$ is independent of the position $x_i(t)$, and therefore we set $c_\varepsilon := vol(B(x_i(t), \varepsilon))$.

Volume of the Ball Under 2-Norm The volume of the m-ball (under the metric $d = d_2$, i.e., the 2-norm), is explicitly given by

$$c_\varepsilon = vol(B(x_i(t), \varepsilon)) = \frac{\sqrt{\pi^m} \varepsilon^m}{\Gamma(1 + \frac{m}{2})},$$

where $\Gamma(r) = \int_0^{+\infty} s^{r-1} e^{-s} ds, \ r > 0$, is the Euler's gamma function (which can be thought of as an extension of the factorial function to fractional arguments).

Cost Functional
The agent is interested in a safe evacuation from the building. This means that she is interested in the minimal exit time that avoids the presence of a huge crowd around her. The exit time depends on the chosen route and the evolution of congestion on

that path. Agent i minimizes the cost

$$J_i = g(x_i(T)) + \int_0^T h_1(G_{I,\varepsilon}(x_i(t)))|u_i(t)|^2 + h_2(G_{I,\varepsilon}(x_i(t)))\, dt,$$

subject to constraints specified below, where h_i is a positive increasing function, with $h_2(0) = 0$. $T > 0$ is the exit time at one of the exits. The final exit cost is represented by g which can be written as $c_3 + \tilde{g}(x)$ where $c_3 > 0$ captures the initial response time of an agent (without congestion). The cost The*initial response time* of agents plays an important role in practice. When it comes to calibration of the model, we observe that the reaction time of people could be incorporated in part through c_3. This would be necessary to better reflect the real-life evacuation which was much higher in our experiments.

Boundary Conditions

In order to complete the description, we need to specify the boundary conditions: There is no agent on the walls, obstacles, and at the exits. The boundary (also called frontier) of D is denoted by ∂D. Roughly speaking $\partial D = clD \backslash intD$ where clD is the closure of the domain D and $intD$ is the interior of the domain D with respect to Euclidean distance. It is the set of points in the closure of D, not belonging to the interior of D. In this setup, the boundary condition is with respect to the normal vector. The fact that no agent should be on the walls, obstacles, and at the exits is translated into $u_i = 0$ at the frontier ∂D of D except at the exits where $u_i = k(x, \rho)$, with $k(x, \rho) \neq 0$.

Then, the best-response problem of agent i given the strategies of the others is the following long-term optimization problem:

$$(P1): \begin{cases} \sup_{u_i} \ -J_i(u_1, \ldots, u_I) \text{ subject to} \\ \dot{x}_i = u_i, \ \text{ on } (0, T) \times D \\ x_i(0) \in D, \\ \text{Boundary constraints:} \\ u_i = 0 \text{ on } \partial D, \ \text{ and} \\ u_i = k(x, \rho) \text{ at Exits.} \end{cases}$$

Note that agent i's problem depends on the positions (and control actions) of all the other agents on her path and around her. This creates an interdependency between the agents and creates a dynamic interactive decision-making problem. We refer this microscopic description of the problem to as a microscopic differential game. This is a state-dependent aggregative differential game with local additive aggregative structure $G_{I,\varepsilon}$.

Interpretation

Agent i located at $x_i(t)$ at time $t \in (0, T)$ has a certain number of other agents around her within a ball of radius ε. This is captured by $G_{I,\varepsilon}$. When ε is sufficiently

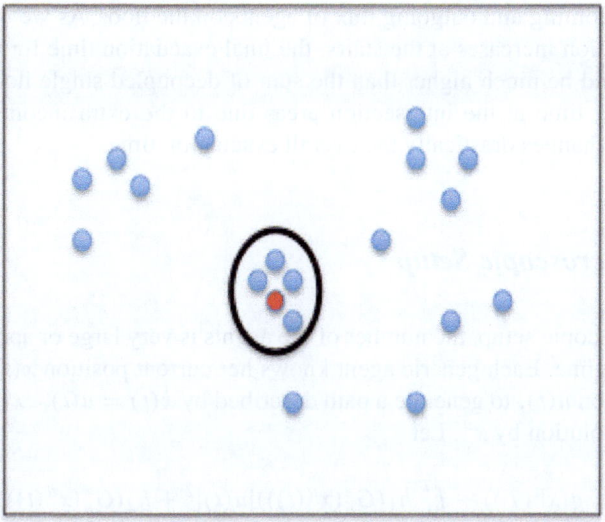

Fig. 1.2 Congestion measure $G_{I,\varepsilon}$ within a radius of ε =1 meter. There are four other agents in the ball

large, one counts all the agents in the building. When ε is small, one counts only the agents located near x_i. Figure 1.2 represents an example of $G_{I,\varepsilon}$ measured within a radius of ε =1 meter. The term $h_1(G_{I,\varepsilon}(x_i))|u_i|^2$ is a good cost model: The agent has a tendency to march faster to one of closest exists when there is no congestion, and agents slow down when there is a congestion on their path or find another alternative exit door. Mathematically speaking, if $G_{I,\varepsilon}(x_i)$ is very high, then $h_1(G_{I,\varepsilon}(x_i))$ is high and consequently u_i should be small. If the congestion term $G_{I,\varepsilon}(x_i)$ is very small, then $h_1(G_{I,\varepsilon}(x_i))$ is smaller and u_i can be increased to speed up the exit time.

Remark 1.1 It is important to note that by adding an extra floor (say the floor $K+1$) to the building changes the dispersion of the flow to the existing behavior in a nonlinear way depending on the shape of the building. In particular, there are significant differences between the single-level building and the multilevel building evacuation. The multilevel building evacuation is not a straightforward extension of the single-level one. One of the main differences between the single-level and multilevel building evacuation is that the multilevel building evacuation framework offers the possibility of representing the interactions between horizontal and vertical components which is not possible in the single-level case. In the single-level case, the vertical interactions are missing, and the agents are looking for the exit and the congestion increases near the exits.

In the multilevel building case, an agent has the choice and possibility to go to the upper level or the lower level, and she will make this decision based her available local information. Moreover, there is an increasing number of obstacles due to the incoming and outgoing flux of agents on the floor. As we will see below, when congestion increases at the stairs, the final evacuation time for the multilevel building would be much higher than the sum of decoupled single floors and stairs. The queueing time at the intersection areas due to the extra incoming flux from upper levels changes drastically the overall evacuation time.

1.2.2 Macroscopic Setup

In the macroscopic setup, the number of the agents is very large or approximated by an infinite regime. Each generic agent knows her current position $x(t)$ and chooses a control action $u(t)$, to generate a path described by $\dot{x}(t) = u(t), \;\; x(0) = x_0 \in D$. Denote this solution by x^u. Let

$$J(u, G_\varepsilon^*) = g(x^u(T)) + \int_0^T h_1(G_\varepsilon^*(x^u(t)))|u(t)|^2 + h_2(G_\varepsilon^*(x^u(t)))\, dt. \qquad (1.2)$$

The best-response problem of a generic agent is

$$(P2) \begin{cases} \sup_u \; -J(u, G_\varepsilon^*) \text{ subject to} \\ \dot{x}(t) = u(t), \;\; \text{on } (0,T)\times D, \\ x(0) \sim \rho_0(.) \text{ is given}, \\ G_\varepsilon^*(.) \text{ is given}, \\ \text{boundary constraints:} \\ u = 0 \text{ on } \partial D, \text{ and} \\ u = k(x,\rho) \text{ at exits}, \end{cases}$$

where G_ε^* is the equilibrium number of agents in the m-ball as I becomes very large. Note, however, that the generic agent does not influence G_ε^*. The evolution of G_ε^* can be obtained from the equilibrium strategies (if any) of the other agents in the building. The main difficulty here is that $G_\varepsilon^*(x)$ is a sectorized local mean-field term, meaning that the number of agents in the ball at $y \neq x$ may be different from the one located at x.

1.2.3 Evacuation Design and Planning

We consider a building designer who wishes to anticipate and to optimize the global evacuation time of the building in emergency situation. The designer has the global state of the system [106] and aims to indicate the optimal routes for evacuation of all the agents. Let the global cost functional be

$$J_0(u) = \int_{x\in D} g(x)\rho^u(T, dx) + \int_0^T \int_{x\in D} \{h_1(G_\varepsilon(., x))|u|^2(.) \\ + h_2(G_\varepsilon(.))\}\rho^u(dt, dx).$$

Further, let the congestion measure be

$$G_\varepsilon(t, x) = \frac{1}{c_\varepsilon} E_Y\{\mathbb{1}_{Y\in B(x(t),\varepsilon)}\} = \frac{1}{c_\varepsilon}\mathbb{P}_Y(Y \in B(x, \varepsilon)),$$

where E_Y denotes the expectation with respect to the distribution of $Y \sim \rho^u(t, dy)$.

The macroscopic optimization problem of the designer is given by

$$(P3) \begin{cases} \sup_u \ -J_0(u) \ \text{subject to} \\ G_\varepsilon(t, x^u(t)) = \frac{1}{c_\varepsilon}\int_{y\in B(x^u(t),\varepsilon)} \rho^u(t, dy), \\ \rho^u_t + div_x(\rho^u u) = 0, \\ \rho_0(.) \text{ is given }, \\ \text{boundary constraints:} \\ u = 0 \text{ on } \partial D, \text{ and} \\ u = k(x, \rho) \text{ at Exits.} \end{cases} \quad (1.3)$$

In contrast to (P1) and (P2), the problem (P3) is not a fixed-point problem. It is a mean-field-type control problem as the probability distribution ρ depends on u. The main feature of (P3) is that the functional $J_0(u)$ is nonlinear in ρ.

Remark 1.2 Problem (P3) can be reformulated as follows:

$$\begin{aligned} &\inf_u J_0(u) = E[g(X(T)) + \int_0^T h_1[G_\varepsilon(t, X(t))]|u|^2 + h_2[G_\varepsilon(t, X(t))]\, dt], \\ &\text{Law}(X(t)) = \rho^u(t, .), \\ &\text{Law}(X(T)) = \rho^u(T, .), \end{aligned}$$

where E denotes the expectation operator. This reformulation can also be seen as a stochastic optimization problem with initial-final probability measure constraint. Thus, the global optimization of the first-order mean-field system leads to a stochastic optimal control problem of mean-field type [460, 463].

1.2.4 Transition from Micro to Macro

We explain how the microscopic model (P1) and the macroscopic model (P2) are related. When the positions are initially independently distributed according to a smooth probability density function ρ_0, then, as I goes to infinity, the local mean-field term $G_{I,\varepsilon}(t, x(t))$ converges to $G_\varepsilon(t, x(t))$, and hence, by continuity, the microscopic cost converges to the macroscopic one. The error gap depends on how fast $h_1(G_{I,\varepsilon})$ goes to $h_1(G_\varepsilon)$ as I increases.

1.2.5 Evolution of Number of Remaining Agents per Floor

Given an initial distribution ρ_0 of all agents in the building, we forecast the time evolution of $\rho(t, .)$ through the transport equation. We would like to know more details on the flows per floor. What are the dynamics of ρ_k, F_k and G_k for floor k? To answer to this question, the building is decomposed as

$$D = (\cup_{k=0}^{K} D_k) \cup (\cup_{k=0}^{K-1} I_k^+) \cup (\cup_{k=1}^{K} I_k^-),$$

where:

- I_k^+ is the set of stairs and elevators between k and $(k+1)$-th floor. This is represented by a weak blue color entry on floor k and weak red (light coral) color in $k+1$.

$$I_k^+ = I_k^{+,d} \cup I_k^{+,a},$$

with $I^{+,d}$ being the flow departing from k to $k+1$, $I_k^{+,a}$ the flow arriving to k from $k+1$.
- I_k^- is the set of stairs and elevators between k and $(k-1)$-th floor. This is represented by a blue color entry on floor k and red color in $k-1$.

$$I_k^- = I_k^{-,d} \cup I_k^{-,a},$$

with $I^{-,d}$ being the flow departing from k to $k-1$, $I_k^{-,a}$ the flow arriving to k from $k-1$.

Let $\rho_k = \rho \mathbb{1}_{x \in D_k}$. The flow dynamics n_k on floor k is therefore given by

$$\dot{n}_k = a_k - d_k,$$

where the arrival term is

$$\begin{aligned} a_k(t) = &\sum_j \mathbb{1}_{\{x_j(t) \in \text{red line door}\}} \\ &+ \sum_k \mathbb{1}_{\{x_k(t) \in \text{light coral line door}\}} \end{aligned} \tag{1.4}$$

and the departure term is

$$\begin{aligned} d_k(t) &= \sum_j \mathbb{1}_{\{x_j(t) \in \text{blue line door}\}} \\ &\quad + \sum_k \mathbb{1}_{\{x_k(t) \in \text{weak blue line door}\}} \\ &\quad + \sum_k \mathbb{1}_{\{x_k(t) \in \text{green exit door}\}}. \end{aligned} \tag{1.5}$$

In a single-level building, there is no flux a_k coming from other floors (since there is no other floor). Thus, $\dot{n} = -d$ which is monotonically decreasing. In a building with two or more floors, the flux n may oscillate as illustrated below in Sect. 1.4.

1.3 Analysis

In this section, we provide detailed analyses of the three problems (P1), (P2), (P3).

1.3.1 Games with Atomic Decision-Makers

1.3.1.1 Well-Posedness of Problem P1

An admissible strategy u_i of agent i is an m-dimensional measurable and square integrable process over $[0, T]$. Given $(u_1, \ldots, u_{i-1}, u_{i+1}, \ldots, u_I)$ a strategy profile of the other agents, with $u_j,\ j \neq i$, square integrable, the best response problem of agent i, problem (P1) has at least one solution in the space of measurable and integrable function over $[0, T] \times D \to \mathbb{R}^m$.

Based on the existence of the best response strategy, we discuss below construction of a Nash equilibrium of the dynamic game with I agents. A (Nash) equilibrium of the dynamic game is a strategy profile $(u_1^*, \ldots, u_I^*)$ such that for every agent i,

$$J_i(u^*) \leq J_i(u_1^*, \ldots, u_{i-1}^*, u_i, u_{i+1}^*, \ldots, u_I^*),$$

$$\text{for all alternative strategies } u_i \in L^2([0, T] \times D,\ \mathbb{R}^m).$$

Below we provide optimality systems for finding Nash equilibria.

1.3.1.2 Pontryagin Maximum Principle

Necessary conditions for open-loop Nash equilibria can be obtained by using Pontryagin maximum principle (see Chap. 4 in Volume I). We introduce the non-optimized Hamiltonian as

$$H_i^0(x_i, u_i, G_{I,\varepsilon}, p_i) = -h_1(G_{I,\varepsilon}(x_i))|u_i|^2 - h_2(G_{I,\varepsilon}(x_i)) + p_i.u_i,$$

where p is the adjoint variable. The Pontryagin maximum principle yields

$$\begin{cases} \dot{p}_i = -H^0_{i,x_i}, & t < T, \ i \in \mathscr{I} \\ p_i(T) = -g_{x_i}(x_i(T)), & i \in \mathscr{I} \\ \dot{x}_i = H^0_{i,p_i}, & 0 < t < T, \ i \in \mathscr{I} \\ x_i(0), & i \in \mathscr{I}, \end{cases} \tag{1.6}$$

where H^0_{i,x_i} is evaluated at the equilibrium control strategy. This is a system of backward-forward ordinary differential equations. I equations in p_i are backward in time, and I equations in x are forward in time.

1.3.1.3 Dynamic Programming Principle

As in [125], we use the dynamic programming principle to find such equilibria of the game. If there exists a continuously differentiable solution v such that

$$\begin{cases} v_{i,t} + H_i(x_i, v_{i,x_i}, G_{I,\varepsilon}(x_i)) = 0, \ \text{ on } (0,T) \times D, \\ v_i(T, x_i) = g(x_i), \ \text{ on } \{T\} \times D, \\ G_{I,\varepsilon}(x_i) = \frac{1}{c_\varepsilon} \frac{\sum_{j \neq i} \mathbb{1}_{\{d(x_j, x_i) \leq \varepsilon\}}}{I-1}, \\ u_i(t, x_i, G) = 0 \ \text{ on } (0,T) \times \partial D, \\ u_i(t, x_i, G) = k(t, x_i, G) \ \text{ on Exits}, \end{cases}$$

then the control strategy vector $(u_1^*, \ldots, u_I^*)$ is a Nash equilibrium with

$$u_i^* = \frac{v_{i,x_i}}{2h_1(G_{I,\varepsilon}(x_i))},$$

and the equilibrium payoff vector is given by $(v_1, \ldots, v_I)$, where H_i is the Legendre–Fenchel transform of the function $h_1(G_{I,\varepsilon}(x_i))|u_i|^2 + h_2(G_{I,\varepsilon}(x_i))$.

The optimal strategy of agent i is in state-and-local-congestion feedback form $u(t, x_i, G_{I,\varepsilon}(x_i))$ which has an aggregative structure. This means that the key decision parameter of i at time t is her position and the number of agents around her. Note however that this equilibrium system may easily become complex in terms of number of partial differential equations (PDEs) and constraints. One way to reduce the complexity is to classify agents by their intrinsic characteristics.

1.3.2 Games with Macroscopic Setup

One Dimension: Crowd on a Line

While there are no pedestrians in a one-dimensional scenario, utilizing a 1-D analysis can assist us in comprehending the relevant parameters when extrapolated to higher dimensions, and thus we begin by establishing a one-dimensional context.

If each agent follows the dynamics $\dot{x}(t) = u^*(x(t), G_\varepsilon(t, x(t)))$,, then the distribution of the state of agents solves a continuity (advection) equation $\rho_t + (u^*\rho)_x = 0,\ 0 < t < T$ with the initial distribution of states $\rho(0, .) = \rho_0(.)$. We integrate the advection equation to arrive at a transport equation[1] $F_t + u^*.F_x = 0$ where $F(t, x(t)) = \int_{(-\infty, x(t))} \rho(t, ds)$ is the cumulative distribution function. In the one-dimensional case, the ball becomes the interval $[x(t) - \varepsilon, x(t) + \varepsilon]$ with a length of 2ε. One can easily express $G_\varepsilon(t, x)$ in terms of the probability distribution ρ.

$$G_\varepsilon(t, x(t)) = \frac{1}{2\varepsilon} \int_{x(t)-\varepsilon}^{x(t)+\varepsilon} \rho(t, ds) = \frac{F(t, x(t) + \varepsilon) - F(t, x(t) - \varepsilon)}{2\varepsilon}. \quad (1.7)$$

If the distribution of the positions admits a density $\rho(t, ds) = \rho(t, s)ds$, that is integrable over $[0, T] \times D$,, then the congestion measure G is differentiable with respect to x and the partial derivative $G_{\varepsilon,x}(t, x) = \partial_x G_\varepsilon(t, x) = \frac{\rho(t,x+\varepsilon)-\rho(t,x-\varepsilon)}{2\varepsilon}$, is close to $\rho_x(t, x)$ when ε is very small. Note that this partial derivative function will be used in the maximum principle.

Two Dimensions: 2D

Let us now move to a two-dimensional scenario for each floor, which can effectively represent various real-life situations by examining (x_1, x_2, h), where we maintain a constant height, denoted as h, for all individuals. Thus, we examine $(x_1(t), x_2(t))$ per floor in the building.

Let the cumulative distribution of $(x_1(t), x_2(t))$ be

$$F(t, x(t)) = \int_{-\infty}^{x_1(t)} \int_{-\infty}^{x_2(t)} \rho(t, ds_1 ds_2).$$

Then, the function F solves the transport equation $F_t + u^*.F_x = 0$, and the cross derivative of F is $F_{x_1 x_2}(t, x) = \rho(t, x)$.

Proposition 1.3 *The probability of the random variable to be in the rectangle* $[x_1 - \varepsilon, x_1 + \varepsilon] \times [x_2 - \varepsilon, x_2 + \varepsilon]$ *is*

[1] For a constant action u the explicit solution of the transport equation is given by $F(t, x) = F_0(x - ut)$.

$$
\begin{aligned}
4\varepsilon^2 G(t,x) &= P(Y \in [x_1-\varepsilon, x_1+\varepsilon] \times [x_2-\varepsilon, x_2+\varepsilon]) \\
&= F(t, x_1+\varepsilon, x_2+\varepsilon) - F(t, x_1-\varepsilon, x_2+\varepsilon) \\
&\quad - F(t, x_1+\varepsilon, x_2-\varepsilon) + F(t, x_1-\varepsilon, x_2-\varepsilon).
\end{aligned} \tag{1.8}
$$

Proof of Proposition 1.3 To compute the probability of the rectangle, we write the domain $(-\infty, x_1+\varepsilon) \times (-\infty, x_2+\varepsilon)$ as union of four mutually exclusive events: $S_1 \cup S_2 \cup S_3 \cup S_4$, where

$$
\begin{aligned}
S_1 &= [x_1-\varepsilon, x_1+\varepsilon] \times [x_2-\varepsilon, x_2+\varepsilon], \\
S_2 &= (-\infty, x_1-\varepsilon] \times [x_2-\varepsilon, x_2+\varepsilon], \\
S_3 &= (-\infty, x_1-\varepsilon] \times (-\infty, x_2-\varepsilon], \\
S_4 &= [x_1-\varepsilon, x_2+\varepsilon] \times (-\infty, x_2-\varepsilon].
\end{aligned} \tag{1.9}
$$

The addition rule of the probability measure for mutually exclusive set implies

$$P(Y \in S_1 \cup S_2 \cup S_3 \cup S_4) \tag{1.10}$$

$$= P(Y \in S_1) + P(Y \in S_2) + P(Y \in S_3) + P(Y \in S_4) \tag{1.11}$$

$$= F(t, x_1+\varepsilon, x_2+\varepsilon). \tag{1.12}$$

$$P(Y \in S_3 \cup S_4) = P(Y \in S_3) + P(Y \in S_4) = F(t, x_1+\varepsilon, x_2-\varepsilon). \tag{1.13}$$

This implies that

$$P(Y \in S_1 \cup S_2) = P(Y \in S_1) + P(Y \in S_2) \tag{1.14}$$

$$= F(t, x_1+\varepsilon, x_2+\varepsilon) - F(t, x_1+\varepsilon, x_2-\varepsilon). \tag{1.15}$$

Similarly,

$$P(Y \in S_2 \cup S_3) = P(Y \in S_2) + P(Y \in S_3) \tag{1.16}$$

$$= F(t, x_1-\varepsilon, x_2+\varepsilon), \tag{1.17}$$

and

$$P(Y \in S_3) = F(t, x_1-\varepsilon, x_2-\varepsilon). \tag{1.18}$$

This implies that

$$P(Y \in S_2) = F(t, x_1-\varepsilon, x_2+\varepsilon) - F(t, x_1-\varepsilon, x_2-\varepsilon). \tag{1.19}$$

Thus,

$$P(Y \in S_1) = P(Y \in S_1 \cup S_2) - P(Y \in S_2) \tag{1.20}$$
$$= F(t, x_1 + \varepsilon, x_2 + \varepsilon) - F(t, x_1 + \varepsilon, x_2 - \varepsilon)$$
$$-F(t, x_1 - \varepsilon, x_2 + \varepsilon) + F(t, x_1 - \varepsilon, x_2 - \varepsilon).$$

This completes the proof. ■

Note that $G(t, x)$ admits partial derivatives given by

$$4\varepsilon^2 G_{x_j}(t, x) \tag{1.21}$$
$$= F_{x_j}(t, x_1 + \varepsilon, x_2 + \varepsilon) - F_{x_j}(t, x_1 - \varepsilon, x_2 + \varepsilon)$$
$$-F_{x_j}(t, x_1 + \varepsilon, x_2 - \varepsilon) + F_{x_j}(t, x_1 - \varepsilon, x_2 - \varepsilon). \tag{1.22}$$

This computation will be used for the open-loop optimality principle in (1.27). Under 2-norm, one gets

$$\pi\varepsilon^2 G(t, x) = P(Y \in B_2(x, \varepsilon)) \tag{1.23}$$
$$= \int_0^{\varepsilon} \int_0^{2\pi} \rho(t, x_1 + r\cos\theta, x_2 + r\sin\theta) r dr d\theta$$
$$= \int_0^{\varepsilon} \int_0^{2\pi} \tilde{\rho}(t, r, \theta) dr d\theta, \tag{1.24}$$

where

$$\tilde{\rho}(r, \theta) = r\rho(t, x_1 + r\cos\theta, x_2 + r\sin\theta).$$

Three Dimensions: 3D

We now examine the three-dimensional case. Let the cumulative distribution of $(x_1(t), x_2(t), x_3(t))$ be

$$F(t, x(t)) = \int_{-\infty}^{x_1(t)} \int_{-\infty}^{x_2(t)} \int_{-\infty}^{x_3(t)} \rho(t, ds_1 ds_2 ds_3).$$

Then, F solves the transport equation $F_t + u^*.F_x = 0$ and the cross derivative of F is $F_{x_1x_2x_3}(t, x) = \rho(t, x)$.

Proposition 1.4 *The probability of the cube $\prod_{j=1}^3 [x_j - \varepsilon, x_j + \varepsilon]$ under ρ is*

$$8\varepsilon^3 G(t, x) = P(Y \in \prod_{j=1}^3 [x_j - \varepsilon, x_j + \varepsilon]) =$$
$$F(t, x_1 + \varepsilon, x_2 + \varepsilon, x_3 + \varepsilon) - F(t, x_1 - \varepsilon, x_2 + \varepsilon, x_3 + \varepsilon)$$
$$+F(t, x_1 - \varepsilon, x_2 - \varepsilon, x_3 + \varepsilon) - F(t, x_1 + \varepsilon, x_2 - \varepsilon, x_3 + \varepsilon)$$
$$+F(t, x_1 + \varepsilon, x_2 - \varepsilon, x_3 - \varepsilon) - F(t, x_1 + \varepsilon, x_2 + \varepsilon, x_3 - \varepsilon)$$
$$+F(t, x_1 - \varepsilon, x_2 + \varepsilon, x_3 - \varepsilon) - F(t, x_1 - \varepsilon, x_2 - \varepsilon, x_3 - \varepsilon).$$

Proof of Proposition 1.4 Introduce the following three functions

$$f_3(t,s_1,s_2,x_3) := \int_{-\infty}^{x_3} \rho(t,s_1,s_2,s_3)ds_3,$$

$$f_{23}(t,s_1,x_2,x_3) := \int_{-\infty}^{x_2}\int_{-\infty}^{x_3} \rho(t,s_1,s_2,s_3)ds_2ds_3,$$

$$\begin{aligned} f_{123}(t,x_1,x_2,x_3) &:= \int_{-\infty}^{x_1}\int_{-\infty}^{x_2}\int_{-\infty}^{x_3} \rho(t,s_1,s_2,s_3)ds_1ds_2ds_3 \\ &= F(t,x_1,x_2,x_3). \end{aligned}$$

It follows that

$$f_{23}(t,s_1,x_2,x_3) = \int_{-\infty}^{x_2} f_3(t,s_1,s_2,x_3)ds_2,$$

$$f_{123}(t,x_1,x_2,x_3) = \int_{-\infty}^{x_1} f_{23}(t,s_1,x_2,x_3)ds_1.$$

Thus,

$$8\varepsilon^3 G(t,x) = P(Y \in \prod_{j=1}^{3}[x_j-\varepsilon, x_j+\varepsilon]) \tag{1.25}$$

$$\begin{aligned}
&= \int_{x_1-\varepsilon}^{x_1+\varepsilon}\int_{x_2-\varepsilon}^{x_2+\varepsilon}\int_{x_3-\varepsilon}^{x_3+\varepsilon} \rho(t,s_1,s_2,s_3)ds_1ds_2ds_3 \\
&= \int_{x_1-\varepsilon}^{x_1+\varepsilon}\int_{x_2-\varepsilon}^{x_2+\varepsilon} f_3(t,s_1,s_2,x_3+\varepsilon)ds_1ds_2 \\
&\quad - \int_{x_1-\varepsilon}^{x_1+\varepsilon}\int_{x_2-\varepsilon}^{x_2+\varepsilon} f_3(t,s_1,s_2,x_3-\varepsilon)ds_1ds_2 \\
&= \int_{x_1-\varepsilon}^{x_1+\varepsilon} f_{23}(t,s_1,x_2+\varepsilon,x_3+\varepsilon)\,ds_1 - \int_{x_1-\varepsilon}^{x_1+\varepsilon} f_{23}(t,s_1,x_2-\varepsilon,x_3+\varepsilon)\,ds_1 \\
&\quad + \int_{x_1-\varepsilon}^{x_1+\varepsilon} f_{23}(t,s_1,x_2-\varepsilon,x_3-\varepsilon)ds_1 - \int_{x_1-\varepsilon}^{x_1+\varepsilon} f_{23}(t,s_1,x_2+\varepsilon,x_3-\varepsilon)ds_1 \\
&= F(t,x_1+\varepsilon,x_2+\varepsilon,x_3+\varepsilon) - F(t,x_1-\varepsilon,x_2+\varepsilon,x_3+\varepsilon) \\
&\quad +F(t,x_1-\varepsilon,x_2-\varepsilon,x_3+\varepsilon) - F(t,x_1+\varepsilon,x_2-\varepsilon,x_3+\varepsilon) \\
&\quad +F(t,x_1+\varepsilon,x_2-\varepsilon,x_3-\varepsilon) - F(t,x_1+\varepsilon,x_2+\varepsilon,x_3-\varepsilon) \\
&\quad +F(t,x_1-\varepsilon,x_2+\varepsilon,x_3-\varepsilon) - F(t,x_1-\varepsilon,x_2-\varepsilon,x_3-\varepsilon).
\end{aligned}$$

This completes the proof. ■

Clearly, $G(t, x_1, x_2, x_3)$ admits partial derivatives with respect to x_j. This calculation will be used for the open-loop optimality principle in (1.27).

Arbitrary Dimension
For completeness, we provide below the change of variables for 2-norm m-ball. Introduce the m-dimensional Euclidean space, the radial and angular coordinates $\theta_1, \theta_2, \ldots, \theta_{m-1}$ where θ_{m-1} ranges over $[0, 2\pi)$ radians (or over [0, 360) degrees), and the other angles range over $[0, \pi]$ radians (or over [0, 180] degrees). The relation with the Cartesian coordinates is

$$\begin{aligned}
s_1 &= x_1 + r\cos(\theta_1)\\
s_2 &= x_2 + r\sin(\theta_1)\cos(\theta_2)\\
s_3 &= x_3 + r\sin(\theta_1)\sin(\theta_2)\cos(\theta_3)\\
&\vdots\\
s_{m-1} &= x_{m-1} + r\sin(\theta_1)\cdots\sin(\theta_{m-2})\cos(\theta_{m-1})\\
s_m &= x_m + r\sin(\theta_1)\cdots\sin(\theta_{m-2})\sin(\theta_{m-1}).
\end{aligned}$$

The volume element is

$$dV = \sin^{m-2}(\theta_1)\sin^{m-3}(\theta_2)\cdots\sin(\theta_{m-2})\,d\theta_1\,d\theta_2\cdots d\theta_{m-1}.$$

$$\begin{aligned}
&\int_{B(x,\varepsilon)} \rho ds \qquad (1.26)\\
&= \int_0^{\varepsilon}\int_{\theta_1=0}^{2\pi}\int_{\theta_2=0}^{\pi}\cdots\int_{\theta_{m-1}=0}^{\pi} \rho(t, x_1 + r\cos(\theta_1), \ldots, x_m\\
&\quad + r\sin(\theta_1)\cdots\sin(\theta_{m-1}))dV.
\end{aligned}$$

In the hypercube case with Cartesian coordinates, the probability can be expressed as

$$2^m\varepsilon^m G(t, x(t)) = \sum_{\substack{b_1,\ldots b_m\\ b_j\in\{\pm\varepsilon\}}} \left(\prod_{j=1}^{m} sign(b_j)\right) F(t, x_1 + b_1, \ldots, x_m + b_m).$$

In the m-dimensional spherical case,

$$\frac{\sqrt{\pi^m}\varepsilon^m}{\Gamma(1+\frac{m}{2})}G(t,x) = \int_{B(x,\varepsilon)} \rho(t,s)ds$$
$$= \int_0^{\varepsilon}\int_{\theta_1=0}^{2\pi}\int_{\theta_2=0}^{\pi}\cdots\int_{\theta_{m-1}=0}^{\pi} \rho(t, x_1 + r\cos(\theta_1), \ldots, x_m$$
$$+ r\sin(\theta_1)\cdots\sin(\theta_{m-1}))dV.$$

In the multidimensional case, the continuity equation is $\rho_t + div_x(u^*\rho) = 0,\ 0 < t < T$ where the operator div_x denotes the divergence, and the congestion measure is $G_\varepsilon(t, x(t)) = \frac{1}{c_\varepsilon}\int_{B(x(t),\varepsilon)} \rho(t, ds)$, which admits a partial derivative with respect to the vector x_i given by

$$c_\varepsilon G_{\varepsilon,x_i}(t,x) := c_\varepsilon \partial_{x_i} G_\varepsilon(t,x)$$
$$= \int_{\sum_{j\neq i} s_j^2 \le \varepsilon^2} \rho(t,\tilde{s}_+)\prod_{j\neq i} ds_j - \int_{\sum_{j\neq i} s_j^2 \le \varepsilon^2} \rho(t,\tilde{s}_-)\prod_{j\neq i} ds_j,$$

where

$$\tilde{s}_+ := s_1, \ldots, s_{i-1}, x_i + \sqrt{\tau_i}, s_{i+1}, \ldots, s_m,$$

$$\tilde{s}_- := s_1, \ldots, s_{i-1}, x_i - \sqrt{\tau_i}, s_{i+1}, \ldots, s_m,$$

$$\tau_i = \sqrt{\varepsilon^2 - \sum_{j\neq i} s_j^2}.$$

The consequence of the above calculation can be summarized as follows: If the positions of the agents can be represented by a probability density function, then, G_ε is differentiable in x, and $\partial_x G_\varepsilon$ is well defined.

Well-Posedness of Problem P2

A best response of a generic agent to the congestion trajectory G_ε is a strategy u^* such that

$$J(u^*, G_\varepsilon) = \inf_u J(u, G_\varepsilon),$$

where $G_\varepsilon(.,.)$ is a congestion measure trajectory of the agents inside the building. Then, the payoff functional $J(u, G_\varepsilon)$ is continuous, strictly convex in u and coercive in u. Hence, J admits a minimum over $[0, T] \times D$. This means that the best response problem (P2) of a generic agent has at least one solution in the space of functions over $[0, T] \times D$ satisfying

$$\int_0^T u^2(t,x)\,\rho(t,x)dxdt < +\infty.$$

Definition 1.5 A mean-field equilibrium of the macroscopic game is a triplet $(u^*, v^*, G^*_\varepsilon)$ such that for every generic agent, $v^* = J(u^*, G^*_\varepsilon) \leq J(u, G^*_\varepsilon),\ \forall\, u$ and G^*_ε coincides with the distribution of agents around a position.

Below we derive backward-forward optimality systems toward finding mean-field equilibria.

Pontryagin Maximum Principle for P2

We introduce the non-optimized Hamiltonian in macroscopic setting as $H^0(x, u, G, p) = -h_1(G_\varepsilon(x))|u|^2 - h_2(G_\varepsilon(x)) + p.u$, where p is the adjoint variable. Pontryagin maximum principle yields

$$\begin{cases} \dot{p} = -H^0_x, \\ p(T) = -g_x(x(T)), \\ \dot{x} = \frac{p}{2h_1(G_\varepsilon(x))},\ 0 < t \leq T, \\ x(0) \in D. \end{cases} \tag{1.27}$$

The set of control actions is a convex subset of $\mathbb{R}^m$ and has a nonempty interior, and the drift maps and instantaneous cost are differentiable in u. The Hamiltonian $H^0(.,., G_\varepsilon, p(t))$ is concave in (x, u) for almost everywhere (a.e.) $t \in [0, T]$. Then, for convex terminal cost function $g,$, u^* is an optimal response if $H^0(x^*(t), u^*, G^*_\varepsilon, p^*(t)) = \max_u H^0(x^*(t), u, G^*_\varepsilon, p^*(t))$.

Below we provide a sufficient condition for optimality for P2 using dynamic programming principle (Bellman's approach) in finite dimension.

Dynamic Programming Principle for P2

We introduce the (optimized) Hamiltonian as

$$H(x, p, G_\varepsilon) = \sup_u\{-h_1(G_\varepsilon(x))|u|^2 - h_2(G_\varepsilon(x)) + pu\}, \tag{1.28}$$

where H is the Legendre transform of the instantaneous cost. The Hamiltonian can be explicitly computed as

$$H(x, p, G_\varepsilon) = \frac{|p|^2}{4h_1(G_\varepsilon(x))} - h_2(G_\varepsilon(x)), \tag{1.29}$$

and the optimal strategy is in mean-field feedback form

$$u^* = \frac{p}{2h_1(G_\varepsilon(x))} = H_p(x, p, G_\varepsilon(x)). \tag{1.30}$$

The dynamic programming principle for P2 leads to the following optimality system:

$$(mfs)\begin{cases} v_t + H(x, v_x, G_\varepsilon(x)) = 0, \ \text{ on } (0,T)\times D \\ v(T,x) = -g(x), \ \text{ on } D \\ \rho_t + div_x(\rho H_p) = 0, \\ \rho(0,.) = \rho_0(.) \text{ on } D \\ u = 0, \ \ \rho = 0 \text{ on } \partial D \\ u = k(x,\rho), \ \ \text{at exits.} \end{cases} \tag{1.31}$$

This is a *first-order mean-field game system* under boundary constraints. Next, we discuss the existence of mean-field equilibria from the system (mfs). Let:

- h_i be smooth and strictly increasing functions, $h_1 > 0, h_2 \geq 0$.
- The initial distribution of positions be L^2 and absolutely continuous (with respect to the Lebesgue measure), and G_ε is bounded away from zero.
- g is continuous and bounded.

Then, the congestion measure G_ε is bounded by $\frac{1}{vol(B)} = \frac{1}{c_\varepsilon}$, and hence the two functions h_i are bounded, and their maximum values are $h_i(\frac{1}{vol(B)})$. Moreover, the instantaneous payoff is strictly concave in u (quadratic in u) and coercive uniformly in the position x. The function H is a smooth and strictly convex function in p (quadratic in p)

$$\lim_{|p|\to+\infty} \frac{H(x,p,G_\varepsilon)}{|p|} = +\infty. \tag{1.32}$$

Then, the system (mfs) has at least one solution. Next we provide an explicit representation of the solution that would reduce drastically the complexity of the numerics.

1.3.2.1 Explicit Solution of HJB Equation

Given the congestion measure $G_\varepsilon(.)$, we now look into solving the congestion-dependent Hamilton–Jacobi–Bellman (HJB) $v_t + H(x, v_x, G_\varepsilon(x)) = 0$. For constant h_2, one arrives at

$$(*)\ v_t + \frac{|v_x|^2}{4h_1(G_\varepsilon(x))} - h_2 = 0.$$

If $h_1 = constant$,, there is an explicit solution given by Hopf–Lax formula. If $h_1(G_\varepsilon(x)) > 0$ and nonconstant, (*) has a viscosity solution, and we use Mané potential approach to get an explicit representation of the solution to the partial differential equation (*). To do so, we split the system into two simpler ones.

$$\begin{cases} (s1) & \frac{|w_x|^2}{4h_1(G_\varepsilon(x))} - h_2 = \alpha \\ (s2) & w_t + \alpha = 0 \\ . & \end{cases}$$

The solution of (*) is obtained by summing up the solutions of (s1) and (s2). The viscosity solution of (s1) starting from x to the exit y is a sort of a generalized shortest path that takes into account the local congestion term $G_\varepsilon(x)$. (s1) is reduced to the Eikonal equation when there is no congestion. We denote this solution by $S^\alpha(x, y)$. The critical value $\underline{\alpha}$ should be carefully chosen to get a global viscosity solution for all $\alpha > \underline{\alpha}$. The solution of (s2) has the form $\alpha(T-t)$. Thus, the solution of (*) has the following representation

$$v(t, x) = \inf{}_y \sup_{\alpha > \underline{\alpha}} \{g(y) + S^\alpha(x, y) - \alpha(T - t)\}. \tag{1.33}$$

1.3.3 Design and Control of Crowd Dynamics

The state is an infinite dimensional (deterministic) object given by $\rho_t^u = -div_x(\rho^u u) =: b(u, \rho)$ and the running payoff is $-f = -\int_y [h_1(G_\varepsilon(y))u^2 + h_2(G_\varepsilon(y))]\rho(t, dy)$. We can therefore apply the dynamic programming principle on the space of measures (which is of infinite dimension) to obtain

$$0 = V_t(\rho) + \mathscr{H}(\rho, V_\rho), \tag{1.34}$$

where V_ρ is the Gateaux derivative with respect to the probability measure ρ, and

$$\mathscr{H}(\rho, V_\rho) = \int_y [\frac{|\partial_y V_\rho|^2}{4h_1(G_\varepsilon(y))} - h_2(G_\varepsilon(y))]\rho(t, dy)$$

with final value $V(T, \rho) = \langle -g, \rho(T, .)\rangle$, and the optimal control action $\bar{u} = \frac{\partial_y V_\rho}{2h_1(G_\varepsilon(y))}$.

We refer the reader to chapter on Master Adjoint Systems (MASS) in Volume I of the book [125]. The MASS [460, 464] in the localized congestion measure is given by the following result:

Proposition 1.6 *The function $v^*(t, y) = V_\rho$ solves the partial differential equation*

$$v_t^* + \hat{\mathscr{H}}(y, v_y^*) + \int \hat{\mathscr{H}}_\rho \rho(t, d\tilde{y}) = 0,$$

where $\hat{\mathscr{H}}(y, v_y^*) = \frac{|v_y^*|^2}{4h_1(G_\varepsilon(y))} - h_2(G_\varepsilon(y))$. The final function at time T is $v^*(T, y) = -g(y)$.

Note that the optimal control action of (P3) is

$$\bar{u} = \frac{v_y^*}{2h_1(G_\varepsilon(y))}.$$

The expression of the optimal control for (P3) differs from the one of (P2) when $v_y^* = \partial_y V_\rho$ differs from v_y. This is in part due to the extra term $+\int \hat{\mathcal{H}}_\rho \rho(t, d\tilde{y})$ in Proposition 1.6.

Proposition 1.7 *The weak derivative of the function $v^*(t, x) = V_\rho$ provides a first-order adjoint equation as*

$$p_t = -\hat{\mathcal{H}}_x(x, v_x^*) - \int \hat{\mathcal{H}}_{y\rho}\rho(dy),$$

and $p(T) = -g_y(x(T))$.

Proof This is a weak derivative of the function $v^*(t, y) = V_\rho$:

$$v_{t,y}^* = -\partial_y[\hat{\mathcal{H}}(y, v_y^*)] = -\hat{\mathcal{H}}_y(y, v_y^*) - \int \hat{\mathcal{H}}_{\tilde{y}\rho}\, \rho(d\tilde{y}),$$

and $v_y^*(T, y) = -g_y(y)$. Set $p = v_y^*$ to obtain the desired result. ■

The gap between the global optimum payoff [106] and expected of the total equilibrium payoff is $V(0, \rho_0) - \mathbb{E}v(0, X_0) \geq 0$.

1.3.4 Constrained Control Action

It is important to model the displacement constraints of an agent. The control action u_i cannot be any element $\mathbb{R}^m$. It is restricted to a set $U_i \subset \mathbb{R}^m$. One can consider, for example, the control action constraint $|u_i| \leq R$ to capture the fact that the velocity/speed is limited, and the maximum norm of the velocity is R. Therefore, the game problems to be analyzed are under the non-empty and convex set U_i. The structure of the equilibrium strategy is given by $u_i^* = \text{proj}_{U_i}\left[\frac{p_i}{2h_1(G_{I,\varepsilon}(x_i))}\right]$, where $\text{proj}_{U_i}[y]$ denotes the projection operator of the vector y onto the set U_i.

1.4 Implementation

In this section, we develop numerical simulation and a validation framework for the mean-field system. The function h_1 models the congestion measure in the neighborhood of the agent, while function h_2 models the potential energy and the response time of that agent. For the implementation, we choose $h_1(G_\varepsilon) = (c_0 + c_1 G_\varepsilon)^{3/2}$

Fig. 1.3 ERB C1 map

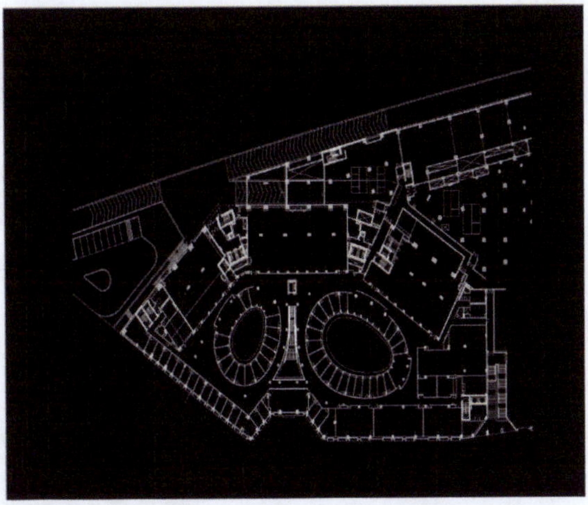

with $c_0 = c_1 = 1$ and h_2 =constant. The number c_3 is typically the reaction time of the people (when there is alarm for example) is set to 40 time units.

1.4.1 Evacuating a Small-Scale Multilevel Building

As a first example, we consider an evacuation scenario of the research and teaching building C1–B1 at New York University. The initial distribution of position is piecewise constant and takes two values 0 and 2 agents/m^2. We consider a slight modification of the map (geometry) of the ground and first floors (Fig. 1.3). The higher floors from 2 to 10 will be evacuated using the bridge on floor 2. There is one main exit door in the ground floor. Figure 1.4 represents the initial setup of the two lower floors. Figure 1.5 represents the initial density of the agents in the building. An agent is included in a three-dimensional box. Its barycenter is represented by small circle on the projected map in 2D. The exit door is in green-color code in the ground floor. The first floor stairs I_1^- are in blue color in Fig. 1.4.

To go down, agents can use two types of colors:

- blue with code (0, 0, 1) or #0000FF placed at floor 1 to go to the ground floor and
- red with code (1, 0, 0) placed at the ground floor for those coming from the first floor.

Agents can move in the white areas. Agents cannot penetrate through the walls. Agents from the upper floors move down via the stairs to a lower floor. Agents can use the stairs in blue in $I_1^{-,d}$ to go down, and those agents are represented in red to arrive at the ground floor. The agents in the ground floor can exit through the exit door in green color with code (0, 1, 0).

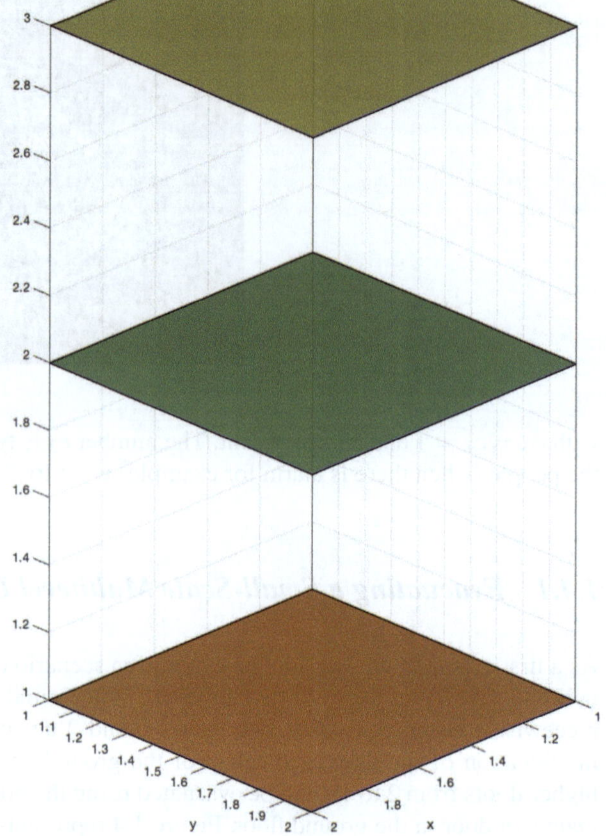

Fig. 1.4 Initial setup of the two lower floors

1.4.1.1 Numerical Methods

Since we have an explicit representation of the Bellman equation (without the diffusion term), the numerical method to solve the backward-forward system with boundary conditions is reduced to the following steps:

- We use the representation (1.33) to compute the value function from the local mean-field $G_\varepsilon(x)$. This determines u^*.
- We solve the transport equation to obtain the number of remaining agents in the building using the congestion measure G.

To numerically solve the transport equation, we need a scheme that preserves positivity (to be normalized into a probability distribution). We use the weak formulation of the Fokker–Planck equation to discretize the system. The system starts with the initial distribution ρ_0 of agents and compute G_0. Let N be a

Fig. 1.5 Initial distribution of agents

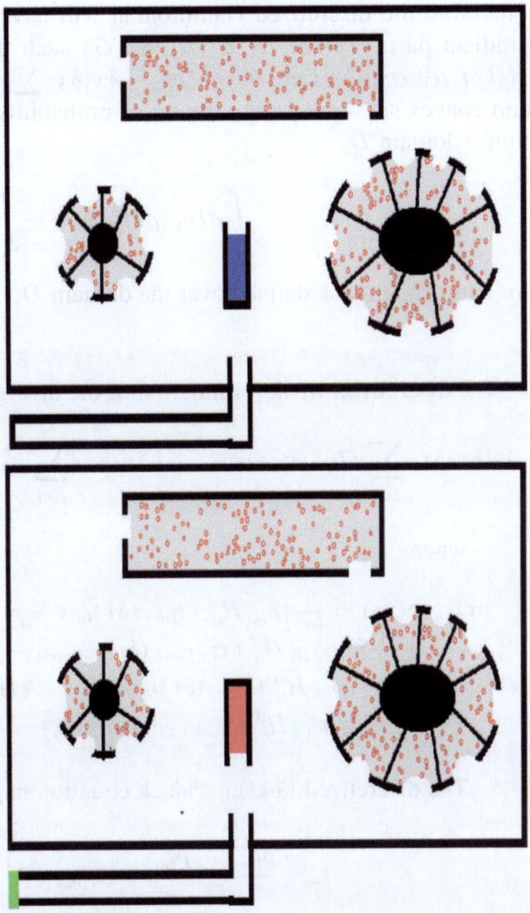

positive integer and the time step size be $\delta t = \frac{T}{N}$. The discrete time space is $\mathcal{T}_d = \{0, \delta t, 2\delta t, \ldots, N\delta t\}$. Let D_d be a grid on the domain and x_{lm} denote a generic point in D_d. The values of v, ρ and G at $(k\delta t, x_{lm})$ are denoted by $v^k_{lm}, \rho^k_{lm}, \; G^k_{lm}$. Introduce the finite difference operators

$$(\delta_1 \rho^k)_{lm} = \frac{\rho^k_{l+1,m} - \rho^k_{l,m}}{\delta x_1},$$

$$(\delta_2 \rho)_{lm} = \frac{\rho^k_{l,m+1} - \rho^k_{l,m}}{\delta x_2}.$$

The four possible points are collected as

$$(\delta \rho^k)_{l,m} = ((\delta_1 \rho^k)_{lm}, (\delta_1 \rho^k)_{l-1,m}, (\delta_2 \rho)_{lm}, (\delta_2 \rho)_{l,m-1}) \in \mathbb{R}^4.$$

Therefore the discretized Hamiltonian will have a four-dimensional vector in the gradient part: $H^d(k\delta t, \alpha_1, \alpha_2, \alpha_3, \alpha_4, G)$ such that $H^d(k\delta t, \alpha_1, \alpha_1, \alpha_2, \alpha_2, G) = H(k\delta t, \alpha_1, \alpha_2, G)$. Let $\mathscr{M} := \{\rho_{lm}^k \mid \delta x_1 \delta x_2 \sum_{l,m} \rho_{lm}^k = 1, \ \rho_{lm}^k \geq 0\}$ be compact and convex set representing the set of probability measures of the discretized (and finite) domain D_d.

$$\int div_x(\rho H_p)\phi = -\int \rho H_p \phi_x$$

for a test function ϕ defined over the domain D.

The discretized weak formulation of the divergence term is

$$\delta x_1 \delta x_2 \sum_{l,m} O_{l,m}(v,\rho)\phi_{lm} = -\delta x_1 \delta x_2 \sum_{l,m} \rho_{lm}^k \langle (\delta\phi)_{lm}, H_\alpha^d(x_{lm}, (\delta v)_{l,m})\rangle,$$

where

$$\begin{aligned} O_{l,m}(v,\rho) = &\tfrac{1}{\delta x_1}[\rho_{lm} H_{\alpha_1}^d(x_{lm}, (\delta v)_{lm}) - \rho_{l-1,m} H_{\alpha_1}^d(x_{l-1,m}, (\delta v)_{l-1,m})] \\ &+ \tfrac{1}{\delta x_1}[\rho_{l+1,m} H_{\alpha_2}^d(x_{l+1,m}, (\delta v)_{l+1,m}) - \rho_{l,m} H_{\alpha_2}^d(x_{l,m}, (\delta v)_{l,m})] \\ &+ \tfrac{1}{\delta x_2}[\rho_{lm} H_{\alpha_3}^d(x_{lm}, (\delta v)_{lm}) - \rho_{l,m-1} H_{\alpha_3}^d(x_{l,m-1}, (\delta v)_{l,m-1})] \\ &+ \tfrac{1}{\delta x_2}[\rho_{l,m+1} H_{\alpha_4}^d(x_{l,m+1}, (\delta v)_{l,m+1}) - \rho_{l,m} H_{\alpha_4}^d(x_{l,m}, (\delta v)_{l,m})]. \end{aligned}$$

The discretized Fokker–Planck equation in ρ is given by

$$\frac{\rho_{lm}^{k+1} - \rho_{lm}^k}{\delta t} + O_{l,m}(v^{k+1}, \rho^k) = 0.$$

At iteration k, we solve the weak formulation for the Fokker–Planck equation based on the expression of $v_k(t,x)$ for the entire trajectory which was computed from G_{k-1}. We iterate these steps until the global error is small.

1.4.1.2 Simulation

For the simulation setup, we have used the computer system with Intel(R) Xeon(R) central processing unit (CPU) E5-1650 v3–3.50 GHz 3.50 GHz with random access memory of 32.0 GB and using 64-bit operating system. The exit time is calculated by simulating the design compartment by zones, resolutions, and floors when agents are initially distributed according to ρ_0. The simulation considers 500 agents divided into 2 refuge zones: the ground floor refuge zone and the second floor refuge zone. Here the Eikonal equation becomes congestion-and-time dependent partial

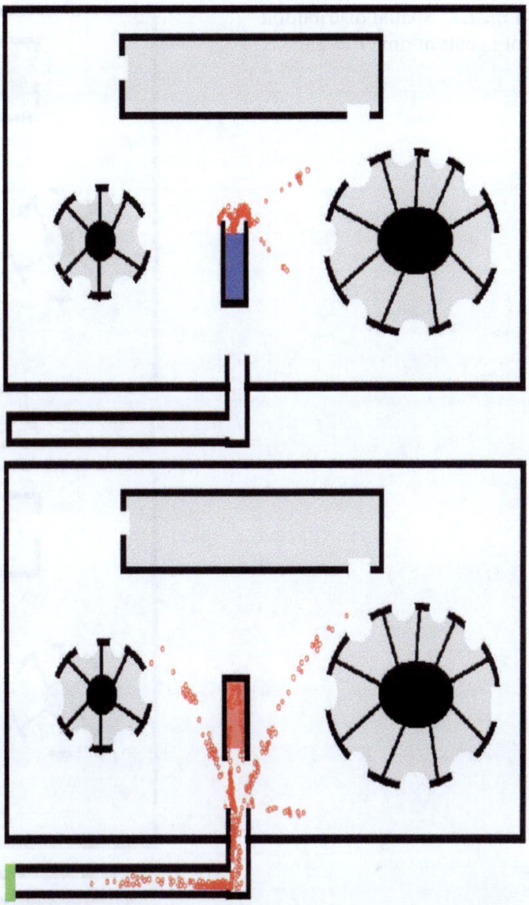

Fig. 1.6 Spatial distribution of agents at time $t = 5$

differential equation. These two features change the behavior of the agents. In particular, we observe that the repulsion forces effects at congested areas are reduced in the mean-field game model. This is in contrast to particle models where repulsion forces and physics laws apply. Intelligent agents, who can think and optimize, have a tendency to slow down and do not push back the others. This means that the repulsive term should be very small in order to better capture cost minimizing behavior. Figure 1.6 plots the distribution at $t = 5$. Figure 1.7 represents the distribution of agents' states at $t = 13$. Figure 1.8 plots the evolution of the number of remaining agents in the building. We observe, in particular, that the number of agents in ground floor starts increasing, because the flow coming from first floor until certain time threshold and then decrease when agents start to exit. Figure 1.9 plots the evolution of the number of agents who have been evacuated safely. The plot has a typical shape of a cumulative distribution function.

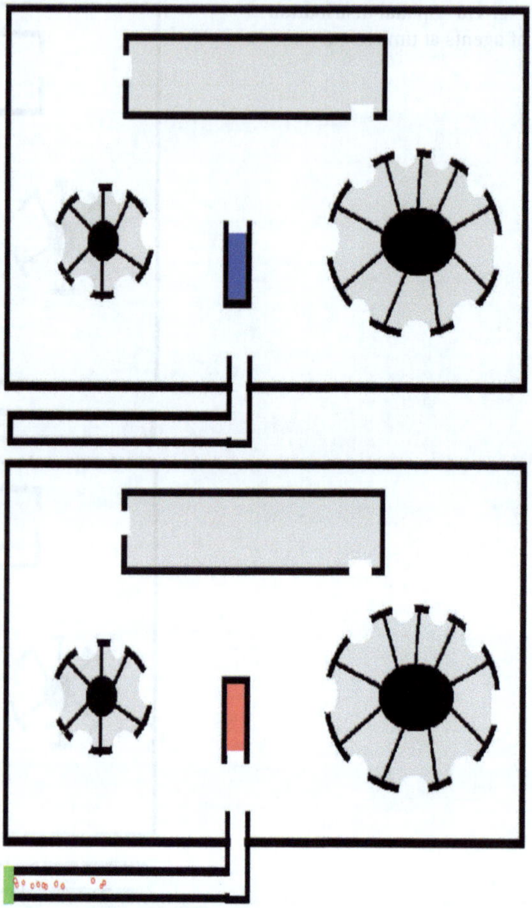

Fig. 1.7 Spatial distribution of agents at time $t = 13$

1.4.1.3 Verification and Validation

As a test, we use the experimental data obtained from real evacuations conducted between August 2014 and November 2015 in building C1–B1 to improve coordination among various components and to ensure the readiness of the emergency process in times of crisis. Table 1.3 and Fig. 1.10 report the results of the exit times. Although the total number of agents was slightly different on different dates, the evacuation time corroborates the proposed mean-field approach even for relatively small number of agents. When the response time of the agents is 40 seconds and the simulation time units is 30 times faster than the physical time, we obtain similar time frame between the simulation and the experimental evacuation time. It is important to notice that human response time c_3 needs to be considered in the simulation in order to better capture the evacuation process observed in practice.

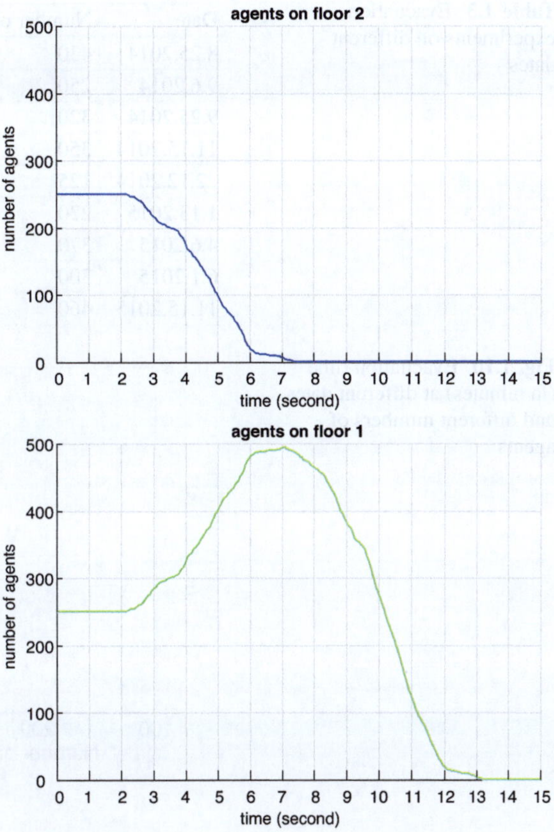

Fig. 1.8 Evolution of the number of remaining agents

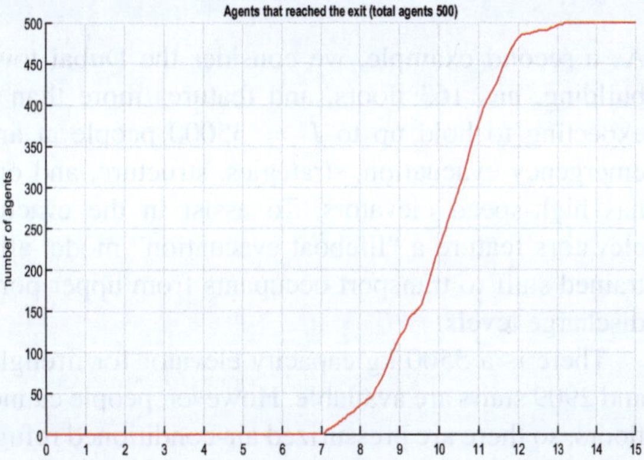

Fig. 1.9 Number of agents evacuated from building vs. time

Table 1.3 Evacuation experiments on different dates

Date	Number of agents inside	Exit time duration
8.25.2014	430	5
9.6.2014	250	3.5
9.25.2014	320	4
11.15.2014	350	4.2
22.12.2014	125	2.5
1.13.2015	220	3
4.6.2015	370	4.8
6.1.2015	300	4
11.15.2015	460	5.5

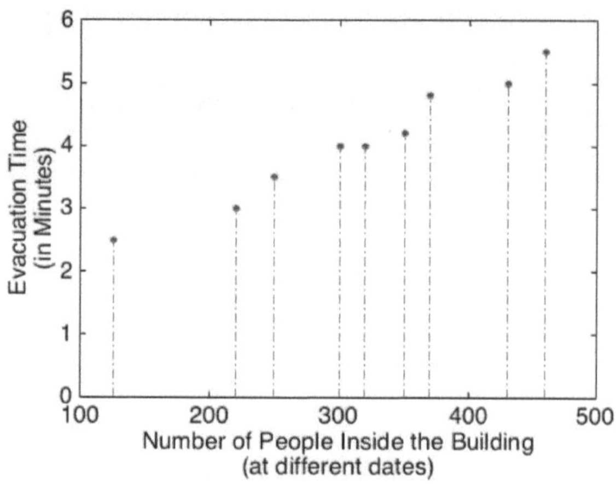

Fig. 1.10 Evacuation time (in minutes) at different dates and different numbers of agents

1.4.2 Evacuation of the Current World's Tallest Building

As a second example, we consider the Dubai tower which as the world's tallest building, has 163 floors, and features more than $K = 150$ available floors and expecting to hold up to $I = 35000$ people at any one time. When it comes to emergency evacuation, strategies, structure, and operations are critical. The tower has high-speed elevators. To assist in the evacuation of occupants, designated elevators feature a "lifeboat evacuation" mode, allowing fire brigade personnel or trained staff to transport occupants from upper portions of the tower to designated discharge levels.

There is a 5500 kg capacity elevator for firefighters and building service work, and 2909 stairs are available. However, people cannot be expected to walk down 163 floors, so there are pressurized air-conditioned refuge areas, approximately every 25 floors where they could wait safely or rest on the way down. In terms of simulated evacuation, we therefore consider interconnection between branches every 25 floors. Figure 1.11 illustrates a sample floor architecture. The areas of refuge are separated from the main structure of the building by a 2-hour fire-resistant construction and are pressurized to minimize the migration of smoke into the compartment. While the

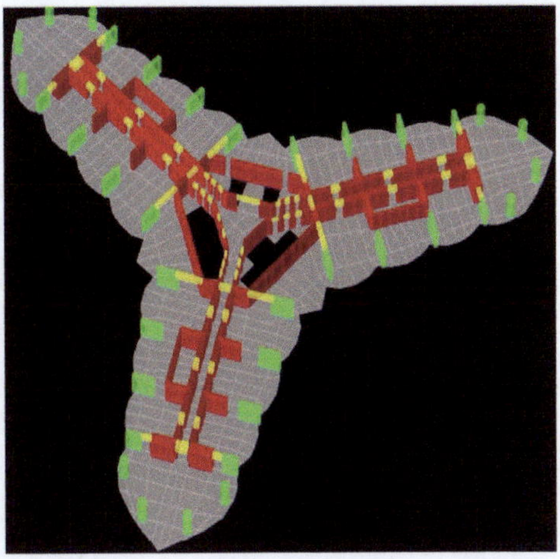

Fig. 1.11 A typical floor design of the tower. Inspired from *Hymenocallis*

design is based on a defend-in-place strategy, where the structure is geared toward isolating the problem and protecting the residents within, there is a possibility that an escalated event may require a full building evacuation. There are 58 elevators of normal situations in the building. A full building evacuation uses only ten enhanced elevators. In case of full evacuation, all the refuge areas at the very high floors need to be evacuated using these ten high-speed elevators if available. The building has three wings, and within the buttress core, the elevators are placed at the center. Floor area is around 309,473 m^2. As already mentioned, the tower stairwell has a total of 2909 steps leading up to the 160th floor.

1.4.2.1 Simulation and Experiments

The simulation considers grid with 35,000 people divided into 6 refuge-zone floors. Except the ground floor group all the others are temporary and need to be relocated in case of full evacuation of the building. For simulation, the special elevator designed for evacuation spent 3*s* per floor and the stairs took 1mn in the absence of congestion. However, in the presence of congestion, the queueing times at stairs and elevator may increase. We use the same computer system as in the previous experiment. We examine the behaviors of the agents from ground floor to floor 25 given that a flow of people will be coming from floor 37 to 25. In contrast to standard evacuation scheme, the tower has the specificity to be evacuated at different vertical directions. For example, an agent located on the 12th floor has the option to go to one of the exit doors located on the 25th floor or to go down at the one of the exits in the ground floor. Such a structural feature is not possible in a single-level building, and it is specific to this tower. To go down, agents can use two types of colors:

- blue with code (0, 0, 1) or #0000FF placed at floor k to go to $k - 1$ and
- red with code (1, 0, 0) placed at floor $k - 1$ for those coming from floor k.

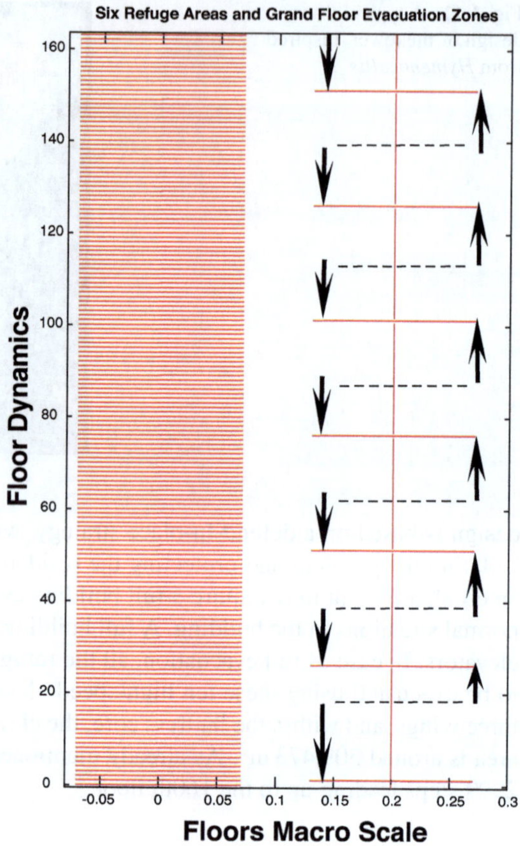

Fig. 1.12 The entire tower macro dynamics. Directions of the network of floor vector fields

To go up, agents can use two types of colors:

- weak-blue with code (0.5, 0.5, 1) placed at floor k and
- weak-red (light coral) with code (1, 0.5, 0.5) placed at floor $k+1$.

Agents can move in the pink and white areas. Agents cannot penetrate through the walls. Agents can use the stairs in weak-blue in $I_k^{+,d}$ to go up. Agents can use the stairs in blue in $I_k^{-,d}$ to go down. If on the same floor, agents can exit using one of the exits in green color with (0, 1, 0). Figure 1.12 represents the macro dynamics of the inter-floor vector fields. We see that people have a tendency to go (up or down) to the "closest and less congested" refuge areas.

Figure 1.13 represents the initial distribution of people.

Figures 1.14 and 1.15 plot the floors configuration at $t = 20$.

Figures 1.16, 1.17, 1.18, 1.19, 1.20, and 1.21 represent the number of remaining agents in the building. Figure 1.22 plots the total number of people who have been evacuated vs. time. We see that the number of agents who have been evacuated remains at zero until the first agent arrives at the exit and then increases

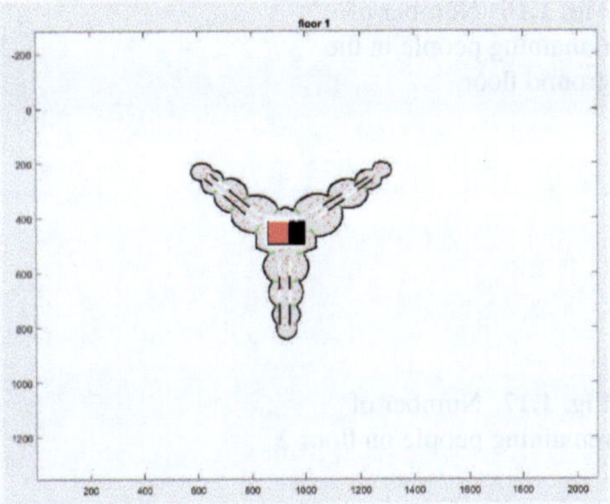

Fig. 1.13 Typical ground floor with several exit doors. Initial distribution of people

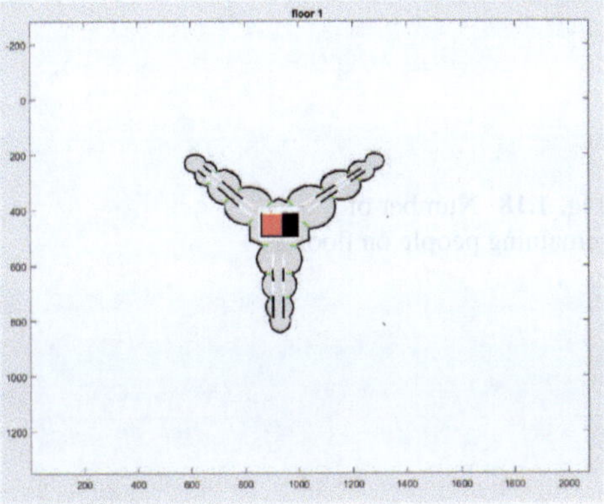

Fig. 1.14 Ground floor: distribution of people at $t = 20$

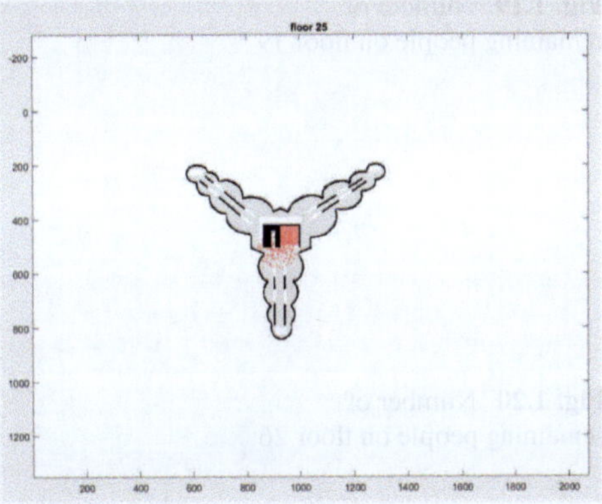

Fig. 1.15 Floor 25: distribution of agents at $t = 20$

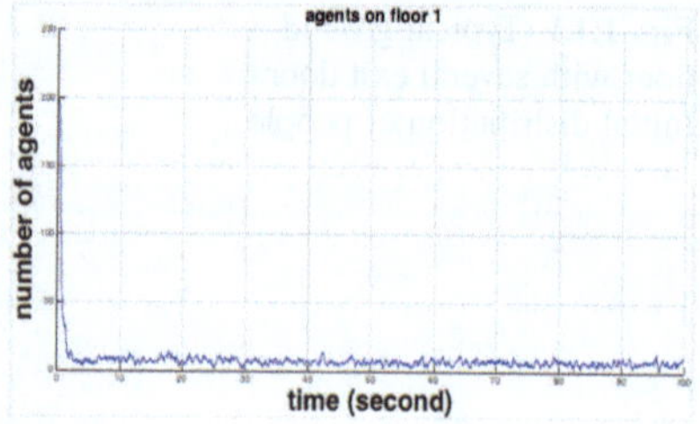

Fig. 1.16 Number of remaining people in the ground floor

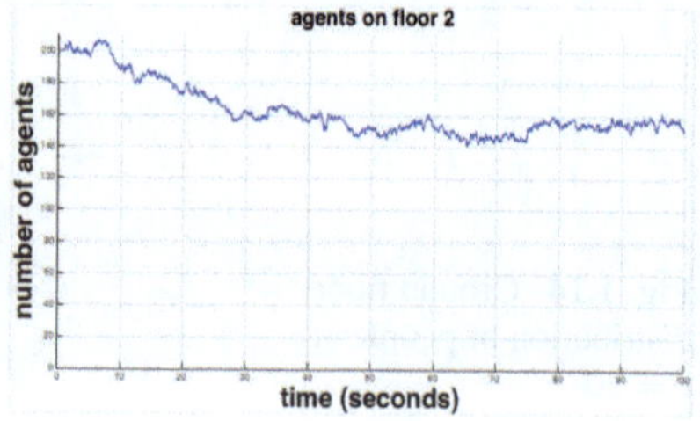

Fig. 1.17 Number of remaining people on floor 2

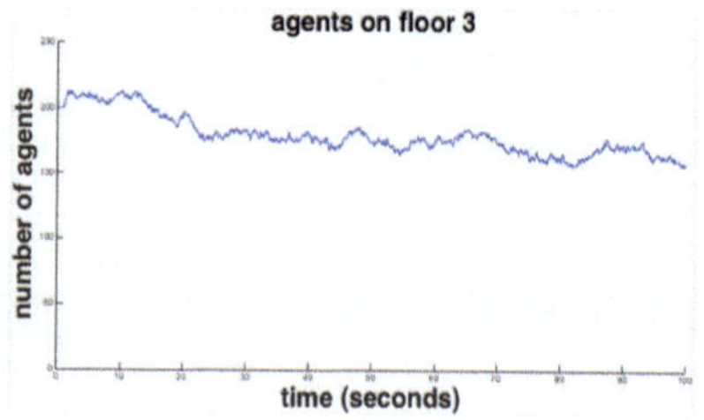

Fig. 1.18 Number of remaining people on floor 3

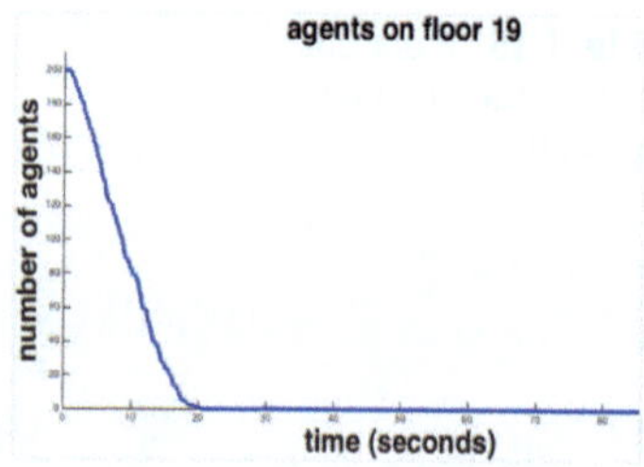

Fig. 1.19 Number of remaining people on floor 19

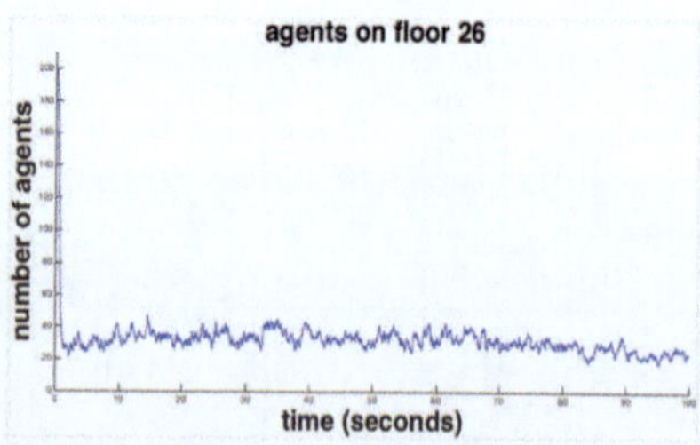

Fig. 1.20 Number of remaining people on floor 26

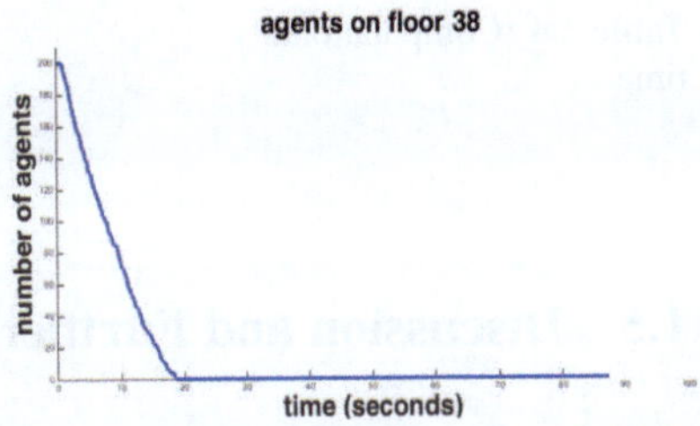

Fig. 1.21 Floor 38: number of remaining people vs. time

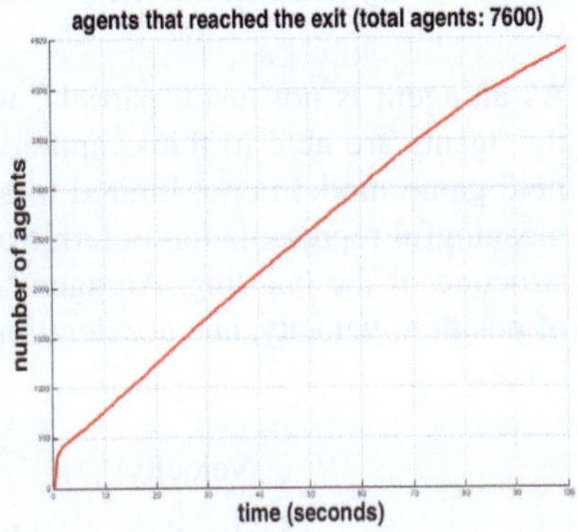

Fig. 1.22 Total number of people who have been evacuated vs. time

progressively. After 100 units of simulation time (without adding the 40 seconds of response time) more than 60% have exited in a scenario with several exit doors on ground floor. Similarly, at $t = 100$, there is no remaining agent on floors 38, 37, as well as on floors 20 and 21. Due to the interconnections between floors through the stairs, these agents are still in the building but are getting closer to the exits (ground floor) or the refuge areas at floor 25. The floors that are closer to the exit floors have their remaining agents augmented by the incoming flow. The system may oscillate, because the incoming flow is not necessarily balanced with the outgoing flow. When the incoming flow is getting smaller (compared to the outgoing flow), the number of remaining agents decreases progressively toward zero, and all the agents will be evacuated. As we observe, the dynamics of this multilevel building cannot be retrieved simply from the single-level analysis. Thus, the analysis of the ground floor or evacuation of a hall has a limited view in the multilevel context. If the number of stairs and doors are limited, the routing options are reduced, and congestion occurs quickly at the stairs and doors (and queueing delay increases) which may slow down the evacuation process. In this particular simulation case, we noticed that the distribution of agents on the floors and the spatial configuration of the floors (presence of obstacle or not) influence the exit time of all the agents in the multilevel building. As the system dimension increases, the number of agents on the floors in the middle will oscillate, and this is governed by the difference between the incoming and outgoing flux in each floor. The running time of the algorithm was affected by the system dimension/size. In our experiment, we used a velocity of move similar to that of a human being (see Table 1.4 and Fig. 1.22).

Table 1.4 Computational time

Number of agent	Computational time
500 agents	10 min 7 sec
1000 agents	23 min 55 sec

1.5 Discussion and Further Improvement

1.5.1 Physical Meaning and/or Interpretation

As an agent is not just a particle, we have considered in (P1–3) the cases where the agents are able to think, anticipate, and optimize their objectives. The mean-field game model is not limited to a theoretical mathematical formula. It has also meaningful representation in terms of the physics and mechanics of the people and structure of the building. An agent's physical dynamics can be rewritten in terms of position, velocity, and acceleration involving the different congestion-dependent forces.

$$\text{Velocity:} \qquad \dot{x} = u(x, G_\varepsilon(x)) \tag{1.35}$$

$$\text{Acceleration: } \dot{u} = H_p\{H_{px} + H_{pp}v_{xt} + H_{pG}G_t\}. \tag{1.36}$$

In the microscopic model, the dynamics of an agent i is

$$\dot{x}_i = u_i \in U_i \subset \mathbb{R}^m, \tag{1.37}$$

$$\dot{u}_i = u_i\{H_{i,px} + H_{i,pp}v_{i,xt} + H_{i,pG}G_t\} \tag{1.38}$$

$$i \in \{1, 2, \ldots, I\}.$$

When people are marching, they possess a kinetic energy which is generatived by the motion. Usually, it captures the work needed to accelerate a body of a given mass from rest to its stated velocity. Then, the objective function J_i can be reinterpreted as the kinetic energy—potential energy.

1.5.2 Stochastic Disturbance

In order to capture measurement errors, we introduce a small noise to the direction/velocity. Let x_i be the stochastic state of agent i at time t satisfying

$$dx_i = u_i dt + \sigma dB_i,$$

where B_i is a Brownian motion, independent across the agents. σB_i should be carefully chosen and restricted, so that agents do not diffuse in all directions. Problem (P1) becomes

$$(SP1): \begin{cases} \sup_{u_i} -\mathbb{E}J_i(u_1,\ldots,u_I) \text{ subject to} \\ dx_i = u_i dt + \sigma dB_i, \text{ on } (0,T)\times D \\ x_i(0) \in D, \\ \text{boundary constraints:} \\ u_i = 0 \text{ on } \partial D, \text{ and} \\ u_i = k(x,\rho) \text{ at Exits.} \end{cases}$$

1.5.2.1 Second-Order Mean-Field Game

Taking the risk-neutral cost as

$$\mathbb{E}J(u, G_\varepsilon^*) = \mathbb{E}g(x(T)) + \mathbb{E}\int_0^T \{h_1(G_\varepsilon^*(x(t)))|u(t)|^2 + h_2(G_\varepsilon^*(x(t)))\}\, dt,$$

, the corresponding stochastic mean-field game problem is

$$(SP2): \begin{cases} \sup_u -\mathbb{E}J(u, G_\varepsilon^*) \text{ subject to} \\ dx(t) = u(t)dt + \sigma dB(t), \\ x(0) \sim \rho_0(.) \text{ is given }, \\ G_\varepsilon^*(t, x(t)) = \frac{1}{c_\varepsilon}\int_{B(x(t),\varepsilon)} \rho^*(t, ds), \\ \rho_t^* + div_x(\rho^* u^*) - \frac{1}{2}\partial_{xx}[\sigma^2\rho^*] = 0, \\ \rho^* = 0 \text{ on } \partial D, \\ u = 0 \text{ on } \partial D, \\ \text{Law}(x^*) = \rho^*. \end{cases} \tag{1.39}$$

Problem (SP2) is a fixed-point problem. The functional $\mathbb{E}J(u, G_\varepsilon^*)$ is linear in the probability measure ρ.

The mean-field equilibrium system is

$$\begin{aligned} & v_t + \frac{|v_x|^2}{4h_1(G_\varepsilon(x))} + \frac{\sigma^2}{2}\Delta v = h_2(G_\varepsilon(x)) \\ & v(T,x) = -g(x), \\ & \rho_t + div_x(\rho u) - \tfrac{1}{2}\partial_{xx}[\sigma^2\rho] = 0, \\ & \rho(0,x) = \rho_0(x). \end{aligned}$$

1.5.2.2 Design and Stochastic Mean-Field-Type Optimization

The mean-field-type control problem of the designer is

$$(SP3): \begin{cases} \sup_{u,G_\varepsilon} \ -J_0(u) \ \text{subject to} \\ dx^u = udt + \sigma dB, \\ G_\varepsilon(t, x^u(t)) = \frac{1}{c_\varepsilon}\int_{B(x^u(t),\varepsilon)} \rho^u(t, ds), \\ \rho^u_t + div_x(\rho^u u) - \frac{1}{2}\partial_{xx}[\sigma^2\rho^u] = 0, \\ \rho^u = 0 \text{ on } \partial D, \\ u = 0 \ \text{ on } \partial D \end{cases}. \tag{1.40}$$

The state ρ is an infinite dimensional object given by $\rho^u_t = -div_x(\rho^u u) + \frac{\sigma^2}{2}\Delta\rho =: \tilde{b}(u,\rho)$, and the running payoff is $-f = -\int_y [h_1(G_\varepsilon(y))u^2 + h_2(G_\varepsilon(y))]\rho(t,dy)$. Let $\tilde{V}(t,\rho)$ be the value of the mean-field-type optimization problem starting from ρ at time t. We can apply the dynamic programming principle (on the space of measures) to obtain

$$0 = \tilde{V}_t(t,\rho) + \tilde{\mathscr{H}}(\rho, \tilde{V}_\rho) + \int \rho\frac{\sigma^2}{2}\Delta_y\tilde{V}_\rho, \tag{1.41}$$

where

$$\tilde{\mathscr{H}}(\rho, \tilde{V}_\rho) = \int_y [\frac{|\partial_y \tilde{V}_\rho|^2}{4h_1(G_\varepsilon(y))} - h_2(G_\varepsilon(y))]\rho(t,dy),$$

with final value $\tilde{V}(T,\rho) = \langle -g, \rho(T,.)\rangle$, and the optimal control action $\bar{u} = \frac{\partial_y\tilde{V}_\rho}{2h_1(G_\varepsilon(y))}$.

The function $\tilde{v}^*(t,y) = \tilde{V}_\rho(t,\rho)(y)$ solves the partial differential equation

$$\tilde{v}^*_t + \tilde{\mathscr{H}}(y, \tilde{v}^*_y) + \frac{\sigma^2}{2}\Delta_y\tilde{v}^* + \int \tilde{\mathscr{H}}_\rho\rho(t,d\tilde{y}) = 0,$$

where $\tilde{\mathscr{H}}(y,\tilde{v}^*_y) = \frac{|\tilde{v}^*_y|^2}{4h_1(G_\varepsilon(y))} - h_2(G_\varepsilon(y))$. The final payoff function is $\tilde{v}^*(T,y) = -g(y)$.

It is important to note that for deterministic first-order dynamics, an explicit representation of the first-order Bellman equation was used via Mané potential in Sect. 1.3.2.1, and this had simplified drastically the complexity of the numerical computations. Unfortunately, we do not have an explicit solution for the second-order Bellman system. Therefore, the explicit solution remains an open issue for the stochastic extension of the individual dynamics provided in Sect. 1.5.2 and to problems SP1–SP3.

Fig. 1.23 Illustration of smoke and crowd

1.5.3 Smoke and Crowd

Crowd congestion at the stairs, doors, and evacuation paths could be quite risky. Another high risk is when smoke is involved. The smoke dynamics is coupled with the cost as it affects the travel time as illustrated in Fig. 1.23.

$$\begin{cases} & x \in D \subset \mathbb{R}^3, \\ v_t + \frac{|v_x|^2}{2c(G_{\varepsilon,\tilde{\varepsilon}})} = f, & (0,T) \times D, \\ m_t + (m\frac{v_x}{c(G_{\varepsilon,\tilde{\varepsilon}})})_x = 0, & (0,T) \times D, \\ G_\varepsilon(x,m) = \frac{1}{vol(B_\varepsilon)} \int_{y\in B(x,\varepsilon)} m(t,dy) & \\ G_{\tilde{\varepsilon}}(x,\tilde{m}) = \frac{1}{vol(B_{\tilde{\varepsilon}})} \int_{y\in B(x,\tilde{\varepsilon})} \tilde{m}(t,dy) & \\ m(0,x) = m_0(x), & \{0\} \times D \\ v_x = 0 = m & \text{at } \partial D, \\ v_x = k(x,m), & \text{at exits}, \end{cases} \tag{1.42}$$

where G_ε is the congestion measure around an agent within a range of size ε, $c(G_{\varepsilon,\tilde{\varepsilon}}) = h_1(G_\varepsilon)\tilde{h}_1(G_{\tilde{\varepsilon}})$, $\tilde{h}_1$ is a nondecreasing function, $\tilde{m}(t,.)$ is the distribution of smoke inside the building at time t, $G_{\tilde{\varepsilon}}$ is a local smoke congestion measure per area.

1.5.4 Co-intelligence of Human–Machine Intelligence Agents

Some of aspects of co-intelligence can be conceptualized through the synergistic enhancement of capabilities. In the context of interactions between human agents and machine intelligence (MI) agents, we model the co-intelligent system's intelligence I_{co} as a function $f(I_H, I_A, \text{Interaction})$ that yields results exceeding individual capacities (I_H, I_A). Another perspective assesses performance, where co-intelligent systems demonstrate superior outcomes $P_{co}(\text{Task})$ on various, diversified, multi-culture tasks compared to individual agents. Extending this to multiple agents via cooperative mean-field-type game (MFTG) theory, co-intelligence arises when the intelligence value of the grand coalition $v(N)$ surpasses the sum of individual values $\sum_{i \in N} v(\{i\})$, with synergistic co-intelligence indicated by the superadditivity of the characteristic function, $v(S \cup T) > v(S) + v(T)$ for disjoint coalitions S and T, mathematically formalizing how collaboration amplifies the collective intelligence or holistic intelligence or problem-solving effectiveness of the system. Human–MI co-intelligence represents a paradigm shift from unilateral machine intelligence automation to shared intentionality and co-learning. In multilingual societies, communication barriers often limit the effectiveness of machine intelligence tools. This paper addresses the problem of enabling nearly real-time, multilingual, audio-to-audio interaction between humans and machine intelligence agents, using a MFTG that naturally captures the risk awareness, distributed, and adaptive nature of such interactions.

A first step toward collective intelligence and co-intelligence is co-learning and mutual learning from other intelligence from various domains, nature, species, etc. Co-intelligence, distinct from individual intelligence quotient, reflects a distributed, emergent intelligence arising from cooperative co-learning dynamics within interconnected agents. Grounded in holistic interdependence and co-creative processes, it characterizes systems that exhibit collective resonance, synergistic decision-making, and alignment with broader informational gradients. From the viewpoint of cooperative mean-field-type game theory, co-intelligence emerges when agents optimize shared objectives under informational coupling and strategic feedback, producing globally beneficial equilibria. Its antithesis, co-stupidity, arises from misaligned incentives and incoherent agent interactions. Both phenomena manifest across scales from individuals to institutions shaped by systemic interagent dynamics and structural information flows. Here we consider $I = 2{,}000{,}000$ human agents indexed by $i \in \{1, \ldots, I\}$. Each human agent i is characterized by a state $X_i(t) \in \mathbb{R}^d$ that includes their position, and a set of audio messages in their mother tongue $\ell_i \in \{1, \ldots, 7164\}$. The control action $\alpha_i(t)$ corresponds to the velocity or motion intention of the agent. Each human agent i is supported by a personal group aa_i of audio–MI agents. The audio–MI agents in aa_i operate in the mother tongue ℓ_i of human i, and their task is twofold:

1. Collaborate among themselves and with other audio–MI agents deployed across Hajj sites.

2. Communicate with their associated human agent i using personalized, fractional-diffusion transformer-based audio encoders.

An audio–MI agent member of aa_i who receives new instruction/message from organizing authorities in a certain language in audio will translate in the language spoken by of the human agent i. This output audio updates the message to i as part of its internal state. The team aa_i handles all input and output messages received by i. If the instructions and guiding messages on sites are currently limited to 200 languages, it will be extended to 7164 languages and in audio, so that it does not exclude audio-literate people who do not speak, read, or write in the 200 languages. MI will learn from Hajj smartphone interactions with pilgrims. From these audio–MI agents, each pilgrim can get personalized help for scheduling, transportation, congestion areas, and evacuation guidance. The collective dynamics involve the position and velocity updates for all human agents, the interagent influence through localized congestion-dependent interactions, and audio-based communication governed by a textless audio diffusion-transformer model across language embeddings. This includes communication with audio-literate people directly in their mother tongues. The distribution ρ_t evolves over the joint space of positions and audio messages. The decision made by audio–MI agents in aa_i is the output audio spoken to agent i and affect the internal state of i. Therefore, the quality of team aa_i output can potentially affect i if i decides to take into account when changing paths or closing directions during the evacuation process. The decisions made of the audio–MI agents in aa_i are the queries–keys–values–weights–biases of the audio-transformer such that they a high-quality audio-to-audio communication in the mother tongue of the corresponding human agent. $X_i = (x_i, x_{l_i}, y_{l_i})$, x_{l_i} is the transformed audio into spectrogram, wavelet scalogram that will be processed by the audio transformer. The internal input-output state of audio–MI agent in aa_i is $(x_{l_i}(t), y_{l_i}(t))$. Here X is a random matrix for multiple reasons: The initial positions are random, the output of the audio–MI is random, and initial message is picked from random event. However, we do not have an infinite number of human agents here. We build an augmented state ρ of the entire interactive system as follows. Let $\rho(t, dX)$ be the probability measure of the entire random matrix $X(t)$ of agents, where $X(t) \in \mathbb{R}^{I \times d}$. The evolution of $\rho(t, X)$ follows the Kolmogorov equation, which is given by

$$\partial_t \rho(t, X) + \sum_{i=1}^{I} \sum_{j=1}^{d} \frac{\partial}{\partial X_{ij}} \left(\rho(t, X) \cdot b_{ij}\left(t, X, \rho(t), \boldsymbol{\alpha}^*(t, X, \rho(t))\right)\right) = 0.$$

Let $V_i(t, \rho)$ be the value function for agent i. The Hamilton–Jacobi–Bellman (HJB) equation in the Wasserstein space is

$$
\begin{aligned}
& i \in \mathscr{I}, \\
& \quad \partial_t V_i(t,\rho) + \int H_i(t, X, \rho, \nabla_\rho V_i(t,\rho)(X))\, d\rho(X) = 0, \\
& \quad V_i(T,\rho) = \int g_i(X,\rho)\, d\rho(X). \\
& \quad H_i(t, X, \rho, P) := \inf_{\alpha_i \in A_i} \left\{ \langle P, B(X, \rho, (\alpha_j)_j)\rangle + L_i(X, (\alpha_j)_j, \rho) \right\},
\end{aligned} \tag{1.43}
$$

where $\langle A, B\rangle$ denotes the canonical inner product of matrices $trace(AB')$ and H_i is the integrand Hamiltonian. The equilibrium cost is $V_i(0, \rho_0)$, and the equilibrium strategies are in state-and-mean-field-type feedback form $\alpha_i(t, X, \rho)$. Assuming the value function of human agent i, $V_i(t, \rho)$ is differentiable with respect to the measure ρ, we define

$$
U_i(t, X, \rho) := \frac{\delta V_i}{\delta \rho}(X).
$$

Then $U_i(t, X, \rho)$ satisfies the following Master Adjoint SyStem (MASS):

$$
\begin{aligned}
& i \in \mathscr{I}, \\
& \partial_t U_i(t, X, \rho) + H_i(t, X, \rho, \nabla_x U_i(t, x, \rho)) \\
& \qquad + \int \tfrac{\delta H_i}{\delta \rho}(t, y, \rho, \nabla_x U_i(t, y, \rho))(X)\, d\rho(y) = 0, \\
& U_i(T, X, \rho) = g_i(X, \rho) + \int \tfrac{\delta g_i}{\delta \rho}(y, \rho)(X) d\rho(y).
\end{aligned} \tag{1.44}
$$

Note that, in general, the function $U_i(t, X, \rho)$ is not the equilibrium cost.

Hajj Evacuation with 2 Million People

We consider a mimic of a real-world scenario in which 2 million pilgrims attending the Hajj in Mecca are assisted in an emergency evacuation by a decentralized network of audio–MI agents. Each human agent has at least one mother tongue from among the 7164 spoken languages. Each is supported by a group aa_i of audio–MI agents using fractional diffusion transformer models to encode and decode messages in the appropriate language. At time t, the state $X(t)$ is a matrix. Its row i is $X_i(t)$ dedicated to agent i and includes their spatial location, current motion, and linguistic communication content. The velocity control $\alpha_i(t)$ adjusts movement based on safety, crowd localized congestion, risk awareness, and coordinated audio instructions. Audio–MI agents communicate with their assigned human and collaborate with nearby MI agents using decentralized learning, real-time audio embedding, and emergent equilibrium strategies derived from MFTG and MASS formulations [460, 464]. The cost function of i is $J_i = L_i(X, (\alpha_j)_j, \rho) =$

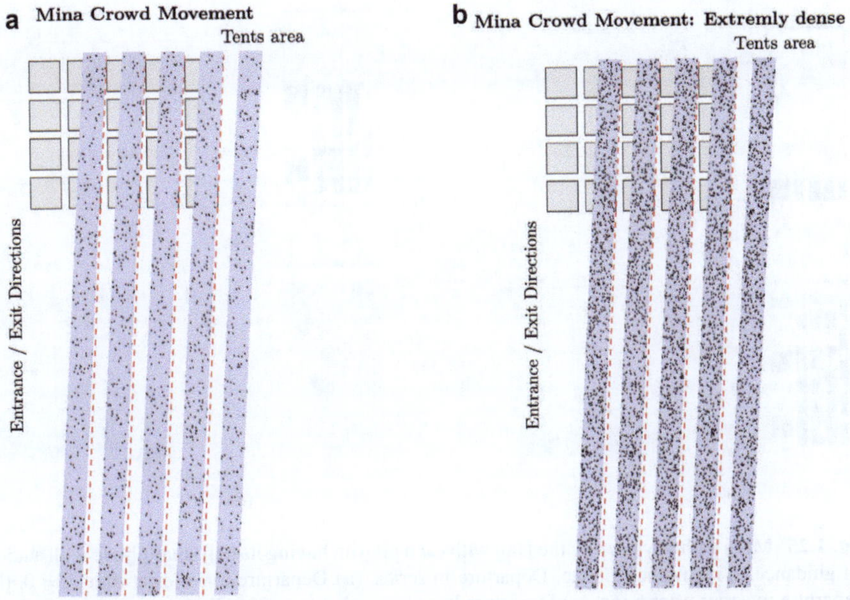

Fig. 1.24 Mina pathways during the Hajj with each pilgrim having its hajj smartphone with audio–MI guidance in its mother tongue. (**a**) Regular number of people around each person. (**b**) Extremely dense case: huge crowd around

$h_i(l_i, X_i, \rho, MIerror^2)\left|\alpha_i\right|^2 + h_{aa_i}(l_i, y_i, \rho)$ where h_{aa_i} is a MI-risk measure associated with human agent i. $h_i(l_i, X_i, \rho, MIerror^2)$ is a localized congestion measure that counts the number of people around the physical position of i. Here we do not recommend to model with respect to an average position (barycenter) as one cannot bring the 2 million people to the same evacuation area. An example of localized measure is $\frac{1}{(I-1)vol(B_{\varepsilon_i})}\sum_{j=1,j\neq i}^{I} \mathbb{I}_{x_j\in B_{\varepsilon_i}(x_i)}$ where B_{ε_i} is the ball of radius ε_i. Intuitively, if the number of people (per volume) around the position x_i is extreme, then the control action which is the velocity here will be decreased, so that the cost to move out of it is higher. We illustrate it the Mina crowd in Fig. 1.24a, b. In the Mina case, the goal is to be at a less crowded area and find a path that avoids huge crowd around. Such a crowd avoidance strategy could save the agent when a panic and a pushing behavior occur during these walking journeys. This crowd avoidance strategy is less risky and safer but has a limited flow occupancy per line and therefore may take a longer time to reach Mina on time. However, if it is too congested, people cannot move or move extremely slowly, so crowd avoidance strategy will win in terms of travel time too.

We simulate the gap areas between pathways are filled in red to clearly distinguish non-movable zones from the paths. Figure 1.25c, d mimic the evacuation process.

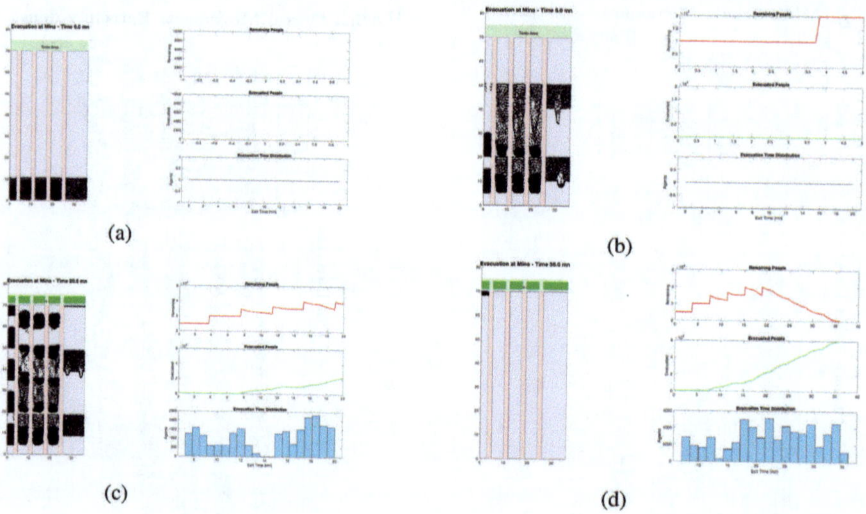

Fig. 1.25 Mina pathways during the Hajj with each pilgrim having its hajj smartphone with audio–MI guidance in its mother tongue. Departure in series. (**a**) Departure in series at time $t = 0$. (**b**) Departure in series after 5 mn. (**c**) Departure in series at time $t = 20$. (**d**) Departure in series after 36 mn

Evacuation of Burj Khalifa and MI-Guided Directions

We consider an evacuation strategy for the Burj Khalifa that uniquely incorporates sky bridges and designated refuge areas. While a vertical evacuation might superficially resemble a horizontal multi-compartment evacuation, critical distinctions exist. The potential hazard associated with a fall from the 148th floor, for instance, fundamentally differs from the risk of injury at ground level. Vertical evacuation can be modeled as a network of pedestrian movement influenced by machine intelligence recommendation flows, where the network represents the routes chosen by individuals. In this context, penalizing an agent based on a global barycenter or mean position of all agents is not necessarily effective, as such a centroid may not physically accommodate the entire crowd. Instead, we employ a localized congestion measure that quantifies the number of individuals in an agent's immediate vicinity. High local density slows movement, while unimpeded movement occurs in sparse areas toward the nearest available exit, subject to flow constraints and obstacles. Evacuation routes include stairs leading to refuge areas on the 25th, 50th, 100th, 125th, and 150th floors, or directly to ground-level exits. Individuals in refuge areas are subsequently evacuated to the ground floor. A key feature of our approach is the coupling of this localized congestion measure with a risk assessment informed by recommendations from MI agents using MFTG. Figure 1.26a, b mimic the evacuation process of a multilevel building.

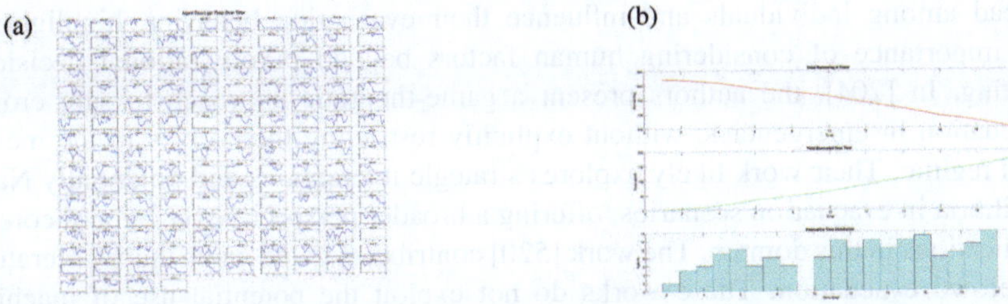

Fig. 1.26 Evacuation of a multilevel building with 163 floors. **(a)** Evolution of the number of people per floor. **(b)** Burj Khalifa evacuation including refuge areas and sky bridges

1.6 Notes

Evacuation models are often employed in engineering analyses within the context of performance based design. The flexibility of the proposed mean-field-based tool and its relatively easy use may help to perform safety design assessments that takes into consideration congestion. This theoretical framework combines a powerful decongestion technique with flexible control over population and behavior to deliver better results. The framework developed here can be easily extended to a nonlocal mean-field dependence, i.e., replacing $G_\varepsilon(t, x(t))$ by the total number of remaining agents in the building. This is simply achieved by increasing the range with ε. Other extensions of the framework include the heterogeneity of agents, evacuation of disabled agents, and design optimal allocation of helpers in the building for a faster and safer evacuation.

Some the challenges that arise in this context are the existence of solution and analysis of the congestion-dependent optimality system, robustness properties when some of the doors are not open, and some of the evacuation elevators are out of service, other solution concepts such as *reverse or inverse* Stackelberg situations in which designer, architect, and evacuation planners anticipate their decisions for evacuation and safety policies and people react to those designs and guidelines and decide on their route to exit [15, 54, 182, 266, 283, 449, 452, 454, 474, 482, 484, 486, 487].

Mean-field-type games have been used to study evacuation problems. A continuum theory for the flow of pedestrians has been presented in [236]. The works in [41, 127, 133–136, 416] examine mean-field-related games of evacuation in multilevel buildings. The authors in [10] address the complex problem of mass evacuation planning by employing mean field game theory. Their work proposes a framework to model and optimize the strategic decisions of a large population during an evacuation. By considering the interactions between evacuees, the approach allows for the derivation of macroscopic evacuation strategies from individual behaviors. This contrasts with the work in [305], who focus on the psychological aspects of crowd evacuation, specifically the role of recurrent emotional contagion within a cyber-physical society. They introduce a model that captures how emotions spread among individuals and influence their evacuation behavior, highlighting the importance of considering human factors beyond purely rational decision-making. In [204], the authors present a game-theoretic approach to safe crowd evacuation in emergencies, without explicitly restricting themselves to the mean-field regime. Their work likely explores strategic interactions and potentially Nash equilibria in evacuation scenarios, offering a broader perspective on game-theoretic applications in this domain. The work [520] contributes to the game theory literature on crowd evacuation. These works do not exploit the potential use of machine intelligence agents to guide or to improve synergistic interaction between human and machine during the evacuation process. For more details on crowd, pedestrian, and evacuation models, see the works in [43, 49, 57, 71, 80, 81, 85, 99, 101, 102, 114, 115, 124, 126, 137, 138, 203, 242, 244, 247, 282, 297, 310, 314, 324, 329, 331, 333, 334, 340, 355, 372, 386, 405, 406, 417, 498, 502, 503, 531, 537, 546–548] and the references therein. Recent advances in infinite population MFTGs can be found below. Correlated equilibria with progressive strategies now offer a richer solution framework than traditional Nash equilibria [68], while finite-state ergodic Markovian settings provide asymptotic insights into long-run behavior [108]. Learning in bounded rationality regimes has gained traction through equilibrium learning algorithms [148] and variational inequality formulations supporting decentralized independent learning [521]. Inverse problems have taken center stage: State anomalies are detected from strategic data [306], boundary observations recover stationary game dynamics [123], Cauchy data resolve hidden states [307], and multipopulation cost structures are shown to be uniquely identifiable [370]. Linear-quadratic models continue to guide theoretical insights, including extensions to discounting [380] and optimization under extended dynamics [60]. Customized applications to energy grid regulation via virtual power plants [536] and systemic risk modeling in large banking networks [94]. Robustness and approximation in discrete-time games under discounted cost criteria have been addressed in [28]. Inverse problems remain central, with progress in identifying unknown dynamics via Carleman estimates under common noise [298] and learning surrogate potentials from data using Gaussian processes to regularize ill-posedness [535]. Singular controls under uncertainty extend games to ambiguous risk settings, relevant for robust financial design [164]. Stochastic control with delays and jump diffusions further reflects real-world complexity in agent dynamics [295]. Structural innovations include games with variable entry times for modeling speculative bubbles [415], deterministic jump dynamics with mixed variational structures [144], and second-order models with controlled diffusion [238]. Numerical methods have also matured, with mesh-independent schemes advancing computational tractability in second-order potential population games [308]. Equilibrium refinements have been explored, notably through coarse correlated equilibria in linear-quadratic settings with applications to emission abatement [79]. In aerospace systems, games have been applied to missile interception with optimal pursuit strategies [35], while financial models now include excess return-driven fund manager interactions [201] and sustainable investment switching [9]. Distributed optimization in power systems via games demonstrates scalable grid control [519]. Progress has been made on global well-posedness to degenerate master equations [40] and Lipschitz stability in inverse coefficient recovery [239]. Nonlocal spatial dynamics and resource accumulation expand classical models [191], and critical mass blowup behavior in decreasing cost games reveals delicate solution structures [105]. Partial observation in stochastic linear-quadratic games [294], last-iterate convergence in regularized graphon settings [140], and hybrid deep learning methods for finite-horizon problems [543] contribute to robust numerical and learning-based solutions. Risk-sensitive games have been addressed via differential dynamic programming [411] and have been deployed for IoT cybersecurity [385]. Finite element methods have been adapted to time-dependent games with non-differentiable Hamiltonians [348], while kinetic-type games with non-separable local dynamics offer new analytical challenges [19]. Models involving Poisson jumps and heterogeneous populations of dominating and negligible agents deepen stochastic formulations [235, 514], complemented by frameworks for nonzero sum games under partial information [342]. Opinion dynamics in social systems have been modeled via large-scale games incorporating structured group behavior [371], and ergodic Choquard games exhibit mass concentration phenomena [65]. On infrastructure and sustainability fronts, optimal control of EV fleets [226], UAV trajectory coordination [404], and tourism modeling via Jacobi processes [525] uses game theory for policy design. Topological bifurcations [315], sticky transition behaviors in network games [66], and the turnpike effect in group behaviors [493] offer novel structural insights. Strategic firm competition games reveal the existence of multiple equilibria due to complementarities [122], and weak-strong uniqueness results enhance mathematical rigor in solution theory [165]. Advances in approximating multi-population discrete-time games under various payoff criteria further improve connections to finite-agent systems [509]. [219] develops second-order Bellman–Isaacs equations on the Wasserstein space for stochastic differential games, while [545] models asset penalization with green premia using a mean-field game framework. Convergence of large-population Stackelberg games is analyzed in [381]. Asset pricing with habit formation is explored in [174], and collective coordination is addressed through synchronization games in [229]. [132] characterizes zero-sum Dynkin games under common noise. Agency problems with moral hazard and accident risk are modeled in [325]. The limits of mean-field modeling in path switching and herd behavior are emphasized in [362], while [316] investigates delayed games with singular controls. A reinforcement-learning-based UAV swarm interception framework is proposed in [542], and governance dynamics between cooperation and competition are analyzed in [104]. Optimal regulation under state-dependent costs is studied in [512]. Scalable mean-field Q-learning is introduced in [532], and entropic minimax games are developed via best-response flows in [285]. Linear-quadratic control under partial information is treated in [163], and Stackelberg strategies with terminal perturbation in [100]. Delayed networked games are examined in [52], and infinite-dimensional mean-field control is advanced in [62]. Reinforcement learning in continuous time is applied in [507], and feedback control under partial observation in [346]. [225] addresses optimal stopping under common noise and non-Markov dynamics. Robust reinsurance and investment with model uncertainty are formulated in [202].

Mean-field variational inference representations are analyzed in [194]. Constrained terminal-state games with common noise are developed in [234], and multi-group LQ differential games are discussed in [25, 42] which proposes a discrete scheme for price-formation games, and [21] explores neural network models [466, 478] for epidemic scenario control.

Chapter 2
Signal Quality-Based CSMA Networks

In the preceding Chap. 1, our focus was on examining the evacuation procedures for multilevel building structures. We explored how the uncoordinated utilization of staircases and exits can result in the dispersion of evacuees and impact the evacuation duration for every individual within a multilevel building. This concept of utilizing resources, akin to the way stairs, movable areas, and exit doors are utilized during evacuations, bears a resemblance to a challenge encountered in wireless networks. Here, the resources in question pertain to channels, spectrum, and antennas.

This chapter presents signal-to-interference-plus noise ratio-based carrier sense multiple access (CSMA) protocols when the network size grows. It starts with modeling the backoff stage as an interactive Markov decision process. Alternative strategies to the IEEE 802.11 exponential backoff are proposed. As the network size grows, a mean-field dynamics emerges (see [125]). The mean-field limit is a Kolmogorov forward equation whose evolution provides an idea about proportion of users per stage of the backoff. Each generic user interacts with all other users connecting around in the same access point. Therefore, each generic user is facing an evolving mean-field vector. The chapter explains how mean-field strategies can be used to reduce collisions and to stabilize the interactive system [460, 464]. See also Chap. 9 for similar methods.

The rest of the chapter is structured as follows. Section 2.1 presents some random access control method as a game. Section 2.2 presents cognitive carrier sense multiple access networks as a game. Mean-field asymptotic regimes are analyzed in Sect. 2.3. Section 2.4 presents a proof of convergence to a mean-field limit dynamics when the number of users grows.

T. Başar et al., *Mean-Field-Type Game Theory II*, Static & Dynamic Game Theory: Foundations & Applications, https://doi.org/10.1007/978-3-032-07031-9_2

2.1 Random Access Control

What Is Aloha?

Aloha is a word in the Hawaiian language. It is mostly used in the sense of "hello." Aloha is a multiple access protocol at the datalink layer and proposes how multiple terminals access the medium without interference or collision. The protocol was developed by Norman Abramson and his colleagues at the University of Hawaii.

The Aloha protocol allows every system to send a frame if it is ready to send. But when a collision occurs, the node will wait for a random amount time and then send the frame again. The process continues until the node has sent all the frames. Since the nodes send their frames without sensing the medium, there is a high probability for collisions to occur. The maximum success rate or throughput that can be achieved with Aloha protocol is only $\frac{1}{2e}$.

The major problem with multiple access is allocating the channel between the users:

- Nodes do not know when other nodes will have data to send.
- Need to coordinate the transmissions.

The advantages of the unslotted Aloha are simplicity and possibility of unequal length packets.

What Is Slotted Aloha?

Time is divided into "slots" of one packet duration, e.g., fixed size packets:

- When a node has a packet to send, it waits until the start of the next slot to send it, which requires synchronization.
- If no other nodes attempt transmission during that slot, the transmission is successful:
  - Otherwise "collision."
  - Collided packets are retransmitted after a random delay.

The success of one terminal's transmission can affect the success of others. If two devices transmit at the same time on the same channel, they can collide, resulting in data loss and wasted resources. Therefore, each terminal's decision depends on what it anticipates other terminals will do, creating an interdependent relationship similar to strategic games where each player's action affects the outcome for all players. This is clearly a strategic decision-making problem.

Under the slotted Aloha, the capacity is slightly improved. The asymptotic capacity of Slotted Aloha is

$$\frac{1}{e} = \lim_{I\to\infty} I.\frac{1}{I}\left(1-\frac{1}{I}\right)^{I-1}.$$

What Is CSMA?

Carrier-sense multiple access (CSMA) is a probabilistic medium access control (MAC) protocol in which a node verifies the absence of other traffic before transmitting on a shared transmission medium, such as an electrical bus, or a band of the electromagnetic spectrum.

Carrier Sense describes the fact that a transmitter uses feedback from a receiver that detects a carrier wave before trying to send. That is, it tries to detect the presence of an encoded signal from another station before attempting to transmit. If a carrier is sensed, the station waits for the transmission in progress to end before initiating its own transmission.

Multiple Access describes the fact that multiple stations send and receive on the medium. Transmissions by one node are generally received by all other stations using the medium.

Below simple modifications of CSMA are presented.

p-Persistent CSMA

Let $p \in (0, 1)$. In a p-persistent CSMA, when the sender is ready to send data, it checks continually if the medium is busy. If the medium becomes idle, the sender transmits a frame with a probability p. If the station chooses not to transmit (the probability of this event is $1 - p$), the sender waits until the next available time slot and transmits again with the same probability p. This process repeats until the frame is sent or some other sender stops transmitting. In the latter case, the sender monitors the channel and, when idle, transmits with a probability p, and so on. p-persistent CSMA is used in carrier-sense multiple access with collision avoidance (CSMA/CA) systems including WiFi and other packet radio systems.

CSMA/CD

Carrier-sense multiple access with collision detection (CSMA/CD) is used to improve CSMA performance by terminating transmission as soon as a collision is detected and reducing the probability of a second collision on retry.

CSMA/CA

Carrier-sense multiple access with collision avoidance (CSMA/CA) is used to improve the performance of CSMA by attempting to be less "greedy" on the channel. If the channel is sensed busy before transmission, then the transmission is deferred for a "random" interval. This reduces the probability of collisions on the channel.

Table 2.1 Aloha game

Resource	Medium
Decision-Makers	Users
Actions	Transmit or Wait
Mixed strategy j.	$(x_j, 1 - x_j), \ 0 \leq x_j \leq 1$
Availability.	$p \in (0, 1)$ is the probability that the medium is available for access
Payoff of j	$u_j(x) = px_j \prod_{i \neq j}(1 - x_i)$
Symmetry	This game is clearly a symmetric game. However, symmetric equilibria are inefficient and do not provide successful transmissions to the medium.
Asymmetry	Asymmetric equilibria such as (1, 0), (0, 1), ..., (0, 1), in which one of the users transmits and all the others wait is a Pareto optimal solution and also a global optimum.
Asymmetry	These asymmetric equilibria are unfair in the sense that only one of the users gets access to the medium, and the others wait.
Asymmetry	In presence of three or more users the other equilibria such as (1, 0), (1, 0), ... (1, 0), (0, 1), ..., (0, 1), i.e., where at least two users decide to transmit, provide a bad equilibrium payoff for everyone as it leads to collision and a waste of slot opportunity.
Equilibria	There is an infinite number of risk-neutral equilibria in this case .
Energy consumption	In order to capture the waste of energy in the retransmissions of the data, an energy consumption cost will be introduced.
New Payoff of j	$u_j(x) = px_j \prod_{i \neq j}(1 - x_i) - cx_j, \ \ c \in (0, 1), p \in (0, 1), c < p$
Energy-based Aloha	By doing so, the strategy profiles in which two or more users transmit will not constitute an equilibria as a deviating user will get 0 which is better than $-c$.
Equilibria	Thanks to the introduction of energy cost, linked to the remaining battery of the device, the number of risk-neutral equilibria is sufficiently reduced. The exact number of risk-neutral equilibria in pure and mixed strategies is $2^I - 1$ where I is the total number of users
Random number	In practice, the number of actively interacting users in the network is a random variable [470]

Aloha Protocol as a Game

Table 2.1 summarizes an Aloha game and the risk-neutral equilibrium structure.

2.2 IEEE 802.11 CSMA-Based Cognitive Networks

IEEE 802.11 CSMA-based cognitive networks can be described as a game where multiple nodes strategically sense and access the wireless medium, aiming to optimize their own transmission opportunities while avoiding interference and achieving efficient spectrum utilization. Next we introduce the backoff process in IEEE 802.11.

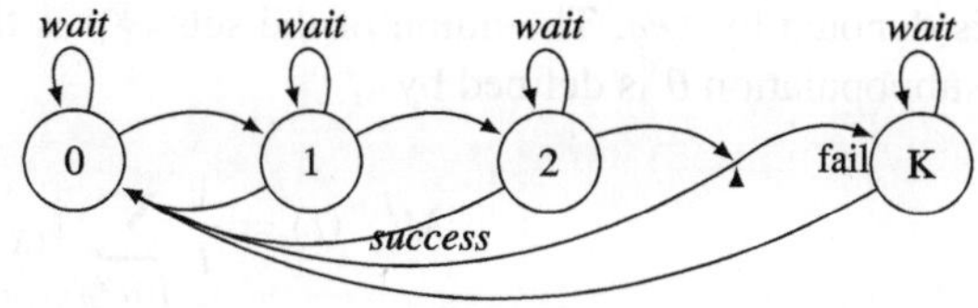

Fig. 2.1 Modeling the backoff: State transitions of the interaction Markov decision process

Backoff in IEEE 802.11:
The backoff process in IEEE 802.11 is governed by a Markov decision process (see [125]) if the duration of per-stage backoff is taken into account:

- Every node with backoff state k attempts a transmission with probability $u_k(.) \in (0, 1)$ at every time slot t.
- If it succeeds, the backoff state k of the node changes to 0.
- Otherwise, the state k changes to $(k + 1)$ modulo $(K + 1)$, where $K \geq 1$ is the maximum backoff state.

The state transitions of the backoff are displayed in Fig. 2.1.

When the number of nodes is large, the validity of the decoupling assumption (nodes independence at the asymptotic) should be justified, not just by a simple fixed point method but a deep study of the stability of the ordinary differential equations. This is because the existence and uniqueness of a rest point do not imply that the dynamics converges to this fixed point. We illustrate this statement via a cycle limit behavior in a heterogeneous CSMA system under control parameters. To construct a cycling behavior of the mean-field limit, we propose several states and classes. We show that the mean-field limit depends on the control parameters and the uncontrolled mean-field limit system can be suboptimal depending on the performance metric. Hence, the control parameters give new insights and help in understanding the behavior of the mean-field limit.

2.3 Mean-Field Asymptotics

We consider cognitive network with two classes of users: primary users and secondary users (the classes are based on the differentiation of contention window or arbitration inter-frame space). The primary users have high priority in the sense that they have reserved time slots for them, but also they can attempt a transmission in the other time slots (the common time slots). Both classes of users can access the channel during common slots. The backoff states of the secondary users are suspended during the reserved time slots. Denote by I the total size of the population of users/nodes. The population is divided into several subpopulations. Let I_θ be the size of subpopulation θ. Hence, $\sum_\theta I_\theta = I$. The set of users in subpopulation θ

is denoted by $\mathscr{P}_\theta$. The union of the sets $\mathscr{P}_\theta$ is the set of users. The mean field of subpopulation θ is defined by

$$M_k^{I,\theta}(t) = \frac{1}{I} \sum_{j \in \mathscr{P}_\theta} \mathbb{1}_{\{X_j^I(t)=k\}},$$

where $X_j^I(t)$ is the backoff state of user j at time t and $\theta \in \{H, L\}$. The per-class decision processes are u_k^θ, $k \leq K^\theta$.

Controlled Drift Limit for Two Classes

If there is no differentiation in the service, the process $M_k^{I,\theta}(It)$ converges in probability to $m_k^\theta(t)$, which is the solution of the system of ordinary differential equations (Kolmogorov equations):

$$\frac{dm_0^\theta}{dt}(t) = \bar{v}^\theta(t)(1-\gamma(t)) - v_0^\theta m_0^\theta(t) + v_{K^\theta}^\theta m_{K^\theta}^\theta(t)\gamma(t)$$

$$\frac{dm_k^\theta}{dt}(t) = v_{k-1}^\theta m_{k-1}^\theta(t)\gamma(t) - v_k^\theta m_k^\theta(t), \quad k = 1, \ldots, K^\theta,$$

where $v_k^\theta = \lim_I (I u_k^{I,\theta})$, $\bar{v}^\theta(t) = \sum_{k=0}^{K^\theta} v_k^\theta m_k^\theta(t)$, and $\gamma(t) = 1 - e^{-\sum_\theta \bar{v}^\theta}$ for $\theta \in \{H, L\}$.

Here $\bar{v}^\theta(t)$ is the mean-field limit of the average attempt rate and $\gamma(t)$ is that of the collision probability.

Unstable Behavior

In this case, the population of each class is $I^H = I_L = 640$. The numbers of backoff stages are equal, $K^H + 1 = K^L + 1 = 21$. The attempt probabilities are

$$(u_0^H, u_1^H, \ldots, u_{20}^H) = (\frac{1}{2400}, \frac{1}{480}, \frac{m}{40}, \ldots, \frac{m^{19}}{40})$$

and

$$(u_0^L, u_1^L, \ldots, u_{20}^L) = (\frac{1}{3840}, \frac{1}{64}, \frac{1}{64}, \ldots, \frac{1}{64}),$$

where $m = 4/5$.

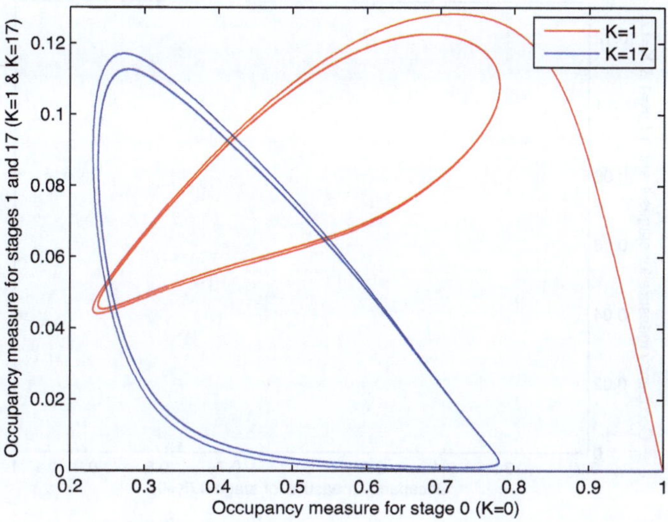

Fig. 2.2 Unstable mean-field

Figure 2.2 illustrates a limit cycle. The dynamics does not converge to a point. It goes to a set.

Stable Configuration
By setting the value of m to $m = 1/2$, we obtain a stable point as illustrated in Fig. 2.3.

2.4 Mean-Field Convergence

In this subsection, we provide a detailed proof of mean-field convergence to deterministic dynamics under specific time-scaling techniques.

Proposition 2.4.0.1 *The process* $M^I_{(\theta,y_\theta)}(It)$ *converges in probability to* $m_{\theta,y_\theta}(t)$ *which is the solution of the system of ODEs:*

$$\begin{cases} \dot{m}_{\theta,0} = \omega_{\bar{s}}\left[\bar{u}_\theta(t)(1-\gamma(t)) - u_{\theta,0}(t)m_{\theta,0} + u_{\theta,K_\theta}(t)m_{\theta,K_\theta}\gamma(t)\right] \\ \qquad +(1-\omega_{\bar{s}})\left(-u_{\theta,0}(t)m_{\theta,0} + u_{\theta,K_\theta}(t)m_{\theta,K_\theta}\right) \\ \dot{m}_{\theta,y_\theta} = \omega_{\bar{s}}\left[u_{\theta,y_\theta-1}(t)m_{\theta,y_\theta-1}\gamma(t) - u_{\theta,y_\theta}(t)m_{\theta,y_\theta}\right] \\ \qquad +(1-\omega_{\bar{s}})\left(u_{\theta,y_\theta-1}(t)m_{\theta,y_\theta-1} - u_{\theta,y_\theta}(t)m_{\theta,y_\theta}\right), \\ \qquad y_\theta \in \{1,\ldots,K_\theta\}, \theta \in \Theta, \end{cases}$$

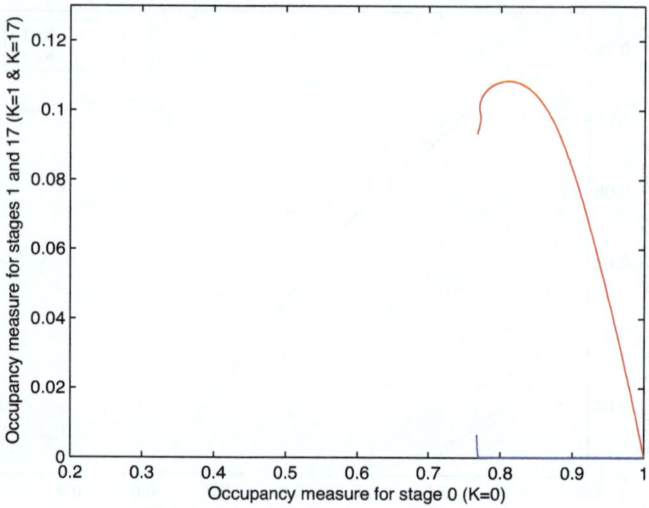

Fig. 2.3 Stable mean field

where $u_{\theta,y_\theta}(t)$ *is the strategy of a player from class* θ *at time* t *in backoff state* y_θ, $\bar{u}_\theta(t) = \sum_{y_\theta=0}^{K_\theta} u_{\theta,y_\theta}(t)m_{\theta,y_\theta}(t)$ *and* $\gamma(t) = 1 - e^{-\sum_\theta \bar{u}_\theta(t)}$ *for* $\theta \in \Theta$.

Here $\bar{u}_\theta(t)$ is the mean-field limit of the average attempt rate, $\gamma(t)$ is the blocking probability, and $\omega_{\bar{s}}$ is the probability of having a good resource state.

The equations of the system of ODEs can be intuitively understood. For each channel configuration, the first term and second term on the right-hand side are, respectively, the inflow caused by blocking probability in the $y_\theta - 1$ backoff stage and the outflow caused by attempts in the y_θ backoff stage times the probability of the corresponding resource state.

To prove the mean-field convergence, we use the following steps: To simplify the notation, we will suppress the dependency on s at many places:

- First, compute the blocking probability for a generic player in each class: Show that this is the probability that he/she transmits times the probability that *at least* one of the others transmits in the same slot in the same interference range (same class or other classes).
- Denote by $1-\xi_x^{I,\theta}(u, m, s)$ the probability that *at least* one of the others transmits in the same slot in the same interference range when the generic user backoff is at level x. Then, the transition probabilities $q_{x,x'}^{I,\theta}(u, m, s)$ of the backoff state of a generic user are given by

$$\begin{cases}
\text{for } x = 0: \\
q^I_{0,1}(t, u^I(t), m(t), \bar{s}) = u^I_0(t)(1 - \xi^I_0(t, u^I(t), m(t), \bar{s})) \\
q^I_{0,0}(t, u^I(t), m(t), \bar{s}) = 1 - u^I_0(t) + u^I_0(t)\xi^I_0(t, u^I(t), m(t), \bar{s}) \\
x \geq 2, q^I_{0,x}(t, u^I(t), m(t), \bar{s}) = 0 \\
\\
\text{for } x \in \{1, \ldots, K-1\}: \\
q^I_{x,x+1}(t, u^I(t), m(t), \bar{s}) = u^I_x(t)(1 - \xi^I_x(t, u^I(t), m(t), \bar{s})) \\
q^I_{x,x}(t, u^I(t), m(t), \bar{s}) = 1 - u^I_x(t) \\
q^I_{x,0}(t, u^I(t), m(t), \bar{s}) = u^I_x(t)\xi^I_x(t, u^I(t), m(t), \bar{s}) \\
x' \notin \{0, x, x+1\}, q^I_{x,x'}(t, u^I(t), m(t), \bar{s}, \bar{s}) = 0 \\
\\
\text{for } x = K: \\
q^I_{K,0}(t, u^I(t), m(t), \bar{s}) = u^I_K(t) \\
q^I_{K,K}(t, u^I(t), m(t), \bar{s}) = (1 - u^I_K(t)) \\
x \notin \{0, K\},\ q^I_{K,x}(t, u^I(t), m(t), \bar{s}) = 0 \\
u^I_x(t) = \frac{u^I_x(t)}{\phi(I)} \\
\xi^I_x(t, u^I(t), m(t), \bar{s}) = (1 - u^I_x(t))^{Im_x(t)-1} \prod_{x' \neq x} (1 - u^I_{x'}(t))^{Im_{x'}(t)}.
\end{cases}$$

- We take the function ϕ such that $\frac{\phi(I)}{I}$ converges to some $\epsilon_1 > 0$. For example, the family of sequences

$$\phi_{\epsilon_1,\epsilon_2,\epsilon_3}(I) = \epsilon_1 I + \epsilon_2 \log(I) + \epsilon_3,\ \epsilon_1 > 0,\ \epsilon_2, \epsilon_3 \geq 0$$

is appropriated. Using the fact that $(1 - \frac{u_{\theta,y_\theta}}{I})^I \longrightarrow e^{-u_{\theta,y_\theta}}$, we can prove that $\xi^{I,\theta}_x(u, m, s)$ has a limit when I goes to infinity. For $u^I = \frac{u}{\phi_{\epsilon_1,\epsilon_2,\epsilon_3}(I)}$, the limit is given by

$$e^{-\sum_x u_x(t)m_x(t)} = e^{-\sum_{\theta\in\Theta}\sum_{y_\theta} u_{\theta;y_\theta}(t)m_{\theta,y_\theta}(t)}.$$

Define block-diagonal the matrix

$$Q^I(u, m, s) = diag[Q^{I,\theta}(u, m, s)]_{\theta\in\Theta};$$

It is easy to see that $I[q^I_{x,x'}(u, m, s) - \mathbb{1}_{\{x=x'\}}]$ has a limit when I goes to infinity. In addition, the resulting function is Lipschitz. We deduce that $Im[Q^I(u, m, s) - I]$ has a limit which is denoted by $f(u, m, s)$. The vector $f(u, m, \bar{s})$ is given by

$$\begin{cases}
f_{\theta,0} = \bar{u}_\theta(t)(1-\gamma(t)) - u_{\theta,0}(t)m_{\theta,0}(t) + u_{\theta,K_\theta}(t)m_{\theta,K_\theta}(t)\gamma(t) \\
f_{\theta,y_\theta} = u_{\theta,y_\theta-1}(t)m_{\theta,y_\theta-1}(t)\gamma(t) - u_{\theta,y_\theta}(t)m_{\theta,y_\theta}(t), \\
y_\theta \in \{1, \ldots, K_\theta\}, \theta \in \Theta.
\end{cases}$$

The vector $\tilde{f}$ is obtained by taking the expectation over the channel state distribution.

- Following similar lines as in Kurtz's theorem, we show that $M^I(\lfloor It \rfloor)$ converges in probability to $m[u, m_0](t)$ and for any $T' > 0$ and $\epsilon > 0$,

$$\mathbb{P}\left(\sup_{t\in[0,T']} \left| M^I(It) - m[u, m_0](t) \right| > \epsilon \middle| M^I(0)=m_0, u^I(t)=u \right)$$

goes to 0 when I goes to infinity, where

$$\frac{d}{dt} m_{\theta,y_\theta}(t) = \tilde{f}_{\theta,y_\theta}(u(t), m(t)), \; m(0) = m_0.$$

Since $M^I(.)$ is a discrete Markov decision process over a finite set, we denote by Λ_I the set of all possible jumps of $M^I(t)$. An element of Λ_I has the form $m' - m$ where $M^I(t) = m$. Then, for a given s, $M^I(t)$ can be written as a function of the jumps:

$$M^I(t) = M^I(0) + \sum_{k=0}^{t-1} \sum_{\lambda_I \in \Lambda_I} \lambda_I \mathbb{1}_{\{M^I(k+1)-M^I(k)=\lambda_I\}}.$$

Let $L_{M^I(k),\lambda_I}$ be the probability to have a jump λ_I from $M^I(k)$. Then, $M^I(t)$ can be rewritten as

$$\begin{aligned} M^I(t) &= M^I(0) + \sum_{k=0}^{t-1} \sum_{\lambda_I \in \Lambda_I} \lambda_I \mathbb{1}_{\{M^I(k+1)-M^I(k)=\lambda_I\}} \\ &\quad + \sum_{k=0}^{t-1} \sum_{\lambda_I \in \Lambda_I} \lambda_I L_{M^I(k),\lambda_I} - \sum_{k=0}^{t-1} \sum_{\lambda_I \in \Lambda_I} \lambda_I L_{M^I(k),\lambda_I} \\ &= M^I(0) + \sum_{k=0}^{t-1} \varsigma^I(k) + \sum_{k=0}^{t-1} f^I(u, M^I(k)), \end{aligned}$$

where

$$\varsigma^I(k) = \sum_{\lambda_I \in \Lambda_I} \lambda_I [\mathbb{1}_{\{M^I(k+1)-M^I(k)=\lambda_I\}} - L_{M^I(k),\lambda_I}]$$

and $f^I(u, m, s)$ is the expected drift under the strategy u. Let $\mathscr{F}_t^I = \sigma(M^I(0), \ldots, M^I(t))$. It is clear that the controlled process $z^I(t) = \sum_{k=0}^{t-1} \varsigma^I(k)$ is a martingale with respect to $\mathscr{F}_t^I$.

- Substituting t by $\lfloor It \rfloor$, we get

$$M^I(\llcorner It \lrcorner) = M^I(0) + z^I(\llcorner It \lrcorner) + \sum_{k=0}^{\llcorner It \lrcorner - 1} f^I(u, M^I(k)).$$

- Using the fact that $If^I(u, m, s) = Im[Q^I(u, m, s) - I]$ converges to f where the function $m \longmapsto f(u, m, s)$ is locality Lipschitz in m and u,, we have that the Cauchy problem associated with the ordinary differential equation (ODE)

$$\left\{ \dot{m}(t) = f(u(t), m(t)),\ m(0) = m_0 \in \Delta(\mathscr{X}),\right.$$

has a unique solution which is $m(t) := m[u, m_0](t)$.

Next, we evaluate the gap between $M^I(\llcorner It \lrcorner)$ and $m[u, m_0](t)$.

$$\begin{aligned} M^I(\llcorner It \lrcorner) - m(t) &= M^I(0) - m_0 + z^I(\llcorner It \lrcorner) \\ &\quad + \sum_{k=0}^{\llcorner It \lrcorner - 1} f^I(u, M^I(k)) - \int_0^t f(u, m(t'))dt' \\ &= M^I(0) - m_0 + z^I(\llcorner It \lrcorner) + B_{I,t}, \end{aligned}$$

where

$$B_{I,t} = \sum_{k=0}^{\llcorner It \lrcorner - 1} f^I(u, M^I(k)) - \int_0^t f(u, m(t'))dt'.$$

The stochastic term $B_{I,t}$ can be decomposed in three parts:

$$B_{I,t} = \sum_{j=1}^{3} B^j_{I,t},$$

where

$$B^1_{I,t} = \sum_{k=0}^{\llcorner It \lrcorner - 1} f^I(u, M^I(k)) - \sum_{k=0}^{\llcorner It \lrcorner - 1} \frac{1}{I} f(u, M^I(k)),$$

$$B^2_{I,t} = \sum_{k=0}^{\llcorner It \lrcorner - 1} \frac{1}{I} f(u, M^I(k)) - \sum_{k=0}^{\llcorner It \lrcorner - 1} \frac{1}{I} f(u, m(\frac{k}{I})),$$

$$B^3_{I,t} = \sum_{k=0}^{\llcorner It \lrcorner - 1} \frac{1}{I} f(u, m(\frac{k}{I})) - \int_0^t f(u, m(t'))\, dt'.$$

We estimate each of the processes $B^j_{I,t}$.

- **Bound for the first term $B^1_{I,t}$.** Let $\bar{\epsilon} > 0$. Using the fact If^I converges to f, for I sufficiency large, one has

$$\| If^I(u, M^I(k)) - f(u, M^I(k)) \| \leq \bar{\epsilon}.$$

Thus,

$$\begin{aligned}
\| B^1_{I,t} \| &= \| \sum_{k=0}^{\lfloor It \rfloor -1} f^I(u, M^I(k)) - \sum_{k=0}^{\lfloor It \rfloor -1} \frac{1}{I} f(u, M^I(k)) \| \\
&\leq \sum_{k=0}^{\lfloor It \rfloor -1} \| f^I(u, M^I(k)) - \frac{1}{I} f(u, M^I(k)) \| \\
&= \frac{1}{I} \sum_{k=0}^{\lfloor It \rfloor -1} \| If^I(u, M^I(k)) - f(u, M^I(k)) \| \leq \frac{1}{I} \bar{\epsilon} \lfloor It \rfloor.
\end{aligned}$$

Hence, $\| B^1_{I,t} \| \leq \bar{\epsilon} \frac{\lfloor It \rfloor}{I}, \ \forall t$.

- **Bound for the third term $B^3_{I,t}$.** To estimate the term $B^3_{I,t}$,, we combine Lipschitz property of the function f relatively to m. Recall that f is c_0-Lipschiz in m if $\forall (m, m'),\ \| f(u, m) - f(u, m') \| \leq c_0 \| m - m' \|$. This implies most linear growth, i.e., there exists $c'_0 > 0$ such that $\| f(u, m) \| \leq c'_0 (1 + \| m \|)$, and the Cauchy problem starting at m_0 has a unique solution.

$$\begin{aligned}
\| B^3_{I,t} \| &= \| \sum_{k=0}^{\lfloor It \rfloor -1} \frac{1}{I} f(u, m(\frac{k}{I})) - \int_0^t f(u, m(t'))\, dt' \| \\
&\leq \| - \int_{\frac{\lfloor It \rfloor}{I}}^t f(u, m(t'))\, dt' + \sum_{k=0}^{\lfloor It \rfloor -1} \frac{1}{I} f(u, m(\frac{k}{I})) \\
&\quad - \int_{\frac{k}{I}}^{\frac{k+1}{I}} f(u, m(t'))\, dt' \| \\
&\leq \| \int_{\frac{\lfloor It \rfloor -1}{I}}^t f(u, m(t'))\, dt' \| + \sum_{k=0}^{\lfloor It \rfloor -1} \| \frac{1}{I} f(u, m(\frac{k}{I})) \\
&\quad - \int_{\frac{k}{I}}^{\frac{k+1}{I}} f(u, m(t'))\, dt' \| .
\end{aligned}$$

There is $c_3(T) > 0$ such that the first part is less than $\frac{c_3(T)}{I}$ by Lebesgue integrability of f. The term in the sum of the second part is less than $\frac{c'_3(T)}{I^2}$. By summation, one gets

$$\| B^3_{I,t} \| \le \frac{c_3'(T)}{I^2} \times \lfloor It \rfloor + \frac{c_3(T)}{I} \longrightarrow 0.$$

- **Bound for the second term $B^2_{I,t}$.** We use the Lipschitz property to estimate the norm of $B^2_{I,t}$.

$$\begin{aligned} \| B^2_{I,t} \| &= \| \textstyle\sum_{k=0}^{\lfloor It \rfloor -1} \frac{1}{I} f(u, M^I(k)) - \sum_{k=0}^{\lfloor It \rfloor -1} \frac{1}{I} f(u, m(\frac{k}{I})) \| \\ &\le \textstyle\sum_{k=0}^{\lfloor It \rfloor -1} \frac{1}{I} \| f(u, M^I(k)) - f(u, m(\frac{k}{I})) \| \\ &\le \textstyle\frac{c_0}{I} \sum_{k=0}^{\lfloor It \rfloor -1} \| M^I(k) - m(\frac{k}{I}) \| . \end{aligned}$$

- Consider the term $\| M^I(\lfloor It \rfloor) - m(t) \|$.

$$\begin{aligned} \| M^I(\lfloor It \rfloor) - m(t) \| &= \| M^I(\lfloor It \rfloor) - m(\tfrac{\lfloor It \rfloor}{I}) + m(\tfrac{\lfloor It \rfloor}{I}) - m(t) \| \\ &\ge \| M^I(\lfloor It \rfloor) - m(\tfrac{\lfloor It \rfloor}{I}) \| - \| m(\tfrac{\lfloor It \rfloor}{I}) - m(t) \| . \end{aligned}$$

The expression $\| m(\frac{\lfloor It \rfloor}{I}) - m(t) \|$ is in order of $\frac{c_2(T)}{I}$. Suppose that $M^I(0) \longrightarrow m_0$ in probability. Then, for I sufficiently large, $\forall t \in [0, T]$, one has

$$M^I(\lfloor It \rfloor) - m(\lfloor It \rfloor) = M^I(0) - m_0 + z^I(\lfloor It \rfloor) + B_{I,t}.$$

The norm can be bounded as

$$\begin{aligned} \| M^I(\lfloor It \rfloor) - m(t) \| &\le \bar{\epsilon} + \| z^I(\lfloor It \rfloor) \| + \| B_{n,t} \| \\ &\le 4\bar{\epsilon} + \| z^I(\lfloor It \rfloor) \| + \frac{c_0}{I} \sum_{k=0}^{\lfloor It \rfloor -1} \| M^I(k) - m(\frac{k}{I}) \| . \end{aligned}$$

Next, we estimate the norm of the process $z^I(t)$ by using Burkholder inequality for zero-mean martingales.

$$\begin{aligned} \mathbb{E}\left(\sup_{t \in [0,T]} \| z^I(\lfloor It \rfloor) \|^2 \right) &= \mathbb{E}\left(\sup_{[0,T]} \| \sum_{k=0}^{\lfloor It \rfloor -1} \varsigma^I(k) \|^2 \right) \\ &\le c_1(T) \sum_{k=0}^{\lfloor It \rfloor -1} \mathbb{E} \| \varsigma^I(k) \|^2 . \end{aligned} \tag{2.1}$$

We check that for our access control problem if each player strategy u^I is in order of $O(\frac{1}{I})$,, then we can distinguish two classes: (i) number of jumps after a successful transmission and (ii) number of jumps after a collision.

$$\mathbb{E}\| \ \varsigma^I(k) \ \|^2 = \mathbb{E} \ \|\sum_{\lambda_I \in \Lambda_I} \lambda_I [\mathbb{1}_{\{M^I(k+1)-M^I(k)=\lambda_I\}} - L_{M^I(k),\lambda_I}] \ \|$$
$$\leq \|\sum_{\lambda_I \in \Lambda_I} \lambda_I [\mathbb{1}_{\{M^I(k+1)-M^I(k)=\lambda_I\}} \ \|^2 + \|\sum_{\lambda_I \in \Lambda_I} L_{M^I(k),\lambda_I} \ \|^2$$
$$+ \|\sum_{\lambda_I \in \Lambda_I} \lambda_I [\mathbb{1}_{\{M^I(k+1)-M^I(k)=\lambda_I\}} \ \|^2 + \| \ f^I(M^I(k)) \ \|^2 .$$

By taking the conditional expectation relative to $\mathcal{F}_t^I$,, each of the terms is in order of $\frac{c_1'(T)}{I^2}$. By combining with (2.1), we get

$$\mathbb{E}\left(\sup_{t\in[0,T]} \ \| \ z^I(\llcorner It \lrcorner) \ \|^2 \right) \leq c_1(T)\frac{c_1'(T)}{I^2} \llcorner IT \lrcorner. \tag{2.2}$$

This means that on a sample path where the random process $\sup_{t\in[0,T]} \ \| \ z^I(\llcorner It \lrcorner) \ \| \leq \bar{\epsilon}$, one gets

$$\| \ M^I(t) - m(\frac{t}{I}) \ \| \leq 5\bar{\epsilon} + \frac{c_0}{I}\sum_{k=0}^{t-1} \| \ M^I(k) - m(\frac{k}{I}) \ \| . \tag{2.3}$$

By taking $\Pi_t = \| \ M^I(t) - m(\frac{t}{I})) \ \| \geq 0, \epsilon = 5\bar{\epsilon} > 0, \bar{\mu}_k = \frac{c_0}{I} > 0$ one has $\Pi_t \leq \epsilon + \sum_{k=0}^{t-1} \bar{\mu}_k \Pi_k$. By the (discrete) Grönwall inequality, we get $\Pi_t \leq \epsilon e^{\sum_{k=0}^{t-1} \bar{\mu}_k} = \epsilon e^{\frac{c_0}{I} t}$.

- We deduce that

$$\mathbb{P}\left(\sup_{t\in[0,IT]} \ \| \ z^I(\llcorner It \lrcorner) \ \| \leq \bar{\epsilon}\right)$$
$$\leq \mathbb{P}\left(\sup_{t\in[0,IT]} \ \| \ M^I(t) - m(\tfrac{t}{I}) \ \| \leq \epsilon + \tfrac{c_0}{I}\sum_{k=0}^{IT-1} \ \| \ M^I(k) - m(\tfrac{k}{I}) \ \|\right)$$
$$\leq \mathbb{P}\left(\sup_{t\in[0,IT]} \ \| \ M^I(t) - m(\tfrac{t}{I}) \ \| \leq \epsilon e^{c_0 \frac{\llcorner IT \lrcorner}{I}}\right). \tag{2.4}$$

By (2.2), we know that $\mathbb{P}\left(\sup_{t\in[0,IT]} \ \| \ z^I(\llcorner It \lrcorner) \ \|^2 \leq \bar{\epsilon}\right) \longrightarrow 1$ when $I \longrightarrow +\infty$. Hence,

$$\mathbb{P}\left(\sup_{t\in[0,IT]} \ \| \ M^I(t) - m(\frac{t}{I}) \ \| > \epsilon e^{c_0 \frac{\llcorner IT \lrcorner}{I}} \right) \longrightarrow 0.$$

This completes the proof. ■

2.5 A Queueing Mean-Field Game in Wireless Networks

This section discusses queueing mean-field games in wireless networks.

Random access control algorithms have played an increasingly important role in the development of wired and wireless networks, but the performance and stability of these algorithms, such as slotted-Aloha, carrier-sense multiple access (CSMA), are still open problems. Distributed medium access control, starting from the first version of Abramson's Aloha to the most recent algorithms used in IEEE802.11, has enabled a rapid growth of wireless networks. They aim at efficiently and fairly sharing a resource among users, even though each user must decide independently (eventually after receiving some messages or listening, sensing channels) when and how to attempt to use the resource. Random access algorithms have generated a lot of research interest, especially recently in attempts to use multi-hop wireless networks to provide high-speed access to the internet with low-cost and low-energy consumption.

In this subsection, we restrict our attention to wireless networks where the resources are receivers, base station or access points and where users interact because of interference, i.e., interfering users cannot transmit simultaneously. There is a collision if another user (mobile) transmits at a higher power level at the same range of the receiver. Motivated by the interest of evolving dense networks, evolutionary game theory was found to provide an appropriate framework to study such networks. We provide asymptotic analysis of spatial random access based on mean-field interaction where interfering users share resources placed on a undirected graph using an Aloha-type access control. We consider several classes of users. Users are spatially distributed, and the arrival processes around each vertex are assumed to be independent.

The interactions between the users are *nonreciprocal* in the sense that the set of users causing interference differ from the set of those suffering from these interferences.

We consider a spatial nonreciprocal random access model described as follows:

Let us define a system composed of an undirected graph $G = (V, E)$ with finite vertices $V = \{1, 2, \ldots, K\}$, and E be the set of edges, and many classes of users distributed in 3D. For each vertex $i \in V$,, the one-hop neighborhood of i is denoted by

$$\mathscr{I}_i = \{i\} \cup \{j \in V,\ (i, j) \in E\}.$$

The set $\mathscr{I}_i$ corresponds to the set of 1-adjacent nodes to i. Time is slotted. During each time slot t, a random number Λ_t of users arrive in the system. Denote by Λ_t^i the number of users arriving at vertex i at time t. We assume that arrival process $\{\Lambda_t^i\}_t$ is i.i.d in time with expectation $\mathbb{E}(\Lambda_t^i) = \lambda_i$. Denote by W_t^i the random variable representing the number of users at vertex i at time t. A user at vertex i requires service with some probability. The user transmits with some probability $\frac{u_i}{\sum_{j \in \mathcal{I}_i} W_t^j}$. If B_t^i is the total number of users requiring service at time t, then the random variable B_t^i follows a binomial law with parameters W_t^i and $\frac{u_i}{\sum_{j \in \mathcal{I}_i} W_t^j}$.

$$B_t^i \sim \text{Bin}\left(W_t^i, \frac{u_i}{\sum_{j \in \mathcal{I}_i} W_t^j} \right).$$

There is a successful transmission at vertex i in a given time slot t if only one of the users transmits at i and no other user in $\mathcal{I}_i$ transmits at time t.

2.5.1 Nonreciprocal Interference

We describe below a wireless Aloha-like network with three intersecting ranges. All sources (players) have the same actions set (but their decisions can be different): transmit or wait. The number of transmitters with which a given transmitter interacts may vary from one transmitter to another. We do not need the reciprocity property: If transmitter Tx1 interacts with transmitter Tx2, we do not require the converse to hold. We provide below an example of multiple access and interference control to illustrate this nonreciprocity. We consider local interactions between transmitters; for each transmitter, there corresponds a receiver. We shall say that a transmitter Tx1 is subject to an interaction (interference) from transmitter Tx2 if the transmission from Tx2 overlaps that from Tx1 and provided that *the receiver of the transmission from Tx1 is within interference range of transmitter Tx2.*

Consider the example depicted in Fig. 2.4. It contains three receivers or destinations represented by squares, with ranges $\mathcal{I}_j,\ j \in \{1, 2, 3\}$. A transmission of a source i within a distance r of the receiver Bj causes interference to a transmission from a source $j \neq i$ to receiver Bj but also i causes interference to any transmitter located in the an intersection range of coverage (the destinations of i and j can be different). We see that sources in area $R_1 := \mathcal{I}_1 \backslash (R_{12} \cup R_{13} \cup R_{123})$ cause no interference to any transmitter from $R_{12} \cup R_{13} \cup R_{123}$ transmitting to B_2 or B_3. The transmitters located in area R_1 cause no interference to any transmitter located in areas R_2 or R_3. But the transmission from sources in R_1 suffers from interference from source located in R_{12}, R_{13}, R_{123}. The sources in R_1 interfere with each other at their common destination R_1. Thus, the interference is not reciprocal. A transmitter in areas $R_{12} \cup R_{13} \cup R_{23} \cup R_{123}$ plays simultaneously several one-shot games with a unidimensional action (transmit or not) and receives a vector of payoffs. Only the payoff obtained at the receiver in which the mobile transmits can be nonzero.

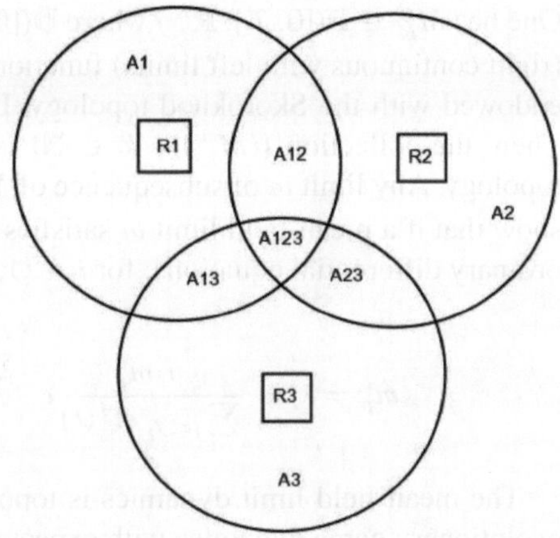

Fig. 2.4 Nonreciprocal interference control

2.5.1.1 Discrete-Time Evolution

One can write the evolution of W introduced above as

$$W_{t+1}^i = W_t^i - \mathbb{1}_{\{B_t^i=1\}} \prod_{j \in \mathscr{I}_i \backslash \{i\}} \mathbb{1}_{\{B_t^j=0\}} + \Lambda_t^i, \; i \in V.$$

Given $(W_t^1, \ldots, W_t^K)$,, the random variables $B_t^1, \ldots, B_t^K$ satisfy the decoupling conditions. Moreover, $(W_t^1, \ldots, W_t^K) \in \mathbb{N}^K$ is an irreducible Markov chain.

2.5.1.2 Mean-Field Limit

Let us study at the expected change in W in one slot when all users use the strategy profile $u_1, \ldots, u_K$. Define

$$f(u_1, \ldots, u_K, w) = \mathbb{E}[W_{t+1} - W_t \mid W_t = w]. \tag{2.5}$$

The drift is given by

$$f_i = \begin{cases} \lambda_i - \frac{u_i w^i}{\sum_{j \in \mathscr{I}_i} w^j}(1 - \frac{u_i}{\sum_{j \in \mathscr{I}_i} w^j})^{w^i-1} \prod_{j \in \mathscr{I}_i \backslash \{i\}} (1 - \frac{u_j}{\sum_{k \in \mathscr{I}_j} w^k})^{w^j} \text{ if } w^i > 0, \\ \lambda_i \text{ if } w^I = 0. \end{cases} \tag{2.6}$$

For any sequence (w_t^i) such that $w_t^i \longrightarrow +\infty$,, define the process

$$M_k^{w_t} = \frac{W(\lfloor k \sum_i w_t^i \rfloor)}{\sum_i w_t^i}.$$

One has $M_k^w \in \mathbb{D}([0, T], \mathbb{R}^K)$ where $\mathbb{D}([0, T], \mathbb{R}^K)$ is the space of cadlag or RCLL (right continuous with left limits) functions from the compact interval $[0, T]$ to $\mathbb{R}^K$ endowed with the Skorokhod topology. Let $(w_t^i)_t$ be a sequence going to infinity. Then, the collection $\{(M_k^w)_k,\ k \in \mathbb{N}\}$ is a compact closed set in the Skorokhod topology. Any limit m of subsequence of M_k^w is Lipschitz continuous. We can easily show that if a mean-field limit m satisfies $\forall i,\ \sum_{j\in\mathcal{I}_i} m^j(0) > 0$, then m solves the ordinary differential equations: for $i \in \{1, 2, \ldots, K\}$

$$\dot{m}_t^i = \lambda_i - \frac{u_i m_t^i}{\sum_{j\in\mathcal{I}_i} m^j(t)} e^{-\sum_{k\in\mathcal{I}_i\setminus\{i\}} \frac{u_k m^k(t)}{\sum_{j\in\mathcal{I}_k} m^j(t)}} e^{-\frac{u_i m_t^I}{\sum_{j\in\mathcal{I}_i} m_t^j}}.$$

The mean-field limit dynamics is topologically equivalent to the ray-projection evolutionary game dynamics with expected rate transition

$$\lambda_i - \frac{u_i m_t^i}{\sum_{j\in\mathcal{I}_i} m_t^j} e^{-\sum_{k\in\mathcal{I}_i} \frac{u_k m_t^k}{\sum_{j\in\mathcal{I}_k} m_t^j}}.$$

The rate region is non-convex and is given by

$$\mathcal{C} := \left\{ (r_1, \ldots, r_K) \in [0, 1]^K \mid r_i \le u_i \prod_{j\in\mathcal{I}_i\setminus\{i\}} (1 - u_j),\ u_i \in [0, 1], \right.$$
$$\left. i \in \{1, 2, \ldots, K\} \right\}.$$

2.5.2 *Medium Access Control with Queues*

The above problem can be extended to multiple queues as follows:

- There are I secondary users who are the potential participants. There are m channels which are resources. A resource can be available or not.
- When available and depending on its queue, each user can select a list of channels. A choice of user j is a subset $a_j \subset \{1, 2, \ldots, m\}$. It includes also the empty subset in which case $a_j = \{\}$ is interpreted as the waiting strategy. The action of j is own queue-dependent and constrained:

$$\sharp a_j = |a_j| \le q_j.$$

The constrained set of action of j is therefore

$$\mathscr{A}_j(q_j) := \{a_j \mid a_j \subset \{1, 2, \ldots, m\},\ \sharp a_j \leq q_j\}.$$

Note that $\mathscr{A}_j(q_j)$ is empty when $q_j = 0$.

- The instantaneous reward for user j at channel c is

$$r_{j,c}(a_1, \ldots, a_I) = \begin{cases} \alpha_{j,c} & \text{if } c \in a_j,\ \ c \notin a_i \ \ i \neq j, \\ 0 & \text{otherwise} \end{cases}$$

where $\alpha_{j,c} = \mathbb{1}_{\{\text{availability},\ SNR_{j,c} \geq \beta_{j,c}, q_j > 0\}}$ is the indicator for success condition. The total number of successes of user j at the corresponding round is

$$r_j = \sum_{c \in a_j} r_{j,c}, \quad \sharp a_j \leq q_j.$$

- User j has its own queue of packets to manage. Let $AQ_{j,t}$ be the number of packets user j receives at time t. Then the queue size dynamics of user j is

$$q_{j,t+1} = \max\{0, q_{j,t} - r_{j,t}\} + AQ_{j,t+1}, \tag{2.7}$$

$$q_{j,0} \geq 0. \tag{2.8}$$

The term $\max\{0, q_{j,t} - r_{j,t}\}$ says that it is counted in queue size when the condition $q_{j,t} \geq r_{j,t}$, is met. User j cannot have more successes than the number of packets sent by user j, and hence, user j cannot have more successes than its own queue size at the corresponding round.

The analysis of such interactions with queues is an open issue. Below we illustrate in Tables 2.2, 2.3, 2.4, 2.5, 2.6, 2.7 the stochastic game for two channels and two users with non-saturated queues. The entries with ()* are state-based equilibria and state-based global optima [106].

Table 2.2 Each user has at least two packets $q_1 \geq 2, q_2 \geq 2$

	f_1	f_2	$both$	$Wait$
f_1	$0, 0$	$(\alpha_1, \alpha_2)^*$	$0, \alpha_2$	$\alpha_1, 0$
f_2	$(\alpha_2, \alpha_1)^*$	$0, 0$	$0, \alpha_1$	$\alpha_2, 0$
$both$	$\alpha_2, 0$	$\alpha_1, 0$	$0, 0$	$(\alpha_1 + \alpha_2, 0)^*$
$Wait$	$0, \alpha_1$	$0, \alpha_2$	$(0, \alpha_1 + \alpha_2)^*$	$0, 0$

Table 2.3 Each user has exactly one packet $q_1 = 1 = q_2$.

	f_1	f_2	$Wait$
f_1	$0, 0$	$(\alpha_1, \alpha_2)^*$	$\alpha_1, 0$
f_2	$(\alpha_2, \alpha_1)^*$	$0, 0$	$\alpha_2, 0$
$Wait$	$0, \alpha_1$	$0, \alpha_2$	$0, 0$

Table 2.4 One of the users has at least two packets and the other has one packet $[q_1 \geq 2, q_2 = 1]$ or $[q_1 = 1, q_2 \geq 2]$

	f_1	f_2	$Wait$
f_1	$0, 0$	$(\alpha_1, \alpha_2)^*$	$\alpha_1, 0$
f_2	$(\alpha_2, \alpha_1)^*$	$0, 0$	$\alpha_2, 0$
$both$	$\alpha_2, 0$	$\alpha_1, 0$	$(\alpha_1 + \alpha_2, 0)^*$
$Wait$	$0, \alpha_1$	$0, \alpha_2$	$0, 0$

	f_1	f_2	$both$	$Wait$
f_1	$0, 0$	$(\alpha_1, \alpha_2)^*$	$(0, \alpha_2)$	$\alpha_1, 0$
f_2	$(\alpha_2, \alpha_1)^*$	$0, 0$	$(0, \alpha_1)$	$\alpha_2, 0$
$Wait$	$0, \alpha_1$	$0, \alpha_2$	$(0, \alpha_1 + \alpha_2)^*$	$0, 0$

Table 2.5 One of the users has at least two packets and the other has no packet $[q_1 \geq 2, q_2 = 0]$ or $[q_1 = 0, q_2 \geq 2]$

	$Wait$
f_1	$\alpha_1, 0$
f_2	$\alpha_2, 0$
$both$	$(\alpha_1 + \alpha_2, 0)^*$
$Wait$	$0, 0$

	f_1	f_2	$both$	$Wait$
$Wait$	$0, \alpha_1$	$0, \alpha_2$	$(0, \alpha_1 + \alpha_2)^*$	$0, 0$

Table 2.6 One of the users has one packet and the other has no packet $[q_1 = 1, q_2 = 0]$ or $[q_1 = 0, q_2 = 1]$

	$Wait$
f_1	$(\alpha_1, 0)$
f_2	$(\alpha_2, 0)$
$Wait$	$0, 0$

	f_1	f_2	$Wait$
$Wait$	$(0, \alpha_1)$	$(0, \alpha_2)$	$0, 0$

Table 2.7 Both queues are empty $[q_1 = 0, q_2 = 0]$

	$Wait$
$Wait$	$(0, 0)^*$

2.6 Notes

Currently, the number active mobile users within a certain range of a wireless access point could be relatively large but is still finite. It scales with the number of stations and the extension of the network over the geographical area. The game-theoretic formulation proposed in this chapter considers finite number of interacting mobile users. Hence, the asymptotic analysis should be interpreted as a simplified or approximated regime, and it does not capture the scenario observed in practice. This theoretical analysis can be used to determine the capacity region limit of the current technology. It can also be used to determine the stability of the system which is crucial for the throughput and other quantities-of-interest formulas; see [12, 15, 253, 263, 265, 266, 283, 317–319, 421, 427, 428, 432, 433, 435, 436, 449–451, 451, 452, 454, 467, 468, 470, 470, 472–474, 476–478, 481, 484, 488–490, 550, 551] and the references therein.

Chapter 3
Power Savings, Mobility, and Data Queues

In this chapter, we study power savings, user mobility, and data queues. Section 3.1 examines dynamic power control and allocation problems in wireless environment. Section 3.1.1 introduces a dynamic power control problem. Dynamic power allocation problems are presented in Sect. 3.1.2. Section 3.2 provides a very simple but detailed and meaningful model of the network characteristics that influence the performances in large-scale regime in the presence or not of the licensed users. In Sect. 3.2.1, we define the key parameters of the model and the performance metrics. In Sect. 3.2.8, we present spatiotemporal game theoretic formulations in cognitive radio networks. Section 3.2.9 presents a game that can be seen as a reaction to the interference field. In Sect. 3.2.9.1, an asymptotic mean-field stochastic game is proposed. A deterministic version of the problem is examined in Sect. 3.2.9.1. Section 3.3 focuses on mobility and power savings in wireless networks. Section 3.4 examines anti-jamming mean-field games. In Sect. 3.4.1, we present the basic cognitive radio model and describe the key parameters of the performance metrics. In Sect. 3.4.2.1, we present global optimization and micro game problems between secondary users. In Sect. 3.4.3, we formulate the hierarchical mean-field game and develop an algorithm for power savings and probability of success optimization. In Sect. 3.4.5, we explain how to extend the basic model to more realistic cognitive radio networks scenarios based on SINR-based CSMA-like access with multiple primary users and multiple secondary users.

3.1 Dynamic Power Allocation

This section examines dynamic power control and allocation problems in a wireless environment. The section is structured as follows. Section 3.1.1 introduces a dynamic power control problem. Dynamic power allocation problems are presented in Sect. 3.1.2

T. Başar et al., *Mean-Field-Type Game Theory II*, Static & Dynamic Game Theory: Foundations & Applications, https://doi.org/10.1007/978-3-032-07031-9_3

3.1.1 Dynamic Power Control

We consider power management in wireless networks with the following data:

- $I \geq 2$ interacting mobile users over a base station.
- independent channel states whose dynamics are driven by log-normal noises

We examine the mean-field limit when the size of the population goes to infinity, and we derive explicit optimal power control solutions for quadratic cost functions. We denote by $x_j^I(t),\ j \in \{1, 2, \ldots, I\}$, the log-normal driven channel states with the explicit individual state evolution:

$$\begin{array}{l} j \in \{1, 2, \ldots, I\}, \\ dx_j^I = -a_j^I(x_j^I + b_j^I)dt + \sigma_j^I dB_j, \quad t > 0, \\ x_j^I(0), \text{given}. \end{array} \tag{3.1}$$

where a_j^I, b_j^I, σ_j^I are real numbers and B_j are mutually independent Wiener processes. We further have the control evolution as the additive power adjustment:

$$dp_j^I = u_j^I dt,$$

where u_j^I is power adjustment parameter which plays the role of a control strategy of player j. We consider the log-normal fading channel, i.e., $e^{x_j^I(t)}$, and in this setting, the received power from j is $e^{x_j^I(t)} p_j^I(t)$. Then, the signal-to-noise plus interference ratio (SINR) of user j can be written in the following form:

$$\text{SINR}_j := \frac{e^{x_{j,t}^I} p_j^I(t)}{N_0 + \sum_{i \neq j} e^{x_i^I(t)} \mu_{ij}^I p_i^I(t)}. \tag{3.2}$$

The term N_0 above represents a background noise, and the factor μ_{ji}^I can be normalized to $\frac{\alpha}{I}$, due to linear increase of length of the CDMA spreading sequence. It also denotes the squared cross-correlation between the signature sequences of users j and i.

In order to fulfill the requirement of user j, the SINR of j should be above a certain target level $\gamma_{th,j}^I$. We address the question of guaranteeing a certain quality-of-service (QoS) to each mobile as follows:

$$\log_2(1 + \text{SINR}_j) \geq \bar{w}_j^I \implies \text{SINR}_j \geq 2^{\bar{w}_j^I} - 1 =: \gamma_{th,j}^I.$$

This inequality is equivalent to

$$\left[e^{x_j^I(t)} p_j^I(t) - \gamma_{th,j}^I \left(N_0 + \sum_{i \neq j} \mu_{ij}^I e^{x_i^I(t)} p_i^I(t) \right) \right] \geq 0$$

Instead of imposing this inequality, we consider the quadratic function (in p)

$$\left[e^{x_j^I(t)}p_j^I(t) - \gamma_{th,j}^I\left(N_0 + \sum_{i\neq j}\mu_{ij}^I e^{x_i^I(t)}p_i^I(t)\right)\right]^2.$$

which can be seen as a sort of balance between energy saving and quality of service satisfaction. In addition to the QoS requirement, we should ask whether the constraint is satisfied with the lowest energy consumption. Toward that end, we add the cost of energy consumption of mobile j in the form $\wp_j^I\left(u_j^I\right)^2$. This can be also obtained by considering a constraint on $\mathbb{E}\left[\left(u_j^I\right)^2\right]$. We denote by R the matrix of cost weights where

$$R = \begin{pmatrix} \wp_1 & 0 & \dots & 0 \\ 0 & \wp_2 & \dots & 0 \\ \dots & \dots & \ddots & \vdots \\ 0 & 0 & 0 & \wp_I \end{pmatrix}, \wp_j > 0$$

The long-term payoff of j is given by

$$\mathbb{E}\int_0^\infty e^{-\beta t}\left[-\sum_j\left(\gamma_{th,j}^I I_{-j}^I - e^{x_j^I(t)}p_j^I(t)\right)^2 - u'Ru\right]dt,$$

where $I_{-j}^I = N_0 + \frac{\alpha}{I}\sum_{i\neq j}e^{x_i^I(t)}p_i^I(t)$, and β are a discount factor. We write the instantaneous cost function as $r^I(x,p,u) = \langle p, C(x)p\rangle + 2\langle D(x), p\rangle + \langle u, Ru\rangle + E$ obtained after expanding and collecting the polynomial terms in p.

$$\begin{array}{l}\sum_j\left(\gamma_{th,j}^I I_{-j}^I - e^{x_j^I(t)}p_j^I(t)\right)^2 \\ = \sum_j\left(\gamma_{th,j}^I I_{-j}^I\right)^2 + \sum_j\left(e^{x_j^I(t)}p_j^I(t)\right)^2 - 2\sum_j\gamma_{th,j}^I I_{-j}^I e^{x_j^I(t)}p_j^I(t),\end{array} \tag{3.3}$$

with

$$\begin{array}{l} j\in\{1,2,\ldots,I\}, \\ \left(I_{-j}^I\right)^2 = \left(N_0 + \frac{\alpha}{I}\sum_{i\neq j}e^{x_i^I(t)}p_i^I(t)\right)^2 \\ = \frac{\alpha^2}{I^2}\sum_{i\neq j}e^{2x_i^I(t)}(p_i^I(t))^2 \\ +\left(\frac{\alpha^2}{I^2}\sum_{i\neq j}\sum_{i'\neq j}e^{x_i^I(t)}p_i^I(t)e^{x_{i'}^I(t)}p_{i'}^I(t)\right. \\ +2N_0\frac{\alpha}{I}\sum_{i\neq j}e^{x_i^I(t)}p_i^I(t) + N_0^2. \end{array} \tag{3.4}$$

By inserting the last expression, we arrive at

$$
\begin{aligned}
&\sum_j \left(\gamma_{th,j}^I I_{-j}^I - e^{x_j^I(t)} p_j^I(t)\right)^2 \\
&= \sum_j \frac{\alpha^2\left(\gamma_{th,j}^I\right)^2}{I^2} \sum_{i\neq j} e^{2x_i^I(t)}\left(p_i^I(t)\right)^2 + \sum_j \frac{\alpha^2(\gamma_{th,j}^I)^2}{I^2} \sum_{i\neq j}\sum_{i'\neq j} e^{x_i^I(t)} p_i^I(t) e^{x_{i'}^I(t)} p_{i'}^I(t) \\
&+ \sum_j 2N_0 \frac{\alpha\left(\gamma_{th,j}^I\right)^2}{I} \sum_{i\neq j} e^{x_i^I(t)} p_i^I(t) + \sum_j N_0^2\left(\gamma_{th,j}^I\right)^2 + \sum_j \left(e^{x_j^I(t)} p_j^I(t)\right)^2 \\
&-2\sum_j \gamma_{th,j}^I N_0 e^{x_j^I(t)} p_j^I(t) - 2\sum_j \gamma_{th,j}^I e^{x_j^I(t)} p_j^I(t) \frac{\alpha}{I} \sum_{i\neq j} e^{x_i^I(t)} p_i^I(t)
\end{aligned}
\tag{3.5}
$$

It follows that

$$
\begin{aligned}
&j \in \{1, 2, \ldots, I\}, \\
&C(x) = E_1'\left(\sum_j v_j v_j'\right) E_1 \\
&D(x) = \left(\sum_j \gamma_{th,j}^I N_0 v_j'\right) E_1 \\
&E = \sum_j N_0^2 \left(\gamma_{th,j}^I\right)^2
\end{aligned}
\tag{3.6}
$$

where

$$
\begin{aligned}
&E_1 = \mathrm{diag}\left(e^{x_j^I(t)}, j \in \{1, 2, \ldots, I\}\right), \\
&j \in \{1, 2, \ldots, I\}, \\
&v_j(k) = \frac{\alpha \gamma_{th,j}^I}{I} \mathbb{I}_{\{k\neq j\}} - \mathbb{I}_{\{k=j\}} \\
&k \in \{1, 2, \ldots, I\},
\end{aligned}
\tag{3.7}
$$

where *diag* denotes the diagonal matrix.

The centralized power control problem is then given by the following dynamic global optimization problem [106]

$$
\bar{V}(x, p) = \mathrm{Sup}_{u^I=(u_1^I,\ldots,u_I^I)} \mathbb{E}\left(\int_0^{+\infty} -e^{-\beta t} r^I(x_t^I, p_t^I, u_t^I)\, dt \mid x_0^I = x, p_0 = p\right)
$$

such that

$$
(ID) \begin{cases}
dx_j^I = -a_j^I(x_j^I + b_j^I)dt + \sigma_j^I dB_j, \quad t \geq t_0, \\
x_j^I(0), \\
dp_j^I = u_j^I dt, \\
p_j^I(0)
\end{cases}
$$

For large I, the exact socially optimal solution (in particular the Pareto optimality property) requires centralized information, and this leads to high computational complexity and high information exchange cost [460, 463, 478]. Using mean-field limit, one can compute the optimal social cost offline; the resulting set of strategies asymptotically achieves the social optimum as I tends to infinity. We can use the framework of linear-quadratic control to solve the problem. The Hamilton–Jacobi–Bellman optimality equation [460, 464] is given

$$\beta\bar{V} = \frac{1}{2}\sum_j \sigma_j^2 \partial^2_{x_j x_j}\bar{V} + \inf_u\left[r(x,p,u) + \langle \partial_x \bar{V}, f\rangle + \langle \partial_p \bar{V}, u\rangle\right] \quad (3.8)$$

$$= \frac{1}{2}\sum_j \sigma_j^2 \partial^2_{x_j x_j}\bar{V} - \sum_j a_j(x_j + b_j)\partial_{x_j}\bar{V} + \langle p, C(x)p\rangle$$

$$+ 2\langle D(x), p\rangle + \frac{1}{4}\langle \partial_p \bar{V}, R^{-1}\partial_p \bar{V}\rangle$$

and the optimal control is

$$u^* = -\frac{1}{2}R^{-1}\partial_p \bar{V}.$$

We now determine the optimal value $\bar{V}$. We write $\bar{V}$ as a quadratic function in p; i.e., for each x, there exists a symmetric matrix $\tilde{K}(x)$, a vectorial function $\tilde{S}(x)$ and scalar function $\tilde{q}$ such that

$$\bar{V}(x,p) = \langle p, \tilde{K}(x)p\rangle + 2\langle p, \tilde{S}(x)\rangle + \tilde{q}(x)$$

By reincorporating these functions, we obtain the following partial differential systems

$$\beta\tilde{K} = \sum_j f_j \partial_{x_j}\tilde{K} - \tilde{K}R^{-1}\tilde{K} + C + \frac{1}{2}\sum_j \sigma_j^2 \frac{\partial^2}{\partial x_j^2}\tilde{K}, \quad (3.9)$$

$$\beta\tilde{S} = \sum_j f_j \partial_{x_j}\tilde{S} - \tilde{K}R^{-1}\tilde{S} + D + \frac{1}{2}\sum_j \sigma_j^2 \frac{\partial^2}{\partial x_j^2}\tilde{S} \quad (3.10)$$

$$\beta\tilde{q} = \langle f, \partial_x \tilde{q}\rangle - \langle \tilde{S}, R^{-1}\tilde{S}\rangle + \frac{1}{2}\sum_j \sigma_j^2 \frac{\partial^2}{\partial x_j^2}\tilde{q} \quad (3.11)$$

We conclude that $u^* = -R^{-1}\left[\tilde{K}(x)p + \tilde{S}(x)\right]$, i.e.,

$$u_j^*(t) = -\frac{K_{jj}(x)p_j(t)}{\wp_j} - \frac{1}{\wp_j}\sum_{j\neq i}\tilde{K}_{ji}(x)p_i(t) - \frac{1}{\wp_j}\tilde{S}_j(x).$$

Note that in the centralized scheme, each optimal control depends not only on its power levels but also on the weighted sum of the received powers of the other mobiles. One can find an explicit solution to the PDE in $\tilde{K}$ by setting $\tilde{k}_{ij} = Ke^{x_i}e^{x_j}$. The PDE in K becomes

$$(\beta - \sum_j f_j)K = -\tilde{K}R^{-1}\tilde{K} + c + \frac{1}{2}\sum_j \sigma_j^2 K, \tag{3.12}$$

which is a standard quadratic equation, which admits a unique positive solution as c is positive definite.

We now examine the noncooperative decentralized power control problem. When using HJB optimality conditions, one arrives at a system of $3I$ partial differential equations to solve.

In order to derive closed-form expressions, we use a mean-field asymptotic approach. We assume that the channel states are *asymptotically i.i.d*, and their dynamics are given by (ID) with $a_j^I \longrightarrow a$, $b_j^I \longrightarrow b$, $\sigma_j^I \longrightarrow \sigma$, $\wp_j^I \longrightarrow \wp$.

It then follows that the process

$$\gamma\left(N_0 + \frac{\alpha}{I}\sum_{i \neq j} z_i(t)\right)$$

converges weakly to $m^*(t)$ under admissible controls that satisfy the asymptotic indistinguishability property. The individual dynamic optimization problem reads

$$\bar{V}_j(x_j, p) = \text{Sup}_{u_j}\, \mathbb{E}\left(\int_0^{+\infty} e^{-\beta t} r_j(x_j(t), p_j(t), u_j(t), m^*(t))\, dt \mid x_{j,0} = x_j,\, p_0 = p\right)$$

such that

$$\begin{cases} dx_j = -(ax_j + b)dt + \sigma dB_j, \ t \geq 0, \\ \qquad dp_j = u_j dt, \end{cases}$$

where $r_j = -|z_j(t) - m^*(t)|^2 - \wp u_j(t)^2$, with $\mathbb{E}(z_{j,0}^2) < \infty$, $z_j(t) = e^{x_j(t)} p_j(t)$.

The optimal control is determined by

$$u_j^* = -\frac{b}{\wp}\left(\bar{K} z_j + \bar{S}\right)$$

where $\bar{K}$ is positive and $\bar{S}$ is a bounded and continuous solution of a system of algebraic equation involving $m*$

Note that the optimal power control of mobile j at time t depends only its individual state $x_j(t)$, its power p_j, and its function $\bar{S}$, which depends on the mean-field limit $m^*(t)$.

It is natural to determine the gap between the global optimum payoff [106] and payoff obtained at decentralized dynamic game solutions. This gap is known as *price of anarchy* or inefficiency of anarchy. In the context of mean-field games, the sum of individual payoffs at the worse equilibrium can be infinite. Hence, we define the expected payoff of the infinite population or consider the limiting of the price of anarchy with I mobiles in which the global social welfare is normalized. In the latter case, the gap can be arbitrarily large. We refer to [29, 30, 78, 231–233, 478, 533, 534] for more details.

3.1.2 Mean-Field Power Allocation Games

We extend the power control model into a power allocation over several channels.

Received Signals
We now consider networks which can be modeled as a parallel multiple access channels, which consist of several orthogonal multiple access channels. The sets of transmitters and channels are, respectively, denoted by $\mathscr{I}$ and $\mathscr{L}$. The cardinality of $\mathscr{I}$ is $I \geq 2$, and the cardinality of $\mathscr{L}$ is L. The $L-$dimensional vector of received signals is denoted by $y = (y_1, \ldots, y_L)$ and defined by $y(t) = \sum_{j=1}^{I} H_j(t)s_j(t) + n(t)$, where t is the time index (the associated rate depends if fast fading or slow fading channels are considered), $\forall j \in \mathscr{I}$, $H_j(t)$ is the channel transfer matrix from the transmitter j to the receiver, $s_j(t)$ is the vector of symbols transmitted by transmitter j at time t, and the vector $n(t)$ represents the noise observed at the receiver, which is assumed to be centered, Gaussian, and distributed as $\mathbb{E}\left(nn^{\dagger}\right) = \text{diag}\left(N_{0,1}, \ldots, N_{0,L}\right)$. We will exclusively deal with the scenario where $\forall j \in \mathscr{I}$, the matrix H_j is an m-dimensional diagonal matrix, i.e., $H_j(t) = \text{diag}\left(h_{j,1}(t), \ldots, h_{j,L}(t)\right)$, whose entries $h_{j,\ell}(t)$, $\forall \ell \in \mathscr{L}$, are i.i.d. complex Gaussian random variables with independent real and imaginary parts, each with zero-mean and variance $N_{0,\ell}$. This scenario models, for instance, the case of intersymbol interference channels which have been diagonalized in the frequency domain by using a pre-cyclic prefix and the Fourier transform, or the case of the uplink of a network with several access receivers operating over nonoverlapping flat-fading channels. We let $h(t) = (h_1(t), \ldots, h_I(t))$ where $h_j(t) = (h_{j,1}(t), \ldots, h_{j,L}(t))$. The vector of transmitted symbols $s_j(t)$, $\forall j \in \mathscr{I}$ is characterized in terms of power by the covariance matrix $p_j(t) = \mathbb{E}\left(s_j(t)s_j^{\dagger}(t)\right) = \text{diag}\left(p_{j,1}(t), \ldots, p_{j,L}(t)\right)$. As a matter of fact, $p_{j,\ell}$, $\forall (j, \ell) \in \mathscr{I} \times \mathscr{L}$, represents the transmit power allocated by the transmitter j over the AP ℓ. Now, since transmitters are power-limited, we have that

$$\forall j \in \mathscr{I}, \quad \sum_{\ell=1}^{L} p_{j,\ell} \leq p_{j,\max}. \tag{3.13}$$

We define a power allocation vector for transmitter $j \in \mathscr{I}$ as a vector $p_j = \left(p_{j,1}, \ldots, p_{j,L}\right)$ with nonnegative entries satisfying (3.13). We denote by $p =$

$(p_1, \ldots, p_I)$. Finally, the band of channel $\ell \in \mathscr{L}$ will be denoted by W_ℓ and the one of the system by $W = \sum_{\ell=1}^{L} W_\ell$.

Channel Dynamics

We use the change of variables $g^I_{j,\ell} := e^{x^I_{j,\ell}}$. The variable $x^I_{j,\ell}$ will be called the individual (channel) state of transmitter j, and the super vector $x^I(t) = \big(x^I_1(t), \ldots, x^I_I(t)\big)$ with $x^I_j = (x^I_{j,1}, \ldots x^I_{j,L})$ will be called the state of the system or network. We assume that the state $x^I_{j,\ell}(t)$ follows the corresponding evolution or dynamics:

$$dx^I_{j,\ell} = -a^I_{j,\ell}(x^I_{j,\ell} + b^I_{j,\ell})dt + \sigma^I_{j,\ell} dB_{j,\ell},\ t \geq 0,\ j \in \mathscr{I} \tag{3.14}$$

where $a^I_{j,\ell}, b^I_{j,\ell}, \sigma^I_{j,\ell}$ are real and $B_{j,\ell,t}$ are mutually independent Wiener processes. This model is very well justified in the literature. In particular, let us point out that the deterministic part allows one to account for the path loss and shadowing effects and the random part can also model rapid channel variations or channel uncertainty (when channel gains are estimated).

Individual State Evolution

Introduce

$$A^I_j = \text{diag}(a^I_{j,\ell}, \ell \in \mathscr{L}) = \begin{pmatrix} a^I_{j,1} & 0 & \ldots & 0 \\ 0 & a^I_{j,2} & \ldots & 0 \\ \ldots & \ldots & \ddots & \vdots \\ 0 & 0 & 0 & a^I_{j,L} \end{pmatrix} \tag{3.15}$$

Similarly, let $B^I_j = \text{diag}\big(a^I_{j,\ell} b^I_{j,\ell}, \ell \in \mathscr{L}\big)$, $\Gamma^I_j = \text{diag}\big(\sigma^I_{j,\ell},\ \ell \in \mathscr{L}\big)$. Then the individual state evolution is described by $dx^I_j = -\big[A^I_j x^I_j + B^I_j\big]dt + \Gamma^I_j dB_{j,t}$.

System State Evolution

The system state evolution is obtained by combining all the individual state evolutions. Let $A^I = \text{diag}\big(A^I_1, \ldots, A^I_I\big)$ be the $IL \times IL$ dimensional matrix:

$$\begin{pmatrix} A^I_1 & 0 & \ldots & 0 \\ 0 & A^I_2 & \ldots & 0 \\ \ldots & \ldots & \ddots & \vdots \\ 0 & 0 & 0 & A^I_I \end{pmatrix}. \tag{3.16}$$

Similarly, we introduce the matrices B^I and Γ^I. The system state evolution is then given by

$$dx^I = -\big[A^I x^I + B^I\big]dt + \Gamma^I dB. \tag{3.17}$$

Control Signal of Each Transmitter
Usually, by control, it is meant in this context finding a vector-valued function $p_j(t)$ for transmitter j, which represents the radio frequency signal emitted by the transmitter. Here, we introduce an auxiliary function $u^I_{j,\ell}(t)$, which is a signal controlling the transmitter and represents the power amplifier supply. At each time t, transmitter j adjusts its power allocation scheme at each link l by using the control signal $u^I_{j,\ell}(t)$. This control signal is linked to the transmit power used for this link by the relation

$$dp^I_{j,\ell} = u^I_{j,\ell}dt. \tag{3.18}$$

We do not place any restriction on the domain in which U^I will be in. By defining a cost function associated with each control, it is clear that any optimal control u which may depend on the state will be almost surely in a bounded set. We restrict our attention to this simple class of strategies. Note that, in general differential games, there are more general classes of strategies (non-anticipative strategies with/out delay) under which the analysis can be carried out. Finally, we introduce $dp^I_j = u^I_j dt$ where u^I_j is the vector with component l equal to $u^I_{j,\ell}(t)$.

Dynamic Power Allocation Problem in the Finite Case
The joint power control-allocation game can be described in a strategic form. The set of players is $\mathscr{I}$, which is the set of transmitters. The strategy of a transmitter is the control-allocation vector $u^I_j(t) = (u^I_{j,1}(t), \ldots, u^I_{j,L}(t))$. The objective of a transmitter is to reach a transmission rate or signal-to-interference-plus-noise ratio (SINR) target with minimum effort in terms of energy consumed by the power supply. When the receiver implements single-user decoding, the SINR of transmitter j on channel ℓ is given by

$$\mathrm{SINR}^I_{j,\ell}(t) = \frac{p^I_{j,\ell}(t) g^I_{j,\ell}(t)}{N_{0,\ell} + \sum_{i \in \mathscr{I}\setminus\{j\}} p^I_{i,\ell}(t) g^I_{i,\ell}(t)}. \tag{3.19}$$

The constraint for transmitter j is to ensure a minimum requirement for the quality of service in terms of rate:

$$\forall j,\ \forall l,\ \forall t, \log_2\left(1 + \mathrm{SINR}^I_{j,\ell}(t)\right) \geq \bar{w}^I_{j,\ell} \tag{3.20}$$

which is equivalent to

$$\mathrm{SINR}^I_{j,\ell}(t) \geq 2^{\bar{w}^I_{j,\ell}} - 1 =: \gamma^I_{j,\ell}. \tag{3.21}$$

This inequality is further equivalent to

$$\left[e^{x^I_{j,\ell}} p^I_{j,\ell} - \gamma^I_{j,\ell} \left(N_{0,\ell} + \sum_{i \neq j} e^{x^I_{i,\ell}} p^I_{i,\ell} \right) \right] \geq 0. \tag{3.22}$$

In practice, this amounts to minimizing the following quadratic term in the power vectors: $\sum_{\ell\in\mathcal{L}}\left[e^{x^I_{j,\ell}(t)}p^I_{j,\ell}(t)-\gamma^I_{j,\ell}\left(N_{0,\ell}+\sum_{i\neq j}e^{x^I_{i,\ell}(t)}p^I_{i,\ell}(t)\right)\right]^2$. The objective of transmitter j is to minimize the above quantity with a minimum amount of energy which is translated by unit cost matrix $R_j=\text{diag}(r^I_{j1},\ldots,r^I_{jL})$. The payoff function for transmitter j is therefore defined by

$$\Phi_{j,\beta}(u)=\mathbb{E}\int_0^\infty e^{-\beta t}\left[-\sum_{l=1}^L\left(\gamma^I_{j,\ell}I^I_{-j,\ell,t}-e^{x^I_{j,\ell,t}}p^I_{j,\ell,t}\right)^2-\langle u^I_j(t),R_ju^I_j(t)\rangle\right]dt, \tag{3.23}$$

where β is a positive parameter allowing the transmitters to evaluate the current and future payoffs differently, $I^I_{-j,\ell}(t)=N_{0,\ell}+\sum_{i\neq j}e^{x^I_{i,\ell}(t)}p^I_{i,\ell}(t)$, and $u^I_j=(u^I_{j,1},\ldots,u^I_{j,L})$.

Optimal Centralized Policies

The centralized power allocation problem is then given by the following dynamic optimization problem

$$\bar{V}^I(x,p)=\sup_{u^I}\mathbb{E}\left(\int_0^{+\infty}-e^{-\beta t}\mathcal{R}_c(x^I_t,p^I_t,u^I_t)\,dt|x^I_0=x,p^I_0=p\right)$$

such that the individual state dynamics (ID) of all the transmitters are satisfied.

$$(ID)\begin{cases}dx^I_{j,\ell}=-a^I_{j,\ell}(x^I_{j,\ell}+b^I_{j,\ell})dt+\sigma^I_{j,\ell}dB_{j,\ell},\\ dp^I_{j,\ell}=u^I_{j,\ell}dt\\ \forall j\in\mathcal{I},\ \forall\ell\in\mathcal{L},\end{cases}$$

where $u^I(t)=\left(u^I_1(t),\ldots,u^I_I(t)\right)$, $\mathcal{R}_c(x,p,u)=\langle p,A_1(x)p\rangle+2\langle A_2(x),p\rangle+A_3+\langle u,Ru\rangle$ is the total sum of payoffs of all the transmitters, obtained after expanding and collecting the polynomial terms in p. The initial vector x_0 is an element belonging to the set of feasible values along the optimal state trajectory at time $t_0=0$. We assume that the vector (x,p) is observed perfectly. The information structure of the agents is given by (i) model knowledge and (ii) perfect state observation.

The basic objective is to maximize the total payoff function (the sum of all the transmitter payoffs).

For large I, the exact globally optimal solution requires centralized information, and this leads to high computational complexity. Using mean-field limit under suitable assumptions, one can compute the optimal total payoff offline; the resulting set of strategies would asymptotically be not that far from the globally optimizing one as I tends to infinity.

The centralized problem can be formulated as

$$\begin{cases} \sup_{u^I} \mathbb{E} \int_0^{+\infty} e^{-\beta t} \left[-\sum_{j=1}^{I} \sum_{l=1}^{L} \left(\gamma_{j,\ell}^I I_{-j,\ell}^I - e^{x_{j,\ell,t}^I} p_{j,\ell,t}^I \right)^2 - \langle u_t^I, R.u_t^I \rangle \right] dt \\ dx_t^I = -[A^I x_t^I + B^I] dt + \Gamma^I dB_t \\ dp_t^I = u_t^I dt \end{cases}$$

where $R = \text{diag}(R_j)_j$.

Since the problem is time invariant and defined over an infinite horizon, the optimizing control u will be a function of (x, p) only. This problem is then in the framework of linear-quadratic (LQ) stochastic optimal control. Hence, one can obtain

Proposition 3.1 *A set of controls $u_* = \eta(x, p)$ constitutes an optimal solution to the centralized problem if there exists a continuously differentiable function $\bar{V}^I$ such that the Hamilton–Jacobi–Bellman-Fleming (HJBF) equation given by*

$$\beta \bar{V}^I(x, p) = \frac{1}{2} \sum_{\ell} \sum_{j} \sigma_{j,\ell}^{I,2} \frac{\partial^2}{\partial x_{j,\ell}^2} \bar{V}^I \tag{3.24}$$

$$+ \inf_u \left[r(x, p, u) + \langle \partial_x \bar{V}^I, f \rangle + \langle \partial_p \bar{V}^I, u \rangle \right]$$

$$= \frac{1}{2} \sum_{\ell} \sum_{j} \sigma_{j,\ell}^{I,2} \frac{\partial^2}{\partial x_{j,\ell}^2} \bar{V}^I + \left(r(x, p, \eta) + \langle \partial_x \bar{V}^I, f \rangle + \langle \partial_p \bar{V}^I, \eta \rangle \right)$$

is satisfied.

Carrying out the minimization, this implies that if there exists a regular function $\bar{V}^I$ such that the Hamilton–Jacobi–Bellman (HJB) optimality equation

$$\beta \bar{V}^I = \frac{1}{2} \sum_{j,\ell} \sigma_{j,\ell}^{I,2} \frac{\partial^2}{\partial x_{j,\ell}^2} \bar{V}^I + \sum_{j,\ell} a_{j,\ell}^I (x_{j,\ell} + b_{j,\ell}^I) \frac{\partial}{\partial x_{j,\ell}} \bar{V}^I +$$

$$\langle p, A_1(x) p \rangle + 2 \langle A_2(x), p \rangle + A_3 + \frac{1}{4} \langle \partial_p \bar{V}^I, R^{-1} \partial_p \bar{V}^I \rangle$$

is satisfied, then the optimal control is

$$u_*^I = -\frac{1}{2} R^{-1} \partial_p \bar{V}^I. \tag{3.25}$$

To determine the optimal value $\bar{V}^I$, we write it as a quadratic function in p; i.e., there exists a symmetric matrix $\tilde{K}(x)$, a vectorial function $\tilde{S}(x)$ and scalar $\tilde{q}$ such that

$$\bar{V}^I(x, p) = \langle p, \tilde{K}(x) p \rangle + 2 \langle p, \tilde{S}(x) \rangle + \tilde{q}(x).$$

We conclude that $u^I_* = -R^{-1}\big[\tilde{K}(x^I)P^I + \tilde{S}(x^I)\big]$

$$u^I_{*,j\ell t} = -\frac{\tilde{K}_{jj\ell}(x^I)p^I_{j\ell t}}{r^I_{j\ell}} - \frac{1}{r^I_{j\ell}}\sum_{j\neq i}\tilde{K}_{ji\ell}(x^I)p^I_{i\ell}(t) - \frac{1}{r^I_{j,\ell}}\tilde{S}_{j\ell}(x^I).$$

This extends the result presented in Sect. 3.1.1.

3.1.3 Decentralized Selfish Policies

In this subsection, we focus on the decentralized power allocation problem. Each user solves its long-term payoff under individual state evolution constraint. We assume that each user knows the payoff functions $\Phi_\beta = (\Phi_{j,\beta})_{j\in\mathcal{I}}$ as well as the state dynamics.

$$\begin{cases} \sup_{u_j} \ \Phi_{j,\beta}(u) \\ \text{s.t} \\ dx^I_j = -[A^I_j x^I_j + B^I_j]dt + \Gamma^I_j dB_j \\ dp^I_j = u^I_j dt \end{cases}$$

We define a feedback Nash equilibrium of the dynamic power allocation game. If the transmitters information structure is reduced to the initial profile x, p at time 0 and the trajectory generated, the strategies become functions of (x, p) and the current time t. A strategy u_j can be written as $u_j(t, x, p)$. Denote by $\mathcal{U}_j$ be the set of such strategies. Under this class of strategies, a feedback Nash equilibrium is a configuration in which none of the transmitters improves its long-term payoff by unilaterally changing its strategy.

Determination of Feedback Nash Equilibria

A sufficiency condition for a feedback Nash equilibrium η is characterized as following: if there exists regular functions $\{V^I_j\}_j$ such that the systems of HJB equations—for all $j \in \mathcal{I}$,

$$\beta V^I_j(x,p) = \frac{1}{2}\sum_\ell\sum_j \sigma^{I,2}_{j,\ell}\partial^2_{x^2_{j,\ell}} V^I_j + \inf{}_u\Big[r^I_j(x,p,\eta^u_j) + \langle\partial_x V^I_j, f\rangle + \langle\partial_p V^I_j, u\rangle\Big] \quad (3.26)$$

$$= \frac{1}{2}\sum_\ell\sum_j \sigma^{I,2}_{j,\ell}\partial^2_{x^2_{j,\ell}} V^I_j + [\tilde{r}^I_j(x,p,\eta) + \langle\partial_x V^I_j, f\rangle + \langle\partial_p V^I_j, \eta\rangle] \quad (3.27)$$

where $\eta_j^u = (\eta_1, \ldots, \eta_{j-1}, u_j, \eta_{j+1}, \ldots, \eta_I)$. The HJB equations in order to derive the equations in x. Thus, one gets a system of $3IL$ differential equations to solve to get the best response correspondence.

Large System Analysis

We now examine the noncooperative decentralized mean-field power allocation problem. When using HJB optimality conditions, one gets a system of $3IL$ partial differential equations to solve. Moreover, fixed point equations and closed-loop parameters are in these equilibrium equations. In order to reduce the number of unknown parameters of these equations and to derive closed form expressions, we use a mean-field asymptotic approach. We assume that the channel states evolutions described by (ID) satisfy asymptotically: $a_{j,\ell}^I \longrightarrow a, b_{j,\ell}^I \longrightarrow b_\ell, \sigma_{j,\ell}^I \longrightarrow \sigma_\ell, r_{j,\ell}^I \longrightarrow r_\ell$. For each link ℓ, define the random measure $M_\ell^I = \frac{1}{I}\sum_{j=1}^I \delta_{z_{j,\ell}^I}$ with marginal $M_{\ell,t}^I = \frac{1}{I}\sum_{j=1}^I \delta_{z_{j,\ell,t}^I}$ where $z_{j,\ell,t}^I = e^{x_{j,\ell,t}^I} p_{j,\ell,t}^I$. The long-term performance metric of the transmitter j is a function of the random measure $(M_{\ell,t}^I)_\ell$ and $z_j(t)^I$, i.e., $\phi\big(z_j(t)^I, M^I(t)\big)$ where ϕ is continuous differentiable function (a quadratic function). The weak convergence of the occupancy measure to some deterministic measure $m^*(t)$ which evolves in time, combined with the weak convergence of the pair of processes $\big(z_{j,t}^I, M^I(t)\big)$ determines the mean-field convergence of the performance metric of user j.

Under independent Wiener processes of individual state dynamics, we know that if the initial occupancy measure $M^I(0)$ is weakly convergent and the process $\big(z_{j,\ell}^I(0)\big)_{j,\ell}$ are exchangeable, the mean-field convergence holds. Using the asymptotic exchangeable assumptions of the Wiener processes $B_{j,\ell}$, we conclude that the individual state evolution generated a random process $z_{j,\ell,t}^I$ generates the process $M^I(t)$ which has the mean-field convergence property.

Proposition 3.2 *If the process $\big(z_j^I(t)\big)_j$ is asymptotically exchangeable, then $M^I(t)$, converges in distribution to a measure $m^*(t)$.*

Moreover, the pair of processes $\big(z_j^I(t), M^I(t)\big)$ converges in law. Using this result, one has that the asymptotic of the pair $\big(z_j^I(t), M^I(t)\big)$ is well-defined.

Note that the asymptotic exchangeability is a more general assumption in the sense that it does not require symmetry assumption as it is widely used the literature.

Using the above result, we have the mean-field convergence of the process

$$\gamma_\ell\left(N_{0,\ell} + \frac{\alpha}{I-1}\sum_{i\neq j} z_{i,\ell,t}\right)$$

to $m_{\ell,t}^*$. For a given mean-field limit trajectory $\{m^*(t)\}_t$, the individual dynamic optimization problem reads

$$V_j(x_j, p_j, m^*) =$$

$$\sup_{u_j} \mathbb{E}\left(\int_0^{+\infty} e^{-\beta t}\tilde{r}_j(x_j(t), p_j(t), u_j(t), m^*(t))\, dt \mid x_{j,0} = x_j, p_{j,0} = P_j, m_0^* = m^*\right)$$

such that

$$\begin{cases} dx_{j,\ell} = -(a_\ell x_{j,\ell} + b_\ell)dt + \sigma_\ell dB_{j,\ell},\ t \geq 0,\ \ell \in \mathcal{L}, \\ dp_{j,\ell} = u_{j,\ell}dt,\ \ell \in \mathcal{L} \end{cases}$$

where $\tilde{r}_j = -\sum_{\ell\in\mathcal{L}}\left(|z_{j,\ell}(t) - m_\ell^*(t)|^2 + r_\ell u_{j,\ell}^2(t)\right)$, with $\mathbb{E}(z_{j,\ell}^2(0)) < \infty$, $z_{j,\ell}(t) = e^{x_{j,\ell}(t)} p_{j,\ell}(t)$.

Note that the optimal power allocation of user j at time t depends only on its individual state $x_j(t)$, and the function $\bar{S}$ which depends on the mean-field limit $m^*(t)$.

The mean-field limit solves the following fixed-point system

$$\begin{array}{l} \ell \in \mathcal{L}, \\ m_\ell^*(t) = \gamma_\ell\left(N_{0,\ell} + \alpha\mathbb{E}(z_\ell(0))\right) \\ +\gamma_\ell\alpha\int_0^t\left\{\mathbb{E}\left(u_\ell^* e^{x_\ell(t)}\right) + a_\ell\mathbb{E}\left(x_\ell^* e^{x_\ell(t)}\right) + \left(b_\ell + \sigma_\ell^2\right)\mathbb{E}\left(e^{x_\ell(t)}\right)\right\} dt'. \end{array} \tag{3.28}$$

As u_ℓ^* is a function of m_ℓ^*, it is a fixed-point integral in m_ℓ^*.

Mean-Field Price of Anarchy

It is natural to bring up the question of the gap between the global optimum payoff and the worst payoff obtained at the decentralized dynamic power allocation solutions. This gap is known as *price of anarchy* or inefficiency of anarchy. In the context of mean-field games, the sum of individual payoffs at the worst equilibrium can be infinite. Hence, we need to define the expected payoff of the infinite population or to consider the limit as I goes to infinity of the price of anarchy in which the global social welfare is normalized [460, 464].

We define the *mean-field price of anarchy* (MFPoA) as follows:

$$\text{MFPoA} = \| \liminf_{I\to\infty} \frac{1}{I}\sum_{j=1}^{I}\tilde{r}_j^I(GO) - \tilde{r}_j(u_*, m^*(t)) \| \tag{3.29}$$

which measures the gap between the limit of global optimum payoff per individual $\lim_I \frac{1}{I}\sum_{j=1}^{I}\tilde{r}_j^I(GO)$ (the $\underline{\lim}$ is taken when the limit does not exists) and the payoff at the mean field equilibrium. The reason to consider the global optimum payoff per individual instead of the global optimum payoff is that the term $\sum_{j=1}^{I}\tilde{r}_j^I(GO)$ can be infinite when I grows to infinity.

The mean-field price of anarchy is in order of $O\left(\alpha\frac{L}{\beta}\max_{\ell\in\mathcal{L}}\theta_\ell\right)$ where θ_ℓ is the Lipschitz constant of ϕ_ℓ.

The proof of this result follows from the mean-field convergence and the fact that the integrand is Lipschitz continuous with Lipschitz constant less than $\alpha L \max_\ell \gamma_\ell$.

3.1.4 Notes

Gaussian distribution is widely used as the law of channel state. One of the reasons of that choice is that the Gaussian is the worst case for the error estimation, namely, the quadratic gap between $s(t)$ and $y(t)$. Once we move from this objective by choosing, for example, $\tilde{r}_j^I(x_t^I, p_t^I, u_t^I)$, the Gaussian distribution may not be a minimizer. To model this case, we introduce *Nature* as a player who chooses the distribution of the individual state (ID) at each time with a fixed expectation and fixed variance. It is now clear that Nature needs not to follow a Gaussian distribution. A (mixed) strategy of Nature is to choose a distribution among all the possible distributions with fixed first two moments. In order to compute at the worst case for the users, we say that Nature minimizes the users' payoff among all the distributions with zero expectation and variance equal to Γ. This leads a distributionally robust power allocation problem.

The number active mobile users within a certain range of a station, relay, or access point could be relatively large but is still finite. It scales with the number of stations and the extension of the network. Currently, we do not have a communication network with infinite number of interacting mobile users interfering at the same station, relay, or access point. Locally, the number of users interfering with each other is finite. The game-theoretic formulation considers also finite number of interacting mobile users. Here, the asymptotic analysis should be interpreted as a simplified or approximated regime, and it does not capture the scenario observed in practice. For more details, see [13, 16, 96, 97, 129, 142, 190, 210, 211, 261, 267, 284, 393, 424, 426, 448, 475], [26, 44, 46, 208, 209, 215, 216, 221, 270, 286, 357, 358, 365, 394, 480, 489], [6, 14, 39, 45, 50, 131, 150, 151, 283, 400, 451, 459, 461, 471, 472, 476, 483], [7, 20, 138, 154, 222, 252, 255, 278, 330, 418, 419, 453, 469, 478, 485, 528] and the references therein.

3.2 Interference Field in Wireless Networks

In Sect. 3.1, we have discussed how using power control one can reduce the interference created by multiple interacting mobile users. This section presents some key mean-field terms in wireless communication networks. Most of the performance metrics are based on the quality of experience of the users, which includes the quality of service and the outage probability of transmissions. The signal quality is influenced by the so-called interference coming from the other transmitter-receiver pairs, obstacles, and blocking objects. We derive a mean-field term which is an interference field across a map that combines the interference on the location, received positions, and angles. The interference field is strongly connected to the performance of the networks and to the satisfaction of the users.

Interference Management

Wireless technology is proliferating rapidly, and the vision of pervasive wireless networking, computing, and communications offers the promise of many industrial and individual benefits. This explosion of wireless applications creates an ever-increasing demand for more radio spectrum. However, most easily usable spectrum bands have been allocated, although many measurements, nevertheless, show that at any specific time and location; these spectrum bands are significantly underutilized, and lot of energy is wasted.

The presence of high-priority primary users and the requirement that the cognitive users should cause only limited interference with them define an access paradigm called cognitive access. Cognitive access control is relevant to both the cognitive radio and the coexistence community. Cognitive access control protocols play an important role to exploit the spectrum opportunities, manage the interference to primary users, and coordinate the spectrum access among cognitive users.

Spectrum Access as a Game

In the absence of primary users, the secondary or cognitive users can opportunistically share the spectrum in a distributed and autonomous manner. However, simultaneous transmissions will generate an aggregate interference that could be potentially high and thus reduces user performance. The spectrum access problem is clearly *strategic decision-making problem*, also called *game*, where the decision involves when to transmit and with which transmit power. The decision policy is based on location, queue, remaining energy, channel state, and resource availability.

We have studied carrier-sense multiple access protocols in Chap. 2. In this section, we study multiple access problems in cognitive radio networks. We specifically examine the large-scale interaction that takes place among secondary users as a reaction to an evolving interference called interference field.

Cognitive Access Game

Cognitive access control coupled with state estimation and detection for identifying and exploiting instantaneous spectrum opportunities is one of the central challenges. In addition, the opportunistic spectrum access needs to take into consideration energy constraints of sensing, exploration, exploitation, and transmission. Then the question is how to develop cognitive sensing and access strategies that learn from observations and offer improved fairness and performance over time without central controllers or dedicated communication/control channels.

Interference Management

Interference and spectrum management have been widely studied from a game-theoretic perspective. However, only a few works are available for optimization and games in large-scale networks. When the number of users becomes sufficiently large, important structure of the finite game outcome can be obtained from an infinite population game called mean-field stochastic game. In it, each user reacts to the interference created by the other users. Since interference evolves in time and depends on the position of the receiver, this creates an interference vector which we call interference field, in reference to the vector field in vector calculus. The interference field is an assignment of an interference to each point in a subset of an appropriate Euclidean space [125].

We use mean-field game to transform the large-scale interaction into a condensed form where many opportunities of exploiting underutilized dimensions appear. In the asymptotic regime, each user interacts with a large representative subset of the entire population, i.e., is reacting to the interference "field" which evolves in time. We provide a mean-field framework which analyzes the evolution over time of the underutilized resources (space, frequency, energy).

Discrete Time Model We propose a game-theoretic model in discrete time. There are two main reasons for proposing a difference game model. First, data measurements and acquisition in cognitive networks are done in discrete time. Therefore, the state, energy, and movement of cognitive users can be collected only in discrete time, and the time step between two measurements should be sufficient for short transmission and communication. Second, the proposed model does not need to solve complex partial integrodifferential equations (PIDE).

This section provides a simple but detailed and meaningful model of the network characteristics that influence the performances in the large-scale regime in the presence or not of the licensed users.

In Sect. 3.2.1, we introduce the key parameters of the model and the performance metrics. In Sect. 3.2.8, we present spatiotemporal game theoretic formulations in cognitive radio networks. Section 3.2.9 presents a game that can be seen as a reaction to the interference field. In Sect. 3.2.9.1, an asymptotic mean-field stochastic game is proposed. A deterministic version is examined in Sect. 3.2.9.1.

Notations

Below we summarize the notations that will be used in this section.

$\mathcal{I}$	Set of users
I	Cardinality of $\mathcal{I}$
$h_{jr,t}$	Channel state from transmitter j to receiver r at slot t
$B_{j,t}$	Brownian motion associated to j at slot t
$E_{j,t}$	Remaining energy of user j at slot t
E_t	Vector of energy state at slot t
x_t	Location of transmitters at slot t
q_t	Queue length vector
s_t	(h_t, E_t, x_t, q_t)
$\gamma_j(s_t, p_t, y_t)$	SINR of user j at slot t
$\mathbb{E}$	Mathematical expectation operator

3.2.1 Wireless Model

Consider a large network with $I \geq 2$ cognitive users and one primary user (licensed user). The set of cognitive users is denoted by $\mathcal{I} = \{1, 2, \ldots, I\}$. To each user, we associate four important types of dynamics: channel, mobility, energy, and queue dynamics.

- The channel dynamics models the channel state evolution. Our channel model generates an ergodic process and captures the behavior of time-varying Gaussian channels.
- The mobility dynamics captures a cognitive user mobility. Based on the mobility pattern, we define a path-loss model between a transmitter and a receiver to capture small interference from cognitive user very far from the corresponding receiver.
- Energy dynamics models the evolution of the remaining energy. A cognitive user can transmit only when sufficient amount of energy is available.
- The queue dynamics captures the size of the queue length of a generic cognitive user. In contrast to the classical spectrum access game where each user is assumed to have always some data to send (saturated queues), here, non-saturated queues are in taken into consideration. If the queue is empty, the cognitive user will use "null" power and save energy. It is important to note that the queue dynamics are correlated, because they depend on the success probability which is mainly driven by the interference field.

The quality of cognitive wireless communication suffers from strong variations of the received signal power due to signal shadowing and multipath fading (spatiotemporal). The shadowing of the signal is influenced by obstacles in the propagation path (buildings, bridges, trees, etc.) and results in attenuation over the entire signal bandwidth. The shadowed areas are larger for low elevations than for high elevations. Multipath fading occurs, because the signal is received not only via the direct path but also after having been reflected from objects in the surroundings. Due to their different propagation distances, multipath signals can add destructively, resulting in a deep fade. We describe below these effects via simple and basic dynamics that include channel dynamics, path loss, and mobility pattern.

3.2.2 Channel Model

We study a very basic channel model described by a discrete time stochastic equation

$$h_{jr,t+1} = h_{jr,t} + \frac{\varsigma}{2}\left(\mu_h - h_{jr,t}\right) + \sigma_h\left(B_{j,t+1} - B_{j,t}\right), \tag{3.30}$$

where $B_{j,t}$ is a Brownian motion independent across the users and independent across time, $h_{jr,t}$ is the channel state from j to r at time t. ς, μ_h, and σ_h are positive real numbers.

Below we comment on the channel model. For generic Gaussian channel, the Brownian dynamics of h_t is sometimes described by

$$h_{t+1} - h_t = \sigma_h\left(B_{t+1} - B_t\right) \tag{3.31}$$

which can be seen as a particular case of (3.30) for $\varsigma = 0$. Below we point out the limitations of the channel model (3.31) and justify the proposed channel model (3.30) for $\varsigma > 0$. Let $\bar{g} = h^2$. For small time scales, one can use Itô's rule in (3.31), and we get

$$d\bar{g}_t = \frac{2\sigma_h^2}{2}dt + \sigma_h\partial_h\bar{g}_t dB_t = \sigma_h^2 dt + 2\sigma_h h_t dB_t \tag{3.32}$$

Note that the variance of h_t goes to infinity with time, and this can be interpreted as the channel gain becoming arbitrarily large, which is not a realistic model. In order to avoid that, we can use the following Ornstein–Uhlenbeck dynamics

$$dh_t = \frac{\varsigma}{2}(\mu_h - h_t)dt + \sigma_h dB_t \tag{3.33}$$

where μ_h and σ_h are nonnegative real numbers and ς is normalized to be 1. Then, the stationary distribution of h_t is Gaussian with mean μ_h and variance σ_h^2. These parameters are finite.

The probability density function $m_h(h,t)$ of the Ornstein–Uhlenbeck process satisfies the Fokker–Planck–Kolmogorov equation

$$\frac{\partial m_h}{\partial t} = \frac{\varsigma}{2}\frac{\partial}{\partial h}[(h - \mu_h)m_h] + \frac{\sigma_h^2}{2}\frac{\partial^2 m_h}{\partial h^2}.$$

The general solution of this equation has a time-varying Gaussian distribution. For $\varsigma = 1$, the stationary solution of this equation is the limit for time tending to infinity and is a Gaussian distribution with mean μ_h and variance σ_h^2,

$$m_{h,st}(h) = \sqrt{\frac{1}{2\pi\sigma_h^2}} e^{-\frac{1}{2\sigma_h^2}(h-\mu_h)^2}.$$

Interestingly, such a density is learnable using the individual mobility patterns.

Another channel model that takes into consideration the Gaussian structure is the *log-normal channel*. The well-known log-normal channel can be modeled as $g_t = e^{\eta_t}$ where η_t is the solution of a linear stochastic differential equation (SDE) with finite asymptotic mean and variance.

3.2.3 Mobility Model

Different mobility models (random walk, Levy flight, random waypoint) of wireless nodes and human mobility have been studied. Here we consider a stochastic pattern of the users given by stochastic difference equation. User j's mobility is

$$x_{j,t+1} = x_{j,t} + \delta.v(t, x_{j,t}) + \sigma_x w_{j,t+1} \tag{3.34}$$

where $w_{j,t+1}$ is a stochastic process with zero mean and finite variance, v is the drift function, and $\sigma_x > 0$.

How to Identify the Drift of the Mobility Pattern in Practice?
In this mobility model, one of the key functions to be identified is the drift function v. One can determine the function v from large data or sample of users' movements. Given different realized paths, one can compute the conditional expected changes in one time slot, i.e., $\mathbb{E}\left(\frac{x_{j,t+1}-x_{j,t}}{\delta} \mid x_t\right)$, from the experiments. The quantity $\mathbb{E}\left(\frac{x_{j,t+1}-x_{j,t}}{\delta} \mid x_t\right)$ is the conditional expectation of v knowing the positions x.

3.2.4 Path-Loss Model

Given a user transmitter j located at position x_j and a receiver r located at position y_r, we define a path-loss function $l : \mathbb{R}_+ \longrightarrow \mathbb{R}_+$, given by

$$l(d_{jr}) = l\left(\left|x_j - y_r\right|\right) = \left(\epsilon^2 + \left|x_j - y_r\right|^2\right)^{\alpha/2} \tag{3.35}$$

where $d_{jr} = \left| x_j - y_r \right|$ is the Euclidean distance from j to r, $\epsilon > 0$ and $\alpha \geq 2$ is the path-loss exponent. The virtual received power at receiver r is then given by

$$\frac{p_j |h_{jr}|^2}{l(d_{jr})}.$$

The proposed path-loss model has the advantage that the function l is bounded away from 0 even if x and y coincide, i.e., $l(0) = |\epsilon|^{\alpha} > 0$. Note that if $\epsilon = 0$, then the denominator $l(d)$ may be zero and one gets infinity if the numerator is nonzero.

3.2.5 *Remaining Energy Dynamics*

The battery state of each user is modeled as a stochastic process which depends on the power consumption. A generic player starts initially with some energy budget $E_0 > 0$ (full battery level). Its evolution is described by

$$E_{j,t+1} = E_{j,t} - \delta . p_{j,t} + \sigma_e E_{j,t} \left(\tilde{B}_{t+1} - \tilde{B}_t \right), \; j \in \mathcal{I}. \tag{3.36}$$

where the remaining energy budget $E_{j,t}$ is subject to positivity constraint, $p_{j,t} \geq 0$ is the power consumption of user j at time t, and $\sigma_e > 0$ is a constant. The presence of the noise term is to capture the stochastic nature of the anticipated prediction and the fact that under renewable energy (such as wind or solar-powered system), the battery may be recharged if the weather conditions allow. Moreover, $p_{j,t} = 0$ if $E_{j,t} \leq \epsilon_E$, where ϵ_E is a fixed minimum energy threshold for the device to be usable.

3.2.6 *Queue Dynamics*

Let $A_{j,k}$ be the arrival of data of user j at time k and $D_j \mathbb{1}_{\text{success}}(s_t, p_t, y_t)$ the departure process conditioned on successful reception. Then, the evolution of queue length can be modeled as

$$q_{j,t+1} = \left[q_{j,t} - D_j \mathbb{1}_{\text{success}}(s_t, p_t, y_t) \right]_+ + A_{j,t+1},$$

where $[z]_+ = \max(0, z)$, $s_t = (h_t, E_t, x_t, q_t)$ is the state of the transmitters at slot k, p_t is the power profile, and q_t is the queue length vector. A success transmission is partially conditioned to a good SINR that we describe below. If $q_j = 0$, then user j has no data to send, and therefore the transmit power of j needs to 0.

3.2.7 SINR Model

The signal-to-interference-plus-noise ratio (SINR) of user j transmitting to a receiver located at $y_{r,t}$ at slot t is

$$\gamma_j(s_t, p_t, y_t) = \frac{\frac{p_{j,t}|h_{jr,t}|^2}{l(d_{jr,t})}}{N_0(y_{r,t}) + \sum_{i \neq j} \frac{p_{i,t}|h_{ir,t}|^2}{l(d_{ir,t})}}$$

where $N_0(y_{r,t}) > 0$ is the background noise at receiver r. This basic SINR model captures both the distance between the transmitter and the receiver and the channel randomness.

There are several analyses in the literature that are dedicated to extreme cases, namely, high SINR and low SINR regimes. Here, we do not consider any specific high SINR and low SINR regimes and we explain why.

On the High-SINR Assumption There are many reasons for not considering the high SINR regime. The first reason is that if the SINR is high, then, in many practical cases, it means that the signal is good (satisfactory) and hence transmissions are successful. Therefore, there is no reason to consume more power, which would lead to higher energy costs. The second reason for not considering the high SINR regime is that the term "high" is not precisely quantified. How high should SINR be? Hence one can conclude that a statement like "SINR tends to infinity" is not practical.

On the Low-SINR Assumption Similarly, the low SINR regime assumption carries out with it many unanswered questions. If the strength of the signal is too low/bad, say, less than a certain threshold β, then transmissions would be unsuccessful. Therefore, no power control would be needed (in many applications) as observed by many studies.

For all these reasons, we do not impose such extreme regime approximations on the model adopted.

3.2.8 Game Formulations

From an information-theoretic viewpoint, the spectrum utilization efficiency of a cognitive radio can be measured by the achievable capacity, which is in turn a function of its received signal-to-interference-plus-noise ratio (SINR) as well as its occupied spectrum bandwidth. In a distributed cognitive radio network, each cognitive radio decides on its transmission power based on the sensed radio environment. Its decision not only impacts its own achievable performance but also affects that of at least its neighboring cognitive radios via interference. Hence, radio resource allocation is an interactive decision-making process, which can be suitably modeled as a game. The set of players is $\mathcal{J}$. In each time slot t, each player $j \in \mathcal{J}$

chooses a transmit power $p_{j,t}$ based on its own state and associated receiver in order to optimize its performance which is a function of the SINR γ_j of that user. The system is subject to stochastic changes such as channels, queue lengths, and mobility. Therefore, a stochastic long-term game model is more appropriate than one-shot game model. Below, we describe the stochastic game that captures the interference problem in large-scale cognitive radio networks.

3.2.9 *Reaction to the Interference Field*

We formulate the problem of spectrum access and interference management in a mean-field setting. We show that the problem can be seen as a reaction to the interference field in the sense that each cognitive user will be facing to the availability of the spectrum and an evolving deterministic sequence which is the interference field [125, 254].

We propose a mean-field stochastic game framework in which each cognitive user interacts with a large representative subset of the entire population, i.e., is reacting to the interference "field" which evolves in time.

In the context of large population of cognitive users, the interference created at a receiver located at y at time t is

$$I_t(y) := \int_s \frac{|h|^2 p_t^*(s, y)}{l(|x - y|)} m_t(ds),$$

where $p_t^*(s, y)$ is the transmit power which is dependent of the state s and the location y of the receiver. The term m_t represents the distribution of states, that is, *mean-field*. Given the evolution of interference I_t, each cognitive user reacts to the interference by adjusting its transmit power depending on its own queue size, location, and remaining energy. The adjustment of the transmit powers together with the new channel state and queue size determines a new interference $I_{t+1}(y)$.

The instantaneous payoff R is a positive function of the SINR and hence a function of the individual state of the user, the transmit power of that user, and the interference field. This means that the evolution of the aggregative term I_t is the key parameter that drives the strategic aspect and the performance of a user. For this reason, we refer to this problem as a *reaction to the interference field*. The access probability and density of success can be expressed as functions of the interference field.

3.2.9.1 Mean-Field Stochastic Game

The mean-field approach provides a natural way of defining and computing macroscopic properties of dynamic wireless networks by averaging over all spatial-temporal patterns for the cognitive users. Modeling wireless networks in terms of *mean-field interaction* seems particularly relevant for large-scale interacting cognitive networks. The idea is to treat such a network as a collection of interacting cognitive users characterized by dynamic random processes in the entire Euclidean plane or space and analyzing it in a stochastic optimization, strategic decision-making, and probabilistic way. In particular, the energy state, channel gains, and the locations of the network elements are seen as realizations of some random processes. A key property in this approach is the asymptotic indistinguishability per class property, that is, a class of power control strategies with the property that asymptotically, the law of any collection of the states becomes invariant under any permutation of the indices of the users within the same class/type.

In the cognitive setting, we have two classes of users (primary users and cognitive users), and for each class, the asymptotic indistinguishability property holds. Since the underlying random model satisfies the asymptotic indistinguishability per class property, the stochastic control and large deviation analysis provide a way of estimating spatial mean field which often capture the key dependencies of the network performance characteristics (access probability, connectivity, probability of success, capacity, energy efficiency, etc.) as functions of a relatively small number of parameters. Typically, these are the densities of the underlying controlled random processes. As we have seen, the dependency in the density can be significantly reduced by exploiting the structure of the problem and considering interference field instead of the whole distribution [460, 464].

The main advantage of the mean-field game approach in wireless networks is that it describes not only the optimal power management but also the evolutions of density of users, interference, and SINR as solutions of controlled difference equations.

In the mean-field stochastic game, the instantaneous payoff function of cognitive user is

$$R: \quad (s, y, p, I) \longmapsto R(s, y, p, I) \in \mathbb{R}.$$

Each cognitive user optimizes its long-term payoff. We assume that $R(s_t, y_t, p_t, I_t)$ is positive and increasing in the transmit power p_t. As a consequence, the future payoff decreases with the state as well as the remaining energy decreases with the consumption. Any interior optimal response to the interference field maximizes a sum of R and v_{t+1}.

On a Game with One-and-a-Half Player

The mean-field stochastic game with state payoff given by $R(s_t, y_t, p_t, I_t)$ can be interpreted as a one-and-a-half player game. It is not a two-player game, because the interference cannot be separated from the individual decision. It is not a single-player game, either, because the interference field is not totally controlled by a single transmitter, and effect of transmitter-receiver pair will be negligible at some spatial position.

Why One-and-a-Half Player?
Each generic transmitter-receiver pair is somehow part of the field I_t and has the characteristics s_t, y_t, p_t. These altogether describe completely the dynamics of the system and the outcome of the game. Therefore, it is a one-and-a-half player game.

This interpretation may appear to be surprising, since in many cases, a player is seen as a decision-maker. Then, what is the meaning of "half-player"?

By one-and-a-half player, we mean that this is not a one-player interaction (we have more than one characteristic), and it is less than two independent characteristics.

How Useful Is the One-and-a-Half Player Game?
The interpretation above allows us to understand the complexity reduction of the mean-field stochastic game. It says that the key factors are composed of individual characteristics and the interference field, which simplifies drastically the analysis of the complex systems with many transmitter-receiver pairs if one can solve the resulting mean-field systems. The one-and-a-half player game represents the mean-field stochastic game driven by the interference field.

Strategies and Payoffs
A history for a generic player at time t is a vector $h_t = (s_0, y_0, p_0, I_0, \ldots, I_{t-1}, s_t, y_t)$. Based on its own history, h_t, the generic player makes a decision at time t. The collection of the functions that map histories to the action space constitutes a pure (behavioral) strategy for the long-term game. The initial state and the strategy determine together a probability measure over the horizon $\mathcal{T}$. Notice that, in order to play the game, the only information requirement to the user is the interference field $I_t(y)$. The expected payoff of the player is the one induced by the initial state and the strategy:

$$F_T = \mathbb{E}\left(g(s_T) + \sum_{t=0}^{T-1} R(s_t, y_t, p_t, I_t(y_t)) \mid s_0, y_0, I_0 \right)$$

Mean-Field Equilibrium
We define a mean-field equilibrium as a pair of sequence of strategies $\{p_t^*\}_t$ and interference $\{I_t^*\}_t$ such that $\{p_t^*\}_t$ is an optimal response to the long-term payoff generated with $\{I_t^*\}_t$ and $\{I_t^*\}_t$ is consistent with the transmit power $p_t^*\}$. This leads to a fixed-point equation between the two processes which can be reduced to a fixed point between the equilibrium payoff $\{v_t^*\}$ and the interference $\{I_t^*\}_t$.

Reacting to a Deterministic "field" Each user interacts with a large representative subset of the entire population, i.e., is reacting to the interference "field" which evolves (forward) in time. In the large population regime, each user is facing to a deterministic and time-varying object which is the limiting scaled interference. To play the mean-field game, we then need to understand only how the interference field will evolve, since all the states of the other users are transformed in the aggregative term; we do not need the knowledge of each of them separately and build the strategy based on the interference field.

Structure of the Optimal Transmit Power

Let $\tilde{A}_{j,t}$ be the random variable representing the acceptance condition of the cognitive user j at time t. Clearly, the optimal transmit power $p_t^*(s, y) = 0$ if there is no energy or no data in the queue. The dependence of the optimal power in terms of queue size is very important, since the cognitive user will be able to save energy if its queue is empty. This allows one to take into consideration non-saturated queue analysis. In the same line, we have $p_t^*(s, y) = 0$ if $\tilde{A}_{j,t} = 0$. Notice that even if the instantaneous payoff is nondecreasing in the transmit power, the strategy consisting of transmitting with the maximum power at each step t is not optimal in the long term. This is because the remaining energy (or battery lifetime) goes down rapidly (for T sufficiently large enough). Thus, the optimal power which is a maximizer of

$$R(s_t, y_t, p_t, I_t^*) + \mathbb{E}v_{t+1}(h_{t+1}, E_t - p_t + w_t, x_{t+1}, q_{t+1}, y_t),$$

balances the instantaneous payoff and the future payoff.

Performance

We provide a generic expression of the access probability of cognitive users. Given the position of the primary receiver at $y_{PU,t}$, and that of the j-th secondary user located at $x_{j,t}$ at time t, the access probability of the latter at time t is

$$\mathbb{P}\left(\tilde{A}_{j,t} = 1\right) = \mathbb{P}\left(\frac{p^*(s_{j,t}, y_{PU,t})|h_{jPU,t}|^2}{l(\|x_{j,t} - y_{PU,t}\|)} \leq \bar{\gamma}\right) \tag{3.37}$$

where $\tilde{A}_{j,t}$ is the acceptance condition of user j and $\bar{\gamma} > 0$ is a fixed threshold.

The coverage probability of cognitive user for a given receiver position is defined as the probability of the signal-to-interference-plus-noise ratio to be greater than a certain threshold. This is a key performance metric, since it is involved in the throughput, outage probability, and queue dynamics. The performance R takes into consideration the process $\tilde{A}_{j,t}$ and hence the access conditions for cognitive users.

Let $v_t(s, y)$ be the continuation equilibrium payoff starting from the individual state s and receiver located at y at time t. Then v_t and I_t satisfy the fixed-point equations called mean-field system:

$$v_{T+1}(s, y) = \bar{g}(s, y), s = (h, E, x, q)$$

$$v_t(s_t, y) = R(s_t, y, p_t^*, I_t)$$

$$+\mathbb{E}v_{t+1}(h_{t+1}, E_t - p_t^* + w_t, x_{t+1}, q_{t+1}, y),$$

$$I_t(y) := \int \frac{|h|^2 p_t^*(s, y)}{l(\left|x - y\right|)} m_t(ds)$$

$$\bar{p}_t^* = \bar{m}_{E^*,t} - \bar{m}_{E^*,t+1},$$

$$I_0 \text{ fixed.}$$

Mean-Field Game

We now examine a deterministic mean-field game by fixing the variance of the above state processes to be zero. The deterministic mean-field system is

$$v_{T+1}(s, y) = \bar{g}(s, y), \; s = (h, E, x, q)$$

$$v_t(h_t, E_t, x_t, q_t, y) = R(h_t, E_t, x_t, q_t, y, p_t^*, I_t)$$
$$+ v_{t+1}(h_{t+1}, E - p_t^*, x_{t+1}, q_{t+1}, y),$$

$$I_t(y) := \int \frac{|h|^2 p_t^*(s,y)}{l(\big|x-y\big|)} m_t(ds)$$

$$\bar{p}_t^* = \bar{m}_{E^*,t} - \bar{m}_{E^*,t+1},$$

$$I_0 \text{ fixed.}$$

As a consequence, we deduce that

$$v_t(.., E, y) - v_{t+1}(.., E - p_t^*, y) = R(s, y, p_t^*, I_t) \geq 0$$

Hence, the equilibrium payoff is increasing with initial energy E. Using the same relation, one gets that the value function decreases with time.

Next, we aim to understand the behavior of the optimal strategy as a function of the available energy. In a one-shot game we know that the strategy when active decreases the available energy. The question is whether such property holds for the mean-field difference game.

Let p be the optimal strategy (when active) associated with E and p' the one associated with E'. Assume that $E \leq E'$. At the final stage, we know that Bellman terminal condition that $p \leq p'$. Now, at stage $T - 1$, the remaining energies satisfy $E \leq E'$; thus, the best response yields $p_{\max} \leq p'_{\max}$, i.e., $p \leq p'$.

The intuition above can be generalized using the so-called S-modular game. A game is called supermodular (submodular) if it satisfies the increasing (decreasing) difference property. Here we look at in terms of E, p and I.

Using S-modularity, one has

$$v_t(.., E, y) - v_t(.., E', y) = R(s, y, p_t, I_t) - R(s, y, p'_t, I_t) \leq 0.$$

This shows that the action p increases with E. Note that if R does not necessarily increase with p, the above structural result may not hold. However, the fact that the energy goes down with consumption and time remains can be exploited to bound the optimal power.

Hierarchical Mean-Field Game

In the reaction to interference game between primary user and cognitive users, there must be an equation to express the dynamic optimization problem of the primary user and each of the cognitive users. Suppose now that the primary user's state and strategy can be detected by the active cognitive users before they make their decisions. Then, each cognitive user reacts to the primary user strategy and the cumulative interference which is a combination of I_t and $\frac{p_{PU,t}|h_{PUr,t}|^2}{l(d_{PUr,t})}$. Hence, each cognitive user solves the system given in Sect. 3.2.9.1. Denote by $(p^*_{SU,t}(p_{PU}), m^*_t(p_{PU}))$ a solution. Then, the primary user anticipates and optimizes its long-term payoff in which the term $(p^*_{SU,t}(p_{PU}), m^*_t(p_{PU}))$ is plugged in the payoff evaluation.

3.2.10 Notes

The instantaneous payoff function proposed in this section is generic and is not explicit. We have restricted ourselves to functions that are positive and increasing in the transmit power of the corresponding user. Thus, the proposed approach can be used in many wireless situations in which this property is preserved. Most of the utility or payoff functions proposed in the literature are for a fixed coding/decoding scheme, specific interference management, and modulation techniques that are imposed to the users. Recent advances in that direction suggest a strategic component of these features by the users leading to the so-called *strategic information theory*. This will of course modify the performance bounds as well as the strategic behavior of the users depending of the information available to the users. The mean-field game framework can be applied in strategic information theory-based payoffs if the above conditions are met. However, the optimal interference and spectrum management under novel payoffs remain open. In this section, we have presented spectrum access and interference management problems in large-scale networks using mean-field stochastic games. We have formulated the problem as a strategic decision-making problem between cognitive users subject to interference limitations to licensed users. By exploiting the structure of the game, we have shown that the problem can be reduced to a *reaction to interference field* which simplifies drastically the analysis of the problem. We conducted equilibrium analysis in the asymptotic regime, where a mean-field system consisting of the evolution of the interference (forward in time) and an equilibrium payoff evolution (backward in time) is provided. Finally, we argued that the framework can be adapted to deterministic systems. The mobility of the users will be discussed in the next section, Sect. 3.3.

3.3 Mobility and Power Savings in Wireless Networks

We consider in this section large-scale wireless network with arbitrarily varying channels and mobility of users. Each user has its own battery state which decreases stochastically with the power consumption and is driven by a reflected Brownian motion. Using mean-field game framework, we describe the evolution of signal-to-interference ratio, the optimal power, and the energy cost. The evolution of density of battery state and users' distribution allow us to compute the probability of success, density of success per unit of area, as well as the Shannon rate of a generic user in the wireless network.

The mean-field approach provides a natural way of defining and computing macroscopic properties of dynamic wireless networks by averaging over all spatial-temporal patterns for the nodes. Modeling wireless networks in terms of *mean-field stochastic games* seems particularly relevant for large-scale interacting networks. The idea entails treating such a network as a collection of interacting nodes characterized by dynamic random processes in the whole Euclidean plane or space and analyzing it via probability theory, optimization, and game theory. In particular, the channel gains and the locations of the network elements are seen as the realizations of some random processes. When the underlying random model satisfies asymptotic indistinguishability per class property, the stochastic control and large deviation analysis provide a way of estimating spatial mean field which often capture the key dependencies of the network performance characteristics (connectivity, probability of success, capacity, energy efficiency, etc.) as functions of a relatively small number of parameters. Typically, these are the densities of the underlying controlled random processes.

3.3.1 Wireless Model

In recent years, many different organizations all over the world have been involved in activities of wireless communications. The quality of this kind of communication suffers from strong variations of the received signal power due to signal shadowing and multipath fading (spatiotemporal). The shadowing of the signal by obstacles in the propagation path (buildings, bridges, trees, etc.) results in attenuation over the entire signal bandwidth. The shadowed areas are larger for low elevations than for high elevations. Multipath fading occurs, because the signal is received not only via the direct path but also after having been reflected from objects in the surroundings. Due to their different propagation distances, multipath signals can add destructively, resulting in a deep fade. We propose mathematical models for channel state and mobility based on stochastic equations that capture the stochastic nature of the network.

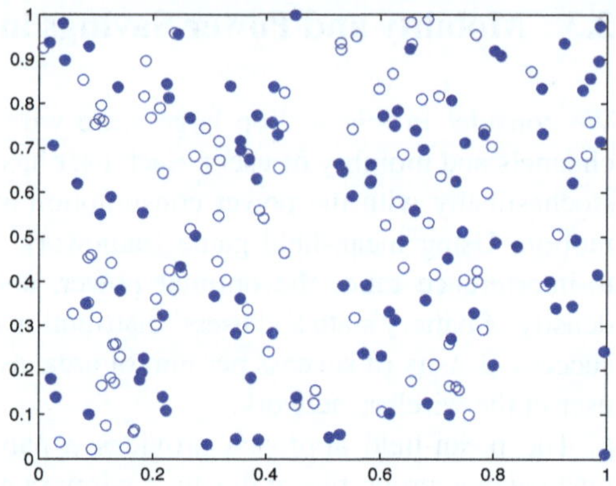

Fig. 3.1 A typical scenario of large-scale networks

We consider large-scale wireless network distributed over an Euclidean space (see Fig. 3.1) which we model as a collection of nodes, located in some domain $\mathscr{D}$, which can in turn be transmitters or receivers (depending on the network considered, nodes may be mobile users, base stations in a cellular network, access points of 802.11 mesh, long-term evolution, etc.). The set of users is denoted by $\mathscr{I}$. Below we describe the battery management, the mobility of users, and the signal-to-interference-plus-noise ratio (SINR) which include the channel gains, the locations of the users, and the transmit powers.

3.3.2 Battery Management

We propose a stochastic modeling of the battery state. Each user j has its own battery state represented by

$$db_{j,t}=-p_{j,t}dt+\sigma_t dB_{j,t}, \tag{3.38}$$

where $p_{j,t}$ is the power transmitted by j at time t, and $\sigma_t > 0$ is the variance of the noise at time t. The battery state $E_{j,t}$ is subject to a positivity constraint. $B_{j,t}$ denotes a standard Brownian motion (also called Wiener process). We assume that the process $B_{j,t}$ is independent across the players and across time. This basic model captures the scenario where the user starts with a total budget of energy which decreases during the course of the game due to consumption. The uncertainty (of Brownian type) captures the stochastic nature of the battery.

Note that the battery state processes $E_{j,t}$ may not be independent. They can be correlated via the power management strategy $p_{j,t}$ which we will be adjusted depending on the interference and the battery state. The fundamental energy-constraint question of the user is when to be active and how much power to use in view of the battery lifetime. Figure 3.2 illustrates a typical example of the battery state decay.

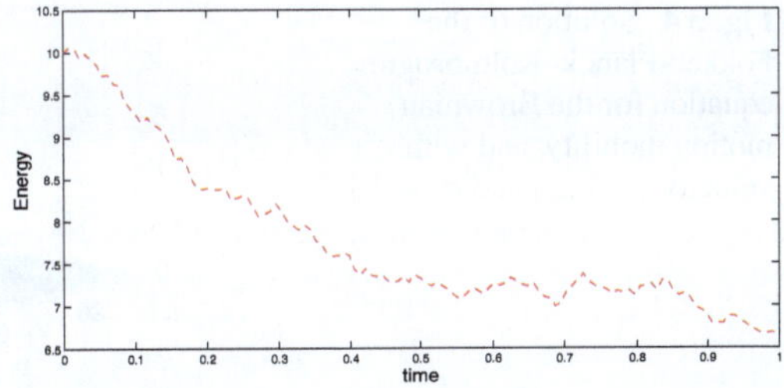

Fig. 3.2 Battery decay model driven by a Brownian motion

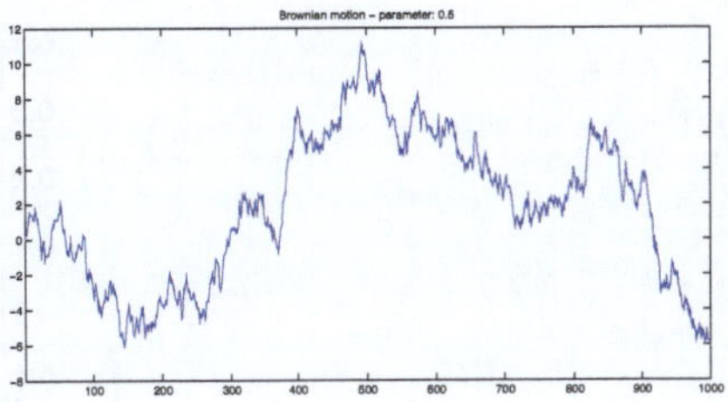

Fig. 3.3 Brownian motion (Bm)

3.3.3 *Mobility of Users*

At each time t, each user j is located at some position $x_{j,t}$ which changes according to a mobility pattern. An example of mobility could be Brownian motion mobility or fractional Brownian motion. We consider Brownian mobility in the form $dx = \tilde{v}(t,x)dt + \tilde{\sigma} dB_t$ where $\tilde{v} \in \mathbb{R}^k,\ k \in \{1,2\}$ is the velocity and $\tilde{\sigma} > 0$ (Fig. 3.3).

The density of the above mobility model can be characterized by a partial differential equation (PDE) as known as the Fokker–Planck–Kolmogorov equation (FPK):

$$\partial_t \nu_t(x) + \partial_x(\tilde{v}(t,x)\nu_t(x)) - \frac{\tilde{\sigma}^2}{2}\partial^2_{xx}\nu_t(x) = 0. \tag{3.39}$$

Recall that a Gaussian with mean $\bar{m}$ and variance σ^2 has a density given by

$$\nu_0(x) = \frac{1}{\sqrt{2\pi\sigma^2}} e^{-\frac{(x-\bar{m})^2}{2\sigma^2}}.$$

For $\tilde{v} = 0$, the fundamental solution of (3.39) is explicitly given by a Gaussian density with (time-varying) variance $\sigma^2 t$

$$\nu_t(x) = \frac{1}{\sqrt{2\pi\sigma^2 t}} \exp\left(-\frac{x^2}{2\sigma^2 t}\right).$$

In Fig. 3.4, we plot the evolution of the density, the density captured at time $t = 10$, and the evolution of the second moment starting from the point 0.

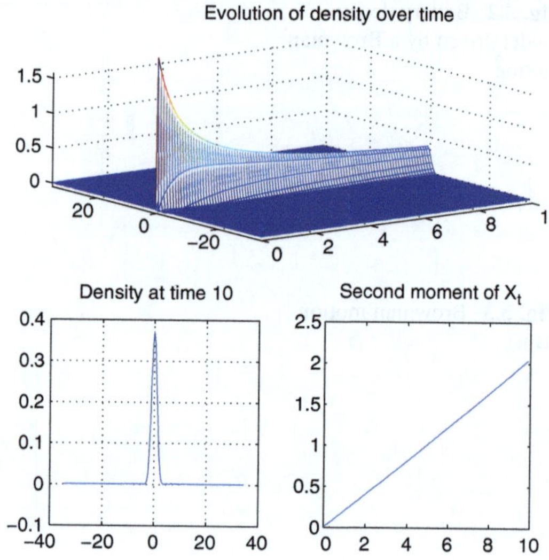

Fig. 3.4 Solution of the Fokker–Planck–Kolmorogrov equation for the Brownian motion mobility, and with velocity $\epsilon \sin(x_t)$ and $\sigma = 0.3$

The distance of the transmitter to the receiver plays an important role. We denote by

$$d_{ij,t} = \| x_{j,t} - y_{i,t} \|$$

the usual Euclidean distance from transmitter j to the receiver i located at position y_i.

SINR

At a given time, several nodes transmit simultaneously, each toward its own receiver. Each transmitter-receiver pair requires its own wireless link. The signal received from the link transmitter may be jammed by the signals received from the other transmitters. Even in the classical scenario where the signal power radiated from a point decays in an isotropic way with Euclidean distance, the geometry of the locations of the nodes plays a key role, since it determines the signal-to-interference-and-noise ratio (SINR) at each receiver and hence the possibility of establishing simultaneously this collection of links at a given bit rate. The interference seen by a receiver is the sum of the signal powers received from all transmitters, except its own transmitter. The power received from transmitter (node) j by the receiver of node i at time t is then equal to $p_{j,t}g_{ij,t}/l(d_{ji,t})$ where $g_{jj,t}$ is the channel gain which we assume to be independent and identically distributed across the nodes and across time. An important special case consists in taking

$$l(d) = \left(A(d^2 + \epsilon_0^2)\right)^{\alpha/2}, \ A > 0, \ \alpha > 2, \epsilon_0 > 0.$$

Each user is characterized by its battery state $b_{j,t}$, its power $p_{j,t}$, its channel gain $g_{ij,t}$ and its location (distance) $d_{ij,t}$. All the parameters are stochastic and may change in time t. We refer to *battery-state channel gain location* of a user as its state characteristic.

The signal-to-interference-plus-noise ratio (SINR) is given by

$$\mathrm{SINR}_{j,t} := \frac{p_{j,t} g_{jj,t}/l(d_{ji,t})}{N_0 + I_{j,t}},$$

where $I_{j,t}$ is the scaled interference from the other users and N_0 is a background noise.

We say that transmitter j covers its receiver i in the reference time slot t if $\mathrm{SINR}_{j,t} \geq \underline{\gamma}$, where $\underline{\gamma}$ is a SINR threshold. The threshold $\underline{\gamma}$ characterizes the minimum requirement and is chosen such that $\underline{\gamma} < \frac{\bar{p}\bar{g}}{N_0}$, for some power $\bar{p} > 0$ and some $\bar{g} > 0$. This means that when the received power is small (the channel gain is too small or power is too small or both are small) or when the interference is very high, the minimum requirement may not be satisfied. On the other hand, if interference is relatively small and channel is good, the good location of user and associated power can be found depending on the battery state (if no energy is available, the user cannot transmit).

Remark 3.3 The virtual power $p_{j,t} g_{kj,t}$ (the virtual received power at receiver k from user j) is clearly not an i.i.d, random variable. As we will see later, the optimal power will be a function of the battery state, the mean interference at the receiver of j and the channel gain and the location of the user. So, the optimal power is a response to a mean-field term which depends on the reaction of the other users, thus *interdependence*. We clarify this statement in Sect. 3.3.4 where we provide explicitly the optimal power.

Instantaneous Payoff

Based on the SINR, we construct an instantaneous payoff function of user j as a tradeoff between probability of success and power consumption and maintenance cost $c(s_{j,t}, p_{j,t})$. An example of a cost function would be $c(s_j, p_j) = \lambda p_j + \tilde{\lambda}$ where $\tilde{\lambda}$ is a fixed energy cost (independent of s_j) and λp_j is the power consumption cost with $\lambda > 0$. The instantaneous payoff function is then

$$r_{j,t} = \mathbb{1}_{\{\mathrm{SINR}_{j,t} \geq \underline{\gamma}\}} - c(s_{j,t}, p_{j,t}).$$

Using the relation $\mathbb{E}(\mathbb{1}_A(X_t)) = \mathbb{P}(X_t \in A)$, we deduce that when there is no cost term ($c \equiv 0$), the expected value of $r_{j,t}$ is the probability of success. Each user j optimizes its long-term utility by choosing its power management $p_{j,t}$ during its sojourn time in the network.

In the mean-field regime, the optimal response $p_t^* = p^*[t, b_t, g_t, x_t, I(m_t)]$ to the interference $I(m_t)$ is obtained by solving a system of coupled partial differential equations (PDEs): The first PDE is a Hamilton–Jacobi–Bellman equation which captures the evolution of the optimal utility, and the second PDE is the evolution of the mean-field $m_t := m(t, b, g, x)$. Below we summarize the key parameters of the stochastic game in the network.

Players	Users
User's state	$(b_{j,t}, g_{j,t}, x_{j,t}),\ b_{j,t} \geq 0$
Actions	power $p_{j,t} \geq 0$
Instantaneous payoff	$r_{j,t}$
Mass	m_t

For example, if (b_t, g_t, x_t) denotes the random variable representing the joint characteristic of a generic user, i.e., *battery state-channel gain-location*, then the evolution of the law of (b_t, g_t, x_t) is the solution of a (multidimensional) Fokker–Planck–Kolmogorov (FPK) equation. The marginal density of battery state for a fixed (g, x) is the solution of

$$\partial_t \bar{m}_t(b) - \partial_b \left(p_t^*(b, g, x, I(m))\bar{m}_t(b)\right) = \frac{\sigma_t^2}{2}\partial_{bb}^2 \bar{m}_t(b) \tag{3.40}$$

$$\bar{m}(t = 0, b) = \rho(b) \tag{3.41}$$

where $\rho(.)$ is the initial density of energy state and ∂_y denotes the partial differential operator.

The FPK equation has many advantages in terms of the wireless network characteristics: It generates all the moments by integration and using Itô's formula. In contrast of the classical approaches used in the wireless literature, the mean-field approach provides not only the first or second moments but also the evolution of the density distribution of users, their battery, and their channel gains for any time. Using the complementary PDE (3.47), we characterize the evolution of the cumulative function of optimal power management in the domain as well as the evolution of the outage probability which is a key performance metric of the network.

Stopping Time

The first time of the remaining energy to be less than ϵ can be chosen as a stopping time of the game. Typically, the battery state is initialized at $B_0 > \epsilon > 0$ at time 0. We define

$$T_\epsilon = \inf\{t \mid t \geq 0,\ b_t \leq \epsilon\}.$$

Note that the user may sometimes want to be in sleep mode: When the interference is very high and the system is congested around its location, a user may prefer to go into sleep/inactive mode in order to save energy and wake up later. This is taken into account in the strategy output $p_{j,t}(.)$. The value of $p_{j,t}$ will be 0 if user j is in sleep mode. The power chosen at time t should be between 0 and the maximum available power at the time. Without loss of generality, one can restrict $p_{j,t}$ to lie in the interval $[0, \xi(x_{j,t})]$ for a certain monotone and positive function ξ.

Best Receiver Association
Optimize over y_j. Find the best receiver (not necessarily the closest) for the user j to connect to at time t depending on its location and the receiver distributions.

Total Power Minimization Under QoS
Minimize $\int_0^T p_{j,t}dt$ such that $SINR_{j,t} \geq \underline{\gamma}$ if the user j is active.

3.3.4 The Micro Game

We now consider a network with size $I < +\infty$. In order to emphasize the fact that the network is finite, we add an index "I" to the parameters introduced below. The problem formulated is a stochastic differential game given by

- $\mathcal{I} = \{1, 2, \ldots, I\}$ is the set of users.
- Let $0 < T < +\infty$. For $t \in [0, T] =: \mathcal{T}$, the control of j corresponds to $p^I_{j,t} \in \mathbb{R}$, i.e., the transmitted power of j.
- The instantaneous payoff of user j at time t is defined by $r_{j,t} : \mathcal{S} \times \mathbb{R}^2 \longrightarrow \mathbb{R}$,

$$r_{j,t}(s^I_{j,t}, p^I_{j,t}, I^I_{j,t}) = \mathbb{1}_{\{\text{SINR}_{j,t} \geq \underline{\gamma}\}} - c_t(s^I_{j,t}, p^I_{j,t}).$$

- Each individual state $s^I_{j,t}$ evolves according to the controlled SDE (3.38), the channel gain, and mobility pattern.

Let $\mathcal{F}^I_t = \sigma(s^I_{t'}, B_{t'}, t' \leq t), t \in \mathcal{T}$, be a sigma-algebra generated by state processes and the standard Brownian motion. The filtration $\mathcal{F}^I_t$ plays an important role in the construction of a strategy at time t. Next, we define key concepts for the control of an user.

- An open-loop control for user j at time t is function of time
- A state-feedback control for user j at time t is function of $s^I_t = (s^I_{1,t}, \ldots, s^I_{I,t})$ and time.
- An own-state-feedback control for user j at time t is a function of $s^I_{j,t}$ and time.
- An aggregate-based and own-state feedback control at time t is a function of $s^I_{j,t}$, the current mean-field term $M^I_t = \frac{1}{I}\sum_{j\in\mathcal{I}} \delta_{s^I_{j,t}}$ and time, where δ denotes the Dirac measure.

Evolution of the Interference

The interference is a stochastic process, since the power is based on the state $s_{j,t} = (b_{j,t}, g_{j,t}, x_{j,t})$. We provide an explicit and exact formula for the interference $I_{j,t}$ using stochastic calculus tools.

We recall first Itô's formula for SDE:

Let $dx_t = \tilde{f}(t, x_t)dt + \tilde{\sigma}(t, x_t)dB_t$, and $l(t, x)$ be a real-valued function, defined for $t \in [0, T]$ and $x \in \mathbb{R}^I$, with continuous partial derivatives $\partial_t l(t, x), \partial_x l(t, x), \partial^2_{xx} l(t, x)$

If $y_t = l(t, x_t)$ then

$$dy_t = \left[\frac{\partial}{\partial t} l(t, x_t) + \langle \tilde{f}(t, x_t), \partial_x l(t, x_t)\rangle + \frac{1}{2} tr\left[\tilde{\sigma}\partial^2_{xx} l \tilde{\sigma}'\right]\right] dt + (\partial_x l)'\tilde{\sigma} dB_t \tag{3.42}$$

where $\partial_x l(t, x_t)$ is the gradient with the respect to the vector x, $\partial^2_{xx} l$ is the Hessian matrix, and σ_t' is the transposition of σ_t.

By applying Itô's formula, we obtain the SDE satisfied by the interference $I_{j,t}(s_t^I, y_t^I)$.

Microscopic Stochastic Interaction Between Users

We say that a collection of mappings $\{p^I_{j,t}(.)\}_{j,t}$ constitutes an *own-state-and-aggregate-based feedback (Nash) equilibrium* if for every user j, $p^I_{j,t}$ maximizes its long-term payoff $R^I_{j,T}(., p^I_{-j})$ where $\{p^I_{i,t}(.)\}_{i \neq j, t \leq T}$.

Similarly, we say that $\{p^I_{j,t}(.)\}_{j,t}$ is own-state feedback (Nash) equilibrium if it is a (Nash) equilibrium and the control is of own-state feedback type.

Clearly, in general, there is no reason for own-state feedback control to be optimal because of imperfect observation of the other states. However, due to the limitations in terms observations, these types of controls are easily implementable. One can use own-state-and-aggregate-based feedback strategy if the environment allows the user to observe the statistics of the aggregative term $\bar{I}_t^I$ at time t via mean-field dynamics.

In the mean-field setting, when $I \longrightarrow +\infty$., we will discuss how to compute $\bar{I}_t^I$, using the Fokker–Planck–Kolmogorov (FPK) equation. Below we state a general methodology for finding own-state-and-aggregate-based feedback strategy which is a best response to $\bar{I}_t^I(y)$.

Let η_t be the information pattern at time t. Then, a feedback equilibrium is defined as follows:

Definition 3.4 A strategy profile in the form $p^I_{j,t} = u^I_{j,t}(t, \eta_t)$ constitutes a feedback Nash equilibrium if there exist functions $v_j^I(t, s_1^I, \ldots, s_I^I)$ defined on $\mathcal{T} \times \mathcal{S}^I$ and satisfying the following relations for each $j \in \mathcal{I}$:

$$\begin{cases} v_j^I(T, s^I) = q(s^I), \\ v_j^I(t, s^I) = q(s_{j,T}^*) + \int_{t' \in [t,T]} r_{j,t'}(s_{j,t}^*, u_{j,t'}^*(\eta_{t'}), \bar{I}_{j,t}^*)\, dt' \\ \quad \geq q(s_T^{*,(j)}) + \int_{t' \in [t,T]} r_{j,t'}(s_{j,t'}^{*,(j)}, a_{j,t'}(\eta_{t'}), \bar{I}_{j,t}^*)\, dt', \end{cases}$$

for all $a_{j,.}(.)$, $s^I \in \mathbb{R}^I$, where on the interval $[t, T]$,

$$\begin{cases} ds_{j,t'}^* = -u_{j,t'}^*(\eta_{t'})dt + \sigma_t dB_{j,t}, \\ s_{j,0}^* = s_j^I,\ j \in \mathcal{I} \\ ds_{t'}^{*,(j)} = -a_{j,t'}(\eta_{t'})dt + \sigma_t dB_{j,t}, \\ s_{j,0}^{*,(j)} = s_j^I,\ j \in \mathcal{I}. \end{cases}$$

The following lemma provides a way of finding a feedback Nash equilibrium under regular payoff and perfect state observation (of the full vector $s^I = (s_1^I, \ldots, s_I^I)$).

Lemma 3.5 *A strategy profile $p_{j,t} = u_{j,t}(s^I) \in \mathbb{R}$, $j \in \mathcal{I}, t \in \mathcal{T}$ provides a feedback Nash equilibrium if there exist continuously differentiable functions v_j^I : $\mathcal{T} \times \mathcal{S}^I \longrightarrow \mathbb{R}$, $j \in \mathcal{I}$ satisfying the following set of partial differential equations called Hamilton–Jacobi–Bellman equations:*

$$\begin{cases} v_j^I\big(T, s^I\big) = q\big(s^I\big) \\ 0 = \frac{\partial}{\partial t} v_j\big(t, s^I\big) + \sup_{a_j'(.)} \Big\{ r_{j,t}\big(s_j^I, a_{j,t}'(s^I), \bar{I}_{j,t}^I(s^I)\big) \\ \quad + \langle \partial_{s^I} v_j^I(t, s^I), f(s^I, a_{j,t}'(s^I)) \rangle \Big\} \\ \quad + \frac{1}{2}\sigma_t^2 \partial_{s^I s^I}^2 v_j^I \\ = r_{j,t}\big(s_j^I, u_{j,t}, \bar{I}_t^I\big) + \langle \partial_{s^I} v_j^I(t, s^I), f(s^I, u_{j,t}(s^I)) \rangle \\ \quad + \frac{1}{2}\sigma_t^2 \partial_{s^I s^I}^2 v_j^I \end{cases}$$

Since the instantaneous payoff is discontinuous in p, it will be difficult to exploit useful information from this lemma regarding our problem of success probability. Moreover, it will be very difficult to have access to all the states (battery, channel, position) of all the transmitters and receivers in a large-scale wireless network. Therefore, we propose to look at the asymptotic case. Below we show how to use mean-field game to overcome the curse of dimensionality.

Micro to Macro

In accordance to the above, we now study the case where $I \longrightarrow +\infty$. The key element in the convergence to mean field will be the aggregate interference term (scaled interference) defined by

$$\bar{I}_{j,t}^I(y_j) := \frac{1}{I-1} I_{j,t}^I = \frac{1}{I-1} \sum_{k \neq j} p_{k,t} g_{kj,t} / l(d_{kj}).$$

Next, we discuss how to prove that the aggregative term $\bar{I}_t^I$ converges to $\bar{I}_t$ when $I \longrightarrow +\infty$?

To prove weak convergence, we use the sufficiency conditions called *asymptotic indistinguishability per class property* which in this context is the following: Under the power control strategies, asymptotically, the law of any collection of the states is invariant by permutation of the index of the users. In particular, at the infinite population, the index of users does not matter; only its class (group/type) and its state (battery state, channel gains per receiver, position) matter.

Suppose that the control $p_{j,t}^I(.,.) \longrightarrow p_t(.,)$. The collection of the state system $(s_1^I, \ldots, s_I^I)$ satisfies the asymptotic indistinguishability per class property when I goes to infinity. In particular, homogeneous control law preserves the indistinguishability property.

Let $\delta_{s_{j,t}^I}$ be the Dirac measure over $s = (b, g, x)$. Let $p_t(b, g, x)$ be the power control policy of a generic user at time t with state $s = (b, g, x)$. One has

$$\bar{I}_{j,t}^I(y_j) = \frac{1}{I-1}\sum_{k\neq j} p_{k,t} g_{kj,t}/l(d_{kj}) \tag{3.43}$$

$$= \int_{s\in\mathcal{S}} \frac{p_t(b,g,x)g}{l(\parallel x - y_j \parallel)}\left(\frac{1}{I-1}\sum_{k\neq j}^{I}\delta_{s_{k,t}^I}\right)(ds) \tag{3.44}$$

$$= -\frac{1}{I-1}\frac{p_{k,t}g_{kj,t}}{l(d_{kj,t})} + \frac{I}{I-1}\int_{\mathcal{S}}\frac{p_t(b,g,x)g}{l(\parallel x - y_j \parallel)} M_t^I(ds)$$

where $M_t^I = \frac{1}{I}\sum_{k=1}^I \delta_{s_{k,t}^I}$.

This shows the following result:

Lemma 3.6 *The interference at any point of the receiver y_j has a kernel given by*

$$K(s, y_j) = \frac{p_t(b,g,x)g}{l(\parallel x - y_j \parallel)},$$

$$\bar{I}_{j,t}^I(y_j) = -\frac{1}{I-1}\frac{p_{k,t}g_{kj,t}}{l(d_{kj,t})} + \frac{I}{I-1}\int_s K(s, y_j) M_t^I(ds).$$

To prove the (weak) convergence of $\bar{I}_{j,t}^I(y_j)$, it suffices to prove that the *mean-field object* M_t^I converges weakly (i.e., in distribution).

Lemma 3.7 *If M_0^I converges in probability to $\rho_0(.)$ when I goes infinity and $p_{j,t}^I$ preserves the asymptotic indistinguishability per class property, then M_t^I converges in distribution to m_t. Moreover, the law of m_t (renamed $m_t(.)$) is solution of the Fokker–Planck–Kolmogorov equation:*

$$\begin{aligned} &\partial_t m_t(s) + \partial_b(p_t^*(.))m_t(s)) + \partial_x(\tilde{v}(t,x))m_t(s)) \\ &= \frac{\sigma_t^2}{2}\partial_{bb}^2 m_t(s) + \frac{\bar{\sigma}^2}{2}\partial_{gg}^2 m_t(s) + \frac{\tilde{\sigma}^2}{2}\partial_{xx}^2 m_t(s), \\ &m_0(s) = \rho_0(s),\ \forall s \in \mathcal{S}, \end{aligned} \tag{3.45}$$

where $p_t^*(.)$ is the optimal control strategy which is a function of $t, s, \bar{m}_t$ and the value function v_t.

Corollary 3.8 *Under the same assumptions as in Proposition 3.32, when $I \longrightarrow +\infty$,*

$$\bar{I}_{j,t}^I(y) \longrightarrow \int_{s\in\mathcal{S}} K(s,y)m_t(ds).$$

Proof The result follows from Proposition 3.32 and the fact that $\frac{1}{I-1}\frac{p_{k,t}g_{kj,t}}{l(d_{kj})}$ converges weakly to 0 and $\frac{I}{I-1} \longrightarrow 1$ when $I \longrightarrow +\infty$. ■

Continuum of Users

We now consider a stylized game with continuum of users in a homogeneous population. We define a cell associated with a point y as $\{x \mid \text{SINR}_t(x,y) \geq \underline{\gamma}\}$. This leads to the question of equilibrium cell formation.

Here we consider the one-shot game in each user chooses its action in a non-empty compact set $\mathbb{S}_b = [0, B_0] \subset \mathbb{R}$. Let $\Delta(\mathbb{S}_b)$ be the set of probability distribution on $\mathbb{S}$. We say that $m \in \Delta(\mathbb{S}_b)$ is a static equilibrium if

$$\text{supp}(m) \subseteq \arg\max_{p\in\mathbb{S}_b} \left[\mathbb{P}_s\left(\text{SINR}(s,p,I(m)) \geq \underline{\gamma}\right) - \bar{c}(p)\right]$$

Assumption C1: $\bar{c}$ Is Convex and Continuous with Respect to p

Assume C1 holds. Then, the game has at least one Nash equilibrium in pure strategies. Consider a converging subsequence $p_j^{\phi(I)}$ of symmetric pure equilibrium (power) $\left(p_j^{\phi(I)}\right)_{j\in\mathcal{I}}$ of the game with I users, $\phi(I) \longrightarrow +\infty$. Then, any limiting subsequence of $\frac{1}{I}\sum_{i=1}^{I}\delta_{p_i^I}$ leads to a static equilibrium.

Dynamic Energy Management and Mean-Field Response

In the asymptotic population regime (massively dense case), each user adjusts his power management dynamically in response to the interference caused by the mass. We formulate the mean-field response problem as

Mean-Field Response Problem

Let B be a one-dimensional Brownian motion defined on $(\Omega; \mathcal{F}; \mathbb{P})$, where $\mathcal{F}$ is the natural filtration generated by B and x_0 is a random variable

(continued)

independent to B, having density $m_0(x)$. Maximize the objective function

$$\sup_{\{p_t\}_{t\in\mathcal{T}}} \mathbb{E}\left(q(x_T) + \int_0^T \left[\mathbb{1}_{\{\text{SINR}(s_t,p_t,I(m_t^*,y))\geq\underline{\gamma}\}} - c_t(s_t,p_t)\right] dt \mid s_0\right)$$

subject to the dynamics of s_t is given by

$$\begin{aligned} db_t &= -p_t dt + \sigma_t dB_t^1,\ t \in (0,T], \\ b_0 &\in \mathbb{R}, \\ dx_t &= \tilde{v}(t,x)dt + \tilde{\sigma} dB_t^2,\ t \in (0,T], \\ x_0 &\in L^2(\mathbb{R}^m), \\ dh &= \bar{\sigma} dB_t^3, \\ g_0 &= |h_0|^2, \end{aligned} \tag{3.46}$$

where m_t^* is the optimal mean-field trajectory and the Brownian motions $B_t^k, k \in \{1,2,3\}$ are assumed to be independent.

The distribution of the optimal mean-field trajectory is obtained by solving the Fokker–Planck–Kolmogorov forward equation given by (3.40). In this equation, one has $p^*(.)$. Thus, we need to find the generic expression of an optimal control.

Lemma 3.9 *If $F_t(.)$ is the cumulative distribution function of the marginal mean-field limit (which is stochastic) with respect g, x, then, $F_t(.)$ is solution of the PDE:*

$$\partial_t F_t(b) - p_t^* \partial_b F_t(b) - \frac{\sigma_t^2}{2} \partial_{bb}^2 F_t(b) = 0. \tag{3.47}$$

Proof This is a direct integration of the FPK equation. ■

This equation is important, since it characterizes the evolution of the optimal distribution of battery state management $F_t(b) = \mathbb{P}\left(b_t^* \leq b\right)$ depending the optimal dynamic power.

Shannon Capacity as a Payoff

We define the (Shannon) throughput (bit-rate) of the channel from transmitter x_j to its receiver y_j to be

$$Thp_j = w \log\left(1 + \text{SINR}_{j,t}\right),\ w > 0. \tag{3.48}$$

Now, the payoff function is a difference between bit rate and energy cost. This is captured by the payoff

$$r_{j,t} = w \log\left(1 + \text{SINR}_{j,t}\right) - c(s_{j,t}, p_{j,t}).$$

Closed-Form Expression
We provide here the explicit form of the optimal control in response to the interference. We compute the static equilibrium for the Shannon capacity. Let $m \in \Delta(\mathbb{S}_b)$ is a static equilibrium. Then,

$$\text{supp}(m) \subseteq \arg\max_{p\in\mathbb{S}_b}[\log(1+\text{SINR}(s,p,I(m))) - c(s,p)]$$

The expression of the optimal power is obtained using the following lemma:

Lemma 3.10 *Assume that the function* $-c(s_{j,t}, p_{j,t})$ *is concave in* p_j. *Then, the optimal control is determined by*

$$\frac{a_t}{1+SINR_{j,t}} - \partial_b v_t(s) = \partial_p c(s,p),\ g \neq 0$$

where $a_t := \frac{g_t/l(d_t)}{N_0+I_t}$. *In addition, if* $c(.) = 0$, *then the optimal control is*

$$p_t^*(s) = \left[\frac{1}{\partial_b v_t(s)} - \frac{1}{a_t}\right]_+ \mathbb{1}_{\{g\neq 0\}} \tag{3.49}$$

$$= \left[\frac{1}{\partial_b v_t(s)} - \frac{(N_0+I_t(m_t,y))l(\|x-y\|)}{g}\right]_+ \mathbb{1}_{\{g\neq 0\}}$$

Remark 3.11 As we can observe, the optimal control does not consist to transmit with the maximum power at step. This is because transmitting with the maximum power gives better throughput today, but the remaining energy for the future goes down (if there is no direct harvesting from external sources). Thus, using the optimal strategy of "static problem" does not necessarily lead to optimal policy for the dynamic long-term problem as it is well-known in dynamic optimization. This example shows that we have to take into consideration the future when deciding on the power allocation today. Here the future is captured in the value function v_t starting from time t.

The mean-field equilibrium is solution of the backward/forward system:

$$\begin{aligned}
&v_T(s) = q(s),\ s\in\mathcal{S},\\
&\partial_t v_t(s) - p_t^*\partial_b v_t(s) + \partial_g v_t(s) + \partial_x v_t(s)\\
&+\log(1+p_t^*\tfrac{g_t/l(d_t)}{N_0+I_t})+\\
&\tfrac{\sigma_t^2}{2}\partial_{bb}^2 v_t(s) + \tfrac{\bar{\sigma}^2}{2}\partial_{gg}^2 v_t(s) + \tfrac{\tilde{\sigma}^2}{2}\partial_{xx}^2 v_t(s) = 0,\\
&\partial_t m_t(x) - p_t^*\partial_b m_t(x)\partial_x(\tilde{v}(t,x)v_t(x))\\
&-\tfrac{\sigma_t^2}{2}\partial_{bb}^2 m_t(s) - \tfrac{\bar{\sigma}^2}{2}\partial_{gg}^2 m_t(s) - \tfrac{\tilde{\sigma}^2}{2}\partial_{xx}^2 m_t(s) = 0,\\
&m_0(s) = \rho_0(s),\ s\in\mathcal{S}.
\end{aligned} \tag{3.50}$$

3.3.5 *Hierarchical Mean-Field Games for Cognitive Radio Networks*

In this subsection, we study mobility effects and power savings in cognitive radio networks using mean-field games. We extend the idea of one-and-a-half player game developed in Sect. 3.2 for discrete time and augment with macro-players, called primary users.

We consider two types of users: finitely many primary users and infinitely many secondary users. Each user has to sense the channel when he or she is active and is subject to a long-term energy constraint. We formulate the interaction between primary and secondary users as a hierarchical mean-field game. We provide explicit formulas and algorithmic power management for both primary and secondary users. Complete characterizations of the optimal distribution of energy and probability of success are given.

Wireless technology is proliferating rapidly, and the vision of pervasive wireless computing and communications offers the promise of many industrial and individual benefits. While consumer devices such as cell phones, personal data assistants, and laptops receive a lot of attention, the impact of wireless technology is much broader, e.g., through sensor networks for safety applications, vehicular networks, smart grid control, home automation, medical wearable and embedded wireless devices, and entertainment systems. This explosion of wireless applications creates an ever-increasing demand for more radio spectrum. However, most easily usable spectrum bands have already been allocated, although many measurements, nevertheless, show that at any specific time and location, these spectrum bands are significantly *underutilized* and considerable amount of energy is wasted.

Cognitive radios offer the promise of being a disruptive technology innovation that will enable the future wireless world. Cognitive radios are fully programmable wireless devices that can *sense*, *learn* their environment, and dynamically adapt their transmission waveforms, energy management policies, channel access methods, spectrum uses, and networking protocols as needed for good network and application performance.

Cognitive radio aims at exploiting these voids in order to accommodate extra radio devices referred to as *secondary users* (SUs) in a network whose first function is to serve the primary users (PUs), e.g., licensed to use the spectrum. The key requirement is that the primary users have to be as little affected as possible by the presence of secondary users.

The protocols are carrier sensing-based mechanisms for secondary users to exploit the spectrum left over by primary users. Each primary transmitter has a dynamic protection zone. If a secondary user located in this zone transmits at the same time, he/she interferes with the primary user. Any secondary user located in this protection zone is hence called a contender of this transmitter. In the general setting, the union of all primary transmitter protection zones does not cover the whole space. Thus, one can accommodate secondary users in the remaining space to better utilize the spatial resource under energy constraints.

This subsection concentrates on exploiting mobility and energy consumption by considering cognitive radio in a wireless network with users distributed in the Euclidean plane. In order to save energy, it is important to control the transmission power and/or the sensing threshold of secondary users in order to further limit the interference they cause to primary users. The users are energy limited, and each secondary user decides its transmission power depending on the primary signal strength and its remaining battery state.

It is not true that licensed solutions avoid interference, making them more robust than unlicensed solutions:

In reality, any wireless network, whether licensed or unlicensed, faces the risk of interference. Many claims of interference in unlicensed bands are largely overstated, and technology implementations mitigate remaining interference issues.

Licensed solutions are not more secure than unlicensed solutions:

In many practical scenarios, security is unrelated to whether a wireless system operates on licensed or unlicensed spectrum. In fact, licensed solutions that operate on fixed frequencies are often susceptible to malicious interference. A well-designed unlicensed solution offers greater protection against intentional interference or jamming.

Licensed spectrum may not ensure a long-term guarantee of availability:

Spectrum licenses last for only a fixed time (window) with no guarantee that they will be available in the future. Given the widespread commercial use of unlicensed spectrum, its future is near certain.

The licensed solutions can transmit at higher powers. However, they may not have automatic cost advantage resulting from the need for less network infrastructure. This is because licensed solutions have other hidden costs that increase the total cost of ownership, while a well-designed unlicensed solution reduces both capital and operational costs.

Primary-Secondary System

We start with a very basic model which we refer to as *model 1*. We consider two types of transmitter-receiver pairs (users): *primary* and *secondary users* distributed over a domain $\mathscr{D}$ in a Euclidean space. The physical layer setting considers a signal model with one primary user (a transmitter-receiver pair) and n secondary users. Interference is treated as noise, and the transmission is successful if the signal-to-interference-plus-noise ratio (SINR) is larger than a certain threshold (constant).

For the MAC layer setting, we use carrier sense SINR-based model. When a user is active, he/she senses the medium to check whether there is any primary contender or not. The tagged secondary user transmits only if it sees the network free of primary contenders. To do so, a user has to choose a transmit power depending on the state of the battery (if no energy is available, to user cannot transmit). Whenever he/she is active, a primary user can access the channel. Then, a primary user has to design its transmit power. For secondary users, because of interference, not every transmission attempt is successful (Fig. 3.5).

Our goal here is to address the following:

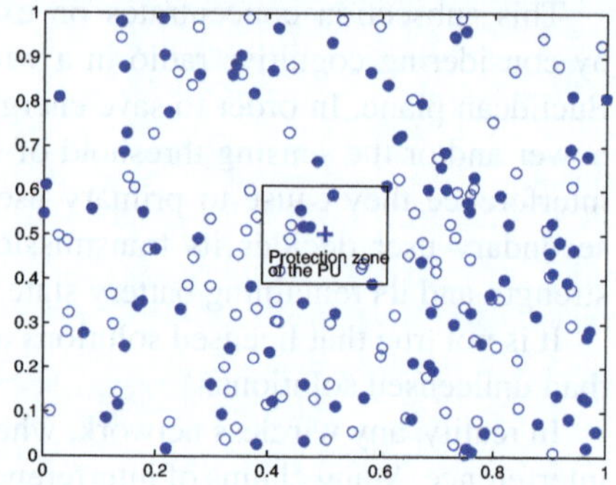

Fig. 3.5 Typical cognitive radio scenario

1. Optimal energy consumption.
2. The effects of mobility in energy consumption.
3. Optimal use of spectrum holes maintaining protection of PUs (quality of service)

Mobility of Users

The mobility of the PU is described by

$$dx_{PU,t} = v_{PU,t}(x_{PU,t})dt + \sigma_{PU} dB_t \tag{3.51}$$

where $x_{PU,t} \in \mathscr{D}$ for all $t \in \mathscr{T}$, σ_{PU} is a positive constant and the mobility of jth SU is described by

$$dx_{j,t} = v_{SU,t}(x_{j,t}, x_{PU,t})dt + \sigma_{SU} dB_{j,t} \tag{3.52}$$

where $x_{j,t} \in \mathscr{D}$ for $j \in \mathscr{I}_{SU}$ and $t \in \mathscr{T}$, σ_{SU} is constant and $v(.)$ is a velocity function.

The processes $B_{j,t}$ are standard Brownian motions (also called Wiener process) in $\mathbb{R}$, independent across the users and across time and defined on probability space $(\Omega, \mathscr{F}, \mathbb{P})$. We assume that for any user j, $B_{j,0}$ is independent of the initial distribution of users. The mobility process in (3.52) is understood as a classical Itô's stochastic differential equation (SDE).

We are also interested in optimizing the energy (battery) usage of all the users, from a initial value of E_0. We consider a model with renewable energy (e.g., solar-powered devices). In that case, a stochastic model is well justified.

Power Consumption

The battery state of each user is modeled as a stochastic process which depends on the power consumption. A generic player starts initially with energy $E_0 > 0$ (full battery level). Then, the energy usage dynamics is described by Itô's SDE

$$dE_{j,t} = -p_{j,t}dt + \sigma_E dB_{j,t}, \ j \in \mathscr{I}. \tag{3.53}$$

with constraint $E_{j,t} \geq 0$, for all $t \in \mathcal{T}$, where $p_{j,t} \geq 0$ is the power consumption of user j at time t and $\sigma_E > 0$ is a constant. The presence of the noise term is to capture the stochastic nature of the anticipated prediction and the fact that under renewable energy, the battery may be recharged if the weather conditions allow. Moreover, $p_{j,t} = 0$ if $E_{j,t} \leq \epsilon_E$, where ϵ_E is a fixed minimum energy threshold for the device to be usable. We will provide the distribution of energy in Sect. 3.4.2 using the Fokker–Planck–Kolmogorov equation.

Virtual Received Power

The virtual received power from a user i located at $x_{i,t}$ at a receiver located at $y_{j,t}$ is defined as

$$r_{i,j,t} = \frac{p_{j,t} g_{i,j,t}}{l(\|x_{i,t} - y_{j,t}\|)} \tag{3.54}$$

for $(i, j) \in \mathcal{I}^2$ where $p_{j,t}$ is the transmit power, $g_{i,j,t}$ is the channel gain (i.i.d random process), $l(\|x_{i,t} - y_{j,t}\|)$ is such that $l(d) = c\big(d^2 + \epsilon^2\big)^{\alpha/2}$ where $\alpha \geq 2$ is the pathloss exponent, $c > 0$. The term ϵ is the component of a hidden constant dimension, that is, every receiver location has components $\big(y^1_{j,t}, y^2_{j,t}, \epsilon\big)$ and every sender has components $\big(x^1_{j,t}, x^2_{j,t}, 0\big)$. This way we avoid a vanishing denominator.

We define the state of user i at time t as $s_{i,t} = \big(E_{i,t}, x_{i,t}, g_{i,\cdot,t}\big) \in \mathcal{S} = \mathcal{R}_+ \times \mathcal{D} \times \mathcal{R}_+$.

Scaled SINR

We now define the *scaled signal to interference plus noise ratio (sSINR)* from SU i to receiver j at time t as

$$SSINR_{i,j,t} := \frac{r_{i,j,t}}{N_0(y_{j,t}) + \frac{1}{I-1}\sum_{i' \neq i} r_{i',j,t} + r_{PU,j,t}} \tag{3.55}$$

for $(i, j) \in \mathcal{I}^2_{SU}$ where $N_0 : \mathcal{D} \longrightarrow \mathbb{R}_+$, $N_0(y_{j,t}) > 0$ is the ambient noise at the receiver location and $r_{PU,j,t}$ is the interference generated by the PU to the receiver located at $y_{j,t}$.

A transmission is successful from SU i to receiver j if

1. sSINR$_{i,j,t} \geq \beta_{SU}$, where β_{SU} is a fixed minimum quality of the signal, and
2. $r_{i,PU,t} \leq \gamma$, where γ is a fixed quantity, that is, the PU should not get more than a certain amount of noise from the transmission of SU.

The sSINR from PU to SU j is defined as

$$sSINR_{PU,j,t} = \frac{r_{PU,j,t}}{N_0(y_{j,t}) + \frac{1}{I}\sum_{i'=1}^{I} r_{i',j,t}} \tag{3.56}$$

Note that the scaled interference form is widely used in code division multiple access (CDMA) networks.

We say that SU i belongs to the primary user's protection zone if the virtual received power is greater than some threshold γ. The access condition is then $A_{i,t} = 1$, where $A_{i,t} = \mathbb{1}_{\{\frac{p_{i,t} g_{i,PU,t}}{l(\|x_{i,t}-y_{PU,t}\|)} \leq \gamma\}}$.

3.3.6 Asymptotics

First note that the only quantity in $sSINR_{PU,j,t}$ (3.56) which impacts in I is $\frac{1}{I}\sum_{i'=1}^{I} r_{i',j,t}$. Considering indistinguishable SU nodes, we can model the system with a mean-field interaction model.

The scaled interference from SUs to a receiver j located at $y = y_{j,t}$ at time t is

$$\frac{1}{I}\sum_{i=1}^{I} r_{i,j,t} \quad = \frac{1}{I}\sum_{i=1}^{I} \frac{p_{j,t} g_{i,j,t}}{l(\|x_{i,t}-y\|)} \tag{3.57}$$

$$= \int_{\mathcal{S}} \frac{\bar{p}_{SU,t}(s, y_{i,t}, t)\bar{g}(x_{i,t}, y)}{l(\|x_{i,t} - y\|)} M_t^I(ds) \xrightarrow{I \to +\infty} \int_{\mathcal{S}} K_t(s, y) m_t(ds) \tag{3.58}$$

where $K : \mathcal{S} \times D \to \mathbb{R}^+$ is the kernel $\bar{p}.\bar{g}/l$ and M_t^I is the occupancy measure of the $n = I$ SUs. $\bar{p}_{SU}$ is the optimal control for SUs at time t, $\bar{p}_{SU} : \mathcal{S} \times D \times \mathcal{T} \to \mathbb{R}^+$. $\bar{g}_t$ is a positive i.i.d. random process defined by $\bar{g} = |h|^2$.

Hence, the following holds:

Lemma 3.12 *Under the descriptions above, and assuming that kernel K is m-integrable, the scaled interference from the SUs converges weakly to*

$$\bar{I}_t(y) := \int_{\mathcal{S}} K_t(s, y) m_t(ds)$$

The proof follows from the mean-field convergence result provided in Appendix, Volume I of this book. ■

We have the dynamics of h_t given by

$$dh_t = \sigma_G dB_t, \tag{3.59}$$

and using Itô's rule, we get

$$d\bar{g}_t = \sigma_G^2 dt + 2\sigma_G h_t dB_t. \tag{3.60}$$

The dynamics of s_t are described by

$$ds_t = \begin{pmatrix} -p_t \\ v_t \\ \sigma_G^2 \end{pmatrix} dt + \begin{pmatrix} \sigma_{PU} & 0 & 0 \\ 0 & \sigma_E & 0 \\ 0 & 0 & 2\sigma_G h_t \end{pmatrix} dB_t = f dt + \sigma dB_t \tag{3.61}$$

Next define the continuous virtual received power from user j by a receiver located at y, $\bar{r} : \mathscr{S} \times D \times \mathscr{T} \to \mathbb{R}^+$:

$$\bar{r}_{\cdot}(s_j, y, t) = \frac{\bar{p}_{\cdot}(s_j, y, t)\bar{g}(x_j, y)}{l(\|x_j - y\|)} \tag{3.62}$$

where $\cdot$ can be PU or SU, and x_j is the location of user j.

Note that if $\bar{p}_{\cdot,t}$ is integrable with respect to the measure m_t, then $\bar{r}_{\cdot,t}$ is integrable and the resulting integral is the scaled interference $\bar{I}_{SU} : D \times \mathscr{T} \to \mathbb{R}$:

$$\bar{I}_{SU}(y, t) = \int_{s \in \mathscr{S}} \bar{r}_{SU}(s, y, t) m_t(ds) \tag{3.63}$$

and for PU,

$$\bar{I}_{PU}(y, t) = \int_{s \in \mathscr{S}} \bar{r}_{PU,t}(s, y, t)\delta_{s_{PU}}(ds) \tag{3.64}$$

Now we define the *mean SINR, (mSINR)* for PU and for SUs, given a receiver at location y :

$$mSINR_{PU}(y, t) = \frac{\bar{r}_{PU}(s_{PU}, y, t)}{N_0(y) + \bar{I}_{SU}(y, t)} \tag{3.65}$$

$$mSINR_x(y, t) = \frac{\bar{r}_{SU}(s_x, y, t)}{N_0(y) + \bar{I}_{SU}(y, t) + \bar{I}_{PU}(y, t)} \tag{3.66}$$

for $x \in D$.

Performance of a Generic User

We start with the access probability. Given the position of the primary receiver, and that of the i-th SU located at $x_t = x$ at time t, the *access probability* of the latter at time t is

$$\mathbb{P}\left(A_{i,t} = 1 | x, y\right) = \mathbb{E}(A_{i,t} | x, y) \tag{3.67}$$

$$= \mathbb{P}\left(\frac{\bar{p}(s_{SU}, y_{PU}, t)\bar{g}(x, y_{PU}, t)}{l(\|x - y_{PU,t}\|)} \leq \gamma \right) \tag{3.68}$$

Recall that the distribution of the ratio between two random variables X_t and Y_t with continuous density is given by

$$\bar{f}_{Z_t}(z) = \bar{f}_{X_t/Y_t}(z) = \int_{\mathbb{R}} |y| \, \bar{f}_{(X_t,Y_t)}(zy, y) \, dy$$

The above reasoning now leads to the following lemma:

Lemma 3.13 *The access probability for a given power $\bar{p} > 0$ is*

$$\mathbb{P}\left(\frac{\bar{g}(x, y_{PU}, t)}{l(\|x - y_{PU,t}\|)} \geq \frac{\gamma}{\bar{p}}\right) = 1 - F_{Z_t}\left(\frac{\gamma}{\bar{p}}\right)$$

where the random variable $Z_t = \frac{\bar{g}(x, y_{PU}, t)}{l(\|x - y_{PU,t}\|)}$, and F_{Z_t} is the cumulative distribution of Z_t which is ratio distribution.

Note that the denominator may not be Gaussian due to the path-loss exponent. However, using Itô's rule in the small noise regime, one can approximate with time-varying Gaussian-driven processes.

Coverage Probability

The coverage probability of a user i who transmits to receiver at $y_{i,t}$ at time t, is the probability that the $mSINR_{i,j,t}$ of the tagged user is larger than β_{SU}. Given the location of the primary receiver, the coverage probability of the primary transmission at time t is the probability that the $mSINR_{PU,PU,t}$ of the tagged user is larger than β_{PU}.

Lemma 3.14 *The coverage probability of PU at time t with a receiver located y is*

$$C_{PU}(y, t) = 1 - F_{Z_t}\left[\frac{\beta_{PU}}{\bar{p}}\left(N_0(y) + \bar{I}_{SU}(y, t)\right)\right]$$

Hence, the density of coverage of PU at time t in a protection $\mathscr{A}'$ is

$$\int_{y \in \mathscr{A}'} C_{PU}(y, t) m_{x,t}(dy).$$

Proof Let $\bar{p} > 0$ be a power the PU. Using the expression of sSINR for PU, the measure of the event $\{mSINR_{PU,PU,t} \geq \beta_{PU}\}$ is equivalent to the measure event

$$\frac{\bar{g}(x_{PU}, y_{PU}, t)}{l(\|x_{PU} - y_{PU,t}\|)} \geq \frac{\beta_{PU}}{\bar{p}}\left(N_0(y) + \bar{I}_{SU}(y, t)\right).$$

The term $\bar{I}_{SU}(y, t)$ is the mathematical expectation of a random variable. Thus, the right-hand side term $\frac{\beta_{PU}}{\bar{p}}\left(N_0(y) + \bar{I}_{SU}(y, t)\right)$ is nonrandom. Hence,

$$C_{PU}(y,t) = \mathbb{P}\left(mSINR_{PU,PU,t} \geq \beta_{PU}\right) \tag{3.69}$$

$$= 1 - F_{Z_t}\left[\frac{\beta_{PU}}{\bar{p}}\left(N_0(y) + \bar{I}_{SU}(y,t)\right)\right]. \tag{3.70}$$

■

When the realization of the state of PU and the receiver are given, a similar formula to the above can be used, and the coverage probability is given by

$\mathrm{COP}(s_{PU}, y, t) =$

$$1 - F_{Z_t}\left[\frac{\beta_{SU}}{\bar{p}}\left(N_0(y) + \bar{I}_{SU}(y,t) + \frac{\bar{p}_{PU} g_{PU,y}}{l(\| x_{PU} - y \|)}\right)\right]$$

For SU, we cannot use the above reasoning, because now, the denominator is random due to the interference of the primary user (if active) even if $\bar{p}$ is deterministic. However, we use the distribution of the difference of the random variables, which is stated in the following lemma.

Lemma 3.15 *In the presence of PU, the coverage probability of a generic SU at a receiver located at* y_{SU} *is*

$$1 - F_{\bar{Z}_t}\left(\beta_{SU}\left(N_0(y_{SU}) + \bar{I}_{SU}(y_{SU}, t)\right)\right)$$

where $\bar{Z}_t$ *is the random variable*

$$\frac{\bar{p}_{SU,t}\bar{g}(x_{SU}, y_{SU}, t)}{l(\|x_{SU} - y_{SU,t}\|)} - \beta_{SU}\frac{\bar{p}_{PU,t}\bar{g}(x_{PU}, y_{SU}, t)}{l(\|x_{PU} - y_{SU,t}\|)}.$$

Proof The coverage inequality event for fixed transmit power gives

$$\frac{\bar{p}_{SU,t}\bar{g}(x_{SU}, y_{SU}, t)}{l(\|x_{SU} - y_{SU,t}\|)} - \beta_{SU}\frac{\bar{p}_{PU,t}\bar{g}(x_{PU}, y_{SU}, t)}{l(\|x_{PU} - y_{SU,t}\|)}$$

$$\geq \beta_{SU}\left(N_0(y_{SU}) + \bar{I}_{SU}(y_{SU}, t)\right).$$

We then take the probability of that event by observing that the right-hand side is deterministic.

The probability density of $\bar{Z}_t$ is obtained by convolution of the two density functions. ■

The density of interference per area $\mathscr{A} \subset \mathscr{D}$ at time t is

$$d_{SU,t}(\mathscr{A}) = \int_{y \in \mathscr{A}} \bar{I}_{SU}(y,t) m_{x,t}(dy) \tag{3.71}$$

where $m_{x,t}$ is the marginal pdf of x. This density allows us to quantify how much interference is around the protection zone of the PU.

The distribution of the remaining energy at time t in area $\mathscr{A} \subseteq \mathscr{D}$ is

$$\mu_{E,t}(\mathscr{A}) := \int_{x \in \mathscr{A}} \int_{g \in \mathbb{R}^+} m_t(E, dx, dg) \tag{3.72}$$

and the distribution of users which have a remaining energy at least $\bar{E}$ is

$$\int_{[0,\bar{E}]} m_{E,t}(de) = F_{E,t}(\bar{E}) \tag{3.73}$$

We note that $(t, \bar{E}) \in \mathscr{T} \times \mathbb{R}^+ \longmapsto F_{E,t}(\bar{E})$ is solution of the following partial differential equation.

$$\partial_t \mu_t(\bar{E}) - \bar{p}(., t)\partial_{\bar{E}} \mu_t(\bar{E}) = \frac{1}{2}\partial^2_{\bar{E},\bar{E}} \mu_t(\bar{E}), \tag{3.74}$$

$$\mu_0(\bar{E}) = F_0(\bar{E}), \tag{3.75}$$

where $\bar{p}(., t)$ is the control of SU and F_0 is a given cumulative function.

We numerically examine four different types of control policies:

- Constant power consumption $\bar{p}(., t) = p_0$,
- Uniform power consumption $\bar{p}(., t) = E_0/T$,
- Adaptive with remaining energy and remaining time $\bar{p}(., t) = \frac{E_t}{1+T-t}$,
- Optimal control $\bar{p}(., t) = \bar{p}^*(., t)$.

Power Saving-Based Payoffs

The payoff function of PU (which will be maximized) is

$$\mathbb{E}\left(q_{PU,T}(s_{PU,T}) - \int_0^T p_{PU,t} dt\right) \tag{3.76}$$

where $q_{PU,T}$ is some terminal value. The payoff function of SU (which will be maximized) is

$$\mathbb{E}\left(q_{SU,T}(s_{i,T}) + \int_0^T \mathbb{1}_{\{mSINR_{i,j,t} \geq \beta_{SU}\}}\right) \tag{3.77}$$

where $q_{SU,T}$ is some terminal value.

The measure m_t satisfies the Fokker–Planck–Kolmogorov equation

$$\partial_t m_t(s) + \mathrm{div}_s(f m_t(s)) = \frac{1}{2}\mathrm{trace}(\sigma\sigma' \partial^2_{ss} m_t), \tag{3.78}$$

where div denotes the divergence operator and ∂^2_{ss} is the second derivative operator with respect to s.

Global Optimization

Global optimization of the probability of success for SUs given a receiver at location y_t is the optimization problem:

$$\int_0^T \int_{x\in D} m_{x,t}(dx)dt$$

$$\mathbb{P}(mSINR_x(y_t,t)\geq\beta_{SU}\cap\bar{r}_{SU}(s_x,y_{PU,t},t)\leq\gamma) \quad (3.79)$$

$$\text{s.t. } \partial_t m_{x,t} + \partial_x(vm_{x,t}) = \frac{\sigma_x^2}{2}\partial^2_{xx}m_{x,t} \quad (3.80)$$

where $m_{x,t}$ is the marginal pdf of m_t with respect to x.

Using Hamilton–Jacobi–Bellman's equation for mean-field-type global optimization, we have that the probability of success constitutes an optimal solution if there exists a function $v(m)$ which satisfies

$$\partial_t v + \int \sup_{p_{SU}}\{\mathbb{P}(mSINR_x(y_t,t)\geq\beta_{SU}\cap\bar{r}_{SU}(s_x,y_{PU,t},t)\leq\gamma) \quad (3.81)$$

$$+\langle\partial_m v, -\partial_x(vm_{x,t}) + \frac{\sigma_x^2}{2}\partial^2_{xx}m_{x,t}\rangle\} = 0. \quad (3.82)$$

Using simple algebraic manipulations, we get from

$$\mathbb{P}(mSINR_x(y_t,t)\geq\beta_{SU}\cap\bar{r}_{SU}(s_x,y_{PU,t},t)\leq\gamma) \quad (3.83)$$

$$= \mathbb{P}\left(\frac{\bar{r}_{SU}(s_x,y_t,t)}{N_0(y_t)+\bar{I}_{SU}(y_t,t)+\bar{I}_{PU}(y_t,t)} \geq \beta_{SU} \cap \frac{\bar{p}_{SU}(s_x,y_{PU,t},t)g(x_j,y_{PU,t})}{l(\|x_j-y_{PU,t}\|)} \leq \gamma\right), \quad (3.84)$$

the following condition on $\bar{p}_{SU}$

$$\bar{p}_{SU} \leq \gamma\frac{l(\|x-y_{PU,t}\|)}{g(x,y_{PU,t})} =: \bar{b} \quad (3.85)$$

$$\bar{p}_{SU} \geq \mathbb{1}_{\{g\neq 0\}}\beta_{SU}\frac{(N_0(y_t)+\bar{I}_{SU}(y_t,t)+\bar{I}_{PU}(y_t,t))l(\|x-y_t\|)}{g(x,y_t)} =: \bar{a}. \quad (3.86)$$

So we have $\bar{a}\leq\bar{p}_{SU}\leq\bar{b}$, and if $\bar{a}\leq\bar{b}$ then $\bar{p}^*_{SU}=\bar{a}$, otherwise $\bar{p}^*_{SU}=0$.

Hierarchical Mean-Field Game

In the language of *mean-field stochastic games*, the above cognitive radio problem corresponds to the case of one atomic player (PUs) and many nonatomic players (SUs). When the major player is in active period, its problem reduces to an energy management problem depending on the network characteristics, the density of minor players around the primary transmitter-receiver. We formulate this as hierarchical (Stackelberg) mean-field response problem where any minor player in the neighborhood of the major player will have the required current state and control feedback. A generic minor player will respond consequently.

Stackelberg Mean-Field Response Problem for PU
Let B be a three-dimensional Brownian Motion defined on $(\Omega; \mathcal{F}; \mathbb{P})$, where $\mathcal{F}$ be the natural filtration generated by B and s_0 be any random variable independent to B having density $m_0(s)$. Maximize the function

$$\sup_{\{\bar{p}_{PU,t}\}_{t\in\mathcal{T}}} \mathbb{E}\left(q_0(s_{PU,T}) + \int_0^T b_{PU}(s_{PU,t}, \bar{p}_{PU,t}, I_t^*)\, dt\right)$$

and the state dynamics of primary user $s_{PU,t}$ is given by

$$ds_{PU,t} = f(s_{PU,t})dt + \sigma dB_t,\ t \in \mathcal{T}\backslash\{0\}, \tag{3.87}$$

$$s_{PU,0} \in \mathbb{R}^3,\ t \in \mathcal{T}\backslash\{0\}. \tag{3.88}$$

and $m_t^* \in \mathcal{MR}_+[p_{pu}]$ is the optimal reaction mean-field trajectory of secondary users, and $I_t^* \in \mathcal{I}[p_{PU}] = \int_{\mathcal{S}} K_t(s,y)m_t^*[p_{pu}](ds)$.

What Is New in the Hierarchical Mean-Field Game?
In the feedback response of mean-field stochastic game capturing the interaction between PU and SUs, there must be an equation to express the dynamic optimization problem of the PU and each of the SUs. Usually this involves one equation per user. Since we have modeled SUs as having similar state behaviors, there is one equation per class. This equation is generally a Bellman equation, since the corresponding long-term payoff optimization falls within the framework of stochastic control of mean-field type. Hence, the Bellman equations will be used to compute optimal feedback energy management policy and the value equation for each SU given the current policy of the PU.

An equation is also needed to express mean-field behavior of each class. The dynamics of the distribution is governed by a Kolmogorov forward equation. In this Kolmogorov forward equation, the optimal power management behaviors of the SUs appear as data, since it is the infinite collection of individual behaviors that are aggregated and constitute collective behavior by consistency and this for any admissible policy of the PU. Thus, the modeling of behavior of a group of SUs (subpopulation) naturally leads to a B.K. (Bellman and Kolmogorov) system of equations. In the absence of a major player around its location, the problem of a generic minor player is to solve a mean-field (Nash) system provided in Algorithm 1 for $p_{PU} = 0$.

In the presence of a major player, the problem of a generic minor player entails solving a mean-field (Nash) system given the state and control of the major player at current time (see Algorithm 1).

The importance of this approach is that the mean-field games formalism involves *the density of players on the state space entering in the Bellman equation*. Thus, the mean-field equilibrium is defined by a B-K system in which the Bellman equations are doubly coupled: For any policy of the PU, individual behaviors are given by the Kolmogorov forward equation, and, at the same time, the distribution of SUs in the state space enters in the Bellman equation. This means that SUs can incorporate into their preferences the density of states/actions of other SUs at the anticipated equilibrium given the control of the PU. Therefore, each SU can construct his strategy by taking account of the anticipated distribution of strategies and of the actions of other SU and PU. The details are provided in Algorithm 1 to compute the optimal power and the evolution of the interference.

3.3.7 Optimal Power Consumption for the PU

Consider a PU transmitting to a receiver located $y_t = y$. The energy minimization problem is given by

$$\min \quad \int_0^T \bar{p}_{PU}(s_{PU}, y, t)dt \tag{3.89}$$

$$\text{s.t. } mSINR_{PU}(y,t) \geq \beta_{PU}, \tag{3.90}$$

Lemma 3.16 *If the cumulated scaled interference limit is independent of the policy of PU, the optimal response to interference is given by*

$$\bar{p}^*_{PU}(t) = \mathbb{1}_{\{E_{PU,t}\geq \epsilon_E\}}\mathbb{1}_{\{\bar{g}(x_{PU,t},y)\neq 0\}}\cdot \tag{3.91}$$

$$\frac{\beta_{PU} l(\|x_{PU,t}-y\|)(N_0(y)+\bar{I}_{SU}(y,t))}{\bar{g}(x_{PU,t},y)} \tag{3.92}$$

Proof The quality of satisfaction for PU gives

$$\frac{(\bar{p}_{PU}(s_{PU}, y, t)\bar{g}(x_{PU,t}, y))/l(\|x_{PU,t} - y\|)}{N_0(y) + \bar{I}_{SU}(y,t)} = \beta_{PU} \tag{3.93}$$

Note that $N_0(y)$ and $\bar{I}_{SU}(y,t)$ are deterministic quantities. We get

$$\bar{p}^*_{PU}(t) = \mathbb{1}_{\{E_{PU,t}\geq \epsilon_E\}}\mathbb{1}_{\{\bar{g}(x_{PU,t},y)\neq 0\}}\cdot \tag{3.94}$$

$$\frac{\beta_{PU} l(\|x_{PU,t}-y\|)(N_0(y)+\bar{I}_{SU}(y,t))}{\bar{g}(x_{PU,t},y)} \tag{3.95}$$

■

Now, the PU estimates $\bar{I}_{SU}(y,t)$ in order to save power follow the steps below:

- For $t \in \mathcal{T}$
  - If PU is active at time t then estimate the interference from SUs, $\hat{I}_{SU}(y,t)$.

– if $\bar{g}(x_{PU,t}, y)\neq 0$ & remaining energy $E_{PU,t}\geq\epsilon_E$ then E $p^*_{PU}(t) = \frac{\beta_{PU} l(\|x_{PU,t}-y\|)(N_0(y)+\hat{I}_{SU}(y,t))}{\bar{g}(x_{PU,t},y)}$ else $p^*_{PU}(t) = 0$

To estimate the interference from SUs, $\hat{I}_{SU}(y, t)$, we use the following steps:

- Solve FPK equation (3.78) to get $\hat{m}_t$
- Compute $\hat{I}_{SU}(y, t) = \int_{\mathscr{S}} K_t(s, y)\hat{m}_t(ds)$

Algorithm 1 Algorithm for SUs: power saving and success probability

```
For t ∈ 𝒯 and each SU j
At each time t, when active: sensing phase before transmission phase
if PU is Active and j in the protection of PU then
    Update the state and control fed back (s_{PU,t}, p_{PU,t}).
    Compute the optimal power using MFS − SU(s_{PU,t}, p_{PU,t}) (Stackelberg-MFS-SU)
    Set a_t = 1_{g≠0} β_SU (N_0(y) + Ī_SU(y,t) + Ī_PU(y,t)) l(‖x−y‖)/g
    if ∂_E v_t = 0 then
        p*_{SU,t} ∈ [a_t, +∞)
    else if ∂_E v_t > 0 then
        if a_t < 1/(∂_E v_t) then
            choose p*_{SU,t} = a_t.
        else if a_t ≥ 1/(∂_E v_t) then
            p*_{SU,t} = 0 (do not transmit)
        end if
    else if ∂_E v_t < 0, then
        PU is Inactive or j is not in the protection of PU.
        choose p*_{SU,t} = p_max 1_{E_SU(t) ≥ ε_E}.
    end if
else
    Solve MFS − SU(s_{PU,t}, 0) (Nash-MFS-SU)
end if
```

Stackelberg-MFS-SU

$$\begin{cases} \partial_t(v_{SU}) + \sup_{\bar{p}_{SU}}\{b_{SU}(s, \bar{p}_{SU}, \bar{p}_{PU}) + f_{SU}\nabla_s v_{SU}\} \\ + \frac{1}{2}\text{trace}(\sigma\sigma'\partial^2_{ss} v_{SU}) = 0 \\ \partial_t(m) + \text{div}_s(f_{SU}[p_{PU}]m) = \frac{1}{2}\text{trace}(\sigma\sigma'\partial^2_{ss} m) \end{cases} \tag{3.96}$$

The final condition $v_{SU}(T, s) = q(s)$ is given. The initial distribution of secondary users $m(0, s) = m_0(s)$ is also given. Hence, we have a backward-forward system for each fixed policy of PU. This system is referred as Stackelberg-MFS-SU. The solutions of this system (eventually normalized) are needed in order to compute the optimal energy management of PU.

If the Stackelberg-MFS-SU system has unique solution for a given $\bar{p}_{PU}$, (denoted by $(\bar{p}^*_{SU}[\bar{p}_{PU}], v^*[\bar{p}_{PU}], m^*[\bar{p}_{PU}])$), then the PU can anticipate the feedback control of SUs and solve

$$\begin{cases} \partial_t(v_{PU}) + \sup_{\bar{p}_{PU}}\{b_{PU}(s, \bar{p}_{SU}, \bar{p}_{PU}, m^*) + f_{PU}\nabla_s v_{PU}\} \\ + \frac{1}{2}\text{trace}(\sigma\sigma'\partial^2_{ss}v_{PU}) = 0 \\ \text{s.t. } (\bar{p}_{SU}, m^*) \in \mathcal{SR}(\bar{p}_{PU}) := \{(\bar{p}^*_{SU}[p_{PU}], m^*[p_{PU}])\} \end{cases} \tag{3.97}$$

The case where $\mathcal{SR}$ has multiple elements is not considered here.

Multiple Receivers for Each PU
We now extend the model 1 to multiple receivers, which we *call model* 2. We consider the extended scenario where each PU can choose multiple receivers around its protection zone. We assume that the power dedicated to each receiver of the PU is managed by virtual machines that we call *virtual battery managers* (VBMs). Each VBM computes the power that is required for the proposed task/service. This allows one to do parallel computing in order to save energy. The virtual power recommended by the virtual machine may not be compatible all the time; therefore, the transmitter of PU has a scheduler which validates the proposed request of the VBM. The positive requests then result in a power allocation scheme across receivers.

Multiple PUs: One Receiver per PU
We next discuss extension of model 1 to multiple PUs, which we call model 3. In many practical scenarios, one can consider the following: A PU is located at the center of its protection zone, and many SUs are trying to access the unused spectrum. Each SU uses carrier sensing in order to guarantee that not too much interference is created to the PU. We are interested in its access probability and its probability of success under energy consumption constraint. For that, we need to define the access indicator random variable of a SU taking account the other SUs and all the Pus, and the access indicator random variable of a PU taking account the other PUs. Then let

$$\mathbb{1}_{a(k,k')} := \mathbb{1}_{\{t_k \le t_{k'}\}} + \mathbb{1}_{\{t_k > t_{k'}\}}\mathbb{1}_{\{r_{k,k',t} \le \gamma\}}\mathbb{1}_{\{r_{k',k,t} \le \gamma\}}, \tag{3.98}$$

where $t_k \in \mathcal{T}_a := [0, 1]$ is the access request time of user k. $\mathbb{1}_{\{k,k'\}}$ reads "if user k requested access before user k' then it gets the medium (indicator is equal to 1), else he can get the medium only if the interference received by k' is not too high (less or equal γ)." Now, given a PU i, the access indicator random variable is

$$\prod_{i' \in PU_a \setminus \{i\}} \mathbb{1}_{\{a(i,i')\}} \tag{3.99}$$

where PU_a is the set of active PUs. Given a SU j, the access indicator random variable is

$$\left(\prod_{i\in PU_a} \mathbb{1}_{\{r_{j,i,t}\leq\gamma\}}\mathbb{1}_{a(j,i)}\right)\prod_{j'\in SU_a\backslash\{i\}} \mathbb{1}_{\{a(j,j')\}} \quad (3.100)$$

where SU_a is the set of active SUs.

Two PUs

We start with two PUs case, denoted by P_1 and P_2.

Access Probability

We compute its access probability and probability of success under energy consumption constraint. Compared to model 1, the *access probability* of active PU i is now given by

$$\mathbb{E}\left[\prod_{\{i'\in PU_a\backslash\{i\}\}} \mathbb{1}_{\{a(i,i')\}}\right] \quad (3.101)$$

The probability of success of SU j is

$$\mathbb{E}\left[\left(\prod_{i\in PU_a} \mathbb{1}_{\{r_{j,i,t}\leq\gamma\}}\mathbb{1}_{a(j,i)}\right)\prod_{j'\in SU_a\backslash\{i\}} \mathbb{1}_{a(j,j')}\right]. \quad (3.102)$$

Probability of Success

The transmission scheme treats interference as a noise and a transmission with a given power policy of SU is successful if he/she is allowed to transmit and its sSINR is greater than β_{SU} where in the interference terms we count only the users who are allowed to transmit.

Using the same limiting arguments as before, we get $mSINR_{P_1}$ for a receiver located at y

$$mSINR_{P_1}(y,t) = \frac{\bar{r}_{P_1}(s_{P_1}, y, t)}{N_0(y) + \bar{I}_{SU}(y,t) + \bar{r}_{P_2}(s_{P_2}, y, t)} \quad (3.103)$$

We now have a random variable (r.v.). in the denominator. Note that r_{P_1} and r_{P_2} are i.i.d. Then,

$$\mathbb{P}(\bar{r}_{P_1} - \beta\bar{r}_{P_2} \geq \beta(N_0 + \bar{I}_{SU})) = \mathbb{P}(P_1 - P_2 \geq \tilde{\beta}) = \mathbb{P}(R \geq \tilde{\beta}) \quad (3.104)$$

We use the characteristic function to describe the r.v. R

$$\mathbb{E}(e^{iR}) = \mathbb{E}(e^{iP_1}e^{-iP_2}) = \mathbb{E}(e^{iP_1})\mathbb{E}(e^{iP_2}) \tag{3.105}$$

From the characteristic function, we derive the expression for $\bar{r}_{P_1}$.

3.3.8 Notes

In this section, we have studied mean-field models for cognitive radio networks under mobility and energy constraints. Using recent developments in the field of *mean-field stochastic games*, we were able to characterize the complete evolution of access probability, success probability, throughput, density of energy, and optimal power for both primary and secondary users whose mobilities are driven by Brownian motions. These performance metrics are fundamental in the emerging cognitive radio networks.

The results here enable one to deal with large number of nodes under indistinguishability per class property for the power allocation scheme. Based on coupled partial differential equations that describe the hierarchical mean-field equilibrium, we have provided algorithm for optimal energy management.

The analysis and results in this section lead to new questions:

- *Can we extend this mobility model to more realistic observed human-based mobility?* The performance metrics under Levy process (flight, motion, stable process, etc.) and anomalous diffusion are open issues.
- *Multiple antennas-based large cognitive radio networks* with different classes of users, each class having its initial distribution and its own mobility scheme. This allows one to study heterogeneity of mobility in a given map sectionized by area (mobility model for city center is different than the one of suburban). It would be important to study what is the impact of this on the optimal use of spectrum holes in cognitive MIMO networks.
- *Queue dynamics* In this section, we have not examined the evolution of the queues. The model can be completed with queue dynamics for a user which described a reflected SDE from

$$d\tilde{q}_{j,}t = (\lambda_t(s_t) - \mu_t(\tilde{q}_t, s_t, \bar{p}_t))\, dt + \tilde{\sigma}(\tilde{q}_t, s_t, \bar{p}_t) dB_{j,t}$$

The term $\lambda_t(s_t)$ is the expected value of arrival process, μ_t is the probability to have successful transmission (departure), and $\tilde{\sigma}(.)dB_{j,t}$ is a noise term. In addition, the queue dynamics is subjected to a positivity constraint. Then the question of queue stability and capacity region arise for the queueing-based hierarchical mean-field game system.

3.4 Anti-jamming Mean-Field Games

Reaction to the interference field was presented in Sect. 3.2 as a game with one-and-half player. This representation does not hold when it comes to multiple types of decision-makers, some being atomic and some others less atomic. In this section, we study anti-jamming problems in cognitive radio networks using mean-field games. We consider three types of users: jammers, primary users, and secondary users. When active, each secondary transmitter-receiver uses carrier sensing and is subject to long-term energy constraint. We formulate the interaction between primary and large number of secondary users as a hierarchical mean-field difference game. This mean field framework allows one to describe the evolution of the density distribution and the associated performance metrics using coupled partial differential equations. We provide explicit formulas and algorithmic power management for both primary and secondary users based on attacker strategies. A complete characterization of the optimal distribution of energy and the Shannon-like information rate is given.

Wireless technology is proliferating rapidly, and the vision of pervasive wireless computing and communications offers the promise of many industrial and individual benefits. While consumer devices such as cell phones, personal data assistants, and laptops receive a lot of attention, the impact of wireless technology is much broader, e.g., through sensor networks for safety applications, vehicular networks, smart grid control, home automation, medical wearable and embedded wireless devices, and entertainment systems. This explosion of wireless applications creates an ever-increasing demand for more radio spectrum. However, most easily usable spectrum bands have been allocated, although many measurements, nevertheless, show that at any specific time and location, these spectrum bands are significantly *underutilized* and lot of energy is wasted.

This manifests itself through voids in either energy, time, space, or spectrum.

Cognitive radios offer the promise of being a disruptive technology innovation that will enable the future wireless world. Cognitive radios are fully programmable wireless devices that can *sense*, *learn* their environment and dynamically adapt their transmission waveform, energy management policy, channel access method, spectrum use, and networking protocols as needed for good network and application performance.

Cognitive radio aims at exploiting these voids in order to accommodate extra radio devices referred to as *secondary users* (SUs) in a network whose first function is to serve the primary users (PUs), those licensed to use the spectrum. The key requirement is that the primary users have to be as little affected as possible by the presence of secondary users.

The protocols are carrier sensing-based mechanism for secondary users to exploit the spectrum left over by primary user. Each primary transmitter has a dynamic protection zone. If a user located in this zone transmits at the same time, he/she interferes with the primary user. Any user located in this protection zone is hence called a contender of this transmitter. In the general setting, the union of all primary transmitter protection zones does not cover the whole space. Thus, one can accommodate secondary users in the remaining space to better utilize the spatial resource under energy constraints.

Due to the intrinsic feature of dynamic spectrum access, cognitive radio networks are very vulnerable to malicious attacks. First, in the opportunity-based spectrum access, secondary users do not own the spectrum band, and hence their access to that band cannot be protected from adversaries. Second, the spectrum is highly dynamic in nature, and the traditional security enhancing mechanisms are not directly applicable, since they only fit in a static spectrum environment and needed to be adapted in particular in large-scale networks. Moreover, some cognitive radio networks may work in a distributed fashion, which makes it more difficult to fight against malicious attacks than in a centralized system. As cognitive radio networks benefit from technology evolution to be capable of utilizing spectrum adaptively and intelligently, the same technologies can also be exploited by malicious attackers to launch more complicated and unpredictable attacks with even greater damage. Therefore, ensuring security is critical for the wide deployment of cognitive radio networks.

Although several approaches have been shown to be able to improve spectrum utilization or bring economic gains for the PUs, most of them are based on the assumption that the users (secondary) only aim at maximizing the spectrum utilization, either in a cooperative way where all users are coordinated by the same network controller and serve a common goal or in a selfish or egoistic manner where the autonomous secondary users want to maximize their own payoff. However, such an assumption does not hold when the secondary users are in a hostile environment, where there exist malicious attackers whose objective is to cause damage to the legitimate users and prevent the spectrum from being utilized efficiently under energy consumption cost. Therefore, how to secure spectrum sharing is of critical importance to the wide deployment of the cognitive radio technology. Malicious attackers can launch various types of attacks in different layers of a cognitive radio network.

Licensed solutions are not more secure than unlicensed solutions. In many practical scenarios, security is unrelated to whether a wireless system operates on licensed or unlicensed spectrum. In fact, licensed solutions (PUs) that operate on fixed frequencies are often susceptible to malicious interference. A well-designed unlicensed (SUs) solution offers greater protection against intentional interference or jamming.

The objective of a jammer is to cause the most damage to the SUs with the limited jamming capability and energy constraints. Given the limited jamming capability, the attackers can adopt an attacking strategy that targets as many data channels as possible to reduce the gain of the SUs by transmitting data.

The present section proposes a feedback strategy against the attackers and concentrates on exploiting mobility and energy consumption by considering cognitive radio in a wireless network with users distributed in the Euclidean plane. In order to save energy it is important to control the transmission power and/or the sensing threshold of secondary in order to further limit the interference they cause to primary users (PUs). The users are energy limited and each secondary user decides on their transmission power depending on the primary signal strength and its remaining battery state.

The objective is to contribute to the understanding of jammers' behavior and the impact of mobility, channel uncertainty, and energy constraints that will become more acute with the increased deployment of renewable resources.

We introduce a battery-state power management in the large-scale network. The transmitted power is to be adapted in response to the localized interference, state, and powers of the primary users, the jammer's strategy, the channel gains, and the remaining energy for the future. Due to these additional assumptions, the wireless network is *dynamic and interactive*. For example, the evolution of the optimal energy management (we set the power $p = 0$ for user in sleep mode or silent mode) will depend on the optimal trajectory of the density distribution of battery state as well as the channel gains, the jammer data, and spatial distribution of the user at the current time.

Stochastic control and game theory provide the natural frameworks for the design of defense, since the attackers will try every means to prevent legitimate users from efficiently utilizing the spectrum resources, and the interactions between the mutually distrustful parties can be modeled as a noncooperative/cooperative game. Dynamic games, e.g., differential games, stochastic games, and sequential games, can be used to help derive the optimal defense strategies that accommodate both the environment dynamics, the mobility, the queue dynamics, and the adaptive attacking strategy. The interaction between attackers and defenders can be seen as pursuit evasion and attacker tracking in multiband networks. Evolutionary game theory is also appropriate to examine evolutionarily stable strategies that will overwhelm possible attacks considered as mutants. Evolutionary network formation games, coalitional games, and graphical games can help legitimate regular nodes form evolutive optimal partitions or networks to defend against multiple distributed attackers in a large area. Establishing reputation systems can identify malicious attackers that provide false feedback, huge power usage, create feedback from fake identities, acquire new identities, start over with a clear reputation (whitewashing), etc. Similarly, robustness against evolving coalition of the malicious nodes can be studied.

In Sect. 3.4.1, we present the basic cognitive radio model and describe the key parameters of the performance metrics. We then present global optimization and micro game problems between SUs. Section 3.4.2 focuses on asymptotics of the interference terms. In Sect. 3.4.3, we formulate the hierarchical mean-field game and develop algorithm for power saving and probability of success optimization. In Sect. 3.4.5, we discuss how to extend the basic model to more realistic cognitive radio networks scenarios based on SINR-based CSMA-like access with multiple PUs and multiple SUs.

We summarize some of the notations in Table 3.1.

Table 3.1 Summary of notations

Symbol	Meaning
$\mathscr{I}_{\text{Jam}}$	Set of potential jammers
$\mathscr{I}_{PU}$	Set of potential primary users
$\mathscr{I}_{SU}$	Set of potential secondary users
n	Cardinality of $\mathscr{I}_{SU}$
$\mathscr{I} := \mathscr{I}_{\text{Jam}} \bigcup \mathscr{I}_{SU} \bigcup \mathscr{I}_{PU}$	Set of potential users
$p_{j,t}$	Control of user j at time t
$E_{j,t}$	Remaining energy
$s_{j,t}$	State of user j at time t
M_t^I	Mean-field term
$b(.)$	Instantaneous payoff function
$\bar{p}(s, y, t)$	Generic feedback control at time t
$\bar{r}(s, y, t)$	Virtual received power at time t
$\mathscr{T} = [0, T],\ T > 0$	Time horizon
$\mathscr{T}_I$	$\{0, \delta_I, 2\delta_I, \ldots\}$
R_T	Long-term payoff
$v_{SU}(T, s)$	Value function of a generic SU

3.4.1 Malicious Behaviors in Cognitive Radio

We start with a very basic model which we refer to *model 1*. We consider two types of transmitter-receiver pairs (users): *primary* and *secondary users* distributed over a domain $\mathscr{D}$ in Euclidean space. The physical layer setting considers a signal model with one primary user (a transmitter-receiver pair), one jammer, and I secondary users. In total, we have $I + 2$ nodes. Interference is treated as noise, and the transmission is successful if the signal-to-interference-plus-noise ratio (SINR) is larger than a certain threshold (constant) and the access conditions are met.

For MAC layer setting, we use carrier sense SINR-based model. When a user is active, he/she senses the medium to check whether there is any primary contender. The tagged secondary user transmits only if it sees the network free of primary contenders. To do so, a user has to choose a transmit power depending on the state of the battery (if no energy is available, the user cannot transmit). Whenever he/she is active, a primary user can access the channel. For secondary users, because of interference, not every transmission attempt is successful.

Due to the limitation of the number of antennas and/or the total power, we assume that the attackers can jam limited number of channels in each time slot. Then, the objective of the jammers is to cause the most damage to the SUs with the limited jamming capability and energy constraints. Given the limited jamming capability, the attackers can adopt an attacking strategy that targets at as many data channels as possible to reduce the gain of the SUs by transmitting data (Fig. 3.6).

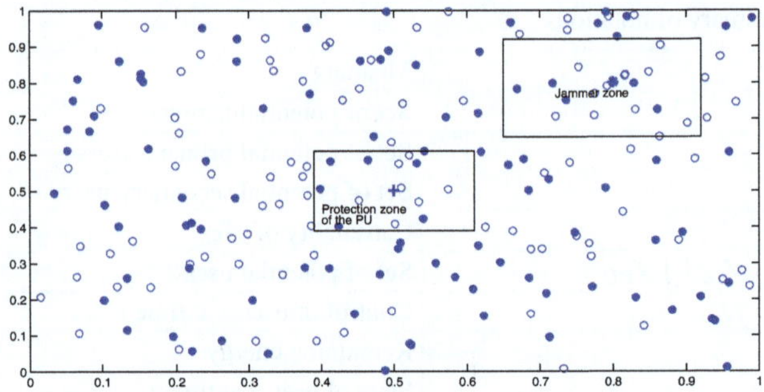

Fig. 3.6 Typical cognitive radio jamming scenario under consideration. Star is a jammer. Black nodes are active SUs, white nodes are inactive SUs, and the plus sign is a PU

Mobility of Users

Let $\delta_I > 0$ be the duration of time step. The mobility of the PU is described by

$$\tilde{x}_{PU,k+1} = \tilde{x}_{PU,k} + \tilde{v}_{PU}(\tilde{x}_{PU,k})\delta_I + \sigma_{PU}\left(\tilde{B}_{k+1} - \tilde{B}_k\right)\sqrt{\delta_I} \tag{3.106}$$

where $\tilde{B}_k$ is a standard Brownian motion, where $\tilde{x}_{PU,k} \in \mathscr{D}$ for all $k \in \mathbb{Z}_+$ and σ_{PU} is positive constant, $v_{PU}(.)$ is a velocity function which we assume to be smooth and uniformly Lipschitz function. When $\delta_I \longrightarrow 0$, the mobility of the PU is described by

$$dx_{PU,t} = v_{PU}(x_{PU,t})dt + \sigma_{PU}dB_t \tag{3.107}$$

where $x_{PU,t} \in \mathscr{D}$ for all $t \in \mathscr{T}$ and σ_{PU} is positive constant, and the mobility of jth SU by

$$dx_{j,t} = v_{SU}(x_{j,t}, x_{PU,t})dt + \sigma_{SU}dB_{j,t}, \tag{3.108}$$

where $x_{j,t} \in \mathscr{D}$ for $j \in \mathscr{I}_{SU}$ and $t \in \mathscr{T}$ and σ_{SU} is a positive constant, $v_{SU}(.)$ is a velocity function which we assumed to be smooth and uniformly Lipschitz.

The processes $B_{j,t}$ are standard Brownian motion (also called Wiener process) in $\mathbb{R}$, independent across the users and across time and defined over a measure space $(\Omega, \mathscr{F}, \mathbb{P})$. We assume that for any user j, $B_{j,0}$ is independent of the initial distribution of users. The mobility process in (3.108) is understood as classical Itô's stochastic differential equation (SDE). The conditions on the coefficients ensure that the trajectory of the SDE is well-posed for given initial point. An example of generic Brownian mobility is illustrated in Fig. 3.7.

When $\delta_I \longrightarrow 0$, the density of the above mobility model can be characterized by a FPK:

$$\partial_t \nu_t(x) + \partial_x(v_{SU}(x)\nu_t(x)) - \frac{\sigma_{SU}^2}{2}\partial_{xx}^2 \nu_t(x) = 0. \tag{3.109}$$

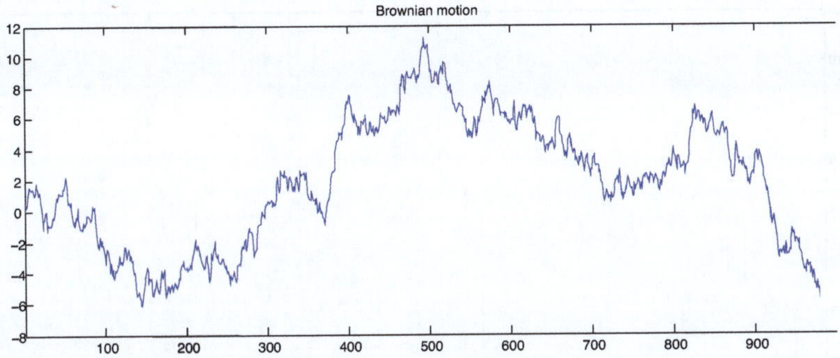

Fig. 3.7 A generic Brownian mobility

Recall that a Gaussian with mean $\bar{m}$ and variance σ_{SU}^2 has a density given by $\nu_0(x) = \frac{1}{\sqrt{2\pi\sigma_{SU}^2}} e^{-\frac{(x-\bar{m})^2}{\sigma_{SU}^2}}$. For $v_{SU} = 0$, the fundamental solution of (3.109) is explicitly given by a Gaussian density with (time-varying) variance $\sigma_{SU}^2 t$

$$\nu_t(x) = \frac{1}{\sqrt{2\pi\sigma_{SU}^2 t}} \exp\left(-\frac{x^2}{2\sigma_{SU}^2 t}\right).$$

We consider a jammer which has a substantial effect of the system performance. The mobility of the Jammer is described by

$$\tilde{x}_{Jam,k+1} = \tilde{x}_{Jam,k} + \tilde{v}_{\mathrm{Jam}}(m, \tilde{x}_{Jam,k})\delta_I + \sigma_{\mathrm{Jam}}\left(\tilde{B}_{k+1} - \tilde{B}_k\right)\sqrt{\delta_I} \qquad (3.110)$$

and the limiting as δ_I goes to zero is given by

$$dx_{\mathrm{Jam}} = v_{\mathrm{Jam}}(m, x_{\mathrm{Jam}})dt + \sigma_{\mathrm{Jam}} dB. \qquad (3.111)$$

We are also interested in optimizing the energy (battery) usage of all the users, from a initial value of $E_0 > 0$. We consider a model with renewable energy (e.g., solar-powered devices). In that case, a stochastic model is well justified.

Power Consumption

The battery state of each user is modeled as a stochastic process which depends on the power consumption. A generic player starts initially with energy $E_0 > 0$ (full battery level).

$$\tilde{E}_{j,k+1} = \tilde{E}_{j,k} - p_{j,k}\delta_I + \sigma_e\left(\tilde{B}_{k+1} - \tilde{B}_k\right)\sqrt{\delta_I}, \ j \in \mathcal{I}. \qquad (3.112)$$

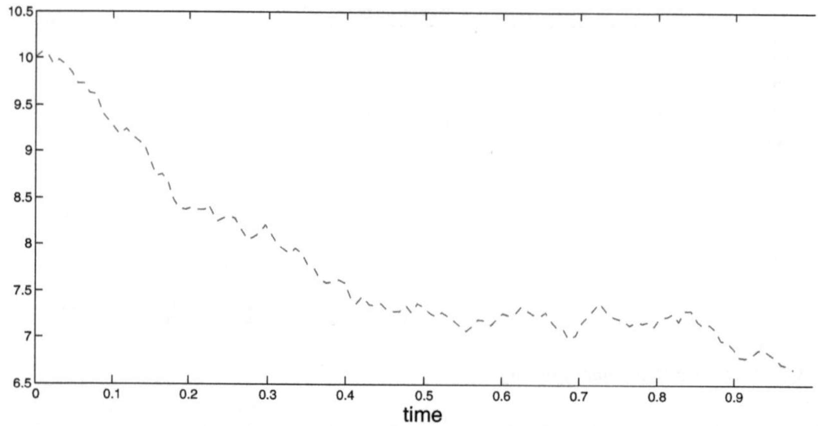

Fig. 3.8 Evolution of remaining energy

The energy $\tilde{E}_{j,k}$ is subject to positivity constraint. Similarly, when $\delta_I \longrightarrow 0$, the energy usage dynamics is described by Itô's SDE

$$dE_{j,t} = -p_{j,t}dt + \sigma_e dB_{j,t}, \ j \in \mathcal{J}. \tag{3.113}$$

with constraint $E_{j,t} \geq 0$, for all $t \in \mathcal{T}$, where $p_{j,t} \geq 0$ is the power consumption of user j at time t and $\sigma_e > 0$ is constant. The presence of noise term is to capture the stochastic nature of the anticipated prediction, and the fact that under renewable energy, the battery may be recharged if the weather conditions allow. Moreover, $p_{j,t} = 0$ if $E_{j,t} \leq \epsilon_E$, where ϵ_E is a fixed minimum energy threshold for the device to be usable. We will provide the distribution of energy using FPK equation. A typical trajectory of energy consumption with small noise is depicted in Fig. 3.8.

The variance of (3.113) becomes very big when $t \longrightarrow +\infty$, while the game stopped at T. To control this, we modify the dynamics (3.113) to be

$$dE_{j,t} = -p_{j,t}dt + \sigma_e E_{j,t} dB_{j,t}, \ j \in \mathcal{J}. \tag{3.114}$$

Note that we can restrict the instantaneous $p_{j,t}$ in a certain interval $[0, p_{\max}]$ or a remaining energy-dependent constraint $p_{j,t} \leq \delta E_{j,t}$. Then, all the power solutions provided below need to be projected in these sets.

For the case where a user can transmit to multiple receivers Rxs, the remaining will be written as

$$\tilde{E}_{j,k+1} = \tilde{E}_{j,k} - \delta_I \left[\sum_{r \in Rx(k)} p_{jr,k} \right] + \sigma_e \tilde{E}_{j,k} \left(\tilde{B}_{k+1} - \tilde{B}_k \right) \sqrt{\delta_I},$$

where $p_{jr,k}$ is the power allocated to receiver r, and the limiting is

$$dE_{j,t} = -\left[\sum_{r \in Rx(k)} p_{jr,t}\right] dt + \sigma_e E_{j,t}\, dB_{j,t} \tag{3.115}$$

Virtual Received Power

The virtual received power from a user i located at $x_{i,t}$ at a receiver located at $y_{j,t}$ is defined as

$$r_{i,j,t} := \frac{p_{j,t} g_{ij,t}}{l(\|x_{i,t} - y_{j,t}\|)} \tag{3.116}$$

for $(i, j) \in \mathcal{I}^2$ where $p_{j,t}$ is the transmit power, $g_{i,j,t}$ is the channel gain (positive random process), $l(\|x_{i,t} - y_{j,t}\|)$ is such that $l(d) = c(d^2 + \epsilon^2)^{\alpha/2}$ where $\alpha \geq 2$ is the pathloss exponent, $c > 0$. The term ϵ is the component of a hidden constant dimension; that is, every receiver location has components $(y^1_{j,t}, y^2_{j,t}, \epsilon)$, and every sender has components in the form $\left(x^1_{j,t}, x^2_{j,t}, 0\right)$. Then, we avoid a vanishing denominator.

We define the state of the user i at time t as $s_{i,t} := (E_{i,t}, x_{i,t}, g_{i,\cdot,t}) \in \mathcal{S} = \mathbb{R}_+ \times \mathcal{D} \times \mathbb{R}_+$.

Scaled SINR

We now define the *scaled signal to interference plus noise ratio (sSINR)* from SU i to receiver j at time t as

$$\text{sSINR}_{i,j,t} := \frac{r_{i,j,t}}{N_0(y_{j,t}) + \frac{1}{I-1}\sum_{i' \neq i} r_{i',j,t} + r_{PU,j,t} + r_{Jam,j,t}} \tag{3.117}$$

for $(i, j) \in \mathcal{I}^2_{SU}$ where $N_0 : \mathcal{D} \longrightarrow \mathbb{R}_+$, $N_0(y_{j,t}) > 0$ is the ambient noise at the receiver location and $r_{PU,j,t}$ is the interference generated by the primary user to the receiver located at $y_{j,t}$. The term $\frac{1}{I-1}$ in the denominator is to take into consideration the number of other SUs than i.

A transmission is successful from SU i to receiver j if

1. $\text{sSINR}_{i,j,t} \geq \beta_{SU}$, where β_{SU} is a fixed minimum quality of the signal, and
2. $r_{i,PU,t} \leq \gamma$, where γ is a fixed quantity; that is, the primary user should not get more than certain amount of noise from the transmission.

The sSINR from primary user to a receiver j is defined as

$$\text{sSINR}_{PU,j,t} := \frac{r_{PU,j,t}}{N_0(y_{j,t}) + \frac{1}{I}\sum_{i'=1}^{I} r_{i',j,t} + r_{Jam,j,t}} \tag{3.118}$$

where the factor $\frac{1}{I}$ averages with the total number of SUs. Note that the scaled interference form is widely used in code-division multiple access (CDMA) networks.

We say that a secondary user i belongs to the primary user's protection zone if the virtual received power is greater than some threshold $\gamma > 0$. The access condition in the presence of PU is then $A_{i,t} = 1$ where $A_{i,t} = \mathbb{1}_{\left\{\frac{p_{i,t}g_{i,PU,t}}{l(\|x_{i,t}-y_{PU,t}\|)} \leq \gamma\right\}}$.

3.4.2 Asymptotics

First note that the only quantity in $\text{sSINR}_{PU,j,t}$ (3.118) which impacts I is $\frac{1}{I}\sum_{i'=1}^{I} r_{i',j,t}$. Considering indistinguishable SU nodes, we can model the system with a mean-field interaction model.

The scaled interference from SUs to a receiver j located at $y_{j,t} = y$ at time t is

$$\frac{1}{I}\sum_{i=1}^{I} r_{i,j,t} = \frac{1}{I}\sum_{i=1}^{I} \frac{p_{j,t}g_{i,j,t}}{l(\|x_{i,t} - y\|)}$$

$$= \int_{\mathscr{S}} \frac{\bar{p}_{SU,t}(s, y, t)\bar{g}(x, y, t)}{l(\|x - y\|)} M_t^I(ds) \xrightarrow{I \to +\infty} \int_{\mathscr{S}} K_t(s, y) m_t(ds) \quad (3.119)$$

where $K_t : \mathscr{S} \times \mathscr{D} \to \mathbb{R}_+$ is the kernel $\bar{p}.\bar{g}/l$, $M_t^I(\cdot, x, \cdot) = \frac{1}{I}\sum_{i=1}^{I} \delta_{x - x_{i,t}}$ is the occupancy measure of the n SUs. For details in the limit, we refer the reader to [56]. $\bar{p}_{SU}$ is the control for SUs at time t, $\bar{p}_{SU} : \mathscr{S} \times \mathscr{D} \times \mathscr{T} \to \mathbb{R}_+$. $\bar{g}_t$ is a positive random process, defined by $\bar{g}_t = |h_t|^2$.

Hence, the following holds:

Lemma 3.17 *Under the above assumptions, the scaled interference from the SUs converges weakly to*

$$\bar{I}_t(y) := \int_{\mathscr{S}} K_t(s, y) m_t(ds).$$

The dynamics of h_t are given by

$$dh_t = \sigma_h dB_t \quad (3.120)$$

and using Itô's rule, we get

$$d\bar{g}_t = \frac{2\sigma_h^2}{2} dt + \sigma_h \partial_h \bar{g}_t dB_t = \sigma_h^2 dt + 2\sigma_h h_t dB_t \quad (3.121)$$

Then, the dynamics of s_t are described by

$$ds_t = \begin{pmatrix} -p_t \\ v_t \\ \sigma_h^2 \end{pmatrix} dt + \begin{pmatrix} \sigma_{PU} & 0 & 0 \\ 0 & \sigma_E & 0 \\ 0 & 0 & 2\sigma_G h_t \end{pmatrix} dB_t \quad (3.122)$$

$$= f dt + \sigma dB_t \tag{3.123}$$

Note that the variance of h_t goes to infinite with time, and this can be interpreted as the channel becoming better and better. In order to avoid that, we can use the following Ornstein–Uhlenbeck dynamics

$$dh_t = \frac{1}{2}(\mu_h - h_t)dt + \sigma_h dB_t \tag{3.124}$$

Then, the stationary distribution of h_t is Gaussian with mean μ_h and variance σ_h^2. In this case,

$$dg_t = h_t(\mu_h - h_t)dt + 2\sigma_h h_t dB_t \tag{3.125}$$

The probability density function $m_h(h, t)$ of the Ornstein–Uhlenbeck process satisfies the FPK equation:

$$\frac{\partial m_h}{\partial t} = \frac{1}{2}\frac{\partial}{\partial h}[(h - \mu_h)m_h] + \frac{\sigma^2}{2}\frac{\partial^2 m_h}{\partial h^2}$$

The general solution of this equation, taking $\mu_h = 0$ for simplicity, and with initial condition ξ_0, is

$$m_h(h, t) = \sqrt{\frac{1}{2\pi\sigma^2(1 - e^{-t})}} \exp\left\{\frac{-1}{2\sigma^2}\left[\frac{(h - \xi_0 e^{-\frac{1}{2}t})^2}{1 - e^{-t}}\right]\right\}$$

The stationary solution of this equation is the limit as time tends to infinity which is a Gaussian distribution with mean μ_h and variance σ^2,

$$m_{h,st}(h) = \sqrt{\frac{1}{2\pi\sigma^2}}\, e^{-\frac{1}{2\sigma^2}(h-\mu_h)^2}.$$

Another channel model is the *log-normal channel* which can be modeled as $g_t = e^{\eta_t}$ where η_t is solution of a linear SDE.

Now, we define the continuous virtual received power from user j by a receiver located at y, $\bar{r} : \mathscr{S} \times \mathscr{D} \times \mathscr{T} \to \mathbb{R}_+$

$$\bar{r}_\cdot(s_j, y, t) := \frac{\bar{p}_\cdot(s_j, y, t)\bar{g}(x_j, y)}{l(\|x_j - y\|)} \tag{3.126}$$

where $\cdot$ can be PU, Jammer or SU, and x_j is the location of user j.

Note that if $\bar{p}_{\cdot,t}$ is integrable w.r.t., the measure m_t, then $\bar{r}_{\cdot,t}$ is integrable, and the resulting integral is the scaled interference $\bar{I}_{SU} : \mathscr{D} \times \mathscr{T} \to \mathbb{R}$

$$\bar{I}_{SU}(y,t) = \int_{s\in\mathscr{S}} \bar{r}_{SU}(s,y,t)m_t(ds) \tag{3.127}$$

and for PU,

$$\bar{I}_{PU}(y,t) = \int_{s\in\mathscr{S}} \bar{r}_{PU,t}(s,y,t)\delta_{s_{PU}}(ds) \tag{3.128}$$

and for jammer,

$$\bar{I}_{\mathrm{Jam}}(y,t) = \int_{s\in\mathscr{S}} \bar{r}_{Jam,t}(s,y,t)\delta_{s_{\mathrm{Jam}}}(ds) \tag{3.129}$$

Now we define the *mean SINR, (mSINR)* for PU, jammer, and SUs, given a receiver at location y

$$mSINR_{PU}(y,t) := \frac{\bar{r}_{PU}(s_{PU},y,t)}{N_0(y)+\bar{I}_{SU}(y,t)+\bar{I}_{\mathrm{Jam}}(s_{\mathrm{Jam}},y,t)}, \tag{3.130}$$

$$mSINR_{\mathrm{Jam}}(y,t) := \frac{\bar{r}_{\mathrm{Jam}}(s_{\mathrm{Jam}},y,t)}{N_0(y)+\bar{I}_{SU}(y,t)+\bar{I}_{PU}(y,t)}, \tag{3.131}$$

$$mSINR_x(y,t) := \frac{\bar{r}_{SU}(s_x,y,t)}{N_0(y)+\bar{I}_{SU}(y,t)+\bar{I}_{PU}(y,t)+\bar{I}_{\mathrm{Jam}}(y,t)} \tag{3.132}$$

for $x \in \mathscr{D}$.

Performance of a Generic User

Given the position of the primary receiver at y_{PU}, and that of the i-th secondary user located at $x_t = x$ at time t, the *access probability* of the latter at time t is

$$\mathbb{P}\left(A_{i,t}=1\right) = \mathbb{E}(A_{i,t}) = \mathbb{P}\left(\frac{\bar{p}(s_{SU},y_{PU},t)\bar{g}(x,y_{PU},t)}{l(\|x-y_{PU,t}\|)} \le \gamma\right) \tag{3.133}$$

Lemma 3.18 *The access probability of SU (resp., Jammer) at time t for a given power $\bar{p} > 0$ is*

$$\mathbb{P}\left(\frac{\bar{g}(x,y_{PU},t)}{l(\|x-y_{PU,t}\|)} \le \frac{\gamma}{\bar{p}}\right) = F_{Z_t}(\frac{\gamma}{\bar{p}})$$

where F_{Z_t} is the cumulative distribution of $Z_t = \frac{\bar{g}(x,y_{PU},t)}{l(\|x-y_{PU,t}\|)}$, which is ratio distribution.

Proof Given $\bar{p} > 0$, we write that $A_{i,t} = 1$ which is $\frac{\bar{g}(x,y_{PU},t)}{l(\|x-y_{PU,t}\|)} \le \frac{\gamma}{\bar{p}}$ which gives the cumulative function of Z_t at the point $\frac{\gamma}{\bar{p}}$. ■

In order to compute F_{Z_t}, we use the ratio distribution formula. Recall that the distribution of the ratio between two random variables X_t and Y_t with continuous density is given by

$$\bar{f}_{Z_t}(z) = \bar{f}_{X_t/Y_t}(z) = \int_{\mathbb{R}} |y|\, \bar{f}_{(X_t,Y_t)}(zy, y)\, dy$$

The ratio distribution of two independent random Gaussian variables with parameters $\mathcal{N}(\mu'_x, \sigma'^2_x), \mathcal{N}(\mu'_y, \sigma'^2_y)$ is a well-known result and is explicitly given by

$$pz(z) = \frac{b'(z)\cdot c'(z)}{a'^3(z)} \frac{1}{\sqrt{2\pi}\sigma'_x\sigma'_y}\left[2\Phi\left(\frac{b'(z)}{a'(z)}\right) - 1\right] + \frac{1}{a'^2(z)\cdot \pi\sigma'_x\sigma'_y} e^{-\frac{1}{2}\left(\frac{\mu'^2_x}{\sigma'^2_x}+\frac{\mu'^2_y}{\sigma'^2_y}\right)} \tag{3.134}$$

where

$$a'(z) = \sqrt{\frac{1}{\sigma'^2_x}z^2 + \frac{1}{\sigma'^2_y}}, \tag{3.135}$$

$$b'(z) = \frac{\mu'_x}{\sigma'^2_x}z + \frac{\mu'_y}{\sigma'^2_y}, \tag{3.136}$$

$$c'(z) = e^{\frac{1}{2}\frac{b'^2(z)}{a'^2(z)} - \frac{1}{2}\left(\frac{\mu'^2_x}{\sigma'^2_x}+\frac{\mu'^2_y}{\sigma'^2_y}\right)}, \tag{3.137}$$

$$\Phi(z) = \int_{-\infty}^{z} \frac{1}{\sqrt{2\pi}} e^{-\frac{1}{2}u^2}\, du \tag{3.138}$$

Note that the denominator of Z_t may not be Gaussian due the path-loss exponent. However using Itô's rule in the small noise regime, one can approximate processes driven by time-varying Gaussian processes.

Coverage Probability

The coverage probability of user i who transmits to receiver at $y_{i,t}$ at time t, is the probability that the $mSINR_{ij,t}$ of the tagged user is larger than β_{SU}. Given the location of the primary receiver, the coverage probability of the primary transmission at time t is the probability that the $mSINR_{PU,PU,t}$ of the tagged user is larger than β_{PU}.

Lemma 3.19 *The coverage probability at time t is*

$$\mathbb{P}\left\{\frac{\bar{g}(x_{PU}, y_{PU}, t)}{l(\|x_{PU}-y_{PU,t}\|)} - \frac{\beta_{PU}}{\bar{p}}\bar{I}_{Jam}(s_{Jam}, y, t) \geq \frac{\beta_{PU}}{\bar{p}}\left(N_0(y)+\bar{I}_{SU}(y,t)\right)\right\}.$$

In particular, in the absence of Jammer, the coverage probability of PU at time t with a receiver located y is

$$C_{PU}(y,t)=1-F_{Z_t}\left[\frac{\beta_{PU}}{\bar{p}}\left(N_0(y)+\bar{I}_{SU}(y,t)\right)\right]$$

Hence, the density of coverage of PU at time t in a protection zone $\mathscr{A}'$ is

$$\int_{y\in\mathscr{A}'} C_{PU}(y,t)m_{x,t}(dy).$$

Proof Let $\bar{p}>0$ be the power of PU. Using the expression of sSINR for PU, the measure of the event $\{mSINR_{PU,PU,t}\geq\beta_{PU}\}$ is equivalent to the measure event

$$\frac{\bar{g}(x_{PU},y_{PU},t)}{l(\|x_{PU}-y_{PU,t}\|)}-\frac{\beta_{PU}}{\bar{p}}\bar{I}_{\text{Jam}}(s_{\text{Jam}},y,t)\geq\frac{\beta_{PU}}{\bar{p}}\left(N_0(y)+\bar{I}_{SU}(y,t)\right).$$

The term $\bar{I}_{SU}(y,t)$ is the expectation of a random variable.

Now, if the jammer is offline, the right-hand side term, i.e., $\frac{\beta_{PU}}{\bar{p}}\left(N_0(y)+\bar{I}_{SU}(y,t)\right)$ is nonrandom. Hence,

$$\begin{aligned}C_{PU}(y,t)&=\mathbb{P}\left(mSINR_{PU,PU,t}\geq\beta_{PU}\right)\\&=1-F_{Z_t}\left[\frac{\beta_{PU}}{\bar{p}}\left(N_0(y)+\bar{I}_{SU}(y,t)\right)\right].\end{aligned}$$

■

When the realization of the state of PU, jammer and the receiver are given (fixed), a similar formula as above can be used, and the coverage probability of SU is given by

COP$(s_{SU},y,t)=$

$$1-F_{Z_t}\left[\frac{\beta_{SU}}{\bar{p}}\left(N_0(y)+\bar{I}_{SU}(y,t)+\bar{I}_{\text{Jam}}(y,t)+\frac{\bar{p}_{PU}g_{PU,y}}{l(\|x_{PU}-y\|)}\right)\right]$$

For SU, we cannot use the above reasoning, because now, the denominator is random due to the interference from the primary user (if active) and the one from jammer even if $\bar{p}$ is deterministic. However, we use distribution of difference of random variables which is stated in the following lemma.

Lemma 3.20 *In the presence of PU and jammer, the coverage probability of a generic SU at a receiver located at y_{SU} is*

$$1-F_{\bar{Z}_t}\left(\beta_{SU}\left(N_0(y_{SU})+\bar{I}_{SU}(y_{SU},t)\right)\right)$$

where $\bar{Z}_t$ is the random variable

$$\frac{\bar{p}_{SU,t}\bar{g}(x_{SU}, y_{SU}, t)}{l(\|x_{SU} - y_{SU,t}\|)} - \beta_{SU}\frac{\bar{p}_{PU,t}\bar{g}(x_{PU}, y_{SU}, t)}{l(\|x_{PU} - y_{SU,t}\|)}$$

$$- \beta_{SU}\frac{\bar{p}_{Jam,t}\bar{g}(x_{Jam}, y_{SU}, t)}{l(\|x_{Jam} - y_{SU,t}\|)}.$$

Proof The coverage inequality event for fixed transmit power gives

$$\frac{\bar{p}_{SU,t}\bar{g}(x_{SU}, y_{SU}, t)}{l(\|x_{SU} - y_{SU,t}\|)} - \beta_{SU}\frac{\bar{p}_{PU,t}\bar{g}(x_{PU}, y_{SU}, t)}{l(\|x_{PU} - y_{SU,t}\|)}.$$

$$- \beta_{SU}\frac{\bar{p}_{Jam,t}\bar{g}(x_{\mathrm{Jam}}, y_{SU}, t)}{l(\|x_{\mathrm{Jam}} - y_{SU,t}\|)} \geq \beta_{SU}\left(N_0(y_{SU}) + \bar{I}_{SU}(y_{SU}, t)\right).$$

We then take the probability of that event by observing that the right-hand side is deterministic. ■

The probability density of $\bar{Z}_t$ obtained by convolution of the density functions. *Density of interference per area $\mathscr{A} \subset \mathscr{D}$* at time t is

$$d_{SU,t}(\mathscr{A}) = \int_{y\in\mathscr{A}} \bar{I}_{SU}(y, t)m_{x,t}(dy) \tag{3.139}$$

where $m_{x,t}$ is the marginal pdf with respect to x. This density allows one to quantify how much interference is around the protection zone of the PU.

Proposition 3.21 *Consider SDE (3.113). The distribution of remaining energy is a weak solution of the following FPK forward equation:*

$$\partial_t m_{E,t}(\bar{E}) = \partial_{\bar{E}}\left[\bar{p}(.)m_{E,t}(\bar{E})\right] + \frac{\sigma_e^2}{2}\partial^2_{\bar{E}\bar{E}}m_{E,t}(\bar{E}), \tag{3.140}$$

$$m_{E,0}(\bar{E}) = \nu_0(\bar{E}). \tag{3.141}$$

For time-independent $\bar{p}$, the stationary solution is given by

$$m_{st}(\bar{E}) = -C_0 D\exp\left(-D\int_{\xi\in[0,\bar{E}]}\bar{p}(\xi)d\xi\right) = C_0' e^{-\Phi(\bar{E})} \tag{3.142}$$

where C_0 is a normalization constant and $D := 2/\sigma_e^2$ and Φ are the potential.

Proof The statements follow solving a linear second-order ODE. ■

Proposition 3.22 *Consider SDE (3.114). The distribution of remaining energy is a weak solution of the following Fokker–Planck–Kolmogorov forward equation:*

$$\partial_t m_{E,t}(\bar{E}) = \partial_{\bar{E}}\left[\bar{p}(.)m_{E,t}(\bar{E})\right] + \frac{\sigma_e^2}{2}\partial^2_{\bar{E}\bar{E}}\left[\bar{E}^2 m_{E,t}(\bar{E})\right], \tag{3.143}$$

$$m_{E,0}(\bar{E}) = \nu_0(\bar{E}). \tag{3.144}$$

For time-independent $\bar{p}$, the stationary solution is given by

$$m_{st}(\bar{E}) = -C_1\frac{D}{\bar{E}^2}\exp\left(-D\int_{[0,\bar{E}]}\frac{\bar{p}(\xi)}{\xi^2}d\xi\right) = C_1' e^{-\Psi(\bar{E})} \tag{3.145}$$

where C_1 is a normalization constant and $D := 2/\sigma_e^2$ and Ψ are the potential.

Proof The statement follows solving a linear second-order ODE. ■

The distribution of remaining energy at time in area $\mathscr{A} \subseteq \mathscr{D}$ is

$$\mu_{E,t}(\mathscr{A}) := \int_{x\in\mathscr{A}}\int_{g\in\mathbb{R}_+} m_t(E, dx, dg) \tag{3.146}$$

and the distribution of users which have a remaining energy at least $\bar{E}$ is

$$\int_{e\in[0,\bar{E}]} m_{E,t}(de) = F_{E,t}(\bar{E}). \tag{3.147}$$

Proposition 3.23 *Consider SDE (3.113). Then, the function $\mu(t,\bar{E}) \in \mathscr{T} \times \mathbb{R}_+ \longmapsto F_{E,t}(\bar{E})$ is solution of the following partial differential equation:*

$$\partial_t \mu_t(\bar{E}) - \bar{p}(.,t)\partial_{\bar{E}}\mu_t(\bar{E}) = \frac{\sigma_e^2}{2}\partial^2_{\bar{E},\bar{E}}\mu_t(\bar{E}), \tag{3.148}$$

$$\mu_0(\bar{E}) = F_0(\bar{E}). \tag{3.149}$$

where $\bar{p}(.,t)$ is the control of secondary user and F_0 is a given cumulative function. Similarly, under SDE (3.114), the cumulative distribution of energy is solution of the PDE

$$\partial_t \mu_t(\bar{E}) - \bar{p}(.,t)\partial_{\bar{E}}\mu_t(\bar{E}) = \frac{\sigma_e^2}{2}\partial_{\bar{E}}\left[\bar{E}^2\partial_{\bar{E}}\mu_t\right], \tag{3.150}$$

$$\mu_0(\bar{E}) = F_0(\bar{E}). \tag{3.151}$$

For time-independent $\bar{p}$, the stationary solutions are given by

$$\mu_{st}(\bar{E}) = \int_{[0,\bar{E}]} m_{st}(d\xi). \tag{3.152}$$

Proof These statements follow by integration of the Fokker–Planck–Kolmogorov forward equation. ■

The total throughput is the mean number of successful transmissions in the network per time slot which is

$$\frac{1}{T}\int_0^T \int_{(s,y)\in\mathcal{S}\times\mathcal{D}} \mathbb{P}(mSINR(s,y,t) \geq \beta) m_t(ds) m_{x,t}(dy)\, dt.$$

We have provided the generic analytical expressions of the preceding analysis for deterministic power of SUs and PU. However during the interaction between the users, correlation and randomness will impact these quantities. Thus, we need to examine these performance metrics under optimal behavior of the users. As we will see the optimization per user will be interdependent random process between the powers, hence independence cannot be assumed.

Power Saving-Based Payoffs
The payoff function of PU (which will be maximized) is

$$\mathbb{E}\left(\bar{v}_{PU,T}(s_{PU,T}) - \int_0^T p_{PU,t} dt\right) \tag{3.153}$$

where $\bar{v}_{PU,T}$ is some terminal cost. The payoff function of SU (which will be maximized) is

$$\mathbb{E}\left(\bar{v}_{SU,T}(s_{i,T}) + \int_0^T \mathbb{1}_{\{mSINR_{i,j,t}\geq\beta_{SU}\}}\, dt\right) \tag{3.154}$$

where $\bar{v}_{SU,T}$ is some terminal cost.

The payoff function of jammer is

$$\mathbb{E}\left(\bar{v}_{Jam,T}(s_{i,T}) - \int_0^T \mathbb{1}_{\{mSINR_{y,t}\geq\beta_{SU}\}}\, dt\right) \tag{3.155}$$

where $\bar{v}_{Jam,T}$ is some terminal cost.

3.4.2.1 Global Optimization and Micro Game Formulations

We know that m_t satisfies the FPK equation

$$\partial_t m_t(s) + \text{div}_s(f m_t(s)) = \frac{1}{2}\text{trace}(\partial^2_{ss}[\sigma\sigma' m_t]), \tag{3.156}$$

where div denotes the divergence operator and ∂^2_{ss} is second derivative operator with the respect to s. Our goal is *Global* optimization of the probability of success for SUs given a receiver at location y_t,

$$\int_0^T \int_{x\in\mathscr{D}} m_{x,t}(dx)dt$$

$$\mathbb{P}(mSINR_x(y_t,t)\geq\beta_{SU} \cap \bar{r}_{SU}(s_x, y_{PU,t}, t)\leq\gamma) \tag{3.157}$$

$$\text{such that } \partial_t m_{x,t} + \partial_x(v m_{x,t}) = \frac{\sigma_x^2}{2}\partial^2_{xx} m_{x,t} \tag{3.158}$$

where $m_{x,t}$ is the marginal pdf of m_t with respect to x.

Using HJB equation, we have that the probability of success constitutes an optimal solution if there exists a function $v(s)$ which satisfies

$$\partial_t v + \sup_{p_{SU}} \{\mathbb{P}(mSINR_x(y_t,t) \geq \beta_{SU} \cap \bar{r}_{SU}(s_x, y_{PU,t}, t) \leq \gamma) \tag{3.159}$$

$$+\langle \partial_m v, -\partial_x(v m_{x,t}) + \frac{\sigma_x^2}{2}\partial^2_{xx} m_{x,t}\rangle\} = 0 \tag{3.160}$$

Proposition 3.24 *Let*

$$\bar{a} := \mathbb{1}_{\{g\neq 0\}}\beta_{SU}\frac{(N_0(y_t)+\bar{I}_{SU}(y_t,t)+\bar{I}_{PU}(y_t,t))l(\|x-y_t\|)}{\bar{g}(x,y_t)} \tag{3.161}$$

and

$$\bar{b} := \gamma\frac{l(\|x-y_{PU,t}\|)}{\bar{g}(x,y_{PU,t})} \tag{3.162}$$

If $\bar{a} \leq \bar{b}$, then the energy-efficient optimal feedback control is given by $\bar{p}^*(.,t) = \bar{a}$. Otherwise, $\bar{p}^*(.,t) = 0$

Proof Using simple algebraic manipulations, we get from

$$\mathbb{P}(mSINR_{SU}(y_t,t) \geq \beta_{SU} \cap \bar{r}_{SU}(s_x, y_{PU,t}, t) \leq \gamma) \tag{3.163}$$

$$= \mathbb{P}\left(\frac{\bar{r}_{SU}(s_{SU},y_t,t)}{N_0(y_t)+\bar{I}_{SU}(y_t,t)+\bar{I}_{PU}(y_t,t)} \geq \beta_{SU} \cap \frac{\bar{p}_{SU}(s_{SU},y_{PU,t},t)g(x,y_{PU,t})}{l(\|x-y_{PU,t}\|)} \leq \gamma\right) \tag{3.164}$$

the following condition on $\bar{p}_{SU}$

$$\bar{a} \leq \bar{p}_{SU} \leq \bar{b} \tag{3.165}$$

This means that if nonzero, the optimal power $\bar{p}_{SU}$ should belong to $[\bar{a}, \bar{b}]$, i.e., if $\bar{a} \leq \bar{b}$, then $\bar{p}^*_{SU} = \bar{a}$, otherwise $\bar{p}^*_{SU} = 0$. ■

Micro Game in the Absence of PU and Jammer

We consider here secondary users of group size $I < +\infty$. In order to emphasize the fact that the network is finite, we add an index "I" to the parameters. The problem formulated above is a stochastic difference game formulated as follows:

- $\mathcal{I} = \{1, 2, \ldots, I\}$ is the set of SUs. $\delta_I > 0$ is the slot length.
- Let $0 < T < +\infty$. For $k \in \left\{0, 1, \ldots, \lfloor \frac{T}{\delta_I} \rfloor\right\} =: \mathcal{T}_I$, the control of j corresponds to $p^I_{j,k} \in \mathbb{R}_+$, i.e., the transmitted power of SU j.
- The instantaneous payoff of user j at time t is defined by $b_{j,k} : \mathcal{S} \times \mathbb{R}^2 \longrightarrow \mathbb{R}$,

$$b_j\left(s^I_{j,k}, p^I_{j,k}, I^I_{j,k}\right) = \mathbb{1}_{\{\text{sSINR}_{j,k} \geq \beta_{SU}\}}$$

- Each individual state $s^I_{j,t}$ evolves according to the controlled difference equations.

Let $\mathcal{F}^I_k = \sigma\left(s^I_{k'}, B_{k'}, k' \leq k\right)$, be a sigma-algebra generated by state processes and the standard Brownian motions. The filtration $\mathcal{F}^I_k$ plays an important role in the construction of a strategy at time step k. Next, we define key concepts for the control of a user.

- An open-loop control for user j at time t is function of time that takes values in a given control set.
- A state feedback control for user j at time t is function of $s^I_k = \left(s^I_{1,k}, \ldots, s^I_{n,k}\right)$ and time.
- An own-state-feedback control for user j at time k is a function of $s^I_{j,k}$ and time.
- An aggregate-based and own-state feedback control at time k is a function $s^I_{j,k}$, the current mean-field term $M^I_k = \frac{1}{I}\sum_{j=1}^{I} \delta_{s^I_{j,k}}$ and time, where δ denotes the Dirac measure.

The Microscopic Stochastic Interaction Between the Users

As discussed in volume I, a strategy in the difference game is a collection of mapping that depends on the past observations, the information patterns, and the probabilistic state space. Due to the fact that the network is complex and limited in terms of state observations, we restrict our attention to simple classes of strategies. This is a strategic decision-making problem. Hence, the state dynamics cannot be assumed to be independent processes.

Where the interdependence between the individual states comes from? Because the users will be reacting to what they receive as their payoffs which depend on the other states, their controls may change over time, with implicit dependence of the other users' decisions and states. This is one of the main reasons why the classical law of large numbers (LLN) cannot be directly applied to the above interaction problem when I is large. We will clarify how to deal with this via *asymptotic indistinguishability per class* argument.

We say that a collection of mapping $\{p^I_{j,k}(.)\}_{j,k}$ constitutes an *own-state-and-aggregate-based feedback (Nash) equilibrium* if for every user j, $p^I_{j,k}$ maximizes the expected value of the cumulative payoff $R^I_{j,T}(., p^I_{-j})$ where $p^I_{-j} = \{p^I_{j',k}(.)\}_{j'\neq j, k\leq T}$. Similarly, we say that $\{p^I_{j,k}(.)\}_{j,k}$ is own-state feedback (Nash) equilibrium if it is a (Nash) equilibrium and the control is of own-state feedback type.

Clearly, in general, there is no reason for own-state feedback control to be optimal (resp., best response) because of imperfect observation of the other states. However, due to the limitations in terms of observations, these type of controls are easily implementable. One can use own-state-and-aggregate-based feedback strategy if the environment allows the user to observe the statistics of the aggregative term $\bar{I}^I_k$ at time k via mean-field dynamics. We will explain the density can be anticipated and estimated by solving (numerically) FPK equation.

In the mean-field limit setting, when $I \longrightarrow +\infty$., we will discuss how to compute $\bar{I}^I_t$, using FPK equation. Below we state a general methodology for finding own-state-and-aggregate-based feedback strategy which is response to the scaled interference $\bar{I}^I_t(y)$.

Let η_k be the information pattern at time k. Then, a feedback equilibrium is defined as follows:

Definition 3.25 A strategy profile in the form $p^I_{j,k} = p^I_{j,k}(k, \eta_k)$ constitutes a feedback Nash equilibrium solution if there exist functions $v^I_j(k, s^I_1, \ldots, s^I_I)$ defined on $\mathcal{T}_I \times \mathcal{S}^I$ and satisfying the following relations for each $j \in \mathcal{I}$:

$$\begin{cases} v^I_j(T, s^I) = \bar{v}_{SU}(s^I), \\ \bar{v}_{SU}(s^*_{j,T}) + \sum_{t'\in\mathcal{T}_I} b_{j,t'}(s^*_{j,t}, p^*_{j,t'}(\eta_{t'}), \bar{I}^*_{j,t'})\,\delta_I \\ \geq \bar{v}_{SU}(s^{*,(j)}_T) + \sum_{t'\in\mathcal{T}_I} b_{j,t'}(s^{*,(j)}_{j,t'}, a_{j,t'}(\eta_{t'}), \bar{I}^*_{j,t'})\,\delta_I, \end{cases}$$

for all $a_{j,.}(.)$, $s^I \in \mathbb{R}^I$, where on the discrete set $\mathcal{T}_I$,

$$\begin{cases} E^*_{j,t'+\delta_I} = E^*_{j,t'} - \delta_I p^*_{j,t'}(\eta_{t'}) + \sigma_e(B_{j,t+\delta_I} - B_{j,t}))\sqrt{\delta_I}, \\ s^*_{j,0} = s^I_j,\ j \in \mathcal{I} \\ s^{*,(j)}_{t'+\delta_I} - s^{*,(j)}_{t'} = f_{j,t'}(p_{t'}, \eta_{t'})\delta_I + \sigma(B_{j,t'+\delta_I} - B_{j,t'}))\sqrt{\delta_I}, \\ s^{*,(j)}_{j,0} = s^I_j,\ j \in \mathcal{I}. \end{cases}$$

where $f_{j,t}$ is the drift of the state of j.

Next, we discuss how to find a feedback equilibrium under regular payoff and perfect state observation (of the full vector $s^I = (s^I_1, \ldots, s^I_I)$).

A strategy profile $p_{j,k} = p_{j,k}(s^I) \in \mathbb{R},\ j \in \mathcal{I}, k \in \mathcal{T}_I$ provides a feedback Nash equilibrium if there exist value functions $v^I_j : \mathcal{T}_I \times \mathcal{S}^I \longrightarrow \mathbb{R},\ j \in \mathcal{I}_{SU}$ satisfying the following Bellman system:

$$\begin{cases} v_j^I(T, s^I) = \bar{v}_{SU}(s^I), \\ v_j^I(k\delta_I, s^I) = \sup_{a'_j(.)} \Big\{ b_j(s_j^I, a'_{j,k}(s^I), \bar{I}_{j,k}^I(s^I)) \\ \quad + \mathbb{E}\left[v_j^I((k+1)\delta_I, s_{k+1}^I) \mid s_k^I = s^I \right] \Big\} \end{cases}$$

When $\delta_I \longrightarrow 0$, one gets the HJB system.

A strategy profile $p_{j,t} = p_{j,t}(s^I) \in \mathbb{R},\ j \in \mathcal{I}_{SU}, t \in \mathcal{T}$ provides a feedback Nash equilibrium if there exist continuously differentiable functions $v_j^I\ :\ \mathcal{T} \times \mathcal{S}^I \longrightarrow \mathbb{R},\ j \in \mathcal{I}_{SU}$ satisfying the following set of equations:

$$\begin{cases} v_j^I(T, s^I) = \bar{v}(s^I) \\ 0 = \frac{\partial}{\partial t} v_j(t, s^I) + \sup_{a'_j(.)} \Big\{ b_{j,t}(s_j^I, a'_{j,t}(s^I), \bar{I}_{j,t}^I(s^I)) + \langle \partial_{s^I} v_j^I(t, s^I), \\ \quad f(s^I, a'_{j,t}(s^I)) \rangle \Big\} + \frac{1}{2}\sigma^2 \partial^2_{s^I s^I} v_j^I \\ = b_{j,t}(s_j^I, p_{j,t}, \bar{I}_t^I) + \langle \partial_{s^I} v_j^I(t, s^I), f(s^I, p_{j,t}(s^I)) \rangle + \frac{1}{2}\sigma^2 \partial^2_{s^I s^I} v_j^I \end{cases}$$

Since the instantaneous payoff is discontinuous in p, it will be difficult to exploit useful information from this result regarding our problem of success probability. Moreover, it will be very difficult to observe all the states (battery, channel, location, queue) of all the transmitter-receiver pairs in a large-scale wireless network. Therefore, we propose to look at the own-state feedback controls and the asymptotic case. In the subsection below, we show how to use mean field game in order to reduce the curse of dimensionality problem.

3.4.3 *Hierarchical Mean-Field Game*

A Toy Example Let us illustrate a Stackelberg game using a very simple example where the location of the users is fixed and energy consumption costs are taken into account. There are two users: PU and SU. Each user has three choices: $0 = p^{(0)} < p^{(1)} < p^{(2)}$ (transmit powers). To each transmit power, we associate a cost: $c_0 = 0 < c_1 < c_2 < 1$. The payoff matrix is represented in the following table where PU is column player and SU is the row player. The entries of the table are the payoffs of SU and PU:

SU \ PU	$p^{(0)}$	$p^{(1)}$	$p^{(2)}$
$p^{(0)}$	$(0, 0)$	$(0, 1 - c_1)$	$(0, 1 - c_2)$
$p^{(1)}$	$(1 - c_1, 0)$	$(-c_1, -c_1)$	$(-c_1, 1 - c_2)$
$p^{(2)}$	$(1 - c_2, 0)$	$(1 - c_2, -c_1)$	$(-c_2, -c_2)$

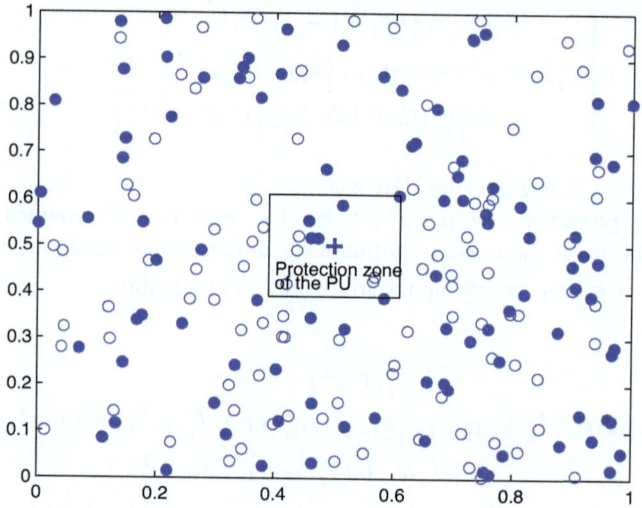

Fig. 3.9 Typical cognitive radio scenario under consideration. Black nodes are active SUs, white nodes are inactive SUs, and the plus sign is a PU

PU announces its action first, and SU reacts to it subsequently. If PU chooses the power $p^{(0)} = 0$ (does not transmit), then the optimal response for SU is to choose $p^{(1)}$, i.e., $p^*_{SU}(p^{(0)}) = p^{(1)}$. In that case, PU gets nothing (0) and SU gets $1 - c_1 > 0$.

Similarly, if $p_{PU} = p^{(1)}$, then $p^*_{SU}(p^{(1)}) = p^{(2)}$. In this case, PU gets 0, and SU gets $1 - c_2 > 0$.

Finally if $p_{PU} = p^{(2)}$, then $p^*_{SU}(p^{(2)}) = p^{(0)} = 0$. Then, PU gets $1 - c_2 > 0$, and SU gets 0.

Clearly, the choice of PU will be $p^{(2)}$ and PU gets $1 - c_2 > 0$. The Stackelberg solution with PU as leader is then given by $(p_{PU}, p_{SU}) = (p^{(2)}, p^{(0)})$.

This example shows that in order to compute the optimal power of PU, the primary user needs to check all possible responses of the SU (assuming that the entries of the table are known by the PU). This does not cover the possibility of nonunique response by SU.

Below, we formulate the analogue of this process for the mean-field game in cognitive radio networks (Fig. 3.9).

Hierarchy in Mean-Field Interaction

We describe now the hierarchical game between PU and SUs. The dynamic game is played as follows:

- Start with a initial energy distribution, location distributions of transmitter-receiver pairs, and channel state distribution.
- At each stage t,

  - PU has its own protection zone. Based on its current state and receiver, PU chooses an appropriate transmit power to meet the feasibility requirement.
  - Each SU uses carrier sensing and, depending on its current state, reacts to the PU and chooses a transmit power to meet the access requirement. If the access requirement is satisfied, the SU can access the channel, and its transmission is successful if the received SINR exceed is certain threshold.
  - The state of all users is changed stochastically according to their dynamics.

- The PU has access priority (licensed user) and has a protection zone. The PU focuses on long-term energy consumption cost.
- The objective of each SU is twofold: opportunistically access channel while maintaining PU protection and optimize the number of successful transmission under battery state constraints.

In order to rigorously define the dynamic hierarchical game, we need to specify the notions of histories, controls, strategies, and long-term payoffs.

In the language of *mean-field stochastic games*, the above cognitive radio problem corresponds to the case of one atomic player (PU) and many nonatomic players (SUs).

A generic nonatomic player will respond subsequently.

Stackelberg Mean-Field Problem for PU

Let B be a four-dimensional Brownian Motion defined on $(\Omega; \mathcal{F}; \mathbb{P})$, where $\mathcal{F}$ is the natural filtration generated by B and s_0 is a random variable independent to B having density $m_0(s)$. PU solves the optimization problem:

$$\sup_{\{\bar{p}_{PU,t}\}_{t\in\mathcal{T}}} \mathbb{E}\left(\bar{v}_{PU}(s_{PU,T}) + \int_0^T b_{PU}(s_{PU,t}, \bar{p}_{PU,t}, I_t^*)\, dt\right)$$

with its state dynamics $s_{PU,t}$ is given by

$$ds_{PU,t} = f_{PU}(s_{PU,t}, \bar{p}_{PU,t})dt + \sigma dB_t,\ t \in \mathcal{T}\backslash\{0\}, \tag{3.166}$$

$$s_{PU,0} \in \mathbb{R}^3,\ t \in \mathcal{T}\backslash\{0\}. \tag{3.167}$$

and $m_t^* \in \mathcal{MR}_+[p_{PU}]$ is the optimal reaction mean-field trajectory of SUs described by the FPK and

$$I_t^* \in \mathcal{I}[p_{PU}] = \left\{\int_{\mathcal{S}} K_t(s, y) m_t^*[p_{PU}](ds),\ m_t^* \in \mathcal{MR}_+[p_{PU}]\right\}.$$

Hierarchy in Mean-Field Game
We provide Algorithms 2, 3, 4 to compute the optimal power and the evolution of the interference.

Nonuniqueness Condition
If there exist two different measures that generate the same interference, then the same value can be obtained by different probability measures: Let $y \in \mathcal{D}$ and Γ be the mapping $m \longmapsto \int K(s, y)m(ds)$. The function Γ associates to each measure the resulting interference at the point y. Then, $\Gamma(m^1) = \Gamma(m^2)$ implies that $m^1 - m^2$ is orthogonal to K in $L^2(\mathbb{R}^d)$. Since $m^1 - m^2$ is a signed measure which belongs to the Hilbert space, the orthogonal set is not reduced to the singleton $\{0\}$.

Optimal Power Consumption for the PU
Consider a PU transmitting to a receiver located $y_t = y$. The cost of energy minimization problem is given by

$$\inf \quad \int_0^T \bar{p}_{PU}(s_{PU}, y, t)dt \tag{3.168}$$

$$\text{such that } mSINR_{PU}(y, t) \geq \beta_{PU}, \tag{3.169}$$

Lemma 3.26 *If the cumulated scaled interference limit is independent of the policy of PU, the optimal response to interference is given by*

$$\bar{p}^*_{PU}(t) = \mathbb{1}_{\{E_{PU,t} \geq \epsilon_E\}} \mathbb{1}_{\{\bar{g}(x_{PU,t}, y) \neq 0\}} \cdot \frac{\beta_{PU} l(\|x_{PU,t} - y\|)(N_0(y) + \bar{I}_{SU}(y,t))}{\bar{g}(x_{PU,t}, y)} \tag{3.170}$$

Proof The quality of satisfaction for PU gives

$$\frac{(\bar{p}_{PU}(s_{PU}, y, t)\bar{g}(x_{PU,t}, y))/l(\|x_{PU,t} - y\|)}{N_0(y) + \bar{I}_{SU}(y, t)} = \beta_{PU} \tag{3.171}$$

Note that $N_0(y)$ and $\bar{I}_{SU}(y, t)$ are deterministic quantities. We get

$$\bar{p}^*_{PU}(t) = \mathbb{1}_{\{E_{PU,t} \geq \epsilon_E\}} \mathbb{1}_{\{\bar{g}(x_{PU,t}, y) \neq 0\}} \cdot \frac{\beta_{PU} l(\|x_{PU,t} - y\|)(N_0(y) + \bar{I}_{SU}(y,t))}{\bar{g}(x_{PU,t}, y)} \tag{3.172}$$

■

Now the PU estimates $\bar{I}_{SU}(y, t)$ in order to save power according to Algorithm 2.

Algorithm 3 computes the interference field around the PU.
Algorithm 4 computes the optimal power of SU.

Algorithm 2 Power saving algorithm for PU

for $t \in \mathcal{T}$ **do**
  if PU is active at time t **then**
    Estimate the interference from SUs, $\hat{I}_{SU}(y, t)$, using Algorithm 3.
    if $\bar{g}(x_{PU,t}, y) \neq 0$ **and** remaining energy $E_{PU,t} \geq \epsilon_E$ **then**
      $p^*_{PU}(t) = \frac{\beta_{PU} l(\|x_{PU,t} - y\|)(N_0(y) + \hat{I}_{SU}(y,t))}{\bar{g}(x_{PU,t}, y)}$
    else
      $p^*_{PU}(t) = 0$
    end if
  end if
end for

Algorithm 3 Estimated interference from SUs to PU

1: Solve FPK equation (3.156) to get $\hat{m}_t$
2: Compute $\hat{I}_{SU}(y, t) = \int_{\mathcal{S}} K_t(s, y)\hat{m}_t(ds)$

Algorithm 4 Algorithm for SUs: power saving and success probability

For $t \in \mathcal{T}$ and each SU j
At each time t, when active: sensing phase before transmission phase
if PU is active and j in the protection of PU **then**
  Update the state and control feedback $(s_{PU,t}, p_{PU,t})$.
  Compute the optimal power using $MFS - SU(s_{PU,t}, p_{PU,t})$ (Stackelberg-MFS-SU)
  Set $a_t = \mathbb{1}_{\{g \neq 0\}} \beta_{SU}(N_0(y) + \bar{I}_{SU}(y, t) + \bar{I}_{PU}(y, t)) \frac{l(\|x-y\|)}{\bar{g}}$
  if $\partial_E v_t = 0$ **then**
    choose $p^*_{SU,t} \in [a_t \mathbb{1}_{\{E_{SU}(t) \geq \epsilon_E\}}, \bar{b}_t)$.
  else if $\partial_E v_t > 0$ **then**
    if $a_t < \frac{1}{\partial_E v_t}$ **then**
      $p^*_{SU,t} = a_t \mathbb{1}_{\{E_{SU}(t) \geq \epsilon_E\}} \mathbb{1}_{\{a_t < \bar{b}_t\}}$.
    else if $a_t \geq \frac{1}{\partial_E v_t}$ **then**
      $p^*_{SU,t} = 0$ (do not transmit).
    end if
  else if $\partial_E v_t < 0$, **then**
    PU is Inactive or j is not in the protection of PU.
    $p^*_{SU,t} = p_{\max} \mathbb{1}_{\{E_{SU}(t) \geq \epsilon_E\}}$.
  end if
else
  Solve $MFS - SU(s_{PU,t}, 0)$ (Nash-MFS-SU)
end if

Stackelberg-MFS-SU Given PU's control strategy, we write a a value function and mean-field equation associated with SUs.

$$\begin{cases} \partial_t(v_{SU}) + \sup_{\bar{p}_{SU}}\{b_{SU}(s, \bar{p}_{SU}, \bar{p}_{PU}) + f_{SU}\nabla_s v_{SU}\} \\ + \frac{1}{2}\text{trace}(\sigma\sigma'\partial^2_{ss} v_{SU}) = 0 \\ \partial_t(m) + \text{div}_s(f_{SU}[p_{PU}]m) = \frac{1}{2}\text{trace}(\partial^2_{ss}[\sigma\sigma' m]) \end{cases} \tag{3.173}$$

In this system, the first equation is a backward Hamilton–Jacobi–Bellman equation. The final condition $v_{SU}(T, s) = \bar{v}(s)$ is given. The second equation of the system is Fokker–Planck–Kolmogorov forward equation given the initial distribution of secondary users $m(0, s) = m_0(s)$. Hence, we have a backward-forward system for each fixed policy of PU. This system is referred as Stackelberg-MFS-SU. The solutions of this system (eventually normalized) are needed in order to compute the optimal energy management of PU.

If the Stackelberg-MFS-SU system has unique solution for a given $\bar{p}_{PU}$, (denoted by $(\bar{p}^*_{SU}[\bar{p}_{PU}], v^*[\bar{p}_{PU}], m^*[\bar{p}_{PU}])$), then the PU can anticipate the feedback control of SUs and solve

$$\begin{cases} \partial_t v_{PU} + \sup_{\bar{p}_{PU}}\{b_{PU}(s, \bar{p}_{SU}, \bar{p}_{PU}, m^*) + f_{PU}\nabla_s v_{PU}\} \\ + \frac{1}{2}\text{trace}(\sigma\sigma'\partial^2_{ss} v_{PU}) = 0, \\ \text{s.t. } (\bar{p}_{SU}, m^*) \in \mathscr{SR}(\bar{p}_{PU}) := \{(\bar{p}^*_{SU}[p_{PU}], m^*[p_{PU}])\} \end{cases} \tag{3.174}$$

The case where $\mathscr{SR}$ has multiple elements is not considered in this work.

Under the assumptions of Sect. 3.4.1,

(i) Algorithm 4 provides a mean-field equilibrium given the control of PU.
(ii) The coupled Algorithms 3, 2, and 4 provide a hierarchical mean-field solution of the cognitive radio network.

Hamiltonian of PU

Consider $b_{PU,y} = 1\!\!1_{\{mSINR \geq \beta_{PU}\}}$. The Hamiltonian of PU is defined by

$$H_{PU,y}(s, \nabla_s v, I_{SU}) = \sup_{\bar{p}_{PU}}\{b_{PU} + \langle f_{PU}, \nabla_s v\rangle\}$$

By taking into consideration the reaction of the SUs, the Hamiltonian becomes

$$\tilde{H}_{PU,y}(s, \nabla_s v, I_{SU}[.]) = \sup_{\bar{p}_{PU}, I_{SU} \in \mathscr{MR}[p_{PU}]}\{b_{PU} + \langle f_{PU}, \nabla_s v\rangle\}$$

Proposition 3.27

$$H_{PU,y}(s, \nabla_s v_{PU}, I_{SU}) = \begin{cases} 1 - \bar{a}\nabla_E v \textit{ if } E_{PU,t} \geq \epsilon_E,\ \bar{a} < \frac{1}{\partial_E v_{PU}(s)} \\ 0 \qquad\qquad \textit{otherwise} \end{cases}$$

$$\textit{where } \bar{a} = \frac{\beta_{PU} l(\left|x_{PU} - y\right|)}{\bar{g}(x_{PU}, y)}(N_0(y) + I_{SU}(y))$$

Proof The proof follows by maximization of the function $b_{PU,y} - p_{PU}\partial_E v_{PU}$ and the positivity condition for $1 - p^*\partial_E v_{PU}$. ■

Note that the Hamiltonian of PU does not depend on $\nabla_g v$ and $\nabla_x v$. This is because the drift of mobility dynamics (resp., channel gains) does not depend on the transmit power. However, the optimal feedback strategy depends on the location, the channel gain, the initial energy, and the mean field.

Hamiltonian of a Generic SU
The Hamiltonian of SU is defined by

$$H_{SU,y}(s, \nabla_s v_{SU}, \bar{p}_{PU}, I_{SU}) = \sup_{\bar{p}_{SU}}\{b_{SU} + \langle f_{SU}, \nabla_s v_{SU}\rangle\}$$

Proposition 3.28

$$H_{SU,y}(s, \nabla_s v_{SU}, \bar{p}_{PU}, I)$$
$$= \begin{cases} 1 - \bar{a}'\nabla_E v_{SU} \textit{ if } E_{SU} \geq \epsilon_E,\ \bar{a}' < \min\left(\frac{1}{\partial_E v_{SU}(s)}, \frac{\gamma l(\left|x_{SU} - y\right|)}{\bar{g}(x_{SU}, y)}\right) \\ 0 \qquad\qquad \textit{otherwise} \end{cases}$$

$$\textit{where } \bar{a}' = \frac{\beta_{SU} l(\left|x_{SU} - y\right|)}{\bar{g}(x_{SU}, y)}(N_0(y) + \bar{r}(x_{PU}, y) + I_{SU}(y))$$

Proof The proof is similar to Proposition 3.27 subject to the access conditions and success conditions for SU given $\bar{p}_{PU}$. ■

Note that the above Hamiltonians are not smooth. The nonzero part of the Hamiltonian is monotone in I.

Approximations and Error Bounds
The hierarchical mean-field solutions are difficult to compute explicitly, and there are no effective numerical scheme either. We propose an estimation of the mean-field system and provide error bounds induced by these approximations. The basic idea is to start with first moment approximation.

Conditional Density

Since the mobility dynamics and channel gains do not depend on control, we can decouple the associated density from m. However, the energy dynamics depends on E, x, g due to the control. Therefore, we are interested in estimating the conditional density $\hat{m}_t(E|x, g) = m_t(E|\ x, g)$ We approximate the conditional distribution of energy by a Gaussian with time-varying mean $\hat{\mu}_t(x, g)$ and variance $\hat{\sigma}_t^2(x, g)$. These two functions evolve according to ODEs.

First-Order Approximation

We will replace the interference term $\int_s K_t(s, y)m_t(ds)$ by

$$\hat{I}_t = \int_s \frac{(\tau_1(x, g, t)E + \tau_2(x, g, t))g}{l()}\hat{m}_t(dE|x, g)m_{x,t}(dx)m_{g,t}(dg)$$

where the coefficients $\tau_j(x, g, t)$ do not depend on E, and the power is approximated by $\tau_1(x, g, t)E + \tau_2(x, g, t))$.

$$\begin{aligned} H \quad &= \sup_p\{\mathbb{1}_{\{mSINR \geq \beta_{SU}\}}(p) - p\partial_E v\} \\ &= \begin{cases} 1 - p^*\partial_R v \text{ if } p^* \in [a, +\infty) \text{ and } a \leq b \\ 0 \qquad\qquad \text{otherwise} \end{cases} \\ &= (1 - a\partial_E v)\mathbb{1}_{\{a \leq b,\ a < \frac{1}{\partial_E v}\}} \end{aligned}$$

Note that, in the case of $p^* = 1/\partial_E v$, we have $H = 0$ for all time, so we are not optimizing at all.

We study the deviation of the Hamiltonian from I_{SU} to $\hat{I}$. The corresponding Hamiltonian values will be denoted by H and $\hat{H}$, respectively.

Next we evaluate the error $H - \hat{H}$:

$$\begin{aligned} H - \hat{H} &= (1 - a\partial_E v)\mathbb{1}_{\{a \leq b\}} - (1 - \hat{a}\partial_E v)\mathbb{1}_{\{\hat{a} \leq b\}} \\ &= \begin{cases} (a - \hat{a})\partial_E v & \text{if } a \leq b, \hat{a} \leq b \\ 1 - a\partial_E v_t & \text{if } a \leq b, \hat{a} > b \\ -1 + \hat{a}\partial_E v_t & \text{if } a > b, \hat{a} \leq b \\ 0 & \text{otherwise} \end{cases} \end{aligned}$$

From

$$v(s, t) = arg \max_p \mathbb{E}R(s(T; t, s))$$

where $\partial_t v + H(v, \nabla_s v, s) + \frac{\sigma^2}{2}\partial_{ss}^2 v = 0$, we can estimate the error

$$\begin{aligned}\mathbb{E}(v(\hat{s}_T, T) - v(s_t, t)) &= \mathbb{E}\left[\int_t^T dv(t', \hat{s}_{t'})\right]\\ &= \mathbb{E}\left[\int_t^T (\partial_t v + \hat{H} + d\sigma^2 \partial_{ss}^2 v) dt\right]\\ &= \mathbb{E}\left[\int_t^T (\hat{H} - H)(t', \hat{s}_{t'})\, dt'\right]\end{aligned}$$

We conclude that the indicator function is appropriated for approximation, since a small error will propagate throught the gap $\int_t^T (1 - a\partial_E v)\mathbb{P}\left(a \leq b, \hat{a} > b\right)$. This error propagation is in part due to the non-smoothness of the indicator function.

3.4.4 Anti-jamming

We now turn to the anti-jamming hierarchical game.

Anti-jamming Stackelberg Strategy of PU

$$\sup_{\bar{p}_{PU}} \inf_{\bar{p}_{\text{Jam}}} \mathbb{E}\left(\bar{v}_{PU}(s_{PU,T}) + \int_0^T b_{PU}(s_{PU,t}, \bar{p}_{PU,t}, s_{Jam,t}, \bar{p}_{Jam,t}, I_t^*)\, dt\right)$$

and the state dynamics of PU $s_{PU,t}$ is given by

$$ds_{PU,t} = f_{PU}(s_{PU,t}, \bar{p}_{PU,t})dt + \sigma dB_t, \quad t \in \mathcal{T}\backslash\{0\},$$
$$s_{PU,0} \in \mathbb{R}^3,$$

The state of the Jammer is

$$ds_{Jam,t} = f_{\text{Jam}}(s_{\text{Jam},t}, \bar{p}_{\text{Jam},t})dt + \sigma dB_t, \quad t \in \mathcal{T}\backslash\{0\},$$
$$s_{Jam,0} \in \mathbb{R}^3,$$

Each generic SU's defense strategy is formalized by solving the following problem:

$$\sup_{\bar{p}_{SU}} \inf_{\bar{p}_{\text{Jam}}} \mathbb{E}\left(\bar{v}_{SU}(s_{SU,T}) + \int_0^T b_{SU}(s_{SU,t}, \bar{p}_{SU,t}, s_{PU,t}, \bar{p}_{PU,t}, s_{Jam,t}, \bar{p}_{Jam,t}, I_t^*)\, dt\right)$$

(continued)

where $m_t^* \in \mathcal{MR}_+[p_{PU}]$ is the minimax-robust mean-field trajectory of SUs, and

$$I_t^* \in \mathcal{I}[p_{PU}] = \left\{ \int_{\mathcal{S}} K_t(s, y) m_t^*[p_{PU}](ds),\ m_t^* \in \mathcal{MR}_+[p_{PU}] \right\}.$$

We introduce the Issaac's condition:

$$\sup_{\bar{p}_{SU}} \inf_{\bar{p}_{\text{Jam}}} \tilde{H}_{SU} = \inf_{\bar{p}_{\text{Jam}}} \sup_{\bar{p}_{SU}} \tilde{H}_{SU},$$

where

$$\tilde{H}_{SU} := b_{SU} + \langle f_{SU}, \nabla_{s_{SU}} v \rangle + \langle f_{\text{Jam}}, \nabla_{s_{\text{Jam}}} v \rangle$$

Sion's minimax theorem [396] is a generalization of von Neumann's minimax theorem. It states

Lemma 3.29 ([396]) *Let $\mathcal{P}_{SU}$ be a non-empty compact convex subset of a linear topological space and $\mathcal{P}_{Jam}$ a non-empty convex subset of a linear topological space. If $\tilde{H}_{SU}$ is a real-valued function upper semicontinuous and quasi-concave on $\mathcal{P}_{SU}$, for all $\bar{p}_{Jam} \in \mathcal{P}_{Jam}$ and is lower semicontinuous and quasi-convex on $\mathcal{P}_{Jam}$, for all $\bar{p}_{SU} \in \mathcal{P}_{SU}$, then*

$$\sup_{\bar{p}_{SU}} \inf_{\bar{p}_{Jam}} \tilde{H}_{SU} = \inf_{\bar{p}_{Jam}} \sup_{\bar{p}_{SU}} \tilde{H}_{SU}.$$

Next, we have the following result:

Proposition 3.30 *Given PU's data, the game with Shannon information rate as an instantaneous payoff has a value and a saddle point. Moreover, the Jammer's optimal strategy is given by*

$$\omega_2 = -\frac{\varphi_3}{\varphi_2} + \frac{\varphi_1 + \sqrt{\varphi_1^2 + 4\frac{1}{(-\partial_{E_{Jam}} v_t)}\varphi_1\varphi_2}}{2\varphi_2}$$

where

$$\varphi_3 = N_0(y) + \bar{I}_{SU}(s, y, t) + \bar{I}_{PU}(s, y, t),$$

$$\varphi_2 = \frac{\bar{g}_{Jam}(s, y, t)}{l\left(\left|x_{Jam} - y\right|\right)},$$

$$\varphi_1 = \bar{r}(s_{SU}, y, t),$$

and the representative SU's defense strategy is given by

$$\omega_1 = \left[\frac{1}{\partial_{E_{SU}} v_t} - \frac{1}{\Psi_t(\omega_2)}\right]_+,$$

$$\Psi_t(\omega_2) = \frac{\frac{\bar{g}}{l}}{N_0 + \bar{I}_{SU} + \bar{I}_{PU} + \omega_2 \frac{\bar{g}_{Jam}(s,y,t)}{l(\|x_{Jam} - y\|)}}.$$

Proof Consider the minimization of the function

$$\omega_2 \longmapsto \log\left(1 + \frac{\varphi_1}{\varphi_3 + \varphi_2\omega_2}\right) - \omega_1 \partial_{E_{SU}} v - \omega_2 \partial_{E_{\text{Jam}}} v$$

By convexity, this leads to $\varphi_3 + \varphi_2\omega_2 > 0$ as a solution of the following equation

$$z^2 + \varphi_1 z + \frac{\varphi_1\varphi_2}{\partial_{E_{\text{Jam}}} v} = 0.$$

Similarly, the function

$$\omega_1 \longmapsto \log\left(1 + \omega_1 \Psi(\omega_2)\right) - \omega_1 \partial_{E_{SU}} v - \omega_2 \partial_{E_{\text{Jam}}} v$$

is concave. Then, the KKT conditions provide the desired result. Thus, a saddle point is obtained by solving the fixed point system between ω_1 and ω_2. ■

3.4.5 Multiple Receivers for Each Primary User

In this subsection, we extend model 1 to multiple receivers, which we refer to as *Model 2*. This is the scenario where each PU can choose multiple receivers around its protection zone. We can think this problem as having multiple virtual PUs, one for each receiver. Then the hierarchical reasoning we used in model 1 can be generalized to this case. We assume that the power dedicated to each receiver of the PU is managed by virtual machines that we call *virtual battery managers* (VBMs). Each VBM computes the power that is required for the proposed task/service. This allows one to do parallel computing in order to save energy. The virtual power recommended by the virtual machine may not be compatible at all the time; therefore, the transmitter of PU has a scheduler which validate or not the proposed request of the VBM. The positive requests are then the resulting power allocation scheme over the receivers.

Figure 3.10 illustrates the power allocation mechanism: VBMs interact with the Battery Scheduler of PU which in turn interact with transmitter's power. allocation policy at time t. Denote by $RxPU = \{Rx_1, \ldots, Rx_I\}$ the set of receivers of PU

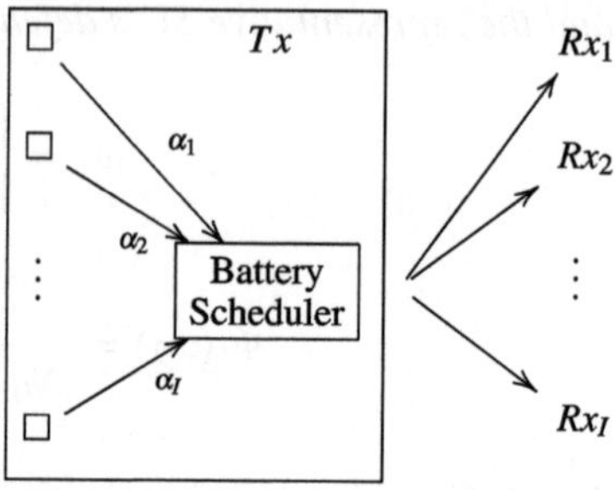

Fig. 3.10 Power allocation mechanism for multiple virtual PUs

and $(\alpha_1, \ldots, \alpha_I)$ the power recommended by VBMs. Then, the allocated power is $p_{PU,k} = \alpha_k$ if the scheduler validates the input by VBM k and $p_{PU,k} = 0$ otherwise. The figure below represents the VBMs inputs.

Naturally the question now is the existence and computation (if any) of PU's optimal response strategy.

Now $\bar{p}_{PU} = (\bar{p}_{PU,r})_r$ is the transmit power vector of PU. $\bar{p}_{SU,r}$ is the power associated with r-th receiver of PU.

$$b_{PU}(s_{PU}, \bar{p}_{SU}, \bar{p}_{PU}, m^*) = \sum_{r \in RxPU} b_{PU,k}(s_{PU}, \bar{p}_{SU}, \bar{p}_{PU,r}, m^*)$$

and the energy dynamics is modified to be

$$E_{k+1} = E_k - \delta_I \left[\sum_{r \in Rx(k)} p_{jr,k} \right] + \sigma_e E_k \left(\tilde{B}_{k+1} - \tilde{B}_k \right) \sqrt{\delta_I}$$

and the limiting is

$$dE = -\left(\sum_{r \in RxPU} \bar{p}_{PU,r} \right) dt + \sigma_e E \, dB_t \tag{3.175}$$

What is the optimal power allocation for PU? Recall that for a classical static multiple access channel power allocation, the splitting power is obtained by the water-filling operator using Karush-Kuhn-Tucker conditions.

Proposition 3.31 *The optimal power allocation of PU under the long-term Shannon rate is given by*

$$p^*_{PU,r} = \left[\frac{1}{\partial_E v_{PU}(s, y)} - \frac{1}{a_{PU,r}} \right]_+,$$

where

$$a_{PU,r} = \frac{\frac{\bar{g}(x,y_r)}{l(\|x-y_r\|)}}{N_0(y) + \bar{I}_{SU}(y_r,t)}.$$

Proof This follows from a direct computation of the optimum of the concave function (with respect to the own policy)

$$\sum_{r \in RxPU} \log(1 + mSINR(y_r)) - p_r \partial_E v_{PU}(s, y)$$

■

3.4.5.1 Jamming at Multiple Receivers

Multiple PUs: One Receiver per PU

Here we extend model 1 to multiple PUs, which we refer to as Model 3. In many practical scenarios, one can consider the following: A PU is located at the center of its protection zone, and many SUs are trying to access the unused spectrum. Each SU uses carrier sensing in order to guarantee that not too much interference is created to the PU. We are interested in its access probability and its probability of success under energy consumption constraint. For that, we need to define the access indicator random variable of a SU taking into account the other SUs and all the PUs and the access indicator random variable of a PU taking into account the other PUs. Then let

$$\mathbb{1}_{a(k,k')} := \mathbb{1}_{\{t_k \leq t_{k'}\}} + \mathbb{1}_{\{t_k > t_{k'}\}} \mathbb{1}_{\{r_{k,k',t} \leq \gamma\}} \mathbb{1}_{\{r_{k',k,t} \leq \gamma\}} \tag{3.176}$$

where $t_k \in \mathscr{T}_a := [0, 1]$ is the access request time of user k. $\mathbb{1}_{\{k,k'\}}$ reads "if user k requested access before user k' then it gets the medium (indicator is equal to 1), else he can get the medium only if the interference received by k' is not too high (less or equal than γ)." Now, given a PU i, the access indicator random variable is

$$\prod_{i' \in PU_a \setminus \{i\}} \mathbb{1}_{a(i,i')} \tag{3.177}$$

where PU_a is the set of active PUs. Given a SU j, the access indicator random variable is

$$\left(\prod_{i \in PU_a} \mathbb{1}_{\{r_{j,i,t} \leq \gamma\}} \mathbb{1}_{\{a(j,i)\}} \right) \prod_{j' \in SU_a \setminus \{i\}} \mathbb{1}_{a(j,j')} \tag{3.178}$$

where SU_a is the set of active SUs.

Two PUs

We start with two PUs denoted by P_1 and P_2.

Access Probability
We compute access probability and probability of success under energy consumption constraint. Compared to model 1, the *access probability* of active PU i is changed to

$$\mathbb{E}\left[\prod_{\{i' \in PU_a \setminus \{i\}\}} \mathbb{1}_{a(i,i')}\right] \tag{3.179}$$

Probability of Success
The probability of success of SU j is

$$\mathbb{E}\left[\left(\prod_{i \in PU_a} \mathbb{1}_{\{r_{j,i,t} \leq \gamma\}} \mathbb{1}_{a(j,i)}\right) \prod_{j' \in SU_a \setminus \{i\}} \mathbb{1}_{a(j,j')}\right]. \tag{3.180}$$

The transmission scheme treats interference as a noise, and a transmission with a given power policy of SU is successful if he/she is allowed to transmit and its sSINR is greater than β_{SU} where in the interference terms we count only the users which are allowed to transmit.

Using the same limiting arguments as before, we get $mSINR_{P_1}$ for a receiver located at y

$$mSINR_{P_1}(y,t) = \frac{\bar{r}_{P_1}(s_{P_1}, y, t)}{N_0(y) + \bar{I}_{SU}(y,t) + \bar{r}_{P_2}(s_{P_2}, y, t)} \tag{3.181}$$

We now have a random variable in the denominator. Note that r_{P_1} and r_{P_2} are independent. Then, the characteristic function can be used to describe the expression

$$\mathbb{P}(\bar{r}_{P_1} - \beta \bar{r}_{P_2} \geq \beta(N_0 + \bar{I}_{SU})) \tag{3.182}$$

Multiple PUs with Multiple Receivers and Multiple SUs
We now extend the model 2 to multiple PUs case, called *model 4*. Each PU has multiple receivers in this range. In order to exploit efficiently, the carrier sensing each receiver can send a beacon to a transmitter.

Queue-Aware Cognitive Radio Networks
Model 4 can be completed with queue dynamics

$$\tilde{q}_{j,k+1} = \tilde{q}_{j,k} - \delta_I \mathscr{P}_{\text{success}}(s_k, p_k, y_k) + \tilde{A}_{j,k+1} + \sigma_q(.)\left(\tilde{B}_{j,k+1} - \tilde{B}_{j,k}\right)\sqrt{\delta_I},$$

with $\lambda_t(s_t)\delta_I$ the conditional expectation of the arrival process $\tilde{A}_{j,k+1}$

This equation says that the evolution of the queue is described by the difference between arrival and departure of packets plus noise where the departure is the success probability.

Each limiting own-queue dynamics will be represented by a reflected SDE in the form

$$dq_{j,t} = (\lambda_t(s_t) - \mu_t(q_t, s_t, \bar{p}_t, y_t))\, dt + \sigma_q(q_t, s_t, \bar{p}_t) dB_{j,t}.$$

The term $\lambda_t(s_t)$ is the expected value of arrival process, μ_t is the probability to have successful transmission (departure of packet), and $\sigma_q(.)dB_{j,t}$ is a noise term. In addition, the queue dynamics are subject to a positivity constraint: $\tilde{q}_{j,t} \geq 0$. Then the question of queue stability and capacity region arise for the queueing-based hierarchical mean-field game system.

A user augmented state is now given by $\underline{s}_{j,k} = (E_{j,k}, x_{j,k}, g_{j,.,k}, q_{j,k})$ described as

$$\begin{cases} k \in \mathbb{Z}_+, \\ \textbf{Own} - \textbf{energy management} \\ E_{j,k+1} = E_{j,k} - \delta_I p_{j,k} + \sigma_e E_{j,k} \left(B_{j,k+1} - B_{j,k}\right) \sqrt{\delta_I} \\ \textbf{Mobility dynamics} \\ x_{j,k+1} = x_{j,k} + \delta_I \tilde{v}_{\theta_j}(x_{j,k}) + \sigma_x \left(B_{j,k+1} - B_{j,k}\right) \sqrt{\delta_I} \\ \theta_j \in \{PU, SU\} \\ \textbf{Own} - \textbf{channel state dynamics} \\ h_{jr,k+1} = h_{jr,k} + \frac{\delta_I}{2} \left(\mu_h - h_{jr,k}\right) + \sigma_h \left(B_{j,k+1} - B_{j,k}\right) \sqrt{\delta_I} \\ \textbf{Own} - \textbf{queue length evolution} \\ q_{j,k+1} = q_{j,k} - \delta_I \mathscr{P}_{\text{success}}(s_k, p_k, y_k) + A_{j,k+1} + \sigma_q \left(B_{j,k+1} - B_{j,k}\right) \sqrt{\delta_I}, \\ \textbf{Initial conditions} \\ s_{j,0} = (E_{j,0}, x_{j,0}, g_{j,.,0}, q_{j,0}),\ j \in \mathscr{I}, \\ \textbf{Boundary conditions} \\ q_{j,k} \geq 0,\ x_{j,k} \in \mathscr{X},\ E_{j,k} \geq 0 \end{cases}$$

Proposition 3.32 *Suppose that (i) M_0^I converges in probability to $\rho_0(.)$ when I goes infinity, (ii) $p_{j,k}^I$ preserves the asymptotic indistinguishability per class property, and (iii) the law of initial measure ρ_0 has a continuous density bounded by $c'e^{-d'\|\underline{s}\|^2}$ for some positive constants c', d'. Then M_k^I converges in distribution to m_k. Moreover, the law of m_t (renamed $m_t(.)$) is solution to the Fokker–Planck–Kolmogorov equation*

$$\partial_t m_t(\underline{s}) + \partial_{\underline{s}}(f_{SU}(p_t^*(.), \underline{s}) m_t(\underline{s})) = \partial_{\underline{s}\underline{s}}^2 [\sigma_{SU} \sigma_{SU}^* m_t(\underline{s})],$$
$$m_0(\underline{s}) = \rho_0(\underline{s}),\ \forall \underline{s} = (E, x, g, q) \tag{3.183}$$

where $p_t^*(.)$ will be the optimal control which is a function of t, $\underline{s}$, $\bar{m}_t$ and the value function v_t.

Proof The proof follows the same lines as above with two classes: PU and SU.

For the weak convergence, the mean-field process under control laws that preserve the asymptotic indistinguishability per class property, we use [438] which is mainly based on de Finetti's theorem. For the derivation of Fokker–Planck–Kolmogorov (FPK), we use distribution theory and integrate by parts. Combining together, we get the evolution of the mass by (3.183). See Volume I for more details. ■

3.4.6 Notes

We have studied mean-field models for cognitive radio networks under mobility and energy constraints. Using recent developments in the field of mean-field stochastic games, we have characterized the evolution of the access probability, success probability, throughput, and density of energy, and optimal power for both primary and secondary users moves according to dynamics driven by Brownian motions. These performance metrics are fundamental in the emerging cognitive radio networks [3, 5, 120, 291, 301, 392, 442, 478, 508, 524]. The result presented in this section allows one to deal with large number of nodes under indistinguishability per class property for the power allocation scheme. Based on coupled partial differential equations that describe the hierarchical mean-field equilibrium, we have provided an algorithm for finding an optimal energy management strategy. These results open new questions:

- *Can we extend the mobility model to more realistic observed human-based mobility?* One possibility is to investigate the performance metrics under Lévy process (flight, motion, stable process, etc.) and anomalous diffusion.
- *Multiple antennas-based large cognitive radio networks* which different classes of users, each class having its initial distribution and its own mobility scheme. This allows one to study heterogeneity of mobility in a given map sectionized by area (the mobility model for downtown is different from the one of suburban areas). It would be important to know the impact of this on the optimal use of spectrum holes in cognitive MIMO networks.
- *Mean field for small number of users?* Last but not the least, the framework developed here is not suitable for the case when there is a small number of secondary users. While cognitive radio networks have been widely studied for the case of a small number of users, it would be an interesting direction

(continued)

to explore the so-called non-asymptotic mean-field game approach to this problem, i.e., a mathematical modeling of spectrum allocation under mobility and energy limitations with a small number of secondary users [6, 120, 424, 440, 478]. We refer the reader to Volume I of the book for non-asymptotic mean-field error bounds.

3.5 Battery-State-Dependent Power Management

Energy consumption is of importance for wireless devices, their users, and their stochastic environment. With the increasing proliferation of communication devices, the power management strategy becomes an important issue to be addressed as well as the payoff optimization problem. This section addresses the power management problem using a mean-field stochastic game framework. Based on the earlier Chap. 2 and Sect. 3.2, we extend the modeling and include the details of battery management.

We consider large number of mobile devices controlling their transmission power to a receiver. Each mobile implements a power control strategy where the transmission power is allowed to depend on the energy level (state) of its battery. The available action (reachable base stations and power levels) depends on the state of the battery. Given the remaining energy in its battery, the mobile has to choose an available the right power level to optimize its long-term payoff. One of the important element for each mobile is its throughput which can be characterized as a function of the SINR at the base station. A second important element for the mobile device is power management or power saving. The mobile has to control both the power consumption and the time when the batteries are replaced. Each battery lifetime game corresponds to a stochastic game with finite horizon (absorbing state of battery when the energy is very small). The goal is to find jointly the power consumption policies such that all the mobile devices achieve as high payoff as possible with communication quality guarantee (e.g., QoS requirement) but also to control the battery state. When the batteries are recharged dynamically with different types of alternative energy such as renewable energies (solar, wind, etc.), the battery transition state becomes irreducible Markov decision process under each policy depending on an exogenous parameter which characterizes the weather conditions (e.g., a good weather will correspond to a sunny moment for the solar-power systems and to a windy slot for wind-powered systems). In this case, the interaction becomes a stochastic game with infinite horizon, and we shall consider the long-term average payoff (discounted or not).

When there is no alternative renewable energy, the battery needs to be replaced or recharged after some time. Thus, it is sufficient to analyze the dynamic game for one cycle. We examine the finite horizon case with total payoff until the absorbing state.

In this section, we present a mean-field approach to energy management. We formulate a mean-field stochastic game model for battery-state-dependent power management.

The remainder of the section is structured as follows. In the next section, we present the model description and the mean-field modeling of power management. Then, we focus on the characterization of equilibria and equilibrium payoffs in Sect. 3.5.2. In Sect. 3.5.3, we analyze the stationary mean-field case.

3.5.1 Energy Game Model

We consider I mobile devices with individual battery-state-dependent transmissions. Each battery state evolution is modeled by a Markov decision process which describes the remaining energy level for the corresponding mobile. The battery state of mobile device j is then influenced by its remaining energy level and its own power consumption policy. This influence is taken into consideration via a decreasing transition probability in the power consumption. The more the power consumption is, the less the remaining energy will be.

It is important to note that the Markov process of each mobile device is controlled by itself, and the battery life time of mobile j depends only its own strategy. We assume that each player starts at the highest energy state or level. We denote by $\mathcal{X} = \{0, 1, 2, \ldots, |\mathcal{X}| - 1\}$ the set of states of the battery and by $\mathcal{A}$ the set of all the possible power levels. Starting from the highest state, each player will visit the different levels in the decreasing order until reaching the last state 0. The state 0 corresponds to the *Empty state* of the battery, and the state $|\mathcal{X}| - 1$ corresponds to the *Full state*. The other states are intermediary states of the battery. When the battery of mobile j is at state 0, there is no power level for mobile j, i.e., zero power. We denote by $\mathcal{A}(x)$ the set of power levels available at state $x \in \mathcal{X}$. Then, this says that $\mathcal{A}(0) = \{0\}$. We assume $\mathcal{A}(x)$ is a subset of $\mathcal{A}(x+1)$, and the maximum power level in $\mathcal{A}(x)$ is strictly less than the maximum power level in $\mathcal{A}(x+1)$.

In order to define the Markov decision process for the power management of each mobile, we need the controlled transition probabilities. Let $q_t(x'|x, a)$ be the probability of moving from x to x' when power a is used at time t in state x. The support of $q_t(.|x, a)$ is reduced to $\{x, x-1\}$ for any power level a and any state $x \neq 0$. Thus, we need to specify only $q_t(x-1|x, a)$ because

$$q_t(x|x, a) = 1 - q_t(x-1|x, a), \; x \geq 1, \; a \in \mathcal{A}(x). \tag{3.184}$$

We assume that $q_t(x-1|x, a)$ is nondecreasing in a. This captures the fact the more the power is consumed less the remaining energy is.

Most of the time, the performance of each mobile depends on the transmit power of the others. When I is very large, it is not possible for a mobile to observe the actions of all the other mobiles in the absence of mediation and coordination between the mobiles. Here, we introduce a mean-field approach for this regime. We define the key object: Let

$$M_t^I(x) = \frac{1}{I}\sum_{j=1}^{I} \mathbb{1}_{\{x_{j,t}^I = x\}},\quad x \in \mathscr{X},$$

where $\mathbb{1}_{\{.\}}$ denotes the indicator function. The term $M_t^I(x)$ represents the fraction of players in state x at time t. The vector $(M_t^I(x))_{x\in\mathscr{X}}$ is the mean field at time t when the population of mobiles is of size I.

Note that the vector $(M_t^I(x))_{x\in\mathscr{X}}$ satisfies

$$\forall\, x \in \mathscr{X},\ M_t^I(x) \geq 0, \text{ and } \sum_{x\in\mathscr{X}} M_t^I(x) = 1.$$

Thus, $M_t^I \in \Delta(\mathscr{X})$ the set of probability measure over $\mathscr{X}$.

Similarly, we define

$$\tilde{M}_t^I(x,a) = \frac{1}{I}\sum_{j=1}^{I} \mathbb{1}_{\left\{\left(x_{j,t}^I, a_{j,t}^I\right) = (x,a)\right\}},\quad x \in \mathscr{X}.$$

We define the instantaneous payoff function of a generic mobile j as a function of its current state, its current strategy, and an aggregative term which is function of the mean field.

$$r_{j,t}^I(x_{j,t}^I, a_{j,t}^I, \tilde{M}_t^I) = r_t\left(a_{j,t}^I,\ \xi\left(\sum_{x,a} a\tilde{M}_t^I(x,a)\right)\right)$$

The term $\xi\left(\sum_{x,a} a\tilde{M}_t^I(x,a)\right)$ can be expressed as $\xi\left(\sum_a a\bar{M}_t^I(a)\right)$ where $\bar{M}_t^I(a) = \sum_{x\in\mathscr{X}} \tilde{M}_t^I(x,a)$. This last formulation allows us to capture several different SINR-based payoff functions. When r is the quotient and ξ is the identity function, the $\text{SINR}_j\left(a_{j,t}^I, \tilde{M}_t^I\right)$ is given by

$$\text{SINR}_j\left(a_{j,t}^I, \tilde{M}_t^I\right) = \frac{a_{j,t}^I}{\sigma^2 + \alpha\sum_{a\in\mathscr{A}} a\bar{M}_t^I(a) - \frac{\alpha}{I}a_{j,t}^I}.$$

The factor σ^2 is the variance of the noise, and α represents a load factor.

3.5.1.1 Example of Payoff Functions

Several interesting instantaneous payoff functions can be written in this form:

- $r^I_{j,t} = \frac{f_t(\text{SINR}_j)}{c_t\left(x^I_{j,t}, a^I_{j,t}\right)}$ is the obtained gain per cost unit.
- $r^I_{j,t} = \frac{c_t\left(x^I_{j,t}, a^I_{j,t}\right)}{f_t(\text{SINR}_j)}$ is the power consumption per gain.
- $r^I_{j,t} = f_t(\text{SINR}_j) - c_t\left(x^I_{j,t}, a^I_{j,t}\right)$ is the benefit measurement.

where f_t is a well-chosen function. For example, the Shannon rate-based payoff f_t corresponds to the logarithm function and the bit-error-rate function $y \longmapsto (1 - e^{-\beta y})^L,\ \beta > 0,\ L > 0$ The function c_t is for the cost of power consumption, a nondecreasing function. Such time-dependent payoff functions allow us to model scenarios where the objective of the mobile changes in time. In some frame, the mobile takes care of energy consumption cost, and some other frames it takes care of its throughput only.

3.5.1.2 On the Mean-Field Object

The goal now is to describe the evolution of the mean-field $\tilde{M}^I_t$ in function of the strategies of the players and the individual battery states. For sufficiency conditions of convergence of the process $\tilde{M}^I_t$, we refer to [444] which shows that under regularity assumptions and finite second moment conditions, the mean-field limit can be characterized by differential equations. When the second moment is not necessarily finite, a stochastic mean-field convergence is provided. The noisy mean field is characterized by Itô's SDE or a discrete jump process. We refer to appendix of volume I for the mean-field convergence.

Note that the convergence to a mean-field limit does not necessarily imply the convergence into a stationary regime. A counterexample is provided in [444] where a limit cycle of M^I_t is observed in a scenario with unique stationary distribution. The phenomenon is well-known in evolutionary game dynamics (deterministic or stochastic) where limit cycles arise in rock-scissor-paper games, Shapley's game, etc.

The mean-field convergence depends mainly on how the interaction between the mobiles scales with the size of the system. Under the *indistinguishability property* (also called exchangeability property), we can show a weak convergence to a random measure. We refer to [198, 326, 366, 413] for more details on the mean-field convergence. See also the discussions in [438, 439] for controlled mean-field processes. Note that if a mobile can precompute (offline) the mean field, the strategy of a mobile device can be a function of $\tilde{M}^I_t$. For example, the following strategy

$$u^I_j(a|x) = \frac{\tilde{M}^I_t(x,a)}{M^I_t(x)} \mathbb{1}_{\{M^I_t(x)>0\}} = \frac{\tilde{M}^I_t(x,a)}{\sum_{a' \in \mathcal{A}(x)} \tilde{M}^I_t(x,a')} \mathbb{1}_{\{M^I_t(x)>0\}} \quad (3.185)$$

can be adopted if the mean-field process is observed.

We assume that the maximum lifetime of a single battery is $T < \infty$. Then, the cumulative payoff of a mobile starting from $t_0 = 0$ is up to the lifetime of the battery under its own strategy, i.e., the first time the battery state process enters into $\{x = 0\}$ under the power consumption strategy adopted by mobile j :

$$\tau_j(u_j^I, x) = \inf\left\{t > 0,\ x_{j,t}^I = 0 \mid x_{j,0}^I = x, u_j^I\right\}$$

A private history of mobile j is the collection of (i) all the power consumptions used by j, (ii) all the previous states of its own battery, and (iii) all the mean-field trajectory and/or the kernel of the mean field.

A (behavioral) strategy u_j^I of a mobile j a collection of mapping from the set of own histories up to date to the set probability distributions over $\mathcal{A}\left(x_{j,t}^I\right)$. The mean-field trajectory and a strategy of the mobile induce a probability over the play until absorbing at state 0. We write $\mathbb{E}_{u^I,x,\tilde{m}_0}$ to denote the expectation under the induced probability measure. Let

$$F_{j,\tau_j}^I\left(u_j^I, \tilde{M}^I\right) = \mathbb{E}_{u^I,x,\tilde{m}_0}\left(\sum_{t' \leq \tau_j} r(a_{j,t'}^I, \tilde{M}_{t'}^I) \mid x_{j,0}^I = |\mathcal{X}| - 1, \tilde{M}_0^I = m_0\right)$$

be the cumulative payoff of mobile j. Note that one can consider the discounted payoff by using recharging system of batteries.

3.5.1.3 Best-Response to Mean-Field Trajectory

A reaction set to the mean-field object $\tilde{M}^I$ can then be defined

$$\mathcal{BR}^I(\tilde{M}^I) = \arg\max_{v_j^I} F_{j,T}^I\left(v_j^I, \tilde{M}^I\right),$$

where v_j^I describes the set of strategies of mobile j. An element $u_j^I \in \mathcal{BR}^I(\tilde{M}^I)$ is a best response to the mean-field $\tilde{M}^I$.

When I goes to infinity, we define the following: A mean-field equilibrium (u^*, m^*) is such that (i) $u^* \in \mathcal{BR}(\tilde{m}^*)$ and (ii) u^* produces the mean-field $\tilde{m}^*$, i.e., the mean-field limit under u^* gives m^* :

$$m^* = \langle m^*, \mathcal{L}(.|u^*, \tilde{m}^*)\rangle.$$

The infinite population power management problem can be formulated as

$$\begin{cases} \sup_v \ F_{j,T}(v, \tilde{m}) \\ \tilde{m}_{t+1} \sim \mathscr{L}_t(.|u_t, \tilde{m}_t), \\ x_{j,t+1} \sim q_t(.|x_{j,t}, v_t) \end{cases}$$

We now formulate a Bellman equation to solve the power management problem

$$\begin{cases} V(\tilde{m}) = \sup_v \ F_{j,T}(v, \tilde{m}) \\ V_{T+1}(x, \tilde{m}) = 0 \\ V_t(x, \tilde{m}_t) = \max_{a'}\{r_t(a', \tilde{m}_t) + \sum_{x' \in \mathscr{X}} q_t(x'|x, a')V_{t+1}(x', \tilde{m}_{t+1}) \\ \tilde{m}_{t+1} \sim \mathscr{L}_t(.|u_t, \tilde{m}_t) \end{cases}$$

A stationary mean-field equilibrium is then a solution of

$$\begin{cases} u_x^* \in \arg\max_{v \in \mathscr{A}(x)} \left\{r_t(v, \tilde{m}^*) + \sum_{x' \in \mathscr{X}} q_t(x'|x, v_t)V_{t+1}(x', \tilde{m}^*)\right\} \\ \tilde{m}^* \sim \mathscr{L}_t(.|u^*, \tilde{m}^*) \end{cases}$$

3.5.2 Convergence and Mean-Field Best Response

The next proposition characterizes limiting payoffs and the subgame perfect mean-field equilibrium under the mean-field convergence at the initial conditions.

Proposition 3.33 *Assume that $\tilde{M}^I(0) \longrightarrow \tilde{m}_0$. Then, the following holds*

- *Convergence of payoff functions:*

$$\lim_{I \longrightarrow \infty} F^I_{j,T}\left(v, \tilde{M}^I_0\right) = F_{j,T}(v, \tilde{m}_0).$$

- *Characterization of limiting optimal strategies: For all $I' \in \mathbb{N}$, there exists $I > I'$ such that*

$$\mathscr{BR}^I(\tilde{M}^I_0) \subseteq \mathscr{BR}^\infty(\tilde{m}_0)$$

where

$$V^I_t\left(x, \tilde{m}_t\right) = \max_v \left\{r_t\left(v, \tilde{m}^I_t\right) + \sum_{x' \in \mathscr{X}} q_t(x'|x, v)V^I_{t+1}\left(x', \tilde{m}^I_{t+1}\right)\right\},$$

and

$$\mathscr{BR}^I_t\left(\tilde{m}^I_t\right) = \left\{v, \ V^I_t\left(x, \tilde{m}^I_t\right) = r_t\left(v, \tilde{m}^I_t\right) + \sum_{x'} q_t(x'|x, v)V^I_{t+1}(x', \tilde{m}^I_{t+1})\right\}$$

is the set of actions reaching the maximum and

$$\mathscr{BR}_t^{\infty}(\tilde{m}_t) = \left\{ v,\ V_t^{\infty}(x, \tilde{m}_t^*) = r_t(v, \tilde{m}_t^*) + \sum_{x'} q_t(x'|x, v) V_{t+1}^{\infty}(x', \tilde{m}_{t+1}^*) \right\}$$

$$\tilde{m}_{t+1}^* = \langle \tilde{m}_t^*, \mathscr{L}(.|u_t^*, \tilde{m}_t^*) \rangle.$$

Moreover, the pair $(u_t^*, \tilde{m}_t^*)$ *is subgame perfect mean-field Nash equilibrium in the sense that for any history (starting from the same* $\tilde{m}_0$*), the continuation strategy generates a mean-field equilibrium of the remaining game, and strategy* u_t^* *combined with* $q_t(.|.)$ *produces the mean-field equilibrium* $\tilde{m}_{t+1}^*$ *to which each of the mobile devices reacts and* u_{t+1}^* *is optimal via the mean-field dynamic programming principle.*

- *In particular, if* $q_t(.) = q(.), r_t(.) = r(.)$ *a stationary mean-field limit solves*

$$\tilde{m}^* = \langle \tilde{m}^*, \mathscr{L}(.|u^*, \tilde{m}^*) \rangle$$

and

$$u^* \in \mathscr{BR}^{\infty}(\tilde{m}^*),$$

i.e.,

$$V_t^{\infty}(x, \tilde{m}^*) = r(u^*, \tilde{m}^*) + \sum_{x'} q(x'|x, u^*) V_{t+1}^{\infty}(x', \mathscr{L}(.|u^*, \tilde{m}^*, .).\tilde{m}^*).$$

Proof Let $\mathscr{F}_t = \sigma\left(x^I(t'), a^I(t')\right)_{0 \leq t' \leq t}$ be the filtration associated with the process $\{X^I\}$, and $\{a^I\}$ up to time t. The evolution of the system is described by the transition probabilities q_t which generates $\mathscr{L}_t$. We first remark that the sequence $\{F_{j,T}^I(v, \tilde{M}_0^I)\}_{n \geq 0}$ is bounded by the infinite norm of r. Assume that the transition probabilities are regular enough: $\tilde{M}^I(0) \longrightarrow \tilde{m}_0$. Then, for an increasing mapping $\phi : \mathbb{N} \longrightarrow \mathbb{N}$, one has

$$\lim_I F_{\beta}^{\phi(I)}(\tilde{M}^{\phi(I)}(0)) = F(\tilde{m}_0)$$

Let $(u_*^I, \tilde{m}_*^I)$ be the optimal pair of process with starting population profile $\tilde{m}_0^I$. The process $(u_*^I, \tilde{m}_*^I)$ is tight, because $\mathscr{A}$ is finite. There is a convergence in law. Denote by \$ the law of the limit. Then, the equality $F(\tilde{m}_0) = V(\tilde{m}_0)$ holds \$−almost surely. Thus, $\lim_{\phi(I) \longrightarrow +\infty} F_T^{\phi(I)}(\tilde{m}_0) = F(\tilde{m}_0)$ and the mapping

$$(u, m) \longmapsto F(u, m)$$

is continuous. A similar reasoning yields the convergence of the optimal strategies and the inclusions.

The statement of subgame perfect mean-field equilibrium follows immediately from the Bellman equation. Any backward solution of the Bellman equation which is consistent with the mean-field limit $\tilde{m}_t$ is a subgame perfect mean-field equilibrium. Now, assuming that the payoff and the transition are time-homogeneous, the same analysis holds for the stationary equation. This leads to a stationary mean-field equilibrium. This completes the proof. ■

3.5.3 *Stationary Mean Field*

We now examine the payoffs under stationary strategies of the mobile devices. A stationary strategy of mobile j is a strategy which depends only the value of the current individual state.

$$\bar{u}_j : (I, x) \in \mathbb{N} \times \mathscr{X} \longrightarrow u_j^I(x) \in \Delta(\mathscr{A}(x))$$

Under this class of strategies, one can compute explicitly the sojourn time of the controlled Markov process. The first time the Markov decision process enters state 0, i.e., the lifetime of the battery starting from the state $|\mathscr{X}| - 1$, is the total sum of the sojourn time of the MDP at all the intermediary states. Then,

$$\tau_j(u_j^I, x) = 1 + \sum_{x'} \tau_j(u_j^I, x')q(x'|x, u_j^I)$$

which can be written as

$$\tau_j\big(u_j^I, x\big) = 1 + \tau_j\big(u_j^I, x\big)q\big(x|x, u_j^I(x)\big) + \tau_j(u_j^I, x-1)\big(1 - q(x-1|x, u_j^I(x))\big)$$

Using the relation

$$\tau_j\big(u_j^I, x\big) = \sum_{1 \le x' \le x} \tau_j(u_j^I, x') - \tau_j\big(u_j^I(x'-1), x'-1\big)$$

with $\tau_j\big(u_j^I, 0\big) = 0$, we can deduce that

$$\tau_j\big(u_j^I, x\big) = \sum_{x' \in \mathscr{X} \backslash \{0\}, x' \le x} \frac{1}{1 - q\big(x'|x', u_j^I(x')\big)}$$

and the expected sojourn time $x' \neq 0$ is

$$\frac{1}{1 - q(x'|x', u_j(x'))}.$$

Hence, an explicit expression for the payoff at the limit is

$$\sum_{x' \in \mathscr{X}} \frac{1}{1 - q(x'|x', u_j(x'))} r(u_j(x'), \tilde{m})$$

A strategy in state x of the battery is said aggressive if it is the maximum power available at that state x, i.e., $a_{\max}(x) = \max \mathscr{A}(x)$. We denote the aggressive strategy by $u^*_{\max}$.

An aggressive strategy may increase the performance of mobile j but decrease the battery lifetime. Under the aggressive strategy, the support of the mean-field limit $\tilde{m}_{\max}$ belong to $\{(0,0), (1, a_{\max}(1)), \ldots, (|\mathscr{X}|-1, a_{\max}(|\mathscr{X}|-1))\}$. Next we provide a sufficient condition for the aggressive strategy and its mean-field limit to be a mean-field equilibrium.

Proposition 3.34 (Aggressive Mean-Field Equilibrium) *A sufficient condition for a consistent pair $(u^*_{\max}, \tilde{m}^*_{\max})$ to a mean-field equilibrium is*

$$\sum_{x \in \mathscr{X}} \frac{1}{1 - q(x|a_{\max}(x), x)} r(a_{\max}(x), \tilde{m}_{\max}) \geq \sum_{x \in \mathscr{X}} \frac{1}{1 - q(x|a', x)} r(a', \tilde{m}_{\max})$$

$$\forall a' \in \mathscr{A}(x)$$

Proof Suppose $u^*_{\max}$ produces $\tilde{m}^*_{\max}$ with support in the set $\{(0,0), (1, a_{\max}(1)), \ldots, (|\mathscr{X}|-1, a_{\max}(|\mathscr{X}|-1))\}$ and the pair $(u^*_{\max}, m^*_{\max})$ satisfies that the payoff

$$\sum_{x \in \mathscr{X}} \frac{1}{1 - q(x|a_{\max}(x), x)} r\left(a_{\max}(x), \tilde{m}^*_{\max}\right)$$

is greater than the payoff of any other pure stationary strategy

$$\sum_{x \in \mathscr{X}} \frac{1}{1 - q(x|a', x)} r\left(a', \tilde{m}^*_{\max}\right), a' \in \mathscr{A}(x).$$

By a standard mixed extension, the inequality holds for any alternative mixed stationary strategy $v \in \Delta(\mathscr{A}(x))$. Then, it follows that $u^*_{\max}$ is a mean-field best response to $\tilde{m}^*_{\max}$ and satisfies the Bellman equation (in mixed stationary strategies). Combining with the fact that the consistency relationship $m^*_{\max} = \langle m^*_{\max}, \mathscr{L}(.|\ u^*_{\max}, \tilde{m}^*_{\max})\rangle$ is satisfied, we deduce that the pair is a mean-field equilibrium. ■

A strategy in state x of the battery is said *passive* if it is the lowest nonzero power available at that state x, i.e., $a_{\min}(x) = \min\left(\mathscr{A}(x)\backslash\{0\}\right)$. Under our assumption of $\mathscr{A}(x) \subset \mathscr{A}(x+1)$, one has that

$$\forall x \neq 0, a_{\min}(x) = \min\left(\mathscr{A}(1)\backslash\{0\}\right).$$

A passive strategy may result in lower performance for a mobile, but the passive strategy has the advantage of saving energy for future interactions. Under the passive strategy, the support of the mean-field limit $\tilde{m}_{\min}$ belongs to $\{(0,0), (1, a_{\min}(1)), \ldots, (|\mathscr{X}|-1, a_{\min}(1))\}$. We denote the passive strategy by $u^*_{\min}$.

Next we provide a sufficient condition for the passive strategy and its mean-field limit to constitute a mean-field equilibrium.

Proposition 3.35 (Passive Mean-Field Equilibrium) *A sufficient condition for a consistent pair* $(u^*_{\min}, \tilde{m}^*_{\min})$ *to be a mean-field equilibrium is*

$$\sum_{x\in\mathscr{X}} \frac{1}{1-q\big(x|a_{\min}(1),x\big)} r\big(a_{\min}(1), \tilde{m}^*_{\min}\big) \geq \sum_{x\in\mathscr{X}} \frac{1}{1-q(x|a',x)} r\big(a', \tilde{m}^*_{\min}\big)$$

$$\forall a' \in \mathscr{A}(x)$$

The proof follows the same lines as in the previous proposition. Suppose that $u^*_{\min}$ produces $\tilde{m}^*_{\min}$ and the pair satisfies the inequality. Then, it follows that $u^*_{\min}$ is a mean-field best response to $\tilde{m}_{\min}$ and satisfies the Bellman equation. Hence, it is a mean-field equilibrium.

3.5.4 Notes

We have examined the case of infinite population in [224, 434] using the framework of evolutionary games. Characterization of weakened evolutionarily state states (in the sense of occupation measures) has been obtained. However, the evolution of the occupation measures has not been studied in the above references. Note that the evolution of the mean-field limit will play an important role and the system may not be in stationary regime even when the size is infinite. Considering an aggregative interference model, we transformed the standard signal-to-interference-plus-noise ratio (SINR)-based payoff in terms of a mean-field term. The convergence of this mean-field term is discussed in appendix as well as in the recent advances in [444]. In the infinite population limit, we establish a connection with the works in [64, 254, 478]. Explicit conditions for mean-field best-response as well as sufficiency conditions for stationary mean-field equilibria for aggressive power management (transmit at the maximum power available at the corresponding state) and passive power management (transmit at the lowest nonzero power at each state of the battery) are given [464].

Chapter 4
User-Base Station Association in Next-Generation Networks

The classical optimal transportation problem entails finding the most cost-effective way of moving masses from one set of locations to another, minimizing its transportation cost. The formulation of this problem and its solution have been useful to understand various mathematical, economic, and control theory phenomena, such as Witsenhausen's counterexample in stochastic control theory, the principal-agent problem in microeconomic theory, and location and planning problems. The main difficulty is that one has to deal with optimization in infinite dimensions (the space of probability measures with transport constraints).

In this chapter, we incorporate the effect of network congestion to the optimal transportation problem, and we present expression for its solution. Here, the congestion term is the key mean-field term [460, 464]. As an application, we focus on mobile association problems in next-generation cellular networks: the determination of the cells corresponding to each (macro, small, femto, pico)base station, i.e., the locations at which mobile terminals prefer to connect to a given base station rather than to others. Two types of problems can be considered: a global optimization of the total power needed by the mobile terminals of the whole network (global optimum [106]) and an individual optimization, in which each mobile terminal chooses the base station that leads to the minimization of its power usage. The analysis of this chapter combines a mean-field-type game theory and optimal transportation to characterize both solutions.

Optimal transportation problems have their origins in planning problems (optimally transporting coal from mines to steel factories). In the classical optimal transportation problem, a central planner has to find a transportation plan between two nonnegative probability measures so as to minimize the overall cost. Such problems have been studied in recent years for their various applications in partial differential equations, control theory, probability theory, microeconomic theory, and congestion management. Typical applications of these problems in engineering are optimal allocations of resources or assignment problems.

T. Başar et al., *Mean-Field-Type Game Theory II*, Static & Dynamic Game Theory: Foundations & Applications, https://doi.org/10.1007/978-3-032-07031-9_4

The most common cost functions in optimal transportation depend basically on some distance between two probability measures (Monge–Kontorovich distance, Lévy–Prokhorov metric, Wasserstein distance, total variation distance). However, in many situations, the decision of which assignment to choose depends not only on the distance but also even more heavily on the congestion the assignment generates, i.e., there is a mean-field effect [460, 464].

One of these cases, which is the main focus of this chapter, is the mobile association problem in next-generation cellular networks, which entails determination of cells corresponding to each base station.

> The cell of a base station corresponds to the locations at which mobile terminals (users) prefer to connect to a given base station rather than to others.

Within this context, two types of problems can be considered: a global optimization of the total power needed to satisfy a certain throughput by the mobile terminals of the whole network and an individual optimization, in which each mobile terminal chooses the base station trying to minimize the power consumption while being satisfied in terms of throughput.

In future wireless networks, mobile terminals are expected to be able to access multiple radio access technologies, capable of choosing the wireless access technology to use and the access point to which to connect. Within this decentralized context, we study the mobile association problem, where we determine the cells corresponding to each base station, i.e., the locations at which mobile terminals prefer to connect to a given base station rather than to other base stations. We analyze the case where mobile terminals within a cell share the same spectrum, and consequently, mobile terminals' decisions on which base station to connect to affect the decision process of other mobile terminals in the network due to network congestion. From these interactions, mobile terminals learn their optimal access point (self-learning), where the individual optimality depends upon the context (type of service, bit error rate, delay).

From a mean-field game theory perspective, in a decentralized scenario, the assignment problem entails each mobile terminal (player) choosing the base station that best serves its needs.

This is an anonymous game since each player's utility depends on how many of his peers choose which strategy (which base station to connect). It is also related to what we have called *reaction to the mean field* [460, 464].

Optimal transport theory is a general framework that allows one to better understand the deployment of large-scale wireless networks in a more efficient way. This explains also the cell formation and user distribution in the network. We determine the spatial locations at which mobile terminals would prefer to connect to a given base station rather than to other base stations in the network. This situation arises when the decision to connect to a given base station instead of others is left to the users. For comparison, we obtain the spatial locations which are more convenient from a centralized point of view. Obviously, in both approaches, the optimality will depend upon the context. In both considered cases, our aim is minimization of the total power of the network, which can be considered as an energy-efficient objective, while maintaining a certain level of throughput for each user connected to the network. We propose the theory of optimal transportation as the main tool of modelization of these mobile association problems. We characterize these mobile associations under different policies and give illustrative examples of this technique.

The remainder of this chapter is organized as follows: In Sect. 4.1, we give some basic concepts and results in optimal transportation. Then we incorporate congestion in this setting and provide closed-form expressions for its solution. Section 4.2 outlines the problem formulation of minimizing the total network power under quality of service constraints. We address the problem for the downlink case. Two different policies are considered: round robin scheduling policy (also known as time fair allocation policy) and rate fair allocation policy. Round robin scheduling policy and rate fair allocation policy are studied in detail in Sects. 4.3 and 4.4, respectively. In Sect. 4.4.2, we study the price of anarchy or performance gap between the centralized and decentralized scenarios. In Sect. 4.4.3, we provide numerical examples for one-dimensional and two-dimensional mobile terminal deployment distribution functions.

4.1 Basics in Optimal Transportation

Optimal transportation, also called the theory of mass transportation, goes back to the original works by Monge in 1781 [335, 336] and later in 1942 by Kantorovich [257]. The work of Brenier in 1987 [72] has renewed the interest in the subject, and since then, many different works have been written on this topic (see, e.g., the recent books of Villani [125, 500, 501] and references therein).

The original problem of Monge was "What is the optimal way to move piles of sand to fill up given holes of the same total volume?"

The mathematical framework for this problem is the following: We consider a domain (connected open subset of a finite-dimensional vector space), denoted by Ω, two probability measures μ and ν on that domain. Monge's problem can be stated as follows: Given two probability measures, μ and ν, and a constant $p \geq 1$, we consider the minimization problem (denoted by $M_p(\mu, \nu)$):

(continued)

$$M_p(\mu, \nu) := \inf \left\{ \left(\int_{\Omega} |x - T(x)|^p \, d\mu(x) \right)^{1/p} : \right.$$

$$\left. T : \Omega \to \Omega \text{ Borel } T\#\mu = \nu \right\}. \quad (4.1)$$

In the original problem, Monge considered that the cost of moving a commodity from position x to a position y depends on the distance $|x - y|$, i.e., the cost function $c \equiv c(|x-y|)$. The cost of moving a commodity from position x through T to its new position $T(x)$ will then be given by $c(|x - T(x)|)$.

$$\text{Min} \int_{\Omega} c(|x - T(x)|) \, d\mu(x), \text{ such that } \quad T\#\mu = \nu, \quad (4.2)$$

where μ and ν are probability measures and $T : \Omega \to \Omega$ is an integrable function. This problem is known as Monge's problem in optimal transportation.

Density Case

The particular case where the measure μ admits a density f will be of interest. Here, f could represent the density function of the sandpiles in a one-dimensional space, and g could represent the density function of holes in one-dimensional space. A function T which relates both probability density functions is called a transport map if the following condition is satisfied:

$$\int_A F(y)g(y) \, dy = \int_{\{x \in X : T(x) \in A\}} F(T(x)) f(x) \, dx, \quad (4.3)$$

for all continuous function F, where X is the support.[1] of the probability density function f and we denote this condition (following the optimal transportation notation) as

$$T\#\mu = \nu. \quad (4.4)$$

Condition 4.3 is an equation of conservation of information.

Relaxation

New approaches to formulate and to tackle the Monge problem (4.2) have led to applications in many different fields, such as partial differential equations, microeconomics, and control theory.

[1] The support of a function f is the closure of the set of points where the function is not zero, i.e., support $(f) = \overline{\{t \, : \, f(t) \neq 0\}}$.

Instead of optimizing over the set of maps T, one can enlarge the set as follows:

The relaxed formulation of Monge's problem (denoted by $W_p(\mu, \nu)$), also Kantorovich's problem or Monge–Kantorovich's problem, can be stated as follows:

$$W_p(\mu, \nu) := \inf_{\gamma \in \Pi(\mu,\nu)} \left\{ \left(\int_{\Omega \times \Omega} |x - y|^p \, d\gamma(x, y) \right)^{1/p} \right\}, \tag{4.5}$$

where $\Pi(\mu, \nu)$ is the set of probability measures such that $\pi_1(\gamma) = \mu$ and $\pi_2(\gamma) = \nu$ where π_1 is the projection on the first component and π_2 is the projection on the second component. $W_p(\mu, \nu)$ is also a metric between the measures μ and ν called Wasserstein metric.

Kantorovich's problem with cost function c reads

$$\mathrm{Min}_{g \in \Pi(\mu,\nu)} \iint_{\Omega \times \Omega} c(x, y) \, dg(x, y), \tag{4.6}$$

where

$$\Pi(\mu, \nu) = \{g \; : \; \pi_1 \# g = \mu \; \text{ and } \; \pi_2 \# g = \nu\}, \tag{4.7}$$

denotes the ensemble of transport plans g, $\pi_1(x, y)$ stands for the projection on the first axis x, and $\pi_2(x, y)$ stands for the projection on the second axis y.

Through this chapter, we will see how to incorporate network congestion in this problem and how its solution can be applied to cellular networks. In uplink transmissions in cellular networks, we have mobile terminals sending information to base stations, and we are interested in finding out the optimal way of "moving" information from mobile terminals to base stations.

Note that, in communication systems, there exists packet loss, so, in general, this condition may never be satisfied, but considering an estimation of the packet loss (for example, by sending a standard packet test), this condition can be satisfied by modifying the reception measure ν. If we cannot obtain a good estimation of this reception measure, we can consider it in its current form as a conservative policy.

Monge's problem is not only a difficult problem because of the highly nonlinear structure of its objective function, but it also presents several limitations in its formulation. Examples of the limitations in Monge's formulation can be found in [501]. As a simple example, consider the interval $\Omega = [0, 2]$, the probability

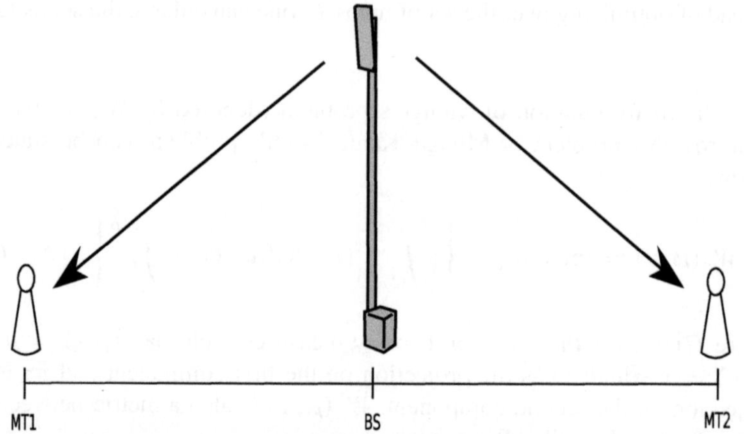

Fig. 4.1 Difficulty with Monge's formulation

function $\mu = \delta_1$ (the Dirac delta function concentrated at 1),[2] and the probability function $\nu = \frac{1}{2}(\delta_0 + \delta_2)$ (the semisum of two Dirac delta functions: one concentrated at 0 and the other one concentrated at 2). Since in Monge's formulation, there is no splitting of mass, i.e., everything that is transmitted from one location x has to go to another location $T(x)$, this simple problem does not have a transport map (see Figs. 4.1 and 4.3). This limitation is due in part to the original considered problem, but, as we will see, this limitation is overcome in Kantorovich's formulation. Kantorovich considered another formulation of this problem in [257].

Transport plan: Kantorovich noticed that the problem of transportation from one location to another can be seen as "graphs of functions" (called transport plans) in the product space (see Fig. 4.2).

In Fig. 4.2, Kantorovich considered "graphs" where the projection in the first axis coincides with the mobile terminal position ($MT_1 = 3.5$, $MT_2 = 5$, and $MT_3 = 6.5$) and the second axis coincides with the base station position ($BS_1 = 4$ and $BS_2 = 6$).

The idea is to minimize the objective function over the space of graphs $g = (\mathrm{Id} \times T)$ in the product space where Id denotes the identity.

---

[2] The Dirac delta function concentrated at point a can be thought of as a function such that

$$\delta_a(x) = \begin{cases} +\infty, & x = a \\ 0, & x \neq a \end{cases} \tag{4.8}$$

constrained to satisfy the identity

$$\int_{-\infty}^{+\infty} \delta_a(x)\, dx = 1. \tag{4.9}$$

This is a heuristic definition. However, the Dirac delta function can be rigorously defined either as a distribution or as a measure. For a rigorous definition, see [63, 73, 74].

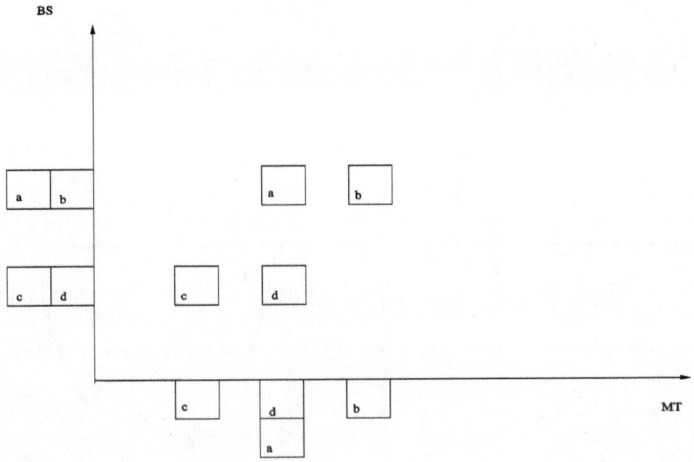

Fig. 4.2 Kantorovich's formulation

The relationship between Monge's and Kantorovich's problems is that every transport map T of Monge's problem determines a transport plan $g_T = (\mathrm{Id} \times T)\#\mu$ in Kantorovich's problem with the same cost. However, Kantorovich's problem considers more functions than the ones coming from Monge's problem (which can always be viewed as the product of the identity and the map T), so we can choose from a bigger set $\Pi(\mu, \nu)$. Then, every solution of Kantorovich's problem is a lower bound to Monge's problem, i.e.

$$\begin{aligned} &\mathrm{Min}_{g \in \Pi(\mu,\nu)} \iint_{\Omega \times \Omega} c(x, y)\, dg(x, y) \\ &\leq \mathrm{Min}_{T\#\mu=\nu} \int_{\Omega} c(|x - T(x)|)\, f(x)\, dx. \end{aligned} \tag{4.10}$$

Figures 4.3 and 4.4 illustrate a case where Monge's problem cannot model a simple scenario of two mobile terminals and one base station. Kantorovich's problem, however, can model very general scenarios.

A probability measure μ is absolutely continuous with respect to the Lebesgue measure if the function $F(x) = \mu((-\infty, x])$ is locally an absolutely continuous real function. A function f is an absolutely continuous real function if there exists an integrable function g such that $f(x) = f(a) + \int_a^x g(t)\, dt$.

Theorem 4.1 *Consider the cost function* $c(\left|x - y\right|) = \left|x - y\right|^p$. *Let* μ *and* ν *be probability measures in* Ω *and fix* $p \geq 1$. *We assume that* μ *can be written as* $d\mu = f(x)\, dx$. *Then, the optimal value of Monge's problem coincides with the optimal value of Kantorovich's problem, i.e.,* $M_p(\mu, \nu) = W_p(\mu, \nu)$, *and there exists an optimal transport map from* μ *to* ν, *which is also unique almost everywhere if* $p > 1$.

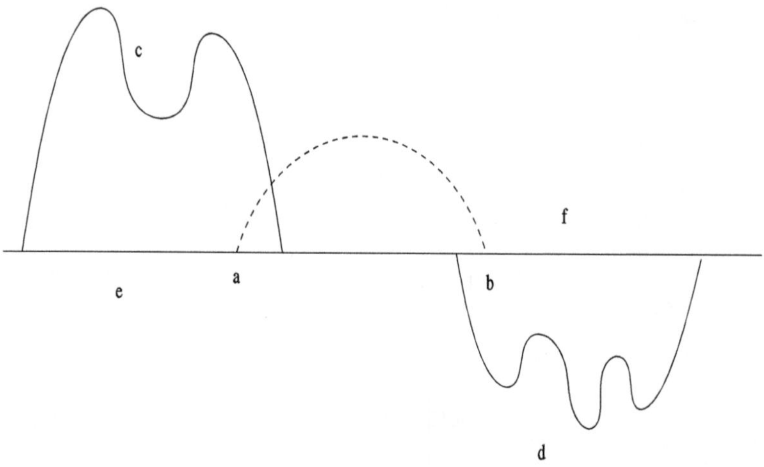

Fig. 4.3 Limitation of Monge's approach

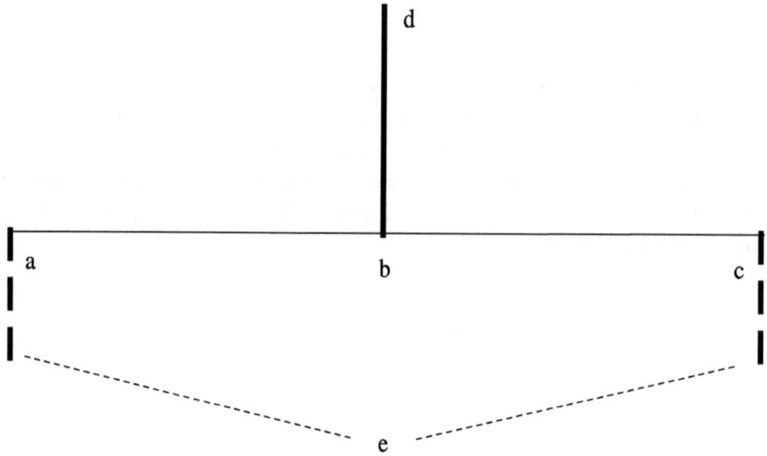

Fig. 4.4 Monge's problem

This result is very difficult to obtain, and it has been proved only recently (see [72] for the case $p = 2$ and the references in [501] for the other cases).

The case that we are interested in can be solved because the image of the transport plan is a discrete finite set.

Congestion-Dependent Cost Function

Thanks to optimal transport theory, it is possible to characterize the partitions considering general settings. Toward this purpose, consider locations $(x_1, y_1)\ldots, (x_K, y_K)$, and the Euclidean distance $d_i(x, y)=\sqrt{(x-x_i)^2+(y-y_i)^2}$, and F a continuous function.

Theorem 4.2 *Consider the problem:*

$$\text{(P1)} \quad \underset{C_i}{\text{Min}} \sum_{i=1}^{K} \iint_{C_i} [F(d_i(x, y)) + s_i \left(\iint_{C_i} m_0(\omega, z)\, d\omega\, dz \right) \Big] m_0(x, y)\, dx\, dy, \tag{4.11}$$

where C_i is the cell partition of Ω. Suppose that s_i are continuously differentiable, nondecreasing, and convex functions. Problem (P1) *admits a solution that satisfies*

$$\text{(S1)} \begin{cases} C_i = \{(x, y) : F(d_i(x, y)) + s_i(N_i) + N_i \cdot s_i'(N_i) \\ \qquad \leq F(d_j(x, y)) + s_j(N_j) + N_j \cdot s_j'(N_j)\} \\ N_i = \iint_{C_i} m_0(\omega, z)\, d\omega\, dz. \end{cases} \tag{4.12}$$

The proof is based on Proposition 3.5 of Crippa et al. [117].

Proof If the probability measure μ can be written as $d\mu = f(x)\, dx$ (i.e., it is absolutely continuous with respect to the Lebesgue measure), then the optimal values of both problems coincide $M_p(\mu, \nu) = W_p(\mu, \nu)$, and there exists an optimal transport map from μ to ν which is unique f-a.e. if $p > 1$. Another important characteristic of this relaxation is that it admits a dual formulation:

$$\begin{aligned} &W_p^p(\mu, \nu) = \sup \left\{ \textstyle\int_\Omega u\, d\mu + \int_\Omega v\, d\nu : \right. \\ &u(x) + v(y) \leq |x - y|^p \quad \text{for } \mu\text{-a.e. } x \text{ and } \nu\text{-a.e. } y \\ &\left. u \in L^1_\mu(\Omega), v \in L^1_\nu(\Omega) \right\}. \end{aligned} \tag{4.13}$$

Moreover, there exists an optimal pair (u, v) for this dual formulation, and when ν is an atomic probability measure (it can be written as $\nu = \sum_{i \in \mathbb{N}} b_i \delta_{y_i}$), the dual formulation becomes

$$\begin{aligned} &W_p^p(\mu, \nu) = \sup \left\{ \textstyle\int_\Omega u\, d\mu + \sum_{i \in \mathbb{N}} b_i v(y_i) : \right. \\ &u(x) + v(y_i) \leq |x - y_i|^p \quad \text{for } \mu\text{-a.e. } x \text{ and every } i \in \mathbb{N}, \\ &\left. u \in L^1_\mu(\Omega), v \in L^1_\nu(\Omega) \right\}. \end{aligned} \tag{4.14}$$

There exists another interesting characteristic when one of the measures is absolutely continuous with respect to the Lebesgue measure, and the other measure is an atomic measure. If the probability measure μ can be written as $d\mu = f(x)\, dx$ where f is a nonnegative function, $(y_i)_{i \in \mathbb{N}}$ is a sequence of points in the domain Ω such that $\nu = \sum_{i \in \mathbb{N}} b_i \delta_{y_i}$, $(B_i)_{i \in \mathbb{N}}$ is a partition of the domain such that the map $T(x) = \sum_{i \in \mathbb{N}} y_i \mathbf{1}_{B_i}(x)$ is an optimal transport map from μ to ν, and the pair (u, v) is a solution of dual formulation 4.13, then

$$u(x) = \inf_{i \in \mathbb{N}} (|x - y_i|^p - v(y_i)) = \sum_{i \in \mathbb{N}} (|x - y_i|^p - v(y_i)) \mathbf{1}_{B_i}(x) \text{ for } f\text{-a.e. } x. \tag{4.15}$$

We also have a similar converse characteristic. If $(B_i)_{i\in\mathbb{N}}$ is a partition of Ω, we set $b_i = \int_{B_i} f(x)\,dx$, $\nu = \sum_{i\in\mathbb{N}} b_i\delta_{y_i}$, and $T(x) = \sum_{i\in\mathbb{N}} y_i \mathbf{1}_{B_i}(x)$, and there exists two functions $u \in L^1_\mu(\Omega)$ and $v \in L^1_\nu(\Omega)$ satisfying condition 4.15, then T is optimal for $M_p(\mu, \nu)$ and the pair (u, v) is optimal for dual formulation 4.13.

In order to prove the existence and uniqueness of the solution, we need to consider the following: We denote by S the unit simplex in $\mathbb{R}^k$:

$$S = \left\{ c = (c_1, \ldots, c_k) \in \mathbb{R}^k : c_i \geq 0, \sum_{i=1}^{k} c_i = 1 \right\}. \tag{4.16}$$

From Theorem 4.1, we deduce that

$$\begin{aligned} &\inf_{\substack{(A_i)_{i\in\{1,\ldots,k\}} \\ \text{partition of } \Omega}} \left\{ \sum_{i=1}^k \int_{A_i} \left[|x - x_i|^p + s_i\left(\int_{A_i} f(x)\,dx\right)\right] f(x)\,dx \right\} \\ &= \inf_{c\in S} \left\{ W_p^p\left(f, \sum_{i=1}^k c_i\delta_{x_i}\right) + \sum_{i=1}^k s_i(c_i)c_i \right\}. \end{aligned} \tag{4.17}$$

The function $F : S \to \mathbb{R}$ defined by $F(c_1, \ldots, c_k) = W_p^p\left(\mu, \sum_{i=1}^k c_i\delta_{x_i}\right)$ is continuous and convex.

If the functions s_i are lower semicontinuous, then there exists an optimum, and if, in addition, the maps $\eta_i(t) := t h_i(t)$ are strictly convex, the optimum is unique.

We can then characterize the solution. If $(A_i)_{i=1,\ldots,k}$ is an optimum and s_i are differentiable in $]0, 1]$ and continuous in 0, then the following holds:

$$\begin{aligned} C_i = &\left\{ x \in \Omega \; : |x - x_i|^p + s_i(N_i) + N_i s_i'(N_i) \right. \\ &\left. \leq |x - x_j|^p + s_j(N_j) + N_j s_j'(N_j),\ \forall j \neq i \right\} \end{aligned} \tag{4.18}$$

$$N_i = \int_{C_i} f(x)\,dx. \tag{4.19}$$

Theorem 4.3 *Consider the problem:*

$$\text{(P2)} \quad \underset{C_i}{\text{Min}} \sum_{i=1}^{K} \iint_{C_i} \left[F(d_i(x, y)) \cdot m_i \left(\iint_{C_i} m_0(\omega, z)\,d\omega\,dz \right) \right] m_0(x, y)\,dx\,dy, \tag{4.20}$$

where C_i is the cell partition of Ω. Suppose that m_i are differentiable. Problem (P2) *admits a solution that satisfies*

$$\text{(S2)} \begin{cases} C_i = \{(x, y) : m_i(N_i)F(d_i(x, y))\,m_0(x, y) + U_i(x, y) \\ \qquad \leq m_j(N_j)F(d_j(x, y))\,m_0(x, y) + U_j(x, y)\} \\ U_i = m_i'(N_i) \iint_{C_i} F(d_i(x, y)) m_0(x, y)\,dx\,dy \\ N_i = \iint_{C_i} m_0(\omega, z)\,d\omega\,dz. \end{cases} \tag{4.21}$$

Proof Problem (P2) can be rewritten as follows:

$$\inf\left\{W_{c_i}\left(f\,dx, \sum_{i=1}^{k} c_i\delta_{x_i}\right) : c_i \geq 0, \sum_{i=1}^{k} c_i = 1\right\}, \tag{4.22}$$

where

$$W_{c_i}\left(f\,dx, \textstyle\sum_{i=1}^{k} c_i\delta_{x_i}\right) = \inf_{(A_i)_{i=1}^k}\left\{\textstyle\sum_{i=1}^{k}\int_{A_i} F(|x-x_i|)\cdot m_i(c_i) f(x)\,dx : \right. \\ \left.\textstyle\int_{A_i} f\,dx = c_i,\ (A_i)_{i=1}^k \text{ partition de } \Omega\right\}. \tag{4.23}$$

Let γ be optimal for Kantorovich's problem, and then, one can consider

$$\gamma(x,y) = \sum_{i=1}^{k} \gamma_i(x)\otimes\delta_{x_i}(y), \qquad c_i = \gamma_i(\Omega), \tag{4.24}$$

where γ_i is a positive measure such that $\gamma_i \leq f\,dx$.

The function $F : S \to \mathbb{R}$ defined by $F(c_1,\ldots,c_k) = W_p^p\left(\mu, \sum_{i=1}^k c_i\delta_{x_i}\right)$ is lower semicontinuous. Then, there exists a solution, and it is unique almost surely.

Let $(c_i)_{i=1}^k$ be a solution of (P2) and $(A_i)_{i=1}^k$ the associated optimal partition. Let us fix two indices i_0 and j_0 and a point $x_0 \in A_{i_0}$. Let $\varepsilon \geq 0$. Let us consider the open ball of radius ε and center x_0, which we denote by B_ε. We denote its measure by c_ε. We make a small variation on the optimal partition by taking from A_{i_0} the ball B_ε and adding it to A_{j_0}. Since the partition $(A_i)_{i=1}^k$ is optimal, then

$$\begin{aligned}
&\int_{A_{i_0}} F(|x-x_{i_0}|)h_{i_0}(c_{i_0})f(x)\,dx + \int_{A_{j_0}} F(|x-x_{j_0}|)h_{j_0}(c_{j_0})f(x)\,dx \\
&\leq \int_{A_{i_0}\setminus B_\varepsilon} F(|x-x_{i_0}|)h_{i_0}(c_{i_0}-c_\varepsilon)f(x)\,dx \\
&+ \int_{A_{j_0}\cup B_\varepsilon} F(|x-x_{j_0}|)h_{j_0}(c_{j_0}+c_\varepsilon)f(x)\,dx,
\end{aligned}$$

which is equivalent to

$$\begin{aligned}
&\int_{A_{i_0}} F(|x-x_{i_0}|)(h_{i_0}(c_{i_0}) - h_{i_0}(c_{i_0}-c_\varepsilon))f(x)\,dx \\
&+ \int_{B_\varepsilon} F(|x-x_{i_0}|)h_{i_0}(c_{i_0}-c_\varepsilon)f(x)\,dx \\
&\leq \int_{A_{j_0}} F(|x-x_{i_0}|)(h_{j_0}(c_{j_0}+c_\varepsilon) - h_{j_0}(c_{j_0}))f(x)\,dx \\
&+ \int_{B_\varepsilon} F(|x-x_{j_0}|)h_{j_0}(c_{j_0}+c_\varepsilon)f(x)\,dx.
\end{aligned}$$

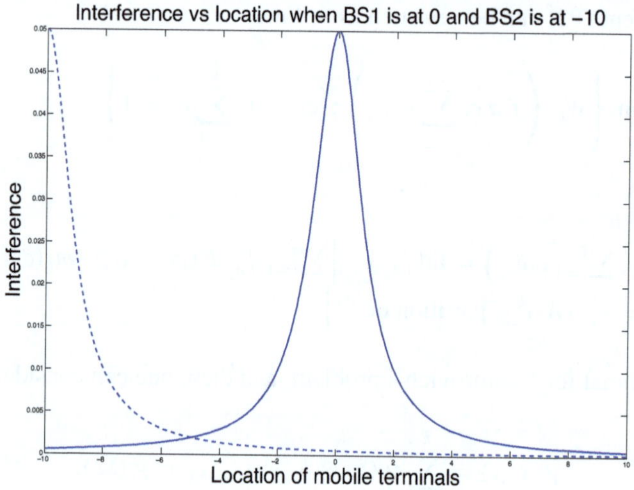

Fig. 4.5 Interference as a function of location of mobile terminals when BS1 is at position 0 (solid line) and BS2 at −10 (dotted line)

Dividing the previous equation by c_ε and taking the limit as $\varepsilon \to 0$, we obtain

$$\int_{A_{i0}} F(|x - x_{i_0}|)h'_{i_0}(c_{i_0}) f(x)\,dx + F(|x - x_{i_0}|)h_{i_0}(c_{i_0}) f(x_0)$$

$$\leq \int_{A_{j0}} F(|x - x_{j_0}|)h'_{j_0}(c_{j_0}) f(x)\,dx + F(|x - x_{j_0}|)h_{j_0}(c_{j_0}) f(x_0). \quad (4.25)$$

Reorganizing the terms, we obtain the desired result. ■

Figure 4.5 represents interference as a function of the location of mobile terminals when BS1 is at position 0 and BS2 at −10. Figure 4.6 represents the SINR.

Figure 4.7 represents the Wardrop equilibrium: thresholds determining the cell boundaries as a function of the location of the base stations. The network is deployed over the interval [−10, 10] (one-dimensional case) presented here vertically. We consider a uniform distribution of mobile terminals, and we find the threshold (solid line) determining the cell boundaries as a function of the base station positions (dotted lines) by changing the position of one of them. BS_1 is fixed at position 0, and we change the position of BS_2 from −10 to +10.

Figure 4.8 represents the worst equilibrium: thresholds determining the cell boundaries (vertical axis) that give the worst equilibrium in terms of the SINR as a function of the location of BS_2 for BS_1 at position 0.

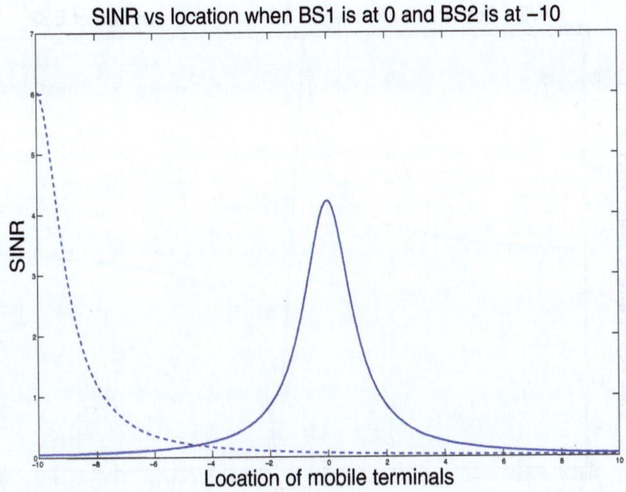

Fig. 4.6 SINR as a function of location of mobile terminals when BS1 is at position 0 (solid line) and BS2 at -10 (dotted line)

Fig. 4.7 Best response to mean field

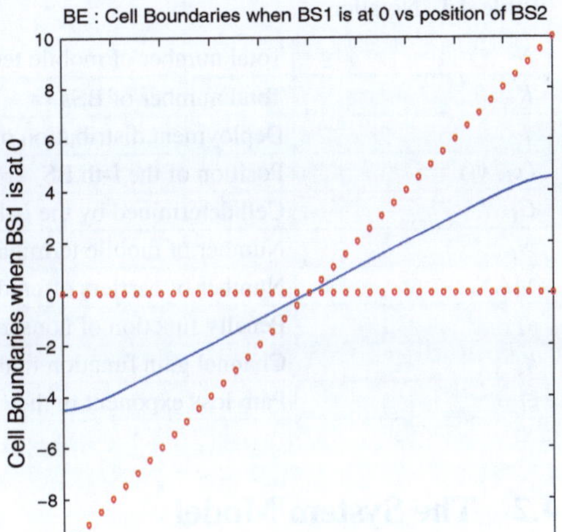

Note that in problem (P1), if the functions $s_i \equiv 0$, the solution of system (S1) becomes the well-known Voronoi cells. In problem (P2), if we have that the functions $h_i \equiv 1$, we find again the Voronoi cells. In general, however, the Voronoi configuration is far from being optimal.

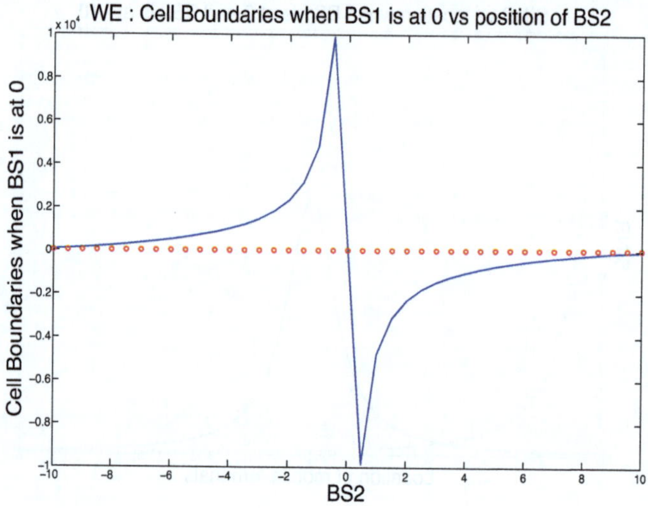

Fig. 4.8 Worst best response to mean field

Table 4.1 Notation

N	Total number of mobile terminals in the network
K	Total number of BSs
f	Deployment distribution of mobile terminals
(x_i, y_i)	Position of the i-th BS
C_i	Cell determined by the i-th BS
N_i	Number of mobile terminals associated with the i-th BS
MAX_i	Number of carriers offered by the i-th BS
κ_i	Penalty function of nonservice
h_i	Channel gain function in the i-th cell
ξ_i	Path loss exponent in the i-th cell

4.2 The System Model

A list of key notation used in this section can be found in Table 4.1. We consider a network deployed in a region, denoted by Ω,[3] over the two-dimensional plane. The mobile terminals (MTs) are distributed according to a given probability density function $m_0(x, y)$. The proportion of mobile terminals in the subregion $\mathcal{A} \subseteq \Omega$ is given by

$$\iint_{\mathcal{A}} m_0(x, y)\, dx\, dy. \tag{4.26}$$

[3] The domain Ω is considered bounded.

The number of mobile terminals in a subregion of the network $A \subseteq \Omega$, denoted by $N(A)$, can be approximated by

$$N(A) = N\left(\iint_A m_0(x, y)\, dx\, dy\right), \tag{4.27}$$

where N is the total number of mobile terminals in the network. Distributions $m_0(x, y)$

1. If the mobile terminals are distributed uniformly over the network, then the probability density function of the mobile terminals is given by

$$m_0(x, y) = \frac{1}{|\Omega|}, \tag{4.28}$$

where $|\Omega|$ is the total area of the network.
2. If the mobile terminals are distributed according to different levels of population density, then the probability density function of the mobile terminals is given by

$$m_0(x, y) = \begin{cases} f_{\text{HD}} \text{ if } (x, y) \text{ is at a high-density region,} \\ f_{\text{ND}} \text{ if } (x, y) \text{ is at a normal density region,} \\ f_{\text{LD}} \text{ if } (x, y) \text{ is at a low-density region,} \end{cases} \tag{4.29}$$

where $f_{\text{HD}} > f_{\text{ND}} > f_{\text{LD}}$ are defined similarly to Eq. 4.28 with constants of normalization K_{HD}, K_{ND}, and K_{LD}, such that $K_{\text{HD}} > K_{\text{ND}} > K_{\text{LD}}$.
3. If the mobile terminals are distributed in a circle with more mobile terminals in the center of the network area and fewer mobile terminals toward the boundary, then we have that the probability density function of the mobile terminals is given by

$$m_0(x, y) = \frac{R_D^2 - (x^2 + y^2)}{K_D}, \tag{4.30}$$

where R_D is the radius of the network and K_D is a coefficient of normalization.

In the network, we consider K base stations (BSs), denoted by BS_i , $i \in \{1, \ldots, K\}$, located at fixed positions (x_i, y_i) , $i \in \{1, \ldots, K\}$. We assume that the neighboring BSs transmit their signals in orthogonal frequency bands. Furthermore, we assume that interference between BSs that are far from each other is negligible. Consequently, instead of considering the SINR (signal-to-interference-plus-noise ratio), we consider as performance measure the SNR (signal-to-noise ratio).

We consider the downlink case (transmission from base stations to mobile terminals) and assume that each BS is going to transmit only to mobile terminals associated with it. We denote by C_i the set of mobiles associated with the i-th BS and by N_i the number of mobiles within that cell, i.e., the cardinal of the set C_i, to be determined.

Since the probability density function of the mobile terminals $m_0(x, y)$ considered in this section is general, instead of considering a particular probability density function, e.g., $\tilde{m}_0(x, y)$, and an average throughput, e.g., $\tilde{\theta}(x, y)$, in each location (x, y), we can consider a constant throughput $\theta > 0$ and redefine the distribution of mobile terminals $m_0(x, y)$ as

$$m_0(x, y) := \frac{\tilde{m}_0(x, y)\tilde{\theta}(x, y)}{\theta}. \tag{4.31}$$

In order to do that, we note that since $m_0(x, y)$ must be a probability density function, we should impose

$$\iint m_0(x, y)\,dx\,dy = 1. \tag{4.32}$$

This equation is equivalent to

$$\iint \frac{\tilde{m}_0(x, y)\tilde{\theta}(x, y)}{\theta}\,dx\,dy = 1. \tag{4.33}$$

Since we are considering a constant throughput $\theta > 0$, this equation is equivalent to

$$\frac{1}{\theta}\iint \tilde{m}_0(x, y)\tilde{\theta}(x, y)\,dx\,dy = 1. \tag{4.34}$$

For this equation to hold, we consider

$$\theta := \iint \tilde{m}_0(x, y)\tilde{\theta}(x, y)\,dx\,dy. \tag{4.35}$$

In consequence, the following equation holds:

$$m_0(x, y)\theta = \tilde{m}_0(x, y)\tilde{\theta}(x, y). \tag{4.36}$$

This last equation simply reflects the fact that mobile terminals with double the demand as compared with others would be considered as two different mobile terminals with the same demand as other mobile terminals at the same location.

If the number of mobile terminals is greater than the maximum number of carriers available in the i-th cell, denoted by MAX_i, we consider a penalization cost function given by

$$\kappa_i(N_i) = \begin{cases} 0 & \text{if } N_i \le \text{MAX}_i, \\ \bar{\kappa}_i(N_i - \text{MAX}_i) & \text{if } N_i > \text{MAX}_i. \end{cases} \tag{4.37}$$

We assume that $\bar{\kappa}_i$ can be either a constant function or at least a nondecreasing function in the number of carriers above the limit MAX_i.[4] We first discuss here the case $N_i \leq \mathrm{MAX}_i$. Study of the general case is in Sect. 4.4.1.

The power transmitted P^T from a base station BS_i to a mobile terminal located at position (x, y) is denoted by

$$P_i^T(x, y) = P_i(x, y). \tag{4.38}$$

The received power P^R at a mobile terminal located at position (x, y) served by a base station BS_i is given by

$$P_i^R(x, y) = P_i(x, y)h_i(x, y), \tag{4.39}$$

where $h_i(x, y)$ is the channel gain between base station BS_i and a mobile terminal located at position (x, y). We shall further assume that the channel gain corresponds to the path loss given by

$$h_i(x, y) = (\sqrt{R^2 + d_i^2(x, y)})^{-\xi}, \tag{4.40}$$

where ξ is the path loss exponent, R is the height of the base station, and $d_i(x, y)$ is the distance between a mobile terminal at position (x, y) and BS_i located at (x_i, y_i), i.e.

$$d_i(x, y) = \sqrt{(x_i - x)^2 + (y_i - y)^2}. \tag{4.41}$$

The SNR received at mobile terminals at position (x, y) in cell C_i is given by

$$\mathrm{SNR}_i(x, y) = \frac{P_i(x, y)h_i(x, y)}{\sigma^2}, \tag{4.42}$$

where σ^2 is the noise power. We assume that the instantaneous mobile throughput is given by the following expression, which is based on Shannon's capacity theorem:

$$\theta_i(x, y) = \log(1 + \mathrm{SNR}_i(x, y)). \tag{4.43}$$

We want to satisfy an average throughput for mobile terminals located at position (x, y) given by $\tilde{\theta}(x, y) > 0$. We shall consider for this objective two policies introduced in [259]:

(A) *Round robin scheduling policy:* where each base station devotes an equal fraction of time for transmission to each mobile terminal corresponding to it

[4] For example, the maximum number of possible carriers in WiMAX is around 2048, and thus, by using this technology, we have $\mathrm{MAX}_i = 2048$.

(B) *Rate fair allocation policy:* where each base station maintains a constant power sent to mobile terminals within its cell but modifies the fraction of time allowed to mobile terminals with different channel gains, so that the average transmission rate demand is satisfied (see Chap. 2 and Sect. 3.2)

For more information on these types of policies in the one-dimensional case, see [259].

4.2.1 Round Robin Scheduling Policy: Global Optimization

In the round robin scheduling policy, a base station devotes an equal fraction of time to transmission to the mobile terminal located within its cell. The set of mobile terminals associated with the i-th base station BS_i, denoted by C_i, and the number of mobile terminals within that cell N_i (which is equal to the cardinal of the set C_i) are to be determined. Since a base station BS_i determines its time of service proportional to the number of mobile terminals within its cell, the throughput of a mobile terminal located at position (x, y) under the round robin scheduling policy is given by

$$\theta_i^{RR}(x, y) = \frac{1}{N_i}\theta_i(x, y), \tag{4.44}$$

where N_i is the number of mobile terminals within the i-th cell and $\theta_i(x, y)$ is the instantaneous mobile throughput at mobile terminals located at positions (x, y) from base station BS_i.

In order to satisfy an instantaneous throughput $\tilde{\theta}(x, y)$, we need to have

$$\theta^{RR}(x, y) = \frac{1}{N_i}\theta_i(x, y) \geq \tilde{\theta}(x, y), \tag{4.45}$$

or equivalently, by replacing in the previous inequality $\theta_i(x, y)$ from Eqs. 4.42 and 4.43

$$\frac{1}{N_i}\log\left(1 + \frac{P_i(x, y)h_i(x, y)}{\sigma^2}\right) \geq \tilde{\theta}(x, y), \tag{4.46}$$

which can be equivalently rewritten as

$$P_i(x, y) \geq \frac{\sigma^2}{h_i(x, y)}\left(2^{N_i\tilde{\theta}(x,y)} - 1\right). \tag{4.47}$$

Since our goal is to minimize the total power of the network, this constraint will be met as equality, and using Eq. 4.40, we obtain

$$P_i(x, y) = \sigma^2\left(2^{N_i\bar{\theta}(x,y)} - 1\right)\left(\left[R^2 + d_i^2(x, y)\right]^{1/2}\right)^{+\xi}. \quad (4.48)$$

From Eq. 4.48, we make the following onbservations:

(a) If the number of mobile terminals increases within a cell, i.e., the base station has more mobile terminals assigned to it, the base station will need to transmit more power to each of the mobile terminals within its cell, to satisfy the throughput constraint in Eq. 4.45.
(b) The function $([R^2 + d_i^2(x, y)]^{1/2})^{+\xi}$ on the right hand side give us the dependence of the power needed to satisfy the throughput constraint in Eq. 4.45 with respect to the distance between the base station and the mobile terminal located at position (x, y).

Our objective is to find the optimal mobile association that minimizes the total power of the network while maintaining the throughput constraint given by inequality 4.45 or equivalently by inequality 4.48. The total power of the network is equal to the sum of the power used in all the cells in the network, i.e.

$$P_{\text{total}} = \sum_{i=1}^{K} P_i^{\text{intra}}, \quad (4.49)$$

$$\text{where} \quad P_i^{\text{intra}} = \iint_{C_i} P_i(x, y) m_0(x, y)\, dx\, dy, \quad (4.50)$$

P_i^{intra} is the intracell power consumption in cell C_i.

The global optimization for the mobile association problem is to determine the cells C_i, $i \in \{1, \ldots, K\}$, to minimize the total power of the network:

$$\text{(RR)} \quad \min \sum_{i=1}^{K} \iint_{C_i} P_i(x, y) m_0(x, y)\, dx\, dy, \quad (4.51)$$

subject to 4.48, where $m_0(x, y)$ is the probability density function of the mobile terminals. We solve this problem in Sect. 4.3.

4.2.2 Rate Fair Allocation Policy: User Optimization

In the rate fair allocation policy, each base station will maintain a constant power sent to mobile terminals within its cell, i.e.

$$P_i(x, y) = P_i \text{ for each mobile terminal at location } (x, y) \text{ inside cell } C_i. \quad (4.52)$$

However, the base station modifies the fraction of time allotted to mobile terminals, set in such a way that the average transmission rate to each mobile terminal with different channel gain is the same, denoted by $\tilde{\theta}(x, y)$, for each mobile terminal located at position (x, y).

Let r_i be the fixed rate of mobile terminals located inside cell C_i. The fraction of time that a mobile terminal located at position $(x, y) \in C_i$ receives positive throughput is

$$\frac{r_i}{\log(1 + \mathrm{SNR}_i(x))}. \tag{4.53}$$

Then, the fixed rate r_i is the solution to equation:

$$\iint_{C_i} \frac{r_i}{\log(1 + \mathrm{SNR}_i(x, y))} m_0(x, y)\, dx\, dy = \beta, \tag{4.54}$$

where β is the fraction of time guaranteed from the regulator to have no interference from the other BSs. Then, the rate

$$r_i = \left(\iint_{C_i} \frac{1}{\log(1 + \mathrm{SNR}_i(x, y))} m_0(x, y)\, dx\, dy \right)^{-1} \beta. \tag{4.55}$$

We study the equilibrium states where each mobile terminal chooses the base station which will serve it, i.e., the situation where given the interactions with other mobile terminals, none of the mobile terminal has any incentive to unilaterally change its strategy. A similar notion of equilibrium has been studied in the context of a large number of small players in road-traffic theory by Wardrop [505]. The Wardrop equilibrium is the analog of a Nash equilibrium for the case of a large number of small players. In the following, we discuss the notion of an equilibrium, which we also call the Wardrop equilibrium by analogy to the notion introduced in [505].

Definition 4.4 Consider the nonatomic game where decision-makers are distributed according to m_0 and each decision-maker chooses one of the K cells. The payoff function associated with cell i is θ_i. The Wardrop equilibrium is given in the context of cellular systems through

$$\text{If } \iint_{C_i} m_0(x, y)\, dx\, dy > 0, \text{ then } \theta_i = \max_{1 \le j \le K} \theta_j(C_j), \tag{4.56a}$$

$$\text{else if } \iint_{C_i} m_0(x, y)\, dx\, dy = 0, \text{ then } \theta_i \le \max_{1 \le j \le K} \theta_j(C_j). \tag{4.56b}$$

As previously mentioned, the Wardrop equilibrium is the analog of a Nash equilibrium in the case of a large number of small players, where, in our case, we consider the mobile terminals as the small players. In this setting, the Wardrop equilibrium indicates that if there is a strictly positive proportion of mobile terminals associated with the i-th base station (the left-hand side condition in (4.56a)), then the throughput that the mobile terminals associated with the i-th base station obtain is the maximum that they would obtain from any other base station (right-hand side consequence in Eq. 4.56a). The second condition indicates that if there is one base station that does not have any mobile terminal associated with it (left-hand side condition in (4.56b), it is because the mobile terminals can obtain a higher throughput by connecting to one of the other base stations (right-hand side consequence in Eq. 4.56b).

We assume that each base station is serving at least one mobile terminal (if that is not the case, we could remove the base stations which are not serving any mobile terminal and find an equivalent situation). Then, the equilibrium is given by the situation when

$$\theta_1 = \theta_2 = \ldots = \theta_K. \tag{4.57}$$

To understand this equilibrium situation, consider as an example the simple case of two base stations: BS_1 and BS_2. Assume that base station BS_1 offers more throughput than base station BS_2. Then, the mobile terminals being served by base station BS_2 will have an incentive to connect to base station BS_1. From (4.48), we previously remarked that the transmitted throughput depends inversely on the number of mobile terminals connected to the base station. As more mobile terminals try to connect to base station BS_1, the throughput of each mobile terminal will diminish until arriving at the equilibrium where both base stations will offer the same throughput.

The condition given by (4.48) is equivalent in our setting to the condition

$$\mathrm{SNR}_1 = \mathrm{SNR}_2 = \ldots = \mathrm{SNR}_K. \tag{4.58}$$

Let us denote by β the SNR offered by the base stations at the equilibrium, i.e.

$$\beta = \mathrm{SNR}_1 = \mathrm{SNR}_2 = \ldots = \mathrm{SNR}_K. \tag{4.59}$$

From (4.42), we know that

$$P_i(x, y) = \sigma^2 \frac{\mathrm{SNR}_i(x, y)}{h_i(x, y)}. \tag{4.60}$$

At the equilibrium, from (4.59) and (4.42), we obtain

$$P_i(x, y) = \sigma^2 \frac{\beta}{h_i(x, y)}. \tag{4.61}$$

Replacing the term $P_i(x, y)$ in (4.50)

$$P_i^{\text{intra}} = \iint_{C_i} \sigma^2 \frac{\beta}{h_i(x, y)} m_0(x, y)\, dx\, dy. \tag{4.62}$$

Replacing the channel gain $h_i(x, y)$ from (4.40), we obtain

$$P_i^{\text{intra}} = \beta \iint_{C_i} \sigma^2\left(\left[R^2 + d_i^2(x, y)\right]^{1/2}\right)^{+\xi} m_0(x, y)\, dx\, dy. \tag{4.63}$$

We want to choose the optimal mobile assignment in order to minimize the total power of the network under the constraint that the mobile terminals have an average throughput of θ, i.e.

$$\underset{C_i}{\text{Min}}\ P_{\text{total}} = \sum_{i=1}^{K} P_i^{\text{intra}}. \tag{4.64}$$

Then, our problem reads

$$\text{(RF)} \quad \underset{C_i}{\text{Min}} \sum_{i=1}^{K} \iint_{C_i} \beta\sigma^2\left(R^2 + d_i^2(x, y)\right)^{\xi/2} m_0(x, y)\, dx\, dy. \tag{4.65}$$

We will solve this problem in Sect. 4.4. Thanks to optimal transport theory, we are able to characterize the partitions considering a general setting. In the following section, we will briefly describe optimal transport theory and motivate the solution of the previously considered mobile association problems.

4.3 Round Robin Scheduling Policy

We continue with problem (4.51). We assume that a service provider wants to minimize the total power of the network while maintaining a certain average throughput of θ to each mobile terminal of the system using the round robin scheduling policy.

We see that this problem is an optimal transportation problem like the one in (P1) with a cost function given by

$$F(d_i(x, y)) = \sigma^2\left(R^2 + d_i(x, y)^2\right)^{\xi/2} \quad \text{and} \tag{4.66}$$

$$m_i(x, y) = \left(2^{N_i\theta} - 1\right). \tag{4.67}$$

From the previous theorem, we can derive an explicit expression for this configuration.

Proposition 4.5 *There exists a unique optimum given by*

$$C_i = \Big\{x_0 \in \Omega \ : d_i(x_0, y_0)^p + h_i(N_i) + N_i h_i'(N_i)$$
$$\leq d_j(x_0, y_0)^p + k_j(N_j) + N_j k_j'(N_j), \ \forall j \neq i\Big\} \tag{4.68}$$

$$N_i = \iint_{C_i} m_0(x_0, y_0)\, dx_0\, dy_0. \tag{4.69}$$

Consider now a specific example illustrating this result:

Example 4.6 Consider a network of $N = 2500$ mobile terminals distributed according to $f(x)$ in $[0, L]$ (for example, with $L = 5.6$ miles for WiMAX radius cell). Assume there are two base stations at position $BS_1 = 0$ and $BS_2 = L$ and $R = 1$. Then, the mobile association threshold (the boundary between both cells, i.e., the location at which the mobile terminals obtain the same throughput by connecting to any of the two base stations) is reduced to find x such that the following equality holds:

$$\big(2^{N_1\theta} - 1\big)\big(1 + x^2\big) f(x) + 2^{N_1\theta}\theta \log 2 \left[x + \frac{x^3}{3}\right]$$

$$= \big(2^{N_2\theta} - 1\big)\big(1 + (1-x)^2\big) f(x) + 2^{N_2\theta}\theta \log 2 \left[\frac{4}{3} - 2x + x^2 - \frac{x^3}{3}\right] \tag{4.70}$$

Notice that this is a fixed point equation on x. If the mobile terminals are distributed uniformly, the optimal solution is given by $C_1 = [0, 1/2\, L)$ and $C_2 = [1/2\, L, L]$, which is the solution that the Voronoi cells would give us, and in that case, the number of mobile terminals connected to each base station would be given by

$$N_1 = N_2 = 1250. \tag{4.71}$$

However, if the deployment distribution of the mobile terminals is more concentrated near BS_2 than BS_1, consider, for example, $f(x) = 2x$, the optimal solution is given by $C_1 = [0, q)$ and $C_2 = [q, L]$ with $q = 0.6027\, L$ and

$$N_1 = 908 \quad \text{and} \quad N_2 = 1592. \tag{4.72}$$

Notice that in the global optimization solution, the number of mobile terminals connected to BS_1 is smaller than the number of mobile terminals connected to BS_2. However, the cell size is bigger.

Figure 4.9 represents the thresholds determining the cell boundaries (vertical axis) of the best equilibrium in terms of the SINR as a function of the location of $BS2$ for BS1 at position 0 when we consider a nonhomogeneous distribution given by $f(x) = (L - x)/2L^2$.

In Fig. 4.10, we plot the equilibrium in the nonhomogeneous case: The network is deployed over the interval $[-10, 10]$ (one-dimensional case) presented here vertically. We find the threshold (solid line) determining the cell boundaries as

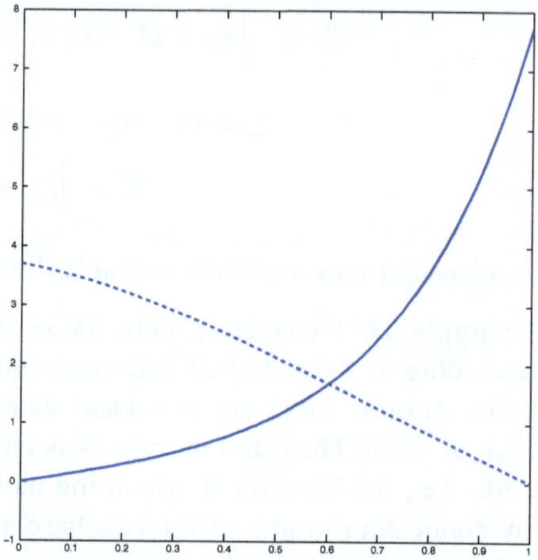

Fig. 4.9 Equilibrium under nonuniform measure

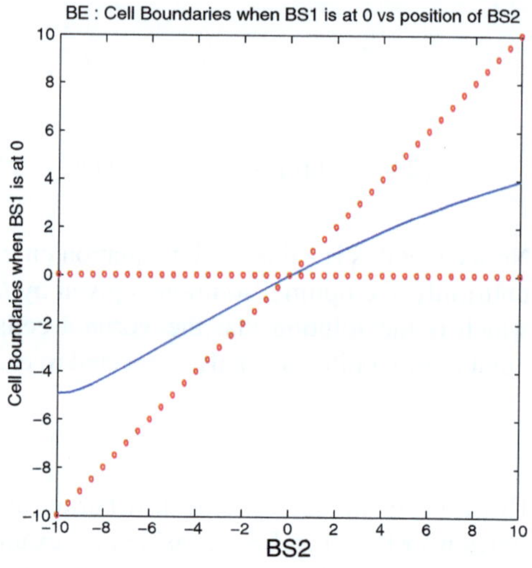

Fig. 4.10 Equilibrium under nonhomogenous measure

a function of the base stations' positions (dotted lines) by changing the position of one of them. BS_1 is fixed at position 0, and we change the position of BS_2 from -10 to $+10$. The deployment distribution of the mobile terminals is given by $f(x) = (L - x)/2L^2$.

4.4 Rate Fair Allocation Policy

Under this policy, we provide the mobile terminals to connect to the base station they prefer in order to minimize their power cost function while maintaining, as quality of service measurement, an average throughput of θ.

Notice that the problem (RF) given in (4.65) is equivalent to (P1) where the functions

$$F(d_i(x,y)) = \left(\sigma^2\left(R^2 + d_i(x,y)^2\right)^{\xi/2}\right), \tag{4.73}$$

and $s_i \equiv 1$. The problem (RF) admits a unique solution given by

Proposition 4.7 *Problem (4.65) admits a unique optimum given by*

$$\begin{aligned} C_i &= \left\{x \in \Omega : \sigma^2\left(R^2 + d_i^2(x_0, y_0)\right)^{\xi/2} \le \sigma^2\left(R^2 + d_j^2(x_0, y_0)\right)^{\xi/2}, \forall j \ne i\right\} \\ N_i &= \iint_{C_i} m_0(x_0, y_0)\, dx_0\, dy_0, \end{aligned} \tag{4.74}$$

which is represented by the Voronoi cells.

4.4.1 Penalization Function

We now focus on a specific penalty function that captures the capacity of the base stations. Note that the penalty function or the case when the number of mobile terminals is greater than the number of carriers available in the cell in the rate fair allocation policy case is equivalent to (P1) where the functions

$$F(d_i(x,y)) = \left(\sigma^2\left(R^2 + d_i(x,y)^2\right)^{\xi/2}\right). \tag{4.75}$$

$$s_i(N_i) = \kappa_i(N_i) = \begin{cases} 0 & \text{if } N_i \le \text{MAX}_i, \\ \bar{\kappa}_i(N_i - \text{MAX}_i) & \text{if } N_i > \text{MAX}_i. \end{cases} \tag{4.76}$$

The problem has then a solution given by

Proposition 4.8 *There exists a unique optimum given by*

$$\begin{cases} C_i = \left\{(x,y) : \sigma^2\left(R^2 + d_i^2(x_0, y_0)\right)^{\xi/2} + s_i(N_i) + N_i \cdot s_i'(N_i) \right. \\ \qquad \left. \le \sigma^2\left(R^2 + d_j^2(x_0, y_0)\right)^{\xi/2} + s_j(N_j) + N_j \cdot s_j'(N_j)\right\} \\ N_i = \iint_{C_i} m_0(\omega, z)\, d\omega\, dz. \end{cases} \tag{4.77}$$

4.4.2 Performance Gap

Consider again the case when the mobile terminals are uniformly distributed on $\Omega = [0, 1]$, but this time, the two antennas are located at coordinates $x_1 = 0$ and $x_2 = 1$. Consider the case when $p = 1$ and

$$s_1(x) = 100 \quad \text{and} \quad s_2(x) = \begin{cases} 0 & \text{for} \quad 0 \leq x \leq 0.999 \\ 1 & \text{for} \quad 0.999 \leq x \leq 1. \end{cases}$$

Then, the equilibrium cell configuration $\left(C_1^E, C_2^E\right)$ is given by

$$C_1^E = \varnothing \text{ and } C_2^E = [0, 1],$$

and the optimum cell configuration $\left(C_1^*, C_2^*\right)$ is

$$C_1^* = [0, 001] \text{ and } C_2^* = [0.001, 1].$$

The optimum is very unfair for mobile terminals living in the first cell C_1^*, who pay $x + 100$, whereas the other mobile terminals just pay the distance from 1.

This is a simple example, but it gives an idea of the performance gap between the centralized and the decentralized scenarios, also known as the price of anarchy.

In Fig. 4.11, we plot the Wardrop equilibrium with multiple base stations: The grid area network is the interval $[-10, 10]$ presented here vertically. We consider a uniform distribution of mobile terminals. We find the threshold (solid and dashed lines) determining the cell boundaries as a function of the base station positions (dotted lines) by changing the position of one of them. BS_1 is fixed at position -10, BS_2 is fixed at position 10, and we change the position of BS_3 from -10 to $+10$.

Fig. 4.11 Multiple BS

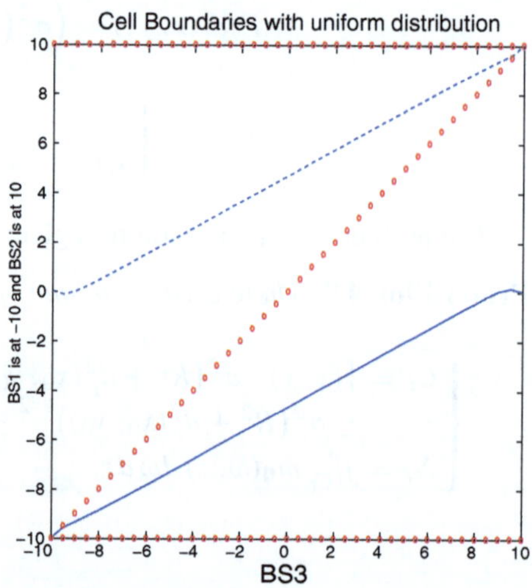

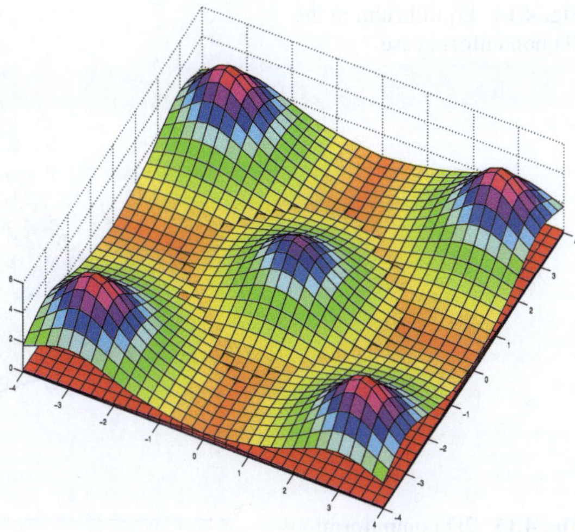

Fig. 4.12 Cell boundaries in 2D

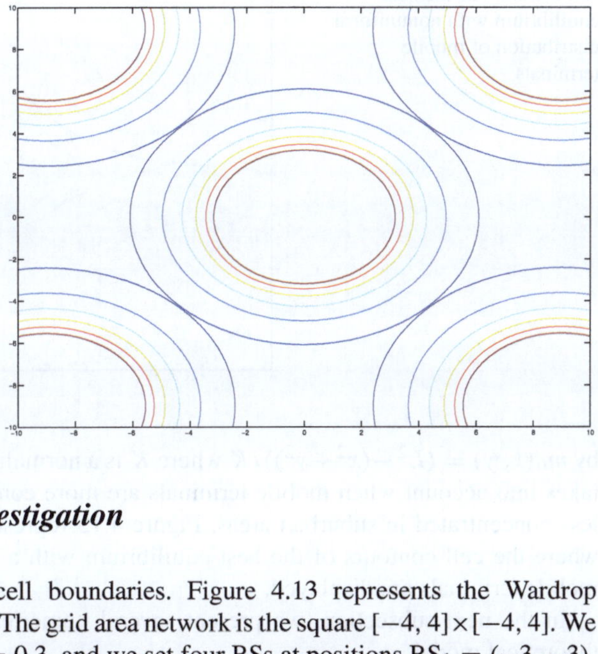

Fig. 4.13 2D case: cell contours of the best equilibrium with uniform distribution of mobile terminals

4.4.3 Numerical Investigation

Figure 4.12 depicts the cell boundaries. Figure 4.13 represents the Wardrop equilibrium in the 2D case: The grid area network is the square $[-4, 4]\times[-4, 4]$. We set the noise parameter $\sigma = 0.3$, and we set four BSs at positions $\mathrm{BS}_1 = (-3, -3)$, $\mathrm{BS}_2 = (3, -3)$, $\mathrm{BS}_3 = (3, 3)$, and $\mathrm{BS}_4 = (-3, 3)$ and one at the origin $\mathrm{BS}_5 = (0, 0)$. We determine the cell boundaries (deep lines) for the uniform distribution of mobile terminals.

Figure 4.14 depicts the Wardrop equilibrium in 2D nonuniform case: The grid area network is the square $[-4, 4] \times [-4, 4]$. We set the noise parameter $\sigma = 0.3$, and we set four BSs at positions $\mathrm{BS}_1 = (-3, -3)$, $\mathrm{BS}_2 = (3, -3)$, $\mathrm{BS}_3 = (3, 3)$, and $\mathrm{BS}_4 = (-3, 3)$ and one at the origin $\mathrm{BS}_5 = (0, 0)$. We determine the cell boundaries (deep lines) for the nonuniform distribution of mobile terminals given

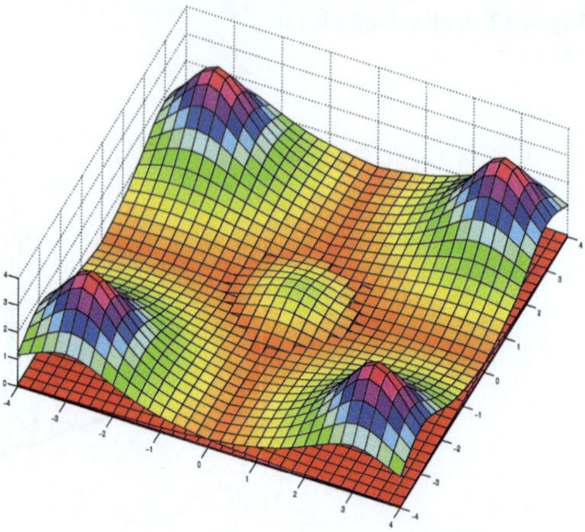

Fig. 4.14 Equilibrium in the 2D nonuniform case

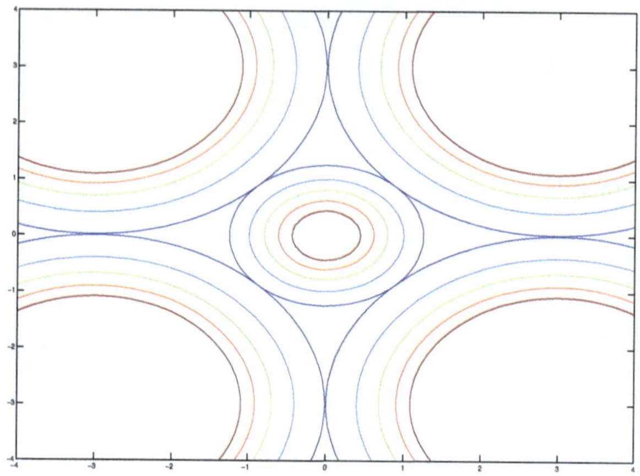

Fig. 4.15 2D nonuniform: cell contours of the best equilibrium with nonuniform distribution of mobile terminals

by $m_0(x, y) = (L^2 - (x^2 + y^2))/K$ where K is a normalization factor. This situation takes into account when mobile terminals are more concentrated in the center and less concentrated in suburban areas. Figure 4.15 represents a 2D nonuniform case where the cell contours of the best equilibrium with a nonuniform distribution of mobile terminals are displayed.

In the next subsection, we present several numerical results that validate our theoretical model.

4.4.4 *One-Dimensional Case*

We first consider the one-dimensional case, and we consider a uniform distribution of mobile terminals in the interval $[-10, 10]$. We set the noise parameter $\sigma = 0.3$. In Fig. 4.7, we fix one base station BS_2 at position 0 and take as parameter the position of base station BS_1. We consider as path loss exponent $\xi = 2$. Red lines show the positions of the BSs. We are able to determine the cell boundary (solid blue curve) from BS_1 and BS_2 at different positions. In Fig. 4.11, we fix two base stations $BS_1 = -10$ and $BS_2 = 10$, and we take as parameter the position of base station BS_3. Red lines show the positions of the BSs. We determine the cell boundary (solid blue curve) from BS_1 and BS_3 and the cell boundary (dotted blue curve) from BS_2 and BS_3.

In the next subsection, we run a simulation of power minimization and the cell formation in LTE-advanced technology.

4.4.5 Green LTE-Advanced Heterogeneous Networks

We consider a geographical area served by a heterogeneous network. The network consists of K spatially and spectrally coexisting tiers, where each tier is distinguished by its transmit power and BS density (e.g., macrocells would typically have a much higher transmit power and lower density, while small cells provide lower power with short range but have higher density). $M = \sum_{k=1}^{K} M_k$ BSs are randomly deployed to satisfy the traffic demand of the customers and covers the total area where M_k is the number of BSs of type k. We denote by (x_j, y_j, z_j) the spatial coordinates of $BS_j (j \in \{1, \ldots, M\})$ in the area. We assume that I users are also randomly placed following a certain user distribution function $f(x, y)$. For instance, the density could follow a uniform distribution with a given user density per km^2 or a normal (Gaussian) distribution corresponding to concentrated users in a hotspot area, and then, the density is reduced as we move away from the center. Orthogonal frequency division multiple access (OFDMA) is the access scheme for the downlink of LTE systems. The available spectrum is divided into resource blocks consisting of 12 adjacent subcarriers. In our framework, we allocate to each user one resource block. In other words, the number of users served by one BS cannot exceed the number of resource blocks denoted N_{RB}. The channel gain for the downlink direction between user i located at $(x, y, 0)$ and BS_j is given by

$$h_j(x, y) = -\kappa - \alpha \log_{10} d_j(x, y),$$

where κ is the pathloss constant, $d_j(x, y) = \left((x - x_j)^2 + (y - y_j)^2 + (0 - z_j)^2\right)^{\frac{1}{2}}$, and α is the pathloss exponent.

The allocated power to satisfy a rate of θ is given by

$$\bar{P}_j(x, y) = C_{\theta, N_0, \kappa} d_j(x, y)^{\frac{\alpha}{10}},$$

where $C_{\theta, N_0, \kappa}$ is positive and depends on θ, N_0, κ. On the other hand, the allocated power per user is

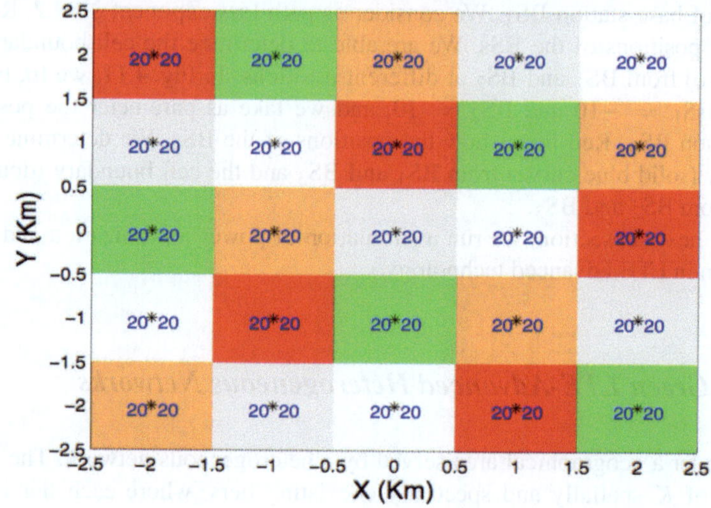

Fig. 4.16 Uniform distribution over $5 \times 5\,\text{km}^2$ for 500 users

$$\frac{\sum_{r=1}^{N_{RB}} P_{j,r}}{N_j} = \bar{P}_j(x, y).$$

Thus, one should minimize the total power $\sum_{j=1}^{M} N_j \int_{\mathscr{C}_j} \bar{P}_j(x, y) f(x, y)\, dxdy$ where $\mathscr{C}_j$ denotes the cell j.

Find the cell association $(\mathscr{C}_j)_j$ such that

$$F(\mathscr{C}_1, \ldots, \mathscr{C}_M) = \sum_{j=1}^{M} N_j \int_{\mathscr{C}_j} \bar{P}_j(x, y) f(x, y)\, dxdy \leq F\big(\tilde{\mathscr{C}}_1, \ldots, \tilde{\mathscr{C}}_M\big), \tag{4.78}$$

where $\big(\tilde{\mathscr{C}}_1, \ldots, \tilde{\mathscr{C}}_M\big)$ is a partition of the domain where the users are distributed according to the probability density f.

Below, we provide numerical solution of problem (4.78) for different distribution f. Figure 4.16 depicts the uniform distribution case over $5 \times 5\,\text{km}^2$ for 500 users. Figure 4.17 depicts the case where users are concentrated in downtown hotspot areas and decrease as we move away from the center (0, 0). Figure 4.18 represents two hotspots where users are concentrated: two neighboring residential areas, 16 stations instead of 25. Figure 4.19 depicts a multitier network: one macrocell and four small cells with 150 users distributed as truncated Gaussian centered with variance 0.5.

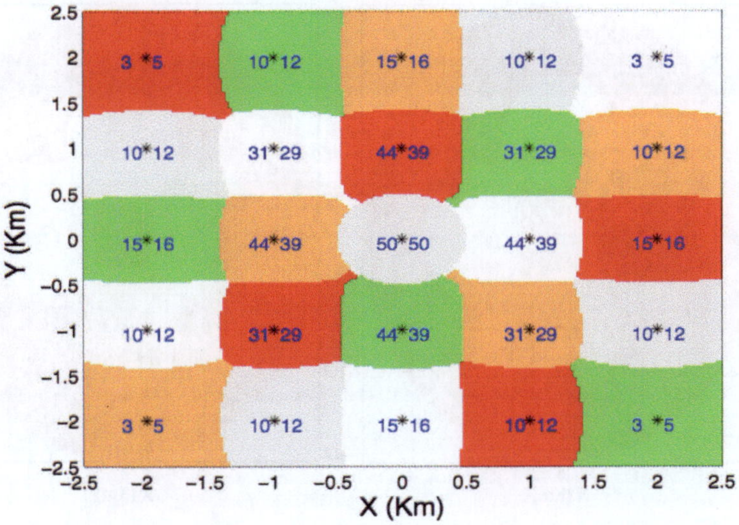

Fig. 4.17 Users are concentrated in downtown hotspot areas and decrease as we move away from the center (0, 0). Truncated Gaussian centered with variance 1.15

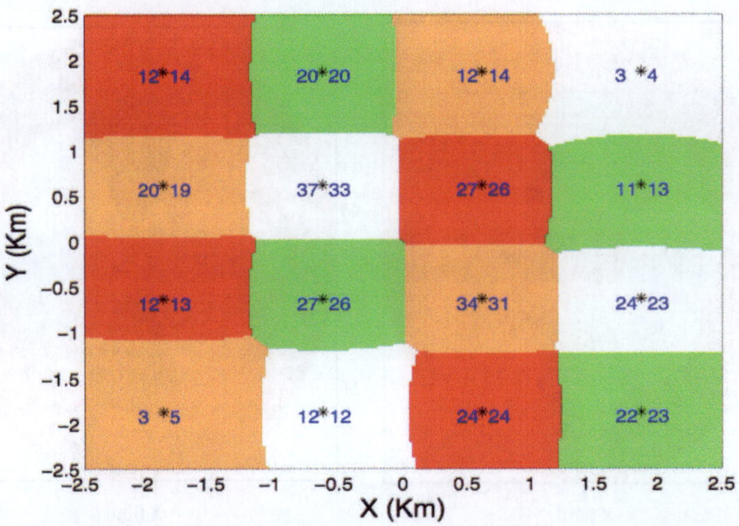

Fig. 4.18 Two hotspots where users are concentrated: two neighboring residential areas. Sixteen stations instead of 25

Figure 4.20 represents another multitier network: one macrocell and eight small cells with 150 users distributed as truncated Gaussian centered with variance 0.5. We observe that the location of the stations provides different shapes and different cell formations.

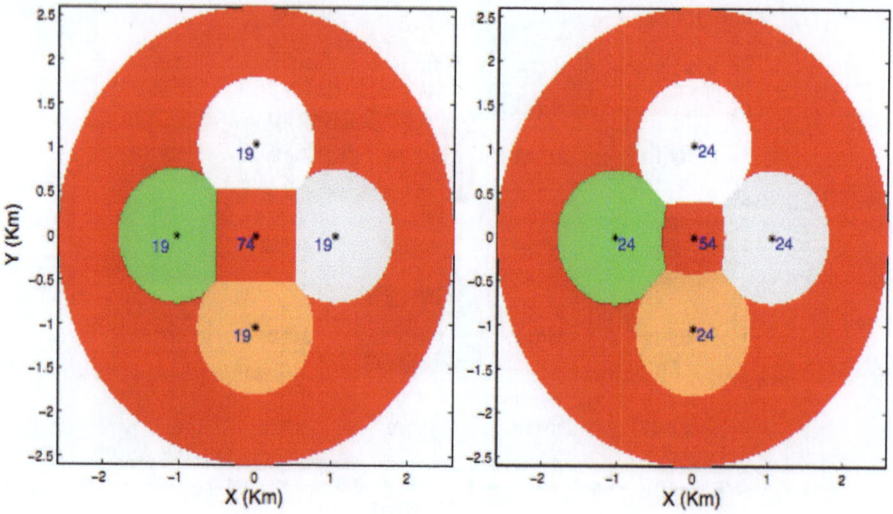

Fig. 4.19 Multitier network: one macrocell and four small cells with 150 users distributed as truncated Gaussian centered with variance 0.5

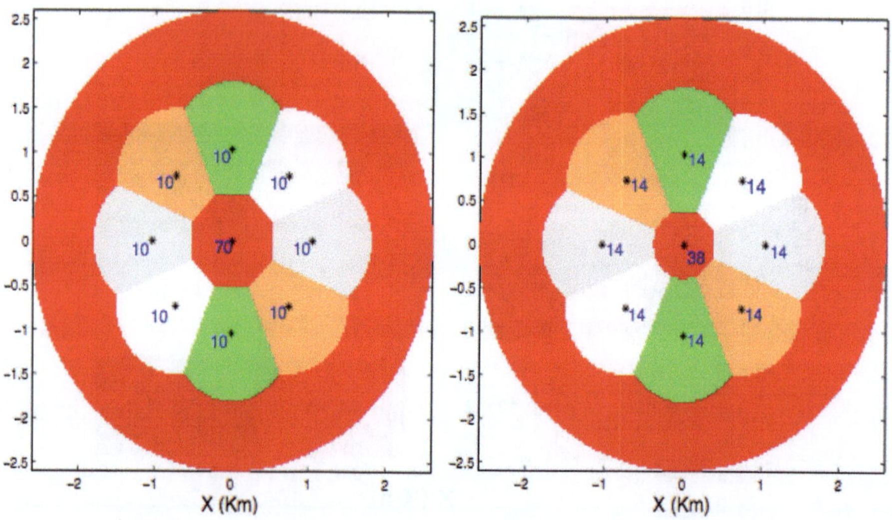

Fig. 4.20 Multitier network: one macrocell and eight small cells with 150 users distributed as truncated Gaussian centered with variance 0.5

4.5 Notes

In this chapter, we have studied the mobile association problem in the downlink scenario. The objective is to determine the spatial locations at which mobile terminals would prefer to connect to a given base station rather than to other base stations in the network if they were offered that possibility (referred to as a decentralized scenario). We examined also the spatial locations which are more convenient from a centralized or from a network operator point of view. In both approaches, the optimality depends upon the context. We have addressed the minimization of the total power of the network, which can be viewed as an energy-efficient objective, while maintaining a certain level of throughput for each user connected to the network. We have discussed an approach using optimal transport theory for the mobile association problems and have characterized these mobile associations under different policies.

The results of this chapter can be extended in several different directions. One of these is to study the price of anarchy between the centralized and decentralized scenario [189, 351, 352, 390–392].

Other works incorporating congestion in optimal transportation are [84, 117]. The work of [84] incorporates the Wardrop equilibrium in the network by assigning probabilities to the used routes.

Starting with the seminal work of Hotelling [230], a large area of research on location games has been developed. In [230], Hotelling introduced the notion of spatial competition in a duopoly situation. Plastria [361] presented an overview of the research on locating one or more new facilities in an environment where competing facilities already exist. Gabszewicz and Thisse [175] provided another general survey on location games. Altman et al. [17] studied the duopoly situation in the uplink scenario of a cellular network where the users are placed on a line segment. The authors realized that, considering the particular cost structure that arises in the cellular context, complex cell shapes arise at equilibrium.

This chapter has focused on a more general situation where a finite number of base stations can compete in one-dimensional and two-dimensional cases without making any assumption on the symmetry of the users' location. In order to do that, we have presented a framework for the mobile association problem using optimal transportation (see the recent books of Villani [500, 501] and references therein). This theory was pioneered by Monge [335, 336] and Kantorovich [257], and it has been proven to be useful in many mathematical, economical, and control theory contexts (see, e.g., [77, 82, 83, 478]). There is a number of works on "optimal transportation"

(continued)

(see [527] and references therein); however, the authors in [527] consider only an optimal selection of routes but do not use the rich theory of optimal transportation. The works on stochastic geometry are similar to our analysis of wireless networks (see, e.g., [32] and references therein), but in our case, we have not considered any particular deployment distribution function.

Part II
Computer Science

Chapter 5
Malware Propagation

In the context of computer networking, malware is a broad term used to describe malicious software or programs specifically designed to damage, disrupt the operation of, steal information from, or gain unauthorized access to computer systems, networks, and devices. The term "malware" is composed of *malicious and software*. Malware can take various forms and may include the following:

- Viruses are self-replicating programs that attach themselves to legitimate files or software. When the infected file is executed, the virus spreads to other files and can cause various forms of damage.
- Worms are standalone malicious programs that can replicate themselves and spread over networks. Unlike viruses, they do not need a host file to infect and can propagate independently.
- Trojans appear to be legitimate or useful software but contain hidden malicious functionality. Users are often tricked into installing them, and once installed, they can perform harmful actions.
- Ransomware encrypts a victim's files or entire system, rendering it unusable. Attackers then demand a ransom to provide the decryption key, allowing the victim to regain access to their data.
- Spyware is designed to secretly collect information about a user's online activities, including keystrokes, browsing history, and login credentials. This information is then transmitted to the attacker.
- Adware displays unwanted advertisements on a user's computer. While not always inherently harmful, it can be considered malware when it disrupts normal computer usage or compromises privacy.
- Rootkits are malicious programs that hide themselves deep within an operating system, making them difficult to detect. They often provide unauthorized access to a system and can be used to conceal other malware.
- A botnet is a network of compromised computers, often controlled by a single entity. These compromised computers, known as bots, can be used for various malicious activities, such as distributed denial-of-service (DDoS) attacks.

T. Başar et al., *Mean-Field-Type Game Theory II*, Static & Dynamic Game Theory: Foundations & Applications, https://doi.org/10.1007/978-3-032-07031-9_5

- Keyloggers record a user's keystrokes without their knowledge, allowing attackers to capture sensitive information like passwords and credit card numbers.
- Backdoors provide a hidden and unauthorized means of accessing a system or network. They are often used by attackers to maintain control over a compromised system.

Malware can be distributed through various means, including email attachments, malicious websites, infected software downloads, and social engineering tactics. Protecting against malware involves using antivirus and anti-malware software, keeping software and operating systems up to date, practicing safe Internet browsing habits, and being cautious when downloading and installing software from untrusted sources. We will demonstrate in this chapter that malware propagation over computer networks can be viewed as a game in various ways.

Malware propagation in opportunistic networks presents unique challenges and characteristics compared to traditional computer networks. Opportunistic networks are decentralized and intermittently connected networks where nodes communicate opportunistically when they come into proximity with each other. Malware can propagate in these networks in several ways:

- Opportunistic networks rely on the physical proximity of nodes for communication. Malware can propagate when an infected device comes into close contact with other devices, allowing the malware to spread through direct wireless connections, Bluetooth, or other short-range communication methods.
- When two nodes in an opportunistic network communicate, they exchange data packets. Malware can exploit this data exchange by attaching itself to legitimate data packets or posing as legitimate traffic. When the infected node communicates with other nodes, it can introduce the malware to the network.
- Malware propagation in opportunistic networks can also leverage social engineering. For example, an infected device may share malicious content with others by portraying it as desirable or useful. This can entice other nodes to accept and propagate the malware unintentionally.
- Opportunistic networks often lack centralized authentication mechanisms due to their decentralized nature. Malware can take advantage of weak or absent authentication to impersonate nodes, leading other nodes to accept its communication and potentially spread the malware further.
- Opportunistic networks frequently employ a store-and-forward mechanism, where nodes store data and forward it to others when opportunities arise. Malware can exploit these buffers to linger within the network, increasing its chances of being propagated to other nodes as they come into contact.
- Opportunistic networks tolerate delays in communication, making them suitable for scenarios like intermittent connectivity in remote areas. Malware can hide in the network for extended periods, increasing its chances of being spread when the network reconnects.
- The topology of opportunistic networks is highly dynamic as nodes continuously join and leave the network. Malware can use this dynamic nature to evade

detection by moving through the network, making it challenging for defenders to track and isolate infected nodes.

- Malware may incentivize nodes to help it propagate by offering rewards or benefits. This can encourage unwitting nodes to participate in the spread of malware, thinking that they are gaining some advantage [460, 464].

In this chapter, we present a framework for mean-field malware propagation in opportunistic wireless networks. Section 5.1 presents a class of malware propagation models. Section 5.2 extends the methodology to regime switching.

5.1 A Malware Propagation Model

Consider a controlled malware propagation in opportunistic networks of two types. The types can represent, for example, different operating systems, different versions of the operating systems, or patched/unpatched version of the same operating system. Different types can be different rates of propagation. We consider $I \geq 2$ mobile nodes that can be infected by a malicious code. There are two infected states: passive and active. Noninfected nodes are susceptible. Then, the set of possible states of a node is $\{P, A, S\}$ (for passive, active, and susceptible), and the set of possible types is $\{\theta_1, \theta_2\}$. The state of the system at time t is $x_t^I = (P_{1,t}, P_{2,t}, A_{1,t}, A_{2,t}, S_t)$, where

$$\begin{aligned} & t \geq 0, \\ & P_t + A_t + S_t = I, \\ & P_{j,t} \geq 0, \textstyle\sum_{j\in\{1,2\}} P_{j,t} = P_t, \\ & A_{j,t} \geq 0, \textstyle\sum_{j\in\{1,2\}} A_{j,t} = A_t, \end{aligned} \tag{5.1}$$

In this example setting, there is no common resource. The occupancy measure is

$$M_\theta^I(t) = \left(P_{\theta,t}/I, A_{\theta,t}/I, S_t/I\right) =: \left(P_{\theta,t}^I, A_{\theta,t}^I, S_t^I\right).$$

At every time step, we want to control the proportion of infected nodes, which is $A_t^I + P_t^I$. There are two fundamental ways to get infected:

1. Caused by a system flaw (e.g., an exploit that could allow arbitrary code execution)
2. Caused by human flaw (e.g., the user is deceived and executes a dangerous piece of code)

We can model this system as a controlled mean-field interaction model [460, 464]. The interaction is simulated using the following rules:

1. A passive node may become susceptible (inoculation) with probability δ_P.

2. A passive node with type θ may opportunistically encounter another passive node of type θ', and both become active. This occurs with a probability proportional to the frequency of other passive nodes at time t. For type θ, the probability is $\lambda\big(P^I_{\theta',t} - \frac{1}{I}\mathbb{1}_{\{\theta=\theta'\}}\big)$. Note that the passive node can decide whether to contact the other passive node or not, and hence, there are two possible actions: $\{m, \bar{m}\}$ (for meet and not meet). Those event will be modeled as a Bernoulli random variable with success (meeting) probability δ_m, which is also denoted as $u(m|P,\theta)$. Here, we model the possibility of getting infected by a system flaw.
3. An active node may become susceptible (inoculation) with probability δ_A.
4. An active node of type θ may become passive with probability $\beta\frac{P^I_{\theta,t}}{h_\theta+P^I_{\theta,t}}$ at time t. Here, it is assumed that, at high concentrations of passive nodes, each active node can infect some maximum number of passive ones per time step. This reflects finite total bandwidth. The parameter $0 \le \beta \le 1$ has the interpretation of the maximum infection rate. The parameter $0 \le h_\theta \le 1$ is the passive node density at which the infection proceeds at half of its maximum rate. Here, we model the possibility of getting infected by a system flaw.
5. A susceptible node may become active with probability δ_S.
6. A susceptible node may become passive in two ways. First, δ_{Sm} is the probability of getting infected by a human flaw. In this case, the susceptible node can "decide" whether to get deceived or not, and hence, there are two possible actions: $\{o, \bar{o}\}$. The stationary strategy in this case will be modeled as a coin toss with probability δ_e. Second, $r\big(P^I_{\theta,t} + P^I_{\theta',t}\big)$ models the probability of encountering a passive node. In this case, the passive node can decide whether to contact the susceptible node or not, and this is modeled in a way analogous to the other two cases.

At every time step, one of the transitions is randomly selected and performed. The number of nodes that do a transition in one time slot is always 0, 1, or 2. In order to control the infected population, each transition has a certain payoff contribution which could be 0 if no infected node is inoculated, $1/I$ if there is a node which is inoculated, and $-1/I$ if one node is infected. In Table 5.1 are the transition probabilities, the contribution to $M^I_{t+1} - M^I_t$, the set of actions, and the contributions to the total payoff.

The intensity, that is, the probability that one arbitrary player does a transition in one time slot, is of the order of $\frac{1}{I}$. The drift, that is, the expected change of M^I in one time step, given the current state of the system, is

$$
\begin{aligned}
f^I_\theta(m) &= \mathbb{E}\big(M^I_{\theta,t+1} - M^I_{\theta,t} | M^I_t = m\big) \\
&= \frac{1}{I}\begin{pmatrix} -p_\theta\delta_P - 2p_\theta\delta_m^2\lambda\frac{Ip_\theta-1}{I} - a_\theta\beta\frac{p_\theta}{h_\theta+p_\theta} + s(\delta_e\delta_{Sm} + \delta_m r(p_\theta + p_{\theta'})) \\ 2p_\theta\delta_m^2\lambda\frac{Ip_\theta-1}{n} - a_\theta\delta_A + a_\theta\beta\frac{p_\theta}{h_\theta+p_\theta} + s\delta_S \\ p_\theta\delta_P + a_\theta\delta_A - s\delta_S - s(\delta_e\delta_{Sm} + \delta_m r(p_\theta + p_{\theta'})) \end{pmatrix},
\end{aligned}
\tag{5.2}
$$

where $m = (p_\theta, p_{\theta'}, a_\theta, a_{\theta'}, s)$. Its limit is

Table 5.1 Probabilities, effects, actions, and payoffs

Case	Transition proba. ($\theta, \theta' \in \{1, 2\}$).	$M^I_{\theta,t+1} - M^I_{\theta,t}$	Actions	Payoff contrib.
1	$P^I_{\theta,t}\delta_P$	$(-1, 0, 1)/n$	Singleton set	$-1/I$
2	$P^I_{\theta,t}\delta^2_m\lambda(P^I_{\theta,t} - \frac{1}{I})$	$(-2, 2, 0)/I$	$\{m, \bar{m}\}$	0
3	$A^I_{\theta,t}\delta_A$	$(0, -1, 1)/n$	Singleton set	$-1/I$
4	$A^I_{\theta,t}\beta\frac{P^I_{\theta,t}}{h_\theta+P^I_{\theta,t}}$	$(-1, 1, 0)/I$	Singleton set	0
5	$S^I_t\delta_S$	$(0, 1, -1)/I$	Singleton set	$1/I$
6	$S^I_t(\delta_e\delta_{Sm} + \delta_m r P^I(t))$	$(1, 0, -1)/n$	$\{o, \bar{o}, m, \bar{m}\}$	$1/I$

$$f(m) = \begin{pmatrix} -p_\theta\delta_P - 2\lambda p^2_\theta\delta^2_m - a_\theta\beta\frac{p_\theta}{h_\theta+p_\theta} + s(\delta_e\delta_{Sm} + \delta_m r(p_\theta + p_{\theta'})) \\ -p_{\theta'}\delta_P - 2\lambda p^2_{\theta'}\delta^2_m - a_{\theta'}\beta\frac{p_{\theta'}}{h_{\theta'}+p_{\theta'}} + s(\delta_e\delta_{Sm} + \delta_m r(p_\theta + p_{\theta'})) \\ 2\lambda p^2_\theta\delta^2_m - a_\theta\delta_A + a_\theta\beta\frac{p_\theta}{h_\theta+p_\theta} + s\delta_S \\ 2\lambda p^2_{\theta'}\delta^2_m - a_{\theta'}\delta_A + a_{\theta'}\beta\frac{p_{\theta'}}{h_{\theta'}+p_{\theta'}} + s\delta_S \\ (p_\theta + p_{\theta'})\delta_P + (a_\theta + a_{\theta'})\delta_A - 2s\delta_S - 2s(\delta_e\delta_{Sm} + \delta_m r(p_\theta + p_{\theta'})) \end{pmatrix}$$

We study here the behaviors of the deterministic mean field and the controlled mean field. The control may stabilize the system.

We first briefly mention the homogeneous mean field. The drift is obtained by computing the expected changes in one time slot:

$$\begin{aligned} f^I(m) &= \mathbb{E}(M^I_{t+1} - M^I_t | M^I_t = m) \\ &= \frac{1}{I}\begin{pmatrix} -p\delta_P - 2p\delta^2_m\lambda\frac{np-1}{n} - a\beta\frac{p}{h+p} + s(\delta_e\delta_{Sm} + \delta_m rp) \\ 2p\delta^2_m\lambda\frac{np-1}{n} - a\delta_A + a\beta\frac{p}{h+p} + s\delta_S \\ p\delta_P + a\delta_A - s\delta_S - s(\delta_e\delta_{Sm} + \delta_m rp), \end{pmatrix} \end{aligned} \tag{5.3}$$

where $m = (p, a, s)$. Then, the limit is

$$f(m) = \begin{pmatrix} -p\delta_P - 2p^2\delta^2_m\lambda - a\beta\frac{p}{h+p} + s(\delta_e\delta_{Sm} + \delta_m rp) \\ 2p^2\delta^2_m\lambda - a\delta_A + a\beta\frac{p}{h+p} + s\delta_S \\ p\delta_P + a\delta_A - s\delta_S - s(\delta_e\delta_{Sm} + \delta_m rp) \end{pmatrix}$$

In all the simulations, we take the following values for the parameters involved:

$$\beta = 0.01, \delta_A = 0.005, \delta_P = 0.0001, \delta_S = 0.0001, \delta_{Sm} = 0.0001.$$

Parameter h (the passive node density at which the infection of active nodes proceeds at half of its maximum value) depends on the stable behavior; that is, $h = 0.3$ is stable, whereas $h = 0.1$ is unstable. Moreover, $\delta_m = 1$, $\delta_e = 1$ means no control.

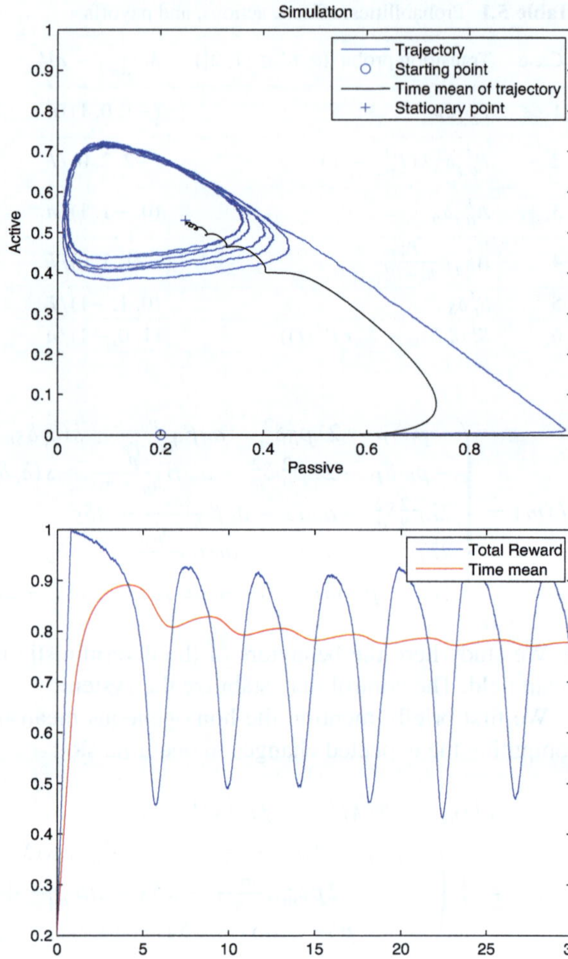

Fig. 5.1 Upper figure: one trajectory with no control. Lower figure: total reward

We investigate the evolution of the system in the following scenarios:

- Trajectory of one run of simulation (Fig. 5.1)
- Mean trajectory of the simulations (Fig. 5.2)
- Mean-field limit trajectory via the ODE (Fig. 5.3)
- Trajectory of the payoff function (both stable and unstable configurations)
- Controlled mean-field limit (Fig. 5.4)
- Trajectory of the heterogeneous malware propagation (Fig. 5.5)
- Optimal control under the mean-field limit (Fig. 5.6, [106])
- The noisy mean field vs. one run of simulation (the stochastic path, Fig. 5.7)

These configurations are analyzed with control as well as without control.

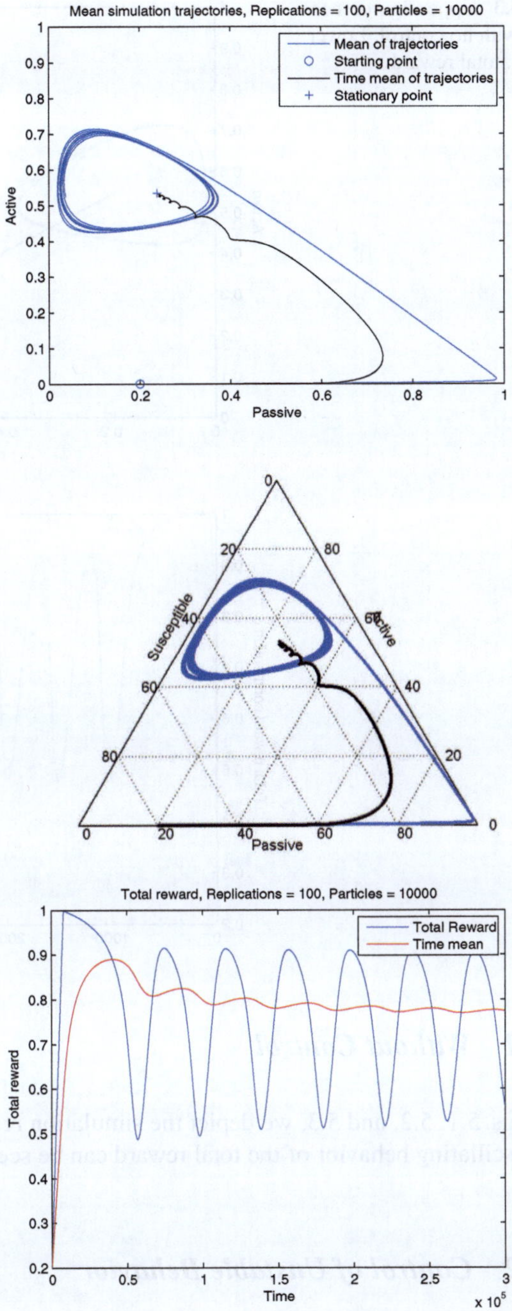

Fig. 5.2 Upper figure: mean of simulation trajectories with no control. Middle figure: ternary plot. Lower figure: total reward

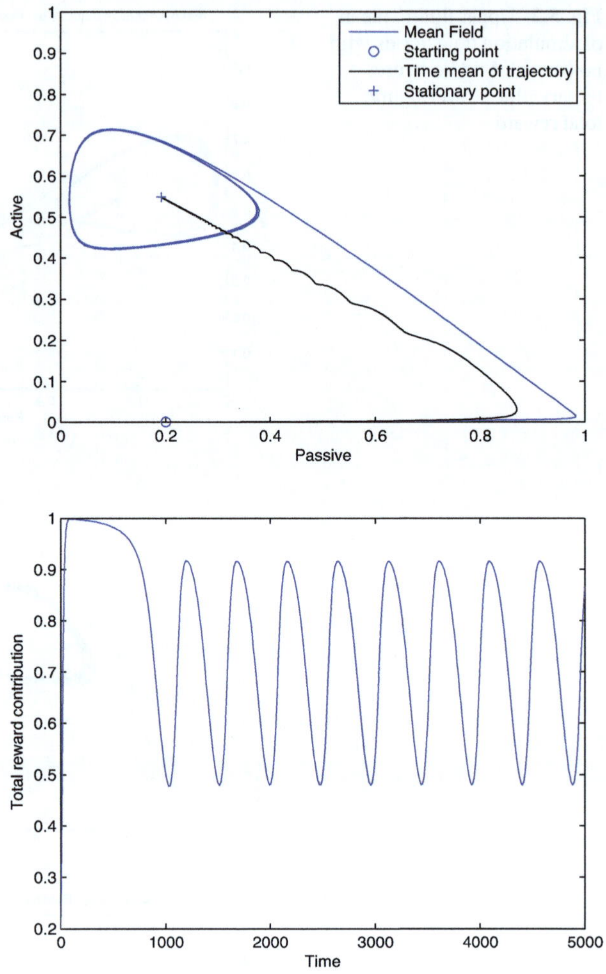

Fig. 5.3 Upper figure: mean field with no control. Lower figure: total reward

5.1.1 *Without Control*

In Figs. 5.1, 5.2, and 5.3, we depict the simulation results for $M(0) = (0.2, 0, 0.8)$. An oscillating behavior of the total reward can be seen.

5.1.2 *Control of Unstable Behavior*

In order to illustrate the case with control parameters, suppose that we want to keep the proportion of infected nodes below 0.9 for all times. One simple way to achieve

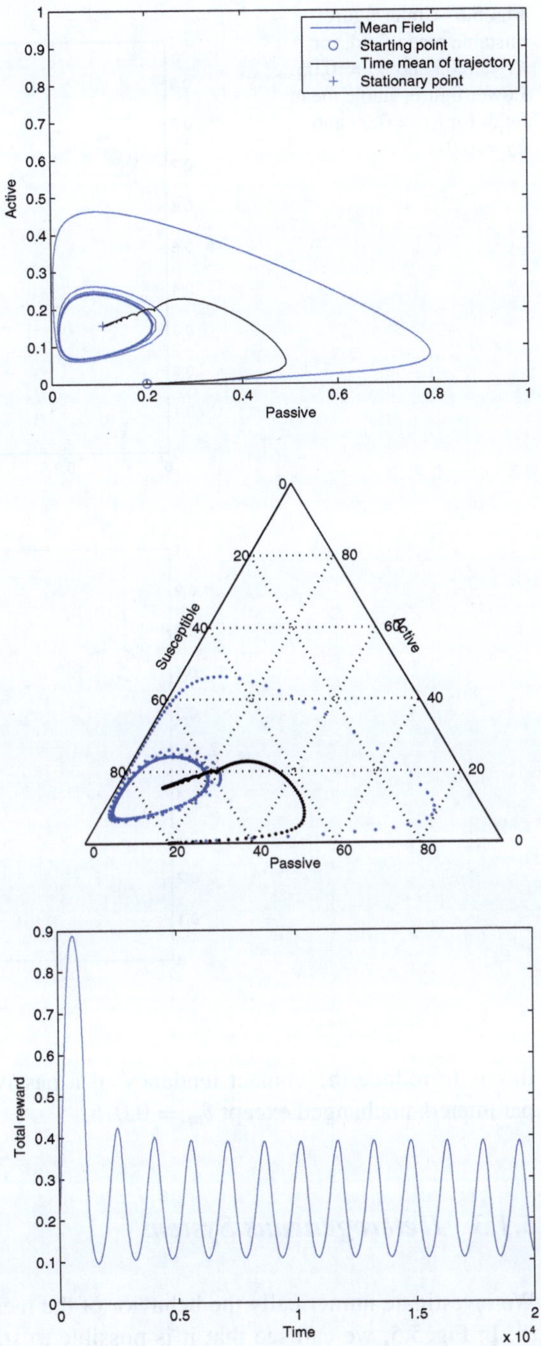

Fig. 5.4 Upper figure: mean field with controlled passive node contact tendency. Middle figure: ternary plot. Lower figure: total reward

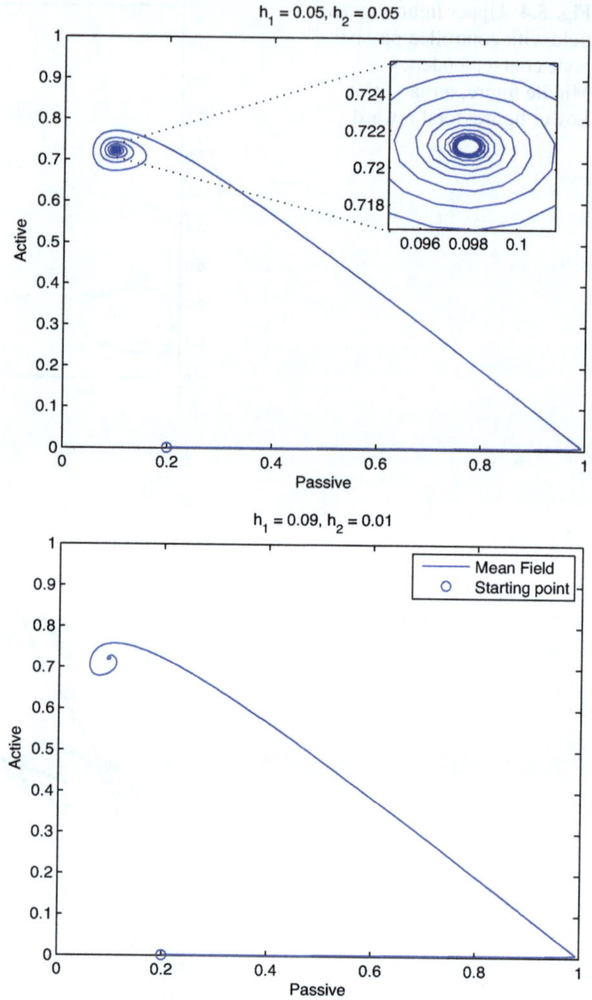

Fig. 5.5 Upper figure: unstable mean field, for $h_1 = 0.05$ and $h_2 = 0.05$. Lower figure: stable mean field, for $h_1 = 0.09$ and $h_2 = 0.01$

this is to reduce the contact tendency of a passive node. In Fig. 5.4, we keep all parameters unchanged except $\delta_m = 0.075$.

5.1.3 Heterogeneous System

We investigate numerically the behavior of the mean-field limit for two types θ and θ'. In Fig. 5.5, we can see that it is possible to stabilize the homogeneous system using the types.

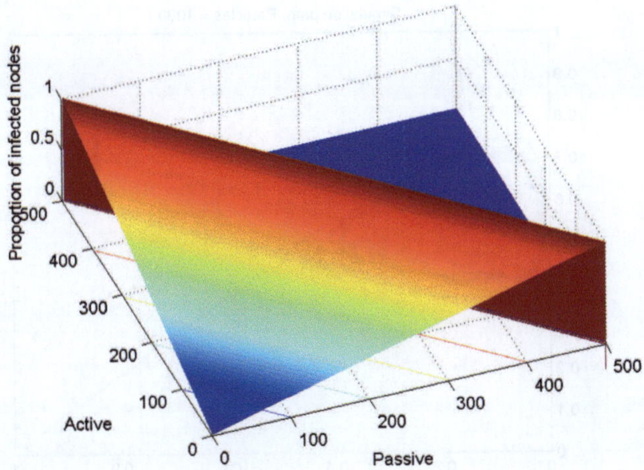

Fig. 5.6 Optimal strategy using backward induction for the homogeneous system

5.1.4 Optimal Strategy for Homogeneous System

Since the payoff function is the same for all the nodes (players), i.e., $r_j(\cdot) = r(\cdot)$, then we obtain a stochastic game with common payoff. Further, the set of actions is the same for all the players. Using backward induction, we get

$$\begin{cases} V_t(x) = \min_{u \in \Delta(\mathscr{A}(x))}\{r(x,u) + \sum_{x'} q(x'|x,u)V_{t+1}(x')\} \\ \forall x \in \mathscr{X},\ t \in \{0, 1, \ldots, T-1\} \\ V_T(x) = x, \forall x \in \mathscr{X} \\ m_{t+1} = m_t L(m_t, u_t) \end{cases}$$

In Fig. 5.6, we plot the optimal strategy obtained by solving the above backward (mean field) Bellman equations.

5.1.5 Stochastic Mean Field vs. One Run of Simulation

In this subsection, we compare one run of simulation with a large I with a noisy mean field, which is a stochastic differential equation built from the drift limit. The variance of noise is obtained from $\tilde{\sigma}^I = \frac{1}{\delta_I} f^I - f$, which its norm is bounded by $\frac{2D\lambda}{I}$. For $I = 1000$, we represent the trajectory of M_t^I and the solution of the stochastic differential equation for

$$d\tilde{m} = f(\tilde{m})dt + \tilde{\sigma}^I d\mathbb{B}_t$$

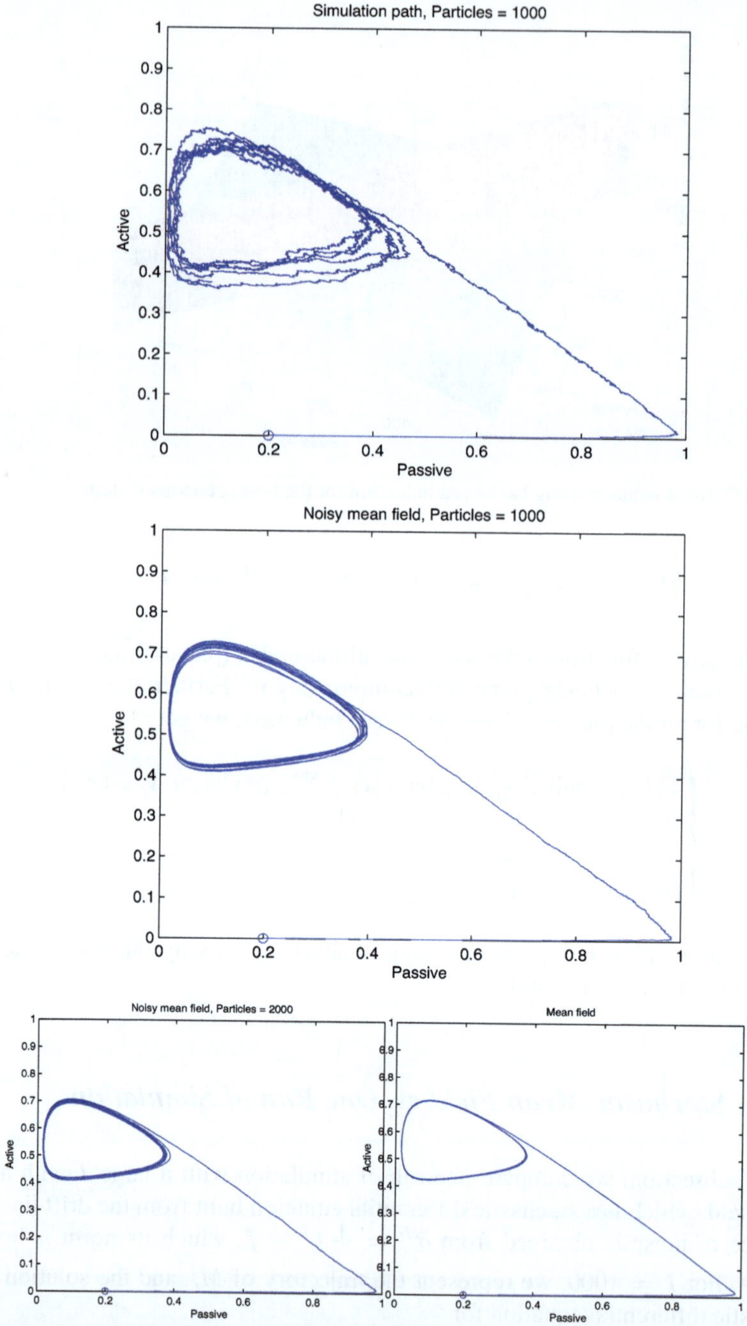

Fig. 5.7 Upper figure: path of one simulation, for $I = 1000$. Middle figure: noisy mean field, for $I = 1000$ and $I = 2000$, respectively. Lower figures: mean field

Note that the smooth version of the last equation can be $\dot{m}^I = \frac{f^I(m^I)}{\delta_I}$. In the leftmost plot of Figure 5.7, we can see the path of one simulation; for $I = 1000$, the rightmost plot is the mean field. In the center, we can see the noisy mean field, for $I = 100$ and $I = 1000$, respectively.

5.2 Malware Propagation with Switching Regimes

We now extend the previous model to include switching regimes. This leads to a hybrid mean-field game. In this example, we have mobile nodes that can be infected by a malicious code. There are two resource states $\mathscr{S} = \{1, 2\}$, corresponding to the two regimes. In state $s = 1$, the value of h is $h_1 = 0.1$. In state 2, $h_2 = 0.3$. The system is stable for $h_2 = 0.3$, and the mean-field dynamics has a limit cycle for h_1 [446]. There are two infected states for each mobile node: passive and active. Noninfected nodes are susceptible. Then, the set of possible individual states of a node is $\{\underline{P}, \underline{A}, \underline{S}\}$ (for passive, active, and susceptible). The state of the system at time t is $X^I(t) = (\underline{P}_s(t), \underline{A}_s(t), \underline{S}_s(t))$, and I is the total number of mobiles in the system (see Table 5.2).

At every time step, we want to control the proportion of infected nodes. We model this system as a controlled mean-field interaction model, with the following transition rules, probabilities, effects $(M_s^I(t+1) - M_s^I(t))$, and payoffs:[1]

The parameter $0 \leq h_s \leq 1$ is the passive node density at which the infection proceeds at half of its maximum rate, where $0 \leq \beta \leq 1$ is the maximum infection rate. δ_m and δ_e are the control parameters (see [446]). In order to control the infected population, each transition has a certain payoff contribution which could be 0 if no infected node is inoculated, $1/I$ if there is a node which is inoculated, and $-1/I$ if one node is infected.

Table 5.2 Transition rules, probabilities, effects, and payoffs

Transition rule	Probability	Effect	Payoff
$\underline{P} \xrightarrow{\delta_{\underline{P}}} \underline{S}$	$\underline{P}_s^I(t)\delta_{\underline{P}}$	$(-1, 0, 1)/I$	$-\frac{1}{I}$
$2\underline{P} \xrightarrow{\lambda} 2\underline{A}$	$\underline{P}_s^I(t)\delta_m^2\lambda\left(\underline{P}_s^I(t) - \frac{1}{I}\right)$	$(-2, 2, 0)/I$	0
$\underline{A} \xrightarrow{\delta_{\underline{A}}} \underline{S}$	$\underline{A}_s^I(t)\delta_{\underline{A}}$	$(0, -1, 1)/I$	$-\frac{1}{I}$
$\underline{A} \xrightarrow{\frac{\beta}{h_s + \underline{P}_s^I(t)}} \underline{P}$	$\underline{A}_s^I(t)\beta\frac{\underline{P}_s^I(t)}{h_s + \underline{P}_s^I(t)}$	$(-1, 1, 0)/I$	0
$\underline{S} \xrightarrow{\delta_{\underline{S}}} \underline{A}$ $\underline{S} \xrightarrow{\delta_{Sm}} \underline{P}$	$\underline{S}^I(t)\delta_{\underline{S}}$	$(0, 1, -1)/I$	$\frac{1}{I}$
$\underline{S} + \underline{P} \xrightarrow{\eta} \underline{P}$	$\underline{S}^I(t)\left(\delta_e\delta_{Sm} + \delta_m\eta\underline{P}^I(t)\right)$	$(1, 0, -1)/I$	$\frac{1}{I}$

[1] $\underline{P} \xrightarrow{\delta_{\underline{P}}} \underline{S}$ means "a passive node may become susceptible (inoculation) with rate $\delta_{\underline{P}}$."

The drift is

$$f^I(m,s) = I\mathbb{E}\big(M_s^I(t+1) - M_s^I(t)|\, M^I(t) = m, s\big) =$$

$$\begin{pmatrix} -\underline{p}_s\delta_{\underline{P}} - 2\underline{p}_s\delta_m^2\lambda\frac{I\underline{p}_s-1}{I} - \underline{a}_s\beta\frac{\underline{p}_s}{h_s+\underline{p}_s} + \underline{s}(\delta_e\delta_{Sm}+\delta_m\eta\underline{p}_s) \\ 2\underline{p}_s\delta_m^2\lambda\frac{n\underline{p}_s-1}{I} - \underline{a}_s\delta_{\underline{A}} + \underline{a}_s\beta\frac{\underline{p}_s}{h_s+\underline{p}_s} + \underline{s}\delta_{\underline{S}} \\ \underline{p}_s\delta_{\underline{P}} + \underline{a}_s\delta_{\underline{A}} - \underline{s}\delta_{\underline{S}} - \underline{s}(\delta_e\delta_{Sm}+\delta_m\eta\underline{p}_s) \end{pmatrix}$$

where $m(s) = (\underline{p}_s, \underline{a}_s, \underline{s}_s)$.

Parameter values are as follows: $\beta = 10^{-2}$, $\delta_{\underline{A}} = 5 \times 10^{-3}$, and $\delta_{\underline{P}} = \delta_{\underline{S}} = \delta_{Sm} = 10^{-4}$. Parameter h_s (the passive node density at which the infection of active nodes proceeds at half of it maximum value in state s) depends the stability of the system. Here, we set $h_1 = 10^{-2}$ in order to obtain an unstable behavior.

At each switch time t_i, a Bernoulli random variable decides the regime ($h = 0.3$ with probability 0.4 and $h = 0.1$ with probability 0.6). Figures 5.8 and 5.9 focus on the hybrid case.

5.3 Corruption Is a Virus

In the corruption case, a mean-field model is proposed in [420]. Corruption is a multifaceted issue that can manifest on different scales, from individual acts of dishonesty to systemic corruption within institutions. The mean-field model of corruption aims to capture this complexity by examining corruption at micro-, meso-, and macroscopic levels. One key aspect of this model is its consideration of the long-term consequences of corruption. By analyzing the state distribution of honest individuals and the dynamics of corruption propagation, researchers can gain insights into strategies for combating corruption over extended periods. The networking aspect of the model is crucial. It examines how corruption spreads within networks, considering factors such as the degree of nodes (i.e., how connected they are) and the number of multi-hop connections. This connectivity plays a pivotal role in shaping the dynamics of corruption, as highly connected nodes can serve as hubs for corruption propagation. The model also explores strategies [460, 464] for combating corruption. It considers both centralized and decentralized approaches to control corruption. Centralized strategies involve top-down interventions or regulations, while decentralized strategies empower individuals or smaller units to take action against corruption.

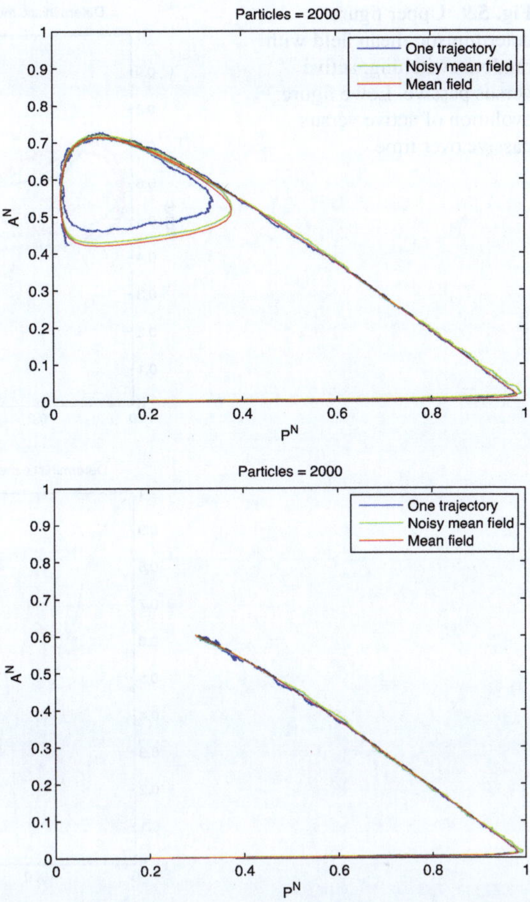

Fig. 5.8 Upper and lower figure: deterministic and noisy mean field without control

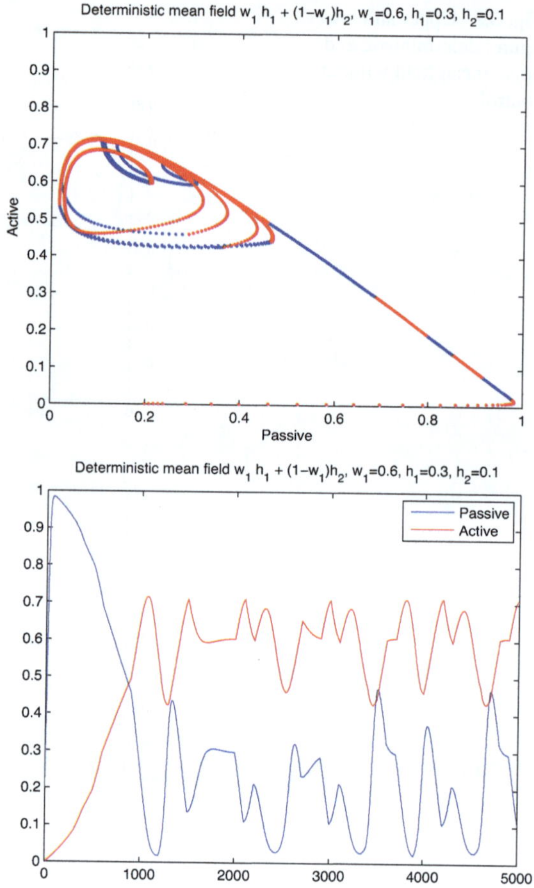

Fig. 5.9 Upper figure: deterministic mean field with random switching: active versus passive. Lowe figure: evolution of active versus passive over time

5.4 Notes

The methodology and results presented in this chapter have been extended to various other settings, including corruption propagation, epidemic propagation, and community building in online media (see [170, 244, 290, 402, 420, 431, 445, 447] and the references therein). Several works apply mean-variance optimization frameworks under dynamic settings with frictions and investment constraints [59, 61], while others propose novel mathematical challenges in management science that leverage such models [58]. In cybersecurity, mean-field game models have been employed to develop botnet defense strategies and understand corruption dynamics in network systems [275, 276].

(continued)

Strategic interactions in oligopoly markets have also been modeled using Cournot and Bertrand competition within mean-field frameworks, offering existence and uniqueness guarantees [197]. In energy economics, applications to fracking and renewable energy competition have been analyzed using mean-field approaches [93], while in control and optimization, fuel follower problems have been studied both in finite-agent and mean-field limits [206]. The modeling of synchronization in coupled oscillators, such as in Kuramoto-type systems, has also been treated in mean-field game formulations [86]. Environmental concerns, particularly carbon emissions regulation, have been addressed using these models to guide electricity production policy [87]. Additionally, timing games and the modeling of bank runs have been formulated via mean-field games [88]. Swarm robotics has benefited from mean-field modeling as a way to abstract large population interactions [152]. More recently, quantum analogs of mean-field games and related stochastic master equations have emerged, extending the theory into quantum probability and dynamics [272–274]. Evolutionary game theory has also been integrated with mean-field ideas to analyze resistance, pressure, and collaboration among agents [271], showcasing the versatility and depth of mean-field modeling across domains.

Chapter 6
Rent-Seeking Games in Cloud Networking

A cloud network is an interconnection of computer systems and resources that are hosted and delivered over the Internet, often through cloud computing services provided by third-party providers. These networks are designed to facilitate the storage, processing, and transmission of data and applications across distributed and virtualized infrastructure, rather than relying solely on on-premises hardware and traditional local area networks (LANs). Key characteristics of a cloud network include virtualization, on-demand services, scalability, resource pooling, self-services, network connectivity, security and compliance, and cost efficiency. Cloud networks often use virtualization technologies to create virtual instances of servers, storage, and networking components. This enables efficient resource allocation and scaling as needed. Users can access and provision computing resources, such as virtual servers, storage, databases, and networking components, on demand. This flexibility allows organizations to adjust their infrastructure to meet changing demands. Cloud networks can scale resources up or down based on demand, which is particularly beneficial for handling fluctuations in user traffic or workloads. This elasticity helps optimize costs and performance. Cloud providers pool computing resources from multiple physical servers and locations. These resources are dynamically allocated to customers as needed, improving resource utilization and efficiency. Users can provision and manage resources through web-based dashboards or APIs without direct involvement from IT administrators, enabling greater autonomy and agility. Cloud networks rely on the Internet and various network technologies to connect users to cloud services and resources. They often include features like load balancing, content delivery networks (CDNs), and global data centers to optimize connectivity and performance. Cloud providers typically offer security measures, compliance certifications, and data encryption to help protect data and ensure regulatory compliance. However, it is also the responsibility of the user to implement proper security practices. Cloud networks often operate on a pay-as-you-go or subscription-based pricing model, which can help organizations reduce upfront capital expenses and align costs with actual

T. Başar et al., *Mean-Field-Type Game Theory II*, Static & Dynamic Game Theory: Foundations & Applications, https://doi.org/10.1007/978-3-032-07031-9_6

usage. Examples of cloud service providers include Amazon Web Services (AWS), Microsoft Azure, and Google Cloud Platform (GCP), which offer a wide range of services and infrastructure components that organizations can use to build and operate their cloud networks. These services include virtual machines, storage solutions, databases, networking tools, and more.

Resource sharing is a fundamental and crucial aspect of cloud networks. It plays a central role in the efficiency, cost-effectiveness, and flexibility of cloud computing environments. Resource sharing allows multiple users and applications to utilize the same physical infrastructure, such as servers, storage devices, and networking equipment. This leads to higher resource utilization rates, reducing wasted capacity and maximizing the efficiency of the hardware. Cloud providers can achieve economies of scale, which often leads to lower costs for customers. Users only pay for the resources they consume, rather than investing in and maintaining dedicated hardware, resulting in cost savings for both small businesses and large enterprises. It enables dynamic scaling of resources based on demand. When one user or application needs more computing power or storage capacity, it can quickly and easily access additional resources from the shared pool. This scalability is essential for handling variable workloads efficiently. Cloud networks can allocate resources rapidly, enabling users to adapt to changing requirements and respond to spikes in demand. This flexibility is particularly valuable in today's fast-paced business environments. Cloud providers can host multiple tenants (users or organizations) on the same physical infrastructure while ensuring isolation and security between them. This multi-tenancy model allows a diverse range of customers to share the same resources without compromising data privacy or security. Sharing resources can lead to reduced energy consumption and a smaller carbon footprint. Cloud providers can optimize data center operations and power consumption, contributing to sustainability efforts. They often have data centers in various geographic regions. Resource sharing across these regions enables users to access resources and services close to their target audience, reducing latency and improving the user experience. It can also improve reliability and redundancy. Data and applications can be distributed across multiple physical locations to enhance fault tolerance and ensure high availability. Resource sharing is integral to the functioning of cloud networks, as it enables providers to efficiently manage and allocate resources while offering customers cost-effective, scalable, and flexible computing solutions.

Rent-seeking games in the context of cloud networking refer to situations where individuals, organizations, or entities attempt to gain economic benefits or advantages by exploiting the structure and dynamics of cloud computing markets and networks without creating corresponding value. In essence, these actors seek to extract rents or profits from the cloud ecosystem without necessarily contributing to the innovation, efficiency, or improvement of cloud services or infrastructure. Cloud resource sharing game is a concept used in the context of cloud computing to model and analyze the allocation of resources (such as computing power, storage, or network bandwidth) among multiple users or entities in a cloud environment. These games are used to study how resources are distributed, allocated, and shared among participants while considering various objectives, constraints, and strategies.

The challenge that operators face is to support more subscribers with higher bandwidth requirements. To meet the bandwidth demand, operators need to seek new technologies to efficiently utilize available network resources, in particular, efficient resource sharing, resource allocation, and flow management. In wireless networks, the problem is particularly relevant due to the inherently limited resources. The success of powerful mobile devices like smartphones or tablets in accessing different kinds of Internet services is not only promoting the growth in traffic volume but also changing the traffic pattern. Certain applications automatically pull or push large amounts of data without the user necessarily being aware. Cloud-based applications transfer more and more data (sometimes even with real-time requirements). This strong growth and continual changes in the traffic characteristics and usage behavior raise questions on how to share limited capacity resources fairly and more efficiently. In order to get a certain level of fairness, the utility function should capture a fair sharing policy.

Cloud networking has some new challenges compared to the classical Open Systems Interconnection (OSI) system. The cloud network layer includes web applications (in the cloud), which makes an additional component compared to the traditional OSI system, where the network is layer 3 and the application is layer 7. Cloud networking is not a separate layer. It goes beyond the boundaries and couples with IaaS (Infrastructure as a Service) and SaaS (Software as a Service). Cloud networking aims at reducing energy consumption cost, leading to a green cloud. However, traditional approaches to energy-efficiency, such as *sleep mode*, cannot be used for the cloud since servers have to be available 24 hours a day and 7 days a week. Another challenge is to enable these different kinds of applications and to apply traffic management that maintains high degrees of resource utilization, a reasonable distribution of capacity, without affecting transport protocol performance and impeding innovation. Such traffic management has to be able to adapt to innovation and be scalable with respect to growing traffic, growing numbers of flows, different application types, etc. One of the challenges in cloud networks is *how to rent efficiently and in a distributed manner, the server space to websites*. These cloud servers would automatically take up as many servers as needed and they charge for bandwidth usage. One of the questions that we address in this chapter is whether and how cloud resource sharing can support demand management evolving in this direction. Resource sharing solutions are very important for data centers, as it is required and implemented at different layers of the cloud network. The resource sharing problem can be formulated as strategic decision-making between the clients.

One of the fundamental challenges in cloud networking is to design efficient and fair solutions. In such systems, a satisfactory solution is an innovative approach that aims to provide all cloud users with a satisfactory payoff anytime, anywhere.

In this chapter, we consider resource sharing problems in cloud networks with generalized fair payoff functions. We formulate the problem as a game. We examine the cloud game with an arbitrary number of cloud users as well as the asymptotic game with an infinite number of users. Exploiting the aggregative structure of the payoff functions, we show that the game has a Nash equilibrium in a wide range of return indices. However, the Nash equilibrium is not an evolutionarily stable strategy in the finite regime. Then, we introduce a myopic mean-field response where each cloud user implements a mean-field-taking strategy. We show that such a *mean-field-taking strategy* is an evolutionarily stable strategy in finite as well as infinite regimes. We provide closed expressions for optimal pricing that lead to efficient resource sharing. As the number of active users grows without bound, we show that the equilibrium strategy converges to a mean-field equilibrium and the optimal prices converge to the optimal price of the mean-field game. Moreover, a very fast mean-field learning algorithm is provided. Then, we address the demand satisfaction problem for which necessary and sufficient conditions for satisfactory solutions are provided.

We consider cloud resource sharing games with continuous action spaces.

- In the finite regime, we provide closed-form expressions of four important game-theoretic solution concepts: Nash equilibrium, mean-field response, evolutionarily stable strategies (finite population version), and satisfactory solutions (that is, each cloud user demand is satisfied anywhere, anytime).
- We show the existence of a Nash equilibrium in some range of parameters and that the Nash equilibrium is stable under best response dynamics in some range of parameters.
- The first surprising result is that the Nash equilibrium is not an evolutionarily stable strategy (finite population version), meaning that it does not optimize the relative payoffs.
- However, we show that the equilibrium is stable in the limiting case. This is due to the fact that in the limiting case, the influence of one generic decision-maker on the total demand is negligible.
- We show that mean-field-taking strategies are evolutionarily stable strategies for a certain range of return indices.
- We provide the expression for optimal pricing as a function of the number of active users.
- Under optimal pricing, no resource is wasted in equilibrium. This means that the efficiency loss tends to zero.

Mean-field games are well adapted for both finite (but large) systems and infinite systems. In contrast to the other classical tools for large-scale systems, the mean-field approach incorporates the dynamics, which allows online demand management (pay-as-you-use scheme). It enables optimization and studies the interactive nature of cloud networking, leading to strategic decision-making problems (so-called games). Recent studies in cloud networking have already adopted a mean-field regime. In the infinite regime, we provide a closed-form expression for the mean-field equilibrium and show that the mean-field equilibrium corresponds to an evolutionarily stable strategy, i.e., it cannot be invaded by a small fraction of deviants. Then, we focus on distributed learning algorithms. We propose a model-based (but still with less information) learning scheme for games with continuous action spaces and large numbers of decision-makers. Each decision-maker will update her learning strategy based on an aggregative term, which is the mean of an increasing function of the actions of the other decision-makers. Each decision-maker will be influenced by the aggregate, and the mean-field behavior is formed from the actions of all decision-makers. Each decision-maker will try to conjecture the aggregative term comprised of the actions of the other decision-makers during each time slot and will respond to the aggregate locally. This drastically simplifies the dimension of the mean-field response system in the asymptotic case.

The chapter is structured as follows. In Sect. 6.1, we present the resource sharing game model. In Sect. 6.2, we examine both equilibrium and stability properties of the game with a finite number of cloud users. Section 6.3 examines the asymptotic game with an infinite number of cloud clients. In Sect. 6.4, we focus on satisfactory solutions. Section 6.6 discusses possible extensions.

6.1 Rent-Seeking in the Cloud Network

Resource-Sharing Mean-Field Game with Finite Number of Clients

We consider resource-sharing mean-field game with a finite (but arbitrary) number of clients in the cloud network. We denote such a game by $\mathcal{G}_I$, where I is the number of clients. The action space of each decision-maker is $\mathcal{A} = \mathbb{R}_+$ which is a non-empty and convex set. Each decision-maker j chooses an action a_j that belongs to the set $\mathcal{A}$. An action may represent a certain demand. All the actions together determine an outcome. Let p_I be the unit price of cloud usage to the clients. The payoff of decision-maker j is given by

$$r_j(a_1, \ldots, a_I) = \left\{ c_I \frac{h(a_j)}{\sum_{i=1}^I h(a_i)} - p_I a_j \right\} \mathbb{I}_{\{\sum_{i=1}^I h(a_i)>0\}},$$

where $\mathbb{I}_S(x)$ denotes the indicator function: It is 1 if $x \in S$ and 0 otherwise. Here, c_I represents the value of the available resource (which can be seen as the capacity of the cloud), and h is a positive and nondecreasing function with $h(0) = 0$. We fix the function h to be x^α, where $\alpha > 0$ denotes a certain return index.

As h is a one-to-one mapping on $\mathbb{R}_+$, one can rewrite the payoff function of decision-maker j as

$$r_j(a_1, \ldots, a_I) = \left\{ c_I \frac{y_j}{\sum_{i=1}^I y_i} - p_I h^{-1}(y_j) \right\} \mathbb{I}_{\{\sum_{i=1}^I y_i>0\}},$$

where $y_j = h(a_j)$. The latter function can be optimized directly on y_j.

This class of games has been studied by Tullock in the 1980s under the name of rent-seeking game, see Tullock [494]. The game $\mathcal{G}_I$ is given by the collection $(\mathcal{I}, \mathcal{A}, (r_j)_{j\in\mathcal{I}})$, where $\mathcal{I} = \{1, 2, \ldots, I\}$, $I \geq 2$. Let

$$m_I := \left(\frac{1}{I}\sum_{i=1}^{I} a_i^\alpha\right)^{\frac{1}{\alpha}}$$

be a weighted α-norm of the actions (this is a particular aggregation procedure). The aggregative term m_I will play the role of the mean-field term in the resource sharing game. The payoff function can be rewritten as

$$r_j(a) := \bar{r}_I(a_j, \tilde{m}_{I,j}) = \frac{c_I}{I}\left(\frac{a_j}{m_I}\right)^\alpha - p_I a_j,$$

where the ratio $\frac{c_I}{I}$ is the theoretical capacity per user in the cloud and

$$\tilde{m}_{I,j}^\alpha = \frac{I}{I-1}\left(m_I^\alpha - \frac{a_j^\alpha}{I}\right)$$

The payoff function of a generic decision-maker j depends only on own-action a_j and the mean-field term m_I. Note that $\tilde{m}_{I,j}^\alpha$ does not depend on a_j. Note that I does not substantially affect the payoff function. However, the number of active users (those who have a nonzero demand) will play an important role in the scaling of the large-scale system.

Resource-Sharing Mean-Field Game with an Infinite Number of Clients
In the current cloud network, the number of clients for clusters grows without bound, and it is becoming increasingly important to dynamically share resources between these large population clients. When the number of active clients is infinite, we analyze the asymptotic game, denoted by $\mathcal{G}_\infty = (\mathcal{A}, \bar{r})$ where $\bar{r} : \mathcal{A}^2 \longrightarrow \mathbb{R}$

$$\bar{r}(a_j, m) = \left\{\frac{c a_j^\alpha}{m^\alpha} - p.a_j\right\} \mathbb{I}_{\{m>0\}}$$

The numbers $c > 0$ and $p > 0$ are scaled asymptotic versions of $\frac{c_I}{I}$ and p_I. The term m can be interpreted as the limiting value of m_I as I goes to infinity. We write $\bar{r}(a_j, m) = p\left(\frac{c}{p}\frac{a_j^\alpha}{m^\alpha} - a_j\right)$. As we will see shortly, one of the important terms in this setting is the ratio $\frac{c}{p}$, which can be interpreted as the inverse of the cost (price) per capacity.

6.2 Finite Number of Active Cloud Clients

In the finite game $\mathcal{G}_I$, we will examine three relevant solution concepts: (Cournot–Nash) equilibrium, aggregate-taking strategies called mean-field response in the finite regime (F-MFR), and evolutionarily stable strategy in the finite regime (F-ESS). We will provide results in terms of optimal reaction to aggregate, consistency, and stability. Similarities and differences between these notions will be discussed.

6.2.1 *Nash Equilibrium*

The best response problem (BR_j) of decision-maker j is

$$(BR_j) : \text{Given } a_{-j} := (a_1, \ldots, a_{j-1}, a_{j+1}, \ldots, a_I), \text{ find } a_j \in \arg\max_{a'_j} r_j(a'_j, a_{-j})$$

The payoff is differentiable outside the origin, and the first derivative with the respect to a_j yields

$$\partial_{a_j} := \frac{c_I}{I}\left[\frac{\alpha a_j^{\alpha-1} m_I^{\alpha} - a_j^{\alpha}\left(\frac{\alpha}{I} a_j^{\alpha-1}\right)}{m_I^{2\alpha}}\right] - p_I$$

$$= \frac{c_I \alpha a_j^{\alpha-1}}{I}\left[\frac{m_I^{\alpha} - \frac{a_j^{\alpha}}{I}}{m_I^{2\alpha}}\right] - p_I,$$

and using the fact that $m_I^{\alpha} - \frac{a_j^{\alpha}}{I} = \sum_{i \neq j} \frac{a_i^{\alpha}}{I}$ is independent of a_j, we obtain the second derivative as

$$\frac{c_I \alpha}{I}\left(m_I^{\alpha} - \frac{a_j^{\alpha}}{I}\right)\left[\frac{(\alpha-1)a_j^{\alpha-2} m_I^{2\alpha} - a_j^{\alpha-1}\frac{2\alpha}{I} a_j^{\alpha-1} m_I^{\alpha}}{m_I^{4\alpha}}\right]$$

which is rewritten as

$$\frac{c_I \alpha}{I}\left(m_I^{\alpha} - \frac{a_j^{\alpha}}{I}\right)\frac{a_j^{\alpha-2}}{m_I^{3\alpha}}\left[(\alpha-1)m_I^{\alpha} - \frac{2\alpha}{I} a_j^{\alpha}\right]$$

An interior solution to the problem (BR_j) is obtained by setting the first-order condition (FOC) to zero and requiring the second-order derivative to be negative. Next, we define a fixed-point of the correspondence $(BR_j)_{j \in \mathcal{I}}$, i.e., a Nash equilibrium.

Definition 6.1 $a = (a_1, \ldots, a_I)$ is a Nash equilibrium profile if no decision-maker has an incentive to deviate unilaterally, i.e, $\forall j \in \mathcal{I},\ a_j \in BR_j(a_{-j})$.

Definition 6.2 A game is symmetric if the action spaces are identical $\mathcal{A}_1 = \ldots = \mathcal{A}_I$ and the payoff functions are invariant under the permutation of clients' actions and indices:

$$r_{\pi(j)}(a_1, \ldots, a_j, \ldots, a_I) = r_j(a_{\pi(1)}, \ldots, a_{\pi(j)}, \ldots, a_{\pi(I)})$$

for any permutation π of $\{1, 2, \ldots, I\}$.

Note that this definition of a symmetric game does not imply that the payoff functions are necessarily symmetric (or indistinguishable):

$$r_j(a_1, \ldots, a_j, \ldots, a_I) = r_{\pi(j)}(a_{\pi(1)}, \ldots, a_{\pi(j)}, \ldots, a_{\pi(I)})$$

for any permutation π of $\{1, 2, \ldots, I\}$. See also symmetric aggregative game in Dubey et al. [143].

The next proposition (whose proof follows from direct computation and hence is not included) provides a closed-form expression for the unique Nash equilibrium for some range of values of the parameter α. It also provides the optimal price p_I^* such that no resource is wasted at equilibrium.

Proposition 6.3 *For the resource sharing game introduced above*

(a) The game is symmetric. All the clients have symmetric strategies at equilibrium whenever the Nash equilibrium exists.

(b) For $0 \leq \alpha \leq 1$, the payoff r_j is concave (outside the origin) with respect to own-action a_j. The best response $BR_j(a_{-j})$ is strictly positive and is given by the root of

$$x^{(\alpha-1)/2}\left(\frac{\alpha c_I}{I p_I} G\right)^{1/2} - \frac{x^\alpha}{I} - G = 0, \quad G := \frac{1}{I}\sum_{i \neq j} a_i^\alpha,$$

and there is a unique equilibrium (hence a symmetric one) obtained from

$$\left(x^{\alpha-1}\frac{\alpha c_I}{I p_I}\frac{I-1}{I}x^\alpha\right)^{\frac{1}{2}} - \frac{x^\alpha}{I} - \frac{I-1}{I}x^\alpha = 0,$$

which is

$$a_{NE}^* = \alpha\frac{(I-1)c_I}{I^2 p_I}.$$

It follows that the total demand $I a_{NE}^*$ at equilibrium is less than $\frac{c_I}{p_I}$, which means that some resources are wasted. The equilibrium payoff is

$$r_j = a_j p_I\left[\frac{G + \frac{a_j^\alpha}{I}}{\alpha G} - 1\right],$$

which is positive for $\alpha \leq 1$.

(c) *For $\alpha > 1$, the activity (participation) of decision-maker j depends mainly on the aggregate of the others. $a_j^* > 0$ only if $G \leq G_*$, and the number of active clients is less than $\frac{\alpha}{\alpha-1}$. If $I > \frac{\alpha}{\alpha-1}$, then $BR_j = 0$.*

(d) *With a cost on participation, the payoff at equilibrium (whenever it exists) is at least* 0.

(f) *By choosing the price $p_I^* = \alpha\frac{(I-1)}{I} < \alpha$, one gets that the total demand at equilibrium is exactly the available capacity of the cloud. Thus, pricing design makes efficient cloud resource sharing. Interestingly, as I goes to infinity, the optimal pricing converges to α.*

(g) *The game is a submodular game with respect to own action and the mean field, i.e., it satisfies a decreasing-difference property: If $a_j > a'_j$, then the difference $\bar{r}_I(a_j, \tilde{m}_{I,j}) - \bar{r}_I(a'_j, \tilde{m}_{I,j})$ is decreasing with m_I.*

Definition 6.4 We say that the cloud renting game is efficient if no resource is wasted, i.e., the equilibrium demand is exactly c_I. Hence, the efficiency ratio is $\frac{Ia^*_{NE}}{c_I}$.

As we can see from (b) of Proposition 6.3, the ratio efficient goes to 1 by setting the price to p_I^*. This type of efficiency loss due to selfishness has been widely seen in the literature of mechanism design and auction theory. We refer the reader to [343] for more recent applications.

Note that the optimal pricing p_I^* depends on the number of active clients in the cloud and the return index α. When the number of active clients varies (for example, due to new entries or exits in the cloud), a new price needs to be set up, which is not convenient. However, for a large cloud, the price can be chosen to be $p = \alpha$ (independently of the number of active clients), and the equilibrium total demand would then be $\frac{(I-1)c_I}{I}$. Hence, the relative loss is $\frac{c_I - \frac{(I-1)c_I}{I}}{c_I} = \frac{1}{I} \longrightarrow 0$. We conclude that $p = \alpha$ is an $\frac{1}{I}$−optimal pricing.

6.2.2 Finite Myopic Mean-Field Response

Since the renting game $\mathcal{G}_I$ has an aggregative structure, it is interesting to evaluate the performance of aggregate-taking cloud users. Each user sees the aggregate m_I as a signal (by ignoring its own effect on the signal). A response to the aggregate in such a game is a mean-field response. Since the reaction ignores the influence of own-action a_j in m_I, such a reaction is called myopic. This leads to an equilibrium concept, which we call finite myopic mean-field response (F-MFR). It corresponds to a mean-field-taking strategy.

Definition 6.5 A pure finite myopic mean-field response is a strategy a^*, such that for every j

$$a_j^* \in \arg\max_{a_j \in \mathcal{A}} \left[\frac{c_I}{I} \left(\frac{a_j}{m_I^*} \right)^\alpha - p_I a_j \right],$$

where $m_I^* = \left(\frac{1}{I} \sum_{i=1}^{I} (a_i^*)^\alpha \right)^{\frac{1}{\alpha}}$.

This is a fixed-point equation since m_I^* contains the action a_j^*.

Proposition 6.6 *For $\alpha \leq 1$, the game $\mathcal{G}_I$ has a F-MFR given by*

$$a_{f-mfr}^* = \frac{\alpha c_I}{I p_I}.$$

For $\alpha > 1$, the game has no F-MFR due to inconsistency.

Note that the F-MFR a_{f-mfr}^* is not a Nash equilibrium (0-Nash equilibrium) of $\mathcal{G}_I$. As we will see below, in the asymptotic regime, a_{f-mfr}^* converges to a mean-field equilibrium. This also indicates that the mean-field equilibrium of the stylized model in the infinite regime is not a Nash equilibrium of the game with a finite number of decision-makers.

Proof Simple differentiation of the payoff function (first and second order) yields to $\frac{\alpha c_I}{I p_I} a^{\alpha-1} \frac{1}{m_I^\alpha} - 1 = 0$, and the second derivative is negative for $\alpha \leq 1$. If $\alpha > 1$, the mean-field-taking strategy leads to unboundedness or to nonparticipation. ■

6.2.3 Finite Evolutionarily Stable Strategy

The notion of evolutionarily stable strategies (ESSs) has been studied in the context of evolutionary games. The original definition was for an infinite population in a pairwise interaction model. However, the idea can be used for a finite population as well. ESS in a finite population (F-ESS) is slightly different than the ESS notion for an infinite population. The reason is that for finite populations, each generic decision-maker has an influence on the mean field, which may result in a non-negligible effect on the relative payoff. The simplest (and weakest) notion is to consider one deviant. The pure strategy $a_i^* = x^*$ is an F-ESS in the mean-field game with a finite number of decision-makers if

$$c_I \frac{(a_i^*)^\alpha}{a_k^\alpha + \sum_{j \notin \{k\}} (a_j^*)^\alpha} - p_I a_i^* \geq c_I \frac{(a_i)^\alpha}{a_i^\alpha + \sum_{j \notin \{i\}} (a_j^*)^\alpha} - p_I a_i \tag{6.1}$$

for all $a_i \in \mathcal{A}$ and for all $k \neq i$.

We note at this point that this definition of F-ESS does not necessarily imply (Nash) equilibrium. As we will see, it will be a mean-field Nash equilibrium in the infinite regime because the effect of $\frac{1}{I-1} a_k^\alpha$ tends to 0.

Following [383], F-ESS is a strategy a^* that maximizes the relative payoff

$$a^* \in \arg\max_{a_j} r_j(a_j, a^*, \ldots, a^*) - r_j(a^*, a_j, a^*, \ldots, a^*).$$

Proposition 6.7 *For $\alpha \leq 1$, the game $\mathcal{G}_I$ has a F-ESS given by $a^*_{f-ess} = \alpha \frac{c_I}{Ip_I}$. Moreover, the F-ESS agrees with the F-MFR. However, for $\alpha > 1$, the two notions do not agree.*

Proof The relative payoff is

$$c_I \frac{a_j^\alpha - (a^*)^\alpha}{a_j^\alpha + (I-1)(a^*)^\alpha} - p_I(a_j - a^*).$$

By differentiating with the respect to a_j, a first-order condition becomes $a^*_{f-ess} = \alpha \frac{c_I}{Ip_I}$. The second-order condition holds for $\alpha \leq 1$. ■

6.2.4 Mean-Field Response Learning Algorithm

As we can see from Proposition 6.3, it is difficult to compute explicitly the best response to the aggregate $m_{I,t-1}$. It requires finding the inverse of the function $\phi : x \longmapsto x^{\frac{1-\alpha}{2}} \left(\frac{x^\alpha}{I} + G\right)$. Hence, a best response learning algorithm cannot be implemented if the best response correspondences are unknown. For the particular case where $\alpha = 1$, one can compute explicitly the best response, and it is given by $br = \left[(\frac{\alpha c_I}{Ip_I} G)^{1/2} - G\right]_+$.

The mean-field response learning algorithm is as follows. It starts at time 0. Each decision-maker j randomly chooses an action $a_{j,0}$. A public signal $m_{I,0}$ is created from these actions and is observed by all decision-makers. At time $t \geq 1$, each decision-maker j responds to the public signal $m_{I,t-1}$ by using the following action:

$$\begin{aligned} a_{j,t} &:= br(m_{I,t-1}) \qquad (6.2)\\ &= \left[\left(\frac{\alpha c_I}{Ip_I}\left(m_{I,t-1} - \frac{a_{j,t-1}}{I}\right)\right)^{1/2} - \left(m_{I,t-1} - \frac{a_{j,t-1}}{I}\right)\right]_+. \end{aligned}$$

Define the convergence time within a certain $\eta-$neighborhood of m^* as the first time the trajectory of the algorithm enters a range within a distance less than η from m^* :

$$T_\eta = \inf\{t > 0 \mid \parallel a_t - a^* \parallel_\infty \leq \eta\}$$

By designing the parameters $c_I, p_I,$ and α, [213, 217] observed the following properties:

- The modified algorithm

$$a_{j,t} = \lambda_{j,t} a_{j,t-1} + (1 - \lambda_{j,t}) br(m_{I,t-1}),$$

where $\lambda_{j,t} > 0$ is a learning rate may converge to the Nash equilibrium for $\lambda_t = \lambda$ sufficiently small.
- For any η, λ sufficiently small, the convergence time T_η is finite, and it is in the order $\log\left(\frac{1}{\eta}\right)$.

We will see below how to accelerate the algorithm whenever convergence occurs. Next we

6.3 Infinite Number of Active Clients

In this section, we study the asymptotic game, which corresponds to the case of an infinite number of decision-makers. In the infinite game, the notion of F-ESS becomes the classical ESS in the sense of Maynard Smith and Price (1973), and the notion of F-MFR becomes a mean-field (Nash) equilibrium.

Definition 6.8 m^* is a pure mean-field (Nash) equilibrium if (i) $m^* \in \mathscr{A}$ and (ii) m^* is consistent and is a best response to itself: $m^* \in \arg\max_{a_j} \bar{r}(a_j, m^*)$

Following Proposition 6.3, we get

Proposition 6.9

(a) The resource sharing game $\mathcal{G}_\infty$ is a symmetric game.
(b) The game $\mathcal{G}_\infty$ is a submodular mean-field game, i.e., $\partial^2_{a_j m}\bar{r} \leq 0$ (at the interior). This means that the incentive to decrease the demand decreases with the level of the mean field.
(c) For $\alpha < 1, m > 0$, the best response to mean field is $br(m) = \left(\frac{\alpha c}{p m^\alpha}\right)^{\frac{1}{1-\alpha}}$, and for $\alpha > 1$, the best response is 0.
(d) For $\alpha \leq 1$, there is a unique mean-field equilibrium given by $m^* = \alpha\frac{c}{p}$, and the optimal pricing is $p^* = \alpha$.
(e) The Nash equilibrium a^*_{NE} can be scaled to converge to the mean-field equilibrium m^* as I grows.

We next discuss the stability of the mean-field equilibrium. We adopt the notion of evolutionarily stable strategy (ESS) introduced by Maynard Smith and Price [403]. The original definition of ESS was for pairwise interaction and bilinear payoffs, whereas we have a nonlinear payoff function.

Definition 6.10 (ESS) Let $\alpha \in [0, 1]$. m^* is an evolutionarily stable strategy (ESS) if it cannot be invaded by a small fraction of deviants, i.e., for any $m \neq m^*$, there exists $1 > \bar{\epsilon}_m > 0$ (which may depend on m), such that for any $\epsilon \in (0, \bar{\epsilon}_m)$, one has

$$\bar{r}(m^*, \varepsilon m + (1-\varepsilon)m^*) > \bar{r}(m, \varepsilon m + (1-\varepsilon)m^*)$$

The parameter $\bar{\varepsilon}_m$ is called invasion barrier.

Next, we show that the mean-field equilibrium is an ESS for $\alpha \in (0, 1)$.

Proposition 6.11 *Let $\alpha \in (0, 1)$. Then, the mean-field equilibrium $m^* = \alpha\frac{c}{p}$ is an evolutionarily stable strategy.*

Proof Let $\alpha \in (0, 1)$, $m \neq m^*$, and $m_\varepsilon := \varepsilon m + (1-\varepsilon)m^* = m^* + \varepsilon(m - m^*)$ for $\varepsilon \neq 1$. It follows that $m^* - m = \frac{m^*-m_\varepsilon}{\varepsilon}$. Let A be the quantity $r(m^*, m_\varepsilon) - r(m, m_\varepsilon)$. Then

$$A = p\left[\frac{c}{p}\left(\frac{(m^*)^\alpha - m^\alpha}{m_\varepsilon^\alpha}\right) - (m^* - m)\right] \tag{6.3}$$

$$= \frac{p}{\alpha m_\varepsilon^\alpha}\left[m^*((m^*)^\alpha - m^\alpha) - \alpha(m^* - m)m_\varepsilon^\alpha\right], \tag{6.4}$$

where we have used the fact that $\frac{c}{p} = \frac{(m^*)}{\alpha}$. The function $x \longmapsto x^\alpha$ is concave in $\mathscr{A}$. Therefore, $(m^*)^\alpha - m^\alpha > \alpha(m^*)^{\alpha-1}(m^* - m)$. By multiplying by m^* both sides, one gets $m^*((m^*)^\alpha - m^\alpha) > \alpha(m^*)^\alpha(m^* - m)$. Thus

$$A > \frac{p}{\alpha m_\varepsilon^\alpha}\left[\alpha(m^*)^\alpha(m^* - m) - \alpha(m^* - m)m_\varepsilon^\alpha\right] \tag{6.5}$$

$$= \frac{p}{m_\varepsilon^\alpha}(m^* - m)\left[(m^*)^\alpha - m_\varepsilon^\alpha\right] \tag{6.6}$$

$$= \frac{p}{m_\varepsilon^\alpha}\frac{m^* - m_\varepsilon}{\varepsilon}\left[(m^*)^\alpha - m_\varepsilon^\alpha\right] \tag{6.7}$$

$$= \frac{p}{\varepsilon m_\varepsilon^\alpha}(m^* - m_\varepsilon)\left[(m^*)^\alpha - m_\varepsilon^\alpha\right] > 0 \tag{6.8}$$

where the last inequality follows because x^α is nondecreasing for $\alpha > 0$ and $m^* \neq m_\varepsilon$. We conclude that $\bar{r}(m^*, \varepsilon m + (1-\varepsilon)m^*) > \bar{r}(m, \varepsilon m + (1-\varepsilon)m^*)$ for any $\varepsilon \in (0, 1)$ and $m \neq m^*$. In particular, the result is true for the case of interests, i.e., for $\varepsilon \in (0, \bar{\varepsilon}_m)$. This completes the proof. ■

Feedback-Free and Derivative-Free Mean-Field Learning

The mean-field response is $a_{j,t+1} \in \arg\max_{a'_j} \bar{r}(a'_j, m_t) = br(m_t)$. In general, this is a multivalued map (correspondence). However, here, the correspondence is a single-valued map, i.e., a classical function, $br = \{f\}$. Thus, Banach–Picard-based mean-field response learning is given by

$$m_{t+1} = f(m_t), \tag{6.9}$$

Note that in Eq. (6.9), only the starting point m_0 is required if the decision-maker knows the structure of f. This means that it is not needed to feed back the mean of the mean field at each step. This is a feedback-free mean-field learning which can be reduced to finding a fixed point of the function f in a feasible set. The mean-field equilibrium can be obtained by finding a root of $g(x) = f(x) - x$.

Definition 6.12 Assume that m_t converges to m^* and let $\eta_t = |m_t - m^*|$. If two positive constants $c_1, o > 0$ exist and $\overline{\lim}_t \frac{\eta_{t+1}}{\eta_t^o} = c_1$, then the sequence $\{m_t\}_t$ is said to converge to m_* with order of convergence o. The number c_1 is called the asymptotic error constant. The cases $o \in \{1, 2, 3, 4\}$ are given special consideration:

(i) If $o = 1$, the convergence of $\{m_t\}_t$ is linear.
(ii) If $o = 2$, the convergence of $\{m_t\}_t$ is quadratic.
(iii) If $o = 3$, the convergence of $\{m_t\}_t$ is cubic.
(iv) If $o = 4$, the convergence of $\{m_t\}_t$ is quartic.

If only the first derivative is available, a quadratic order scheme can be obtained from Newton's method. Now, we have the following two questions:

Is it possible to construct a derivative-free mean-field learning with a quadratic convergence rate?

Without using the derivative, Steffensen has proposed a second-order method with only two evaluations per time slot.

$$(2*) \begin{cases} m_{t+1} = m_t - \frac{g^2(m_t)}{g(m_t+g(m_t))-g(m_t)}, \\ m_0 \end{cases}$$

Is it possible to construct a derivative-free mean-field learning with a quartic convergence rate?

Here, we aim to investigate whether a higher-order learning scheme can be constructed without involving the derivative and with minimal memory (as few previously evaluated values as possible) and a minimal number of evaluations of the function. To construct such a learning algorithm, we start with the secant method given by

$$(1.6*) \begin{cases} m_{t+1} = m_t - \frac{g(m_t)(m_t-m_{t-1})}{g(m_t)-g(m_{t-1})}, \\ m_0 \end{cases}$$

which has the form $m_t = F(m_{t-1}, m_{t-2})$, i.e., a two-step memory scheme. The secant method is known to have a convergence order of 1.6 (hence the name of the system (1.6*)).

We need to remove the memory term $g(m_{t-1})$, i.e., we want a memoryless feedback-free mean-field learning scheme. We introduce the following notation: $\Delta g[a, a'] := \frac{g(a)-g(a')}{a-a'}$, $a \neq a'$, and $b_t = m_t + g(m_t)$. The function g will be evaluated at this point: m_t, $\bar{y}_t$ and b_t, where $\bar{y}_t$ is the analogue of the secant point. Let $\nu(s) = \alpha_0 + \alpha_1(s - m_t) + \alpha_2(s - m_t)^2$ be a polynomial of degree 2, with all the coefficients to be determined. We will determine the unknown coefficients using the following requirement: $\nu(m_t) = g(m_t)$, $\nu(\bar{y}_t) = g(\bar{y}_t)$, and $\nu(b_t) = g(b_t)$. From these relations, we deduce the coefficients:

- $\alpha_0 = g(m_t)$
- $\alpha_1 = \frac{\bar{y}_t \Delta g[b_t, m_t] + m_t \Delta g[\bar{y}_t, m_t] - m_t \Delta g[b_t, m_t] - b_t \Delta g[\bar{y}_t, m_t]}{\bar{y}_t - b_t}$
- $\alpha_2 = \frac{\Delta g[b_t, m_t] - \Delta g[\bar{y}_t, m_t]}{b_t - \bar{y}_t}$

Note that the derivative of ν at the point m_t is exactly the coefficient α_1. We deduce that a plausible derivative-free term of g is $\frac{\bar{y}_t \Delta g[b_t, m_t] + m_t \Delta g[\bar{y}_t, m_t] - m_t \Delta g[b_t, m_t] - b_t \Delta g[\bar{y}_t, m_t]}{\bar{y}_t - b_t}$.

Consider the following scheme:

$$(4*) \begin{cases} \bar{y}_t = m_t - \frac{g(m_t)}{\Delta g[b_t, m_t]}, \\ m_{t+1} = \bar{y}_t - (1 + w_t) \frac{(\bar{y}_t - b_t) g(\bar{y}_t)}{(\bar{y}_t - m_t) \Delta g[b_t, m_t] - (b_t - m_t) \Delta g[\bar{y}_t, m_t])} \\ w_t = 2 \frac{g(\bar{y}_t)}{g(b_t)} \end{cases}$$

Properties 6.13 The proposed speedup scheme (4*)

- Is a derivative-free scheme
- Has three evaluations of the function g per time slot
- Is memoryless
- Is faster than the secant speedup method (superlinear order)
- Is faster than Steffensen's speedup learning method (quadratic order)

Proposition 6.14 *Assume that g has a simple zero in $\mathcal{A}$. Locally, scheme (4*) has convergence order four.*

Proof If g is a sufficiently many-times differentiable function with a simple zero and m_0 is a guess close enough to m^*, then using the Taylor expansion, the error $\varepsilon_{t+1} = m_{t+1} - m^*$ (without absolute value) can be expressed as

$$\varepsilon_{t+1} = c_1 \varepsilon_t^4 + O\left(\varepsilon_t^5\right)$$

where $c_1 = \frac{(1+q_1) q_2 [(5+3q_1) q_2^2 - q_1 (1+q_1) q_3]}{q_1^3}$, $q_o = \frac{g^{(o)}(m^*)}{o!}$. This means that, locally, scheme (4*) has a fourth-order convergence rate. Now, we prove the Taylor expansion:

$$\varepsilon_{t+1} = \varepsilon_t - \frac{g(m_t)}{A_1} - (1 + w_t) \frac{g(\bar{y}_t)}{A_2} \tag{6.10}$$

$$A_1 = \Delta g[b_t, m_t] \tag{6.11}$$

$$A_2 = \frac{(\bar{y}_t - m_t)\Delta g[b_t, m_t] - (b_t - m_t)\Delta g[\bar{y}_t, m_t])}{\bar{y}_t - b_t} \tag{6.12}$$

We use the relation $g(m_t) = \sum_{l=1}^{4} q_l \varepsilon_t^l + O(\varepsilon_t^5)$. Hence

$$\begin{aligned} g(b_t) &= g(m_t + g(m_t)) \\ &= 0 + (\varepsilon_t + g(m_t))c_1 + c_2(\varepsilon_t + g(m_t))^2 \\ &\quad + c_3(\varepsilon_t + g(m_t))^3 + c_4(\varepsilon_t + g(m_t))^4 + O(.), \end{aligned} \tag{6.13}$$

completing the proof. ■

Definition 6.15 (Convergence Time) The convergence time within an $\eta-$neighborhood of m^* is the first time the trajectory enters into this neighborhood, i.e.

$$T_{\eta,\varepsilon_0} := \inf\{t > 0 \mid d(m_t, m^*) \leq \eta,\ d(m_0, m^*) = \varepsilon_0\}$$

Proposition 6.16 *The convergence time* $T_{\eta,\varepsilon_0} = 1 + \lfloor \max(0, T) \rfloor$ *where*

$$T = \frac{1}{\ln(4)} \ln \left[\frac{\ln\left(\frac{1}{\eta |c_2|^{\frac{1}{3}}}\right)}{\ln\left(\frac{1}{\varepsilon_0 |c_2|^{\frac{1}{3}}}\right)} \right],$$

which is in order of $O\left(\log(\log(\frac{1}{\eta}))\right)$.

Proof of Proposition 6.16 Let $t \geq 2$. We reiterate the recursive equation

$$\varepsilon_t \leq c_2 \varepsilon_{t-1}^4 \leq c_2 \left(c_2 \varepsilon_{t-2}^4\right)^4 = c_2^{1+4} \varepsilon_{t-2}^{16} \tag{6.14}$$

$$\leq c_2^{1+4} \left(c_2 \varepsilon_{t-3}^4\right)^{16} = c_2^{1+4+4^2} \varepsilon_{t-3}^{4^3} \leq c_2^{1+4+4^2+\ldots+4^{t-1}} \varepsilon_0^{4^t} \tag{6.15}$$

Recall that for $q \neq 1,\ 1 + q + \ldots + q^{t-1} = \frac{q^t - 1}{q - 1}$. Thus

$$\varepsilon_t \leq c_2^{\frac{4^t-1}{3}} \varepsilon_0^{4^t}.$$

This means that for the convergence time to be within an $\eta-$neighborhood of the mean-field equilibrium, it suffices to choose t satisfying $c_2^{\frac{4^t-1}{3}} \varepsilon_0^{4^t} \leq \eta$. Thus

$$\left(\varepsilon_0 c_2^{\frac{1}{3}}\right)^{4^t} \leq \eta c_2^{\frac{1}{3}}.$$

By taking the logarithm twice (provided that the term is higher than e), one arrives at

$$T = \frac{1}{\ln(4)} \ln\left[\frac{\ln\left(\frac{1}{\eta c_2^{\frac{1}{3}}}\right)}{\ln\left(\frac{1}{\varepsilon_0 c_2^{\frac{1}{3}}}\right)}\right],$$

which is the desired result. ■

6.4 How to Satisfy Client Demand?

We now examine learning satisfactory solutions in cloud resource sharing games with *continuous action spaces*. Our motivations for satisfactory solution seeking are the following: First, we observe that in dynamic interactive systems, most users constantly make decisions which are simply "good enough" rather than the best response or optimal. Simon [395] has adopted the term "satisficing" for this type of decision. As mentioned by Simon himself on page 129 of his paper, *"Evidently, organisms adapt well enough to satisfice"; they do not, in general, "optimize."* Therefore, a satisfactory solution offers an alternative approach and closely models the way cloud users make decisions [349, 407, 408].

Here, a satisfactory strategy is a decision-making strategy that attempts to meet an acceptability threshold. This is contrasted with optimal decision-making or best response strategy, an approach that specifically attempts to find the best option available given the choices of the other users. Following that idea, we define a satisfactory solution as a situation where every user is satisfied, i.e., her payoff is above her satisfaction level.

For each user j, we introduce a satisfaction level s_j^*. The collection

$$\mathcal{G} := (\mathcal{I}, (\mathcal{A}, s_j, s_j^*)_{j\in\mathcal{I}})$$

constitutes a one-shot game in satisfaction form.

Definition 6.17 The demand profile $(a_j)_{j\in\mathcal{I}}$ is a pure satisfactory solution of the cloud resource sharing game if all the users are satisfied, that is

$$s_j(a) = \frac{c_I a_j^\alpha}{\sum_{i=1}^I a_i^\alpha} \geq s_j^*, \ \forall j \in \mathcal{I}.$$

6.4.1 Necessary and Sufficient Condition for Existence

Before going for a pure satisfactory solution seeking, we first need to explore whether the problem is well posed, i.e., the existence of a pure satisfactory solution. The feasibility of the satisfaction problem for a given $s_j^*,\ j \in \mathcal{I}$, is the nonemptiness of the set $\{a = (a_1, \ldots, a_I) \in \mathcal{A}^I \mid s_j(a) \geq s_j^*,\ \forall j\}$. This means that there exists a vector $(\varepsilon_1, \ldots, \varepsilon_I)$, $\varepsilon_j \geq 0$, such that there is a demand profile a that satisfies $\forall j,\ s_j(a) = s_j^* + \varepsilon_j$. Thus, a necessary condition for existence of a satisfactory solution is that such a vector $s^* + \varepsilon$ belongs to the set $s\left(\mathcal{A}^I\right)$, i.e., the range of the vectorial function $s(a) := (s_1(a), \ldots, s_I(a))$. It is not difficult to see that this is also a sufficient condition.

Proposition 6.18 *Suppose that the satisfaction levels are such that* $\sum_{i=1}^I s_i^* < c_I$. *Then, the game has a satisfactory solution and the allocation given by*

$$a_i = (s_i^*)^{\frac{1}{\alpha}},$$

satisfies

$$s_i(a) = \frac{c_I s_i^*}{\sum_{j=1}^I s_j^*} > s_i^*$$

The condition $\sum_{i=1}^I s_i^* < c_I$ translates the fact that the total demand should not exceed the available capacity.

Proof of Proposition 6.16 By assumption, the satisfaction levels s_i^* are such that the condition $\sum_{i=1}^I s_i^* < c_I$ is met. Then we compute $s_i(a)$:

$$s_j(a) = \frac{c_I a_i^\alpha}{\sum_{j=1}^I a_j^\alpha} = \frac{c_I s_i^*}{\sum_{j=1}^I s_j^*} > s_i^*$$

Thus, every user is satisfied. ■

6.4.2 Fully Distributed Algorithm for Satisfactory Solution

In this subsection, we restrict the set $\mathcal{A}$ to the interval $[0, c_I]$. Our goal is to better understand the gaps between the minimum satisfaction requirements of cloud-based applications and the variable performance of dynamic cloud networks and to develop novel solutions to overcome these gaps. To do so, we propose a fully distributed learning algorithm for satisfactory solutions. The only information available to each user is the numerically realized value of its satisfaction $s_{j,t}$ and its own-target s_j^*. The basic Banach–Picard fixed-point iteration is given by

$$a_{j,t+1} = \text{proj}_{\mathcal{A}}\left[a_{j,t}\frac{s_j^*}{s_{j,t}}\right], \tag{6.16}$$

where $\text{proj}_{\mathcal{A}}(x) = \min(c_I, \max(0, x))$. $\text{proj}_{\mathcal{A}}$ is the projection operator over the nonempty and convex set $\mathcal{A}$.

The algorithm is *fully distributed* in the sense that a user need not observe the actions of the others in order to update her strategy iteratively. Denote by f_j the righthand side of (6.16). For the function f_j to be well defined, we need the ratio $\frac{a_j}{s_j}s_j^* = \frac{s_j^*}{c_I}a_j^{1-\alpha}\left(\sum_{i=1}^{I} a_i^{\alpha}\right)$ to have a limit as $a_j \longrightarrow 0$ provided that $\alpha \leq 1$. The set of fixed points of f, to be denoted by $fix(f)$, is contained in the set of satisfactory solutions. It should be clear that if the algorithm converges to some interior point a^*, then a^* is a satisfactory solution. To prove this statement, consider a converging sequence to a^*. Putting together the continuity of the projection map and the continuity of the payoff function, one gets the continuity of $\text{proj}_{\mathcal{A}}\left[a_j\frac{s_j^*}{s_j}\right]$. Taking the limit as t goes to infinity yields $a_j = \text{proj}_{\mathcal{A}}\left[a_j\frac{s_j^*}{s_j}\right]$, i.e., the payoff of user j is $s_j = s_j(a^*) = s_j^*$, which means that every user j is satisfied in the interior steady state.

It is important to investigate the outcomes of the above algorithm since a satisfactory solution can be seen as a global optimum [106] of the game with payoff function $\mathbb{1}_{\{s_j \geq s_j^*\}}$, where $\mathbb{1}_{\{.\}}$ denotes the indicator function. A convergence of the above algorithm exhibits a procedure for finding a global optimum (and hence, the solution would be Pareto optimal).

In the limiting regime with satisfactory level s^*, the algorithm simplifies to $m_{t+1} = \frac{s^*}{c}m_t + \frac{s^*}{c}\varepsilon m_t^{1-\alpha}$, where $\varepsilon > 0$ is sufficiently small, which converges to an interior fixed point.

6.5 Learning Algorithms

For a numerical study, we fix $\alpha = 1$ but allow for different prices in order to observe the users who will not participate in the cloud business due to high cost. To solve the equilibrium problem, we develop a distributed iterative algorithm, which is based on an update method. It is useful to have an iterative algorithm, as the reward function in our case is nonlinear, and we are dealing with a distributed scenario. Each node only needs to do local computations regarding the demand $a_{j,t+1}$ at time $t+1$.

6.5.1 Banach–Picard Learning Algorithm

One of the first iterative schemes for finding the fixed point is the so-called Banach–Picard iterate. It consists of iterating the function many times starting from an arbitrary point a_0 in the domain. $a_{t+1} = f(a_t) = f(f(a_{t-1})) = \ldots = \underbrace{(fofo\ldots of)}_{(t+1)\ times}(a_0)$. Here, f represents the best response. For different initial conditions $a_{j,0}$, we can write the update equation for user j as

$$a_{t+1} = f(a_t) \tag{6.17}$$

This type of equation is called *Banach–Picard update equation*, where $f_j(a_t) = \max\left(0, \left(\sqrt{\frac{c_I}{p_j} D_{-j,t}} - D_{-j,t}\right)\right)$. We can also rewrite $f_j(a_t)$ in terms of the mean action m_t at time t and the action $a_{j,t}$ performed by node j at time t. $f_j(a_t) = \max\left(0, \left(\sqrt{\frac{Ic_I}{p_j}\left(m_t - \frac{a_{j,t}}{I}\right)} - I\left(m_t - \frac{a_{j,t}}{I}\right)\right)\right)$, where $m_t = D_t/I$. It turns out that the update equation in (6.17) does not converge to the optimal solutions. For this reason, we instead use an Ishikawa algorithm [31, 240, 241, 279, 321] given that the function is not monotone and is not a strict contraction.

6.5.2 Ishikawa Learning Algorithm

We slightly modify the update equations and introduce a small parameter $\lambda \in [0, 1]$, where we weight the update between the new and old values. This type of equation is called *Ishikawa Update Equation*, which can be written as

$$a_{t+1} = f(a_t)\lambda + a_t(1 - \lambda) \tag{6.18}$$

Note that if a^* is a fixed point of f, i.e., $f(a^*) = a^*$, then $f(a_k)\lambda + a_k(1 - \lambda) = a^*\lambda + a^*(1 - \lambda) = a^*$. Then, a^* is a fixed point of Eq. (6.18). If f is a Lipschitz, then there exists a learning rate $0 < \bar{\lambda} < 1$ sufficiently small such that for all $0 < \lambda < \bar{\lambda}$, the algorithm converges to a pure Nash equilibrium.

It should be noted that we need the value of $\bar{\lambda}$ dependent on the structure of the reward function and the number of users. In general, it is hard to compute $\bar{\lambda}$, and therefore, in such scenarios, it is better for convergence to use a vanishing λ, which is independent of $\bar{\lambda}$.

For large-scale systems with the same price, it can be useful to convey just the mean instead of the individual demand of each user. In this section, we analyze two different user behaviors. The mean demand can be constructed by summing $c_I\left(\frac{D-a_j}{D^2}\right) = p_j$ over all I and dividing by I:

$$m_s^* = \frac{(I-1)c_I}{2I\sum_{j'=1}^{I} p_{j'}} + \sqrt{\frac{(I-1)^2 c_I^2}{4I^2(\sum_{j'=1}^{I} p_{j'})^2} + \frac{c_I \varepsilon_I}{I^2 \sum_{j'=1}^{I} p_{j'}}}. \tag{6.19}$$

For large-scale systems with the same price and same starting points, Eq. (6.19) can be simplified as $m_s^* \simeq \frac{c}{p}$. Each user tries to maximize its individual reward. The reward for a large-scale system can be written as

$$\bar{r}(a_j, m) = \frac{c a_j}{\bar{m}} - a_j p \tag{6.20}$$

where $\bar{m} = m + \varepsilon$ and $\varepsilon = \lim_{I \longrightarrow +\infty} \frac{\varepsilon_I}{I}$. $\bar{r}$ is linear because the influence of a_j in m is negligible as $I \longrightarrow +\infty$, leading to $\frac{c}{p} = m + \varepsilon$. Hence, $br(.)$ represents the mean-field best response of a user with only the knowledge of $\frac{c}{p} - \varepsilon$, which is

$$br(m) = \begin{cases} M, & m < \frac{c}{p} - \varepsilon \\ 0, & m > \frac{c}{p} - \varepsilon \\ \acute{M}, & m = \frac{c}{p} - \varepsilon \end{cases}$$

where $\acute{M}$ represents any value in $(0, M)$. This means that when $m_t < \frac{c}{p} - \varepsilon$, demand should be increased to some large number $M > \frac{c}{p}$, and when $m_t > \frac{c}{p} - \varepsilon$, demand should be decreased to zero. We can write the update algorithm as

$$m_{j,t+1} = f(m_{j,t})\lambda_t + m_{j,t}(1-\lambda_t) \tag{6.21}$$

Here, we use a vanishing lambda where $\lambda_t = \frac{1}{t+1}$. The ODE for the above iterative algorithm can be written as $\dot{m}_t = f(m_t) - m_t$.

Lemma 6.19 (MF Trajectory) *The gap between the trajectory of the ODE $\dot{m}_t$ and the mean-field equilibrium m^* is $|m_t - m^*|$, which vanishes in the limit, i.e.*

$$\lim_{t \longrightarrow \infty} |m_t - m^*| = 0 \tag{6.22}$$

Proof of Proposition 6.16 For $m_0 \in (0, M)$, $V(m_t) = |m_t - m^*|^2 \geq 0$, $\dot{V}(m) = 2(m_t - m^*)(\dot{m}_t) = 2(m_t - m^*)(f(m_t) - m_t)$. From this, we can argue that $\dot{V}(m) \leq 0$ when $m_t > m^*$ and when $m_t < m^*$ and $V(m^*) = 0$, and hence, m^* is a stable equilibrium point of the ODE. ■

6.6 Discussions

We have studied resource sharing games [212, 214, 263] in cloud networking. We have examined both finite and infinite regimes. We have seen that the equilibrium is not an evolutionarily stable strategy in the finite regime. Using a suitable scaling, the finite myopic mean-field response converges to a mean-field equilibrium, which is also a strict evolutionarily stable strategy. We have provided a necessary and sufficient condition for the existence of a satisfactory solution where each user is satisfied anywhere, anytime. We have proposed an algorithm which is able to perform distributed resource sharing for a certain type of reward function. The algorithm converges to the Nash equilibrium under suitable conditions. It is possible to extend the results of this chapter to other types of reward functions. We have also developed an algorithm for large-scale systems and applied it to finite-scale systems. We discuss below some possible extensions of the game considered.

Resource Sharing in Open Cloud
Assume that a new entry/exit is allowed in the cloud network. A new client will enter (participate) only if its payoff is positive through participation. This requires $\frac{a_{new}^{\alpha}}{I+1} \geq (\alpha - 1)G_I$, and a necessary condition for existence is that the number of active users $I + 1$ should not exceed $\frac{\alpha}{\alpha-1}$.

Resource Sharing with Leadership
Another approach to resource sharing entails partitioning the set of cloud users into different categories and hierarchies. This is captured in the so-called leader-follower games, where the leaders move first and then each of the followers reacts to the aggregate demand of the leaders.

Stochastic Demand
In some situations, demand may depend on a realization of some random events or random variables; for example, user demand may rapidly change depending on news on the web. In that case, the demand will be a function of a random variable w: $a_j = d_j(w)$. One can analyze the equilibria and network stability under such randomness.

6.7 Notes

Foster et al. [168] define the ambiguous cloud as "A large-scale of distributed computing paradigm that is driven by economies of scale, in which a pool of abstracted virtualized, dynamically-scalable, managed computing power, storage, platforms, and services are delivered on demand to external customers over the Internet" [168]. The concept of cloud computing has emerged to satisfy the uneven demand of users wanting cheap computing resources ranging from storage to processing. Every aspect of a computer system can essentially be offered as a service to various users. Thanks

(continued)

to advances in virtualization and scheduling techniques, it is possible to have multiple instances of software running on the same physical system. Amazon's Elastic Compute Cloud (Amazon EC2) [107] is one of the leading providers of cloud computing. In such a situation, but in a different context, Tullock [494] proposed a game-theoretic setting with probabilistic reward in *rent-seeking*. He had noticed that the social loss associated with rent-seeking behavior exceeds a certain threshold. From a cloud networking perspective, this means that a lot of resources may be wasted if the cloud user considers an economic renting [11, 478]. Game-theoretic tools for cloud networking have been recently proposed in [220, 243]. The authors in [243] study cloud resource allocation games. The work in [506] presents a game-theoretic method of fair resource allocation for cloud computing services. They proved the existence of a Nash equilibrium under the feasibility condition of the resource allocation problem. Ma et al. [320] examines a resource allocation for cloud users using the Kelly approach. In [256], Kantere et al. study optimal service pricing for cloud cache services. In [4], Yehuda et al. study the framework of Resource as a Service (RaaS). In [268], Khan et al. study noncooperative, semicooperative, and cooperative game-based grid resource allocation. Rao et al. [368] proposed a game-theoretic approach for the provisioning and operation of the infrastructure under uniform cost models. The authors in [281] addressed the question of whether or not to utilize the cloud for processing, by identifying characteristics of potential cloud beneficiaries and advisable actions to actually gain financial benefits. They proposed a game-theoretic model of an Infrastructure-as-a- Service (IaaS) cloud market, covering the dynamics of pricing and usage. The work in [23] modeled a service provisioning problem in cloud as an action-coupling constrained game and proposed an efficient algorithm for the run time management and allocation of IaaS resources to competing SaaS (Software as a Service). The authors in [23, 268, 281, 368, 382, 491, 506] proposed a Bayesian Nash equilibrium allocation algorithm to solve the resource management problem in cloud computing under heterogeneous distribution of resources and incomplete information. Recently, there has been renewed interest in developing large-scale strategic learning, also called mean-field learning. Large data and a large number of users are frequently met in the context of cloud networking, such as in Amazon's Elastic Compute Cloud (Amazon EC2), which is a web service that provides resizable compute capacity in the cloud. The problem is tackled in [354] using block splitting for large-scale distributed learning. We refer the reader to chapter 1 of [461] for a survey on distributed strategic learning. The work of Possajennikov [363] examines mean-field-taking strategies. In [24], the authors consider a public cloud of more than a hundred thousand servers and many private clouds in addition to the traditional network scale.

Chapter 7
Auctions

An auction is a mechanism or process used to buy and sell goods, services, or assets through competitive bidding. In an auction, multiple participants, often called bidders or buyers, compete to acquire the item or service being auctioned by placing bids. Auctions can be categorized based on the information available to bidders, which refers to the level of information participants have about the item being auctioned, the bids of other participants, and the rules of the auction. In private-value auctions, each bidder has private information on their own valuation of the item. Bidders may have different valuations for the same item based on their unique needs or preferences. Sealed-bid auctions are often used in private-value settings.

In Chap. 6 of Volume I, a non-asymptotic mean-field-type game framework is presented. In this chapter, we use that framework to approximate optimal bidding strategies of auctions with heterogeneous and asymmetric bidders [427, 429, 478, 515–518].

7.1 First-Price Auctions with Asymmetric Bidders

The theory of auctions as games of incomplete information goes back at least to 1961 of the work of Vickrey. A seller has an object to sell. Let us assume she adopted a first-price auction rule, and we have asymmetric bidders with $I \geq 2$ of them for the single item. Each bidder independently submits a single bid without seeing the others' bids. If there is only one bidder with the highest bid, the object is sold to that bidder. The winner pays the amounts corresponding to his bid; that is, the price in this case is the highest bid (or first price bid). If there is more than one highest bidder, the item goes to each of these bidders with equal probability. Bidder j has a valuation v_j of the item. The random variable v_j has a C^1–cumulative distribution function F_j with support $[\underline{v}, \bar{v}]$, where $\underline{v} < \bar{v}$. A strategy of bidder j is a mapping from a valuation to a bid space: $v_j \mapsto b_j(v_j)$. The payoff of bidder j is then

T. Başar et al., *Mean-Field-Type Game Theory II*, Static & Dynamic Game Theory: Foundations & Applications, https://doi.org/10.1007/978-3-032-07031-9_7

$$(v_j - b_j)\mathbb{1}_{\{\max_{j'\neq j} b_{j'}(v_{j'}) < b_j(v_j)\}} - p_j,$$

where p_j is the submission fee.

The information structure of the auction game is as follows:

- Each bidder knows his own valuation (as well as his bid) but not the valuation of the other bidders.
- Each bidder knows the valuation cumulative distributions of the other bidders. The cumulative distribution functions are assumed to be learnable by the bidders and the auctioneer.
- The structure of the game is common knowledge.

When the functions F_j are not all identical, the bidders are said to be asymmetric.

7.1.1 Risk-Neutral Equilibria

We are interested in equilibria, equilibrium payoffs, and the revenue of the seller. The existence of risk-neutral equilibrium of first-price auction games has been widely studied in the literature (see [205, 287, 288, 378, 478]). The risk-neutral payoff of bidder j is

$$(v_j - b)\mathbb{P}\left(\max_{j'\neq j} \tilde{v}_{j'}(b) < v_j(b)\right) - p_j,$$

where the inverse bid of decision-maker j is

$$v_j = \tilde{v}_j(b_j(v_j)), \;\; \tilde{v}_j = b_j^{-1},$$

and b here is a generic variable that represents the bid value. Clearly, no bidder would bid an amount that is greater than his valuation (minus submission fee) because that would lead to a negative payoff. By fixing the bidding strategies of the others, one has attempted to compute the best response correspondence. Any increase in the bid will decrease the gain but increase the probability of winning. This can be viewed as a tradeoff between profit and probability of winning. In order to find an equilibrium under asymmetric valuation distributions, we establish the first-order optimality condition by differentiating the risk-neutral payoff. This results into I ordinary differential equations (ODEs) with $2I$ boundary conditions:

$$v_j'(b) = \frac{F_j(v_j(b))}{F_j'(v_j(b))}\left[\frac{1}{I-1}\sum_{j'=1}^{I}\frac{1}{v_{j'}(b)-b} - \frac{1}{v_j(b)-b}\right] \tag{7.1}$$

$$v_j(\underline{b}) = \underline{v}, \;\; v_j(\bar{b}) = \bar{v}, \tag{7.2}$$

The inverse of the function v is the optimal strategy b. This is clearly intractable even with a small number of bidders. Even with three bidders, it is not clear how the solutions behave with F_j.

Why is this not a standard ODE problem? We cannot apply the standard local existence and uniqueness theorem to the ODE with initial value (lowest bid) $v_j(\underline{b}) = \underline{v}$, because the right-hand-side terms $\frac{1}{v_j(b)-b}$ in the ODEs are unbounded at $\underline{v}$. Further, the equilibrium satisfies $v_j(\bar{b}) = \bar{v}$, but the term $\bar{b}$ is unknown. Due to these difficulties, explicit solutions of (7.1) and (7.2) cannot be obtained. Since obtaining explicit solutions is not possible, one may ask whether it is possible to solve the problem numerically. According to the work in [167], the numerical implementation of systems (7.1) and (7.2) also presents a challenging task. One of the well-known numerical methods involves finding, among the solutions of ODEs together with the initial conditions, the one that satisfies the highest equilibrium bid constraint. Such an approach is known as *forward-shooting method*. However, the forward-shooting method of Marshall et al. [323] does not converge to the solution due to approximation near $\underline{b}$, with the derivative $v'_j(\underline{b})$. It has been shown in [167] that for the special case of power law (i.e., $F(v) = v^\alpha$), a dynamical system approach can be used with the change of variable

$$v_j(b) = bV_j(b),\ b = e^w,$$

In the backward approach, one searches for the value of $\bar{b}$ by solving Eq. (7.1) backward in b subject to the end condition $v_j(\bar{b}) = \bar{v}$ and looking for the value of $\bar{b}$ for which the initial value coincides with $\underline{v}$. However, the standard backward-shooting method is inherently unstable, especially when the bids are near $\underline{v}$. The authors in [166] have shown that the backward-shooting method is unstable even in the symmetric case.

7.1.2 Symmetric Bidders

If all the functions $F_j(v)$ are the same and hence equal to $\bar{m}(v)$, then we know from [553] that the symmetric equilibrium is

$$b(s) = s - \frac{\int_{\underline{v}}^{s} F^{I-1}(x)dx}{F^{I-1}(s)},$$

which is obtained as follows: By symmetry, instead of I ODEs, one now has only one ODE to solve. The ODE is given by

$$v'(b)(v(b) - b) = \frac{G(v(b))}{G'(v(b))}, \tag{7.3}$$

where G is the value distribution of the $I-1$ bidders. Using one-to-one function property and the fact that $(h^{-1})'(x) = \frac{1}{h'(h^{-1}(x))}$, one obtains

$$\frac{1}{h'(x)}(x - h(x))G'(x) = G(x),$$

where $h(x)$ is the bidding strategy. This means that $xG' = h'G + hG' = (hG)'$. By simple integration between the minimum value and v, one gets

$$h(v)G(v) - [hG]_{\underline{v}} = \int_{\underline{v}}^{v} xG'(x)dx = [xG(x)]_{\underline{v}}^{v} - \int_{\underline{v}}^{v} G(x)\, dx.$$

Hence, the optimal bidding strategy in the symmetric case is given by $h(v) = v - \frac{1}{G(v)} \int_{\underline{v}}^{v} G(x)\, dx$.

7.1.3 Error Bound

The non-asymptotic mean-field approach provides a useful error bound for this open problem. For asymmetric distribution, we are able to get a precise error bound when each distribution F_j is close to the arithmetic mean; the equilibrium strategies and payoffs can be approximated in a perturbed range. To do so, we use a non-asymptotic mean-field approach over function space [106]. First, note that the revenue of the seller satisfies the indistinguishability property, since it is, up to a constant, the integral of the product $\prod_{j \in \mathcal{I}} F_j$. We rewrite the function F_j as

$$F_j(v) = \bar{m}(v) + \varepsilon \gamma_j(v),$$

where $\gamma_j(v) = \frac{F_j(v) - \bar{m}(v)}{\varepsilon}$, $\bar{m}(v) = \frac{1}{I} \sum_{j'=1}^{I} F_{j'}(v)$, and $\varepsilon = \max_j \max_{[\underline{v}, \bar{v}]} |F_j(v) - \bar{m}(v)|$

One arrives at

- A good approximation of the asymmetric equilibrium strategies
- An equilibrium payoff with deviation order of $O(\varepsilon^2)$

Example 7.1 The optional second-price auction is currently used in DoubleClick Ad Exchange. Examples of ad exchanges are RightMedia, adBrite, OpenX, and DoubleClick. The process can be described as follows [322, 339]:

- User u visits the webpage w of publisher $p(w)$ that has, say, a single slot for ads.
- Publisher $p(w)$ contacts the exchange E with $(w, P(u), \rho)$, where ρ is the minimum price that p(w) is willing to take for the slot in w and $P(u)$ is the information about user u that $p(w)$ shares with E.

- The exchange E contacts ad networks $adn_1, \ldots, adn_m$ with $(E(w), E(u))$, where $E(w)$ is the information about w provided by E and $E(u)$ is the information about u provided by E. $E(u)$ may be potentially different from $P(u)$.
- Each ad network adn_j returns (b_j, d_j) on behalf of its customers which are the advertisers; b_j is its bid, that is, the maximum it is willing to pay for the slot in page w, and d_j is the ad it wishes to be shown. Each ad network may have multiple advertisers. The ad networks may also choose not to return a bid.
- Exchange E determines a winner j^* for the ad slot among all $(b_j, d_j)'$s and its price c_{j^*}, satisfying $\rho \leq c_{j^*} \leq b_{j^*}$ via an auction (first or second price).
- Exchange E returns winning ad d_{j^*} to publisher p(w) and price c_{j^*} to ad network j^*.
- The publisher $p(w)$ serves webpage w with ad d_{j^*} to user u (the impression of ad d_{j^*}).

Note that from the click of the user to the impression of the ad d_{j^*}, there are many intermediary interactive processes. Auction is one of them and an important one, because it determines the winner ad network. While it is reasonable to consider a large population of users over the Internet, the number of concurrent ad networks remains finite. Therefore, a non-asymptotic mean-field analysis for the revenue is needed. As a user may click several times over webpages, the dynamic auction framework seems more realistic. We examine the dynamic auction in Sect. 7.3.

Remark 7.2 (We Do Not Consider Security Issue) Consider the following auction example. A seller announces an online first-price auction for selling an object (indivisible). Player 1 participates in this auction. She has a valuation 1000 for the item and bids 700 in anticipation of maximizing her payoff. Player 2, her competitor who has valuation 730, breaks the server before the auction is closed and finds out the highest bid by player 1 and places her own bid as 701. Player 2 will win the object, though she has a lower valuation. This example clearly highlights the importance of security in an auction.

7.2 Spiteful Bidders

A player might be losing the auction for a long-term project, and yet she may want to continue to participate in the auction in order to minimize the negative payoff on losing by making her competitor, who would win the auction, pay a high price for the win. This negative dependence of payoff on others' surplus is referred to as spiteful behavior. Below, we show that the non-asymptotic mean-field framework applies to that scenario as well. A spiteful player j maximizes the weighted difference of her own payoff r_j and his competitors' payoffs $r_{j'}$ for all $j' \neq j$. The payoff of a spiteful player j is

$$r_{j,\alpha} := (1-\alpha_j)r_j - \alpha_j \sum_{j'\neq j} r_{j'} \tag{7.4}$$

Obviously, setting α_j equal to zero yields a selfishness (whose payoff equals his exact profit) whereas $\alpha_j = 1$ defines a completely malicious player (jammer) whose only goal is to force the minimization of the profits of the other players. Note that for an altruistic player, we would be considering a payoff in the form $r_{j,\alpha} = (1-\alpha_j)r_j + \alpha_j \sum_{j'\neq j} r_{j'}$.

The payoff of a spiteful player j is

$$\begin{aligned} r_{j,\alpha} = &(1-\alpha_j)(v_j - b_j)\prod_{j'\neq j} F_{j'}(v_{j'}(b_j)) \\ &-\alpha_j \mathbb{E}\left[\max{}_{j'\neq j} v_{j'} |\max{}_{j'\neq j} b_{j'} > b_j(v_j)\right] \\ &+\alpha_j \mathbb{E}\left[\max{}_{j'\neq j} b_{j'} |\max{}_{j'\neq j} b_{j'} > b_j(v_j)\right] \end{aligned} \tag{7.5}$$

As aforementioned, the main difficulty is that the private value distributions are asymmetric. Denote by $b^*_{j,\alpha}(v)$ the equilibrium bid strategy. Even when bidders are selfish ($\forall j,\ \alpha_j = 0$), the above shows that obtaining an explicit expression for $b^*_{j,0}(v)$ is not that easy, if not impossible. However, for symmetric distributions and symmetric coefficients $\alpha_j = \alpha$, the payoff functions reduce to

$$r_{j,\alpha} = (1-\alpha)(v_j - b)F^{I-1}(v) - \alpha\int_{v_j}^{\bar{v}} s(I-1)F^{I-2}(s)f(s)ds$$

$$-\alpha\int_{b_j(v_j)}^{b_j(\bar{v})} (I-1)sF^{I-1}(v(s))f(v(s))v'(s)ds$$

Using the fact that the derivative of $\int_{v(b)}^{\bar{v}} h(s)\, ds$ with the respect to b is given by $-h(v(b))v'(b)$, the first order optimality condition yields to

$$b(v) = v - \frac{(1-\alpha)F(v)}{(I-1)f(v)}b'(v).$$

In particular for uniform distribution $F_j(v) = v$ over $[0, 1]$, the equilibrium strategy for spiteful player satisfies the ODE:

$$b_\alpha(v) = v\left[1 - \frac{(1-\alpha)}{(I-1)}b'(v)\right].$$

It is easy to check that $b(v) = \frac{I-1}{I-\alpha}v$ is a solution of the above ODE. The symmetric equilibrium is increasing and convex in α. We now evaluate the revenue of the seller at equilibrium, for $F_j(v) = \bar{m}(v) = v$:

$$R_\alpha(\bar{m}, \bar{m})) = \mathbb{E}[\max_j b_j(v_j)] = \int b(v) I F^{I-1}(v) f'(v) dv \tag{7.6}$$

$$= \int I\left(\frac{(I-1)}{(I-\alpha)} v\right) v^{I-1}\, dv = \frac{I(I-1)}{(I-\alpha)(I+1)}.$$

Next, we compare this revenue with $R_\alpha(F_1, F_2)$ where

$$F_1(v) = \bar{m}(v) - \varepsilon v(1-v)\left(\frac{1}{2} - v\right),\ F_2(v) = \bar{m}(v) + \varepsilon v(1-v)\left(\frac{1}{2} - v\right)$$

$$R_\alpha(F_1, F_2) = R_\alpha(\bar{m}, \bar{m}) + O(\varepsilon^2).$$

Introducing a numerical scheme for optimal bidding strategies, we first solve the initial-value problem that starts at b_0 near but not equal to $\underline{b}$ so that the denominator does not vanish. ODE starts at $b_0 > \underline{b}$ and moves forward. We fix the starting function to

$$v(b_0) = \underline{v} + (b - b_0) D_1 + \delta_1 D_2 (b - b_0)^{1+\delta_2},$$

where $D_1 = (D_{11}, D_{12})$ is the limit in $\underline{b}$ of the derivative, $D_{11} = \lim_{b \to b_0} \frac{v_1(b) - v_1(\underline{b})}{b - \underline{b}}$, $D_{12} = \lim_{b \to b_0} \frac{v_2(b) - v_2(\underline{b})}{b - \underline{b}}$, δ_i are positive constants, and $D_2 = [D_{21}, D_{22}]$ is the equivalent fractional derivative approximation around $\underline{b}$, $b_0 = \underline{b} + 10^{-\frac{7}{1+\delta_2}}$. To find the solution, we look at the intersection $v_1(b_+) = v_2(b_+) = x$. The solution is $\left(\frac{1}{x} v_1(bx), \frac{1}{x} v_2(bx)\right)$.

Figure 7.1 depicts the inverse optimal bidding strategy for the cumulative distributions $F_1(v) = v^7$ and $F_2(v) = v^8$. The x-axis represents the optimal bidding strategy, and the y-axis is the inverse optimal bidding strategy of the two players.

Figure 7.2 depicts the same for $F_1(v) = v^2$ and $F_2(v) = v^{10}$.

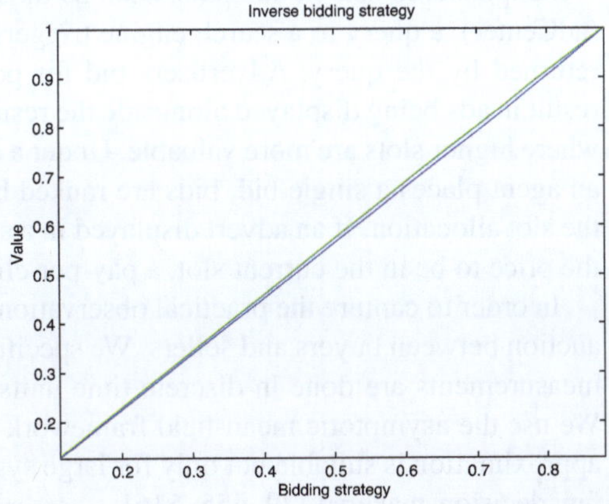

Fig. 7.1 Inverse optimal strategy for the distributions $F_1(v) = v^7$ and $F_2(v) = v^8$. The x-axis represents the optimal bidding strategy, and the y-axis is the inverse optimal bidding strategy of the two players

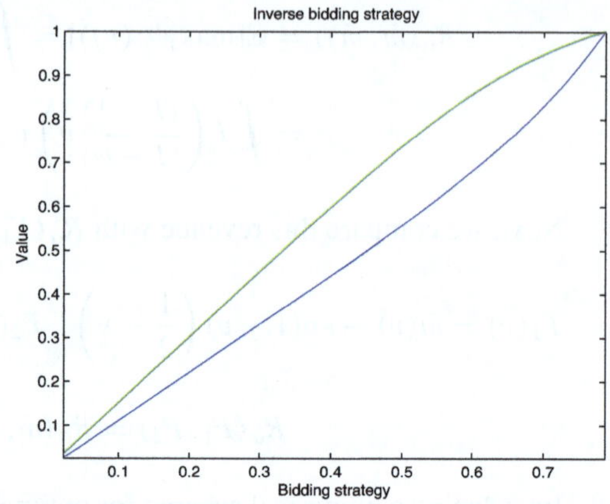

Fig. 7.2 Inverse optimal bidding for the distributions $F_1(v) = v^2$ and $F_2(v) = v^{10}$. The x-axis represents the bidding strategy, and the y-axis is the inverse optimal bidding strategy of the two players

7.3 Dynamic First-Price Auctions with Asymmetric Bidders

We now discuss the extension of the above framework to a dynamic setting. A possible motivation for a dynamic auction is that, currently, a large proportion of Internet users employ search engines or read/watch online media to locate information. For example, search engines are used to obtain an instantaneous selection of offers from various providers operating in cyberspace. Internet auctions, on the other hand, connect potential sellers and buyers from different locations in a virtual auction. Buyers also often get multiple purchase opportunities over time. This is most obvious in the case of today's online auction markets: eBay has auctions for physical goods closing every second, while Google's DoubleClick and Microsoft's Advertising Exchange trade online advertisements at a much faster rate. This creates the possibility for buyers to inter-temporally substitute, adjusting their current bids to account for the option value of waiting for future purchasing opportunities.

In sponsored search auctions, such as those run by Google (AdWords) or Bing (adCenter), a query to a search engine triggers an auction for positions on the page returned by the query. Advertisers bid for positions or slots, and successful bids result in ads being displayed alongside the results to the query. The slots are ordered, where higher slots are more valuable. Under a dynamic auction mechanism for slots, an agent places a single bid, bids are ranked by weight, and ranked bids determine the slot allocation. If an advert displayed in a slot is clicked, then the advertiser pays the price to be in the current slot, a pay-per-click payment model.

In order to capture the practical observation, we model the problem as a dynamic auction between buyers and sellers. We specifically focus on the buyers' side. Since measurements are done in discrete time units, we adopt a model in discrete time. We use the asymptotic mean-field framework presented in Volume I, Chap. 6. This approximation is suitable not only for large systems but also for a small system with few decision-makers [429, 455, 516].

The time space of the interaction is $\mathcal{T} = \{0, 1, \ldots, T-1\}$, where T is the length of the horizon. Since the budget is limited for the entire day or month, we transform the long-term budget constraints into an iterative budget state equation for bidder j :

$$s_{j,t+1} = s_{j,t} - c(s_{j,t}, b_{j,t})\mathbb{1}_{\{b_{j,t} > \max_{j' \neq j} b_{j',t}\}},$$

where $s_{j,t}$ is the remaining budget of bidder j for the corresponding contract frame and $c(s_{j,t}, b_{j,t})$ is the total cost for the bid (if winner). The state $s_{j,t}$ is subject to positivity constraint: $s_{j,t} \geq 0$ almost surely. This clearly limits the action space at time t to the set of bids $b_{j,t}$ such that $c(s_{j,t}, b_{j,t}) \leq s_{j,t}$. Let g be the function $b \mapsto c(s_{j,t}, b)$. The individual state bidder j is the pair $(s_{j,t}, v_{j,t})$. This generates a stochastic game with Markovian transition. Each bidder maximizes her long-term payoff given by

$$R_{j,T} = \mathbb{E}\left[\sum_{t \in \mathcal{T}} (v_{j,t} - b_{j,t})\mathbb{1}_{\{\max_{j' \neq j} b_{j',t}(v_{j,t}) < b_{j,t}(v_{j,t})\}} \mid v_j\right].$$

The history of a bidder j at time t is a collection $(s_{j,t'}, v_{j,t'}, b_{j,t'})_{t' \leq t}$. A bidding strategy $\tau_{j,t}$ of j at time t is a mapping from the history $h_{j,t} \in \mathcal{H}_{j,t}$ to the restricted bid space $[\underline{b}, \bar{b}] \bigcap [0, g^{-1}(s_{j,t})]$.

The game is played as follows. At opportunity $t \in \mathcal{T}$, every player j, $j \in \mathcal{N}$

- Realizes his current value $v_{j,t} \in [\underline{v}, \bar{v}]$ distributed according to $F_{j,t}$
- Submits a bid $b_{j,t} = \tau_{j,t}(v_{j,t}, h_{j,t})$, where $\tau_{j,t} : [\underline{v}, \bar{v}] \times \mathcal{H}_{j,t} \to [\underline{b}, \bar{b}]$ denotes his bidding strategy at auction t
- Updates his information $h_{j,t+1} \in \mathcal{H}_{j,t+1}$, based on the results obtained in auction t. Specifically, he forms a set of beliefs about the distribution of the bidding profile of the players $b_{-j,t+1}$ distributed according to $F_{j',t+1}$, $j' \neq j$

Let $V_{j,0}(s_0, T, \tau_{-j})$ be the value function of bidder j, i.e., it is the supremum, over all possible bidding strategies, of the expectation of the payoff $R_{j,T}$ starting from an initial budget s_0 when the other bidder strategy profile is τ_{-j}.

Based on the standard Bellman optimality criterion, we can readily get the following result.

Theorem 7.3 *Given v_j, the optimal strategy of a player j satisfies*

$$\begin{aligned} b^*_{j,t}(v_j) \in \arg\max\nolimits_{b \leq g^{-1}(s_{j,t})} \mathbb{E}_{\bar{b}_{-j,t}} \Big[&(v_j - b)\mathbb{1}_{\{b > \bar{b}_{-j,t}\}} \\ &+ \mathbb{1}_{\{b > \bar{b}_{-j,t}\}} V_{j,t+1}(s_{j,t} - c(s_{j,t}, b)) \\ &+ \mathbb{1}_{\{b < \bar{b}_{-j,t}\}} V_{j,t+1}(s_{j,t}) \mid v_j \Big], \end{aligned} \tag{7.7}$$

where $\bar{b}_{-j,t} := \max_{j' \neq j} b_{j',t}$ *is the highest bid of the other players (other than* j*). Let* $x_1^* \in \arg\max_{x_1 \in [\bar{b}_{-j,t}, g^{-1}(s_{j,t})]} \left[v_j - x_1 + V_{j,t}(s_{j,t} - c(s_{j,t}, x_1))\right]$. *Let* $W_j := v_j - x_1^* + V_{j,t}(s_{j,t} - c(s_{j,t}, x_1^*))$. *Then, the optimal bidding strategy is*

$$b_{j,t}^*(v_j) = \begin{cases} x_1^* & \text{if } W_{j,t} > V_{j,t+1}(s_{j,t}) \\ x_2^* < \min\left(\bar{b}_{-j,t}, g^{-1}(s_{j,t})\right) & \text{otherwise} \end{cases}$$

where x_2^* *is any real number in* $[0, \min\left(\bar{b}_{-j,t}, g^{-1}(s_{j,t})\right))$. *Further, the value iteration is given by*

$$V_{j,t}(s_{j,t}) = \mathbb{E}_{v_j} \sup_b \mathbb{E}_{\bar{b}_{-j,t}} \Big[(v_j - b)\mathbb{1}_{\{b > \bar{b}_{-j,t}\}} + \mathbb{1}_{\{b > \bar{b}_{-j,t}\}} V_{j,t+1}(s_{j,t} - c(s_{j,t}, b)) + \mathbb{1}_{\{b < \bar{b}_{-j,t}\}} V_{j,t+1}(s_{j,t}) \mid v_j \,\Big]$$

For the symmetric setup (below), we drop the index j.

Theorem 7.4 (Symmetric Beliefs) *Given* v*, the optimal strategy of a generic player satisfies*

$$\begin{aligned} b_t^*(v) \in \arg\max_{b \le g^{-1}(s_t)} \mathbb{E}_{\bar{b}_t} \Big[&(v - b)\mathbb{1}_{\{b > \bar{b}_t\}} \\ &+ \mathbb{1}_{\{b > \bar{b}_t\}} V_{t+1}(s_t - c(s_t, b)) \\ &+ \mathbb{1}_{\{b < \bar{b}_t\}} V_{t+1}(s_t) \mid v\Big], \end{aligned} \tag{7.8}$$

where $\bar{b}_t := \max_{j' \neq j} b_{j',t}$ *is the highest bid of the other players than* j.

Letting $x_1^* \in \arg\max_{x_1 \in [\bar{b}_t, g^{-1}(s_t)]} [v - x_1 + V_t(s_t - c(s_t, x_1))]$, *we have that the optimal bidding strategy is*

$$b_t^*(v) = \begin{cases} x_1^* & \text{if } W_t > V_{t+1}(s_t) \\ x_2^* < \min\left(\bar{b}_t, g^{-1}(s_t)\right) & \text{otherwise} \end{cases}$$

where x_2^* *is any real number in* $[0, \min\left(\bar{b}_{-j,t}, g^{-1}(s_{j,t})\right))$.

The value iteration is given by

$$V_t(s_t) = \mathbb{E}_v \sup_b \mathbb{E}_{\bar{b}_t} \Big[(v - b)\mathbb{1}_{\{b > \bar{b}_t\}} + \mathbb{1}_{\{b > \bar{b}_t\}} V_{t+1}(s_t - c(s_t, b)) + \mathbb{1}_{\{b < \bar{b}_t\}} V_{t+1}(s_t) \mid v\Big]$$

Theorem 7.5 (Non-asymptotic Mean Field) *Let* $F_{j,t}(v) = \bar{m}_t(v) + \varepsilon \gamma_{j,t}(v)$, *where*

$$\gamma_{j,t}(v) = \frac{F_{j,t}(v) - \bar{m}_t(v)}{\varepsilon},$$

$$\bar{m}_t(v) = \frac{1}{I}\sum_{j'=1}^{I} F_{j',t}(v),$$

and

$$\varepsilon = \max_j \sup_{t\in\mathcal{T}} \sup_{[\underline{v},\bar{v}]} |F_{j,t}(v) - \bar{m}_t(v)|,$$

Then, we have the following:

(i) The long-term revenue of the seller is

$$R_0(s_0; F_1, \ldots, F_I) = R_0(s; \bar{m}, \ldots, \bar{m}) + O(\varepsilon^2)$$

for any $I \geq 2$.

(ii) Moreover, if $F_{j,t} = F^I_{j,t}$ depends on I and converges to a common function, which equals $\bar{m}_t(v)$ as I tends to infinity, $\sqrt{I}(F^I_{j,t} - \bar{m}_t(v)) \leq \delta$, then the error gap between the finite regime and the infinite regime in equilibrium is in order of $O(\frac{1}{I})$.

Proof The proof of the statement (i) is a direct extension of the first result in Chap. 6 of Volume I to the time space $\mathcal{T}$. We now prove the statement (ii). Let $\sqrt{I}(F^I_{j,t} - \bar{m}_t(v)) \leq \delta$; then, the error gap at time t reduces to $\delta_{t,\bar{m}_t,\bar{r}}\frac{\delta}{I}$, and hence, the global error in $\mathcal{T}$ is at most

$$\frac{\delta}{I}\left(\sup_{t\in\mathcal{T}}\delta_{t,\bar{m}_t,\bar{r}}\right).$$

■

Note that this is a significant improvement of the classical mean-field approximation since the use of mean-field convergence à la *de Finetti* [277] gives a convergence order of $O(\frac{1}{\sqrt{I}})$. The use of the indistinguishability property of the payoff function helps us to provide a more precise error.

Scaled Auction Problem

We now examine a scaled version of the auction game. Using a scaling factor $\frac{1}{\lambda}$, to the starting state (budget) and horizon, the value

$$W_0^\lambda(s, T) := \lambda V_0\left(\frac{s}{\lambda}, \frac{T}{\lambda}\right),$$

has a certain limit when λ goes to 0. Let $\lim_{\lambda\to 0} W_0^\lambda(s, T) = w(s, T)$. Let $T^* = \inf\{t \geq 0 \mid s_t \leq 0, s_0 > 0\}$. Then, the value w is the solution of the following non-zero sum differential game given by

$$w(s, T) = \sup_{\tau} \int_0^{\inf(T^*,T)} \mathbb{E}_{v,\bar{b}} \left[(v_t - b_t) \mathbb{1}_{\{b_t > \bar{b}_t\}} \right] dt$$

subject to

$$\dot{s}_t = -E[c(s_t, b_t)],$$

$T^* = \inf\{t \geq 0, \mid s_t = 0, s_0 > 0\}$, $w(0, T) = 0$, and $w(s, 0) = 0$. Let $r(b)$ be the instantaneous payoff from the above formulation. Introduce the Hamiltonian

$$H(s, p) = \sup_b [r(b) - pc(s, b)].$$

The value w satisfies the Hamilton–Jacobi–Bellman equation

$$\partial_t w + H(s, \partial_s w) = 0.$$

This payoff also provides an approximation in the asymmetric in the case

$$w(m_1, \ldots, m_I, T) = w(\bar{m}, \ldots, \bar{m}, T) + O(\varepsilon^2),$$

where $\bar{m}$ is the average measure and m_j is the distribution of state of player j.

7.4 Notes

The work in [205] considered repeated second-price auctions under homogeneous valuation distributions and budget constraints. Most auctions involve bidders that are heterogeneous ex ante. For example, ad networks may not have a homogeneous belief about the advertisers' distribution. The non-asymptotic mean-field game approach provides an approximation of the equilibrium bidding and the equilibrium revenue in a heterogeneous auction mechanism [264, 429, 430, 515–518].

Chapter 8
Network of Neural Networks

Machine intelligence (MI) focuses on the creation of models, evolutionary dynamics, and algorithms that enable machines or software to co-learn from data and improve their performance over time. Machine intelligence is therefore an advanced computer science that allows a machine, device, software, program, code, or algorithm to interact intelligently with its environment, which means it can collect data, make decisions, perform actions, and develop strategies to maximize its chances of successfully achieving its preferences and objectives [69, 156, 157, 466]. A neural network is a computational model inspired by the structure and function of the neural units, comprised of interconnected nodes or artificial neural units that process and transform information to solve various machine learning and pattern recognition tasks. A large language model is a type of machine intelligence model that uses a vast amount of data to understand and generate human-like text or natural language responses. The methodology has been extended to large learning models which go beyond texts. A large learning model typically refers to a learning model, often a neural network, that is trained on a substantial amount of data to perform various tasks or make risk-aware predictions with high accuracy. A multimodal learning model is a type of machine intelligence model that can process and integrate information from multiple modes or sources to perform tasks that require a combination of these data types. The emergence of multimodal large learning models (MLLMs) is transforming the field of machine intelligence, enabling machines to comprehend and generate content across various modalities, including text, images, audio, video, music, gestures, sensors, and emotions. These MLLMs have demonstrated exceptional capabilities in tasks such as language understanding, content generation, sentiment analysis, and interactive systems. However, as we delve deeper into the potential of these models, we realize that there is much more to explore. A network of neural networks (NoNNs, [466]) is a system where multiple individual neural networks interact with each other. Each neural network in this network of networks, often called a subnetwork or module, specializes in a specific aspect of the task. It is often used to solve adversarial, cooperative, co-opetitive,

T. Başar et al., *Mean-Field-Type Game Theory II*, Static & Dynamic Game Theory: Foundations & Applications, https://doi.org/10.1007/978-3-032-07031-9_8

competitive, or coalitional strategic learning problems. These subnetworks can communicate and share chosen information to make decisions or perform individual or collective tasks. Risk awareness is vital in NoNNs because it helps identify and mitigate potential weaknesses and uncertainties within the system. It allows for the assessment of the confidence levels of individual subnetworks, enabling more reliable decision-making by weighting their outputs accordingly. Being risk-aware is essential for detecting and defending against adversarial attacks, addressing ethical concerns related to bias and fairness, ensuring regulatory compliance, and making informed decisions about how to fuse the outputs of different subnetworks [466].

This chapter explores a frontier in MLLMs. In this paradigm, we examine a world where multiple agents coexist, each equipped with their own Text–Image–Audio–Video–Music–Gesture–Sensor–Emotion–Voice–Sound–Infrared–Table–Time series–Graph–Hyper–Spectrum–Point Cloud-X-Ray–Inertial measurement unit MLLMs. These agents, driven by their unique preferences and objectives, engage in complex interactions within the framework of mean-field-type game theory. At the heart of this exploration lies the recognition of the diverse nature of agents' information and preferences. Each agent may be risk-aware, co-opetitive, collaborative, cooperative, competitive, partially altruistic, hostile, spiteful, mutually supportive, reciprocal, risk-seeking, ot risk-neutral or exhibit a myriad of other preference profiles. We present some mathematics of large learning models, including large language models and some risk-awareness by means of mean-field-type game theory [106, 457, 458, 460, 463]. We discuss the concept of a network of neural networks, which involves various neural network architectures designed for specific purposes. It covers models like tensor neural network, graph neural network (GNN), graph-tensor neural network (GTNN), tensor-graph neural network (TGN), tensor generative adversarial network (GAN), graph GAN, tensor-graph GAN, graph-tensor GAN, generative pre-trained transformer (GPT), tensor GPT, graph GPT, tensor-graph GPT, graph-tensor GPT, and a tensor-graph neural network with blockchain integration. The chapter also presents the design and analysis aspects of NoNNs, including architecture selection, decision-maker's neural network architecture, shared neural network architecture, strategic learning steps, and the outcomes of NoNN. Additionally, it discusses mean–variance [466] training and design techniques in the context of NoNNs.

The chapter is organized as follows. Section 8.1 presents the functions behind the architectures of large learning models and connects them with mean-field-type game theory. Section 8.2 focuses on transformer-based deep neural networks. Then, mean-field-type transformers (MFTTs) are presented in Sect. 8.3. Section 8.4 presents a network of neural network architecture models. Section 8.5 focuses on the design and analysis of a network of neural networks and multiple interacting agents, including machine intelligence agents.

8.1 Large Learning Models

Generative machine intelligence models are being tested in numerous industries, from language generation to image synthesis to video production. However, underlying these advancements are variational inequality challenges that arise due to the non-convex nature of deep learning architectures. This section addresses a novel approach to handling these challenges by leveraging hidden convexities within the structure of deep learning models, particularly in risk-aware settings, including the expectile risk measure. We explore how reformulating deep learning architectures as a series of games can lead to more efficient outcomes, improved training consistency, and innovative approaches that extend beyond traditional gradient-based methods. The incorporation of game theory within these models allows us to exploit the hidden convexities, allowing for more tractable layer-by-layer learning strategies and enhancing risk-aware model performance.

Deep Learning Architectures as Games

In standard deep learning architectures, the final output is the result of a composition of several layer operators, making the problem highly non-convex both in terms of its parameters and often in its input. This non-convexity poses significant challenges for optimization, as traditional methods typically rely on first-order optimality conditions, such as Fermat's rule, which often lead to local rather than global optima. This limitation is particularly pronounced in deep learning, where the depth of the model exacerbates the complexity of the network. By reformulating the architecture as a series of games, we can begin to exploit hidden convexities within the model. Specifically, by considering each layer as a player in a game, we focus on the interactions between layers rather than treating the entire network as a monolithic optimization problem. In doing so, we exploit certain convexities within the activation functions and matrix-valued softargmax operators. These hidden convexities allow for more efficient equilibrium strategies, sidestepping some of the non-convexity challenges that arise when treating the network holistically. This game-theoretic approach is particularly effective for risk-aware generative machine intelligence models, where the need for robust optimization is paramount to managing uncertainty in the output space.

Sampled Data: A Hierarchical Game Approach

When dealing with finite, sampled data sets, neural networks must look for consistency in their outcomes. Achieving this consistency requires iterating the outcomes of the network repeatedly, a process that can be conceptualized as a feedback loop. In this context, the problem of machine intelligence consistency can be reformulated as a hierarchical game, where each layer in the network relies on information from the previous layer to make decisions about its outputs. This hierarchical information structure reflects how layers in a deep network process data, with each layer transforming the input it receives from the preceding layer.

The consistency game can be described as follows: The layers of the neural network act as the decision-makers, and their decision space is defined by the output at each respective layer. The information structure is based on the input from the preceding layer, which corresponds to the action taken by the previous decision-maker. This forms a fully hierarchical game, specifically a fully Stackelberg game, where each layer operates as a leader, making decisions based on the actions of the preceding layer.

Similarly, the training process can also be framed as a game, where the layers serve as the decision-makers. In this case, the actions correspond to the adjustment of weights and biases. The information structure is determined by the input from the previous layer, leading to a scenario of partial hierarchical information. However, unlike a classical Stackelberg game, the current decision-maker (layer) does not have direct knowledge of the precise action taken by the preceding layer, as it can only observe the resulting input. This creates a unique variation of the Stackelberg structure in the training process. Within this context, the training process of neural networks can be viewed as a game played over the sampled data, where each layer acts based on the information provided by the layer below it. The challenge is to ensure that this iterative process leads to stable, consistent outcomes. By examining the consistency problem as a hierarchical game, we gain new insights into how to stabilize the training process and ensure that the network performs reliably across different samples. This approach also opens up possibilities for more effective optimization techniques that can exploit the structure of the game to improve convergence and performance on real-world data.

Beyond Gradient Descent: Bregman Learning

While gradient descent and its variants have been the cornerstone of earlier deep learning optimization, there are limitations to their applications, especially when dealing with non-convex architectures. To address these limitations, we introduce layer-by-layer Bregman learning, which extends beyond traditional gradient-based methods by incorporating Bregman proximal algorithms that exploit the hidden convexities. These algorithms operate in continuous time, providing a more flexible and potentially more powerful framework for optimization. Bregman learning allows us to define a broader class of discrepancy functions, which can be used to guide the optimization process more effectively than standard Euclidean distances. By leveraging Bregman divergences, we are able to capture more nuanced relationships between data points, leading to improved convergence properties and robustness in training. The continuous-time nature of Bregman proximal algorithms also opens up new avenues for incorporating second-order information into the optimization process, providing a richer understanding and enabling more efficient navigation through non-convex spaces [466].

Transformer-Based Architectures as Games

The advent of transformer-based architectures, particularly in image/audio/text/video processing, has introduced additional challenges for optimization due to the non-convex nature of self-attention mechanisms. Self-attention is not 1-Lipschitz in its parameters (input signal, query, and key), which complicates the application of fixed-point algorithms traditionally used for finding zeroes of operators or variational inequalities. These non-convexities lead to difficulties in finding global optima and ensuring the stability of the model. However, by reformulating transformer-based architectures as games, we can effectively address some of these challenges. The key insight lies in recognizing that the normalization operator used in transformers can be viewed as the subgradient of a norm, which introduces a hidden convexity into the optimization problem. This hidden convexity allows us to define a well-behaved minimization problem at each layer, despite the inherent non-convexity of the self-attention mechanism. By exploiting this structure, we can develop more robust optimization algorithms that are better suited to the unique challenges posed by transformer-based models. The self-attention mechanism is not Lipschitz in the entire space because the cross-interaction between the entries of the sequence data creates a quadratic term, which is not Lipschitz. However, we can still exploit the fact that the signal is normalized before processing the masked or unmasked self-attention mechanism, making it Lipschitz on any non-empty and bounded domain. This property is well exploited by using a normalized distance that maps the entire domain to the unit ball. This careful exploration of deep learning architectures as games reveals hidden convexities that can be leveraged to improve optimization and training stability, particularly in risk-aware generative machine intelligence models. By moving beyond traditional gradient descent and incorporating hierarchical games and Bregman learning, we open up new possibilities for enhancing the performance and robustness of these powerful foundational large learning models. We also present mean-field-type transformers [466] based on different interaction models and opening the door to alternative self-attention mechanisms. This allows us to include some non i.i.d. structures of the data.

Disributionally Robust Training

In addition to the foundational approaches discussed above, we extend our examination to incorporate several advanced training methodologies—mean–variance training, variance-aware training, and distributionally robust training—particularly when the true distribution of the data is unknown. In real-world applications, machine intelligence models often face uncertainty regarding the exact distribution of the underlying data, as they must rely on limited or sampled data sets to make predictions and generate outcomes. In these cases, it becomes crucial to construct an estimated distribution from the available data and adapt the training process accordingly. Mean–variance training offers a way to balance the trade-off between achieving high accuracy (mean performance) and controlling the variability (variance) of the model's output. This is particularly relevant in risk-aware generative machine intelligence, where it is essential not only to maximize the expected performance but also to minimize the uncertainty in the model's predictions. Variance-aware training further sharpens this focus by directly penalizing excessive variance in the model's output, thus improving the robustness of the predictions in the face of noisy or incomplete data. Distributionally robust training provides a powerful framework for managing uncertainty when the exact distribution is unknown. Instead of optimizing for a single estimated distribution, this approach optimizes for the worst-case scenario across a set of plausible distributions. By considering the worst-case performance over this ambiguity set of distributions, we ensure that the model remains robust even under adversarial or highly uncertain conditions. These training methods, when framed within the game-theoretic perspective we propose, further exploit hidden convexities and allow for more effective management of risk and uncertainty in generative large learning models. The combination of these approaches provides a comprehensive toolbox for enhancing model reliability, even when faced with imperfect or incomplete data, a common challenge in many real-world machine intelligence applications.

8.1.1 Three Deep Learning Architecture Problems

We consider a neural network with $L \geq 1$ layers. Let $\{\mathcal{H}_l\}_{0\leq l\leq L}$ be nonzero Hilbert spaces. For every $l \in \{1, \ldots, L\}$, let $W_l : \mathcal{H}_{l-1} \to \mathcal{H}_l$ be a bounded linear operator and consider the family of activation operators $R_l : \mathcal{H}_l \to \mathcal{H}_l$. The deep neural network is defined by the composition of operators:

$$R_L \circ (W_L \cdot +b_L) \circ R_{L-1} \circ (W_{L-1} \cdot +b_{L-1}) \circ \ldots \circ R_1(W_1 \cdot +b_1), \tag{8.1}$$

where W_l is the weight operator and b_l captures the bias parameter. Let $x_0 \in \mathcal{H}_0$, $\{\lambda_t\}_{t\geq 0}$ be a nonnegative sequence, and consider the map defined by

$$\hat{O}_l : \left| \begin{array}{l} \mathcal{H}_{l-1} \to \mathcal{H}_l \\ x \mapsto R_l(W_l x + b_l). \end{array} \right.$$

Iterate layer by layer and by timestep the following:

$$\text{for } t \in \{0, 1, 2, \ldots\} \left| \begin{array}{l} y_{1,t} = \hat{O}_1(x_t), \\ \text{for } l \in \{2, \ldots, L\},\ y_{l,t} = \hat{O}_l(y_{l-1,t}), \\ x_{t+1} = x_t + \lambda_t(y_{L,t} - x_t), \end{array} \right. \tag{8.2}$$

which means that

$$x_{t+1} = x_t + \lambda_t(\hat{O}_L \circ \hat{O}_{L-1} \circ \ldots \circ \hat{O}_1(x_t) - x_t),$$

starting from x_0. The sequence $(\lambda_t)_{t\geq 0}$, with $\lambda_t \geq 0$, is the step-size (learning rate). For consistency of the operation, the output of $\hat{O}_L$ should be in the same space as the input, i.e., $\mathcal{H}_L = \mathcal{H}_0$. As we want to examine situations in which time goes to infinity, the cumulative step-size $\sum_{t=1}^{T} \lambda_t \to +\infty$ as T goes to infinity.

Definition 8.1 The neural network specification is given by the family

$$\mathbb{NN} := (x_0, L, \{R_l, W_l, b_l, \mathscr{H}_l\}_{1 \le l \le L}, \lambda).$$

Problem 8.2 What does a well-trained deep neural network do? To answer this question, we aim to find and characterize the behavior of the neural network for a large class of architectures $\mathbb{NN}$. To this end, we consider the algorithm in (8.2) and examine the possible limit (if any) of $(x_t, y_{1,t}, \ldots, y_{L,t})$ as t goes to infinity.

Problem 8.3 (Training and Design) Given an input–output data set $\{(x_i, d_i = y_{L,i}),\ i \in \{1, \ldots, D\}\}$, find $\mathbb{NN}$ such that

$$\hat{O}_L \circ \hat{O}_{L-1} \circ \ldots \circ \hat{O}_1(x_i) = d_i,\ \forall i \in \{1, \ldots, D\},$$

which is reformulated in a weak sense as evolving equalities: Find a vector $\theta^* := (W^*, b^*) = (W_l^*, b_l^*)_l$ such that

$$r_L \circ \tilde{A}_{L,\theta_L^*} \circ \ldots \circ (r_1 \circ \tilde{A}_{1,\theta_1^*})(x_i) - y_{L,i} = 0, \quad i \in \{1, \ldots, D\}, \tag{8.3}$$

where $\tilde{A}_l = \tilde{A}_{l,\theta_l} : x_{l,i} \mapsto W_l x_{l,i} + b_l$ and $A_{l,i} = A_{l,x_{l,i}} : \theta_l \mapsto W_l x_{l,i} + b_l$, which is linear and nonzero, and $A_{l,i}^\dagger$ is the adjoint operator of $A_{l,i}$. The latter set of equalities is now replaced by variational inequalities: Find a vector $\theta^* := (W^*, b^*) = (W_l^*, b_l^*)_l$ such that the set of variational inequalties

$$\Big\langle (r_L \circ \tilde{A}_{L,\theta_L^*}) \circ \ldots \circ (r_1 \circ \tilde{A}_{1,\theta_1^*})(x_i) - y_{L,i}, A_L^\dagger(\theta_L - \theta_L^*) \Big\rangle \ge 0, \ \ \forall \theta,$$
$$i \in \{1, \ldots, D\}. \tag{8.4}$$

holds.

Find a vector θ^* such that

$$\sum_{i=1}^{D} \omega_i' \Big\langle (r_L \circ \tilde{A}_{L,\theta_L^*}) \circ \ldots \circ (r_1 \circ \tilde{A}_{1,\theta_1^*})(x_i) - y_{L,i}, A_L^\dagger(\theta_L - \theta_L^*) \Big\rangle \ge 0, \ \ \forall \theta, \tag{8.5}$$

where $\omega_i' > 0$, $\sum_{i=1}^{D} \omega_i' = 1$. Moreover, design an algorithm that approximates a θ^*, a solution (if any) to (8.5).

Due to the hierarchical structure of $\mathbb{NN}$, the resulting squared norm of the operator $\hat{O}_L \circ \hat{O}_{L-1} \circ \ldots \circ \hat{O}_1$ is not necessarily convex in the weights W and biases b, and this holds even for simple architectures with piecewise constant activations or piecewise linear activation functions like rectified linear unit (ReLU). The idea here is to exploit the hidden convexity within each layer to propose efficient algorithms for pre-training, training, and analyzing the outcomes of both shallow and deep learning.

Risk awareness is vital in generative machine intelligence because it helps identify and mitigate potential weaknesses and uncertainties within the system. It allows for the assessment of the confidence levels of individual subnetworks, enabling more reliable decision-making by weighting their outputs accordingly. Being risk-aware could help detect and defend against adversarial attacks, addressing ethical concerns related to bias and fairness, ensuring regulatory compliance, and making informed decisions about how to fuse the outputs of different subnetworks.

Problem 8.4 (Risk-Aware Design and Training) The expectile risk-aware training problem in generative machine intelligence is to find L and the operators $\hat{O}_1, \ldots, \hat{O}_L$:

$$\arg\min_{L\geq 1} \arg\min_{\hat{O}_L,\ldots,\hat{O}_1} \mathscr{R}_\alpha[\left|y_L - \hat{O}_L \circ (\hat{O}_{L-1} \circ \ldots \circ \hat{O}_1 x_0)\right|^2], \tag{8.6}$$

where $\mathscr{R}_\alpha$ is a function generating a risk measure such as the α-expectile with $\frac{1}{2} < \alpha < 1$. The expectile will be defined in Sect. 8.1.2.10. The randomness here is the input–output pairs $(x_0, d_L = y_L)$.

Given the structure of the hierarchical information, we can invoke here the Stackelberg solution concept. For $\alpha = \frac{1}{2}$, the α-expectile risk-aware design and training problem reduces to the risk-neutral case: Find L and the operators $\hat{O}_1, \ldots, \hat{O}_L$:

$$\arg\min_{L\geq 1} \arg\min_{\hat{O}_L,\ldots,\hat{O}_1} \int_{x_0}\int_{y_L}[\left|y_L - \hat{O}_L \circ(\hat{O}_{L-1} \circ \ldots \circ \hat{O}_1 x_0)\right|^2] \mathbb{P}(dx_0 dy_L). \tag{8.7}$$

In the above formulation, one can replace $\left|y_L - \hat{O}_L \circ (\hat{O}_{L-1} \circ \ldots \circ \hat{O}_1 x_0)\right|^2$ by a general loss function $Loss(y_L, \hat{O}_L \circ \hat{O}_{L-1} \circ \ldots \circ \hat{O}_1 x_0)$ and then take the risk measure over that. As we can see, the distribution $\mathbb{P}(dx_0 dy_L)$ plays a key role here. In particular, we will discuss the case where the true distribution $\mathbb{P}(dx_0 dy_L)$ is unknown and the learnability under distributional robustness will be introduced and solved using f-divergence and Bregman discrepancy. The training problem for large learning models is inherently non-convex in terms of the parameter space θ. Non-convex optimization landscapes are characterized by multiple local minima, saddle points, and flat regions, making it difficult to guarantee convergence to a global minimum. Currently, there is no general proof of convergence for gradient-based optimization methods like gradient descent, Adam, Nesterov, or even second-order methods like Newton's method when applied to non-convex functions [462]. These algorithms were primarily designed for convex optimization problems, where convergence to a global minimum can be proved under certain conditions (such as strong convexity, coercivity, strong monotonicity, and Lipschitz gradient). In the non-convex setting, especially for large learning models such as GPT, these algorithms may be trapped in suboptimal areas or get stuck in local minima. To fully exploit the internal operations of large learning models—such as the structure of layer normalization, projection to the unit ball, attention mechanisms, multi-head attention, and layer-wise dependencies-reformulating the training problem could lead to more effective solutions. Formulating the training problem of large learning models as a variational inequality instead of a problem of finding zeros of the gradient of the squared error offers significant advantages, particularly in the context of the model's inherent non-convexity. In non-convex optimization, minimization techniques like gradient descent, Secant, Reverse Ishikawa, Newton, Steffensen, Adam, or Nesterov often struggle to avoid local minima, saddle points, and flat regions, which can cause the model to be trapped around suboptimal solutions [150, 482, 486, 487]. By focusing on variational equilibrium rather than zeros of the gradient of the squared error, the original variational inequality formulation aims to find a point where no feasible perturbation in the parameters can provide a descent direction, making it more robust. This approach is particularly beneficial for large learning models like GPT, which exhibit rich internal structures such as multi-head self-attention and layer-wise dependencies. The original variational inequality formulation captures the interdependencies between different layers and components, providing a more holistic approach.

Consider the variational inequality with $F(\theta) = \theta(1-\theta)$, i.e., find all θ^*

$$\langle F(\theta^*), (\theta - \theta^*) \rangle \geq 0, \ \forall \theta,$$

which is the set $\{0, 1\}$. In this case, there is a potential function ϕ such that $\nabla \phi = F$. It suffices to take $\phi = \frac{1}{2}\theta^2 - \frac{1}{3}\theta^3$. Next, we express the variational inequality for finding all zeros of $\nabla \phi$:

$$\langle \nabla \phi(\theta^*), (\theta - \theta^*) \rangle \geq 0, \forall \theta$$

which is the set $\{0, 1\}$. Now, consider the square error $\langle F, F \rangle$, which in this particular case corresponds to F^2. We are attempting to find all values of θ that minimize F^2 and compare them with the values of θ that solve the original variational inequality problem, which involves finding the zeros of $\nabla \phi$. However, it is not a good idea to work with the gradient of F^2 which in this case is $2(\nabla F)F = Hess(\phi)\nabla \phi$. Working with the gradient of the square error F^2 yields

$$\langle (\nabla(F^2))(\theta^*), (\theta - \theta^*) \rangle \geq 0, \ \forall \theta$$

which is

$$\langle (Hess(\phi)\nabla \phi)(\theta^*), (\theta - \theta^*) \rangle \geq 0, \forall \theta,$$

which is the set $\{0, \frac{1}{2}, 1\}$. In other words, the zeroes of the gradient include the local extremum $\frac{1}{2}$. But $\frac{1}{2}$ is not a solution to the original variational inequality problem. It is not a minimizer of F^2.

Through this simple example, we observe that the training problem is better formulated as a variational inequality in the output space, rather than as a finding of zeroes of the gradient of the squared error. Minimizing the squared error norm differs from finding zeros of the gradient of the square error. This is why further stability analysis is required when employing algorithms for solving variational inequalities. It is important to exploit the hidden properties of the intermediary operators, such as projection, proximal, normalization, and ReLU, which are all gradients of some other functions, and use these properties in the training problem instead of limiting our analysis to finding zeroes of the squared error. Note, however, that there is no global potential function for the generative machine intelligence architecture in the general setting [38, 139, 479]. Layer by layer, all hidden potential and all hidden convexities will be useful in the sub-layer games involved.

8.1.2 Sampled Data

We start with a well-designed, well-trained, well-deployed generative machine intelligence model based on the above foundational learning model. This means that the number of layers, L, the activation functions per layer, the weights, and biases per layer are all set. For such a well-trained neural network, we ask the following question: What does a well-trained deep neural network do?

We consider the algorithm in (8.2) and examine the possible limit (if any) of $(x_t, y_{1,t}, \ldots, y_{L,t})$ as t goes to infinity. If the sequence $(x_t, y_{1,t}, \ldots, y_{L,t})_t$ defined in Problem 8.2 does not converge to a point but to a set or a limit cycle, it could be problematic: Two elements in the limit set have different meanings and or two interpretations. For example, if the initial input is a table in a certain format and, after a certain number of iterations, the output is just a noisy image without a specific meaning, there is a model collapse. It is therefore important to examine such long-run outcomes. If the limit is x^*, then we have $y_L^* = x^*$ and $\hat{O}_L \circ \hat{O}_{L-1} \circ \ldots \circ \hat{O}_1(x^*) = x^*$, and therefore, the model can be seen as consistent in the long run.

To answer this question, we characterize the outcomes as fixed points and connect them to constrained Nash games. We introduce below basic notions and two workable cases for the outcomes.

8.1.2.1 Outcomes of $\mathbb{NN}$: Two Computable Cases

We examine two classes of activation functions: averaged nonexpansive and maximally cyclically monotone activation functions.

$\mathbb{NN}$ **Outcomes Under Averaged Nonexpansive Activation Functions**
We introduce the notion of an averaged nonexpansive operator.

Definition 8.5 Let $0 < \gamma \leq 1$. An operator $\hat{O} : \mathcal{H}_0 \to \mathcal{H}_0$ is γ-averaged if $[Id + \frac{1}{\gamma}(\hat{O} - Id)]$ is 1-Lipschitz continuous.

γ-averaged nonexpansive operators play key roles in the asymptotics of $(\hat{O} - Id)(x_t)$ even if the operator $\hat{O}$ has no fixed point [37]. For example, the mapping $\hat{O}(x) = x + e$, with e nonzero, has no fixed point. However, $\hat{O}(x_t) - x_t = e$ which converges to e. When $\hat{O}$ has a fixed point, the difference $\hat{O}(x_t) - x_t$ goes to zero as t goes to infinity where x_t, are the outputs of the algorithm in (8.2). These γ-averaged nonexpansive operators allow us to extend the well-known convergence results of the case of contractive operators under Banach–Picard iterates.

Example 8.6 The softargmax, which is the Boltzmann–Gibbs distribution, is widely used in the architecture of generative machine intelligence. Let $k \geq 2$, $\varepsilon > 0$, $j \in \{1, 2, \ldots, k\}$

$$\text{softargmax: } \mathbb{R}^k \ni z \mapsto \left(\frac{e^{z_1}}{\sum_{j=1}^k e^{z_j}}, \ldots, \frac{e^{z_k}}{\sum_{j=1}^k e^{z_j}} \right) \in \mathbb{R}^k_+.$$

The scaled softargmax

$$\varepsilon \text{ softargmax}\left(\frac{z}{\varepsilon}\right) = \varepsilon \left(\frac{e^{\frac{z_1}{\varepsilon}}}{\sum_{j=1}^k e^{\frac{z_j}{\varepsilon}}}, \ldots, \frac{e^{\frac{z_k}{\varepsilon}}}{\sum_{j=1}^k e^{\frac{z_j}{\varepsilon}}} \right)$$

is 1−averaged. The extension to a real-valued matrix is done row by row or column by column. Thus, $X \mapsto softargmax(X)$ is 1-averaged.

Example 8.7 The rectified linear unit (ReLU) activation function, which is widely used in the architecture of generative machine intelligence, is given by $\text{ReLU}(x) = \max(0, x) = \frac{x+|x|}{2}$ and is $\frac{1}{2}$-averaged. The extension to Hilbert space is done entry-wise. Thus, $X \mapsto \text{ReLU}(X)$ is $\frac{1}{2}$-averaged.

Example 8.8 The sigmoid activation function, which is widely used in the architecture of generative machine intelligence, is given by $\sigma(x) = \frac{1}{1+e^{-x}}$ and is $\frac{5}{8}$-averaged. The extension to Hilbert space is done entry-wise. Thus, $X \mapsto \sigma(X)$ is $\frac{5}{8}$-averaged.

Example 8.9 The scaled sigmoid linear unit (SiLU) activation function, which is widely used in the architecture of generative machine intelligence, is given by $\varepsilon x \sigma(\varepsilon x)$, with $\varepsilon < \frac{1}{2}$, and is $\frac{1+2\varepsilon}{2}$-averaged. The extension to Hilbert space is done entry-wise. Thus, $X \mapsto \sigma(X)$ is $\frac{1+2\varepsilon}{2}$-averaged.

Example 8.10 The unit ball normalization function

$$X \mapsto \hat{\sigma}(X) = \frac{X}{1 + \left| X \right|}$$

defined on Hilbert space is 1-averaged.

Lemma 8.11 *The following three statements are equivalent:*

(a) The operator $\hat{O}$ *is* $\frac{1}{2}$*-averaged.*

(b) $\left|\hat{O}(x) - \hat{O}(y)\right|^2 \leq \langle x - y, \hat{O}(x) - \hat{O}(y)\rangle, \ \forall (x, y).$

(c) $(1 - \mu)Id + \mu\hat{O}$ *is* $\mu/2$*-averaged for all* $\mu \in (0, 2)$.

Proof $\hat{O}$ being $\frac{1}{2}$-averaged is equivalent to $\hat{O} = (Id + M)/2$ with M being 1–Lipschitz. This means that $M = 2\hat{O} - Id$. Hence, the operator $(1-\mu)Id+\mu\hat{O} = (1 - \mu)Id + \mu(Id + M)/2 = (1 - \mu/2)Id + (\mu/2)M$ is $\mu/2$-averaged for all $\mu \in (0, 2)$.

Conversely, if $(1 - \mu)Id + \mu\hat{O}$ is $\mu/2$-averaged for all $\mu \in (0, 2)$, then by choosing $\mu = 1$, one obtains that O is $\frac{1}{2}$-averaged. □

Lemma 8.12 $\hat{O}$ *is* $\frac{1}{2}$*-averaged if* $O = \hat{O}^{-1} - Id$ *is monotone.*

Proof Let $\hat{O}$ be $\frac{1}{2}$-averaged. Then, $\langle x - \hat{O}(x) - (y - \hat{O}(y)), \hat{O}(x) - \hat{O}(y)\rangle \geq 0, \ \forall (x, y)$. Let $\hat{O}(y_1 + x_1) = x_1$ and $\hat{O}(y_2 + x_2) = x_2$ with $y_i = Ox_i$. Set $z_i = y_i + x_i$. Let us compute $\langle y_1 - y_2, x_1 - x_2\rangle = \langle Ox_1 - Ox_2, x_1 - x_2\rangle = \langle z_1 - x_1 - (z_2 - x_2), \hat{O}(z_1) - \hat{O}(z_2)\rangle = \langle z_1 - \hat{O}(z_1) - (z_2 - \hat{O}(z_2)), \hat{O}(z_1) - \hat{O}(z_2)\rangle \geq 0$, hence $\langle Ox_1 - Ox_2, x_1 - x_2\rangle \geq 0$. This means that O is monotone.

Conversely, suppose that O is monotone. We choose $O(\hat{O}(x_i)) + \hat{O}(x_i) \ni x_i, \ i \in \{1, 2\}$ in the graph. Then, $O(\hat{O}(x_i)) \ni x_i - \hat{O}(x_i), \ i \in \{1, 2\}$. By monotonicity $\langle O(\hat{O}(x_1)) - O(\hat{O}(x_2)), \hat{O}(x_1) - \hat{O}(x_2)\rangle \geq 0$, which leads to

$$\langle (x_1 - \hat{O}(x_1)) - (x_2 - \hat{O}(x_2)), \hat{O}(x_1) - \hat{O}(x_2)\rangle \geq 0,$$

which implies that $\hat{O}$ is $\frac{1}{2}$-averaged. □

The $\frac{1}{2}$-averagedness property is important because it is associated with monotone operators, which are crucial in variational inequalities.

Remark 8.13 If $\hat{O}$ is $\frac{1}{2}$-averaged, then there is a monotone operator O such that $\hat{O} = (Id + O)^{-1}$. One can also use the operator $\hat{O}_\lambda = (Id + \lambda O)^{-1}$ approximation.

We now connect outcomes of $\mathbb{NN}$ with game theory. It answers the question of Problem 8.2.

Proposition 8.14 *Consider an* $\mathbb{NN}$ *with activation functions* $r_1, \ldots r_L$ *that are* $\frac{1}{2}$*-averaged. Then, the outcomes of* $\mathbb{NN}$ *(if any) are exactly the constrained Nash equilibria of the nonzero sum game*

$$\mathcal{G}_1 = (\mathcal{L} = \{1, \ldots, L\}, (\mathcal{H}_l, 0, C_l(.))_{\mathcal{L}}),$$

where $C_l(y) = (W_l y_{l-1} + b_l - O_l(y_l)),\ O_l$ *is a monotone operator.*

Proof For each layer l of the deep neural network, the outcome can be written as $r_l(W_l y_{l-1} + b_l) = y_l$ with $y_0 = x_0$. Since r_l is $\frac{1}{2}$-averaged, there is a monotone operator O such that $r_l = (Id + O_l)^{-1}$. It follows that $(W_l y_{l-1} + b_l - y_l) \in O_l(y_l)$. This means that $y_l \in (W_l y_{l-1} + b_l - O_l(y_l)) =: C_l(y)$:

$$\begin{cases} l \in \{1, \dots, L\} \\ \\ Inf_{\{y_l \in C_l(y)\}} 0 \end{cases} \Longrightarrow \begin{cases} y := (y_1, \dots, y_L), \\ y \in C(y) \end{cases} \Longrightarrow \begin{cases} \text{CNE of game:} \\ (\mathcal{L}, (\mathcal{H}_l, 0, C_l(.))_{\mathcal{L}}). \end{cases} \tag{8.8}$$

The latter set is the set of fixed points of the operator C. □

Our interest in averaged operators is due to the weak convergence properties of their Banach–Picard or Mann iterates to their fixed points (if any).

Proposition 8.15 *Let $0 < \gamma \leq 1$, and $\hat{O} = \hat{O}_L \circ \hat{O}_{L-1} \circ \dots \hat{O}_1 : \mathcal{H}_0 \to \mathcal{H}_0$ be an γ-averaged operator such that $\hat{O}$ has at least one fixed point. Let $\{\lambda_t\}_{t\geq 1}$ be a sequence in $[0, \frac{1}{\gamma}]$ such that $\sum_{t\geq 1} \lambda_t(1 - \gamma \cdot \lambda_t) = +\infty$. Assume that $y_0 \in \mathcal{H}_0$, and set $y_{t+1} = y_t + \lambda_t(\hat{O}(y_t) - y_t)$. Then, $\hat{O}(y_t) - y_t$ converges to 0 as t goes to infinity and y_t converges weakly to a point in Fixed($\hat{O}$).*

Proof From the assumptions we note that $\varepsilon_t = \gamma \cdot \lambda_t$ is a sequence in [0, 1] and satisfies $\sum_{t\geq 1} \varepsilon_t(1 - \varepsilon_t) = +\infty$. Set $\widetilde{O} := (1 - \frac{1}{\gamma})Id + \frac{1}{\gamma}\hat{O}$. Then, $x_{t+1} = x_t + \varepsilon_t(\widetilde{O}(x_t) - x_t)$ and Fixed($\widetilde{O}$) = Fixed($\hat{O}$). Let $x^* \in$ Fixed($\hat{O}$), i.e., $\widetilde{O}(x^*) = \hat{O}(x^*) = x^*$. We evaluate $\left|x_{t+1} - x^*\right|$. Indeed

$$\begin{aligned}
&\left|x_{t+1} - x^*\right|^2 = \left|(1 - \varepsilon_t)x_t + \varepsilon_t \widetilde{O}(x_t) - \varepsilon_t x^* - (1 - \varepsilon_t)x^*\right|^2 \\
&= \left|(1 - \varepsilon_t)(x_t - x^*) + \varepsilon_t(\widetilde{O}(x_t) - \hat{O}(x^*))\right|^2 \\
&= (1 - \varepsilon_t)^2\left|x_t - x^*\right|^2 + \varepsilon_t^2\left|\widetilde{O}(x_t) - \widetilde{O}(x^*)\right|^2 \\
&+2\varepsilon_t(1 - \varepsilon_t)\langle x_t - x^*, \widetilde{O}(x_t) - \widetilde{O}(x^*)\rangle \\
&= (1 - \varepsilon_t)^2\left|x_t - x^*\right|^2 + \varepsilon_t^2\left|\widetilde{O}(x_t) - \widetilde{O}(x^*)\right|^2 \\
&-\varepsilon_t(1 - \varepsilon_t)[\left|x_t - \widetilde{O}(x_t)\right|^2 - \left|x_t - x^*\right|^2 - \left|\widetilde{O}(x_t) - \widetilde{O}(x^*)\right|^2] \\
&= [(1 - \varepsilon_t)^2 + \varepsilon_t(1 - \varepsilon_t)]\left|x_t - x^*\right|^2 \\
&+[\varepsilon_t^2 + \varepsilon_t(1 - \varepsilon_t)]\left|\widetilde{O}(x_t) - \widetilde{O}(x^*)\right|^2 - \varepsilon_t(1 - \varepsilon_t)\left|x_t - \widetilde{O}(x_t)\right|^2 \\
&= (1 - \varepsilon_t)]\left|x_t - x^*\right|^2 + \varepsilon_t\left|\widetilde{O}(x_t) - \widetilde{O}(x^*)\right|^2 \\
&-\varepsilon_t(1 - \varepsilon_t)\left|x_t - \widetilde{O}(x_t)\right|^2.
\end{aligned} \tag{8.9}$$

Therefore

$$\begin{aligned}
&\left|x_{t+1}-x^*\right|^2 \\
&\quad = (1-\varepsilon_t)]\left|x_t-x^*\right|^2 + \varepsilon_t\left|\widetilde{O}(x_t)-\widetilde{O}(x^*)\right|^2 - \varepsilon_t(1-\varepsilon_t)\left|x_t-\widetilde{O}(x_t)\right|^2 \\
&\quad \le \left|x_t-x^*\right|^2 - \varepsilon_t(1-\varepsilon_t)\left|x_t-\widetilde{O}(x_t)\right|^2.
\end{aligned} \tag{8.10}$$

By a telescopic sum, we obtain

$$\begin{aligned}
\textstyle\sum_{t=0}^{T}\varepsilon_t(1-\varepsilon_t)\left|x_t-\widetilde{O}(x_t)\right|^2 &\le \textstyle\sum_{t=0}^{T}\left(\left|x_t-x^*\right|^2-\left|x_{t+1}-x^*\right|^2\right) \\
&= \left|x_0-x^*\right|^2-\left|x_{T+1}-x^*\right|^2.
\end{aligned} \tag{8.11}$$

As $\sum_{t=0}^{\infty}\varepsilon_t(1-\varepsilon_t)\left|x_t-\widetilde{O}(x_t)\right|^2 \le \left|x_0-x^*\right|^2 < +\infty$ and $\sum_{t=0}^{\infty}\varepsilon_t(1-\varepsilon_t) = +\infty$, it follows that the sequence $\left|x_t-\widetilde{O}(x_t)\right|^2 \to 0$ as $t\to\infty$. But, $\hat{O}-Id = \gamma(\widetilde{O}-Id)$. Thus, $\left|x_t-\hat{O}(x_t)\right|^2 \to 0$ as $t\to\infty$. □

In high dimensions, we connect some averaged operators to specific cyclic monotone operators, which are useful in deriving potential functions. It characterizes the algorithm of the question of Problem 8.2 as a subdifferential flow.

8.1.2.2 NN Outcomes Under Maximally Cyclically Monotone Activation Operations

The operator $\hat{O}$ is k-cyclically monotone if every $(x_1,\ldots,x_{k+1})$ with $x_{k+1}=x_1$ and every $(u_1,\ldots,u_k)$

$$(x_i,u_i)\in \mathrm{graph}(\hat{O}),\ 1\le i\le k,\ x_{k+1}=x_1$$

implies that

$$\sum_{i=1}^{k}\langle x_{i+1}-x_i, u_i\rangle \le 0.$$

$\hat{O}$ is cyclically monotone if it is k-cyclically monotone for every integer $k\ge 2$.

We next introduce the notion of maximal cyclic monotonicity. A maximally cyclically monotone $\hat{O}$ is cyclically monotone such that there is no other cyclically monotone operator that properly contains the graph of $\hat{O}$.

The next lemma shows that a subdifferential of a proper and convex function is cyclically monotone and it is maximal.

Lemma 8.16 *The subdifferential ∂h is maximally cyclically monotone for every proper and convex function h.*

Proof To see it, we fix an integer $k \geq 2$. For every $1 \leq i \leq k$, consider (x_i, u_i) in the graph of the subdifferential of h. Set $x_{k+1} = x_1$. By the definition of subdifferential, $\langle x_{i+1} - x_i, u_i \rangle \leq h(x_{i+1}) - h(x_i)$. Summing up these inequalities leads to

$$\sum_{i=1}^{k} \langle x_{i+1} - x_i, u_i \rangle \leq \sum_{i=1}^{k} h(x_{i+1}) - h(x_i) = h(x_{k+1}) - h(x_1) = h(x_1) - h(x_1) = 0.$$

□

Example 8.17 Consider the softargmax defined in Example 8.6. One has $\{\text{softargmax}(z)\} = (\partial \phi)(z)$ where

$$\phi(z) = \varepsilon \log \left(\sum_{j=1}^{k} e^{\frac{z_j}{\varepsilon}} \right),$$

which is proper and convex. Hence, the softargmax activation is maximally cyclically monotone.

Example 8.18 The rectified linear unit activation function has a convex antiderivative h : $\text{ReLU}(x) = \partial_x h(x)$ with $h(x) = \frac{1}{2}\text{ReLU}^2(x)$, and hence, it is a maximally cyclically monotone activation operator.

Example 8.19 The sigmoid activation function has a convex antiderivative h : $\sigma(x) = \partial_x h(x)$ with $h(x) = \log(1 + e^x)$, and hence, it is a maximally cyclically monotone activation operator.

The next lemma shows that the converse is actually true. To show it, we construct explicitly a supremum of linear functions from the cyclically monotone inequalities.

Lemma 8.20 *Every maximally cyclically monotone operator $\hat{O}$ is a subdifferential for some proper convex function ϕ.*

Proof Suppose that $\hat{O}$ is a maximally cyclically monotone operator. Then, $graph(\hat{O})$ is non-empty. There is an element (x_0, u_0) in the graph of $\hat{O}$. Let us define the function:

$$\phi(x) = \sup_{k \geq 1} \sup_{(x_i, u_i) \in graph(\hat{O}), 1 \leq i \leq k} \langle x - x_k, u_k \rangle + \sum_{i=0}^{k-1} \langle x_{i+1} - x_i, u_i \rangle.$$

By cyclic monotonicity, $\phi(x_0) = 0$ and ϕ are proper and convex as a supremum of linear maps. Take (x, u) in the graph of $\hat{O}$ and let $-\infty < \eta < \phi(x)$. There are finitely many points (x_i, u_i) in $graph(\hat{O})$, $0 \leq i \leq k$ such that $\langle x - x_k, u_k\rangle + \sum_{i=0}^{k-1}\langle x_{i+1} - x_i, u_i\rangle > \eta$.

We now set $(x_{k+1}, u_{k+1}) = (x, u)$. For every y, $\phi(y) \geq \langle y - x_{k+1}, u_{k+1}\rangle + \sum_{i=0}^{k}\langle x_{i+1} - x_i, u_i\rangle = \langle y - x, u\rangle + \langle x - x_k, u_k\rangle + \sum_{i=0}^{k-1}\langle x_{i+1} - x_i, u_i\rangle > \langle y - x, u\rangle + \eta$.

We deduce that $\phi(y) \geq \langle y - x, u\rangle + \eta$. By letting η go to $\phi(x)$, we obtain that $\phi(y) \geq \langle y - x, u\rangle + \phi(x)$. The latter inequality shows that $u \in \partial\phi$. This implies that the $graph(\hat{O})$ is a subset of $graph(\partial\phi)$. By maximality, we obtain $\hat{O} = \partial\phi$. □

From the relation

$$r_l = [Id + \partial f_l]^{-1} = \left[\partial\left(\frac{1}{2}\Big|\cdot\Big|^2 + f_l\right)\right]^{-1} = \partial\left(\frac{1}{2}\Big|\cdot\Big|^2 + f_l\right)^*,$$

we have that $\phi_l = (\frac{1}{2}\Big|\cdot\Big|^2 + f_l)^*$ and $r_l = \partial\phi_l$, where, as usual, $\psi^*(x) = \sup_y[\langle x, y\rangle - \psi(y)]$ is the Legendre–Fenchel conjugate of ψ.

We now characterize the question of Problem 8.2 with more explicit functions.

Proposition 8.21 *Consider* $\mathbb{NN}$ *with activation functions* $r_1, \ldots r_L$ *that are maximally cyclically monotone. Then, the outcomes of* $\mathbb{NN}$ *(if any) are Nash equilibria of the nonzero sum game*

$$\mathcal{G}_2 = (\mathcal{L} = \{1, \ldots, L\}, (\mathcal{H}_l, g_l)_{\mathcal{L}}),$$

where $g_l(y) = \frac{1}{2}\Big|(W_l y_{l-1} + b_l - y_l)\Big|^2 + f_l(y_l)$.

Proof For each layer l of the deep neural network, the outcome can be written as $r_l(W_l y_{l-1} + b_l) = y_l$ with $y_0 = x_0$. Since r_l is maximally cyclically monotone, there is a proper and convex f_l such that $r_l = (Id + \partial f_l)^{-1}$, i.e., $O_l = \partial f_l$. It follows that $(W_l y_{l-1} + b_l - y_l) \in \partial f_l(y_l)$. By convexity of $\frac{1}{2}\Big|.\Big|^2 + f_l$, the latter inclusion is equivalent to the argmin of $y_l \mapsto \frac{1}{2}\Big|(W_l y_{l-1} + b_l - y_l)\Big|^2 + f_l(y_l)$. This leads to Nash equilibria of the game $\mathcal{G}_2$. □

Remark 8.22 We have seen that the set of outcomes of neural networks in the long run is exactly the set of Nash equilibria of a constrained Nash game. The game is explicitly described in the case of $\frac{1}{2}$-averaged activation operators as well as in the maximally cyclically monotone operators. These characterizations have several advantages. One of the advantages is the use of existing advanced strategic learning algorithms, especially in the case of operators that are not necessarily globally monotone in all the variables but component-wise (layer-wise) monotone. We highlight here the existence of a fixed point. For contraction operators (Lipschitzian with Lipschitz constant strictly less than 1), the existence of a unique fixed point follows from the Banach theorem. For nonexpansive operators, those with a Lipschitz constant is one, a fixed point may not exist, or there may be an infinite number of fixed points. For single-valued operators, one can use the Brouwer–Schauder fixed-point theorem for continuous functions over a convex, closed, bounded, non-empty set into itself. However, here, the space is not bounded (the input signal can be arbitrarily high). Therefore, the Brouwer–Schauder fixed-point theorem is not directly applicable. The closed unit ball is a compact set in finite dimensions, but it is not a compact set in Hilbert spaces (infinite dimensions). Another case of the existence of a fixed-point is Tarski's fixed-point theorem for monotonic operators (sub/super-modularity) over partially ordered sets.

8.1.2.3 Training

We now focus on the training problem given a data set. Given $(x_i, y_{L,i})_{i\in\{1,\ldots,D\}}$, the training problem (limited to weights and biases) is to find $\theta = (W, b) = (W_l, b_l)_l$ such that the neural network $(x_i, L, r, W, b, \{\mathcal{H}_l\}_{1\le l\le L})$ produces an output that matches y_i at each iteration i. This means that

$$(r_L \circ A_{L,i,\theta_L}) \circ \ldots \circ (r_1 \circ A_{1,i,\theta_1})(x_i) = y_{L,i}, \quad t \in \{1, \ldots, D\},$$

where $\theta_l := (W_l, b_l)$ and $A_{l,t,\theta_l} = A_{l,t,(W_l,b_l)} : (W_l, b_l) \mapsto W_l y_{l-1,i} + b_l$ with $y_{l-1,i}$ being the output from layer $l-1$.

Thus, given the data $(x_i, y_{L,i})_{i\in\{1,\ldots,D\}}$, the training problem is to find $\theta^* = (\theta_1^*, \ldots, \theta_L^*)$ such that

$$r_L \circ A_{L,i,\theta_L^*} \circ \ldots \circ (r_1 \circ A_{1,i,\theta_1^*})(x_i) - y_{L,i} = 0, \quad i \in \{1, \ldots, D\}. \tag{8.12}$$

In the sequel, we will denote (for each t) $A_{l,i} := A_{l,i,\theta_l}$ and $A_{l,i}^{\dagger}$ to be its adjoint operator.

8.1.2.4 Training Under Averaged Nonexpansive Activations

We have the following lemma whose proof is immediate.

Lemma 8.23 *Let r_l be a $\frac{1}{2}$-averaged nonexpansive function. Then, for any z, we have*

$$r_l(A_l\theta_l) - z = 0 \Longleftrightarrow A_l\theta_l - (Id + O_l)z = 0,$$

where O_l is a maximally monotone operator.

Proposition 8.24 *Suppose that each r_l is an averaged nonexpansive activation. Then, the solutions (if any) of the training of weights and bias problem (8.12) are* *also solutions of the following variational inclusion: Given an input–output data set* $\{(x^*_{0,i}, y^*_{L,i}),\ i \in \{1, 2, \ldots, D\}$*, find* (W^*, b^*) *such that*

$$0 \in A^{\dagger}_{l,i}[y^*_{l,i} + O_l(y^*_{l,i}) - (W^*_l y^*_{l-1,i} + b^*_l)], \quad i \in \{1, 2, \ldots, D\}. \tag{8.13}$$

These are also Nash equilibria of the nonzero sum game $\mathcal{G}_n$ *given by*

$$\begin{cases} l \in \mathcal{L}, \\ (W^*_l, b^*_l) \in \arg\min_{W_l, b_l} \sum_{i=1}^{D} \omega_{l,i} \left| y^*_{l,i} + O_l(y^*_{l,i}) - (W_l y^*_{l-1,i} + b_l) \right|^2. \end{cases} \tag{8.14}$$

Proof The proof uses Lemma 8.23 to connect the zeros to the minimization. □

Problem 8.3 is on sampled input–output data. One can also formulate the problem not on sampled data but on the exact distribution $\mathbb{P}(dx_0 dy_L)$.

Given $\mathbb{P}(dx_0 dy_L)$, find a vector $\theta^* := (W^*, b^*) = (W^*_l, b^*_l)_l$ such that

$$E\left(\left\langle (r_L \circ \tilde{A}_{L,\theta^*_L}) \circ \ldots \circ (r_1 \circ \tilde{A}_{1,\theta^*_1})(x_0) - y_L, A^{\dagger}_L(\theta_L - \theta^*_L) \right\rangle\right) \geq 0, \ \ \forall \theta. \tag{8.15}$$

Moreover, design an algorithm that approximates a θ^*, a solution (if any) to (8.15).

8.1.2.5 Training Under Maximally Cyclically Monotone Activations

Proposition 8.25 *Suppose that each* r_l *is maximal cyclically monotone activation. Then, the solutions (if any) of the training of weights and bias problem (8.12) are also solutions of the following variational inclusion: Given an input–output data set* $\{(x^*_{0,i}, y^*_{L,i}),\ i \in \{1, 2, \ldots, D\}$*, find* (W^*, b^*) *such that*

$$0 \in A^{\dagger}_{l,i}[y^*_{l,i} + \partial f_l(y^*_{l,i}) - (W^*_l y^*_{l-1,i} + b^*_l)], \quad i \in \{1, 2, \ldots, D\}. \tag{8.16}$$

These are also Nash equilibria of the nonzero sum game $\mathcal{G}_3$ *given by*

$$\begin{cases} l \in \mathcal{L}, \\ (W^*_l, b^*_l) \in \arg\min_{W_l, b_l} \sum_{i=1}^{D} \omega_{l,i} \left| y^*_{l,i} + \partial f_l(y^*_{l,i}) - (W_l y^*_{l-1,i} + b_l) \right|^2. \end{cases} \tag{8.17}$$

Proof Since each of the activation operators r_l is maximal cyclically monotone activation, $r_l = (Id + \partial f_l)^{-1}$, i.e., $O_l = \partial f_l$ for a proper, convex lower semicontinuous function f_l. The training problem is to find $\theta = (W, b) = (W_1, b_1, \ldots, W_L, b_L)$ such that for each $1 \leq i \leq D$,

$$
\begin{cases}
(W_1 x_{0,t}^* + b_1) \in y_{1,i}^* + \partial f_1(y_{1,i}^*),\\
(W_2 x_{1,i}^* + b_2) \in y_{2,i}^* + \partial f_2(y_{2,i}^*),\\
\dots\\
\dots\\
(W_l x_{l-1,i}^* + b_l) \in y_{l,i}^* + \partial f_l(y_{l,i}^*),\\
\dots\\
\dots\\
(W_{L-1} x_{L-2,i}^* + b_{L-1}) \in y_{L-1,i}^* + \partial f_{L-1}(y_{L-1,i}^*),\\
(W_L x_{L-1,i}^* + b_L) \in y_{L,i}^* + \partial f_L(y_{L,i}^*),
\end{cases}
\tag{8.18}
$$

which is rewritten as

$$
\begin{cases}
(W_1^*, b_1^*) \in \arg\min_{W_1, b_1} \left| y_{1,i}^* + \partial f_1(y_{1,i}^*) - (W_1 x_{L,i}^* + b_1)\right|^2,\\
(W_2^*, b_2^*) \in \arg\min_{W_2, b_2} \left| y_{2,i}^* + \partial f_2(y_{2,i}^*) - (W_2 x_{1,i}^* + b_2)\right|^2,\\
\dots\\
\dots\\
(W_l^*, b_l^*) \in \arg\min_{W_l, b_l} \left| y_{l,i}^* + \partial f_l(y_{l,i}^*) - (W_l x_{l-1,i}^* + b_l)\right|^2,\\
\dots\\
\dots\\
(W_L^*, b_L^*) \in \arg\min_{W_L, b_L} \left| y_L^* + \partial f_L(y_{L,i}^*) - (W_L x_{L-1,i}^* + b_L)\right|^2.
\end{cases}
\tag{8.19}
$$

But, this set is exactly the set of Nash equilibria of the nonzero sum game:

$$
\mathcal{G}_3 = \left(\mathcal{L}, \prod_{l=1}^{L} L_2(\mathcal{H}_{l-1}, \mathcal{H}_l) \times \mathcal{H}_l, \left| y_l^* + \partial f_l(y_l^*) - b_l - W_l x_{l-1}^* \right|^2 \right).
$$

Moreover, by Fermat's principle, the last equation in (8.19) implies that

$$
0 \in A_{l,i}^\dagger [y_{l,i}^* + \partial f_l(y_{l,i}^*) - (W_l^* x_{l-1,i}^* + b_l^*)],
$$

which is the first part of the result.

For each $l \in \mathcal{L}$, let $\omega_{l,i} > 0$, $\sum_{i=1}^{D} \omega_{l,i} = 1$. Then, clearly, the system of variational inequalities (8.19) yields

$$
\langle (W_l, b_l) - (W_l^*, b_l^*), \sum_{i=1}^{D} \omega_{l,i} A_{l,i}^\dagger [y_{l,t}^* + \partial f_l(y_{l,i}^*) - (W_l^* y_{l-1,i}^* + b_l^*)] \rangle \geq 0
$$

for all (W_l, b_l). Hence, the set of solutions of the training of weights and bias problem (8.12) is the set of Nash equilibria of the nonzero sum game:

$$\begin{cases} l \in \mathcal{L}, \\ (W_l^*, b_l^*) \in \arg\min_{W_l, b_l} \sum_{i=1}^{D} \omega_{l,i} \left| y_{l,i}^* + \partial f_l(y_{l,i}^*) - (W_l y_{l-1,i}^* + b_l) \right|^2. \end{cases} \tag{8.20}$$

This completes the proof. □

We have the following lemma whose proof is immediate.

Lemma 8.26

(i) For any z, we have

$$A_l^\dagger (r_l(A_l\theta) - z) = \partial_\theta [\phi_l(A_l\theta) - \langle A_l\theta, z\rangle],$$

where $\partial\phi_l = r_l$.

(ii) If there exists θ^* *solving* $r_l(A_l\theta^*) = z$*, then it coincides with the minimizer of* $\theta \mapsto \phi_l(A_l\theta) - \langle A_l\theta, z\rangle$.

Lemma 8.27 *Let L be an integer with* $L \geq 1$*. For every* $l \in \{1, \ldots, L\}$*, let* $\hat{O}_l$ *be* γ*-averaged, and it can be written as* $\hat{O}_l = (Id + \partial f_l)^{-1}$*. Let* $0 < \varepsilon < 1$*. Suppose that there is a solution* θ_l *to the equation* $z_l - \hat{O}_l(\theta_l) = 0$*. Then, the algorithm* $\theta_l^{p+1} = \theta_l^p + \varepsilon(z_l - \hat{O}_l(\theta_l^p))$*, given initial point* θ^0*, converges to a solution to* $\hat{O}_l(\theta_l) = z_l$ *as p goes to infinity.*

Proof By assumption, $Id - \hat{O}_l$ and $\hat{O}_l$ are both minimizers of proximal convex functions and are in particular 1-Lipschitz continuous. We can write $Id - \varepsilon\hat{O}_l$ as an averaged nonexpansive operator as follows:

$$Id - \varepsilon\hat{O}_l = (1-\varepsilon)Id + \varepsilon Id - \varepsilon\hat{O}_l = (1-\varepsilon)Id + \varepsilon(Id - \hat{O}_l),$$

which is ε-averaged. This means that the operator $\tilde{O}_l = Id - \varepsilon\hat{O}_l + \varepsilon z_l$ is ε-averaged. Then, the result on the fixed points of $\tilde{O}_l$ follows Lemma 8.15. □

Lemma 8.28 *Let* A_l *be linear and nonzero and* f_l *be a proper convex function from* $range(A_l)$ *to* $\mathbb{R}$*. Let* $h_l(x) = \inf_{z \in range(A_l)} (f_l^*(z) + \frac{1}{2}\left|z - x\right|^2)$*. Then,* $x \in z + \partial f_l^*(z)$*, i.e.,* $z \in (Id + \partial f_l^*)^{-1}(x)$*. Further*

$$\left|A\right|^2 \leq 1 \implies A_l^\dagger \circ (Id + \partial f_l)^{-1} \circ A_l = (Id + \partial g_l)^{-1},$$

where $g_l(z) = (h_l \circ A_l)^*(z) - \frac{1}{2}\left|z\right|^2$.

Proof From the Fenchel conjugate, one has

$$(Id + \partial f_l)^{-1} = \left[\partial\left(\frac{1}{2}\left|\cdot\right|^2 + f_l\right)\right]^{-1} = \partial\left[\left(\frac{1}{2}\left|\cdot\right|^2 + f_l\right)^*\right] = \partial p_l^*,$$

with $p_l = \frac{1}{2}\left|.\right|^2 + f_l$, i.e., $f_l = p_l - \frac{1}{2}\left|.\right|^2$.

Let $h_l(f_l^*, x) = \inf_z (f_l^*(z) + \frac{1}{2}\left|z - x\right|^2)$. Then

$$\inf_z \left(f_l^*(z) + \frac{1}{2}\left|z - x\right|^2 \right) + \inf_z \left(f_l(z) + \frac{1}{2}\left|z - x\right|^2 \right) = \frac{1}{2}\left|x\right|^2,$$

i.e.,

$h_l(f_l^*, x) + h_l(f_l, x) = \frac{1}{2}\left|x\right|^2$. By differentiating, one obtains

$$\partial(h_l(f_l^*, x)) + \partial(h_l(f_l, x)) = x.$$

Hence

$$A_l^\dagger \circ (Id + \partial f_l)^{-1} \circ A_l = A_l^\dagger \circ \partial p_l^* \circ A_l = \partial(p_l^* \circ A_l).$$

□

Proposition 8.29 (Activation Functions That Have Antiderivatives) *Suppose the training problem has at least one solution. Then, the set of solutions of the training problem coincides with the set of Nash equilibria of the following layer-by-layer nonzero sum game $\mathcal{G}_4$:*

$$\arg\min_{\theta_l} \sum_{i=1}^{D} \omega_{l,i}[\phi_l(A_{l,i}\theta_l) - \langle A_{l,i}\theta_l, y_{l,i}\rangle] \;\; l \in \mathcal{L}, \;\; \omega_{l,i} > 0, \;\; \sum_{i=1}^{D} \omega_{l,i} = 1. \tag{8.21}$$

Moreover, given θ_l^0, the algorithm defined for each i by

$$\theta_l^{p+1} = \theta_l^p - \frac{\gamma}{2\left|A_{l,i}\right|^2} A_{l,i}^\dagger [r_l(A_{l,i}\theta_l^p) - y_{l,i}], \; 0 < \gamma < 1$$

converges to a minimizer θ_l of Problem (8.21)*, as $p \to \infty$.*

Proof Since, for each $i \in \{1, \ldots, D\}$, $r_l(A_{l,i}\theta) = y_{l,i}$ admits a solution $\theta_{l,i}$, then by Lemma 8.26 (i), we have

$$0 \in A_l^\dagger [r_l(A_{l,i}\theta_l) - y_{l,i}] = \partial_\theta[\phi_l(A_{l,i}\theta_l) - \langle A_{l,i}\theta_l, y_{l,i}\rangle], \quad i \in \{1, \ldots, D\},$$

with $\phi_l := (\frac{1}{2}\left| \cdot \right|^2 + f_l)^*$. Therefore, by Lemma 8.26 (ii), $\theta_{l,i}^*$ is a global minimizer of $\phi_l(A_{l,i}\theta_l) - \langle A_{l,i}\theta_l, y_{l,i}\rangle$. This in turn yields that θ_l^* minimizes the weighted function $\sum_{i=1}^D \omega_{l,i}[\phi_{l,i}(A_{l,t}\theta_l) - \langle A_{l,i}\theta_l, y_{l,i}\rangle]$.

Noting that the inclusion $0 \in A^{\dagger}_{l,i}[r_l(A_{l,i}\theta^*_l) - y_{l,i}]$ implies the inclusion $\theta^*_l \in \theta^*_l + A^{\dagger}_{l,i}[r_l(A_{l,i}\theta^*_l) - y_{l,i}]$, we examine the fixed points of the operator $\theta \mapsto \theta + A^{\dagger}_{l,i}[r_l(A_{l,i}\theta) - y_{l,i}]$. Since $\frac{1}{2\left|A_{l,i}\right|^2}(A^{\dagger}_{l,i} \circ r_l \circ A_{l,i})$ is a γ-averaged operator for each i, given $\theta^0_{l,i}$, we can use Lemma 8.27: The iterates

$$\theta_l^{p+1} = \theta_l^p - \frac{\gamma}{2\left|A_{l,i}\right|^2} A^{\dagger}_{l,i}[r_l(A_{l,i}\theta_l^p) - y_{l,i}],\ 0 < \gamma < 1,$$

converge to $\theta_{l,i}$, as $p \to \infty$. The latter can also be retrieved from the convexity and Lipschitz gradient of $\theta \mapsto \phi_l(A_l\theta) - \langle A_l\theta, z\rangle$. This completes the proof. □

Remark 8.30 In general, after a huge training phase, one would like to have weights and bias $\theta = (W, b)$, r_l are time-and-signal input invariant so that the architecture of a deep neural network is fixed for any data set. This means that the operator $A_{l,i}$, which contains $x_{l,i} = y_{l-1,i}$ is signal input-dependent as the input changes. Thus, one would like to find $\theta_{l,i} = \theta_l$ at level l.

Let us consider the formulation in (8.15) of the training problem based only on the distribution $\mathbb{P}(dx_0 dy_L)$. The algorithm is written in a stochastic approximation perspective with state-dependent noise:

$$\theta^{p+1} = \theta^p - \lambda_p B(\theta^p, x_0, y_L)$$

$$= \theta^p - \lambda_p\{E_{\mathbb{P}(dx_0'dy_L')}[B(\theta^p, x_0', y_L')] + B(\theta^p, x_0, y_L) - E_{\mathbb{P}(dx_0'dy_L')}[B(\theta^p, x_0', y_L')]\}.$$

Here, the noise is $B(\theta_i, x_0, y_L) - E_{\mathbb{P}(dx_0'dy_L')}[B(\theta_i, x_0', y_L')]$.

One can also consider a mini-batch stochastic gradient with mini-batch size D :

$$\begin{cases} \theta^{p+1} = \theta^p - \lambda_p E_{\mathbb{P}(dx_0'dy_L')}[B(\theta^p, x_0', y_L')] \\ -\lambda_p\{\frac{1}{D}\sum_{i=1}^{D} B(\theta^p, x_{0,i}, y_{L,i}) - E_{\mathbb{P}(dx_0'dy_L')}[B(\theta^p, x_0', y_L')]\}. \end{cases} \tag{8.22}$$

Remark 8.31 We have the following result:

- Direct gradient descent algorithms and stochastic gradient algorithms have several limitations when the underlying cost function is weakly convex but not strongly convex and not η-smooth and has an unbounded gradient or a non-Lipschitz gradient.
- Let ϕ_l be proper and convex, $x^* \in \arg\min \phi_l$, and given an initial point $x(0) = x_0$, run the gradient flow $\dot{x}(t) = -\nabla\phi_l(x(t))$. Then, the function $t \mapsto \phi_l(x(t))$ decreases. Indeed

$$\frac{d}{dt}\phi_l(x(t)) = \langle -\nabla\phi_l(x(t)), \nabla\phi_l(x(t))\rangle = -\left|\nabla\phi_l(x(t))\right|^2 \le 0.$$

We deduce that $\phi_l(x(t)) \leq \phi_l(x(t'))$ for all $0 \leq t' < t$. Integrating over t', one obtains

$$\phi_l(x(t)) \leq \frac{1}{t}\int_0^t \phi_l(x(t'))dt'. \tag{8.23}$$

We now show that the average regret $\frac{1}{T}\int_0^T(\phi_l(x(t)) - \phi_l(x^*))dt$ vanishes. $\frac{d}{dt}(\frac{1}{2}\left|x(t) - x^*\right|^2) = -\langle\nabla\phi_l(x(t)), x(t) - x^*\rangle = \langle\nabla\phi_l(x(t)), x^* - x(t)\rangle + \phi_l(x(t)) - \phi_l(x(t)) \leq \phi_l(x^*) - \phi_l(x(t))$, where we have used the convexity of ϕ_l.

By integrating over time, $\frac{1}{2}\left|x(t) - x^*\right|^2 - \frac{1}{2}\left|x(0) - x^*\right|^2 \leq \int_0^t(\phi_l(x^*) - \phi_l(x(t')))dt'$.

This implies that the average regret satisfies $\frac{1}{T}\int_0^T(\phi_l(x(t)) - \phi_l(x^*))dt \leq \frac{1}{2T}\left|x(0) - x^*\right|^2$ and by inequality (8.23)

$$0 \leq \phi_l(x(T)) - \phi_l(x^*) \leq \frac{1}{T}\int_0^T(\phi_l(x(t)) - \phi_l(x^*))dt \leq \frac{1}{2T}\left|x(0) - x^*\right|^2$$

for any $T > 0$.

- In this case, $\phi_l(x(t))$ converges, but $x(t)$ does not necessarily converges, and it may oscillate. To reduce oscillations, one can consider the time-average trajectory to break cyclic behavior and or limit cycle. If the first-level time average does not converge, one can use a second-level time average, and so on.
- The time-average trajectory $\frac{1}{T}\int_0^T x(t)dt$ satisfies

$$\begin{aligned}\phi_l\left(\frac{1}{T}\int_0^T x(t)dt\right) - \phi_l(x^*) &\leq \frac{1}{T}\int_0^T(\phi_l(x(t)) - \phi_l(x^*))dt \\ &\leq \frac{1}{2T}\left|x(0) - x^*\right|^2.\end{aligned} \tag{8.24}$$

- For generative machine intelligence outputs, the trajectory of $x(t)$ is extremely important as decisions may be made based on the output $x(T)$. For example, if the output $x(T)$ changes, oscillates, or has different meanings, then the decisions may not be reliable.
- We consider the implicit Euler method gradient descent written as $x_{t+1} - x_t = -\varepsilon\nabla\phi_l(x_{t+1})$, i.e., $\frac{x_{t+1}-x_t}{\varepsilon} + \nabla\phi_l(x_{t+1}) = 0$, which means that x_{t+1} is an extremum of

$$y \longmapsto -\phi_l(y) - \frac{(y - x_t)^2}{2\varepsilon}.$$

The methodology extends to a metric space (with distance d) where the interpolation of the algorithm

$$x_{t+1} \in \arg\max_{y} \left[-\phi_l(y) - \frac{d(y, x_t)^2}{2\varepsilon} \right]$$

converges to a solution of the gradient flow:

$$\dot{x} = -\partial_x \phi_l(x),$$

as ε goes to zero.

These algorithms are improved in terms of speed in the next section. The results do not assume strong convexity as they are not fulfilled by the layer operators.

8.1.2.6 Beyond Gradient Descent

In this section, we work with maximally cyclically monotone activation functions r_l. The quality, distribution, diversity and integrity of the data $\mathscr{D} = \{(x_{0,i}, y_{L,i}) \in \mathscr{H}_0 \times \mathscr{H}_L, 1 \leq i \leq D\}$, for an integer $D \geq 1$, used in the pre-training and training phases are crucial in maintaining the model's output reliability.

In many cases, the true distribution of the underlying sample data $\mathscr{D}$ over $\Omega = \mathscr{H}_0 \times \mathscr{H}_L$ is not available or unknown to the decision-maker. We formulate the problem from the variations $\tilde{m}$ around the data-driven measured or observed distribution m. For convex and proper function f, we define f-divergence as

$$D_f(\tilde{m}||m) = \int_{\omega} f\left(\frac{d\tilde{m}}{dm}\right) dm - f(1).$$

The term $f(1)$ is used for controlling the sign. The f-divergence is not a distance metric as it is not symmetric. However, it can be used to capture discrepancies between two probability measures. For $\rho_l > 0$, the distributional uncertainty set driven by f-divergence ball is

$$B_{\rho_l}(m) = \{\tilde{m}| \int_{\Omega} \tilde{m}(d\omega) = 1, \; D_f(\tilde{m}||m) \leq \rho_l\}.$$

Let $(\Omega, \mathscr{F}, \tilde{m})$ be a measure space, f a measurable function, and ν an absolutely continuous measure, i.e., there exists $\rho(.) > 0$ such that $\nu(dx) = \rho(x)\tilde{m}(dx)$. Let g be a function such that e^g is $\tilde{m}$-integrable. The Legendre–Fenchel transform of the relative entropy yields

$$\sup_{\nu} \left\{ \int g(x)\nu(dx) - \int \log(\rho(x))\nu(dx) \right\} = \log\left(\int_{\Omega} e^{g(\omega)} \tilde{m}(d\omega) \right).$$

From the above relation, we deduce that

$$\sup_\nu \left\{ \int g_l(x)\nu(dx) - \frac{1}{\varepsilon_l}\int \log(\rho(x))\nu(dx) \right\}$$
$$= \sup_\nu \left\{ \int g_l(x)\nu(dx) - \frac{1}{\varepsilon_l}\int \log(\frac{\nu(dx)}{\tilde{m}(dx)})\nu(dx) \right\} = \frac{1}{\varepsilon_l}\log\left(\int_\Omega e^{\varepsilon_l g_l(\omega)}\tilde{m}(d\omega)\right), \tag{8.25}$$

which is the log of the expected value of exponentiated loss. For a very small ε_l, one can connect the latter expression to mean–variance through

$$\frac{1}{\varepsilon_l}\log\left(\int_\Omega e^{\varepsilon_l g_l(\omega)}\tilde{m}(d\omega)\right) = E_{\tilde{m}}g_l + \frac{\varepsilon_l}{2}\text{var}_{\tilde{m}}[g_l] + O(\varepsilon_l^2).$$

It follows from the previous observation that minimizing the log of the expected value of exponentiated loss can be seen as a distributionally robust optimization:

$$\inf_{\theta_l\in\mathscr{H}_l}\frac{1}{\varepsilon_l}\log\left(\int_\Omega e^{\varepsilon_l g_l(\theta,\omega)}\tilde{m}(d\omega)\right)$$
$$= \inf_{\theta_l\in\mathscr{H}_l}\sup_\nu \left\{ \int g_l(\theta,\omega)\nu(dx) - \frac{1}{\varepsilon_l}\int \log\left(\frac{\nu(dx)}{\tilde{m}(dx)}\right)\nu(dx) \right\}, \tag{8.26}$$

with the distributional constraint that the measure μ is absolutely continuous with respect to $\tilde{m}$.

We work with the sample game

$$\mathscr{G}_6 = (\mathscr{L} = \{1,\ldots,L\}, \Omega = (\mathscr{H}_0 \times \mathscr{H}_L), (\Theta_l, g_l)_{\mathscr{L}}),$$

with a given $\omega = (x, y)$, and where $g_l(\theta, x, y) = g_l(\theta,\omega) = \phi_l(A_l\theta) - \langle A_l\theta, y_l\rangle$.

Definition 8.32 Distributionally robust risk-neutral game:

Consider a set $\mathscr{L}$ of decision-makers with $|\mathscr{L}| = L \geq 2$. Each decision-maker $l \in \mathscr{L}$ has its own loss functional $E_{\tilde{m}}g_l$ with $g_l : \prod_{l\in\mathscr{L}}\Theta_l \times \Omega \to \mathbb{R}$ expressed as $g_l(\theta,\omega)$. Further, the distributionally robust risk-neutral game is given by

$$G_{dr} = \left(\mathscr{L}, (\Theta_l \times B_{\rho_l}(m), E_{\tilde{m}}g_l)_{l\in\mathscr{L}}, (\rho_l)_{l\in\mathscr{L}}, m\right).$$

The distributionally robust equilibrium problem of G_{dr} is formulated as follows:

$$l \in \mathscr{L},$$
$$\inf_{\theta_l\in\Theta_l}\sup_{\tilde{m}\in B_{\rho_l}(m)} E_{\tilde{m}}g_l(\theta,\omega). \tag{8.27}$$

Example 8.33 In the rapidly evolving landscape of machine intelligence, it is crucial to recognize and address the unique challenges faced by diverse communities. In many countries in Africa, where audio literacy plays a central role in communication and cultural heritage, the integration of machine intelligence technologies presents both unprecedented opportunities and obstacles. The significance of ensuring that the benefits of machine intelligence reach audio-literate populations cannot be overstated. Despite the growing recognition of machine intelligence's potential in Africa, there exists a notable research gap in tailoring risk assessment frameworks to the specific needs of audio-literate communities. Conventional risk measures often fall short in capturing the nuances of oral traditions and audio-centric communication prevalent in many parts of the continent. This type of analysis with G_{dr} bridges this gap by proposing a data-driven distributionally robust coherent risk assessment framework explicitly designed for machine intelligence applications in Africa. m is the audio data sampled in 7000 dialects on the continent and 2000 local languages in Africa. The true distribution of these local languages is unknown to current machine intelligence systems. The imperative lies not only in acknowledging the existence of this gap but also in actively addressing it. By doing so, we pave the way for responsible and culturally sensitive integration of machine intelligence technologies, ensuring that they empower, rather than marginalize, the diverse and vibrant audio-literate communities across the continent.

We transform the distributionally robust term as follows. The distributionally robust term is given by

$$\begin{aligned} &\sup_{\tilde{m}\in B_{\rho_l}(m)} E_{\tilde{m}} g_l(\theta,\omega) \\ &= \inf_{\lambda_l\geq 0,\mu} E_m \left\{\lambda_l(\rho_l + f(1)) + \mu + \lambda_l f^*\left(\frac{g_l(\theta,\omega)-\mu}{\lambda_l}\right)\right\}. \end{aligned} \tag{8.28}$$

Proposition 8.34 (Triality Theory) *Let h_3 be a real-valued function $(\theta_1,\theta_2,\theta_3)\mapsto h_3(\theta_1,\theta_2,\theta_3)$ defined on $\prod_{k=1}^{3}\Theta_k$. Then, the following inequalities hold:*

$$\begin{aligned} &\sup_{\theta_2\in\Theta_2} inf_{\theta_1\in\Theta_1,\theta_3\in\Theta_3} h_3(\theta_1,\theta_2,\theta_3) \leq \\ &inf_{\theta_3\in\Theta_3}\sup_{\theta_2\in\Theta_2} inf_{\theta_1\in\Theta_1} h_3(\theta_1,\theta_2,\theta_3) \leq \\ &inf_{\theta_1\in\Theta_1,\theta_3\in\Theta_3}\sup_{\theta_2\in\Theta_2} h_3(\theta_1,\theta_2,\theta_3), \end{aligned} \tag{8.29}$$

and similarly

$$\begin{aligned} &\sup_{\theta_1\in\Theta_1,\theta_3\in\Theta_3} inf_{\theta_2\in\Theta_2} h_3(\theta_1,\theta_2,\theta_3) \leq \\ &\sup_{\theta_3\in\Theta_3} inf_{\theta_2\in\Theta_2}\sup_{\theta_1\in\Theta_1} h_3(\theta_1,\theta_2,\theta_3) \leq \\ &inf_{\theta_2\in\Theta_2}\sup_{\theta_1\in\Theta_1,\theta_3\in\Theta_3} h_3(\theta_1,\theta_2,\theta_3). \end{aligned} \tag{8.30}$$

A proof of Proposition 8.34 can be obtained from the duality inequality. ∎

Based on the above, the distributionally robust game can be transformed into the following game $\hat{G}_{rdr}$.

Proposition 8.35 *Reduced distributionally robust risk-neutral game*

Consider a set $\mathcal{L}$ of decision-makers with $|\mathcal{L}|\geq 2$. Each decision-maker $l\in\mathcal{L}$ has its own loss functional $E_m\hat{g}_l$ with $\hat{g}_l:\prod_{j\in\mathcal{L}}\Theta_j\times\mathbb{R}_+\times\mathbb{R}\times\Omega\to\mathbb{R}$ expressed as $\hat{g}_l(\theta,\lambda_l,\mu,\omega)$. The game G_{dr} is equivalent to the reduced distributionally robust risk-neutral game given by

$$\hat{G}_{rdr} = \left(\mathcal{L}, (\Theta_l\times\mathbb{R}_+\times\mathbb{R}, E_m\hat{g}_l)_{l\in\mathcal{L}}, \rho_l, m\right).$$

Further, the reduced distributionally robust equilibrium problem of $\hat{G}_{rdr}$ is formulated as follows:

$$\begin{aligned} & l \in \mathscr{L}, \\ & \inf_{(\theta_l, \lambda_l, \mu) \in \Theta_l \times \mathbb{R}_+ \times \mathbb{R}} E_m \hat{g}_l(\theta, \lambda_l, \mu, \omega). \end{aligned} \tag{8.31}$$

The advantage of the reduced distributionally robust game $\hat{G}_{rdr}$ is that the action space $\Theta_l \times B_{\rho_l}(m)$ is reduced to the space $\Theta_l \times \mathbb{R}_+ \times \mathbb{R}$.

Proof We use triality theory (Proposition 8.34) to derive the following expression:

$$\begin{aligned} & \inf_{\theta_l \in \Theta_l} \sup_{\tilde{m} \in B_{\rho_l}(m)} E_{\tilde{m}} g_l(\theta, \omega) \\ & = \inf_{\theta_l \in \Theta_l} \inf_{\lambda_l \geq 0, \mu} E_m \left\{ \lambda_l(\rho_l + f(1)) + \mu + \lambda_l f^* \left(\frac{g_l(\theta, \omega) - \mu}{\lambda_l} \right) \right\} \\ & = \inf_{(\theta_l, \lambda_l, \mu) \in \Theta_l \times \mathbb{R}_+ \times \mathbb{R}} E_m \hat{g}_l(\theta, \lambda_l, \mu, \omega), \end{aligned} \tag{8.32}$$

where

$$\hat{g}_l(\theta, \lambda_l, \mu, \omega) = \lambda_l(\rho_l + f(1)) + \mu + \lambda_l f^* \left(\frac{g_l(\theta, \omega) - \mu}{\lambda_l} \right).$$

And this completes the proof. □

8.1.2.7 Mean–Variance Loss

We now focus on a game with a mean–variance loss function. This class of games is widely used in risk quantification and belongs to the class of mean-field-type games (MFTGs). The basic ingredients of a one-shot MFTG are as follows:

Definition 8.36 MFTG

- The set of decision-makers $\mathscr{L}$, with cardinality $|\mathscr{L}| = L \geq 2$.
- For every decision-maker $l \in \mathscr{L}$, there is a set of actions $\mathscr{A}_l$ which is non-empty.
- Every decision-maker l has a preference structure that can be represented by an instant performance functional $P_l : \prod_{j \in \mathscr{L}} \mathscr{A}_j \times \mathscr{P}\left(\prod_{j \in \mathscr{L}} \mathscr{A}_j\right) \to \mathbb{R}$, $(a, D_a) \mapsto P_l(a, D_a)$, with $a = (a_1, \ldots, a_L)$. The expected performance functional of the decision-maker l is given by $E[P_l(a, D_a)] := \int_{b \in \prod_{j \in \mathscr{L}} \mathscr{A}_j} P_l(b, D_a) D_a(db)$.

The collection $\mathscr{G} = (\mathscr{L}, (\mathscr{A}_l, E[P_l])_{l \in \mathscr{L}})$ is called a one-shot MFTG in strategic form.

Note that $D \mapsto E[P_l(a, D)]$ is not necessarily linear.

Definition 8.37 Mean-field-type game with mean–variance loss

Consider a set $\mathscr{L}$ of decision-makers with $|\mathscr{L}| \geq 2$. Each decision-maker $l \in \mathscr{L}$ has its own loss functional $\mathrm{mv}_l : \prod_{j \in \mathscr{L}} \Theta_j \times \mathbb{P}(\Omega) \to \mathbb{R}$ expressed as

$$\mathrm{mv}_l(\theta, \tilde{m}) = E_{\tilde{m}}[g_l(\theta, \omega)] + \hat{r}_l \cdot var_{\tilde{m}}[g_l(\theta, \omega)].$$

Here, $\hat{r}_l = \frac{c_l}{\varepsilon + |g_l(x^0, \omega^0)|}$, $c_l > 0, \varepsilon > 0$ with c_l being the risk-sensitivity index of decision-maker l.

The mean–variance game is given by

$$G_{mv} = (\mathscr{L}, (\Theta_l, \mathrm{mv}_l)_{l \in \mathscr{L}}, \tilde{m}) .$$

Further, the mean–variance equilibrium problem of G_{mv} is formulated as follows:

$$\begin{cases} l \in \mathscr{L}, \\ \theta_l \in \arg\min_{\theta_l' \in \Theta_l} \mathrm{mv}_l(\theta_l', \theta_{-l}, \tilde{m}). \end{cases} \tag{8.33}$$

Note that $\tilde{m} \mapsto \mathrm{mv}_l(\theta, \tilde{m})$ is nonlinear. This nonlinearity in the measure $\tilde{m}$ creates an extra challenge in the search for the best-response strategy.

The Mean–Variance as the Best Case of an Expected Loss

Lemma 8.38 *Let mv be a mean–variance trade-off of g, $\hat{\lambda} > 0$. We have*

$$\begin{aligned} mv &= E[g] + \hat{\lambda} \cdot var[g] \\ &= inf_{z \in \mathbb{R}} E\left[\left\{(1 - 2\hat{\lambda} z) g + \hat{\lambda} z^2\right\} + \hat{\lambda} g^2\right] \\ &= inf_{z \in \mathbb{R}} E[\tilde{g}(z)], \end{aligned} \tag{8.34}$$

where $\tilde{g}(z) = (1 - 2z\hat{\lambda}) g + \hat{\lambda} z^2 + \hat{\lambda} g^2$.

Proof We have

$$\begin{aligned} mv &= E[g] + \hat{\lambda} \cdot \mathrm{var}[g] \\ &= E[g] + \hat{\lambda}(E[g^2]) - \hat{\lambda}(E[g])^2 \\ &= E[g] + \hat{\lambda} E[g^2] - \hat{\lambda} \sup_{z \in \mathbb{R}} \left\{2\langle z, E[g]\rangle - z^2\right\} \\ &= E[g] + \hat{\lambda} E[g^2] + \hat{\lambda} \inf_{z \in \mathbb{R}} \left\{-2\langle z, E[g]\rangle + z^2\right\} \\ &= \inf_{z \in \mathbb{R}} \left[E[g] + \hat{\lambda} E[g^2] + \hat{\lambda} \left\{-2\langle z, E[g]\rangle + z^2\right\}\right] \\ &= \inf_{z \in \mathbb{R}} \left\{(1 - 2\hat{\lambda} z) E[g] + \hat{\lambda} z^2\right\} + \hat{\lambda} E[g^2] \\ &= \inf_{z \in \mathbb{R}} E\left[\left\{(1 - 2\hat{\lambda} z) g + \hat{\lambda} z^2\right\} + \hat{\lambda} g^2\right] \\ &= \inf_{z \in \mathbb{R}} E[\tilde{g}(z)], \end{aligned} \tag{8.35}$$

where $\tilde{g}(z) = (1 - 2z\hat{\lambda}) g + \hat{\lambda} z^2 + \hat{\lambda} g^2$. □

Proposition 8.39 *MFTG with measure-linearized mean–variance loss Consider a set $\mathscr{L}$ of decision-makers with $|\mathscr{L}| \geq 2$. Each decision-maker $l \in \mathscr{L}$ has own loss functional $E_{\tilde{m}}\tilde{g}_l(\omega, z_l, \omega)$ with $\tilde{g}_l : \prod_{j\in\mathscr{L}} \Theta_j \times \mathbb{R} \times \Omega \to \mathbb{R}$ expressed as $\tilde{g}_l(\theta, z_l, \omega)$.*

The mean–variance game G_{mv} is equivalent to the reduced mean–variance game, which is given by

$$\tilde{G}_{rmv} = (\mathscr{L}, (\Theta_l \times \mathbb{R}, E_{\tilde{m}}\tilde{g}_l)_{l\in\mathscr{L}}, \tilde{m}).$$

The reduced mean-variance equilibrium problem of $\tilde{G}_{rmv}$ is formulated as follows:

$$\begin{aligned} & l \in \mathscr{L}, \\ & inf_{(\theta_l, z_l)\in\Theta_l\times\mathbb{R}} E_{\tilde{m}}\tilde{g}_l(\theta, z_l, \omega) \end{aligned} \tag{8.36}$$

Proof Based on Lemma 8.38, we transform the mean-variance game G_{mv} as follows:

$$\begin{aligned} & \inf_{\theta_l \in\in \Theta_l} E_{\tilde{m}}[g_l(\theta, \omega)] + \hat{r}_l \cdot var_{\tilde{m}}[g_l(\theta, \omega)] \\ & = \inf_{(\theta_l, z_l)\in\Theta_l\times\mathbb{R}} E_{\tilde{m}}\tilde{g}_l(\theta, z_l, \omega), \end{aligned} \tag{8.37}$$

where

$$\tilde{g}_l(\theta, z_l, \omega) = (1 - 2z_l\hat{r}_l)g_l + \hat{r}_l z_l^2 + \hat{r}_l g_l^2.$$

□

The advantage of $\tilde{G}_{rmv}$ formulation is that $\tilde{m} \mapsto E_{\tilde{m}}\tilde{g}_l(\theta, z_l, \omega)$ is linear. The price to pay is that the action space of the decision-maker l is now augmented from Θ_l to $\Theta_l \times \mathbb{R}$.

8.1.2.8 Distributional Robust Mean Variance of a Loss Functional

The exact probability measure $\tilde{m}$ over Ω is known to the decision-makers. Given a data set with size D, a certain data-driven distribution m is estimated. The decision-makers consider the distributional robustness of their estimation m.

Definition 8.40 Distributionally robust mean-variance MFTG

Consider a set $\mathscr{L}$ of decision-makers with $|\mathscr{L}| \geq 2$. Each decision-maker $l \in \mathscr{L}$ has own loss functional $\text{drmv}_l : (\prod_{j\in\mathscr{L}} \Theta_j) \times \mathbb{P}(\Omega) \to \mathbb{R}$ expressed as $\text{drmv}_l(\theta, \tilde{m}) = \sup_{\tilde{m}\in B_{\rho_l}(m)} \text{mv}_l(\theta, \tilde{m})$. The distributionally robust mean-variance game is given by

$$G_{drmv} = \left(\mathscr{L}, (\Theta_l \times B_{\rho_l}(m), \text{drmv}_l)_{l\in\mathscr{L}}, \rho_l, m\right).$$

The distributionally robust mean–variance equilibrium problem of G_{drmv} is formulated as follows:

$$\begin{aligned} & l \in \mathscr{L}, \\ & \inf_{\theta_l \in \Theta_l} \sup_{\tilde{m} \in B_{\rho_l}(m)} \mathrm{mv}_l(\theta, \tilde{m}). \end{aligned} \tag{8.38}$$

The challenge of G_{drmv} is that the function $\tilde{m} \mapsto \mathrm{mv}_l(\theta, \tilde{m})$ is nonlinear (in the measure) and the distributional uncertainty set $B_{\rho_l}(m)$ is infinite-dimensional (in general).

Proposition 8.41 *Distributionally robust mean–variance MFTG: linearized measure and f-divergence: Consider a set $\mathscr{L}$ of decision-makers with $|\mathscr{L}| \geq 2$. Each decision-maker $l \in \mathscr{L}$ has its own loss functional $E_m\, \hat{\mathrm{mv}}_l(\theta, z_l, \lambda_l, \mu, \omega)$ with*

$$\hat{\mathrm{mv}}_l : \left(\prod_{j \in \mathscr{L}} \Theta_j \right) \times \mathbb{R} \times \mathbb{R}_+ \times \mathbb{R} \times \Omega \to \mathbb{R}.$$

The game G_{drmv} is equivalent to the reduced distributionally robust mean–variance game

$$\hat{G}_{drmv} = \left(\mathscr{L}, (\Theta_l \times \mathbb{R} \times \mathbb{R}_+ \times \mathbb{R},\, E_m \hat{\mathrm{mv}}_l)_{l \in \mathscr{L}}, \rho_l, m \right),$$

whose solution is obtained through

$$\begin{aligned} & l \in \mathscr{L}, \\ & \mathit{inf}_{(\theta_l, z_l, \lambda_l, \mu) \in \Theta_l \times \mathbb{R} \times \mathbb{R}_+ \times \mathbb{R}} E_m \hat{\mathrm{mv}}_l(\theta, z_l, \lambda_l, \mu, \omega). \end{aligned} \tag{8.39}$$

Proof We show that

$$\inf_{\theta_l \in \Theta_l} \sup_{\tilde{m} \in B_{\rho_l}(m)} mv(g_l(\theta, \omega); \tilde{m}||m)$$

is exactly

$$\inf_{(\theta_l, z_l, \lambda_l, \mu) \in \Theta_l \times \mathbb{R} \times \mathbb{R}_+ \times \mathbb{R}} E_m \hat{\mathrm{mv}}_l(\theta, z_l, \lambda_l, \mu, \omega).$$

$$\begin{aligned} & \inf_{\theta_l \in \Theta_l} \sup_{\tilde{m} \in B_{\rho_l}(m)} mv(g_l(\theta, \omega); \tilde{m}||m) \\ & = \inf_{\theta_l \in \Theta_l} \sup_{\tilde{m} \in B_{\rho_l}(m)} E_{\tilde{m}} g_l(\theta, \omega) + \hat{r}_l \mathrm{var}_{\tilde{m}}[g_l(\theta, \omega)] \\ & = \inf_{\theta_l \in \Theta_l} \sup_{\tilde{m} \in B_{\rho_l}(m)} \inf_{z_l \in \mathbb{R}} E_{\tilde{m}} \left[(1 - 2 z_l \hat{r}_l) g_l(\theta, \omega) + \hat{r}_l z_l^2 + \hat{r}_l g_l^2(\theta, \omega) \right] \\ & = \inf_{(\theta_l, z_l) \in \Theta_l \times \mathbb{R}} \sup_{\tilde{m} \in B_{\rho_l}(m)} E_{\tilde{m}} \tilde{g}_l(\theta, z_l, \omega) \\ & = \inf_{(\theta_l, z_l) \in \Theta_l \times \mathbb{R}} \inf_{\lambda_l \geq 0, \mu} E_m \left\{ \lambda_l(\rho_l + f(1)) + \mu + \lambda_l f^* \left(\tfrac{\tilde{g}_l(\theta, z_l, \omega) - \mu}{\lambda_l} \right) \right\} \\ & = \inf_{(\theta_l, z_l, \lambda_l, \mu) \in \Theta_l \times \mathbb{R} \times \mathbb{R}_+ \times \mathbb{R}} E_m \hat{\mathrm{mv}}_l(\theta, z_l, \lambda_l, \mu, \omega), \end{aligned} \tag{8.40}$$

where

$$\hat{\text{mv}}_l(\theta, z_l, \lambda_l, \mu, \omega) = \lambda_l(\rho_l + f(1)) + \mu + \lambda_l f^*\left(\frac{\tilde{g}_l(\theta, z_l, \omega) - \mu}{\lambda_l}\right).$$

$$\tilde{g}_l(\theta, z_l, \omega) = (1 - 2z_l\hat{r}_l)g_l + \hat{r}_l z_l^2 + \hat{r}_l g_l^2,$$

and where we have used (8.32). □

Thanks to the previous transformation, the function $\text{mv}_l(\theta, \tilde{m})$ is transformed into infimization over a linearized function in the measure $\tilde{m}$ resulting in an augmented action space. As a consequence of this step, the action space $\Theta_l \times B_{\rho_l}(m)$ is augmented to be $\Theta_l \times \mathbb{R} \times B_{\rho_l}(m)$ for decision-maker l. As a second step, the action space $\Theta_l \times \mathbb{R} \times B_{\rho_l}(m)$ is reduced to $\Theta_l \times \mathbb{R} \times \mathbb{R}_+ \times \mathbb{R}$ thanks to the Fenchel transformation above.

8.1.2.9 Variance-Aware Bregman Learning

Let b be a strictly convex and twice continuously differentiable function $b : \Theta \rightarrow \mathbb{R}$. The Bregman discrepancy $d_b : \Theta \times \text{reint}(\Theta) \rightarrow \mathbb{R}$ is given by

$$d_b(y, x) = b(y) - b(x) - \langle b_x(x), y - x\rangle,$$

with $\text{reint}(\Theta)$ being the relative interior of Θ.

The Bregman discrepancy $d_b(y, x)$ is the difference between the value of b at the destination point y and the first-order Taylor expansion of b around x. Since b is a convex function, $d_b(y, x) \geq 0$. Moreover, $d_b(y, x) > 0$, $\forall x \neq y$ (by strict convexity of b). Let

$$\hat{\text{mv}}_l(x, y, \lambda, \mu) = E_m\lambda(\rho + f(1)) + \mu + \lambda f^*\left(\frac{\tilde{g}(x, y, \omega) - \mu}{\lambda}\right),$$

where

$$\tilde{g}_l(x, y, \omega) = (1 - 2y\hat{r}_l)g_l(x, y, \omega) + \hat{r}_l y^2 + \hat{r}_l g_l^2(x, y, \omega).$$

Proposition 8.42 (Regret Bound) *Let $s = (x, y, \lambda, \mu)$, and consider the following dynamics*

$$\begin{aligned}
&s(0) = s_{00},\ \dot{s}(0) = s_{01},\ \ddot{s}(0) = s_{02},\\
&\dddot{s} = -\frac{3}{t}\ddot{s} - (e^{\alpha} - \dot{\alpha})(\ddot{s} + \frac{2}{t}\dot{s}) \qquad (8.41)\\
&\quad -\frac{e^{2\alpha+\beta}}{t} b_{ss}^{-1}(s + [t + 2e^{-\alpha}]\dot{s} + te^{-\alpha}\ddot{s})\hat{mv}_{l,s}(t\dot{s} + s).
\end{aligned}$$

Then, we have the following bound:

$$0 \leq \hat{mv}(s(t)) - \hat{mv}(s^*) \leq \frac{c_0}{t}\int_0^t e^{-\beta(t')}dt',$$

where $c_0 = e^{\beta(0)}[\hat{mv}(s(0)) - \hat{mv}(s^*)] + d_b(s^*, s_0 + e^{-\alpha(0)}\dot{s}_0) > 0$ *is the initial error plus the initial Bregman discrepancy.*

Ensemble Variance-Aware Bregman Learning
We now focus on the accuracy bounds over the sample data $\{\tilde{\omega}_k \sim m, k \in \{1, \ldots, D\}\}$ with $D \geq 1$ points.

Proposition 8.43 (Sample Regret Bound) *Let*

$$\hat{mv}(x, y, \lambda, \mu, \omega) = \lambda(\rho + f(1)) + \mu + \lambda f^*\left(\frac{\tilde{g}(x, y, \omega) - \mu}{\lambda}\right),$$

and particle swarm-based stochastic Bregman dynamics given by

$$\begin{array}{l} z(0) = z_{00},\ \dot{z}(0) = z_{01} \\ \frac{d}{dt}[b_z(z + e^{-\alpha}\dot{z})] = -e^{\alpha+\beta}\frac{1}{D}\sum_{k=1}^{D}\hat{mv}_z(x, y, \lambda, \mu, \tilde{\omega}_k), \\ \tilde{\omega}_k \sim m, k \in \{1, \ldots, D\}. \end{array} \tag{8.42}$$

For each sample $(\tilde{\omega}_k)_{k\in\{1,\ldots,D\}} \in \Omega^D$, *and convex function* $\tilde{g}$, *one has*

$$0 \leq \frac{1}{D}\sum_{k=1}^{D}\hat{mv}(z(t), \omega_k) - \hat{mv}(z_D^*, \omega_k) \leq \frac{c_{0,D}}{t}\int_0^t e^{-\beta(s)}ds,$$

where $c_{0,D} = e^{\beta(0)}[\hat{mv}(z_{0,D}) - \hat{mv}(z_D^*)] + d_g(z_D^*, z_{0,D} + e^{-\alpha(0)}\dot{z}_0) > 0$ *is the initial error and* $z_D^* \in \arg\min_z \frac{1}{D}\sum_{k=1}^{D}\hat{mv}(z, \omega_k)$.

Proof The proof follows from the inequalities in Proposition 8.42. □

Remark 8.44 Training when $\hat{O}_L \circ \hat{O}_{L-1} \circ \ldots \circ \hat{O}_1(x) = y$ has no solution: The above analysis was conducted under the assumption of the existence of a solution. Sometimes the output equation may not have an exact solution. For example, the equation $-1 = ReLU(x)$ has no solution. But one can look for an x that will make the difference $ReLU(x) + 1$ as close as possible to zero. In this section, we formulate criteria to reduce the gap when the output data y is not in the range of $\hat{O}_L \circ \hat{O}_{L-1} \circ \ldots \circ \hat{O}_1$. One can, of course, consider $\left|\hat{O}_L \circ \hat{O}_{L-1} \circ \ldots \circ \hat{O}_1(x) - y\right|^2$ and optimize through the parameters inside $\hat{O}_l$.

For example, for ReLU activations, let x be the input signal and y be the last layer output. Then, the following mappings are not convex and are not differentiable:

- (x and y random):

$$(\omega_1, b_1, \omega_2, b_2) \mapsto \mathbb{E}_{x,y}\left|\text{ReLU}(\omega_2(\text{ReLU}(\omega_1 x + b_1)) + b_2) - y\right|^2.$$

-

$$(\omega, b) \mapsto \frac{1}{D}\sum_{t=1}^{D}\left|\text{ReLU}(\omega_2(\text{ReLU}(\omega_1 x_t + b_1)) + b_2) - y_t\right|^2,$$

given a set of input signals and layer outputs $\{(x_i, y_i), i \in \{1, \ldots, D\}\}$, $D \geq 1$.

Still, one can exploit the convexity and the $\frac{1}{2}$-averagedness of $\text{ReLU}(x) = \max(0, x) = \frac{x+|x|}{2}$ to conduct layer-by-layer analysis.

8.1.2.10 Expectile Neural Networks

Expectiles constitute a more generalized concept that allows for a nonlinear partitioning of the data. Instead of dividing the data into two parts based on a fixed proportion, expectiles divide the data such that a certain proportion of the expected values of the data falls below the expectile. The expectile is law-invariant. For any $\frac{1}{2} \leq \alpha < 1$, the α-expectile is a coherent risk measure. Expectiles are often used in risk engineering and financial modeling when dealing with asymmetric and non-normal distributions, where traditional quantiles may not provide an accurate representation of the data. They can be particularly useful in situations where one would like to focus on a specific range of expected outcomes rather than just looking at a fixed percentile.

Let $(\Omega, \mathscr{F}, \mathbb{P})$ be an atomless probability space and $L^q(\Omega, \mathscr{F}, \mathbb{P})$ be the L^q space with $q \geq 1$. Let α-expectile of an integrable random variable g, with $\alpha \in (0, 1)$, be defined by

$$\begin{cases} \rho_\alpha : L^2(\Omega, \mathscr{F}, \mathbb{P}) \to \mathbb{R}, \\ \rho_\alpha(g) \in \arg\min_{l' \in \mathbb{R}} E\{\alpha[(g - l')_+]^2 + (1-\alpha)[(l' - g)_+]^2\} \\ = \arg\min_{l' \in \mathbb{R}} E\{V_\alpha(g - l')\}, \\ V_\alpha : \mathbb{R} \to \mathbb{R}, \\ V_\alpha(x) = \alpha[x_+]^2 + (1-\alpha)[(-x)_+]^2 \\ = \alpha x^2 \mathbb{I}_{\{x \geq 0\}} + (1-\alpha) x^2 \mathbb{I}_{\{x<0\}} = |\alpha - \mathbb{I}_{\{x<0\}}| x^2. \end{cases} \tag{8.43}$$

For $\alpha = \frac{1}{2}$, the above definition yields $\arg\min_{l' \in \mathbb{R}} E[(g - l')^2]$, and we know that $\{E[g]\} = \arg\min_{l' \in \mathbb{R}} E[(g - l')^2]$, which means that $\rho_{\frac{1}{2}}(g)$, the half-expectile of g, coincides with the expected value of g.

Lemma 8.45 *We have the following:*

(a) For each $\alpha \in (0,1)$ and a square integrable $g \in L^2$, the mapping $l' \mapsto E\{V_\alpha(g-l')\}$ is convex.
(b) For each $\alpha \in (0,1)$, and a square integrable $g \in L^2$, the mapping $l' \mapsto E\{V_\alpha(g-l')\}$ is differentiable, the derivative is well defined in L^1, and it is a one-to-one mapping, given by $-2E\{U_\alpha(g-l')\}$ with

$$\begin{cases} U_\alpha : \mathbb{R} \to \mathbb{R}, \\ U_\alpha(x) = \alpha x \mathbb{I}_{\{x \geq 0\}} + (1-\alpha)x\mathbb{I}_{\{x<0\}} = |\alpha - \mathbb{I}_{\{x<0\}}|x. \end{cases} \tag{8.44}$$

(c) Let $\mathbb{U}_\alpha(l') = E\{U_\alpha(g-l')\}$. For each $\alpha \in (0,1)$ and a square integrable $g \in L^2$, the mapping $l' \mapsto E\{V_\alpha(g-l')\}$ has a unique minimizer, and it is given by $\mathbb{U}_\alpha^{-1}(0) = \rho_\alpha(g)$. This means that the α-expectile of the loss g is the number l' uniquely determined by

$$\alpha E[(g-l')_+] = (1-\alpha)E[(l'-g)_+].$$

(d) For each $\frac{1}{2} \leq \alpha < 1$, $\rho_\alpha : L^2 \to \mathbb{R}$ is a coherent risk measure.
(e) For each $0 < \alpha < \frac{1}{2}$, $\tilde{\rho}_\alpha : L^2 \to \mathbb{R}$, defined by $\tilde{\rho}_\alpha(g) = -\rho_\alpha(-g)$, is a coherent risk measure.
(f) For each $0 < \alpha < 1$, $\rho_\alpha : L^1 \to \mathbb{R}$ is law-invariant, i.e., $\mathscr{R}_\alpha(\mathbb{P}_{g_1}) = \mathscr{R}_\alpha(\mathbb{P}_{g_2})$ if $\mathbb{P}_{g_1} = \mathbb{P}_{g_2}$, where $\mathscr{R}_\alpha(\mathbb{P}_g) = \rho_\alpha(g) = \mathbb{U}_\alpha(g)^{-1}(0)$, and $\mathbb{U}_\alpha(g)(l') = E\{U_\alpha(g-l')\} = \int U_\alpha(y-l')\mathbb{P}_g(dy)$ for $l' \in \mathbb{R}$.
(g) $\mu \mapsto \mathscr{R}_\alpha(\mu)$ is a Frechet differentiable.
(h) The α-expectile of a Gaussian random variable satisfies $\mathscr{R}_\alpha(\mathscr{N}(m,\sigma^2)) = m + (\alpha - \frac{1}{2})|\sigma|\sqrt{\frac{8}{\pi}} + O((\alpha-\frac{1}{2})^3)$. For $\frac{1}{2} < \alpha < 1$, set $\alpha - \frac{1}{2} = \frac{\sqrt{\beta}}{2}$. Then

$$\mathscr{R}_{\frac{1+\sqrt{\beta}}{2}}(\mathscr{N}(m,\sigma^2)) = m + |\sigma|\sqrt{\frac{2\beta}{\pi}} + O\left(\frac{\beta\sqrt{\beta}}{8}\right).$$

8.1.2.11 Coherent Risk Quantification

The expectile satisfies the properties of coherent risk measures defined next.

Definition 8.46 A risk measure of a random loss functional assigns a real value to it in order to determine its risk degree. A risk measure R is coherent if it satisfies the following properties:

- Translation principle: $R(g_1 + \lambda) = R(g_1) + \lambda$ for all $\lambda \in \mathbb{R}$.
- Subadditivity: $R(g_1 + g_2) \leq R(g_1) + R(g_2)$ for all g_1, g_2.
- Monotonicity: If $g_1 \leq g_2$, then $R(g_1) \leq R(g_2)$.
- Positive homogeneity: $R(\lambda g) = \lambda R(g)$, for all $\lambda \geq 0$.

Example 8.47 Noncoherent risk measures

- The expected value of g, denoted by $E[g]$ which is also the arg min of $l' \mapsto E[(g-l')^2]$, is a coherent risk measure. It is risk-neutral.
- The variance $var(g)$ is not coherent as $var(g+\lambda) = var(g)$ for all $\lambda \in \mathbb{R}$ and $g \in L^2$. Also, the variance is not positively homogeneous. Similarly, the standard deviation is not coherent.
- The mean–variance, mean–standard deviation, variance-to-mean ratio, mean-to-variance ratio, Sharpe ratio, and dispersion index are not coherent risk measures.
- The quantile and the value at risk are not coherent risk measures.
- Similarly, the exponentiated return and the entropic risk measure given by $RS(g_l) = \frac{1}{\eta_l}\log\left(\frac{1}{\alpha_l}E_m[e^{\eta_l g_l(\theta,m)}]\right)$ are not coherent as they do not satisfy positive homogeneity.

Computing Expectile as a Fixed Point

Let $\alpha \in (0,1)$ and g be an integrable real-valued random variable. Noting that $g - l' = \max(g-l', 0) - \max(-(g-l'), 0)$ for all random variable g. We have that the expectile of g with level α is the unique solution of the equation:

$$\begin{aligned} \rho_\alpha(g) &= E[\max((1+\varepsilon)g - \varepsilon\rho_\alpha(g), g)], \\ \varepsilon &:= \tfrac{2\alpha-1}{1-\alpha}. \end{aligned} \tag{8.45}$$

By setting $\rho_\alpha(g) := x$, one has

$$x = E[\max((1+\varepsilon)g - \varepsilon x, g)] =: F_{g,\varepsilon}(x). \tag{8.46}$$

This means that a fixed-point equation $F_{g,\varepsilon}(x) = x$ is to be solved.

The architecture design problem is to find L and the operators $\hat{O}_1, \ldots, \hat{O}_L$ in

$$\begin{aligned}
&\arg\min\left\{\int_{x_0}\int_{y_1} V_\alpha\left[\left|y_1 - \hat{O}_1 x_0\right|^2\right]\mathbb{P}(dx_0 dy_1) \mid \hat{O}_1\right\}, \\
&\vdots \\
&\arg\min\left\{\int_{y_{l-1}}\int_{y_l} V_\alpha\left[\left|y_l - \hat{O}_l \circ \hat{O}_{l-1} \circ \ldots \circ \hat{O}_1 x_0\right|^2\right]\mathbb{P}(dy_{l-1}dy_l) \mid \hat{O}_1, \ldots, \hat{O}_l\right\}, \\
&\vdots \\
&\arg\min\left\{\int_{y_{L-1}}\int_{y_L} V_\alpha\left[\left|y_L - \hat{O}_L \circ \hat{O}_{L-1} \circ \ldots \circ \hat{O}_1 x_0\right|^2\right]\mathbb{P}(dy_{L-1}dy_L) \mid \right. \\
&\left. \hat{O}_1, \ldots, \hat{O}_L\right\}.
\end{aligned} \tag{8.47}$$

The reduced form of the problem is the following: Fix the number of layers L and the activation functions $r_1, \ldots, r_L$, and find the parameters $\theta_1, \ldots, \theta_L$ in

$$\arg\min\left\{\int_{x_0}\int_{y_1} V_\alpha\left[\left|y_1-\hat{O}_1x_0\right|^2\right]\mathbb{P}(dx_0dy_1)\mid\ \theta_1=(W_1,b_1)\right\},$$

$$\vdots$$

$$\arg\min\left\{\int_{y_{l-1}}\int_{y_l} V_\alpha\left[\left|y_l-\hat{O}_l\circ\hat{O}_{l-1}\circ\ldots\circ\hat{O}_1x_0\right|^2\right]\mathbb{P}(dy_{l-1}dy_l)\mid\theta_k\right.$$
$$\left.=(W_k,b_k),\ k\in\{1,\ldots,l\}\right\}.$$

$$\vdots$$

$$\arg\min\left\{\int_{y_{L-1}}\int_{y_L} V_\alpha\left[\left|y_L-\hat{O}_L\circ\hat{O}_{L-1}\circ\ldots\circ\hat{O}_1x_0\right|^2\right]\mathbb{P}(dy_{L-1}dy_L)\mid\theta_l\right.$$
$$\left.=(W_l,b_l),\ l\in\{1,\ldots,L\}\right\},\tag{8.48}$$

Mini-batch risk-aware training problem is then the following: Given an input–output data set $\{(x_i,d_{L,i})\mid 1\leq i\leq D\}$, with size $D\geq 1$, consider the following cost functional

$$\frac{1}{D}\sum_{i=1}^{D} V_\alpha\left[\left|d_{L,i}-\hat{O}_L\circ\hat{O}_{L-1}\circ\ldots\circ\hat{O}_1x_i\right|^2\right]$$

to be minimized over θ. Bregman learning can be used to solve this problem.

Proposition 8.48 *Suppose that each r_l is averaged nonexpansive activation. Then, the solutions (if any) of the risk-aware training of weights and bias problem are also solutions of the following: Given an input-output data set $\{(x^*_{0,i},y^*_{L,i}),\ i\in\{1,2,\ldots,D\}$, find the set of risk-aware Nash equilibria (W^*,b^*) of the nonzero sum game $\mathcal{G}_n n$ given by*

$$\begin{cases} l\in\mathcal{L},\\ (W_l^*,b_l^*)\in\arg\min_{W_l,b_l}\sum_{i=1}^{D}\omega_{l,i}V_\alpha\left(\left|y^*_{l,i}+O_l(y^*_{l,i})-(W_lx^*_{l-1,i}+b_l)\right|^2\right).\end{cases}\tag{8.49}$$

Proposition 8.49 *Suppose that each r_l is a maximally cyclically monotone activation. Then, (8.49) holds with $O_l=\partial f_l$ for a proper, convex function f_l.*

Proof The result is immediate from Lemma 8.20.

Sample data-independent risk-aware training problem is the minimization of the following cost functional

$$E\left(V_\alpha\left[\left|d_L-\hat{O}_L\circ\hat{O}_{L-1}\circ\ldots\circ\hat{O}_1x_0\right|^2\right]\right)$$

over θ. □

Risk-Aware Nash Solution
The function $\theta \mapsto g_l(\theta) := \sum_{i=1}^{D} \omega_{l,i} V_\alpha(\left| y_{l,i}^* + O_l(y_{l,i}^*) - (W_l y_{l-1,i}^* + b_l)\right|^2)$ is convex as a composition of convex and nondecreasing function V_α and another convex function which is the norm.

Risk-Aware Stackelberg Solution
By exploiting the structure of the game, the reaction at layer l is based on the output of the previous layer $l-1$, making this a Stackelberg game. This also leads to the Stackelberg solution:

$$\begin{aligned}
&\text{Layer 1}: \ \arg\min_{\theta_1} V_\alpha \left[\left| y_1 - \hat{O}_1 x_0 \right|^2 \mid x_0 \right], \\
&\vdots \\
&\text{Layer } l: \ \arg\min_{\theta_l} V_\alpha \left[\left| y_l - \hat{O}_l y_{l-1} \right|^2 \mid y_{l-1} \right], \\
&\vdots \\
&\text{Layer } L: \\
&\arg\min_{\theta_L} V_\alpha \left[\left| y_L - \hat{O}_L y_{L-1} \right|^2 \mid y_{L-1} \right] \}
\end{aligned} \tag{8.50}$$

8.2 Transformer-Based Architectures as Games

8.2.1 Layer Normalization

Layer Normalization as Weighted Double Projection
Consider $x = (x_1, \ldots x_D)'$, composed of $D \geq 2$ elements of $\mathcal{H}^d$, $d \geq 2$. Layer normalization is a concatenation row by row:

$$\mathcal{O}_{l,nn}(x_1, \ldots, x_D) = (\mathcal{O}_{l,nn,1}(x_1), \ldots \mathcal{O}_{l,nn,D}(x_D))'$$

with

$$\mathcal{O}_{l,nn,i}(x_i) = \gamma_{l,i} \frac{\sqrt{d-1}\left(x_i - \frac{\sum_{k=1}^{d} x_{ik}}{d} \mathbb{1} \right)}{\sqrt{\left\langle \left(x_i - \frac{\sum_{k=1}^{d} x_{ik}}{d} \mathbb{1} \right), \left(x_i - \frac{\sum_{k=1}^{d} x_{ik}}{d} \mathbb{1} \right) \right\rangle}} + \beta_{l,i}.$$

We describe the normalization as a composition of three operators: two projection operators followed by an affine operator.

Lemma 8.50 *Let $d \geq 2$. Let $C = \{w \in \mathcal{H}^d,\ \langle w, w\rangle \leq d-1, \langle w, \mathbb{1}\rangle = 0\}$ which is the intersection of the ball with radius $\sqrt{d-1}$ and the hyperplane directed by vector $\mathbb{1}$ with element 1 at each entry, which is orthogonal to $\mathbb{1}$. Note that these two sets are convex and closed, and their intersection is non-empty. Denote the two sets involved by $C_1 = \{w \in \mathbb{R}^k,\ \langle w, \mathbb{1}\rangle = 0\}$ and $C_2 = \{w \in \mathbb{R}^k,\ \langle w, w\rangle \leq d-1\}$, which are closed and convex and have non-empty intersection:*

- *The projection of x into C_1 is given by*

$$Proj_{C_1}(x) = x - \frac{\langle x, \mathbb{1}\rangle}{\langle \mathbb{1}, \mathbb{1}\rangle}\mathbb{1}.$$

- *The projection of x into C_2 is*

$$Proj_{C_2}(x) = x\mathbb{I}_{x\in C_2} + \sqrt{d-1}\frac{x}{\sqrt{\langle x, x\rangle}}\mathbb{I}_{x\notin C_2}.$$

  In particular for the sphere with radius $\sqrt{d-1}$, the projection can be obtained for any nonzero element and is given by

$$\sqrt{d-1}\frac{x}{\sqrt{\langle x, x\rangle}}\mathbb{I}_{x\neq 0}.$$

- *The i-th component of the normalization operation at layer l is given by*

$$\mathcal{O}_{l,nn,i}(x_i) = \gamma_{l,i}(Proj_{C_2} \circ Proj_{C_1})(x_i) + \beta_{l,i} = (A_{l,i} \circ Proj_{C_2} \circ Proj_{C_1})(x_i),$$

  where $A_{l,i} : w_i \mapsto \gamma_{l,i} w_i + \beta_{l,i}$.

Proof Let $e \in \mathcal{H}, e \neq 0$, $\delta \in \mathbb{R}$ and $C_3 = \{w \in \mathcal{H},\ \langle w, e\rangle = \delta\}$. Then, the projection x to C_1 is explicitly given by

$$Proj_{C_3}(x) = x + \frac{\delta - \langle x, e\rangle}{\langle e, e\rangle}e.$$

We apply this result for $\delta = 0$ and $e = \mathbb{1}$, which is nonzero. The second and third assertions are immediate from the projection formula. □

In terms of the practical implementation of this normalization layer in transformer-based foundational large learning models, there are several issues. The first issue is that the zero in the denominator needs to be handled. Another issue is the geometric aspect. In finite dimensions, the hyperplane is of dimension $d-1$, which could lead to a loss of data. It is reasonable to ask whether this projection to the hyperplane is really useful for the quality of the output. For example, take the input matrix to be

$$X_0 = \begin{pmatrix} c_1 & c_1 & \cdots\cdots & c_1 & c_1 \\ c_2 & c_2 & \cdots\cdots & c_2 & c_2 \\ \cdots & \cdots & \cdots\cdots & \cdots & \cdots \\ c_{D-1} & c_{D-1} & \cdots\cdots & c_{D-1} & c_{D-1} \\ c_D & c_D & \cdots\cdots & c_D & c_D \end{pmatrix} \tag{8.51}$$

with $c_i \neq c_j, i \neq j$. The first projection is applied row by row, resulting in the projection of $c_i \mathbb{1}$ to the orthogonal of $\mathbb{1}$ yielding the null vector. There is no chance to reconstruct something similar to X_0 after this operator. Some of the recent implementations of generative machine intelligence have decided to bypass the first projection.

Layer Normalization Operator Is Not Invertible
The function

$$z \mapsto \frac{z}{\sqrt{\langle z, z\rangle}} \mathbb{I}_{z\neq 0}$$

is not invertible. As a consequence, the second projection is not necessarily a good idea, as one cannot retrieve the input signals that are outside the unit sphere.

Layer Normalization Is a Weighted Gradient Flow
The modified normalization operator is given by $\mathcal{O}_{l,nn}(x_1, \ldots, x_D) = (y_1, \ldots, y_D)$, where

$$y_i = \gamma_{l,i} \sqrt{d-1} \frac{\left(x_i - \frac{\sum_{k=1}^d x_{ik}}{d} \mathbb{1}\right)}{\sqrt{\varepsilon_l + \left\langle \left(x_i - \frac{\sum_{k=1}^d x_{ik}}{d}\right), \left(x_i - \frac{\sum_{k=1}^d x_{ik}}{d} \mathbb{1}\right)\right\rangle}} + \beta_{l,i},$$

where $\beta_{l,i} and \gamma_{l,i}$ are weights to be determined and $\varepsilon_l > 0$ is fixed. $\beta_{l,i}$ is a vector in $\mathcal{H}^d$, and $\gamma_{l,i}$ is a diagonal matrix with d diagonal elements in $\mathcal{H}^d$. Here, we divide the denominator by $d-1$ instant of d to an unbiased empirical variance:

Lemma 8.51

- *Let* $\varepsilon_l > 0$. *The mapping* $z \mapsto \sqrt{\varepsilon_l + \langle z, z\rangle}$ *is differentiable with gradient* $\frac{z}{\sqrt{\varepsilon_l + \langle z, z\rangle}}$.
- *The mapping* $\phi_i : x_i \mapsto \sqrt{\frac{\varepsilon_l + \langle (x_i - \frac{\sum_{k=1}^d x_{ik}}{d} \mathbb{1}), (x_i - \frac{\sum_{k=1}^d x_{ik}}{d} \mathbb{1})\rangle}{d-1}}$ *is differentiable with gradient*

$$\frac{\sqrt{d-1}\left(x_i - \frac{\sum_{k=1}^d x_{ik}}{d}\mathbb{1}\right)}{\sqrt{\varepsilon_l + \left\langle \left(x_i - \frac{\sum_{k=1}^d x_{ik}}{d}\mathbb{1}\right), \left(x_i - \frac{\sum_{k=1}^d x_{ik}}{d}\mathbb{1}\right)\right\rangle}}.$$

- *Layer normalization is a weighted gradient flow: We have* $\mathcal{O}_{l,nn}(x_1, \ldots, x_D) = (\mathcal{O}_{l,nn,1}(x_1), \ldots \mathcal{O}_{l,nn,D}(x_D))$ *with* $\mathcal{O}_{l,nn,i}(x_i) = (A_{l,i} \circ (\nabla \phi_i))(x_i)$.

The proof is immediate, following from standard differentiation. ∎

Add & Norm Differs from Norm & Add
The Norm & Add operator is $u_i + \mathcal{O}_{l,nn,i}(x_i)$, which differs from the Add & Norm operator $\mathcal{O}_{l,nn,i}(x_i + u_i)$.

8.2.1.1 Alternative Normalization Operators

One of the key inconveniences of LayerNorm in transformer architectures, such as GPT, arises from its underlying mechanism, which involves the composition of two projections. The first projection is onto a hyperplane that is orthogonal to the vector where all components are equal to 1. Essentially, LayerNorm centers the input by subtracting the mean, ensuring that the data lies on a hyperplane of reduced dimensionality. This projection reduces the degrees of freedom of the data, as it essentially forces the input to conform to a lower-dimensional subspace. This dimensionality reduction can lead to a loss of information and structure within the data. The second projection is onto a sphere by scaling the centered data to have unit variance. Together, these two operations-first a projection onto a hyperplane and then onto a sphere-can cause the input data to lose crucial nuances and relationships, potentially affecting its quality. If these two operations are repeated across multiple layers, as they are in deep transformers, the compounded loss of structure and data quality can result in what is known as *model collapse*, where all input data may converge to a similar form, reducing the model's ability to distinguish between different inputs. Particularly, all constant vectors will be normalized to the same output, while zero vectors are excluded from the normalization process, further contributing to potential collapse. The second significant issue with LayerNorm is its non-invertibility. Unlike transformations such as linear layers, which have clearly defined inverses, LayerNorm irreversibly alters the data by normalizing it, meaning that the original input cannot be reconstructed from the output. This introduces a risk of inconsistency between the input and the output, as the transformation process distorts or compresses the input data in a way that is not recoverable. For example, in a transformer architecture, the output embeddings of each layer depend heavily on the normalized representations from LayerNorm. Since LayerNorm discards absolute magnitude information by normalizing the variance, inputs that were originally distinct may end up having similar normalized representations, causing ambiguities. This can be particularly problematic in tasks that require fine-grained distinctions between token embeddings, such as in autoregressive generation tasks like GPT, where every token influences the next one. If subtle variations between tokens are smoothed out due to LayerNorm, the model might generate inconsistent or incoherent outputs. The inability to reverse this transformation means that errors or distortions introduced by LayerNorm can propagate through the network, potentially leading to degraded performance and reduced ability to capture interesting patterns in the input data.

Alternative Normalization Operator

We now address the invertibility issue by using an alternative normalization operator. Introduce the mapping $\hat{\sigma}_i : x_i \mapsto \frac{x_i}{1+\sqrt{\langle x_i, x_i\rangle}}$. The layer normalization operator is now $\mathcal{O}_{l,nn}(x_1,\ldots,x_D) = (\mathcal{O}_{l,nn,1}(x_1),\ldots\mathcal{O}_{l,nn,D}(x_D))$ with $\mathcal{O}_{l,nn,i}(x_i) = (A_{l,i}\circ\hat{\sigma}_i)(x_i)$. The advantage of this normalization is that it does not reduce the input dimensionality of the object. Moreover, it is invertible in the sense that if $\hat{\sigma}_i(x_i) = w_i$, then $\frac{x_i}{1+\sqrt{\langle x_i, x_i\rangle}} = w_i$.

Taking the norm on both sides yields

$$\frac{\sqrt{\langle x_i, x_i\rangle}}{1+\sqrt{\langle x_i, x_i\rangle}} = \sqrt{\langle w_i, w_i\rangle} = 1 - \frac{1}{1+\sqrt{\langle x_i, x_i\rangle}}.$$

Thus, $x_i = \frac{w_i}{1-\sqrt{\langle w_i, w_i\rangle}} = \hat{\sigma}_i^{-1}(w_i)$ for $\langle w_i, w_i\rangle < 1$.

The question now is to examine this new normalization $x_i \mapsto \frac{x_i}{1+\left|\left|x_i\right|\right|}$. This time, there is no denominator issue, and this also solves the invertibility issue. It is a one-to-one mapping.

Mapping Unbounded to Bounded

Let x_1 and x_2 be two elements of $\mathcal{H}$. Then

$$\begin{aligned} \left|\left|\hat{\sigma}_i(x_1) - \hat{\sigma}_i(x_2)\right|\right| &\le 2, \\ \left|\left|\hat{\sigma}_i(x_i)\right|\right| &\le \min(1, \left|\left|x_i\right|\right|). \end{aligned} \tag{8.52}$$

The normalization can be rewritten as $\hat{\sigma}_i(x_i) = \frac{\left|\left|x_i\right|\right|}{1+\left|\left|x_i\right|\right|}\frac{x_i}{\left|\left|x_i\right|\right|}$, which, for $x_i \neq 0$, is the weighted gradient $\hat{\sigma}_i(x_i) = (\frac{1}{1+\left|\left|x_i\right|\right|})\nabla(\frac{1}{2}\left|\left|x_i\right|\right|^2) = (\frac{\left|\left|x_i\right|\right|}{1+\left|\left|x_i\right|\right|})\nabla(\left|\left|x_i\right|\right|)$.

Nonexpansiveness

The Jacobian is the matrix $\nabla(\hat{\sigma}_i(x_i)) = J(x_i) = \frac{\mathbb{I}}{1+\|x_i\|} - \frac{x_i^\dagger x_i}{(1+\|x_i\|)^2\|x_i\|}$. Since the eigenvalue $\frac{\mathbb{I}}{1+\|x_i\|} \leq 1$ is achieved at $x_i = 0$, we deduce that $\hat{\sigma}_i$ is 1-Lipschitz.

Remark 8.52 The proposed new normalization, defined as $\text{LayerNorm}(z) = \frac{z}{1+\|z\|}$, offers a promising alternative to traditional LayerNorm by addressing several key issues that arise in transformer architectures. First and foremost, this new normalization avoids the two consecutive projections that can degrade data quality in standard LayerNorm. In traditional LayerNorm, data is projected onto a lower-dimensional hyperplane (by removing the mean) and then projected onto a sphere by rescaling based on variance. These projections can cause a loss of critical information and lead to the risk of model collapse, where input vectors of constant values get mapped to the same output and zero vectors are excluded entirely. In contrast, the proposed normalization directly scales the input vector z by dividing it by $1 + \|z\|$, where $\|z\|$ is the norm (or magnitude) of the input. This approach ensures that the magnitude of the input is smoothly adjusted, without the need for projecting onto a hyperplane or sphere. As a result, the original structure and relationships within the data are better preserved, minimizing the risk of information loss or distortions. Another key advantage is that this new normalization is less prone to causing model collapse. Since it scales the input by its 1+ norm, it avoids the issue where all constant vectors map to the same output. By incorporating the norm into the scaling factor, it introduces a natural way to maintain variability in the input, even when the components are identical. Additionally, unlike traditional LayerNorm, this method does not exclude the null vector. When z is zero, the normalization simply returns zero, making the operation well defined even for inputs of zero magnitude, which improves the model's robustness. Moreover, this form of normalization retains crucial information about the input's original magnitude. In traditional LayerNorm, scaling by the variance effectively removes information about the input's magnitude, which can lead to inconsistencies between the input and output, especially in scenarios where magnitude plays an important role in the decision-making process of the model. By dividing by $1 + \|z\|$, the magnitude of the input is controlled without completely eliminating it, allowing the model to retain some notion of the input's relative scale. This can be particularly beneficial in transformer models like GPT, where token embeddings may have significant variation in magnitude, and preserving this information can lead to more precise representations of the input.

An additional major advantage of the normalization $\text{LayerNorm}(z) = \frac{z}{1+\|z\|}$ is that it is fully invertible and represents a one-to-one mapping. This is a significant improvement over traditional LayerNorm, which lacks invertibility due to its reliance on projection steps (onto a hyperplane and then a sphere). With traditional LayerNorm, once the data has been transformed, it is impossible to reconstruct the original input, as some information about the input's absolute magnitude and individual components is lost. This non-invertibility can lead to inconsistencies between the inputs and outputs, making it harder for the model to maintain internal coherence, particularly in deep networks like transformers. In contrast, the proposed normalization allows for a clear and direct inverse function. Given the output $\text{LayerNorm}(z) = \frac{z}{1+\|z\|}$, the original input z can be recovered by simply inverting the normalization process. Specifically, for a given normalized output w, we can recover z as follows: $z = \hat{\sigma}^{-1}(w) = \frac{w}{1-\sqrt{\langle w, w\rangle}}$. This property ensures that no information is irretrievably lost during the normalization process. Since the transformation is smooth and bijective (one-to-one), each unique input corresponds to a unique output, and vice versa. This bijective property guarantees that all the structure of the input is preserved in a reversible manner. In contexts like transformer models, where maintaining precise relationships between inputs and outputs across layers is crucial for tasks like autoregressive language modeling, having a normalization method that is invertible is highly beneficial. Moreover, this invertibility addresses a core limitation of traditional LayerNorm, where normalization can lead to ambiguities or inconsistencies because different inputs may produce the same normalized output (e.g., constant vectors being mapped to the same output). In the new normalization, no such ambiguities arise. Each distinct input vector is mapped to a distinct output, preserving the input's uniqueness, which is critical for models that rely on subtle differences in embeddings to make accurate predictions. Therefore, this new LayerNorm $\hat{\sigma}$ not only preserves the magnitude and structure of the input in a way that traditional LayerNorm does not but also ensures full invertibility and a one-to-one mapping, making it a much more robust choice for architectures like GPT that demand high fidelity in internal data representations.

A Scaled LayerNorm
In finite but large dimensions, one can consider a scaled version such as $\hat{\sigma}_i : x_i \mapsto \frac{x_i}{1+\rho\sqrt{\langle x_i, x_i\rangle}}$, where $\rho > 0$ is a carefully chosen scale. For example, $\rho = \frac{1}{\sqrt{d}}$ can be used. In infinite dimensions, however, there is no such d, and one can choose ρ to depend on the strength of the signal.

8.2.2 *SoftArgMax as a Gradient Flow and Also Argmax*

SoftArgMax Is a Gradient
Following Example 8.6, the log-sum up to $i \geq 2$ mapping $\phi_i(z) = \varepsilon \log\left(\sum_{j=1}^{i} e^{\frac{z_j}{\varepsilon}}\right)$ is differentiable, and its gradient is

$$\text{softargmax}_i : \mathbb{R}^i \ni z \mapsto \left(\frac{e^{\frac{z_1}{\varepsilon}}}{\sum_{j=1}^{i} e^{\frac{z_j}{\varepsilon}}}, \ldots, \frac{e^{\frac{z_i}{\varepsilon}}}{\sum_{j=1}^{i} e^{\frac{z_j}{\varepsilon}}} \right) \in \mathbb{R}_+^i.$$

softargmax_i is a probability distribution over $\{1, \ldots, i\}$. The vector softargmax_i is extended to $\mathbb{R}^D$ by completing with 0, i.e., $\text{softargmax}^*_{ij} = \text{softargmax}_i \mathbb{I}_{j \leq i}$.

SoftArgMax as ArgMax

Consider the entropy function

$$p = (p_1, \ldots, p_i) \mapsto \sum_{j=1}^{i} p_j \log p_j,$$

defined over $\sum_{j=1}^{i} p_j = 1,\ p_j > 0,\ j \in \{1, \ldots, i\}$, the $(i-1)$-simplex of $\mathbb{R}^i$. We use the convention $0 \log 0 = 0$ as $\lim_{x \to 0_+} x \log x = 0$. This entropy is a convex function. Consider the penalized function $z \mapsto \sup_p \{\langle z, p \rangle - \varepsilon \sum_{j=1}^{i} p_j \log p_j\}$; then, p is the argmax which, for a very small ε, can be seen as an argument of the approximate max of the components of z, hence the name "soft" argument maximum or softargmax. The value associated with it is the Fenchel–Legendre conjugate of the entropy.

SoftArgMax Is Not Necessarily an Averaged Operator

The gradient of softargmax is the matrix $\varepsilon \nabla_z \text{softargmax}_D(\frac{z}{\varepsilon}) = \text{diag}(p) - p^\dagger p$ where $p = \text{softargmax}_D(\frac{z}{\varepsilon})$. Each row sums up to zero. To be a contraction, the parameter $\varepsilon > 1$, and to be closer to the softargmax, ε needs to be small enough (closer to zero). These two extremes are not consistent. Thus, softargmax in its current form in the architecture of transformer-based generative machine intelligence is just another normalization (weighting with probabilities), and it does not necessarily focus on the maximum component.

8.2.2.1 Attention Is Not an Averaged Operator

Lemma 8.53 *Let $D \geq 2$, $z_{ij} = \frac{1}{\sqrt{k}} \langle Q x_i, K x_j \rangle$ for $Q, K \in \mathscr{L}(\mathscr{H}^d, \mathscr{H}^k)$ and $z_i = \{z_{ij}\}_{1 \leq j \leq i}$. For every $i \geq 2$, the mapping*

$$x \mapsto \sum_{l=1}^{D} \mathit{softargmax}_{il} \left(\left\{ \frac{1}{\sqrt{k}} \langle Q x_i, K x_j \rangle \right\}_{1 \leq j \leq i} \right) x_l \mathbb{I}_{l \leq i},$$

is not Lipschitz continuous.

Proof The assertion follows from the derivative of the composition and the fact that the quadratic term is not Lipschitz in x.

It suffices to show that $x \mapsto \frac{\sum_{j=1}^{i} e^{\frac{1}{\sqrt{k}}\langle Qx_i, Kx_j\rangle} x_j}{\sum_{l=1}^{i} e^{\frac{1}{\sqrt{k}}\langle Qx_i, Kx_l\rangle}}$ is not Lipschitz.

We examine

$$\frac{e^{\frac{1}{\sqrt{k}}\langle Qx_i, Kx_i\rangle} x_i + \sum_{j=1}^{i-1} e^{\frac{1}{\sqrt{k}}\langle Qx_i, Kx_j\rangle} x_j}{e^{\frac{1}{\sqrt{k}}\langle Qx_i, Kx_i\rangle} + \sum_{l=1}^{i-1} e^{\frac{1}{\sqrt{k}}\langle Qx_i, Kx_l\rangle}} = p_{ii}x_i + \sum_{j=1}^{i-1} p_{ij}x_j.$$

The gradient in x_i leads to the following matrix:

$$\left(\frac{1}{\sqrt{k}}(QK^{\dagger} + KQ^{\dagger})x_i^{\dagger}x_i + \mathbb{I}_{dd}\right)p_{ii} + \sum_{j=1}^{i-1} p_{ij}\frac{1}{\sqrt{k}}Q^{\dagger}Kx_j^{\dagger}x_j$$

$$-\left(\frac{1}{\sqrt{k}}(QK^{\dagger} + KQ^{\dagger})x_i^{\dagger}p_{ii} + \sum_{j=1}^{i-1} p_{ij}\frac{1}{\sqrt{k}}Q^{\dagger}Kx_j^{\dagger}\right)\left(p_{ii}x_i + \sum_{j=1}^{i-1} p_{ij}x_j\right).$$

When $x_i = \frac{1}{\varepsilon}\mathbb{1}_d = x_j$, the probabilities are uniform (on the left diagonal).

$$\left(\frac{1}{\sqrt{k}}(QK^{\dagger} + KQ^{\dagger})x_i^{\dagger}x_i + \mathbb{I}_{dd} + \frac{i-1}{\sqrt{k}}Q^{\dagger}Kx_i^{\dagger}x_i\right)\frac{1}{i}$$

$$-\left(\frac{1}{\sqrt{k}}(QK^{\dagger} + KQ^{\dagger})x_i^{\dagger} + \frac{i-1}{\sqrt{k}}Q^{\dagger}Kx_i^{\dagger}\right)\frac{x_i}{i}$$

can be written as $\frac{1}{i}\mathbb{I}_{dd} + Bx_i^{\dagger}x_i,\ B \neq 0$, whose norm goes to infinity as ε goes to zero. This means that the (matrix) Jacobian is unbounded. □

As a corollary of Lemma 8.53, we have the following.

Corollary 8.54 *The masked self-attention operator* $\mathcal{O}_{att,i} : x \mapsto W(\sum_{j=1}^{i}$ *softargmax*$_{ij}(z_i)Vx_j)$*, with nonzero linear operators* $V \in \mathcal{L}(\mathcal{H}^d, \mathcal{H}^k)$ *and* $W \in \mathcal{L}(\mathcal{H}^k, \mathcal{H}^d)$*, is not an averaged operator.*

Note that all these (masked and unmasked) self-attention operators are not Lipschitz on the entire domain, but they are Lipschitz when restricted to any non-empty, closed, and bounded domain, such as the unit ball.

8.2.2.2 Softargmax of the Attention Mechanism as a Game

From the matrix x from layer l, each player i minimizes

$$a_i = (a_{ij})_{1 \le j \le D} \mapsto g_i(x, a_i) = -\sum_{j=1}^{i} z_{ij} a_{ij} + \sum_{j=1}^{i} a_{ij} \log a_{ij} + \sum_{j=i+1}^{D} a_{ij}^2,$$

defined over

$$\Delta_{i-1}^{*} = \left\{ a_i \in \mathbb{R}^D, \sum_{j=1}^{i} a_{ij} = 1, a_{ij} > 0, j \in \{1, \ldots, i\}, \text{ and } a_{ij} = 0, i < j \le D \right\}.$$

The dependence of the preceding layer decision-makers is via $z_{ij} = \frac{1}{\sqrt{k}} \langle Q x_i, K x_j \rangle$ for $Q, K \in \mathcal{L}(\mathcal{H}^d, \mathcal{H}^k)$. Given x, consider the D-decision-maker non-sum game

$$\hat{G}_{\text{softargmin}} = \left(\mathcal{I} = \{1, \ldots, D\}, (\Delta_{i-1}^{*}, g_i)_{i \in \mathcal{I}}, x \right).$$

This best response of this game for $i \in \mathcal{I}$ is $\{a_i\} = \text{argmin}_{a_i'} g_i(x, a_i') = \{\text{softargmax}_i^{*}(z_i)\}$, which is a probability measure.

Lemma 8.55 *Given the matrix x, the D-decision-maker non-sum game*

$$\hat{G}_{softargmin} = \left(\mathcal{I}, (\Delta_{i-1}^{*}, g_i)_{i \in \mathcal{I}}, x \right)$$

has a unique Nash equilibrium, and it is given by $(softargmax_i^{*}(z_i))_{i \in \mathcal{I}}$.

8.2.2.3 Averaging as a Game

Let $\sum_{j=1}^{D} p_{ij} = 1, p_{ij} \ge 0$, and consider the convex function $y_i \mapsto \sum_{j=1}^{D} p_{ij} \frac{1}{2} \left| y_i - x_j \right|^2$. This function is minimized at $y_i^{*} = \sum_{j=1}^{D} p_{ij} x_j$, which is the convex combination of the elements $x_1, \ldots, x_D$.

8.2.2.4 Self-Attention as a Game

From the previous observation, we see that $y_i^{*} = \sum_{j=1}^{D} \text{softargmax}_{ij}^{*} x_j$ is the minimizer of the convex function $y_i \mapsto \sum_{j=1}^{D} \text{softargmax}_{ij}^{*} \frac{1}{2} \left| x_j - y_i \right|^2$.

Lemma 8.56 *Let h_i be the cost function $h_i : y_i \mapsto \sum_{j=1}^{D} softargmax_{ij}^{*} \frac{1}{2} \left| (WV) x_j - y_i \right|^2$. Given the matrix x, the D-decision-maker non-sum game*

$$\hat{G}_{att} = \left(\mathcal{I}, (\mathcal{H}^d, h_i)_{i \in \mathcal{I}}, x \right)$$

has a unique Nash equilibrium, and it is given by $(\mathscr{O}_{att,i}(x))_{i\in\mathscr{I}}$ *with*

$$\mathscr{O}_{att,i}(x) = (WV)\sum_{j=1}^{D} softargmax^*_{ij} x_j.$$

The full uni-head self-attention operator

$$\mathscr{O}_{att}(x)=(WV)\text{softargmax}\left(x\left(\frac{1}{\sqrt{k}}QK^\dagger\right)x^\dagger\right)x=(M_1)\text{softargmax}(x(M_2)x^\dagger)x,$$

where $M_1 = WV$, $M_2 = \frac{1}{\sqrt{k}}QK^\dagger$.

Definition 8.57 The multi-head self-attention performance evaluation

$$\mathscr{H}^{H(Dd+d^2)} \ni (M_{1h}, M_{2h})_{1\le h\le H} \mapsto \frac{1}{\sqrt{H}}\sum_{h=1}^{H} M_{1h}\text{softargmax}(x(M_{2h})x^\dagger)x$$

where $\langle C_1, C_2\rangle = trace(C_1^\dagger C_2) = trace(C_1 C_2^\dagger)$, where $C_1, C_2. \in \mathscr{L}(\mathscr{H}^d, \mathscr{H}^D)$.

Lemma 8.58 *Let* $\hat{g}_i$ *be the cost function*

$$\hat{g}_i : y_i \mapsto \sum_{j=1}^{D}\sum_{h=1}^{H} softargmax^*_{ij,h}\frac{1}{2H}\left|\sqrt{H}(W_h V_h)x_j - y_i\right|^2,$$

with $softargmax^*_{ij,h} = softargmax^*_{ij,h}(Q_h, K_h)$. *Given the matrix* x, *the D-decision-maker non-sum game*

$$\hat{G}_{multihead} = \left(\mathscr{I}, (\mathscr{H}^d, \hat{g}_i)_{i\in\mathscr{I}}, x\right)$$

has a unique Nash equilibrium, and it is given by $(\mathscr{O}_{att,i}(x))_{i\in\mathscr{I}}$ *with* $\mathscr{O}_{att,i}(x) = \frac{1}{\sqrt{H}}\sum_{h=1}^{H}(W_h V_h)\sum_{j=1}^{D} softargmax^*_{ij,h} x_j$.

Corollary 8.59 *Let* $\tilde{g}_i$ *be the cost function*

$$\tilde{g}_i : y_i \mapsto \sum_{j=1}^{D}\sum_{h=1}^{H} softargmax^*_{ij,h}\frac{1}{2H}\left|\sqrt{H}(W_h V_h)x_j + x_i - y_i\right|^2.$$

Given the matrix x, *the D-decision-maker non-sum game*

$$\hat{G}_{Idplusmultihead} = \left(\mathscr{I}, (\mathscr{H}^d, \tilde{g}_i)_{i\in\mathscr{I}}, x\right)$$

has a unique Nash equilibrium, and it is given by $(Id + \mathscr{O}_{att,i}(x))_{i\in\mathscr{I}}$ *with* $(Id + \mathscr{O}_{Id+att,i})(x) = x_i + \frac{1}{\sqrt{H}}\sum_{h=1}^{H}(W_h V_h)\sum_{j=1}^{D} softargmax^{*}_{ij,h} x_j$.

Remark 8.60

- As the parameter $\frac{1}{\sqrt{k}}$ is not too large, the softargmax operation is not really an approximate maximizer. This is only probabilistic weights assigned to the vectors. The denominator, which is a sum of exponentials, creates some difficulties in practice. For $D = 200{,}000$, the sum over 200,000 exponential terms, we may generate a very big number depending on the signal input. One may consider instead a logarithmic scale: for $j \leq i$, $\chi_{ij} = \log p_{ij} = z_{ij} - \log(\sum_{l=1}^{i} e^{z_{il}})$, i.e.,

$$\chi_{ij} = \frac{1}{\sqrt{k}}\langle Qx_i, Kx_j\rangle - \log\left(\sum_{l=1}^{i} e^{\frac{1}{\sqrt{k}}\langle Qx_i, Kx_l\rangle}\right).$$

- The softargmax in the attention is not needed at this stage of the architecture. We can use a simpler scaling factor, such as sigmoid attention. For each head h, compute the weight matrix

$$S_{ij,h} = \frac{1}{\sqrt{k}}\langle Q_h x_i, K_h x_j\rangle.$$

  One can apply a sigmoid function to each entry of S_h weight $S'_{ij,h} = \sigma(S_{ij,h}) = \frac{e^{S_{ij,h}}}{D+e^{S_{ij,h}}}$. The rows are not normalized to probabilities in this case. However, $x_i + \frac{1}{\sqrt{H}}\sum_{h=1}^{H}(W_h V_h)\sum_{j=1}^{D} S'_{ij,h} x_j$ can still be seen as a game.
- Another attention mechanism is obtained with the function

$$S'_{ij,h} = e^{-\left|Q_h^{\dagger} K_h (x_i - x_j)\right|^2}.$$

- Yet, another attention mechanism is obtained with the function

$$S'_{ij,h} = \frac{\left|Q_h^{\dagger} K_h (x_i - x_j)\right|}{D + \left|Q_h^{\dagger} K_h (x_i - x_j)\right|}.$$

8.2.3 Feedforward Operator

The feedforward operator $\mathscr{O}_{ff} : x \mapsto W_2 r_l(W_1 x + b_1) + b_2 = (A_2 \circ r_l \circ A_1)(x)$ where $W_1 \in\in \mathscr{L}(\mathscr{H}^d, \mathscr{H}^q)$, $b_1 \in \mathscr{H}^q$ and $W_2 \in \mathscr{L}(\mathscr{H}^q, \mathscr{H}^d)$, $b_2 \in \mathscr{H}^d$ and r_l is an entry-wise activation function.

Definition 8.61 The entry-wise activation function in a Hilbert space is given by

$$r_l(x) = \sum_{k=0}^{\infty} r_l(\langle x, e_k\rangle)e_k,$$

where $(e_k)_k$ is an orthonormal basis of the Hilbert space.

As a particular case, for Euclidean spaces, this basis is finite.

Lemma 8.62 *If for each entry k, the map is $x_k \mapsto r_l(x_k)$ γ_l-averaged, then*

$$x = (x_k)_k \mapsto r_l(x) = \sum_{k=0}^{\infty} r_l(\langle x, e_k\rangle)e_k$$

is also γ_l-averaged.

Proof

$$\begin{aligned} r_l(x) &= \textstyle\sum_{k=0}^{\infty} r_l(\langle x, e_k\rangle)e_k \\ &= \textstyle\sum_{k=0}^{\infty}((1-\gamma_l)\langle x, e_k\rangle + \gamma_l \tilde{r}_l(\langle x, e_k\rangle))e_k \\ &= (1-\gamma_l)\textstyle\sum_{k=0}^{\infty}\langle x, e_k\rangle e_k + \gamma_l \sum_{k=0}^{\infty} \tilde{r}_l(\langle x, e_k\rangle)e_k \\ &= ((1-\gamma_l)Id + \gamma_l\tilde{r}_l)(x) \end{aligned} \tag{8.53}$$

with $\tilde{r}_l(x) := \sum_{k=0}^{\infty} \tilde{r}_l(\langle x, e_k\rangle)e_k$. One has $\tilde{r}_l(x) - \tilde{r}_l(y) = \sum_{k=0}^{\infty}(\tilde{r}_l(\langle x, e_k\rangle) - \tilde{r}_l(\langle y, e_k\rangle))e_k$.

Then, $\left|\tilde{r}_l(x) - \tilde{r}_l(y)\right|^2 = \sum_{k=0}^{\infty} \left|\tilde{r}_l(\langle x, e_k\rangle) - \tilde{r}_l(\langle y, e_k\rangle)\right|^2 \leq \sum_{k=0}^{\infty} \left|\langle x, e_k\rangle - \langle y, e_k\rangle\right|^2$.

As $\langle x, e_k\rangle \mapsto \tilde{r}_l(\langle x, e_k\rangle)$ is 1-Lipschitz, we have $\left|\tilde{r}_l(x) - \tilde{r}_l(y)\right|^2 \leq \sum_{k=0}^{\infty} \left|\langle x - y, e_k\rangle\right|^2 \leq \left|x - y\right|^2$. This implies that $\tilde{r}_l(x)$ is 1-Lipschitz. Hence, $r_l = (1-\gamma_l)Id + \gamma_l\tilde{r}_l$ is an γ_l-averaged operator. □

Lemma 8.63 *The map* $(W_1, b_1, W_2, b_2) \mapsto \left|W_2 r_l(W_1 x + b_1) + b_2\right|^2$ *is not necessarily convex even if r_l is convex.*

Proof The non-convexity is because of the multiplicative factor. Take $r_l = ReLU$ (which is a convex function) as an illustration. Then, the feedforward operator $\mathcal{O}_{ff}$ is not convex in the parameter (W_1, b_1, W_2, b_2). □

8.2.3.1 Layer *l* Operator of the Transformer

The full operator at layer l is given by

$$\hat{\mathcal{O}}_l = (Id + \mathcal{O}_{l,ff} \circ \mathcal{O}_{l,nn}) \circ (Id + \mathcal{O}_{l,att} \circ \mathcal{O}_{l,nn}),$$

where

$$\begin{aligned}
&\text{Parameters} : x_0, l, D, d, k, \gamma_{1l,i}, \beta_{1l,i}, H, Q_{lh}, K_{lh}, V_{lh}, W_{lh}, \gamma_{2l,i}, \beta_{2l,i}, \\
&W_{1li}, W_{2li}, b_{1li}, b_{2li}, \varepsilon_l > 0, \\
&\text{Input at} l : \; x_l \in \mathcal{H}^{Dd}, \\
&u_{l,i} = \gamma_{1l,i} \frac{\sqrt{d-1}\left(x_i - \frac{\sum_{k=1}^d x_{ik}}{d} \mathbb{1}\right)}{\sqrt{\varepsilon_l + \left\langle \left(x_i - \frac{\sum_{k=1}^d x_{ik}}{d} \mathbb{1}\right), \left(x_i - \frac{\sum_{k=1}^d x_{ik}}{d} \mathbb{1}\right) \right\rangle}} + \beta_{1l,i}, \\
&\%\% \; \mathcal{O}_{l,nn}(x_l) = u_l = (u_{l,i})'_{1 \le i \le D}, \\
\\
&\tilde{u}_{l,i} = \sum_{h=1}^H W_{lh} (\textstyle\sum_{j=1}^i \text{softargmax}_{ij} (\frac{1}{\sqrt{k}} \langle Q_{lh} u_i, K_{lh} u_j) V_{lh} u_j), \\
&\%\% \; \mathcal{O}_{l,att}(u) = \tilde{u}_l = (\tilde{u}_{l,i})_{1 \le i \le D}, \\
\\
&\hat{u}_l = x_l + \tilde{u}_l = (Id + \mathcal{O}_{l,att} \circ \mathcal{O}_{l,nn})(x_l), \\
\\
&\hat{y}_l = \mathcal{O}_{l,nn}(\hat{u}_l), \\
&\hat{y}_{l,i} = \gamma_{2l,i} \frac{\sqrt{d-1}\left(\hat{u}_{li} - \frac{\sum_{k=1}^d \hat{u}_{lik}}{d} \mathbb{1}\right)}{\sqrt{\varepsilon_l + \left\langle \left(\hat{u}_{li} - \frac{\sum_{k=1}^d \hat{u}_{lik}}{d} \mathbb{1}\right), \left(\hat{u}_{li} - \frac{\sum_{k=1}^d \hat{u}_{lik}}{d} \mathbb{1}\right) \right\rangle}} + \beta_{2l,i}, \\
\\
&\tilde{y}_{li} = W_{2li} r_l (W_{1li} \hat{y}_{li} + b_{1li}) + b_{2li}, \\
&\%\% \; \tilde{y}_l = \mathcal{O}_{l,ff}(\hat{y}_l) = (\tilde{y}_{li})'_{1 \le i \le D}, \\
&y_{li} = \hat{u}_{li} + \tilde{y}_{li}, \\
\\
&\text{Output at} l : y_l \\
&\text{Input at} l + 1 : x_{l+1} = y_l,
\end{aligned} \tag{8.54}$$

and this holds for $1 \le l \le L, \; L \ge 2$.

Lemma 8.64 *The output operator at layer l of the transformer*

$$y_{l-1} = (y_{l-1,i})'_i \mapsto \hat{\mathcal{O}}_l(y_{l-1}) = y_l = (y_{l,i})'_{1 \le i \le D}$$

is not necessarily an averaged operator.

Proof This follows from the quadratic term in the softargmax and the self-attention operator. □

This non-averagedness at layer l creates a big issue in the layer-by-layer fixed-point analysis.

Lemma 8.65 *For a fixed input matrix y_{l-1}, the mapping*

$$(\gamma_{1l,i}, \beta_{1l,i}, Q_{lh}, K_{lh}, V_{lh}, W_{lh}, \gamma_{2l,i}, \beta_{2l,i}, W_{1li}, W_{2li}, b_{1li}, b_{2li})_{1\leq i\leq D}$$

$$\mapsto \left|\hat{\mathcal{O}}_l(y_{l-1})\right|^2 = \left|(Id + \mathcal{O}_{l,ff} \circ \mathcal{O}_{l,nn}) \circ (Id + \mathcal{O}_{l,att} \circ \mathcal{O}_{l,nn})(y_{l-1})\right|^2$$

is non-convex.

Proof Follows from the composition of operators and the multiplicative weights. □

This non-convexity at layer l creates another big issue in the layer-by-layer training.

Sum of Composition of the Operators

Let us consider the operator $E_1 = (Id + \mathcal{O}_{l,att} \circ \mathcal{O}_{l,nn})$. As the normalization occurs before taking the self-attention, $\mathcal{O}_{l,att}$ restricted at the unit ball is of course Lipschitz as it is a continuously differentiable function and the closed unit ball is a convex, closed, bounded, and non-empty set but not compact (in Hilbert spaces). Hence, the Jacobian has a bounded norm in the closed unit ball. We can scale with V to get an averaged operator on the unit ball. This holds not only for the standard normalization $\mathcal{O}_{l,nn} = A_{1l,nn} \circ \text{proj}_{C_2} \circ \text{proj}_{C_1}$ but also for the modified normalization $\mathcal{O}_{l,nn} = A_l \circ \hat{\sigma}$.

The second operator $E_2 = (Id + \mathcal{O}_{l,ff} \circ \mathcal{O}_{l,nn})$ with $\mathcal{O}_{l,ff} = A_{2l,ff} \circ r_l \circ A_{1l,ff}$ can be made averaged by designing the linear operators involved. However, for training, we do not need such restrictions on weights and biases if the activation function of the feedforward is a gradient. For consistency and coherence check, however, we need a well-behaved fixed-point operation of the overall network. The full operator at layer l is

$$\hat{\mathcal{O}}_l = E_2 \circ E_1$$

$$= (Id + A_{2l,ff} \circ r_l \circ A_{1l,ff} \circ A_{2l,nn} \circ \text{proj}_{C_2} \circ \text{proj}_{C_1})\circ$$
$$(Id + \mathcal{O}_{l,att} \circ A_{1l,nn} \circ \text{proj}_{C_2} \circ \text{proj}_{C_1}).$$

or

$$\hat{\mathcal{O}}_l = (Id + A_{2l,ff} \circ r_l \circ A_{1l,ff} \circ A_{2l,nn} \circ \hat{\sigma}_{2,nn}) \circ (Id + \mathcal{O}_{l,att} \circ A_{1l,nn} \circ \hat{\sigma}_{1,nn}).$$

In both cases, each layer operation can be easily decomposed to yield a unique Nash equilibrium of a properly designed hierarchical game.

8.2.3.2 Outcome of the Full Transformer

The output of the full transformer is $\hat{\mathcal{O}}_L \circ \hat{\mathcal{O}}_{L-1} \circ \ldots \circ \hat{\mathcal{O}}_1$.

Iterate layer by layer and by timestep the following:

$$
\begin{array}{|l}
Parameters\\
Input : x_0\\
\text{for } t \in \{0, 1, 2, \ldots\},\\
y_{1,t} = (Id + \mathcal{O}_{1,ff} \circ \mathcal{O}_{1,nn}) \circ (Id + \mathcal{O}_{1,att} \circ \mathcal{O}_{1,nn})(x_t),\\
\text{for } l \in \{2, \ldots, L\},\ y_{l,t} = (Id + \mathcal{O}_{l,ff} \circ \mathcal{O}_{l,nn}) \circ (Id + \mathcal{O}_{l,att} \circ \mathcal{O}_{l,nn})(y_{l-1,t}),\\
x_{t+1} = x_t + \lambda_t(y_{L,t} - x_t),
\end{array}
\tag{8.55}
$$

which means that

$$x_{t+1} = x_t + \lambda_t(\hat{\mathcal{O}}_L \circ \ldots \circ \hat{\mathcal{O}}_1(x_t) - x_t),$$

starting from x_0. For consistency of the operation, the output of $\hat{O}_L$ should be in the same space as the input, i.e., $\mathcal{H}_L = \mathcal{H}_0$.

Definition 8.66 The transformer-based neural network $\mathbb{TNN}$ specification is given by the family

$$(x_0, L, D, d, k, (r_l, \gamma_{1l,i}, \beta_{1l,i}, H, Q_{lh}, K_{lh}, V_{lh}, W_{lh}, \gamma_{2l,i}, \beta_{2l,i},\\ W_{1li}, W_{2li}, b_{1li}, b_{2li})_{1\leq l\leq L, 1\leq i\leq D}, \lambda).$$

Problem 8.67 What does a well-trained transformer-based deep neural network do? To answer this question, we aim to find and characterize the behavior of the neural network for a large class of architectures $\mathbb{TNN}$. To this end, we examine the possible limit (if any) of $(x_t, y_{1,t}, \ldots, y_{L,t})$ as t goes to infinity:

$$
\begin{aligned}
& l = 1,\\
& y_1 = \hat{\mathcal{O}}_1(x^*),\\
& \ldots\\
& l \in \{2, \ldots, L-1\}:\\
& y_l = \hat{\mathcal{O}}_l(y_{l-1}),\\
& \ldots\\
& x^* = y_L = \hat{\mathcal{O}}_L(y_{L-1}),
\end{aligned}
\tag{8.56}
$$

which is rewritten in a two-step operation per layer as follows:

$$
\begin{aligned}
& l = 1, \\
& y_{\frac{1}{2}} = (Id + \mathcal{O}_{1,att} \circ \mathcal{O}_{1,nn})(x^*), \\
& y_1 = (Id + \mathcal{O}_{1,ff} \circ \mathcal{O}_{1,nn})(y_{\frac{1}{2}}), \\
& \dots \\
& l \in \{2, \dots, L-1\} : \\
& y_{l-\frac{1}{2}} = (Id + \mathcal{O}_{l,att} \circ \mathcal{O}_{l,nn})(y_{l-1}), \\
& y_l = (Id + \mathcal{O}_{l,ff} \circ \mathcal{O}_{l,nn})(y_{l-\frac{1}{2}}), \\
& \dots \\
& y_{L-\frac{1}{2}} = (Id + \mathcal{O}_{L,att} \circ \mathcal{O}_{L,nn})(y_{L-1}), \\
& x^* = y_L = (Id + \mathcal{O}_{L,ff} \circ \mathcal{O}_{L,nn})(y_{L-\frac{1}{2}}),
\end{aligned}
\tag{8.57}
$$

which means that $(y_{\frac{1}{2}}, y_1, y_{\frac{3}{2}}, y_2, y_{L-\frac{1}{2}}, y_L)$ is a fixed-point operator:

$$
(\hat{\mathcal{F}}_{\frac{1}{2}}, \hat{\mathcal{F}}_1, \dots, \hat{\mathcal{F}}_{L-\frac{1}{2}}, \hat{\mathcal{F}}_L),
$$

where

$$
\begin{aligned}
& l = 1, \\
& y_{\frac{1}{2}} = \hat{\mathcal{F}}_{\frac{1}{2}}(x^*), \\
& y_1 = \hat{\mathcal{F}}_1(y_{\frac{1}{2}}), \\
& \dots \\
& l \in \{2, \dots, L-1\} : \\
& y_{l-\frac{1}{2}} = \hat{\mathcal{F}}_{l-\frac{1}{2}}(y_{l-1}), \\
& y_l = \hat{\mathcal{F}}_l(y_{l-\frac{1}{2}}), \\
& y_{L-\frac{1}{2}} = \hat{\mathcal{F}}_{L-\frac{1}{2}}(y_{L-1}), \\
& x^* = y_L = \hat{\mathcal{F}}_L(y_{L-\frac{1}{2}}).
\end{aligned}
\tag{8.58}
$$

Proposition 8.68 *The outcomes of transformer-based neural network* $\mathbb{TNN}$ *are exactly the Nash equilibria of a game.*

Proof We have shown above that each operator $\hat{\mathscr{F}}_{l-\frac{1}{2}}$ and $\hat{\mathscr{F}}_l$ is a result of a game. By collecting all these games, we arrive at a hierarchical game where each hierarchical level l is a multi-step sequential game. At the hierarchical level l, each decision-maker has the layer input layer, which is the action input output of the preceding layer decision-makers. If the $\mathbb{TNN}$ has an outcome matrix x^*, it means that it is a fixed point of $\hat{\mathscr{F}}$, which means that one gets a fixed point to $\hat{\mathscr{O}}$. At this fixed-point x^*, each decision-maker is playing its unique best response to preceding input and preceding decision-maker at the same sub-layer. □

Lemma 8.69 (Existence) *Let $F : \mathscr{H}_0 \to \mathscr{H}_0$ be a continuous operator and B_ρ a ball of $\mathscr{H}$ with radius $\rho > 0$. The restricted variational inequality problem of finding $x^* \in B_\rho$, such that $\langle F(x^*), x - x^*\rangle \geq 0,\ \forall \theta \in B_\rho$, has at least one solution.*

This existence result is extremely useful in generative machine intelligence as the normalization operation brings the element to a certain closed ball B_ρ, which is a convex, bounded, and non-empty set of $\mathscr{H}_0$, and the operators involved are continuous (modified normalization, projection, attention, averaging, ReLU, softargmax, and Affine map are all continuous).

Proof The variational inequality restricted to a convex, closed, bounded, and non-empty set is equivalent to $x = \text{proj}_{B_\rho}(x - \gamma F(x)), \gamma > 0$, which admits at least one solution according to the Brouwer–Schauder fixed-point theorem. □

Proposition 8.70 (Existence) *Let $\sigma_{L,nn}$ be the normalization (holonorm) $((\frac{x_i}{1+\left|x_i\right|})_{1\leq i\leq D})'$. The $\mathbb{TNN}$ operator $\sigma_{L,nn} \circ \hat{\mathscr{O}}_L \circ \hat{\mathscr{O}}_{L-1} \circ \ldots \circ \hat{\mathscr{O}}_1$ maps the unit ball into itself and is continuous on $\mathscr{H}^{Dd}$. Moreover, this operator has at least one fixed point.*

Proof Follows from Lemma 8.69. □

The result provides the existence of a consistent outcome up to a normalization or scaling. If $\left|x^*\right| < 1$, then one can re-scale or up-scale using the mapping $\hat{\sigma}^{-1}$.

Next, we focus on the training problem of $\mathbb{TNN}$.

8.2.3.3 Design and Training of the Full Transformer

Problem 8.71 Design: Find L, H, D, d, k and the operators $\hat{\mathscr{O}}_1, \ldots, \hat{\mathscr{O}}_L$:

$$\arg\min_{L,D\geq 1} \arg\min_{\hat{\mathscr{O}}_L,\ldots,\hat{\mathscr{O}}_1} \int_{x_0}\int_{y_L} [\Phi(y_L, \hat{\mathscr{O}}_L \circ \ldots \circ \hat{\mathscr{O}}_1 x_0)]\mathbb{P}(dx_0 dy_L), \tag{8.59}$$

where Φ is a real-valued risk-aware cost function.

The minimization over functional space and the transport map are more involved. As an illustration, let us consider the Monge problem: Given two measures m_0, ν_1. Design: Find $\hat{\mathscr{O}}_1 : \mathscr{H}^d \to \mathscr{H}^d$:

$$\arg\min_{\hat{\mathcal{O}}_1}\{\int_{x_0}[\Phi(x_0,\hat{\mathcal{O}}_1 x_0)]m_0(dx_0),\ \hat{\mathcal{O}}_1\#m_0=\nu_1\}, \tag{8.60}$$

where $\hat{\mathcal{O}}_1\#m_0$ is the push-forward operator induced by $\hat{\mathcal{O}}_1$, i.e., $\nu_1(B) = m_0(\hat{\mathcal{O}}_1^{-1}(B))$ for any measurable subset.

Some special cases of the Monge problem can be solved for initial absolutely continuous measures m_0 through the Kantorovich relaxation:

Given two measures m_0, ν_1, find a joint probability measure $\mathbb{P}(dx_0 dy_1)$ such that

$$\arg\min_{\mathbb{P}}\{\int_{x_0}\int_{y_1}[\Phi(y_1,\hat{\mathcal{O}}_1 x_0)]\mathbb{P}(dx_0 dy_1),\ m_0=\mathbb{P}_x,\nu_1=\mathbb{P}_y\}, \tag{8.61}$$

where $\mathbb{P}_x = m_0$ is the first marginal, $\mathbb{P}_y = \nu_1$ is the second marginal, and m_0 is assumed to be an absolutely continuous measure.

Remark 8.72 A number of observations are in order. First, the absolute continuity for the initial measure means the initial data input. The data input is currently large but still finite. Thus, it is not a density-based data. This means that one needs to get an estimate density from the training data. Second, using mean-field-type team [106], the optimal map $\hat{\mathcal{O}}_1$ depends on the measure (training data), which is problematic for the architectures of generative machine intelligence. Third, the optimal transport map $\hat{\mathcal{O}}_1$ does not have the shape of ReLU, softargmax, masked/unmasked self-attention used at the layers. Can we instead design activation functions that all collectively approximate the optimal transport map? This question is related to the approximation capability of transformer-based neural networks.

We therefore focus on a sub-goal, which is the minimization of parameters:

$$\theta=(\gamma_{1l,i},\beta_{1l,i},Q_{lh},K_{lh},V_{lh},W_{lh},\gamma_{2l,i},\beta_{2l,i},\ W_{1li},W_{2li},b_{1li},b_{2li})_{1\le i\le D,1\le l\le L}$$

However, $\theta\mapsto\int_{x_0}\int_{y_L}[\Phi(y_L,\hat{\mathcal{O}}_L\circ\ldots\circ\hat{\mathcal{O}}_1 x_0)]\mathbb{P}(dx_0 dy_L)$ is non-convex, i.e., the risk-aware minimization over θ is non-convex even if Φ is a convex function. This means that some hidden convexity structures in the operator $(Id+\mathcal{O}_{l,ff}\circ\mathcal{O}_{l,nn})\circ(Id+\mathcal{O}_{l,att}\circ\mathcal{O}_{l,nn})$ need to be exploited.

8.2.4 Extremely Large Data and Asymptotics

As D grows to infinity, the size of the matrix $x=(x_1,\ldots,x_D)^\dagger$ explodes. The normalization layer is done row-wise. We choose a generic row, say i. $LayerNorm(x_i)$ is well defined. The projection to a hyperplane or ball is also well defined.

8.2.4.1 Unmasked Self-Attention

The single-headed unmasked attention mechanism has the following term:

$$\sum_{l=1}^{D} \frac{e^{\frac{1}{\sqrt{k}}\langle Qx_i, Kx_j\rangle}}{\frac{D}{D}\sum_{l=1}^{D} e^{\frac{1}{\sqrt{k}}\langle Qx_i, Kx_l\rangle}} x_j = \frac{1}{D}\sum_{l=1}^{D} \frac{e^{\frac{1}{\sqrt{k}}\langle Qx_i, Kx_j\rangle}}{\frac{1}{D}\sum_{l=1}^{D} e^{\frac{1}{\sqrt{k}}\langle Qx_i, Kx_l\rangle}} x_j$$

$$= \frac{\int e^{\frac{1}{\sqrt{k}}\langle Qx_i, Ky\rangle} y m_D(dy)}{\int e^{\frac{1}{\sqrt{k}}\langle Qx_i, Ky'\rangle} m_D(dy')},$$

where $m_D(dy) = \left(\frac{1}{D}\sum_{j=1}^{D}\delta_{x_j}\right)(dy)$ is the empirical measure of the input data set. The integrability of these exponential components imposes conditions on Q, K, V, W and on the measure m_D. If there is a limiting measure m such that all those integrals are finite, then the multi-headed unmasked self-attention is well defined and is given by

$$x_i + \frac{1}{\sqrt{H}}\sum_{h=1}^{H} W_h V_h \frac{\int e^{\frac{1}{\sqrt{k}}\langle Q_h x_i, K_h y\rangle} y m(dy)}{\int e^{\frac{1}{\sqrt{k}}\langle Q_h x_i, K_h y'\rangle} m(dy')}.$$

Definition 8.73 Given $m \in \mathscr{P}(\mathscr{H}^d)$, the (Boltzmann–Gibbs) mean-field unmasked self-attention is $\mathrm{mf}_{unmasked-att}$:

$$\begin{cases} \mathscr{H}^d \to \mathscr{H}^d \\ x \mapsto x + \frac{1}{\sqrt{H}}\sum_{h=1}^{H} W_h V_h \left(\frac{\int_{\mathscr{H}^d} e^{\frac{1}{\sqrt{k}}\langle Q_h x, K_h y\rangle} y m(dy)}{\int_{\mathscr{H}^d} e^{\frac{1}{\sqrt{k}}\langle Q_h x, K_h y'\rangle} m(dy')} \right). \end{cases} \tag{8.62}$$

8.2.4.2 Masked Self-Attention

As D grows to infinity, the size of the matrix $x = (x_1, \ldots, x_D)^\dagger$ explodes, but now we examine the position of x_i in the list, and $i \leq D$ should be encoded. We can capture it through the variable $s_i = \frac{i}{D}$. This describes the set $\{\frac{1}{D}, \frac{2}{D}, \ldots, \frac{D-1}{D}, \frac{D}{D}\}$. As D goes to infinity, this becomes the non-empty compact interval $[0, 1]$. The normalization layer is done row-wise with the encoding (s_i, x_i). The masked occupancy measure yields

$$\hat{m}_D(ds', dx') = \left[\frac{1}{D}\sum_{j=1}^{D}\delta_{(s_j, x_j)}\right](ds'dx')$$

and is a probability measure over the product space $[0, 1] \times \mathscr{H}^d$. We denote its limiting measure (if any) by $\hat{m}(ds', dx')$, and the second marginal as $m(B) = \int_0^1\int_B \hat{m}(ds', dx')$ will be a probability measure over $\mathscr{H}^d$.

We choose a generic row; say i. $LayerNorm(x_i)$ is well defined for each s_i. The projection to a hyperplane or ball is well defined. The attention mechanism becomes

$$\sum_{j=1}^{i} \frac{e^{\frac{1}{\sqrt{k}}\langle Qx_i, Kx_j\rangle}}{\frac{D}{D}\sum_{l=1}^{i} e^{\frac{1}{\sqrt{k}}\langle Qx_i, Kx_l\rangle}} x_j = \frac{1}{D}\sum_{j=1}^{D} \frac{e^{\frac{1}{\sqrt{k}}\langle Qx_i, Kx_j\rangle}\mathbb{I}_{\frac{j}{D}\leq\frac{i}{D}}}{\frac{1}{D}\sum_{l=1}^{D} e^{\frac{1}{\sqrt{k}}\langle Qx_i, Kx_l\rangle}\mathbb{I}_{\frac{l}{D}\leq\frac{i}{D}}} x_j$$

$$= \frac{\int_0^1\int_{\mathcal{H}^d} e^{\frac{1}{\sqrt{k}}\langle Qx_i, Ky\rangle}\mathbb{I}_{s\leq s_i}\, y\hat{m}_D(dsdy)}{\int_0^1\int_{\mathcal{H}^d} e^{\frac{1}{\sqrt{k}}\langle Qx_i, Ky'\rangle}\mathbb{I}_{s'\leq s_i}\hat{m}_D(ds'dy')}.$$

If there is a limiting measure $\hat{m}$ such that all those integrals are finite, then the one-headed masked self-attention is well defined and is given by

$$\frac{\int_0^1\int_{\mathcal{H}^d} e^{\frac{1}{\sqrt{k}}\langle Qx_i, Ky\rangle}\mathbb{I}_{s\leq s_i}\, y\hat{m}(dsdy)}{\int_0^1\int_{\mathcal{H}^d} e^{\frac{1}{\sqrt{k}}\langle Qx_i, Ky'\rangle}\mathbb{I}_{s'\leq s_i}\hat{m}(ds'dy')}.$$

Thus, the multi-headed masked self-attention is

$$x_i + \frac{1}{\sqrt{H}}\sum_{h=1}^{H}(W_hV_h)\frac{\int_0^1\int_{\mathcal{H}^d} e^{\frac{1}{\sqrt{k}}\langle Q_hx_i, K_hy\rangle}\mathbb{I}_{s\leq s_i}\, y\hat{m}(dsdy)}{\int_0^1\int_{\mathcal{H}^d} e^{\frac{1}{\sqrt{k}}\langle Q_hx_i, K_hy'\rangle}\mathbb{I}_{s'\leq s_i}\hat{m}(ds'dy')}.$$

Definition 8.74 Given $\hat{m} \in \mathcal{P}([0,1]\times\mathcal{H}^d)$, the mean-field-masked self-attention is $\mathrm{mf}_{masked-att}$:

$$\begin{cases} [0,1]\times\mathcal{H}^d \to [0,1]\times\mathcal{H}^d \\ (s,x) \mapsto \left(s, x + \frac{1}{\sqrt{H}}\sum_{h=1}^{H} W_hV_h \frac{\int_0^1\int_{\mathcal{H}^d} e^{\frac{1}{\sqrt{k}}\langle Q_hx, K_hy\rangle}\mathbb{I}_{s'\leq s}\, y\hat{m}(ds'dy)}{\int_0^1\int_{\mathcal{H}^d} e^{\frac{1}{\sqrt{k}}\langle Q_hx, K_hy'\rangle}\mathbb{I}_{s''\leq s}\hat{m}(ds''dy')}\right). \end{cases} \tag{8.63}$$

8.2.4.3 Sigmoid Self-Attention

An alternative is the sigmoid unmasked self-attention given by

$$x_i + \frac{1}{\sqrt{H}}\sum_{h=1}^{H}(W_hV_h)\sum_{j=1}^{D}\frac{e^{\frac{1}{\sqrt{k}}\langle Q_hx_i, K_hx_j\rangle}}{D + e^{\frac{1}{\sqrt{k}}\langle Q_hx_i, K_hx_j\rangle}} x_j$$

$$= x_i + \frac{1}{\sqrt{H}}\sum_{h=1}^{H}(W_hV_h)\frac{1}{D}\sum_{j=1}^{D}\frac{De^{\frac{1}{\sqrt{k}}\langle Q_hx_i, K_hx_j\rangle}}{D + e^{\frac{1}{\sqrt{k}}\langle Q_hx_i, K_hx_j\rangle}} x_j,$$

which is rewritten

$$x_i + \frac{1}{\sqrt{H}} \sum_{h=1}^{H} (W_h V_h) \frac{1}{D} \sum_{j=1}^{D} \frac{e^{\frac{1}{\sqrt{k}}\langle Q_h x_i, K_h x_j \rangle}}{1 + \frac{1}{D} e^{\frac{1}{\sqrt{k}}\langle Q_h x_i, K_h x_j \rangle}} x_j$$

$$= x_i + \frac{1}{\sqrt{H}} \sum_{h=1}^{H} (W_h V_h) \int \frac{e^{\frac{1}{\sqrt{k}}\langle Q_h x_i, K_h y \rangle}}{1 + \frac{1}{D} e^{\frac{1}{\sqrt{k}}\langle Q_h x_i, K_h y \rangle}} y m_D(dy).$$

For large D, the term $\frac{1}{D} e^{\frac{1}{\sqrt{k}}\langle Q_h x_i, K_h y \rangle} \to 0$. Thus, the asymptotic sigmoid unmasked self-attention is

$$x_i + \frac{1}{\sqrt{H}} \sum_{h=1}^{H} W_h V_h \int e^{\frac{1}{\sqrt{k}}\langle Q_h x_i, K_h y \rangle} y m(dy).$$

Definition 8.75 Given $m \in \mathscr{P}(\mathscr{H}^d)$, the mean-field sigmoid unmasked self-attention is $\text{mfSigmoid}_{unmasked-att}$:

$$\begin{cases} \mathscr{H}^d \to \mathscr{H}^d \\ x \mapsto \left(x + \frac{1}{\sqrt{H}} \sum_{h=1}^{H} W_h V_h \int_{\mathscr{H}^d} e^{\frac{1}{\sqrt{k}}\langle Q_h x, K_h y \rangle} y m(dy) \right). \end{cases} \tag{8.64}$$

The sigmoid masked self-attention is

$$x_i + \frac{1}{\sqrt{H}} \sum_{h=1}^{H} (W_h V_h) \sum_{j=1}^{i} \frac{e^{\frac{1}{\sqrt{k}}\langle Q_h x_i, K_h x_j \rangle}}{D + e^{\frac{1}{\sqrt{k}}\langle Q_h x_i, K_h x_j \rangle}} x_j.$$

It follows that the asymptotic sigmoid masked self-attention is

$$x_i + \frac{1}{\sqrt{H}} \sum_{h=1}^{H} W_h V_h \int_0^1 \int_{\mathscr{H}^d} e^{\frac{1}{\sqrt{k}}\langle Q_h x_i, K_h y \rangle} \mathbb{I}_{s \le s_i} y \hat{m}(ds dy).$$

Definition 8.76 Given $\hat{m} \in \mathscr{P}([0, 1] \times \mathscr{H}^d)$, the mean-field sigmoid masked self-attention is $\text{mfSigmoid}_{masked-att}$:

$$\begin{cases} [0, 1] \times \mathscr{H}^d \to [0, 1] \times \mathscr{H}^d \\ (s, x) \mapsto \left(s, x + \frac{1}{\sqrt{H}} \sum_{h=1}^{H} W_h V_h \int_0^1 \int_{\mathscr{H}^d} e^{\frac{1}{\sqrt{k}}\langle Q_h x, K_h y \rangle} \mathbb{I}_{s' \le s} y \hat{m}(ds' dy) \right). \end{cases} \tag{8.65}$$

8.2.4.4 Projected Boltzmann–Gibbs Self-Attention

Let $x_i \in \mathscr{H}^d$ be a nonzero vector.

$$\mathrm{proj}_{x_i^{\perp}}(y_i) := y_i - \left\langle y_i, \frac{x_i}{\langle x_i, x_i\rangle}\right\rangle x_i.$$

Then

$$\begin{array}{l} \left|\mathrm{proj}_{x_i^{\perp}}(y_i)\right|^2 = \left|y_i - \langle y_i, \frac{x_i}{\langle x_i, x_i\rangle}\rangle x_i\right|^2 \\ = \langle y_i - \frac{\langle y_i, x_i\rangle}{\langle x_i, x_i\rangle} x_i, y_i - \frac{\langle y_i, x_i\rangle}{\langle x_i, x_i\rangle} x_i\rangle \\ = \langle y_i - \frac{\langle y_i, x_i\rangle}{\langle x_i, x_i\rangle} x_i, y_i\rangle - \langle y_i - \frac{\langle y_i, x_i\rangle}{\langle x_i, x_i\rangle} x_i, \frac{\langle y_i, x_i\rangle}{\langle x_i, x_i\rangle} x_i\rangle \\ = \langle y_i, y_i\rangle - \frac{(\langle y_i, x_i\rangle)^2}{\langle x_i, x_i\rangle} \leq \langle y_i, y_i\rangle \end{array} \quad (8.66)$$

This means that $\left|\mathrm{proj}_{x_i^{\perp}}(y_i)\right| \leq \left|y_i\right|$. From Lemma 8.50, the operator $\mathrm{proj}_{x_i^{\perp}}(y_i)$ is indeed a projection into the orthogonal space to x_i.

Let $\varepsilon > 0$. The Boltzmann–Gibbs self-attention works with the parameter $\varepsilon > 0$, which allows us to examine its relationship to $\arg\max_{j\in\{1,2,\ldots,D\}} \langle Q_h x_i, K_h x_j\rangle$ as ε goes to zero, i.e., $\frac{1}{\varepsilon}$ goes to infinity. Note that $\frac{1}{\sqrt{k}}$ does not go to infinity.

The projected Boltzmann–Gibbs unmasked self-attention

$$x_i + \mathrm{proj}_{x_i^{\perp}}\left(\frac{1}{\sqrt{H}}\sum_{h=1}^{H} W_h V_h \frac{\sum_{l=1}^{D} e^{\frac{1}{\varepsilon}\langle Q_h x_i, K_h x_j\rangle} x_j}{\sum_{l=1}^{D} e^{\frac{1}{\varepsilon}\langle Q_h x_i, K_h x_l\rangle}}\right),$$

which is

$$x_i + \mathrm{proj}_{x_i^{\perp}}\left(\frac{1}{\sqrt{H}}\sum_{h=1}^{H} W_h V_h \frac{\int e^{\frac{1}{\varepsilon}\langle Q_h x_i, K_h y\rangle} y m_D(dy)}{\int e^{\frac{1}{\varepsilon}\langle Q_h x_i, K_h y'\rangle} m_D(dy')}\right).$$

With H finite and D going to infinity, one has the projected mean-field Boltzmann–Gibbs unmasked self-attention as

$$x_i + \mathrm{proj}_{x_i^{\perp}}\left(\frac{1}{\sqrt{H}}\sum_{h=1}^{H} W_h V_h \frac{\int e^{\frac{1}{\varepsilon}\langle Q_h x_i, K_h y\rangle} y m(dy)}{\int e^{\frac{1}{\varepsilon}\langle Q_h x_i, K_h y'\rangle} m(dy')}\right).$$

Definition 8.77 Given $m \in \mathscr{P}(\mathscr{H}^d)$, the projected mean-field Boltzmann–Gibbs unmasked self-attention is $\mathrm{pmfBG}_{unmaskedatt}$:

$$\begin{cases} \mathscr{H}^d \to \mathscr{H}^d \\ x \mapsto x + \mathrm{proj}_{x^\perp}\left(\frac{1}{\sqrt{H}}\sum_{h=1}^H W_h V_h \frac{\int_{\mathscr{H}^d} e^{\frac{1}{\varepsilon}\langle Q_h x, K_h y\rangle} y m(dy)}{\int_{\mathscr{H}^d} e^{\frac{1}{\varepsilon}\langle Q_h x, K_h y'\rangle} m(dy')}\right). \end{cases} \tag{8.67}$$

Similarly, the projected mean-field Boltzmann–Gibbs masked self-attention is given by

$$x_i + \mathrm{proj}_{x_i^\perp}\left(\frac{1}{\sqrt{H}}\sum_{h=1}^H W_h V_h \frac{\sum_{l=1}^i e^{\frac{1}{\varepsilon}\langle Q_h x_i, K_h x_j\rangle} x_j}{\sum_{l=1}^i e^{\frac{1}{\varepsilon}\langle Q_h x_i, K_h x_l\rangle}}\right).$$

It follows that projected mean-field Boltzmann–Gibbs masked self-attention is

$$x_i + \mathrm{proj}_{x_i^\perp}\left(\frac{1}{\sqrt{H}}\sum_{h=1}^H W_h V_h \frac{\int_0^1\int_{\mathscr{H}^d} e^{\frac{1}{\varepsilon}\langle Q_h x, K_h y\rangle}\mathbb{I}_{s\le s_i} y\hat{m}(dsdy)}{\int_0^1\int_{\mathscr{H}^d} e^{\frac{1}{\varepsilon}\langle Q_h x_i, K_h y'\rangle}\mathbb{I}_{s'\le s_i}\hat{m}(ds'dy')}\right).$$

Definition 8.78 Given $\hat{m} \in \mathscr{P}([0,1]\times\mathscr{H}^d)$, the projected mean-field Boltzmann–Gibbs masked self-attention is $\mathrm{pmfBG}_{masked-att}$:

$$\begin{cases} [0,1]\times\mathscr{H}^d \to [0,1]\times\mathscr{H}^d \\ (s,x) \mapsto \left(s, x + \mathrm{proj}_{x^\perp}\left(\frac{1}{\sqrt{H}}\sum_{h=1}^H W_h V_h \frac{\int_0^1\int_{\mathscr{H}^d} e^{\frac{1}{\varepsilon}\langle Q_h x, K_h y\rangle}\mathbb{I}_{s'\le s} y\hat{m}(ds'dy)}{\int_0^1\int_{\mathscr{H}^d} e^{\frac{1}{\varepsilon}\langle Q_h x, K_h y'\rangle}\mathbb{I}_{s''\le s}\hat{m}(ds''dy')}\right)\right). \end{cases} \tag{8.68}$$

8.2.4.5 Mean-Field Limit Transformer

In all these self-attention operations, the number of heads H can be relatively large (such as $H = 96$) but not infinite. In practice, the data size can be large but still not infinite. At each layer, the number of operations can be large but not infinite. In this context, instead of applying row-by-row probability transformation of this huge matrix, one may think of taking a generic row and applying a mean-field effect. This leads to a generic decision-maker interacting with the mean field at each layer l. For a nonvanishing learning rate, one gets a mean-field difference equation, while the mean field is frozen (because a generic decision-maker does not affect the mean field in the infinite population regime). In this case, the outcomes of $\mathbb{TGN}_{mf}$ are steady states of some difference equation, and the mean field solves a measure push-forward equation. The other operators of the transformer block are done only for each x_i leading to

$$(Id + \mathscr{O}_{l,ff,i} \circ \mathscr{O}_{l,nn,i}) \circ (Id + \mathscr{O}_{l,att,i} \circ \mathscr{O}_{l,nn,i})(y_{l-1,i}, m),$$

there is no other rows to operate, but the measure m is supposed to be learnable to compute the averaging terms of the attention mechanism. Instead of the infinite data $\{(x_i, y_i), i \in \{1, \ldots, \infty\}$, one has

$$\left(x_i, y_i, \tilde{m}(dx'dy') = \lim_{I\to\infty} \left[\frac{1}{I-1} \sum_{j\neq i} \delta_{(x_j,y_j)} \right] (dx'dy') \right)$$

from which we extract the marginal distributions $m = \mathbb{P}_X$ and $\mathbb{P}_Y$.

In the infinite data mean-field limit, the input is (x_i, m), and the same m is transferred at each layer and each generic neuron unit where the computation of the attention mechanism is required. The output is (y_i, m^*).

Definition 8.79 The mean-field transformer-based neural network $\mathbb{TNN}_{mf}$ specification is given by the family

$$(x_i, y_i, \tilde{m}, L, d, k, H, (r_l, \gamma_{1l}, \beta_{1l}, Q_{lh}, K_{lh}, V_{lh}, W_{lh}, \gamma_{2l}, \beta_{2l}, W_{1l}, W_{2l}, b_{1l}, b_{2l})_{1\leq i\leq D}, \lambda).$$

where the pair $(x_i, y_i) \in (\mathscr{H}^d)^2$ is a generic initial draw according to $\tilde{m}$.

The consistency check now requires only a double fixed-point equation: $x_i^* = \hat{\mathscr{O}}_{L,i} \circ \hat{\mathscr{O}}_{L-1,i} \circ \ldots \circ \hat{\mathscr{O}}_{1,i}(x_i^*, m^*)$ and $\mathbb{P}_{x_i^*\ |m^*} = \mathscr{L}(x_i^*\ |m^*) = m^*$. We rename $\hat{\mathscr{O}}_{l,i}$ to be $\hat{\mathscr{F}}_l \circ \hat{\mathscr{F}}_{l-\frac{1}{2}\ |m^*}$, where the explicit dependence on m^* appears in $\hat{\mathscr{F}}_{l-\frac{1}{2}\ |m^*}$, where the mean-field self-attention mechanism is applied, and where $\hat{\mathscr{F}}_l = (Id + \mathscr{O}_{l,ff} \circ \mathscr{O}_{l,nn})$ and $\hat{\mathscr{F}}_{l-\frac{1}{2}\ |m^*} = (Id + \mathscr{O}_{l,att} \circ \mathscr{O}_{l,nn})$ are now defined on $\mathscr{H}^d \times \mathscr{P}(\mathscr{H}^d)$.

If these two fixed-point equations are fulfilled, the sequence

$$\begin{aligned}
& l = 1, \\
& y_{\frac{1}{2}} = \hat{\mathscr{F}}_{\frac{1}{2}}(x^*, m^*), \\
& y_1 = \hat{\mathscr{F}}_1 \circ \hat{\mathscr{F}}_{\frac{1}{2}}(x^*, m^*), \\
& \ldots \\
& l \in \{2, \ldots, L-1\}: \\
& y_{l-\frac{1}{2}} = \hat{\mathscr{F}}_{l-\frac{1}{2}\ |m^*} \circ \hat{\mathscr{F}}_{l-1} \circ \hat{\mathscr{F}}_{l-\frac{3}{2}\ |m^*} \circ \ldots \circ \hat{\mathscr{F}}_1 \circ \hat{\mathscr{F}}_{\frac{1}{2}}(x^*, m^*), \\
& y_l = \hat{\mathscr{F}}_l \circ \hat{\mathscr{F}}_{l-\frac{1}{2}\ |m^*} \circ \ldots \circ \hat{\mathscr{F}}_1 \circ \hat{\mathscr{F}}_{\frac{1}{2}}(x^*, m^*), \\
& \ldots \\
& y_{L-\frac{1}{2}} = \hat{\mathscr{F}}_{L-\frac{1}{2}\ |m^*} \circ \hat{\mathscr{F}}_{L-1} \circ \ldots \circ \hat{\mathscr{F}}_1 \circ \hat{\mathscr{F}}_{\frac{1}{2}}(x^*, m^*), \\
& x^* = y_L = \hat{\mathscr{F}}_L \circ \hat{\mathscr{F}}_{L-\frac{1}{2}\ |m^*} \circ \ldots \circ \hat{\mathscr{F}}_1 \circ \hat{\mathscr{F}}_{\frac{1}{2}}(x^*, m^*),
\end{aligned} \tag{8.69}$$

where the dependence on m^* is in the operator $\hat{\mathscr{F}}_{l-\frac{1}{2}} = \hat{\mathscr{F}}_{l-\frac{1}{2}}|_{m^*}$ and the resulting output

$$(y_{\frac{1}{2}}, y_1, y_{\frac{3}{2}}, y_2, \ldots, y_{L-\frac{1}{2}}, y_L = x_i^*, m^*)$$

are the outcomes of $\mathbb{TGN}_{mf}$.

Proposition 8.80 *The infinite population $\mathbb{TGN}_{mf}$ is a mean-field game, and the outcomes of $\mathbb{TGN}_{mf}$ are mean-field equilibria.*

Proof In an infinite population hierarchical mean-field game, the infinite population mean-field object is frozen. Here, the frozen object is m. Considering a well-designed infinite population $\mathbb{TGN}_{mf}$, the operator $\hat{\mathscr{O}}_l$ at each layer l can be viewed as the best-response function of decision-maker l to (y_{l-1}, m_{l-1}). This implies that the output y_l is an optimal response to the previous layer's outputs and to the frozen mean field. The choice of the layer decision-maker does not affect the mean field, as it is frozen. Thus, each layer functions as a decision-maker that optimizes its action based on the action of the preceding decision-maker and the mean field, as in the best-response dynamics to the mean field. This defines an infinite population hierarchical mean-field game. Given that each layer in the $\mathbb{TGN}_{mf}$ represents a decision-maker optimizing its action based on the infinite population mean field of actions from previous layers, the sequence of outputs $(y_1^*, y_2^*, \ldots, y_L^*)$ and frozen mean-field vector $(m_1^*, m_2^*, \ldots, m_L^*)$ constitutes a mean-field-type Nash equilibrium if it is best response, and the resulting action will reproduce m^* as its probability distribution, i.e., $\mathscr{L}(y_l^*) = m_l^*$ creating a fixed point. This hierarchical mean-field equilibrium of $\mathbb{TGN}_{mf}$ is characterized by each generic decision-maker choosing y_l^* as a response to (y_{l-1}^*, m_{l-1}^*), such that the resulting distribution m_l^* is consistent with the responses of all other generic decision-makers. □

This outcome is also an infinite population mean-field Nash equilibrium of the hierarchical mean-field game. Note that, here, the mean field is an infinite population mean field of actions in the sense that m is a distribution of the other decision-makers' actions. Due to the infinite population assumption, x_i, which is an action of a generic decision-maker, does not influence the population mean-field limit m. We will see that this has strong limitations from an interaction perspective, where each operation may affect the distribution, which is discussed next.

The training problem at this infinite population regime is to move from a signal input distribution to an output distribution subject to the difference equation constraint:

$$\begin{aligned}
&\mathcal{D} = \{\mathbb{P}\}, \mathbb{P} \in \mathcal{P}(\mathcal{H}^d \times \mathcal{H}^d) \\
&\inf_{\theta\in\Theta} E\Phi(y_T, \hat{y}), \\
&\text{such that} \\
&y_0 = x_0 = \hat{x} \sim \mathbb{P}_1 = m_0, \; \hat{y} \sim \mathbb{P}_2 \\
&y_{t+1} = y_t + \lambda_t(\hat{\mathcal{F}}_L \circ \hat{\mathcal{F}}_{L-\frac{1}{2}}|_{m_t} \circ \ldots \circ \hat{\mathcal{F}}_1 \circ \hat{\mathcal{F}}_{\frac{1}{2}, m_t}(y_t) - y_t), \; t > 0 \\
&m_{t+1} = ((1-\lambda_t)Id + \lambda_t \hat{\mathcal{F}}_{L-\frac{1}{2}}|_{m_t} \circ \ldots \circ \hat{\mathcal{F}}_1 \circ \hat{\mathcal{F}}_{\frac{1}{2}, m_t}|_{\theta=\theta^*})\# m_t,
\end{aligned} \tag{8.70}$$

where Φ is a real-valued risk-aware cost function and T is the number of iterations of transformer $\mathbb{TNN}_{mf}$.

A pair (θ, m) solution of (8.70) is also a mean-field Nash equilibrium while restricted to the class of time-and-state-independent controls. Note that generically the mean-field Nash equilibrium here is a function of (y_t, m_t), which means that the restriction of constant control may lose optimality [106].

On the Learnability of the Input Distribution

The distribution m can be obtained as the marginal of $\mathbb{P}(dx_0 dy_L)$ with respect to the first variable. If $\mathbb{P}(dx_0 dy_L)$ is unkown, m can be approximated from m_D, which is the input empirical measure. In that case, the approximated measure m should not be considered as exact. One can use a distributionally robust approach as in Proposition 8.35 to take into consideration all measures around the learned/estimated one.

Beyond Independent and Identically Distributed Data Set

Mean-field convergence goes far beyond independent and identically distributed data set input data. It includes indistinguishability, multi-class indistinguishability, pairwise indistinguishability, k-wise indistinguishability, and asymptotically k-wise indistinguishability. These situations allow us to cover a broader class of data sets in generative machine intelligence, where the i.i.d. assumption does not hold. Details are given in the Appendix of Volume I.

8.3 MFTT: Mean-Field-Type Transformer

Traditional transformers, especially those relying on pairwise attentions such as the Boltzmann–Gibbs, Coulomb, sigmoid, Cucker-Smale, Kuramoto, Krause, and consensus-based masked and unmasked self-attention, face scalability issues as the dimensionality or sequence length increases. MFTTs circumvent this by working with distributions rather than pairwise interactions. This means that MFTTs can efficiently handle high-dimensional data without the need to explicitly consider all pairwise interactions, which significantly reduces computational complexity. MFTTs are inherently designed to work with probability distributions, making them particularly well suited for continuous data or scenarios where the input and output spaces are not discrete. This flexibility allows MFTTs to model more complex data distributions and interactions compared to traditional transformers that are often limited to discrete representations. By incorporating mean-field-type interactions, MFTTs can capture more subtle dependencies in the data. Traditional transformers might struggle to model such dependencies due to their fixed-dimensional representations and distribution-independent attention mechanisms. MFTTs, on the other hand, can adapt their representations based on the distribution of the inputs and outputs, potentially leading to improved performance on tasks that involve risk-awareness and structured data. Mean-field-type game theory often provides insights into the distribution-dependent quantities of interest of systems with several interacting components. In the context of MFTTs, this means that the model can generalize better from observed data distributions to unseen data, improving robustness and adaptability across a range of tasks and data distributions.

In the previous section, the distribution of the data was assumed to be identical and fixed across layers and across time. However, during the operation of generative machine intelligence, the distribution of the new data need not be the same across layers. This leads to a mean-field-type game problem and a mean-field-type transformer. In this setting, we do not have an infinite number of decision-makers per sub-player. There is no n. There is indeed an initial distribution of data μ_0. Then, an initial signal input x_0 is generated according to μ_0. With some abuse of notation, one can consider an $\mathscr{H}^d$-valued random variable x_0 with probability law $\mu_0 = \mu_{x_0} = \mu_{y_0}$. We equip the Hilbert space with the inner product $E(\langle x_1, x_2 \rangle)$, where E denotes the expected value operator. The underlying processes are hence assumed to have finite variances.

The earlier implementations of transformers operated on fixed-dimensional representations, and their attention mechanisms were designed to handle the interactions between all pairs of elements within a sequence. Mean-field-type interactions include not only pairwise interactions but also interactions with the distribution of signal inputs and or outputs of the preceding layer. Mean-field-type transformer (MFTT) extends the traditional transformer architecture by incorporating the concept of mean-field-type game theory, allowing it to work directly with distributions and providing a framework for handling interactions in a more scalable and flexible manner.

8.3.1 Mean-Field-Type Self-Attention Mechanisms

We provide below several examples of mean-field-type self-attention mechanisms.

Example 8.81 In addition to the mean-field-type interactions provided in Sect. 8.2.4 below, the following mean-field-type attention mechanisms without freezing the distributions are defined:

- The mean-field-type unmasked (Boltzmann–Gibbs) self-attention is $\text{mft}_{unmasked-att}$:

$$\begin{cases} \mathscr{H}^d \times \mathscr{P}(\mathscr{H}^d) \to \mathscr{H}^d \\ (x,\mu) \mapsto x + \frac{1}{\sqrt{H}} \sum_{h=1}^{H} W_h V_h \left(\frac{\int_{\mathscr{H}^d} e^{\frac{1}{\sqrt{k}}\langle Q_h x, K_h y\rangle} y \mu(dy)}{\int_{\mathscr{H}^d} e^{\frac{1}{\sqrt{k}}\langle Q_h x, K_h y'\rangle} \mu(dy')} \right). \end{cases} \tag{8.71}$$

- The mean-field-type masked (Boltzmann–Gibbs) self-attention is $\text{mft}_{masked-att}$:

$$\begin{cases} [0,1] \times \mathscr{H}^d \times \mathscr{P}([0,1] \times \mathscr{H}^d) \to [0,1] \times \mathscr{H}^d \\ (s,x,\hat{\mu}) \mapsto \left(s, x + \frac{1}{\sqrt{H}} \sum_{h=1}^{H} W_h V_h \frac{\int_0^1 \int_{\mathscr{H}^d} e^{\frac{1}{\sqrt{k}}\langle Q_h x, K_h y\rangle} \mathbb{I}_{s' \le s}\, y \hat{\mu}(ds'dy)}{\int_0^1 \int_{\mathscr{H}^d} e^{\frac{1}{\sqrt{k}}\langle Q_h x, K_h y'\rangle} \mathbb{I}_{s'' \le s}\, \hat{\mu}(ds''dy')} \right). \end{cases} \tag{8.72}$$

- The mean-field-type unmasked sigmoid self-attention is $\text{mftSigmoid}_{unmasked-att}$:

$$\begin{cases} \mathscr{H}^d \times \mathscr{P}(\mathscr{H}^d) \to \mathscr{H}^d \\ (x,\mu) \mapsto \left(x + \frac{1}{\sqrt{H}} \sum_{h=1}^{H} W_h V_h \int_{\mathscr{H}^d} e^{\frac{1}{\sqrt{k}}\langle Q_h x, K_h y\rangle} y \mu(dy) \right). \end{cases} \tag{8.73}$$

- The mean-field-type masked sigmoid self-attention is $\text{mfSigmoid}_{masked-att}$:

$$\begin{cases} [0,1] \times \mathscr{H}^d \times \mathscr{P}([0,1] \times \mathscr{H}^d) \to [0,1] \times \mathscr{H}^d \\ (s,x,\hat{\mu}) \mapsto \left(s, x + \frac{1}{\sqrt{H}} \sum_{h=1}^{H} W_h V_h \int_0^1 \int_{\mathscr{H}^d} e^{\frac{1}{\sqrt{k}}\langle Q_h x, K_h y\rangle} \mathbb{I}_{s' \le s}\, y \hat{\mu}(ds'dy) \right). \end{cases} \tag{8.74}$$

- The projected mean-field-type unmasked Boltzmann–Gibbs self-attention is $\text{pmftBG}_{unmaskedatt}$:

$$\begin{cases} \mathscr{H}^d \times \mathscr{P}(\mathscr{H}^d) \to \mathscr{H}^d \\ (x,\mu) \mapsto x + \text{proj}_{x^\perp} \left(\frac{1}{\sqrt{H}} \sum_{h=1}^{H} W_h V_h \frac{\int_{\mathscr{H}^d} e^{\frac{1}{\varepsilon}\langle Q_h x, K_h y\rangle} y \mu(dy)}{\int_{\mathscr{H}^d} e^{\frac{1}{\varepsilon}\langle Q_h x, K_h y'\rangle} \mu(dy')} \right). \end{cases} \tag{8.75}$$

- The projected mean-field-type masked Boltzmann–Gibbs self-attention is $\text{pmftBG}_{masked-att}$:

$$\begin{cases} [0,1] \times \mathscr{H}^d \times \mathscr{P}([0,1] \times \mathscr{H}^d) \to [0,1] \times \mathscr{H}^d \\ (s,x,\hat{\mu}) \mapsto \left(s, x + \text{proj}_{x^\perp} \left(\frac{1}{\sqrt{H}} \sum_{h=1}^{H} W_h V_h \frac{\int_0^1 \int_{\mathscr{H}^d} e^{\frac{1}{\varepsilon}\langle Q_h x, K_h y\rangle} \mathbb{I}_{s' \le s}\, y \hat{\mu}(ds'dy)}{\int_0^1 \int_{\mathscr{H}^d} e^{\frac{1}{\varepsilon}\langle Q_h x, K_h y'\rangle} \mathbb{I}_{s'' \le s}\, \hat{\mu}(ds''dy')} \right) \right). \end{cases} \tag{8.76}$$

- (masked) Normalized Distance:

$$\begin{cases} [0,1] \times \mathscr{H}^d \times \mathscr{P}([0,1] \times \mathscr{H}^d) \to [0,1] \times \mathscr{H}^d \\ (s, x, \hat{\mu}) \mapsto (s, x + j_1), \\ j_1 = \text{proj}_{x^\perp}\left(\frac{1}{\sqrt{H}} \sum_{h=1}^H W_h V_h \int_0^1 \int_{\mathscr{H}^d} \phi_h(\Big|(x-y)\Big|) y \mathbb{I}_{s' \le s} \hat{\mu}(ds'dy)\Big)\right). \end{cases} \tag{8.77}$$

where ϕ is a nonnegative continuous bounded function, one can choose for example $\frac{\Big|x\Big|}{1+\Big|x\Big|}$.

- (masked) Vicsek and Projected Cucker–Smale:

$$\begin{cases} [0,1] \times (\mathscr{H}^d)^2 \times \mathscr{P}([0,1] \times (\mathscr{H}^d)^2) \to [0,1] \times (\mathscr{H}^d)^2 \\ (s, x_1, x_2, \hat{\mu}) \mapsto (s, x_1 + x_2, x_2 + j_2), \\ j_2 = \text{proj}_{x_2^\perp}\left(\frac{1}{\sqrt{H}} \sum_{h=1}^H W_h V_h \int_0^1 \int_{\mathscr{H}^d} \phi_h(\Big| Q_h K_h^\dagger (x_1 - y_1)\Big|)(y_2 - x_2) \mathbb{I}_{s' \le s} \hat{\mu}(ds'dy_1dy_2)\Big)\right), \end{cases} \tag{8.78}$$

where ϕ is a smooth nonnegative function vanishing at infinity. Here (x_1, x_2) is interpreted as position-velocity.

- (masked) Projected Newton

$$\begin{cases} [0,1] \times (\mathscr{H}^d)^2 \times \mathscr{P}([0,1] \times (\mathscr{H}^d)^2) \to [0,1] \times (\mathscr{H}^d)^2 \\ (s, x_1, x_2, \hat{\mu}) \mapsto (s, x_1 + x_2, x_2 - j_2), \\ j_2 = \text{proj}_{x_2^\perp}\left(\frac{1}{\sqrt{H}} \sum_{h=1}^H W_h V_h \int_0^1 \int_{\mathscr{H}^d} \nabla V_2(\Big| Q_h K_h^\dagger (x_1 - y_1)\Big|) \mathbb{I}_{s' \le s} \hat{\mu}(ds'dy_1dy_2)\Big)\right), \end{cases} \tag{8.79}$$

where V_2 is a smooth repulsive function. Here, (x_1, x_2) is interpreted as position velocity.

- (masked) Projected Hegselmann–Krause:

$$\begin{cases} [0,1] \times \mathscr{H}^d \times \mathscr{P}([0,1] \times \mathscr{H}^d) \to [0,1] \times \mathscr{H}^d \\ (s, x, \hat{\mu}) \mapsto \left(s, x + \text{proj}_{x^\perp}\left(\frac{1}{\sqrt{H}} \sum_{h=1}^H W_h V_h \frac{\int_0^1 \int_{\mathscr{H}^d} \phi(\Big| Q_h K_h^\dagger (x-y)\Big|) \mathbb{I}_{s' \le s} (y-x) \hat{\mu}(ds'dy)}{\int_0^1 \int_{\mathscr{H}^d} \phi(\Big| Q_h K_h^\dagger (x-y')\Big|) \mathbb{I}_{s'' \le s} \hat{\mu}(ds''dy')}\right)\right). \end{cases} \tag{8.80}$$

- Gradient: $-\nabla V_1(x) - \int_0^1 \int_{\mathscr{H}^d} \nabla V_2(x-y) \mathbb{I}_{s'' \le s} \hat{\mu}(ds''dy')$.
- Kuramoto: Gradient model with V_1, V_2 in one dimension (with sine function).
- (masked) Kernel-based

$$\begin{cases} [0,1] \times \mathscr{H}^d \times \mathscr{P}([0,1] \times \mathscr{H}^d) \to [0,1] \times \mathscr{H}^d \\ (s, x, \hat{\mu}) \mapsto \left(s, x + \text{proj}_{x^\perp}\left(\frac{1}{\sqrt{H}} \sum_{h=1}^H W_h V_h \int_0^1 \int_{\mathscr{H}^d} ker(h; x, y) \mathbb{I}_{s' \le s} \hat{\mu}(ds'dy)\right)\right). \end{cases} \tag{8.81}$$

- Biot–Savart $ker_1(y) = \frac{y^\perp}{\langle y, y \rangle}$, where $ker(x, y) = ker_1(x - y)$.
- Platlak–Keller–Segel: $ker_2(y) = \frac{y}{\langle y, y \rangle}$.

- Coulomb: $ker_3(y) = \nabla\phi(y)$, $\phi(y) = -\log(\left|y\right|)\mathbb{I}_{d=2} + \frac{1}{\left|y\right|^{d-2}}\mathbb{I}_{d\geq 3}$.
- (masked) Higher-order interaction: $(k+1)$-wise interaction

$$\begin{cases} [0,1]\times\mathcal{H}^d \times \mathcal{P}([0,1]\times(\mathcal{H}^d)^k) \to [0,1]\times\mathcal{H}^d \\ (s,x,\hat{\mu}^{\otimes k}) \mapsto (s, x+j_1), \\ j_1 = \text{proj}_{x^\perp}\left(\frac{1}{\sqrt{H}}\sum_{h=1}^H W_h V_h \int_0^1 \int_{\mathcal{H}^d} ker(h;x,y_1,\ldots,y_k)\mathbb{I}_{s'\leq s}\hat{\mu}^{\otimes k}(ds'dy_1\ldots dy_k)\right). \end{cases} \tag{8.82}$$

- (masked) Localized mean-field-type interaction: x interacts with y only when y is a certain neighborhood of x such as $y \in B_\rho(x)$. This is a counting-based measure; it is not a global interaction, but a localized effect in the sense that it is limited to the neighborhood of the action:

$$\begin{cases} [0,1]\times\mathcal{H}^d \times \mathcal{P}([0,1]\times\mathcal{H}^d) \to [0,1]\times\mathcal{H}^d \\ (s,x,\hat{\mu}) \mapsto (s, x+j_1), \\ j_1 = \text{proj}_{x^\perp}\left(\frac{1}{\sqrt{H}}\sum_{h=1}^H W_h V_h \int_0^1 \int_{y\in B_\rho(x)} ker(h;x,y)\mathbb{I}_{s'\leq s}\hat{\mu}(ds'dy)\right). \end{cases} \tag{8.83}$$

Based on these masked self-attention mechanisms of mean-field type, we build MFTTs.

8.3.2 Structure of Mean-Field-Type Transformers

The MFTT operates by iteratively processing data through a sequence of layers, where each layer includes an operator that combines the input with the distribution of inputs (or outputs). Specifically, the operator at each layer, denoted $\hat{\mathcal{O}}_l$, maps an input and a measure (or distribution) over that input to an output and a new measure. Formally, at layer l, the operator $\hat{\mathcal{O}}_l$ is defined as $\hat{\mathcal{O}}_l : \mathcal{H}_{l-1} \times \mathcal{P}(\mathcal{H}_{l-1}) \to \mathcal{H}_l$, where $\mathcal{H}_{l-1}$ represents the space of pre-outputs at the previous layer which serves as pre-inputs at layer l and $\mathcal{P}(\mathcal{H}_{l-1})$ represents the space of probability measures over $\mathcal{H}_{l-1}$.

Given an initial pre-input x_0 with distribution μ_0, one gets the input (x_0, μ_0). The MFTT processes data through the following sequence of operations:

- **Layer 1:** $y_1 = \hat{\mathcal{O}}_1(x_0, \mu_0)$, and $\nu_1 = \hat{\mathcal{O}}_1(. \mid \mu_0)\#\mu_0$. Here, y_1 is the pre-output of the first layer, and ν_1 is the measure induced by applying $\hat{\mathcal{O}}_1$ to μ_0.
- **Subsequent Layers:** At layer l: $y_l = \hat{\mathcal{O}}_l(y_{l-1}, \nu_{l-1})$, and $\nu_l = \hat{\mathcal{O}}_l(. \mid \nu_{l-1})\#\nu_{l-1}$. This process continues until layer L.
- **Final Output:** The output of the MFTT is (y_L, ν_L), which represents the result after processing through all L layers.

Definition 8.82 The mean-field-type transformer (MFTT) is formally defined by the tuple: $(x_0, \mu_0, L, (\mathcal{H}_{l-1} \times \mathcal{P}(\mathcal{H}_{l-1}), \hat{\mathcal{O}}_l, \theta_l)_{l\in\{1,\dots,L\}}, \mathcal{H}_L)$, where $(\hat{\mathcal{O}}_l)_{l\in\{1,\dots,L\}}$ are the sequence of operators applied at each layer.

Let $\mu_{y_{l-\frac{1}{2}}} = \mathbb{P}_{y_{l-\frac{1}{2}}} = \mathcal{L}(y_{l-\frac{1}{2}})$ be the probability law of $y_{l-\frac{1}{2}}$.

$$
\begin{aligned}
&x_0 \sim \mu_0,\\
&l = 1,\\
&y_{\frac{1}{2}} = \hat{\mathcal{F}}_{\frac{1}{2},\mu_{y_0}}(x_0),\\
&y_1 = \hat{\mathcal{F}}_1(y_{\frac{1}{2}}),\\
&\text{compute: } \mu_{y_1}\\
&\dots\\
&l \in \{2,\dots,L-1\}:\\
&y_{l-\frac{1}{2}} = \hat{\mathcal{F}}_{l-\frac{1}{2},\mu_{y_{l-1}}}(y_{l-1}),\\
&y_l = \hat{\mathcal{F}}_l(y_{l-\frac{1}{2}}),\\
&\text{compute: } \mu_{y_l}\\
&\dots\\
&y_{L-\frac{1}{2}} = \hat{\mathcal{F}}_{L-\frac{1}{2},\mu_{y_{L-1}}}(y_{L-1}),\\
&x^* = y_L = \hat{\mathcal{F}}_L(y_{L-\frac{1}{2}}),\\
&\text{compute: } \mu_{y_L}.
\end{aligned}
\tag{8.84}
$$

The mean-field-type fixed-point system yields

$$
\begin{aligned}
&l = 1:\\
&y_{\frac{1}{2}} = \hat{\mathcal{F}}_{\frac{1}{2},\mu^*}(x^*),\\
&y_1 = \hat{\mathcal{F}}_1 \circ \hat{\mathcal{F}}_{\frac{1}{2},\mu^*}(x^*),\\
&\text{compute: } \mu_{y_1}\\
&\dots\\
&l \in \{2,\dots,L-1\}:\\
&y_{l-\frac{1}{2}} = \hat{\mathcal{F}}_{l-\frac{1}{2}\,|\mu_{y_{l-1}}} \circ \hat{\mathcal{F}}_{l-1} \circ \hat{\mathcal{F}}_{l-\frac{3}{2}\,|\mu_{y_{l-2}}}(x^*) \circ \dots \circ \hat{\mathcal{F}}_1 \circ \hat{\mathcal{F}}_{\frac{1}{2},\mu^*}(x^*),\\
&y_l = \hat{\mathcal{F}}_l \circ \hat{\mathcal{F}}_{l-\frac{1}{2}\,|\mu_{y_{l-1}}} \circ \dots \circ \hat{\mathcal{F}}_1 \circ \hat{\mathcal{F}}_{\frac{1}{2},\mu^*}(x^*),\\
&\text{compute: } \mu_{y_l}\\
&\dots\\
&y_{L-\frac{1}{2}} = \hat{\mathcal{F}}_{L-\frac{1}{2}\,|\mu_{y_{L-1}}} \circ \hat{\mathcal{F}}_{L-1} \circ \dots \circ \hat{\mathcal{F}}_1 \circ \hat{\mathcal{F}}_{\frac{1}{2},\mu^*}(x^*),\\
&x^* = y_L = \hat{\mathcal{F}}_L \circ \hat{\mathcal{F}}_{L-\frac{1}{2}\,|\mu_{y_{L-1}}} \circ \dots \circ \hat{\mathcal{F}}_1 \circ \hat{\mathcal{F}}_{\frac{1}{2},\mu^*}(x^*),\\
&\text{compute: } \mu_{y_L}.
\end{aligned}
\tag{8.85}
$$

Iterating the MFTT, the resulting output (if any) provides

$$(x^*, \mu_{x^*}, y_{\frac{1}{2}}, y_1, \mu_{y_1}, y_{\frac{3}{2}}, y_2, \mu_{y_2}, \ldots, y_{L-\frac{1}{2}}, y_L = x^*, \mu_{y_L} = \mu_{x^*})$$

when $\mathscr{H}_L = \mathscr{H}_0$.

We will next show how this outcome is connected to mean-field-type Nash equilibria. Note that, here, the mean field is the distribution of the own action of the decision-maker called the individual mean field.

Lemma 8.83 *Distribution-wise, the mean-field-type transformer-based deep neural network is given by*

$$\begin{aligned}
&\mu_0 \in \mathscr{P}(\mathscr{H}^d)\\
&l = 1,\\
&\nu_{\frac{1}{2}} = \hat{\mathscr{F}}_{\frac{1}{2}, \mu_0} \# \mu_0,\\
&\nu_1 = \hat{\mathscr{F}}_1 \# \nu_{\frac{1}{2}},\\
&\ldots\\
&l \in \{2, \ldots, L-1\}:\\
&\nu_{l-\frac{1}{2}} = \hat{\mathscr{F}}_{l-\frac{1}{2}, \nu_{l-1}} \# \nu_{l-1},\\
&\nu_l = \hat{\mathscr{F}}_l \# \nu_{l-\frac{1}{2}},\\
&\ldots\\
&\nu_{L-\frac{1}{2}} = \hat{\mathscr{F}}_{L-\frac{1}{2}, \nu_{L-1}} \# \nu_{L-1},\\
&\nu_L = \hat{\mathscr{F}}_L \# \nu_{L-\frac{1}{2}}.
\end{aligned} \tag{8.86}$$

The algorithm iterates as follows:

$$\begin{aligned}
\mu_{t+1} = ((1-\lambda_t) Id + \lambda_t \hat{\mathscr{F}}_L \circ \hat{\mathscr{F}}_{L-\frac{1}{2}} {}_{|\nu_{L-1,t}}\\
\circ \ldots \circ \hat{\mathscr{F}}_2 \circ \hat{\mathscr{F}}_{\frac{3}{2}, \nu_{1,t}} \circ \hat{\mathscr{F}}_1 \circ \hat{\mathscr{F}}_{\frac{1}{2}, \mu_t}) \# \mu_t,
\end{aligned} \tag{8.87}$$

starting from $\mu_0 \in \mathscr{P}(\mathscr{H}^d)$.

8.3.3 MFTT as an MFTG with Hierarchies

We demonstrate that any mean-field-type transformer can be represented as a mean-field-type game with multiple hierarchies:

- Initial input (x_0, μ_0) are parameters of the game.

- Each layer l represents a decision-maker.
- The action of the decision-maker l is given by the output y_l.
- The distribution ν_l of y_l is analogous to the distribution of actions in an MFTG.

In the MFTT, the output y_l at each layer l can be interpreted as the action of decision-maker l in an MFTG. This output depends on the previous output y_{l-1} and the distribution ν_{l-1} of the outputs from the preceding layer. Similarly, in an MFTG, each decision-maker's action y_l is influenced by the distribution of actions from previous stages, which corresponds to ν_{l-1} in the MFTT. Each decision-maker in the MFTG reacts to the state actions and distribution of state actions of others, including their own state distribution. In the MFTT, the output y_l at each layer l is determined by ν_{l-1}, the distribution of the previous layer's outputs. Thus, the MFTT's approach to processing distributions directly aligns with the MFTG's framework, where actions are dependent on the distribution of all actions in the game. Here, the MFTG is given by $(\mathcal{L}, (\mathcal{H}_l \times \mathcal{P}(\mathcal{H}_l), g_l)_{l\in\mathcal{L}})$, where $\text{argmin}_z Eg_l(z) = \hat{\mathcal{O}}_l$ with the hierarchical structure of decision l at the hierarchical layer l and the information structure that decision-maker l observes the decision and the distribution of the decision-maker $l-1$.

8.3.3.1 Subgame Perfect Mean-Field-Type Nash Equilibrium

In a well-designed MFTT, the operator $\hat{\mathcal{O}}_l$ at each layer l can be viewed as the best-response function of decision-maker l to the distribution ν_{l-1}. This implies that the output y_l is an optimal response to the distribution of the previous layer's outputs [106]. Thus, each layer functions as a decision-maker that optimizes its action based on the action and distribution of action from the previous layer, as in the best-response dynamics in an MFTG. Given that each layer in the MFTT represents a decision-maker optimizing its action based on the distribution of actions from previous layers, the sequence of outputs $(y_1, y_2, \ldots, y_L)$ and distributions $(\nu_1, \nu_2, \ldots, \nu_L)$ constitutes a mean-field-type Nash equilibrium. This equilibrium is characterized by each decision-maker choosing y_l as a response to ν_{l-1}, such that the resulting distribution ν_l is consistent with the responses of all other decision-makers. Therefore, the MFTT framework models a hierarchical best-response process that results in a mean-field-type Nash equilibrium.

8.3.3.2 Mean-Field-Type Stackelberg Solution

The hierarchical structure of MFTT and of decision-makers across layers can be viewed as a special case of a Stackelberg solution in mean-field-type games. In a multi-layer Stackelberg game, a leader makes a strategic decision first, and then, followers react optimally to this decision. Here, there is one leader per layer. Leader 1, who is decision-maker 1, acts first, then follows leader 2 based on the choice of leader 1, then follows leader 3 based on the choices of leaders 1 and 2, and so on.

This sequential decision-making process inherently influences the equilibrium since the leaders' choices affect the followers' responses, and the equilibrium thus reflects this hierarchical structure.

In an MFTT, each layer represents a decision-maker in a hierarchical sequence where the output of one layer (or decision-maker) serves as the input for the next. The crucial insight here is that while the current layer's pre-output y_l is the best response to the action and distribution (y_{l-1}, ν_{l-1}) of the previous layer's outputs, the decision-making process at each layer does not depend on future layers' choices but rather on the distribution induced by the preceding layer. This structure means that each decision-maker's reaction does not depend on all subsequent decisions and distributions $(y_{l'}, \nu_{l'}, l' \geq l+1)$.

The preceding leaders anticipate the future reactions. That anticipation coincides with the correct reaction due to the structure of the MFTT. This can be seen from the fact $\hat{\mathcal{O}}_l(y, \nu) = \hat{\mathcal{O}}_l(y_{l-1}, \nu_{l-1})$. Thus, the resulting outcome of the MFTT, where each layer's output constitutes a mean-field-type Nash equilibrium, also qualifies as a Stackelberg equilibrium of the hierarchical mean-field-type game.

8.3.3.3 Limitations of MFTT

In order to get a true computational advantage of MFTT over the infinite data size $\mathbb{TNN}$ discussed earlier, the computation of the mean-field-type interaction term $\hat{\mathcal{O}}_l(y_{l-1}, \nu_{l-1})$ should be done in a very efficient way. However, such efficiency needs to be examined in detail depending on the form of interaction (pairwise, three-wise, k-wise, local, localized, global, etc). The efficient computation of quenched measure or McKean–Vlasov measure is still an open issue. One may need another sub-layer dedicated to the computation of the interaction term in the same way as the (finite) self-attention was introduced.

8.3.4 Small Learning Rate Regime

We now examine the dynamics of a transformer-based neural network when λ_t is very small. One can relate this to a pseudo-trajectory of an ordinary differential system or a stochastic differential system.

8.3.4.1 Finite Data Sequence

D be finite. The pseudo-trajectory of the algorithm in (8.2) is captured by the following ODE:

$$\begin{cases} x(0) = x_0 \in \mathscr{H}^d, \\ \dot{x} = (\hat{\mathscr{O}}_L \circ \ldots \circ \hat{\mathscr{O}}_1)(x) - x, \ t > 0 \\ . \end{cases} \tag{8.88}$$

Even if the operator is not globally Lipschitz, it does not create a difficulty in numerous situations. When restricted to any not empty, closed, bounded, convex set of $\mathscr{H}^d$, the operator $\hat{\mathscr{O}}_L \circ \ldots \circ \hat{\mathscr{O}}_1$ is locally Lipschitz on bounded sets, and linear maps involved are all assumed to be bounded. As locally Lipschitz drifts define a unique solution to the ODE for any initial condition, we conclude that there is a unique trajectory for such cases. Note, however, that some of the interaction terms, such as the Coulomb interaction, require a more careful analysis to establish existence due to singularities at zero. As expected, the steady states (if any) of this ODE are the fixed points of the operator $\hat{\mathscr{O}}_L \circ \ldots \circ \hat{\mathscr{O}}_1$.

The training problem to find a time-and-state-independent control action θ such that

$$\begin{cases} \mathscr{D} = \{(\hat{x}_i, \hat{y}_i) \in \mathscr{H}^d \times \mathscr{H}^d, i \in \{1, 2, \ldots, D\}\}, \\ \inf_{\theta \in \Theta} \frac{1}{D} \sum_{i=1}^{D} \Phi(y_i(T), \hat{y}_i), \\ \text{such that} \\ y(0) = \hat{x}, \\ \dot{y} = (\hat{\mathscr{O}}_L \circ \ldots \circ \hat{\mathscr{O}}_1)(y) - y, \ T > t > 0. \end{cases} \tag{8.89}$$

The fact that the control action is restricted to time-and-state-independent control is because of a data-independent architecture at the end of the training process.

8.3.4.2 Continuous Time Mean-Field Limit

This is a stylized network case. In practice, the data size can be large but still not infinite. At each layer, the number of operations can be large but not infinite. In this context, instead of applying softargmax row by row to this huge matrix, one may think of taking a generic row and applying a mean-field effect. This leads to a generic decision-maker interacting with the mean field at each layer l. As the learning rate vanishes, one ends up with a mean-field ODE with the mean field frozen (because a generic decision-maker does not affect the mean field in the infinite population regime). In this case, the outcomes of $\mathbb{TNN}$ are the steady state of some ODEs, and the mean field solves the Fokker–Planck equation. The training problem at this infinite population regime is to move from a single input distribution to an output distribution subject to the ODE constraint:

$$
\begin{aligned}
&\mathscr{D} = \{\mathbb{P}\}, \mathbb{P} \in \mathscr{P}(\mathscr{H}^d \times \mathscr{R}^d) \\
&\inf_{\theta \in \Theta} E\Phi(y(T), \hat{y}), \\
&\text{such that} \\
&y(0) = \hat{x} \sim \mathbb{P}_1 = m_0, \hat{y} \sim \mathbb{P}_2 \\
&\dot{y} = \hat{\mathscr{F}}_L \circ \hat{\mathscr{F}}_{L-\frac{1}{2}}|_m \circ \ldots \circ \hat{\mathscr{F}}_1 \circ \hat{\mathscr{F}}_{\frac{1}{2},m}(y) - y, \ T > t > 0 \\
&\dot{m}(t, y) + div(m(t, y)(\hat{\mathscr{F}}_L \circ \hat{\mathscr{F}}_{L-\frac{1}{2}}|_{m(t,.)} \circ \ldots \circ \hat{\mathscr{F}}_1 \circ \hat{\mathscr{F}}_{\frac{1}{2},m(t,.)}|_{\theta=\theta^*}(y) - y)) = 0,
\end{aligned} \tag{8.90}
$$

where Φ is a real-valued risk-aware cost function.

A pair (θ, m) solution of (8.90) is also a mean-field Nash equilibrium when restricted to the class of time-and-state-independent controls. Note that generically the mean-field Nash equilibrium here is a function of $(y(t), m(t))$, which means that the restriction to constant control may lose optimally.

8.3.4.3 Continuous-Time Mean-Field-Type Transformer

The training problem in the mean-field-type case is

$$
\begin{aligned}
&\mathscr{D} = \{\mathbb{P}\}, \mathbb{P} \in \mathscr{P}(\mathscr{H}^d \times \mathscr{H}^d) \\
&\inf_{\theta \in \Theta} E\Phi(y(T), \hat{y}), \\
&\text{such that} \\
&y(0) = \hat{x} \sim \mathbb{P}_1, \hat{y} \sim \mathbb{P}_2 \\
&\dot{y} = (\hat{\mathscr{F}}_L \circ \hat{\mathscr{F}}_{L-\frac{1}{2}}|_{\nu_{L-1,t}} \circ \ldots \circ \hat{\mathscr{F}}_2 \circ \hat{\mathscr{F}}_{\frac{3}{2},\nu_{1,t}} \circ \hat{\mathscr{F}}_1 \circ \hat{\mathscr{F}}_{\frac{1}{2},\mu_t})(y) - y, \\
&\mu(0, .) = \mathbb{P}_1, \\
&\dot{\mu} + div(\mu(\hat{\mathscr{F}}_L \circ \hat{\mathscr{F}}_{L-\frac{1}{2}}|_{\nu_{L-1}} \circ \ldots \circ \hat{\mathscr{F}}_2 \circ \hat{\mathscr{F}}_{\frac{3}{2},\nu_1} \circ \hat{\mathscr{F}}_1 \circ \hat{\mathscr{F}}_{\frac{1}{2},\mu}(y) - y)) = 0, \\
&l \in \{1, \ldots, L-1\} : \\
&\nu_{l-\frac{1}{2}}(t, .) = \mathscr{F}_{l-\frac{1}{2},\nu_{l-1}(t,.)} \# \nu_{l-1}(t, .), \\
&\nu_l(t, .) = \mathscr{F}_l \# \nu_{l-\frac{1}{2}}(t, .), \\
&\nu_0(t, .) := \mu(t, .).
\end{aligned} \tag{8.91}
$$

The main difference between systems (8.90) and (8.91) is that in the infinite population regime in (8.90), the population mean-field m is frozen, and all the computations at all layers are done with m frozen. In the mean-field-type case (8.91), the mean field is an own-state mean field which is coming from the own-Fokker–Planck equation. In the mean-field-type case, there is a finite number of decision-makers at each layer.

Note that the minimization is not for all the joint distribution $\mathbb{P}$ as in the Wasserstein formulation. Here, the minimization is on the time-and-state-independent control action θ. In this MFTG, each decision-maker at each layer contributes to the mean-field μ in a non-negligible way. It should be clear that a θ that is time-and-state-independent may not be globally optimal. Also, it may not be a global best response. Typically, for adjoint functional p, if the Hamiltonian term

$$\theta \mapsto \langle p, (\hat{\mathscr{F}}_L \circ \hat{\mathscr{F}}_{L-\frac{1}{2}\,|\nu_{L-1,t}} \circ \ldots \circ \hat{\mathscr{F}}_2 \circ \hat{\mathscr{F}}_{\frac{3}{2},\nu_{1,t}} \circ \hat{\mathscr{F}}_1 \circ \hat{\mathscr{F}}_{\frac{1}{2},\mu_t})(y)\rangle$$

admits a minimizer, it may depend on $y, \mu_t, \nu_{1,t}, \ldots, \nu_{L-1,t}$ which is called state-and-mean-field-type control strategy.

8.4 Neural Network Architecture Models

Neural network architecture models are computational frameworks inspired by the interconnected neurons, used for machine learning tasks like pattern recognition and decision-making through layered connections of artificial neurons. In this section, we present some neural network architectures.

8.4.1 Tensor Neural Network

A tensor neural network (TensorNN) generalizes neural networks to higher-order tensor data. Given a tensor $T^{(0)} \in \mathscr{H}^{(d_1 \times d_2 \times \ldots \times d_k)}$, each layer l processes the tensor using a tensor product:

$$\begin{cases} T^{(l+1)} = \sigma_l\left(W^{(l)} \otimes T^{(l)} + b^{(l)}\right), \\ l \in \{1, \ldots, L\}, \end{cases} \tag{8.92}$$

where $\mathscr{H}$ is a Hilbert space, $\otimes$ is the tensor product, $W^{(l)}$ are learnable tensor operators, and $b^{(l)}$ is a bias.

This is an essential tool for engineering tasks involving multi-dimensional data, such as images, videos, and sensor arrays. It extends traditional neural networks to handle tensors of higher orders. It operates on tensor data, where tensors represent multi-dimensional arrays. It uses tensor convolution and pooling operations. The layers process data in a hierarchical and localized manner. It is well suited for structured data with multiple dimensions. It can capture spatial patterns and hierarchies in data:

- TensorNNs can analyze medical images, detecting anomalies in multiple dimensions (e.g., time, depth, and width) for early disease diagnosis.
- TensorNNs can analyze multi-dimensional data from sensors, predicting environmental factors like air quality, water pollution, and climate changes.
- TensorNNs can predict material properties by analyzing complex interactions between multiple attributes, aiding in developing new materials with specific characteristics.

- TensorNNs handle multi-dimensional data (e.g., images, videos, medical scans) more naturally, capturing intricate patterns across multiple dimensions.
- TensorNNs can capture hierarchical features and spatial relationships in data, benefiting applications like image segmentation or video analysis.
- TensorNNs can process local regions in data, which is advantageous for tasks requiring localized understanding and analysis.

Layer $l = 1$ (Tensor Convolution and Pooling)
Tensor Convolution Operation

$$\begin{cases} \text{Input: } X \text{ with dimensions } N \times C \times H \times W \\ \text{Filters: } F_C^{(1)} \text{ filters of size } C \times K^{(1)} \times K^{(1)} \\ \text{Output: } Y^{(1)} \text{ with dimensions } N \times F_C^{(1)} \times (H - K^{(1)} + 1) \times (W - K^{(1)} + 1) \end{cases} \tag{8.93}$$

$$Y_{n,c}^{(1)} = \sigma\left(\sum_{i=1}^{H-K^{(1)}+1} \sum_{j=1}^{W-K^{(1)}+1} \sum_{k=1}^{C} F_{c,k}^{(1)} * X_{n,k,i,j} + b_c^{(1)}\right).$$

Tensor Pooling Operation

$$\begin{cases} \text{Input: } Y^{(1)} \text{ with dimensions } N \times F_C^{(1)} \times (H - K^{(1)} + 1) \times (W - K^{(1)} + 1) \\ \text{Output: } Z^{(1)} \text{ with dimensions } N \times F_C^{(1)} \times (H_{\text{pool}}^{(1)}) \times (W_{\text{pool}}^{(1)}) \end{cases} \tag{8.94}$$

$$Z_{n,c}^{(1)} = \text{pooling}(Y_{n,c}^{(1)}).$$

Layer $l = 2$ to $L - 1$ (Tensor Convolution and Pooling)

1. **Tensor Convolution Operation**

$$\begin{cases} \text{Input: } Z^{(l-1)} \text{ with dimensions } N \times F_C^{(l-1)} \times H_{\text{pool}}^{(l-1)} \times W_{\text{pool}}^{(l-1)} \\ \text{Filters: } F_C^{(l)} \text{ filters of size } F_C^{(l-1)} \times K^{(l)} \times K^{(l)} \\ \text{Output: } Y^{(l)} \text{ with dimensions } N \times F_C^{(l)} \times (H_{\text{pool}}^{(l-1)} - K^{(l)} + 1) \\ \qquad \times (W_{\text{pool}}^{(l-1)} - K^{(l)} + 1) \end{cases} \tag{8.95}$$

$$Y_{n,c}^{(l)} = \sigma_l\left(\sum_{i=1}^{H_{\text{pool}}^{(l-1)}-K^{(l)}+1} \sum_{j=1}^{W_{\text{pool}}^{(l-1)}-K^{(l)}+1} \sum_{k=1}^{F_C^{(l-1)}} F_{c,k}^{(l)} * Z_{n,k,i,j}^{(l-1)} + b_c^{(l)}\right).$$

2. **Tensor Pooling Operation**

$$\begin{cases} \text{Input: } Y^{(l)} \text{ with dimensions } N \times F_C^{(l)} \times (H_{\text{pool}}^{(l-1)} - K^{(l)} + 1) \\ \qquad \times (W_{\text{pool}}^{(l-1)} - K^{(l)} + 1) \\ \text{Output: } Z^{(l)} \text{ with dimensions } N \times F_C^{(l)} \times (H_{\text{pool}}^{(l)}) \times (W_{\text{pool}}^{(l)}) \end{cases} \tag{8.96}$$

$$Z^{(l)}_{n,c} = \text{pooling}(Y^{(l)}_{n,c}).$$

Layer $l = L$ (Fully Connected Layer)
Flattening

$$\begin{cases} \text{Input: } Z^{(L-1)} \text{ with dimensions } N \times F_C^{(L-1)} \times H_{\text{pool}}^{(L-1)} \times W_{\text{pool}}^{(L-1)} \\ \text{Flattened: } z^{(L-1)} \text{ with dimensions } N \times (F_C^{(L-1)} \times H_{\text{pool}}^{(L-1)} \times W_{\text{pool}}^{(L-1)}). \end{cases} \tag{8.97}$$

Fully Connected Layer

$$\begin{cases} \text{Input: } z^{(L-1)} \text{ with dimensions } N \times (F_C^{(L-1)} \times H_{\text{pool}}^{(L-1)} \times W_{\text{pool}}^{(L-1)}) \\ \text{Weight Matrix: } (F_{\text{out}} \times (F_C^{(L-1)} \times H_{\text{pool}}^{(L-1)} \times W_{\text{pool}}^{(L-1)})) \times (F_{\text{in}}^{(\text{out})}) \\ \text{Output: } y_{\text{pred}} \text{ with dimensions } N \times F_{\text{out}} \end{cases} \tag{8.98}$$

$$y_{\text{pred}} = \text{softargmax}(W_{\text{out}} \cdot z^{(L-1)} + b_{\text{out}}).$$

The above can be generalized to the tensors of arbitrary dimension as follows:

Input Layer: Let X be the input tensor of size $N_1 \times N_2 \times \ldots \times N_D$, where $N_1, N_2, \ldots, N_D$ are the dimensions of the tensor. The input features X are given.

Tensor Convolution Layer ($l = 0$ to $L-1$): For layer l, we calculate new tensor representations by applying a tensor convolution operation:

For each position $(i_1, i_2, \ldots, i_D)$, the new representation $H^{(l+1)}(i_1, i_2, \ldots, i_D)$ is computed as follows:

$$H^{(l+1)}(i_1, i_2, \ldots, i_D) =$$

$$\sigma_l \left(\sum_{j_1=1}^{N_1} \sum_{j_2=1}^{N_2} \cdots \sum_{j_D=1}^{N_D} K^{(l)}(i_1, i_2, \ldots, i_D, j_1, j_2, \ldots, j_D) \right.$$
$$\left. \cdot H^{(l)}(j_1, j_2, \ldots, j_D) + b^{(l)} \right),$$

where:

- $H^{(l+1)}(i_1, i_2, \ldots, i_D)$ is the hidden state of the tensor at position $(i_1, i_2, \ldots, i_D)$ in layer $l + 1$.
- σ_l is an activation function (e.g., ReLU).
- $K^{(l)}$ is a learnable tensor kernel of appropriate dimensions.
- $H^{(l)}(j_1, j_2, \ldots, j_D)$ is the hidden state of the tensor at position $(j_1, j_2, \ldots, j_D)$ in layer l.
- $b^{(l)}$ is a bias tensor.

Tensor Pooling Layer ($l = 0$ to $L - 1$)

For layer l, we perform tensor pooling operations to reduce tensor dimensions while retaining important information.

Output Layer: The final layer produces the output predictions. For a classification task, you might apply a softargmax function to the tensor representations:

The output prediction tensor $\hat{Y}$ is given by

$$\hat{Y} = \text{softargmax}(W^{(L)} \cdot H^{(L)} + B^{(L)}),$$

where:

- $\hat{Y}$ is the predicted tensor with label probabilities.
- $W^{(L)}$ is a learnable tensor weight.
- $B^{(L)}$ is a bias tensor.

8.4.2 Graph Neural Network

A graph neural network (GNN) operates on graph-structured data. Let $G = (V, E)$ be a graph with nodes V and edges E. At layer l, a GNN updates node features $X^{(l)} \in \mathbb{R}^{N \times D^{(l)}}$ based on neighboring nodes and edge features:

$$X_i^{(l+1)} = \sigma_l \left(\sum_{j \in \mathcal{N}(i)} W^{(l)} X_j^{(l)} + b^{(l)} \right),$$

where $\mathcal{N}(i)$ is the set of neighbors of node i, $W^{(l)}$ are learnable weights, and σ_l is a nonlinear activation.

GNN is a powerful tool for engineering tasks involving structured data, such as social networks, sensor networks, and molecular graphs. GNN processes data in a graph format, capturing relationships between entities. It operates on graph-structured data and processes node features in a neighborhood aggregation manner. The layers involve graph convolutions and aggregation operations. It utilizes message passing between nodes and can capture complex relationships and dependencies in graph data:

- GNNs can analyze social networks to identify influential nodes, detect communities, and predict links, aiding in understanding social interactions and information spread.
- GNNs can model patient–doctor relationships and disease spread within communities, assisting in early disease detection and healthcare resource allocation.
- GNNs can personalize recommendations based on social connections, enhancing user engagement and promoting collaborative consumption.
- GNNs excel at handling graph-structured data, capturing node relationships and interactions in complex networks.

- GNNs can perform classification tasks on nodes and entire graphs, such as social network node classification or molecular graph classification.
- GNNs use message-passing mechanisms to aggregate information from neighbors, making them suitable for scenarios where information flows in a network.

Layer $l = 0$ (Input Layer)
Input: Graph $G = (V, E)$ with N nodes and M features per node.
Node features: $X^{(0)} \in \mathbb{R}^{N \times M}$.

Layer $l = 1$ to L (Graph Convolution and Aggregation)
Graph Convolution Operation

$$\begin{cases} \text{Input: } X^{(l-1)} \text{ with dimensions } N \times M \\ \text{Weight Matrix: } W^{(l)} \in \mathbb{R}^{M \times F_C^{(l)}} \\ \text{Output: } H^{(l)} \text{ with dimensions } N \times F_C^{(l)} \end{cases} \tag{8.99}$$

$$H_{n,c}^{(l)} = \sigma_l \left(\sum_{j \in \mathcal{N}(n)} \frac{1}{\sqrt{|\mathcal{N}(n)|}} \left(X_n^{(l-1)} \cdot W_c^{(l)} \right) \right).$$

Aggregation Operation

$$\begin{cases} \text{Input: } H^{(l)} \text{ with dimensions } N \times F_C^{(l)} \\ \text{Output: } X^{(l)} \text{ with dimensions } N \times F_C^{(l)} \end{cases} \tag{8.100}$$

$$X_n^{(l)} = \text{aggregation} \left(\left[H_{n,c}^{(l)} \, \forall c \in \{1, \ldots, F_C^{(l)}\} \right] \right).$$

Output Layer (Classification)
Global Pooling Operation

$$\begin{cases} \text{Input: } X^{(L)} \text{ with dimensions } N \times F_C^{(L)} \\ \text{Output: } Z \text{ with dimensions } F_C^{(L)} \end{cases} \tag{8.101}$$

$$Z_c = \text{global\_pooling} \left(\left[X_{n,c}^{(L)} \, \forall n \in \{1, \ldots, N\} \right] \right).$$

Fully Connected Layer

$$\begin{cases} \text{Input: } Z \text{ with dimensions } F_C^{(L)} \\ \text{Weight Matrix: } W_{\text{out}} \in \mathbb{R}^{F_C^{(L)} \times F_{\text{out}}} \\ \text{Output: } y_{\text{pred}} \text{ with dimensions } F_{\text{out}} \end{cases} \tag{8.102}$$

$$y_{\text{pred}} = \text{softargmax}(Z \cdot W_{\text{out}} + b_{\text{out}}).$$

The above description can be generalized to weighted GNNs.

Input Layer

Let X be the input feature matrix of size $N \times F$, where N is the number of nodes and F is the number of input features for each node. The input features X are given.

Graph Convolution Layer ($l = 0$ to $L - 1$)

For layer l, we calculate new node representations by applying a graph convolution operation:

For node i, the new representation $h_i^{(l+1)}$ is computed as follows:

$$h_i^{(l+1)} = \sigma_l \left(\sum_{j \in \mathcal{N}(i)} c_{ij} W^{(l)} h_j^{(l)} + b^{(l)} \right),$$

where:

- $h_i^{(l+1)}$ is the hidden state of node i in layer $l + 1$.
- σ_l is an activation function (e.g., ReLU, GELU).
- $W^{(l)}$ is a learnable weight matrix.
- $h_j^{(l)}$ is the hidden state of node j in layer l.
- $b^{(l)}$ is a bias term.
- $\mathcal{N}(i)$ is the neighborhood of node i.
- $c_{ij} > 0$ is a normalization factor.

Aggregation Layer ($l = 0$ to $L - 1$)

For layer l, we aggregate information from neighboring nodes to obtain the aggregated representations:

For node i, the aggregated representation $\text{Agg}^{(l+1)}(i)$ is calculated as

$$\text{Agg}^{(l+1)}(i) = \frac{1}{|\mathcal{N}(i)|} \sum_{j \in \mathcal{N}(i)} h_j^{(l+1)},$$

where $|\mathcal{N}(i)|$ is the number of neighbors of node i.

Output Layer

The final layer produces the output predictions. For node classification, you might apply a softargmax function to the node representations:

For node i, the output prediction $\hat{y}_i$ is given by

$$\hat{y}_i = \text{softargmax}(W^{(L)} \cdot \text{Agg}^{(L)}(i) + b^{(L)}),$$

where:

- $\hat{y}_i$ is the predicted label probabilities for node i.
- $W^{(L)}$ is a learnable weight matrix.
- $b^{(L)}$ is a bias term.

8.4.3 Graph-Tensor Neural Network

A graph-tensor neural network (GTNN) combines GNN and TensorNN for graph-structured tensor data. At layer l, GTNN updates tensor node features $X^{(l)}$ using graph-structured information and tensor transformations:

$$X_i^{(l+1)} = \sigma_l \left(\sum_{j \in \mathcal{N}(i)} W^{(l)} \otimes X_j^{(l)} + b^{(l)} \right).$$

GTNN bridges the gap between graph-structured data and multi-dimensional tensors. It is ideal for engineering tasks involving heterogeneous data, such as social networks with additional attribute information. It combines graph and tensor processing to handle both structured and relational data. It applies graph convolution on node features and tensor convolution on attributes and incorporates graph aggregation and pooling operations. It is suitable for scenarios where data has both graph relationships and multi-dimensional features and can capture intricate interactions between graph structure and tensor attributes. GTNNs can process both text data and user interactions in social media, detecting sentiments, fake news, and understanding evolving trends. GTNNs can model transportation networks along with population density, optimizing city infrastructure for efficient and sustainable urban development. GTNNs can integrate crime data, geospatial information, and social connections to identify crime hotspots and patterns, assisting law enforcement agencies. The GTNN architecture combines both graph processing and tensor processing capabilities. GTNN can effectively handle data with different modalities (text, audio, video, images, time series, etc.) by incorporating graph structures and multi-dimensional attributes. It allows modeling complex interactions between different modalities and their relationships, making it suitable for capturing the rich and diverse information present in multimodal data. GTNN can integrate the strengths of both graph-based and tensor-based processing to provide a holistic understanding of the data, making it a powerful choice for multimodal learning tasks. GTNNs can seamlessly integrate and process different types of data (graphs, images, text, etc.) in a unified framework, enabling comprehensive analysis of complex multimodal data sets. GTNNs can capture interactions between graph structures and tensor attributes, which is crucial for understanding how different modalities influence each other. GTNNs enable joint processing of attributes and relationships, suitable for scenarios where both the attribute and structural information are important.

Layer $l = 0$ (Input Layer)
The input is a graph $G = (V, E)$ with N nodes and M features per node. The node features: $X^{(0)} \in \mathbb{R}^{N \times M}$.

Layer $l = 1$ to L (Graph-Tensor Convolution and Aggregation)
Graph-Tensor Convolution Operation

$$\begin{cases} \text{Input: } X^{(l-1)} \text{ with dimensions } N \times M \\ \text{Weight Tensor: } \mathscr{W}^{(l)} \in \mathbb{R}^{M \times F_C^{(l)} \times K^{(l)}} \\ \text{Output: } H^{(l)} \text{ with dimensions } N \times F_C^{(l)} \end{cases} \tag{8.103}$$

$$H_{n,c}^{(l)} = \sigma_l \left(\sum_{j \in \mathscr{N}(n)} \frac{1}{\sqrt{|\mathscr{N}(n)|}} \left(X_n^{(l-1)} * \mathscr{W}_c^{(l)} \right) \right).$$

Aggregation Operation

$$\begin{cases} \text{Input: } H^{(l)} \text{ with dimensions } N \times F_C^{(l)} \\ \text{Output: } X^{(l)} \text{ with dimensions } N \times F_C^{(l)} \end{cases} \tag{8.104}$$

$$X_n^{(l)} = \text{aggregation}\left(\left[H_{n,c}^{(l)} \, \forall c \in \{1, \ldots, F_C^{(l)}\} \right] \right).$$

Output Layer (Classification)
Global Pooling Operation

$$\begin{cases} \text{Input: } X^{(L)} \text{ with dimensions } N \times F_C^{(L)} \\ \text{Output: } Z \text{ with dimensions } F_C^{(L)} \end{cases} \tag{8.105}$$

$$Z_c = \text{global\_pooling}\left(\left[X_{n,c}^{(L)} \, \forall n \in \{1, \ldots, N\} \right] \right).$$

Fully Connected Layer

$$\begin{cases} \text{Input: } Z \text{ with dimensions } F_C^{(L)} \\ \text{Weight Matrix: } W_{\text{out}} \in \mathbb{R}^{F_C^{(L)} \times F_{\text{out}}} \\ \text{Output: } y_{\text{pred}} \text{ with dimensions } F_{\text{out}} \end{cases} \tag{8.106}$$

$$y_{\text{pred}} = \text{softargmax}(Z \cdot W_{\text{out}} + b_{\text{out}}).$$

8.4.4 Tensor-Graph Neural Network

A tensor-graph neural network (TGN) extends GNN with tensor operations. At layer l, TGN updates graph node features $X^{(l)}$ using tensor information and graph convolutions:

$$X_i^{(l+1)} = \sigma_l \left(\sum_{j \in \mathcal{N}(i)} W^{(l)} X_j^{(l)} + b^{(l)} \right).$$

TGN is a versatile tool for engineering tasks where both tensor data and graph structures are relevant. It integrates tensor and graph processing to address data with both multi-dimensional features and graph structure.

Layer $l = 0$ (Input Layer)
The input is tensor T with dimensions $N \times M \times P$. A graph: $G = (V, E)$ with N nodes. Node features: $X^{(0)} \in \mathbb{R}^{N \times M}$.

Layer $l = 1$ to L (Tensor Convolution, Graph Aggregation, and Pooling)
Tensor Convolution Operation

$$\begin{cases} \text{Input: } T^{(l-1)} \text{ with dimensions } N \times M \times P \\ \text{Filters: } W^{(l)} \text{ filters of size } K^{(l)} \times K^{(l)} \times Q^{(l)} \\ \text{Output: } T^{(l)} \text{ with dimensions } N \times M \times Q^{(l)} \end{cases} \tag{8.107}$$

$$T_{n,m,q}^{(l)} = \sigma_l \left(\sum_{i=1}^{K^{(l)}} \sum_{j=1}^{K^{(l)}} \sum_{k=1}^{Q^{(l)}} W_{i,j,k}^{(l)} * T_{n,m+i-1,q+k-1}^{(l-1)} + b_q^{(l)} \right).$$

Graph Aggregation Operation

$$\begin{cases} \text{Input: } T^{(l)} \text{ with dimensions } N \times M \times Q^{(l)} \\ \text{Output: } H^{(l)} \text{ with dimensions } N \times Q^{(l)} \end{cases} \tag{8.108}$$

$$H_{n,q}^{(l)} = \sigma_l \left(\sum_{m \in \mathcal{N}(n)} \frac{1}{\sqrt{|\mathcal{N}(n)|}} \left(T_{n,m,q}^{(l)} \right) \right).$$

Pooling Operation

$$\begin{cases} \text{Input: } H^{(l)} \text{ with dimensions } N \times Q^{(l)} \\ \text{Output: } U^{(l)} \text{ with dimensions } N \times R^{(l)} \end{cases} \tag{8.109}$$

$$U_{n,r}^{(l)} = \text{pooling}(H_{n,r}^{(l)}).$$

Output Layer (Classification)
Global Pooling Operation

$$\begin{cases} \text{Input: } U^{(L)} \text{ with dimensions } N \times R^{(L)} \\ \text{Output: } Z \text{ with dimensions } R^{(L)} \end{cases} \tag{8.110}$$

$$Z_r = \text{global\_pooling}\left(\left[U_{n,r}^{(L)}\ \forall n \in \{1, \ldots, N\}\right]\right).$$

Fully Connected Layer

$$\begin{cases} \text{Input: } Z \text{ with dimensions } R^{(L)} \\ \text{Weight Matrix: } W_{\text{out}} \in \mathbb{R}^{R^{(L)} \times F_{\text{out}}} \\ \text{Output: } y_{\text{pred}} \text{ with dimensions } F_{\text{out}} \end{cases} \tag{8.111}$$

$$y_{\text{pred}} = \text{softargmax}(Z \cdot W_{\text{out}} + b_{\text{out}}).$$

These advanced neural network architectures, GNN, TensorNN, GTNN, and TGN, provide powerful tools for optimizing and controlling data in Hilbert spaces. They allow for the incorporation of graph and tensor structures, enabling richer representations and enhanced performance in optimization and control tasks. They enable the modeling and analysis of structured and multi-dimensional data, enhancing the capabilities of optimization, control, and decision-making in various engineering domains. Each architecture is designed to handle different types of data and relationships. The choice of architecture depends on the nature of the data and the specific problem being addressed. These architectures provide flexibility in modeling and learning from diverse types of data, enabling a wide range of applications.

8.4.5 *Tensor Generative Adversarial Network*

Layer $l = 0$ (Input Layer)
Input noise: $Z \in \mathbb{R}^{N \times M \times P}$, where N is batch size, M is dimension of noise, and P is number of channels.

Layer $l = 1$ to L (Generator Layers)
Tensor Convolutional Layer

$$\begin{cases} \text{Input: } Z \text{ with dimensions } N \times M \times P \\ \text{Filters: } W^{(l)} \in \mathbb{R}^{K^{(l)} \times K^{(l)} \times Q^{(l)}} \\ \text{Output: } G^{(l)} \text{ with dimensions } N \times M \times Q^{(l)} \end{cases} \tag{8.112}$$

$$G_{n,m,q}^{(l)} = \sigma_l \left(\sum_{i=1}^{K^{(l)}} \sum_{j=1}^{K^{(l)}} \sum_{k=1}^{P} W_{i,j,k,q}^{(l)} * Z_{n,m+i-1,k} + b_q^{(l)} \right).$$

Layer $l = 1$ to L (Discriminator Layers)
Tensor Convolutional Layer

$$\begin{cases} \text{Input: } X^{(l)} \text{ with dimensions } N \times M \times Q^{(l)} \\ \text{Filters: } W^{(l)} \in \mathbb{R}^{K^{(l)} \times K^{(l)} \times P^{(l)}} \\ \text{Output: } D^{(l)} \text{ with dimensions } N \times M \times P^{(l)} \end{cases} \tag{8.113}$$

$$D_{n,m,p}^{(l)} = \sigma_l \left(\sum_{i=1}^{K^{(l)}} \sum_{j=1}^{K^{(l)}} \sum_{q=1}^{Q^{(l)}} W_{i,j,q,p}^{(l)} * X_{n,m+i-1,q} + b_p^{(l)} \right).$$

Loss Function
The generator (G) and discriminator (D) networks are trained using the minimax adversarial loss:

$$\mathcal{L}_{\text{GAN}}(G, D) = E_{x\sim\text{real}}[\log(D(x))] + E_{z\sim\text{noise}}[\log(1 - D(G(z)))],$$

where x represents real data samples, z represents noise samples, and $D(x)$ and $D(G(z))$ are the discriminator's outputs.

8.4.6 Graph GAN

Layer $l = 0$ (Input Layer)
– Input noise: $Z \in \mathbb{R}^{N \times M \times P}$, where N is batch size, M is dimension of noise, and P is number of channels.

Layer $l = 1$ to L (Generator Layers)
Graph Convolutional Layer

$$\begin{cases} \text{Input: } Z \text{ with dimensions } N \times M \times P \\ \text{Graph Structure: } G = (V, E) \text{ with } N \text{ nodes} \\ \text{Filters: } W^{(l)} \in \mathbb{R}^{K^{(l)} \times Q^{(l)}} \\ \text{Output: } G^{(l)} \text{ with dimensions } N \times Q^{(l)} \end{cases} \tag{8.114}$$

$$G_{n,q}^{(l)} = \sigma_l \left(\sum_{i=1}^{K^{(l)}} \sum_{j \in \mathcal{N}(n)} W_{i,j,q}^{(l)} * Z_{n,i,j} + b_q^{(l)} \right).$$

Layer $l = 1$ to L (Discriminator Layers)
Graph Convolutional Layer

$$\begin{cases} \text{Input: } X^{(l)} \text{ with dimensions } N \times Q^{(l)} \\ \text{Graph Structure: } G = (V, E) \text{ with } N \text{ nodes} \\ \text{Filters: } W^{(l)} \in \mathbb{R}^{K^{(l)} \times P^{(l)}} \\ \text{Output: } D^{(l)} \text{ with dimensions } N \times P^{(l)} \end{cases} \tag{8.115}$$

$$D_{n,p}^{(l)} = \sigma_l \left(\sum_{i=1}^{K^{(l)}} \sum_{j \in \mathcal{N}(n)} W_{i,j,p}^{(l)} * X_{n,i} + b_p^{(l)} \right).$$

Loss Function

The generator (G) and discriminator (D) networks are trained using the minimax adversarial loss:

$$\mathcal{L}_{\text{GAN}}(G, D) = E_{x \sim \text{real}}[\log(D(x))] + E_{z \sim \text{noise}}[\log(1 - D(G(z)))],$$

where x represents real graph samples, z represents noise samples, and $D(x)$ and $D(G(z))$ are the discriminator's outputs.

8.4.7 Tensor-Graph GAN

Layer $l = 0$ (Input Layer)

Input noise: $Z \in \mathbb{R}^{N \times M \times P}$, where N is batch size, M is dimension of noise, and P is number of channels.

Graph: $G = (V, E)$ with N nodes.

Layer $l = 1$ to L (Generator Layers)
Tensor Convolutional Layer

$$\begin{cases} \text{Input: } Z \text{ with dimensions } N \times M \times P \\ \text{Filters: } W^{(l)} \in \mathbb{R}^{K^{(l)} \times K^{(l)} \times Q^{(l)}} \\ \text{Output: } G^{(l)} \text{ with dimensions } N \times M \times Q^{(l)} \end{cases} \tag{8.116}$$

$$G_{n,m,q}^{(l)} = \sigma_l \left(\sum_{i=1}^{K^{(l)}} \sum_{j=1}^{K^{(l)}} \sum_{k=1}^{P} W_{i,j,k,q}^{(l)} * Z_{n,m+i-1,k} + b_q^{(l)} \right).$$

Graph Aggregation Layer

$$\begin{cases} \text{Input: } G^{(l)} \text{ with dimensions } N \times M \times Q^{(l)} \\ \text{Output: } H^{(l)} \text{ with dimensions } N \times Q^{(l)} \end{cases} \tag{8.117}$$

$$H_{n,q}^{(l)} = \sigma_l \left(\sum_{m \in \mathcal{N}(n)} \frac{1}{\sqrt{|\mathcal{N}(n)|}} \left(G_{n,m,q}^{(l)} \right) \right).$$

Layer $l = 1$ to L (Discriminator Layers)
Tensor Convolutional Layer

$$\begin{cases} \text{Input: } X^{(l)} \text{ with dimensions } N \times M \times Q^{(l)} \\ \text{Filters: } W^{(l)} \in \mathbb{R}^{K^{(l)} \times K^{(l)} \times P^{(l)}} \\ \text{Output: } D^{(l)} \text{ with dimensions } N \times M \times P^{(l)} \end{cases} \tag{8.118}$$

$$D_{n,m,p}^{(l)} = \sigma \left(\sum_{i=1}^{K^{(l)}} \sum_{j=1}^{K^{(l)}} \sum_{q=1}^{Q^{(l)}} W_{i,j,q,p}^{(l)} * X_{n,m+i-1,q} + b_p^{(l)} \right).$$

Loss Function
The generator (G) and discriminator (D) networks are trained using the minimax adversarial loss:

$$\mathcal{L}_{\text{GAN}}(G, D) = E_{x \sim \text{real}}[\log(D(x))] + E_{z \sim \text{noise}}[\log(1 - D(G(z)))],$$

where x represents real graph samples, z represents noise samples, and $D(x)$ and $D(G(z))$ are the discriminator's outputs.

8.4.8 Graph-Tensor GAN

Layer $l = 0$ (Input Layer)
Input noise is $Z \in \mathbb{R}^{N \times M \times P}$, where N is batch size, M is dimension of noise, and P is number of channels and the graph $G = (V, E)$ with N nodes.

Layer $l = 1$ to L (Generator Layers)
Graph Convolutional Layer

$$\begin{cases} \text{Input: } Z \text{ with dimensions } N \times M \times P \\ \text{Graph Structure: } G = (V, E) \text{ with } N \text{ nodes} \\ \text{Filters: } W^{(l)} \in \mathbb{R}^{K^{(l)} \times Q^{(l)}} \\ \text{Output: } G^{(l)} \text{ with dimensions } N \times Q^{(l)} \end{cases} \tag{8.119}$$

$$G_{n,q}^{(l)} = \sigma_l \left(\sum_{i=1}^{K^{(l)}} \sum_{j \in \mathcal{N}(n)} W_{i,j,q}^{(l)} * Z_{n,i,j} + b_q^{(l)} \right).$$

Tensor Aggregation Layer

$$\begin{cases} \text{Input: } G^{(l)} \text{ with dimensions } N \times Q^{(l)} \\ \text{Output: } H^{(l)} \text{ with dimensions } N \times Q^{(l)} \end{cases} \tag{8.120}$$

$$H_{n,q}^{(l)} = \sigma_l \left(\sum_{m \in \mathcal{N}(n)} \frac{1}{\sqrt{|\mathcal{N}(n)|}} \left(G_{n,m,q}^{(l)} \right) \right).$$

Layer $l = 1$ to L (Discriminator Layers)
Graph Convolutional Layer

$$\begin{cases} \text{Input: } X^{(l)} \text{ with dimensions } N \times Q^{(l)} \\ \text{Graph Structure: } G = (V, E) \text{ with } N \text{ nodes} \\ \text{Filters: } W^{(l)} \in \mathbb{R}^{K^{(l)} \times P^{(l)}} \\ \text{Output: } D^{(l)} \text{ with dimensions } N \times P^{(l)} \end{cases} \tag{8.121}$$

$$D_{n,p}^{(l)} = \sigma_l \left(\sum_{i=1}^{K^{(l)}} \sum_{j \in \mathcal{N}(n)} W_{i,j,p}^{(l)} * X_{n,i} + b_p^{(l)} \right).$$

8.4.9 GPT

Encoder-only transformers, decoder-only transformers, and encoder–decoder transformers are all compositions of mappings in addition to the above neural network description. Bidirectional encoder representations from transformers, generative pre-trained transformers, use a specific architecture for matrix transformations. The main architecture is given by the composition of the attention layer with the feedforward layer iterated in L blocks. $\mathcal{D} = (\tilde{\sigma}_{ff_L} \circ \tilde{\sigma}_{att_L}) \circ \ldots \circ (\tilde{\sigma}_{ff_1} \circ \tilde{\sigma}_{att_1})$.

An attention layer is defined via an attention multi-head. A single-head attention is given by $\sigma_{att}(x) = \text{softargmax}^*(x\omega_{qk}x')x\omega_{ov}'$, where the autoregressive masking is

$$\text{softargmax}^*(A)_{ij} = \frac{e^{a_{ij}}}{\sum_{i \geq q} e^{a_{iq}}} \mathbb{I}_{\{i \geq j\}},$$

where ω_{qk} is the query–key matrix and ω_{ov} is the output-value matrix. By choosing a set of attention heads, AH, the attention layer, is $\tilde{\sigma}_{att}(x) = x + \sum_{att \in AH} \sigma_{att}(x)$. The feedforward layer $\sigma_{ff_l}(x_l) = x_l + (W_{2l} r_l (W_{1l} x_l + b_{1l}) + b_{2l})$. The full transformer

$$ET(x) = \sigma_e \circ \mathcal{D} \circ \sigma_u(x),$$

where σ_e maps one-hot token acts first via the embedding and σ_u maps the output of the decoder via the unembedding. All the outputs are normalized via

$$\sigma_{n,i}(x)[j] = \beta[j] + \gamma[j]\frac{x_i[j] - \bar{x}_i}{\sqrt{\varepsilon + \nu}},$$

where $\bar{x}_i = \frac{1}{size(x_i)} \sum_j x_{ij}$ and $\nu = \frac{1}{size(x_i)-1} \sum_j (x_{ij} - \bar{x}_i)^2$, γ, β are parameters.
The algorithm becomes

$$\begin{aligned}
&\text{Given: } n, d, x_0 \in \mathbb{R}^{nd}, L \geq 3, \lambda \in \mathbb{R}^T, \\
&\text{for } t \in \{0, 1, 2, \ldots, T-1\} \\
&\qquad l = 1: \quad y_{1,t} = \sigma_{1,n2} \circ \sigma_{1,ff} \circ \sigma_{1,n1} \circ \sigma_{1,att}(x_t) \\
&\qquad \text{for } l \in \{2, \ldots, L\}: \\
&\qquad\qquad y_{l,t} = \sigma_{l,n2} \circ \sigma_{l,ff} \circ \sigma_{l,n1} \circ \sigma_{l,att}(y_{l-1,t}) \\
&\qquad x_{t+1} = x_t + \lambda_t(y_{L,t} - x_t), \\
&x_T \in \mathbb{R}^{nd}.
\end{aligned} \tag{8.122}$$

Let x be a matrix with n rows and d columns. Thus, the row x_i has d entries.

$$\begin{cases}
i \in \{1, \ldots, n\}, h \in \{1, \ldots, H\}, l \in \{1, \ldots, L\} \\
\text{Attention: parametrized by} \\
W_{lh,3i} \in \mathbb{R}^{k_l \cdot d}, \; W_{lh,4i}, W_{lh,5i}, W_{lh,6i} \in \mathbb{R}^{d \cdot k_l}, \\
\text{Given: } x \in \mathbb{R}^{nd} \\
\sigma_{l,att} : \mathbb{R}^{nd} \to \mathbb{R}^{nd}, \; \sigma_{l,n1} : \mathbb{R}^{nd} \to \mathbb{R}^{nd}, \\
\sigma_{l,n1} \text{ parametrized by } \gamma_{li1}, \beta_{li1} \in \mathbb{R}^d, \\
\tilde{u}_{l,i} = \sigma_{l,att,i}(x) \\
= \left(x_i + \sum_{h=1}^H W'_{lh,3i} \sum_{j=1}^i \frac{e^{\frac{\langle W'_{lh,4i} x_i, W'_{lh,5i} x_j \rangle}{\sqrt{k_l}}}}{\sum_{j'=1}^i e^{\frac{\langle W'_{lh,4i} x_i, W'_{lh,5i} x_{j'} \rangle}{\sqrt{k_l}}}} W'_{lh,6i} x_j \right), \\
\bar{\tilde{u}}_{l,i} = \frac{\sum_{j'=1}^d \tilde{u}_{l,i}[j']}{d}, \\
u_{l,i} = \sigma_{l,n1,i}(\tilde{u}_{l,i}) \\
= (\beta_{li1}[j'] + \gamma_{li1}[j'] \sqrt{d-1} \frac{\tilde{u}_{l,i}[j'] - \bar{\tilde{u}}_{l,i}}{\left| \tilde{u}_{l,i} - \bar{\tilde{u}}_{l,i} \right|_2}, \quad j' \in \{1, \ldots, d\}), \\
\\
\text{ReLU Feed Forward: parametrized by} \\
W_{l1i} \in \mathbb{R}^{d \cdot m_l}, W_{l2i} \in \mathbb{R}^{m_l \cdot d}, b_{l,1i} \in \mathbb{R}^{m_l}, b_{l,2i} \in \mathbb{R}^d, \\
\sigma_{l,ff} : \mathbb{R}^{nd} \to \mathbb{R}^{nd}, \; \sigma_{l,n2} : \mathbb{R}^{nd} \to \mathbb{R}^{nd}, \\
\sigma_{l,n2,i} \text{ parametrized by } \gamma_{l,i2}, \beta_{l,i2} \in \mathbb{R}^d, \\
\\
\tilde{y}_{l,i} = \sigma_{l,ff,i}(u_{l,i}) = u_{l,i} + W'_{l2i} ReLU(W'_{l1i} u_{li} + b_{l1i}) + b_{l2i}, \\
\bar{\tilde{y}}_{l,i} = \frac{\sum_{j'=1}^d \tilde{y}_{l,i}[j']}{d}, \\
y_{li} = \sigma_{l,n2,i}(\tilde{y}_{li}) \\
= (\beta_{li2}[j'] + \gamma_{li2}[j'] \sqrt{d-1} \frac{\tilde{y}_{l,i}[j'] - \bar{\tilde{y}}_{l,i}}{\left| \tilde{y}_{l,i} - \bar{\tilde{y}}_{l,i} \right|_2}, \quad j' \in \{1, \ldots, d\}),
\end{cases} \tag{8.123}$$

$$
\left|
\begin{array}{l}
\text{Given: } n, d, H \\
\text{Given: } W_3, W_4, W_5, W_6, \gamma_1, \beta_1 \\
\text{Given: } W_1, W_2, b_1, b_2, \gamma_2, \beta_2 \\
\text{Given: } l, k_l, m_l, x_l \\
\\
\text{for } i \in \{1, \ldots, n\} \\
\qquad \tilde{u}_{l,i} = \sigma_{l,n1,i}(x_{l,i}) \\
\qquad u_{l,i} = \sigma_{l,att,i}(\tilde{u}_l) \\
\\
\qquad \tilde{y}_{l,i} = \sigma_{l,ff,i}(u_{l,i}) \\
\qquad y_{l,i} = \sigma_{l,n2,i}(\tilde{y}_{l,i}) \\
\text{Return: } y_l = (y_{l,i})_{i \in \{1,\ldots,n\}}
\end{array}
\right. . \tag{8.124}
$$

Let

$$
\left|
\begin{array}{l}
\sigma_{l,n1} = (\sigma_{l,n1,i}, i \in \{1, \ldots, n\}), \\
\sigma_{l,att} = (\sigma_{l,att,i}, i \in \{1, \ldots, n\}), \\
\sigma_{l,ff} = (\sigma_{l,ff,i}, i \in \{1, \ldots, n\}), \\
\sigma_{l,n2} = (\sigma_{l,n2,i}, i \in \{1, \ldots, n\}).
\end{array}
\right. \tag{8.125}
$$

The l-th layer operator is $\sigma_{l,ff} \circ \sigma_{l,n2} \circ \sigma_{l,att} \circ \sigma_{l,n1}$.

$$
\begin{aligned}
&\text{Given: } n, d, x_0, L, T, \lambda\\
&\text{for } t \in \{0, 1, \ldots, T\}\\
&\quad l = 1:\\
&\quad \text{for } i \in \{1, 2, \ldots, n\}\\
&\qquad \text{Given: } H, k_1, m_1, h \in \{1, \ldots, H\}\\
&\qquad \text{Given: } W_{1h,3i} \in \mathbb{R}^{k_1\ d},\ W_{1h,4i}, W_{1h,5i},\\
&\qquad\qquad W_{1h,6i} \in \mathbb{R}^{d\ \cdot\ k_1}, \gamma_{1i1}, \beta_{1i1} \in \mathbb{R}^d,\\
&\qquad \text{Given: } W_{11i} \in \mathbb{R}^{d\ \cdot\ m_1}, W_{12i} \in \mathbb{R}^{m_1\ d},\\
&\qquad\qquad b_{1,1i} \in \mathbb{R}^{m_1}, b_{1,2i} \in \mathbb{R}^d,\\
&\qquad\qquad \gamma_{1,i2}, \beta_{1,i2} \in \mathbb{R}^d,\\
&\qquad y_{1,i,t} = \sigma_{1,ff,i} \circ \sigma_{1,n2,i} \circ \sigma_{1,att,i} \circ \sigma_{1,n1,i}(x_t),\\
&\quad \text{Return: } y_{1,t} = (y_{1,i,t}, i \in \{1, 2, \ldots, n\}),\\
&\quad \text{for } l \in \{2, \ldots, L\}:\\
&\qquad \text{for } i \in \{1, 2, \ldots, n\}\\
&\qquad \text{Given: } H, k_l, m_l, h \in \{1, \ldots, H\}\\
&\qquad \text{Given: } W_{lh,3i} \in \mathbb{R}^{k_l\ d},\ W_{lh,4i}, W_{lh,5i},\\
&\qquad\qquad W_{lh,6i} \in \mathbb{R}^{d\ k_l}, \gamma_{li1}, \beta_{li1} \in \mathbb{R}^d,\\
&\qquad \text{Given: } W_{l1i} \in \mathbb{R}^{d\ \cdot\ m_l}, W_{l2i} \in \mathbb{R}^{m_l\ d},\\
&\qquad\qquad b_{l,1i} \in \mathbb{R}^{m_l}, b_{l,2i} \in \mathbb{R}^d,\\
&\qquad\qquad \gamma_{l,i2}, \beta_{l,i2} \in \mathbb{R}^d,\\
&\qquad y_{l,i,t} = \sigma_{l,ff,i} \circ \sigma_{l,n2,i} \circ \sigma_{l,att,i} \circ \sigma_{l,n1,i}(y_{l-1,t})\\
&\quad \text{Return: } y_{l,t} = (y_{l,i,t}, i \in \{1, 2, \ldots, n\}),\\
&\quad x_{t+1} = x_t + \lambda_t(y_{L,t} - x_t),
\end{aligned}
\tag{8.126}
$$

which means that $x_{t+1} = x_t + \lambda_t(\sigma_{L,ff} \circ \sigma_{L,n2} \circ \sigma_{L,att} \circ \sigma_{L,n1} \circ \ldots \circ \sigma_{1,ff} \circ \sigma_{1,n2} \circ \sigma_{1,att} \circ \sigma_{1,n1}(x_t) - x_t)$, starting from x_0.

The full transformer specification is given by $\mathbb{NN} := (x_0, \sigma_{n2}, \sigma_{ff}, \sigma_{n1}, \sigma_{att})$ $(L, d, k, m, H, W_{l,h,3i}, W_{l,h,4i}, W_{l,h,5i}, W_{l,h,6i}, \gamma_{l,i1}, \beta_{l,i1})$, $(W_{l,1i}, W_{l,2i}, b_{l,1i}, b_{l,2i}, \gamma_{l,i2}, \beta_{l,i2}), i \in \{1, \ldots, n\}, h \in \{1, \ldots, H\}, l \in \{1, \ldots, L\}, \{\mathscr{H}_l\}_{1 \le l \le L}, \lambda)$.

8.4.10 Tensor GPT

Layer $l = 0$ (Input Layer)
Input text: $X \in \mathbb{R}^{N \times T \times V}$, where N is batch size, T is sequence length, and V is vocabulary size.

Layer $l = 1$ to L (Transformer Layers)
Tensor Self-Attention Layer

$$\begin{cases} \text{Input: } X \text{ with dimensions } N \times T \times V \\ \text{Output: } H \text{ with dimensions } N \times T \times V \end{cases} \tag{8.127}$$

$$H = \text{MultiheadSelfAttention}(X) + X.$$

Tensor Feedforward Layer

$$\begin{cases} \text{Input: } H \text{ with dimensions } N \times T \times V \\ \text{Output: } Y \text{ with dimensions } N \times T \times V \end{cases} \tag{8.128}$$

$$Y = \text{Feedforward}(H) + H.$$

Layer $l = L + 1$ (Output Layer)
Tensor Linear Layer

$$\begin{cases} \text{Input: } Y \text{ with dimensions } N \times T \times V \\ \text{Output: } \hat{X} \text{ with dimensions } N \times T \times V \end{cases} \tag{8.129}$$

$$\hat{X} = \text{Linear}(Y).$$

Loss Function
The tensor GPT is trained using a suitable loss function, such as the cross-entropy loss, for text generation tasks.

8.4.11 Graph GPT

Layer $l = 0$ (Input Layer)
– Input graph: $G = (V, E)$, where V is the set of nodes and E is the set of edges.

Layer $l = 1$ to L (Graph Transformer Layers)
Graph Self-Attention Layer

$$\begin{cases} \text{Input: } G = (V, E) \\ \text{Output: } G' = (V, E') \end{cases} \tag{8.130}$$

$$G' = \text{GraphSelfAttention}(G) + G.$$

Graph Aggregation Layer

$$\begin{cases} \text{Input: } G' = (V, E') \\ \text{Output: } G'' = (V, E') \end{cases} \tag{8.131}$$

$$G'' = \text{GraphAggregation}(G') + G'.$$

Layer $l = L + 1$ (Output Layer)
Graph Linear Layer

$$\begin{cases} \text{Input: } G'' = (V, E') \\ \text{Output: } \hat{G} = (V, E') \end{cases} \tag{8.132}$$

$$\hat{G} = \text{GraphLinear}(G'').$$

Loss Function
The graph GPT is trained using a suitable loss function for graph generation or analysis tasks.

8.4.12 *Tensor-Graph GPT*

Layer $l = 0$ (Input Layer)
Let $X \in \mathbb{R}^{N \times T \times V}$ be the input text, where N is batch size, T is sequence length, and V is vocabulary size and $G = (V, E)$ be the input graph where V is the set of nodes and E is the set of edges.

Layer $l = 1$ to L (Tensor-Graph Transformer Layers)
Tensor Self-Attention Layer

$$\begin{cases} \text{Input: } X \text{ with dimensions } N \times T \times V \\ \text{Output: } H_x \text{ with dimensions } N \times T \times V \end{cases} \tag{8.133}$$

$$H_x = \text{TensorSelfAttention}(X) + X.$$

Graph Attention Layer

$$\begin{cases} \text{Input: } G = (V, E) \\ \text{Output: } G' = (V, E') \end{cases} \tag{8.134}$$

$$G' = \text{GraphAttention}(G) + G.$$

Tensor-Graph Aggregation Layer

$$\begin{cases} \text{Input: } H_x \text{ with dimensions } N \times T \times V, \text{ and } G' = (V, E') \\ \text{Output: } H_{xg} \text{ with dimensions } N \times T \times V \end{cases} \tag{8.135}$$

$$H_{xg} = \text{TensorGraphAggregation}(H_x, G') + H_x.$$

Layer $l = L + 1$ (Output Layer)
Tensor Linear Layer

$$\begin{cases} \text{Input: } H_{xg} \text{ with dimensions } N \times T \times V \\ \text{Output: } \hat{X} \text{ with dimensions } N \times T \times V \end{cases} \tag{8.136}$$

$$\hat{X} = \text{TensorLinear}(H_{xg}).$$

Loss Function: The tensor-Graph GPT is trained using a suitable loss function for generation or graph analysis tasks.

8.4.13 Graph-Tensor GPT

Layer $l = 0$ (Input Layer)
The input graph: $G = (V, E)$, where V is the set of nodes and E is the set of edges. The input text is $X \in \mathbb{R}^{N \times T \times V}$, where N is batch size, T is sequence length, and V is vocabulary size.

Layer $l = 1$ to L (Graph-Tensor Transformer Layers)
Graph Self-Attention Layer

$$\begin{cases} \text{Input: } G = (V, E) \\ \text{Output: } G' = (V, E') \end{cases} \tag{8.137}$$

$$G' = \text{GraphSelfAttention}(G) + G.$$

Tensor Attention Layer

$$\begin{cases} \text{Input: } X \text{ with dimensions } N \times T \times V \\ \text{Output: } X' \text{ with dimensions } N \times T \times V \end{cases} \tag{8.138}$$

$$X' = \text{TensorAttention}(X) + X$$

Graph-Tensor Aggregation Layer

$$\begin{cases} \text{Input: } G' = (V, E'), \text{ and } X' \text{ with dimensions } N \times T \times V \\ \text{Output: } H_{gx} \text{ with dimensions } N \times T \times V \end{cases} \tag{8.139}$$

$$H_{gx} = \text{GraphTensorAggregation}(G', X') + X'.$$

Layer $l = L + 1$ (Output Layer)
Tensor Linear Layer

$$\begin{cases} \text{Input: } H_{gx} \text{ with dimensions } N \times T \times V \\ \text{Output: } \hat{X} \text{ with dimensions } N \times T \times V \end{cases} \tag{8.140}$$

$$\hat{X} = \text{TensorLinear}(H_{gx}).$$

Loss Function: The graph-tensor GPT is trained using a suitable loss function for graph generation, text generation, or combined tasks.

The equation for a tensor self-attention layer is given by

$$\mathbf{Z} = \text{softargmax}\left(\frac{\mathbf{Q} \cdot \mathbf{K}^T}{\sqrt{d_k}}\right) \cdot \mathbf{V},$$

where

$$\begin{cases} \mathbf{Q}, \mathbf{K}, \mathbf{V} \text{ are the query, key, and value tensors of dimensions } N \times T \times d_k \\ \mathbf{Z} \text{ is the output tensor of the same dimensions as } \mathbf{V} \\ d_k \text{ is the dimension of the key vectors} \\ \text{softargmax}(\cdot) \text{ applies the softargmax operation along the specified dimension} \\ \sqrt{d_k} \text{ is the scaling factor for stabilizing the attention weights.} \end{cases} \tag{8.141}$$

The equation for a multi-head tensor self-attention layer is given by

$$\mathbf{Z} = \text{Concat}(\mathbf{Z}_1, \mathbf{Z}_2, \ldots, \mathbf{Z}_h) \cdot \mathbf{W}_O$$

with

$$\begin{cases} \mathbf{Q}, \mathbf{K}, \mathbf{V} \text{ are the query, key, and value tensors of dimensions } N \times T \times d_k \\ \mathbf{Z}_i = \text{SelfAttention}(\mathbf{Q} \cdot \mathbf{W}_Q^{(i)}, \mathbf{K} \cdot \mathbf{W}_K^{(i)}, \mathbf{V} \cdot \mathbf{W}_V^{(i)}) \\ \mathbf{W}_Q^{(i)}, \mathbf{W}_K^{(i)}, \mathbf{W}_V^{(i)} \text{ are learnable projection matrices for head } i \\ h \text{ is the number of attention heads} \\ \mathbf{W}_O \text{ is the learnable output projection matrix.} e \\ \text{Concat}(\cdot) \text{ concatenates the outputs of all attention heads.} \end{cases} \tag{8.142}$$

The multi-head tensor self-attention layer computes attention scores for multiple heads and then concatenates and projects the outputs. Each head processes the input with its own learnable projection matrices to capture different aspects of the data. The concatenated outputs are linearly transformed by the output projection matrix.

The equation for a feedforward layer at the l-th layer of a tensor neural network is given by

$$\mathbf{Y}^{(l)} = \mathrm{ReLU}(\mathbf{X}^{(l)} \cdot \mathbf{W}^{(l)} + \mathbf{b}^{(l)}) \cdot \mathbf{W}^{(l+1)} + \mathbf{b}^{(l+1)},$$

where

$$\begin{cases} \mathbf{X}^{(l)} \text{ is the input tensor at the } l\text{-th layer of dimensions } N \times d_1 \times d_2 \times \ldots \times d_k \\ \mathbf{Y}^{(l)} \text{ is the output tensor of the same dimensions as } \mathbf{X}^{(l)} \\ \mathbf{W}^{(l)}, \mathbf{W}^{(l+1)} \text{ are learnable weight tensors for the } l\text{-th and } l+1\text{-th layers} \\ \mathbf{b}^{(l)}, \mathbf{b}^{(l+1)} \text{ are bias tensors for the } l\text{-th and } l+1\text{-th layers} \\ \mathrm{ReLU}(\cdot) \text{ applies the rectified linear unit activation function element-wise.} \end{cases} \tag{8.143}$$

The feedforward layer at the l-th layer applies linear transformations followed by element-wise activation to the input tensor $\mathbf{X}^{(l)}$. The first linear combination uses weights $\mathbf{W}^{(l)}$ and biases $\mathbf{b}^{(l)}$, followed by a ReLU activation. The second linear combination using weights $\mathbf{W}^{(l+1)}$ and biases $\mathbf{b}^{(l+1)}$ produces the final output tensor $\mathbf{Y}^{(l)}$.

The equation for a tensor linear layer is given by

$$\hat{\mathbf{T}} = \mathbf{W} \cdot \mathbf{T} + \mathbf{b},$$

where

$$\begin{cases} \mathbf{T} \text{ is the input tensor of dimensions } N \times d_1 \times d_2 \times \ldots \times d_k \\ \hat{\mathbf{T}} \text{ is the output tensor of the same dimensions as } \mathbf{T} \\ \mathbf{W} \text{ is the learnable weight tensor of dimensions } d_0 \times d_1 \times d_2 \times \ldots \times d_k \\ \mathbf{b} \text{ is the bias tensor of dimensions } d_1 \times d_2 \times \ldots \times d_k. \end{cases} \tag{8.144}$$

The operation $\cdot$ represents the tensor contraction, which is similar to the matrix-vector multiplication but extended to tensors of higher order. The result of the tensor linear layer is the transformed tensor $\hat{\mathbf{T}}$ obtained by applying the weight tensor $\mathbf{W}$ and adding the bias tensor $\mathbf{b}$.

8.4.14 Tensor-Graph Neural Network with Blockchain Step

A tensor-graph neural network (TGN) is designed for data with both tensor and graph structures. It combines tensor operations and graph convolutions for powerful data representation:

$$\mathbf{H}_{gx}^{(l+1)} = \sigma\left(\sum_{j\in\mathcal{N}(i)} \mathbf{W}_{gx}^{(l)}\mathbf{H}_{gx,j}^{(l)} + \mathbf{b}_{gx}^{(l)}\right).$$

Blockchain step in TGN: Integrating a blockchain step into TGN involves maintaining a decentralized ledger of transactions. Each node i in the graph may represent a participant in the blockchain network. The tensor-graph features $\mathbf{H}_{gx}^{(l)}$ of each node contain information about their current state:

$$\mathbf{H}_{gx}^{(l+1)} = \sigma\left(\sum_{j\in\mathcal{N}(i)} \mathbf{W}_{gx}^{(l)}\mathbf{H}_{gx,j}^{(l)} + \mathbf{b}_{gx}^{(l)}\right).$$

The blockchain step involves:

- validating and processing transactions from neighboring nodes, updating the local tensor-graph features $\mathbf{H}_{gx}^{(l)}$ accordingly
- propagating validated transactions to neighboring nodes for consensus.

Incorporating a blockchain step into a tensor-graph neural network enhances the model's ability to handle decentralized and distributed data. The blockchain layer enables secure and tamper-resistant data exchange among participants while leveraging the power of tensor and graph representations provided by the TGN architecture.

Decentralized Decision-Maker Architecture with Blockchainized TGNs

Decision-maker i and blockchainized TGN: For each decision-maker i, there exists a blockchainized tensor-graph neural network with L_i layers. The architecture can be described as follows:

$$\mathbf{H}_{gx}^{(l+1,i)} = \sigma\left(\sum_{j\in\mathcal{N}(i)} \mathbf{W}_{gx}^{(l,i)}\mathbf{H}_{gx,j}^{(l,i)} + \mathbf{b}_{gx}^{(l,i)}\right).$$

Blockchain step for decision-maker i: The blockchain step within decision-maker i involves:

Validating and processing transactions from neighboring nodes and updating the local tensor-graph features $\mathbf{H}_{gx}^{(l,i)}$ accordingly.

Propagating validated transactions to neighboring nodes for consensus within decision-maker i's blockchain network.

Decision A_i: The decision A_i made by decision-maker i is based on the final output of their blockchainized TGN:

$$A_i = \text{DecisionFunction}(\mathbf{H}_{gx}^{(L_i,i)}).$$

Collaborative decision-making: The decisions A_i from all decision-makers are aggregated to make a final collaborative decision:

$$\text{FinalDecision} = \text{AggregationFunction}(A_1, A_2, \ldots, A_D).$$

The decentralized decision-maker architecture with blockchainized TGNs enables collaborative decision-making among multiple decision-makers. Each decision-maker's blockchainized TGN processes data and transactions, leading to individual decisions A_i. These decisions are then aggregated to arrive at a final collaborative decision, offering a distributed and consensus-driven approach.

Mean–Variance Game Scenario with Decision-Makers and Blockchainized TGNs

Decision-maker i and blockchainized TGN: For each decision-maker i, there exists a blockchainized tensor-graph neural network with L_i layers. The architecture can be described as follows:

$$\mathbf{H}_{gx}^{(l+1,i)} = \sigma\left(\sum_{j\in\mathcal{N}(i)} \mathbf{W}_{gx}^{(l,i)}\mathbf{H}_{gx,j}^{(l,i)} + \mathbf{b}_{gx}^{(l,i)}\right).$$

Blockchain step for decision-maker i:

The blockchain step within the decision-maker i involves

Validating and processing transactions from neighboring nodes and updating the local tensor-graph features $\mathbf{H}_{gx}^{(l,i)}$ accordingly.

Propagating validated transactions to neighboring nodes for consensus within decision-maker i's blockchain network.

Decision-maker i aims to optimize a mean–variance objective when making decisions based on its blockchainized TGN output at layer L_i:

$$a_i \in \arg\max_{a_i'} \left(\text{mv}(A_i^{(L_i)}))\right),$$

where

$$\begin{cases} A_i^{(L_i)} = \text{Output}(\mathbf{H}_{gx}^{(L_i,i)}) \text{ is the output of the blockchainized TGN at layer } L_i \\ \text{Mean}(\cdot) \text{ and Variance}(\cdot) \text{ compute mean and variance, respectively.} \end{cases} \tag{8.145}$$

Collaborative game dynamics: The decisions a_i from all decision-makers influence the game dynamics, creating a competitive and optimization-driven environment. The game scenario with decision-makers and blockchainized TGNs introduces a competitive aspect where decision-makers aim to optimize their decisions based on a mean–variance objective. The blockchainized TGNs process data, make decisions, and compete in a dynamic environment, showcasing a complex interplay of optimization and collaboration.

8.5 Risk-Quantification of Network of Neural Networks

Given an input–output data set $(x_{ij}, y_{ij})_{j\in\{1,\ldots,D\}}$ associated to decision-maker $i \in \mathcal{I}$, one can compute the sequences

$$\left(\hat{d}_{it}, \hat{\sigma}_{it}^2, \mathrm{mv}_{it}\right)_{t\in\{1,\ldots,D\}},$$

as follows

$$\begin{cases} \hat{d}_{i0} = d_{i0}, \\ \hat{d}_{i,t+1} = \hat{d}_{it} + \mu_t^i(d_{i,t+1} - \hat{d}_{it}), \quad \mu_{it} = \frac{\mu_i}{t+1},\ t \geq 0,\ \mu_i > 0 \\ \hat{\sigma}_{i1}^2 = (d_{i1} - \hat{d}_{i1})^2, \\ \hat{\sigma}_{i,t+1}^2 = \hat{\sigma}_{i,t}^2 + \nu_{it}[(d_{i,t+1} - \hat{d}_{i,t+1})^2 - \hat{\sigma}_{it}^2], \quad \nu_{it} = \frac{\nu_i}{t}, t \geq 1, \nu_i > 0, \\ \mathrm{mv}_{i0} = \hat{d}_{i0}, \\ \mathrm{mv}_{it} = \hat{d}_{it} + \hat{r}_i\hat{\sigma}_{it}^2,\ t \geq 1,\ \hat{r}_i = \frac{r_i}{\varepsilon_i + |d_{i0}|^2},\ r_i > 0. \end{cases} \tag{8.146}$$

We design a network of neural networks that represents a strategic mean–variance game. There $|\mathcal{I}| + 1$ neural networks:

- The systemNN collects the actions from the $|\mathcal{I}|$ decision-makers.
- Each of $|\mathcal{I}|$ decision-makers has its own neural network NN_i for analyzing, computing, understanding, and making its own strategy and for estimating its own payoff.

8.5.1 *Architecture Selection*

We propose a generic architecture for the network of neural networks. The input representation represents the game state or input information $(\tilde{m}, \Omega)$ and simulates/generates a sample data $(\omega_k)_{k\in\{1,\ldots,D\}})$, $D \geq 1$. We create multiple neural networks, each representing a player's strategy and estimated payoff. These neural networks should take the private message and own strategy as input and output the player's chosen strategies and estimated payoff. We also introduce a shared network architecture to capture interactions and dependencies among players' strategies. The shared network systemNN can capture the overall game dynamics and help players adapt their strategies based on each other's actions. We define variance-aware loss functions that guide the networks toward learning mean–variance payoffs and equilibria. systemNN is a shared neural network that returns a numerical value of the mean–variance to each decision-maker via private messages. The private message $\mathrm{mv}_{i,t}$ is returned to i. Decision-maker i does not know $(x_{j,t}, \mathrm{mv}_{j,t})_{j\neq i,t}$. At time t, neural network NN_i send $x_{i,t}$ to systemNN.

8.5.2 Decision-Maker's Neural Network Architecture

- Numerical value of the player's payoff from the shared layer. One-hot encoded vector representing the player's previous action.
- Convert the one-hot encoded previous action into a dense vector representation. Learn embeddings that capture relationships between different actions.
- Numerical value representing the player's received payoff from the shared layer (feedback). This feedback helps the player adjust its strategy based on the impact of its previous actions.
- Concatenate the outputs of the embedding layer and the feedback layer.
- We choose one or more hidden layers to process the concatenated input and some activation functions to introduce nonlinearity.
- Fully connected layer with softargmax activation to produce a probability distribution over possible actions. The output represents the decision-maker's current strategy for choosing actions.

8.5.3 Shared Neural Network Architecture

- Numerical vector representing the state of nature ω. This could include relevant features or information about the current game state. Data to learn m. Strategy choices of all players x_{i0}.
- One or more fully connected (dense) hidden layers to process the concatenated input $(x_i)_{i\in\mathcal{I}}, \tilde{m}$. Use suitable activation functions (e.g., σ) to introduce nonlinearity. These layers capture the interactions between players' strategies and the state of nature.
- Fully connected layer to compute the expected payoff for each player based on their chosen strategies and the state of nature. The output represents the computed payoff value for each player.
- For each player, the shared network sends a private message containing its calculated payoff value. These feedback messages are sent privately to each player's individual strategy neural network for further adjustment.

8.5.3.1 Main Steps of the NN Strategic Learning

Neural network strategic learning

Given: T
for every $i \in \mathcal{I}$
  NN_i initializes $K_i, n_1, \ldots, n_{K_i}, \lambda_i$, and sends $x_{i,0}$
systemNN initializes:
  systemNN receives $x_{i,0}$
  systemNN computes $(mv_{i,0}(x_0))_{i\in\mathcal{I}}$
for $t \in \{0, 1, 2, \ldots, T-1\}$
  for $i \in \mathcal{I}$
  NN_i :
    NN_i receives $\text{mv}_{i,t}$ from systemNN
    $l = 1: \; y_{i1,t} = \sigma_{i1,n_1} \circ \sigma_{i1,n_1-1} \circ \ldots \circ \sigma_{i1,1}(x_{it})$
    for $l \in \{2, \ldots, K_i\}$:
      $y_{il,t} = \sigma_{il,n_l} \circ \sigma_{il,n_l-1} \circ \ldots \circ \sigma_{il,1}(y_{i,l-1,t})$
    $x_{i,t+1} = x_{it} + \lambda_{it}(y_{i,K_i,t} - x_{it})$,
    $x_{i,t+1}$ is sent to the systemNN,
  systemNN:
    systemNN receives $x_{i,t+1}$ from $i \in \mathcal{I}$,
    systemNN computes $(\text{mv}_{i,t+1})_{i\in\mathcal{I}}$
    systemNN sends privately $\text{mv}_{i,t+1}$to $i \in \mathcal{I}$
  Return : $(x_{i,t}, y_{i,K_i,t}, \text{mv}_{i,t})_{t\in\{1,\ldots,T\}}$

(8.147)

with $\sigma_{il,j}(x_i) = \tilde{\sigma}_{il,j}(W_{il}x_i + b_{il})$, W_{il} being a linear bounded operator, and $\mathcal{H}$ being a nonzero Hilbert space.

8.5.4 Risk-Aware Outcomes of the Network of Neural Networks

What do well-trained neural networks do? To answer this question, we aim to find and characterize the asymptotic behavior of the network for a large class of architectures $\mathbb{NN}$. To do so, we examine $\lim_{t\to+\infty} (x_{it}, y_{i1,t}, \ldots, y_{iL,t} = mv_{it})_{i\in\mathcal{I}}$.

The decision-makers are interactive through the (private) signal they receive.

Outcomes of well-trained mean–variance neural networks

- In one dimension, $\sigma_{il,j}(x) = \tilde{\sigma}_{il,j}(W_{il}x_i + b_{il})$. If $\tilde{\sigma}_{il,j}$ is a γ_{il}-averaged operator with $0 < \gamma_{il} < 1$, then there exists a function $f_{il,j}$ such that $\tilde{\sigma}_{il,j} = [Id + \partial f_{il,j}]^{-1}$.
- The asymptotic outcomes of the neural networks are exactly the Nash equilibria of the nonzero sum game:

$$\mathcal{G}_i = (\mathcal{K}_i, (\mathcal{H}_{il},\ f_{il}(\cdot) + \frac{1}{2}\left| \cdot - b_{il} - W_{il}x_{i,l-1}\right|^2)_{l\in\mathcal{K}_i})$$

defined by:

  - The decision-makers are members of $\mathcal{K}_i$.
  - The action space of the decision-maker l is $\mathcal{H}_{il}$.
  - The objective function of decision-maker l output from previous is

$$z_i \mapsto f_{il}(z_i) + \frac{1}{2}\left|z_i - b_{il} - W_{il}x_{i,l-1}\right|^2.$$

- For three or more layers ($K_i \geq 3$), there is a deep neural architecture such that the resulting game $\mathcal{G}_i$ is not a potential game.

8.5.4.1 Mean–Variance Training and Design

For decision-maker i, given an input–output data set $\{(x_{it'}, d_{it'} = y_{iL,t'}),\ t' \in \{1, \ldots, D\}\}$, compute $\{(x_{it'}, mv_{it'} = y_{iL,t'}),\ t' \in \{1, \ldots, D\}\}$. Find $\mathbb{NN}$ such that $\sigma_{iL,k_L} \circ \ldots \circ \sigma_{iL,1} \circ \ldots \circ \sigma_{i1,k_1} \circ \ldots \circ \sigma_{i1,1}(x_{it'}) = mv_{it'},\ \forall t'$, which can be reformulated in a weak sense as a variational inequality.

We let $A_l : (W_l, b_l) \mapsto (W_l x^*_{l-1} + b_l)$, which is a linear operator and let A^*_l be the adjoint of the operator A_l.

Decision-maker i's training of weights and bias problem

Suppose that for each decision-maker i, σ_{il} is γ_{il}-averaged for some $\gamma_{il} \in (0, 1)$, $1 \leq l \leq K_i$. Then, the solutions (if any) of the training of weights and bias problem are also solutions of the following variational inequality: Given an input–output data set $\{(x^*_{i0,t'}, y^*_{iK_i,t'} = \text{mv}_{i,t'}),\ t' \in \{1, 2, \ldots, D\}$, find (W^*_i, b^*_i) (independent of time) such that

$$0 \in A^*_{iK_i,t'}[y^*_{iK_i,t'} + \partial f_{iK_i}(y^*_{K_i,t'}) - (W^*_{iK_i} x^*_{i,K_i-1,t'} + b^*_{i,K_i})],\ \forall t' \in \{1, 2, \ldots, D\}.$$

These are also Nash equilibria of the nonzero sum game defined by

$$\begin{cases} i \in \mathcal{N} \\ l \in \mathcal{K}_i, \\ (W^*_{il}, b^*_{il}) \in \arg\min_{W_{il}, b_{il}} \sum_{t'=1}^{D} \omega_{il,t'} \left| y^*_{il,t'} + \partial f_{il}(y^*_{l,t'}) - (W_{il} x^*_{i,l-1,t'} + b_{il}) \right|^2. \end{cases} \tag{8.148}$$

The dependence on mv_i is via $y^*_{iK_i}$.

The Hidden Convexity of Non-convex Activation Functions

Most of the activation functions used in neural networks are non-convex. However, we have exploited their γ-averagedness structure to derive them as a subdifferential of some functional using Legendre–Fenchel duality. This is connected to Bregman divergence or proximal function. We therefore consider activation functions that can be obtained as a proximal of some other functions. This will allow us to build a class of learning algorithms without imposing strong conditions on the weights [179, 180, 262, 264, 462].

Using the fact $(\partial f_{ik})^{-1} = \partial f^*_{ik}$:

(i) We have

$$A^*_{ik}[\tilde{\sigma}_{ik}(A_{ik}\theta) - y_{ik}] = \partial_\theta[g_{ik}(A_{ik}\theta) - \langle A_{ik}\theta, y_{ik}\rangle],$$

where $\partial g_{ik} := (\frac{1}{2}\left| \cdot \right|^2 + f_{ik})^*$.

(ii) If there exists θ^* solving $\tilde{\sigma}_{ik}(A_{ik}\theta^*) = y_{ik}$, then it coincides with the minimization of

$$\theta \mapsto g_{ik}(A_{ik}\theta) - \langle A_{ik}\theta, y_{ik}\rangle.$$

Decision-maker i's training problem

Suppose the training problem has at least one solution. Then, the set of solutions of the training problem coincides with the set of Nash equilibria of the following layer-by-layer nonzero sum game:

$$\arg\min_{\theta_l} \sum_{t=1}^{D} \omega_{l,t}[g_{il,t}(A_{il,t}\theta_l) - \langle A_{il,t}\theta_l, y_{il,t}\rangle], \quad l \in \mathscr{L}, \; l \in \mathscr{L}, \; \omega_{l,t} > 0,$$

$$\sum_{t=1}^{D} \omega_{l,t} = 1. \tag{8.149}$$

Moreover, given θ_{i0}, the algorithm

$$\theta_{ik,t+1} = \theta_{ik,t} - \frac{\lambda_i}{2\left|A_{ik}\right|^2} A_{ik}^*[\tilde{\sigma}_{ik}(A_{ik}\theta_{ik,t}) - y_{ik}], 0 < \lambda_i < 1,$$

converges to the one of the minimizers of Problem (8.149).

8.5.4.2 Expectile Tensor-Graph Neural Networks

Let $(\Omega, \mathscr{F}, \mathbb{P})$ be an atomless probability space and $L^q(\Omega, \mathscr{F}, \mathbb{P})$ be the L^q space with $q \geq 1$. Let α-expectile of an integrable random variable L, with $\alpha \in (0, 1)$, be defined by

$$\begin{cases} \rho_\alpha : L^2(\Omega, \mathscr{F}, \mathbb{P}) \to \mathbb{R}, \\ \rho_\alpha(L) \in \arg\min_{l \in \mathbb{R}} \; E\{\alpha[(L-l)_+]^2 + (1-\alpha)[(l-L)_+]^2\} \\ = \arg\min_{l \in \mathbb{R}} \; E\{W_\alpha(L-l)\}, \\ W_\alpha : \mathbb{R} \to \mathbb{R}, \\ W_\alpha(x) = \alpha[x_+]^2 + (1-\alpha)[(-x)_+]^2 \\ = \alpha x^2 \mathbb{I}_{\{x \geq 0\}} + (1-\alpha)x^2 \mathbb{I}_{\{x<0\}} = |\alpha - \mathbb{I}_{\{x<0\}}|x^2. \end{cases} \tag{8.150}$$

Input Layer: Let X be the input tensor of size $n_1 \times n_2 \times \ldots \times n_D$, where $n_1, n_2, \ldots, n_D$ are the dimensions of the tensor. The input features X are given.

Layer: For layer l, we calculate new tensor representations by applying a tensor convolution operation:

For each position $(i_1, i_2, \ldots, i_D)$, the new representation $H^{(l+1)}(i_1, i_2, \ldots, i_D)$ is computed as follows:

$$H^{(l+1)}(i_1, i_2, \ldots, i_D) =$$

$$\sigma_l \left(\sum_{j_1=1}^{n_1} \sum_{j_2=1}^{n_2} \cdots \sum_{j_D=1}^{n_D} K^{(l)}(i_1, i_2, \ldots, i_D, j_1, j_2, \ldots, j_D) \right.$$
$$\left. \cdot H^{(l)}(j_1, j_2, \ldots, j_D) + b^{(l)} \right),$$

where $H^{(l+1)}(i_1, i_2, \ldots, i_D)$ is the hidden state of the tensor at position $(i_1, i_2, \ldots, i_D)$ in layer $l+1$, σ_l is an activation function, $K^{(l)}$ is a learnable tensor kernel of appropriate dimensions, $H^{(l)}(j_1, j_2, \ldots, j_D)$ is the hidden state of the tensor at position $(j_1, j_2, \ldots, j_D)$ in layer l, and $b^{(l)}$ is a bias tensor.

Given an input–output data set $\{(x_t, d_{L,t}) \mid 1 \leq t \leq D\}$, with size $D \geq 1$, we consider the following cost functional:

$$\frac{1}{D} \sum_{t=1}^{D} W_\alpha [d_{L,t} - \sigma_L \circ \sigma_{L-1} \circ \ldots \circ \sigma_1 x_t].$$

Find the activation functions $\sigma_1, \ldots, \sigma_L$ in

$$\arg\min\{\int_{x_0} \int_{y_L} W_\alpha [y_L - \sigma_L \circ \sigma_{L-1} \circ \ldots \circ \sigma_1 x_0] \mathbb{P}(dx_0 dy_L) \mid \sigma_1, \ldots, \sigma_L\}.$$

The reduced form of the problem is:
Find the parameters $\theta_1, \ldots, \theta_L$ in

$$\arg\min\{\int_{x_0} \int_{y_L} W_\alpha [y_L - \sigma_L \circ \sigma_{L-1} \circ \ldots \circ \sigma_1 x_0] \mathbb{P}(dx_0 dy_L) \mid \theta_l = (W_l, b_l),$$
$$l \in \{1, \ldots, L\}\}.$$

When multiple agents are involved, the resulting strategic expectile tensor-graph network of neural networks CODIPAS is given by:

Expectile neural network CODIPAS

Given: T
for every $i \in \mathcal{I}$
 NN_i initializes $\alpha_i, \mu_i, K_i, n_1, \ldots, n_{K_i}, \lambda_i,$ and sends $x_{i,0}$
systemNN initializes:
 systemNN receives $x_{i,0}$
 systemNN computes $(\text{W}_{i,\alpha_i,0})_{i\in\mathcal{I}}$
for $t \in \{0, 1, 2, \ldots, T-1\}$
 for $i \in \mathcal{I}$
 NN_i :
  NN $_i$ receives $\text{W}_{i,\alpha_i,t}$ from systemNN
  $\hat{W}_{i,\alpha_i,t} = \hat{W}_{i,\alpha_i,t-1} + \mu_{it}(\text{W}_{i,\alpha_i,t} - \hat{W}_{i,\alpha_i,t-1}),$
  $l = 1: \ y_{i1,t} = \sigma_{i1,n_1} \circ \sigma_{i1,n_1-1} \circ \ldots \circ \sigma_{i1,1}(x_{i,t})$
  for $l \in \{2, \ldots, K_i\}$:
   $y_{il,t} = \sigma_{il,n_l} \circ \sigma_{il,n_l-1} \circ \ldots \circ \sigma_{il,1}(y_{i,l-1,t})$
  $x_{i,t+1} = x_{i,t} + \lambda_{it}(y_{i,K_i,t} - x_{i,t}),$
  $x_{i,t+1}$ is sent to the systemNN,
 systemNN:
  systemNN receives $x_{i,t+1}$ from $i \in \mathcal{I}$,
  systemNN computes $(\text{W}_{i,\alpha_i,t+1})_{i\in\mathcal{I}}$
  systemNN sends privately $\text{W}_{i,\alpha_i,t+1}$ to $i \in \mathcal{I}$
 Return : $(x_{i,t}, y_{i,K_i,t}, \text{W}_{i,\alpha_i,t}, \hat{W}_{i,\alpha_i,t})_{t\in\{1,\ldots,T\}}$
(8.151)

8.5.4.3 Extremile Tensor-Graph Neural Networks

The extremile of order α is

$$E_\alpha(F) \in \arg\min_{l\in\mathbb{R}} E\{k'_\alpha(F(L))\left|L - l\right|^2\}, \tag{8.152}$$

where

$$k_\alpha(x) = (1 - (1 - x)^{s_\alpha})\mathbb{I}_{\{0\le\alpha\le\frac{1}{2}\}} + x^{r_\alpha}\mathbb{I}_{\{\frac{1}{2}<\alpha<1\}}$$

and

$$r_\alpha = \frac{\log(\frac{1}{2})}{\log(\alpha)} = s(1-\alpha).$$

Denote by $y_l = \inf\{y \mid F(y) > 0\}$, $y_u = \sup\{y \mid F(y) < 1\}$, the lower and upper endpoints of the support of F. The extremile is explicitly given by

$$E_\alpha(F) = \begin{cases} \{-\int_{y_l}^{0}\{1-(1-F(y))^{s_\alpha}\}dy + \int_0^{y_u}(1-F(y))^{s_\alpha}\, dy\}\mathbb{I}_{\{0\leq\alpha\leq\frac{1}{2}\}} \\ +\{-\int_{y_l}^{0}(F(y))^{r_\alpha}dy + \int_0^{y_u}(1-F(y))^{r_\alpha}\, dy\}\mathbb{I}_{\{\frac{1}{2}<\alpha\leq 1\}}. \end{cases} \tag{8.153}$$

If the support of F is the entire set of real numbers, then

$$E_\alpha(F) = \int_{-\infty}^{0}(g_\alpha(1-F(y)) - 1)dy + \int_0^{+\infty} g_\alpha(1-F(y))dy,$$

where

$$g_\alpha(x) = x^{s_\alpha}\mathbb{I}_{\{0\leq\alpha\leq\frac{1}{2}\}} + (1-(1-x)^{r_\alpha})\mathbb{I}_{\{\frac{1}{2}<\alpha\leq 1\}}.$$

Let L have a finite absolute first moment, and let L^i be independent observations drawn from the distribution of F of L: Then

$$E_\alpha(F) = \begin{cases} E[\max(L^1,\ldots,L^r)]\mathbb{I}_{\{\alpha=(\frac{1}{2})^{\frac{1}{r}},\ r\geq 1\}}, \\ E[\min(L^1,\ldots,L^s)]\mathbb{I}_{\{\alpha=1-(\frac{1}{2})^{\frac{1}{s}},\ s\geq 1\}}. \end{cases} \tag{8.154}$$

8.5.4.4 Quantile Tensor-Graph Neural Networks

The quantile of order α is

$$E_\alpha(F) \in \arg\min_{l\in\mathbb{R}} E\{k'_\alpha(F(L))\Big|L-l\Big|\}. \tag{8.155}$$

Given an input–output data set $\{(x_i, d_{L,i}) \mid 1 \leq i \leq D\}$, with size $D \geq 1$, the cost functional

$$\frac{1}{D}\sum_{i=1}^{D} U_\alpha[d_{L,i} - \sigma_L \circ \sigma_{L-1} \circ \ldots \circ \sigma_1 x_i]$$

provides quantile-based training weight and bias.

8.6 Diffusion Transformer in Tensor-Graph Networks

This section introduces some basic notions of diffusion models together with transformers. This combination is called transfusion. Diffusion transformers are a generative modeling approach that combines diffusion models with transformer architectures, leveraging the strengths of both for handling structured data like images or text. To describe this process step by step, we begin with the forward diffusion process, which gradually corrupts the data by adding noise over time. This forward process is governed by a stochastic differential equation. Let $x(0)$ denote the initial, clean data at time $t = 0$, and over time, noise B is progressively added, transforming the data into a noisy version by the time $t = T$. The forward SDE is expressed as

$$dx = b(x, t)\, dt + \sqrt{2\varepsilon}\, dB, \quad x(0) \sim \mu(0, .),$$

where x is the data, $b(x, t)$ is the drift term that represents the deterministic part of the process, $\sqrt{2\varepsilon}$ is the diffusion coefficient that controls the noise level, and B is the standard Brownian motion. As time t increases from 0 to T, the data evolves from its clean state $x(0)$ to a highly noisy state $x(T)$, which resembles random Gaussian noise.

The goal of the generative model is to reverse this process, recovering the original data from noise. This is achieved by solving the reverse-time SDE, which effectively denoises the noisy data and retrieves the clean data distribution. The reverse-time SDE is formulated as

$$dx = \left[b(x, t) - 2\varepsilon \nabla_x \log \mu(t, x)\right] dt + \sqrt{2\varepsilon} dB, \quad x(T) \sim \mu(T, .),$$

where $\nabla_x \log \mu(t, x)$ is the gradient of the log-probability of the data at time t. This score function guides the process by indicating how to reverse the noise and restore the data. The starting point of the reverse process is the noisy data at the final time T. This noisy data, denoted by $x(T)$, is typically a sample drawn from a standard Gaussian distribution. The reverse process begins from this noisy state $x(T)$ and aims to recover the clean data $x(0)$ by progressively denoising it over time, moving backward from $t = T$ to $t = 0$. Since the true score function $\nabla_x \log \mu(t, x)$ is unknown, we approximate it by training a neural network. In diffusion transformers, the transformer architecture is used in place of the traditional convolutional network. Transformers use self-attention mechanisms. During the reverse diffusion process, the transformer predicts either the score function $\nabla_x \log \mu(t, x)$ or the denoised data $x(t)$ at each reverse step.

8.6.1 Explicit Solutions to Diffusion Systems

Given a measurable subset S at time t, we would like to learn the mean-field term

$$m(t, S) = \mathbb{P}_{x(t)}(S) = \mathbb{P}(x(t) \in S)$$

and compute specific quantities of interests. We start with the stochastic optimal control problem with $\varepsilon > 0$:

$$(P1) = \begin{cases} \inf_u E[h(T, x_T) + \int_0^T \frac{1}{2}\Big|u(t)\Big|^2 dt], \\ dx = udt + \sqrt{2\varepsilon} dB, \\ x(0) = x_0 \sim m_0 \end{cases}$$

$$(P2) = \begin{cases} v_t + \inf_u \langle u, v_x \rangle + \frac{1}{2}\Big|u\Big|^2 + \varepsilon \text{ trace}(v_{xx}) = 0, \\ v(T, x) = h(T, x), \\ m_t + div(um) - \varepsilon \text{ trace}(m_{xx}) = 0, \\ m(0) = m_0 \end{cases} \tag{8.156}$$

$$(P3) = \begin{cases} v_t - \frac{1}{2}\Big|v_x\Big|^2 + \varepsilon \text{ trace}(v_{xx}) = 0, \\ v(T, x) = h(T, x), \\ m_t - div_x(v_x m) - \varepsilon\text{trace}(m_{xx}) = 0, \\ m(0, dx) = m_0(dx) \end{cases}.$$

We use the Hopf–Cole transformation.

Lemma 8.84 *Let*

$$(HC) = \begin{cases} v_t + b\Big|v_x\Big|^2 + a \;\; trace(v_{xx}) = 0, \\ v(T, x) = h(T, x). \end{cases} \tag{8.157}$$

Then, $v = \frac{a}{b} \log(w)$ *with*

$$(HC2) = \begin{cases} w_t + a \;\; trace(w_{xx}) = 0, \\ w(T, x) = e^{\frac{b}{a} h(T,x)}. \end{cases} \tag{8.158}$$

Proof Use the transformation $w(t, x) = \phi(v(t, x))$. Then, $w_t = \phi'(v)(-b\Big|v_x\Big|^2 - a \text{ trace}(v_{xx}))$, $w_x = \phi'(v)v_x$, and $w_{xx} = \phi''(v)v_x(v_x)' + \phi'(v)v_{xx}$.

This implies that $w_t = -\phi'(v)b\Big|v_x\Big|^2 - a \text{ trace}(w_{xx} - \phi''(v)v_x(v_x)')$, which reduces to

$w_t + a\ \text{trace}(w_{xx}) = 0$ if we choose $-\phi'(v)b + a\phi''(v) = 0$. Hence, the transformation map $\phi(z) = e^{\frac{b}{a}z}$ will reduce the original system into the following system:

$$\begin{cases} w_t + a\ \text{trace}(w_{xx}) = 0, \\ w(T, x) = e^{\frac{b}{a}h(T,x)}, \end{cases} \tag{8.159}$$

which is explicitly solved by Laplace transform, Fourier transform, or separation of variables $a(x)b(t)$. ▮ □

Using the above transformation with $a = \varepsilon, b = -\frac{1}{2}$, we obtain $v(t, x) = -2\varepsilon \log(w(t, x))$. Thus, $v_x(t, x) = -2\varepsilon \frac{w_x(t,x)}{w(t,x)}$. The distribution of x at time t solves

$$(P4) = \begin{cases} v(t, x) = -2\varepsilon \log(w(t, x)), \\ w_t + \varepsilon\ \text{trace}(w_{xx}) = 0, \\ w(T, x) = e^{-\frac{1}{2\varepsilon}h(T,x)}, \\ m_t + 2\varepsilon\ \text{div}_x(\frac{w_x}{w}m) - \varepsilon\ \text{trace}(m_{xx}) = 0, \\ m(0, dx) = m_0(dx). \end{cases} \tag{8.160}$$

Note that $\text{trace}(m_{xx}) = \text{div}_x(\frac{m_x}{m}.m) = \text{div}_x((\log m)_x.m)$. This means that the distribution solves

$$\begin{cases} m_t - div_x((v_x + \varepsilon(\log m)_x)m) = 0, \\ m(0, dx) = m_0(dx). \end{cases} \tag{8.161}$$

As T gets larger, and $\frac{1}{T}h(T, x) \to 0$, we obtain

$$(P5) = \begin{cases} \lambda + \left|U\right|^2 = 2\varepsilon\ \text{trace}(U_x), \\ m_x = -Um \\ \int m(dx) = 1 \end{cases} \tag{8.162}$$

In one dimension, this provides $m(x) = \frac{e^{-U(x)}}{\int e^{-U(y)}dy}$, i.e., the Boltzmann–Gibbs distribution or the softmin of U.

Note that in order to obtain m, we do not need v but v_x. The transformer architecture can be used to estimate a θ that approximate $(\log m)_x$.

8.6.2 *Diffusion Transformer for Risk Quantification*

We start with an open issue with a very local (point-based) density-based optimization.

8.6.2.1 Density-Based Optimization

In the context of infinite population games, we have used nonatomic terms in the quantity of interest in the point mean-field $m(t, x(t))$, i.e., the density at the point $x(t)$. However, here, there is no infinite number of decision-makers in the tensor-graph neural network. We aim to compute $m(t, x(t))$ to examine the output data density $m(t, y)$ at y at time t. There does not exist a stochastic maximum principle for functionals such as $g(t, x(t), m(t, x(t)), u(t, x(t), m(t, x(t))))$, where $x(t)$ is the input data at time t, u is the decision variable at time t, and g is the performance metric (could be a score function or error function). When the function m is twice continuously differentiable, one can apply Itô's formula to $m(t, x(t))$.

In the context of mean-field-type games, one can consider $\frac{m(t,B_\varepsilon(x))}{vol(B_\varepsilon)}$, which is a localized around the entire ball as an approximation of the density at x. One can also consider a generic smooth regularization kernel, which does a convolution with the probability density function. The standard method applies to the regularized forms. The question then is to know the limit as ε goes to zero. As we know, approximating first and then optimizing may differ from optimizing first and then approximating.

8.6.2.2 Time-Reversed Diffusion Process

We consider the input signal diffusion process $(x(t),\ 0 \leq t \leq T)$ with $x(0) \sim m_0$.

$$\begin{cases} dx = bdt + \sqrt{2\varepsilon} dB, \\ x(0) = x_0 \sim m_0, \\ m(t, dx) = \mathscr{L}(x(t)), \\ \\ m_t + div_x\left((b - \varepsilon(\log m)_x)m\right) = 0, \\ m(0, dx) = m_0(dx) \end{cases} . \tag{8.163}$$

Now consider the time-reversed process $(y(t) = x(T - t),\ 0 \leq t \leq T)$, which start with $y(0) = x(T) \sim m(T - 0, dx)$. The probability law of y is $\tilde{m}(t, dy) = \mathscr{L}(y(t)) = \mathscr{L}(x(T - t)) = m(T - t, dy)$. We write the dynamics of $\tilde{m}$ from the Fokker–Planck–Kolmogorov equation in m. The time derivative of $\tilde{m}$ is

$$\begin{aligned} \tilde{m}_t(t, y) = -m_t(T - t, y) &= div_y\left((b(T-t, y) - \varepsilon(\log m(T - t, y))_y)m(T - t, y)\right) \\ &= div_y\left((b(T - t, y) - \varepsilon(\log \tilde{m})_y)\tilde{m}\right). \end{aligned}$$

Putting all together, we arrive at

$$\begin{cases} dx = bdt + \sqrt{2\varepsilon}dB, \\ x(0) = x_0 \sim m_0, \\ m(t, dx) = \mathscr{L}(x(t)), \\ \\ m_t + div_x\left((b - \varepsilon(\log m)_x)m\right) = 0, \\ m(0, dx) = m_0(dx), \\ \\ \tilde{m}_t(t, y) + div_y\left((-b(T-t, y) + \varepsilon(\log \tilde{m}(t, y))_y)\tilde{m}(t, y)\right) = 0, \\ \tilde{m}(0, dy) = \nu_1(dy) = m(T-0, dy). \end{cases} \tag{8.164}$$

Knowing that $-\sigma dB$ and σdB have the same distribution, the process $(y(t),\ 0 \leq t \leq T)$ is diffusion process with drift $\tilde{b}(t, y) = -b(T-t, y) + \varepsilon(\log \tilde{m}(t, y))_y$. This shows a first main result:

$$\tilde{b}(T-t, y) + b(t, y) = \varepsilon(\log m(t, y))_y. \tag{8.165}$$

The associated reversed ODE is

$$\begin{cases} dz = (b(t, z) - \varepsilon(\log m(t, z))_z)dt, \\ z(T) \sim \nu_1. \end{cases} \tag{8.166}$$

The diffusion coefficient, however, needs to be carefully designed to match with shift in the drift: $b(t, y) - \varepsilon(\log m(t, y))_y = (b(t, y) - 2\varepsilon(\log m(t, y))_y) + \varepsilon(\log m(t, y))_y$. Hence, the time-reversed process is given by

$$\begin{cases} dy = (b(t, y) - 2\varepsilon(\log m(t, y))_y)dt + \sqrt{2\varepsilon}dB, \\ y(T) \sim \nu_1. \end{cases} \tag{8.167}$$

The idea is to use this latter equation to sample m_0. To do so, one needs to compute $(\log m(t, y))_y$. Thus, one needs to design a deep neural network to estimate it or to solve the Fokker–Planck–Kolmogorov in an efficient way.

8.6.3 *Mean-Field-Type Diffusion Transformer*

When the drift is of mean-field type as in MFTT

$$\begin{aligned} & b(y, \mu, \nu_1, \ldots, \nu_{L-1}) \\ & = (\hat{\mathscr{F}}_L \circ \hat{\mathscr{F}}_{L-\frac{1}{2}\ |\nu_{L-1}(t,.)} \circ \ldots \circ \hat{\mathscr{F}}_2 \circ \hat{\mathscr{F}}_{\frac{3}{2}, \nu_1(t,.)} \circ \hat{\mathscr{F}}_1 \circ \hat{\mathscr{F}}_{\frac{1}{2}, \mu(t,.)}) \\ & (y) - y, \end{aligned}$$

the ODE with initial stochastic (marginal) data μ_0 is given by

$$\begin{aligned} dy &= b(y, \mu(t,.), \nu_1(t,.), \ldots, \nu_{L-1}(t,.))dt, \\ y_0 &\sim \mu_0(0,.) = \mu_0(.) \end{aligned} \tag{8.168}$$

The backward ODE with terminal stochastic (marginal) data ν_1 is given by

$$\begin{aligned} dz &= (b(z, \mu(t,.), \nu_1(t,.), \ldots, \nu_{L-1}(t,.)) - \varepsilon(\log \mu(t,z))_z)dt, \\ z(T) &\sim \mu(T,.), \end{aligned} \tag{8.169}$$

where

$$\begin{aligned} & l \in \{1, \ldots, L-1\}: \\ & \nu_{l-\frac{1}{2}}(t,.) = \hat{\mathscr{F}}_{l-\frac{1}{2},\nu_{l-1}(t,.)} \# \nu_{l-1}(t,.), \\ & \nu_l(t,.) = \hat{\mathscr{F}}_l \# \nu_{l-\frac{1}{2}}(t,.), \\ & \nu_0(t,.) := \mu(t,.), \end{aligned} \tag{8.170}$$

The mean-field-type diffusion transformer is therefore given by

$$\begin{aligned} dz &= (b(z, \mu(t,.), \nu_1(t,.), \ldots, \nu_{L-1}(t,.) - 2\varepsilon(\log \mu(t,z))_z)dt + \sqrt{2\varepsilon} dB, \\ z(T) &\sim \mu(T,.), \end{aligned} \tag{8.171}$$

In practice, the forward process is used to add noise at each step, and the backward process is used as a denoising process. Using this process, one can get a top-quality image at the end, starting from white noise. However, one needs to be careful with model collapse if the denoising process does not match as indicated in (8.171).

8.7 Notes

In this chapter, we have presented a network of neural networks with a diverse array of neural network architectures tailored to specific applications and challenges. These architectures, ranging from tensor-based networks to graph-based models and their generative adversarial variants, showcase the versatility and adaptability of deep learning in addressing complex problems. The chapter also emphasized the importance of strategic design and analysis in constructing effective NoNN systems, including architecture selection, decision-making mechanisms, and shared architectures. Deep learning [76, 109, 112, 113, 147, 159, 258, 299, 327, 328, 397, 397, 398] has

(continued)

been used to address several problems in computer science, and transformer-based architectures have been widely examined as candidate for backcasting, nowcasting, forecasting, and risk quantification [1, 22, 51, 67, 91, 92, 95, 119, 141, 149, 158, 160, 162, 172, 173, 185, 187, 188, 193, 195, 200, 246, 248–251, 269, 280, 289, 296, 302, 304, 309, 311–313, 332, 341, 345, 347, 350, 356, 369, 389, 399, 409, 412, 492, 504, 510, 511, 513, 522, 526, 529, 530, 538–540, 544, 549]. The connection between deep learning and game theory has been examined in several research papers. The work [462] surveys the interplay between the two fields. On the one hand, game theory has developed different solution concepts. Many of them involve fixed-point arguments and variational inequalities. In a game-theoretic setting, the objective functions do not have to be convex (in the global variable), and there are several strategic learning algorithms for solving game-theoretic problems beyond the class of potential convex games. On the other hand, deep neural network input–output outcomes [55, 360, 495] are fixed points of a composition of activation, attention, and weight operators. The composition of multiple operators is not convex in the weight parameters, which need to be designed during the training phase of the neural network. As a consequence, most of the error bounds, stability, and accuracy analysis provided for *convex loss functions* are not exploitable for the training problem. The convexity in the input signal variable does not necessarily mean or imply convexity in the weight parameters, which also differ from the convexity of the functionals involved in another problem at hand. The works in [110, 111] have established properties of activation functions used in neural networks and used averaged operators to prove convergence in the weak sense. The work in [139] examined large learning model (LLM) properties based on these references to establish connections with the Stackelberg solution and Nash equilibria. Acceleration techniques have been provided in [54, 179, 180, 182, 184, 262, 264]. For more details on GPT and game theory, we refer the reader to the book [466].

Part III
Evolutionary Biology and BioChemistry

Chapter 9
Biochemical Reaction Networks

A reaction network, in the context of chemistry and chemical engineering, is a visual or mathematical representation of the interconnected chemical reactions that occur within a system. It is a way to depict how different chemical species (reactants and products) interact and transform into one another over time. Reaction networks are essential tools for understanding complex chemical processes, such as those in biological systems, industrial chemical reactions, and environmental reactions. Biochemical reaction networks are representations of interconnected chemical reactions that occur within biological systems, encompassing the complex web of molecular interactions in living organisms. Each node or component in the network represents a specific chemical species, which can be an element, a compound, or a molecule involved in the reactions. The edges or links between nodes represent chemical reactions. These reactions describe the conversion of one or more reactants into one or more products, typically following a stoichiometric relationship. Reaction networks often include information about the stoichiometry of each reaction, specifying the molar ratios of reactants and products involved. Some reaction networks indicate the directionality of reactions, distinguishing between forward and reverse reactions. Reaction networks may also include information about the kinetics of reactions [460, 464], including reaction rates and rate constants.

This chapter discusses the application of mean-field game dynamics in chemical reaction networks, transitioning from individual chemical reactions to mean-field modeling, exploring the concept of limit cycles, and introducing biochemical reaction mechanisms. We establish a relationship between biochemical dynamics and mean-field game dynamics. We show that chemical reaction networks can be studied using stochastic mean-field-type game dynamics [443, 460, 463, 499]. Similar methods have been used in Chaps. 2 and 5.

The chapter is organized as follows. Section 9.1 presents the appearance of mean field in chemical reaction networks. Section 9.2 presents biochemical reaction mechanisms. The chapter ends with a concluding remarks with Sect. 9.3.

T. Başar et al., *Mean-Field-Type Game Theory II*, Static & Dynamic Game Theory: Foundations & Applications, https://doi.org/10.1007/978-3-032-07031-9_9

9.1 Mean Field in Chemical Reaction Networks

The connection between Markov chain and mean-field Kolmogorov equation or population dynamics is already established in Chap. 5 of this volume. Stochastic dynamical behavior of a well-stirred mixture of S molecular species that chemically interact through K reaction channels is accurately described by the chemical master equation, which is the stochastic Kolmogorov equation. This stochastic equation presents a contradistinction to earlier works that proposed deterministic macroscopic evolution laws. The derivation clarified the connection between the stochastically correct chemical master equation and the deterministic but often satisfactory reaction rate equation. We work here with constant reaction rates k or reaction rates that are scaled with parameter V, which can be interpreted as volume [430, 443, 464].

Why a Chemical Reaction Network Is a Mean-Field Interaction?

What is a chemical reaction? A chemical reaction is defined as the process, wherein a set of chemical substances interact with each other, which leads to their conversion into other types of forms. The initial substances used in the reaction are collectively called reactants, while the final substances formed after the reaction are known as products. Thus, during a chemical reaction, there is a conversion of the reactants into a single or multiple products. A reactant means a substance or substances that are involved in a chemical reaction. The chemical reactions occur under the appropriate conditions of pressure and temperature in the presence of a catalyst. The catalyst plays a significant role in increasing the rate of a chemical reaction. For example, for the reaction $2A + B \xrightarrow{k_1(.)} \underline{3A}$, the reactants are A and B, $k_1(.)$ the catalyst, and $\underline{3A}$ is the product. A reaction in the system plays the role of an interaction (an encounter) in evolutionary games or population games. In this analogy, there are two types (subpopulations or phenotypes), A and B. This reaction corresponds to an interaction between two players of subpopulation A with one player of subpopulation B.

In general, a chemical reaction network is determined by:

- a finite set of network species, which can correspond to types and/or subpopulations in the context of mean-field interaction
- a finite set of products which can correspond to the possible outcomes
- A finite set of reactions.

We follow the same methodology as in Chap. 5. We examine the large-scale case: all species come in large numbers and their concentrations are determined by the fractions. We will study the dynamics of species, i.e., the dynamics of species concentrations, by using mean-field dynamics.

Table 9.1 Link between mean-field interaction and chemical reaction

Chemical reaction	MF interaction
Molecules	Phenotypes
Species: A, B	Types: A, B
$2A + B \xrightarrow{k_1(.)} \underline{3A}$	3 players: 2 of type A and 1 with type B
Product: 3A	Outcome of the interaction
Composition of the system	Mean field

The following table summarizes the connection between the chemical reaction and some terminology in evolutionary games and/or mean-field interaction (Table 9.1).

In a synthesis reaction, two or more simple substances (reactants) combine together to yield a more complex substance. For example, hydrogen gas together with oxygen forms a more complex product, water. The chemical equation of this reaction is written as follows: $2H_2 + O_2 \xrightarrow{catalyst} 2H_2O$. In the context of evolutionary games, this involves six players.

9.1.1 From Chemical Reaction to Deterministic Mean Field

Consider I_0 players. Each player can have two states $s = 1$ or 2 and an exogenous state $s = 0$, which represents the inactive state. We denote a generic player in state 1 by A and a generic player in state 2 by B. An encounter between the players is represented by a reaction. Let $(m_a, m_b) = (A/I_0, B/I_0)$. Then,

$$\begin{array}{c} 2A + B \longrightarrow 3A \\ A \leftrightarrows \varnothing \\ B \longleftarrow \varnothing \end{array} .$$

The chemical equation $2A + B \longrightarrow 3A$ represents the reaction of A and B to form A. The arrow points in the direction of the reaction from the reactants (substances that react) toward the product or products. In this case, the reactants are A (written as A because each molecule consists of one atom of A) and B and the product is 3A. The coefficient 2 in the A indicates that two molecules of A take part in the reaction, and the 3 in the 3A indicates that three molecules of A are produced. When no number appears, as in the case of A, the number 1 is assumed; one molecule of A takes part in the reaction. The equation shows that two molecules of A react with one molecule of B to form three molecules of A. Table 9.2 described the changes, probability of occurrences of the encounter, effects, actions, and payoffs. Table 9.2 defines also the mean-field Kolmogorov equation detailed in Chap. 5.

Figure 9.1 illustrates the mean field for A and B very large.

Table 9.2 Probabilities, effects, actions, and payoffs. The interpretation is similar to that in Chap. 5

	Transition probability	$M_s^I(t+1) - M_s^I(t)$	Actions	Payoff
1	$k_1 a(a - \frac{1}{I})b$	$(1, -1)/I$	k_1	$c(k_1, a, b)$
2	k_2	$(1, 0)/I$	k_2	$c(k_2, a, b)$
3	$k_3 a$	$(-1, 0)/I$	k_3	$c(k_3, a, b)$
4	k_4	$(0, 1)/I$	k_4	$c(k_4, a, b)$

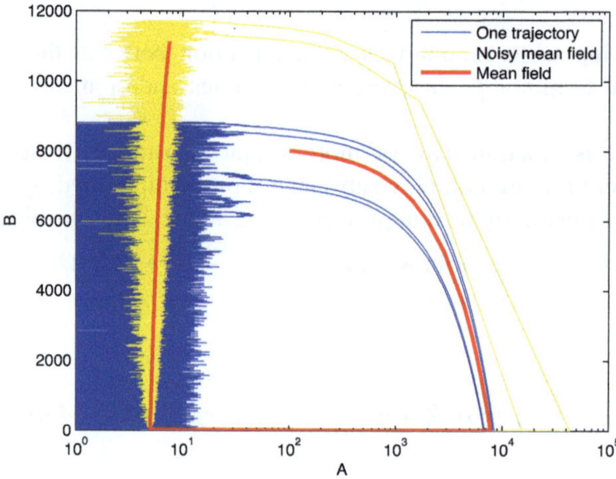

Fig. 9.1 Simulation, noisy mean field, and deterministic mean field

The deterministic mean-field dynamics is obtained by multiplying the effect with the transition probabilities, which leads to the drift. The ordinary differential equations are

$$\frac{d}{dt}a = k_1 a^2 b + k_2 - k_3 a,$$
$$\frac{d}{dt}b = -k_1 a^2 b + k_4. \tag{9.1}$$

As we can see, $\frac{d}{dt}(a+b)$ may not be zero. This says that the proportion may vary in time due to the formation and creation of new molecules of type A or type B. The stationary points of the above system are given by

$$k_1 a^2 b + k_2 - k_3 a = 0, \ k_1 a^2 b = k_4. \tag{9.2}$$

This reduces to

$$k_4 + k_2 = k_3 a, \ k_1 a^2 b = k_4. \tag{9.3}$$

Hence, if $k_i > 0$, then $a_* = \frac{k_4+k_2}{k_3}$, $b_* = \frac{k_4}{k_1}\frac{k_3^2}{(k_4+k_2)^2}$. Next we study the stability of the ODE. The Jacobian of the system is given by

$$\begin{pmatrix} 2ak_1b - k_3 & k_1a^2 \\ -2ak_1b & -k_1a^2 \end{pmatrix}.$$

A sufficiency condition for stability is $2a_*k_1b_* - k_3 - k_1a_*^2 < 0$ and $-(2a_*k_1b_* - k_3)k_1a_*^2 + k_1a^2(2ak_1b) = k_3k_1a_*^2 > 0$. Combining together the conditions reduces to

$$2k_1k_4\frac{k_3}{(k_4+k_2)} < k_3 + k_1\left(\frac{k_4+k_2}{k_3}\right)^2.$$

9.1.2 *Limit Cycle in Chemical Reaction Networks*

We now consider interactions of the following type:

$$\begin{aligned} \varnothing &\longrightarrow A \\ A &\leftrightarrows \varnothing \\ A &\longrightarrow B \\ 2A + B &\longrightarrow 3A. \end{aligned}$$

Table 9.3 provides the corresponding effects, transitions, actions, and payoffs. The ordinary differential equations are

$$\frac{d}{dt}a = k_1 - k_2a - k_3a + k_4a^2b \tag{9.4}$$

$$\frac{d}{dt}b = k_3a - k_4a^2b \tag{9.5}$$

We compute the rest points of the system:

$$\{(a^*, b^*) \big| a^*(k_3 - k_4a^*b^*) = 0, k_1 - (k_2 + k_3)a + k_4(a^*)^2b^* = 0\}.$$

For $\vec{k} = (1, 1, 3, 1)$, we obtain the unique positive rest point

Table 9.3 Transition rates, effects, actions, and payoffs of limit cycle

	Transition rate	$M_s^I(t+1) - M_s^I(t)$	Actions	Payoff
1	k_1	$(1, 0)/I$	k_1	$c(k_1, a, b)$
2	k_2a	$(-1, 0)/I$	k_2	$c(k_2, a, b)$
3	k_3a	$(-1, 1)/I$	k_3	$c(k_3, a, b)$
4	$k_4a(a - 1/I)b$	$(1, -1)/I$	k_4	$c(k_4, a, b)$

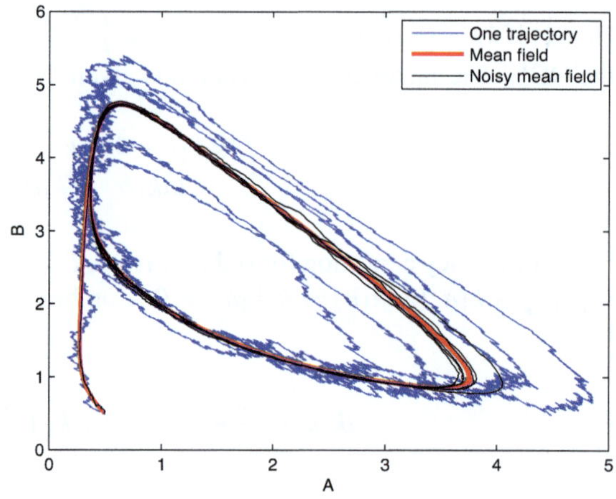

Fig. 9.2 Limit cycle in chemical reaction. One trajectory is obtained from one run of simulation. The mean-field trajectory is from the ODE, and the noisy mean field is when the discrepancy between the ODE and the expected changes is seen as a noise. The steady state (1, 3) is inside the limit cycle and is unstable

$$a^* = 1, \quad b^* = 3$$

Now we examine the stability of the stationary point (1,3). The Jacobian

$$J(a^*, b^*) = \begin{pmatrix} -(k_2 + k_3) + 2k_4ab & k_4a^2 \\ k_3 - 2k_4ab & -k_4a^2 \end{pmatrix}$$

at $(1, 3)$ is

$$J(1, 3) = \begin{pmatrix} 2 & 1 \\ -3 & -1 \end{pmatrix}.$$

The Jacobian has positive eigenvalues, and hence the system is unstable. In Fig. 9.2, we depict the concentrations of A and B. As we can see, there is a limit cycle in the trajectory between A and B.

9.2 Biochemical Reaction Mechanism

We model the processes of transcription and translation of gene regulation as first-order reaction networks involving three species:

R denotes the amount of messenger ribonucleic acid (mRNA). P denotes the amount of protein. D denotes the amount of gene.

A gene can create a molecule of mRNA with rate constant k_1, without destroying itself:

$$G \xrightarrow{\text{with rate } k_1} G + R.$$

A molecule of mRNA can create a protein with rate constant k_2, without destroying itself:

$$R \xrightarrow{\text{with rate } k_2} P + R.$$

A molecule of mRNA and protein can degrade at rate constants k_3 and k_4, respectively:

$$R \xrightarrow{\text{with rate } k_3} \varnothing,$$

$$P \xrightarrow{\text{with rate } k_4} \varnothing.$$

Combing together, one gets

$$G \xrightarrow{\text{with rate } k_1} G + R \tag{9.6}$$

$$R \xrightarrow{\text{with rate } k_2} P + R \tag{9.7}$$

$$R \xrightarrow{\text{with rate } k_3} \varnothing \tag{9.8}$$

$$P \xrightarrow{\text{with rate } k_4} \varnothing \tag{9.9}$$

More precisely, the first two reactions involve catalytic production from a source, and the last two reactions are of degradation type.

Letting g_t, r_t and p_t denote the stochastic processes that specify the levels of gene, mRNA, and protein, respectively, and using our mean-field framework, one gets

$$\begin{aligned} l\mathbb{E}g_t &= \mathbb{E}g_0 = constant \\ \frac{d}{dt}\mathbb{E}r_t &= k_1\mathbb{E}g_t - k_3\mathbb{E}r_t \\ \frac{d}{dt}\mathbb{E}p_t &= k_2\mathbb{E}r_t - k_4\mathbb{E}p_t \ . \end{aligned} \tag{9.10}$$

Figure 9.3 shows one run of simulation, noisy mean field, and deterministic mean field.

Set of Reactions

We consider chemical species $A_1, A_2, \ldots, A_S$ that take part in $R_1, R_2, \ldots, R_K$ different types of chemical reactions. Each reaction R_j is described by

$$R_j : \sum_{s=1}^{S} \alpha_{js} A_s \longrightarrow^{k_j} \sum_{s=1}^{S} \beta_{js} A_s,$$

Fig. 9.3 mRNA, protein, and gene

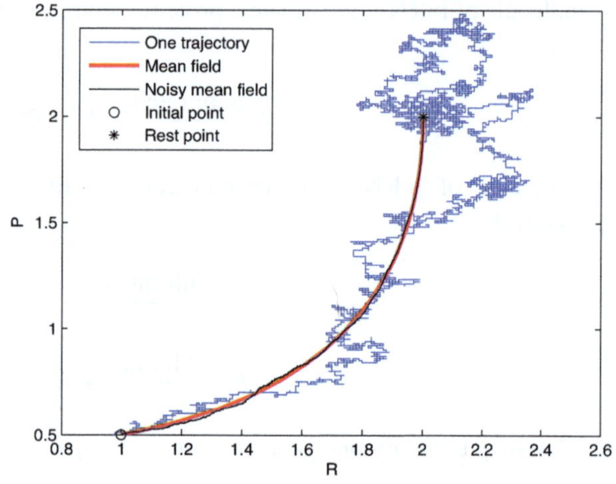

Table 9.4 Transition rates, effects, and payoffs

	Transition rate	Effect	Payoff
1	$k_1 \prod_{s=1}^{S} m^{\alpha_{1,s}}$	$(\beta_{1,s} - \alpha_{1,s})/I, s \in \{1, 2\ldots\}$	$c(k_1, m)$
⋮	⋮	⋮	⋮
K	$k_K \prod_{s=1}^{S} m^{\alpha_{K,s}}$	$(\beta_{K,s} - \alpha_{K,s})/I, s \in \{1, 2\ldots\}$	$c(k_K, m)$

where α_{js} and β_{js} are the numbers of copies of species A_s that participate in the reaction type R_j, and k_j is the reaction rate of j. We are interested in how the concentration of A_s changes.

Effect in R_j: Given $A_s : \alpha_{js} \longrightarrow \beta_{js}$, we define $E_{js} := \beta_{js} - \alpha_{js}$. If $E_{js} < 0$ (resp. > 0), the number of molecules of species A_s decreases (resp. increases) in the reaction of type R_j.

Table 9.4 captures the transitions, effects, and payoffs.

9.3 Notes

Biochemical reaction networks involve several reactions with interacting molecules and species. In this chapter, we have examined biochemical reaction networks and established their mean-field-type game dynamics. By examining each reaction and translating the changes, we have arrive at a description that captures the evolution of concentrations, related to mean-field-type game dynamics. More details can be found in [116, 260, 478, 496, 523].

Chapter 10
COVID-19: Data-Driven Modeling

Data-driven mean-field-type game theory aims to incorporate certain data sets and real measurements into MFTG settings in a closed-loop fashion. Researchers use the measured data to learn and extract additional useful information from the field which the model then takes into consideration. One can thus utilize the model to study emerging dynamics and features and conduct new measurements, intervention measures, and actions. Scientists have developed mean-field-type filters, mean-field-type forecasting, and risk-aware filtering and forecasting based on MFTG theory; they apply these techniques to intelligent transportation systems. In the context of the COronaVIrus Disease 2019, COVID-19 pandemic, data-driven MFTG theory can model interactions wherein the decision-makers are individuals, authorities, institutions, or firms. Each individual has a multilayer interaction. The epidemiological layer consists of the individual's internal state, which is composed of the following factors: age, gender, preexisting health conditions, location, mask and sanitation adherence, and health status. The latter includes many contributing elements: susceptible; exposed; tested (positive or negative) or untested (active or recovered); retested; number of contacts; contact-tracing; hospitalized; recovered; vaccinated with the first, second, or booster dose; up-to-date vaccine; and tested deceased or untested deceased. The individual's economic layer consists of consumption, family revenue (both ex ante and current), employment, and work status (working from home, working part time, working partially from home, and workload), including individual economic supports and benefits from the incentive policies provided. The mobility layer involves the mobility pattern (going to work, marketplaces, grocery stores, hotspots, drug stores, etc.), travel status (a recent trip), transportation usage (utilizing public transport, a personal vehicle, carsharing, cabs, and Uber-like services), and in-city, inter-city, and international movement.

Each authority has multiple layers of interactions that include mitigation policies; testing, hospital, and treatment capacities; and vaccination capabilities. The firms also have various layers that interact with the other categories. One can evaluate the MFTG filter by its global error: measurement/reporting error, estimation error,

T. Başar et al., *Mean-Field-Type Game Theory II*, Static & Dynamic Game Theory: Foundations & Applications, https://doi.org/10.1007/978-3-032-07031-9_10

modeling error, optimization error, computational error, and decision-making error. The development of data-driven MFTG theory is timely due to the rise of high-dimensional systems and data science, both of which encompass a broad range of techniques that range from strategic deep learning and filtering to compressed sensing and useful information extraction from data. The methodology is flexible enough to capture multi-class interactions in epidemic propagation wherein multiple authorities and firms are risk-aware atomic decision-makers and individuals are risk-aware, non-atomic decision-makers. Research in this area has also boosted other scientific endeavors. We continue to advance human-centered machine co-intelligence-driven methods to accelerate knowledge extraction for both COVID-19 and a diverse set of diseases that includes Ebola, malaria, and cancer. These approaches also have applications in other related areas like food security, agriculture, and demand–supply matching in underserved locations.

In this chapter, a class of mean-field-type games with discrete–continuous state spaces is presented. We establish Bellman systems that provide sufficiency conditions for mean-field-type equilibria in state-and-mean-field-type feedback form [460, 463]. We derive unnormalized master adjoint systems (MASS). The methodology is shown to be flexible enough to capture multi-class interaction in epidemic propagation in which multiple authorities are risk-aware atomic decision-makers and individuals are risk-aware non-atomic decision-makers. Based on MASS, we present a data-driven modeling and analytics for mitigating COronaVIrus Disease 2019. The model integrates untested cases, age structure, decision-making, gender, preexisting health conditions, location, testing capacity, hospital capacity, and mobility map on local areas at the district level, in-city, inter-cities, and international. It is shown that the data-driven model can capture most of the reported data on COVID-19 on confirmed cases, deaths, recovered, number of testing, and number of active cases in 66+ countries. The model also explains non-Gaussianity and non-exponential properties in 15+ countries [130, 145, 146, 170, 290, 431, 447, 464, 478]

The chapter is structured as follows. Section 10.1 introduces COronaVIrus Disease 2019. In Sect. 10.2, we present some basic models in continuous time. The data-driven MFTG setup is presented in Sect. 10.3. Section 10.4 presents the COVID-19 data set used. The chapter ends with concluding notes of Sect. 10.7.

10.1 COronaVIrus Disease 2019

COronaVIrus Disease 2019 is one of the severe acute respiratory syndrome-related coronavirus. COVID-19 is a disease caused by the virus SARS-CoV-2. SARS-CoV-2 has infected people across several countries within a few months and caused deaths, economic losses, and food insecurity. As COVID-19 has spread around the globe, authorities have made decisions about local migration, confinement, isolation, quarantine, and healthcare system capabilities. Individuals, depending on the country, culture, social-economic context, age, and season, have also made decisions concerning COVID-19. In this chapter, we introduce an integrated mathematical model and analysis strategies for combatting the propagation of COVID-19 (and similar type of pandemics that may arise in the future) and for examining the economic consequences. For each infected person, one often defines the number of secondary infections that are generated by that patient, which is a random variable. The probability distribution of these random variables is referred to as offspring distribution. The offspring distribution depends on its infectiousness, which can be biological, environmental, behavioral, or combinations of these. The offspring distribution has been estimated from empirical data for outbreaks of several infectious diseases such as tuberculosis, cholera, plaque, smallpox, severe acute respiratory syndromes (SARS), MERS, and Ebola. There are two important sectors that authorities have considered to mitigate COVID-19. One is the public health sector, which has tried to understand the dynamics of COVID-19 (via epidemiological models and efforts) and take measure to flatten the curve of hospitalized COVID-19 patients (mitigation and suppression). In this case, flattening the pandemic curve basically means reducing the number of people who are hospitalized all at once so that the local healthcare system is not overburdened. The other one is the economic policy sector, which has tried to alleviate negative consequences of COVID-19 and by the restrictive policy measures. The economic policies are also used to prevent the amplification and the multiplication of the crisis. Unfortunately, these two sectors and approaches have not been coordinated close enough. The effectiveness of the public health measures may depend on the perception of people in practice and their expectations because each individual also makes its own decisions. Therefore, both decisions, from economic policies by authorities and from country-specific individual preferences, have to be integrated all together to build a reliable model. Several basic compartmental models have been suggested in the literature to simplify the complexity of mathematical modeling of infectious diseases. These include SIR (susceptible-infected-removed), SEIR that is extended SIR with exposed (E) agents, SEIR with quarantine (Q), SEIR with age structure, SEIR with spatial location, SEIRD that is R is divided into recovered (R) or deceased (D), SEIRD with data fitting via parameters, SIRD, SIRD with vaccination, and SIRD with hospitalization. These models had to be modified and expanded to include important and key features of COVID-19 such as context-awareness, decision-making, comorbidities per area, mobility, age, gender, healthcare status, and testing capabilities to capture the evolving COVID-19 data set and to improve the strategies by considering risk-awareness.

We introduce and analyze a data-driven model that integrates some of the basis of economic and pandemic issues. Table 10.1 displays some of the key differences and key features of the data-driven mean-field-type game model.

Data-Driven Mean-Field-Type Game Theory

Below we build upon the MFTG model a data-driven methodology. The data-driven mean-field-type game theory model of the pandemic COVID-19 presented here is built upon the Markov decision process model of epidemics. Despite the simplicity of this model, important features are extracted that can assist potential decisions on the strategy to combat the outbreak, essentially configuring a scenario ranging

Table 10.1 Key features of the data-driven MFTG model

Feature	SIR	SEIR	SEIRD	SEIR-spatial	SEIR-age structured	MFTG
Locality	No	No	No	No	No	Yes
City	No	No	No	No	No	Yes
Sector per city	No	No	No	No	No	Yes
Position	No	No	No	Yes	No	Yes
Infection status	Three states	4	5	4	4	17
Age	No	No	No	No	Yes	Yes
Gender	No	No	No	No	No	Yes
Height	No	No	No	No	No	Yes
Health	No	No	No	No	No	Yes
Family size	No	No	No	No	No	Yes
Economic status	No	No	No	No	No	Yes
Testing strategy	No	No	No	No	No	Yes
Decision-making	No	No	No	No	No	Yes
Untested positive	No	No	No	No	No	Yes
Data-driven	No	No	No	No	No	Yes

from short-term suppression to long-term mitigation depending on the achieved reduction in the hospitalized number, contact number, exposed number, testing number, preexisting health conditions, age, gender, location, mobility pattern, and spatial contact rate. Basically, data-driven MFTG is given as follows:

- Time is continuous.
- Data is reported at discrete-time instants.
- Decision-makers: individuals, firms, and authorities.
- Control actions: individuals' actions include contact/meeting rates and a movement path to go outside for food, pharmacies, or other basic needs. Firms' actions include the production essential, moderate-essential, and less-essential goods and total working hours. An authority at specific location controls the testing actions and migration rates within that area, between different areas of a city, or between cities or the country level.
- Objective: An individual aims to reduce its risk of being hospitalized and to limit its economic consequences. Each firm aims to maximize its revenue while keeping its employees from infections. Each authority aims to reduce the number of deaths under cost–benefit considerations (see Fig. 10.1).

The introduced data-driven MFTG model has the following key features:

- Contextual awareness: This aspect encompasses region-specific factors such as areas, cities, and country-specific characteristics, as well as comorbidity features.
- Comprehensive decision-making and risk-aware cost-benefit analysis: This capability applies to governing bodies, businesses, and individuals, facilitating integrated decision-making while considering the associated risks and benefits.
- Data-driven mobility mapping: The patterns of mobility within various areas, cities, and countries are inherently derived from the data-driven model.

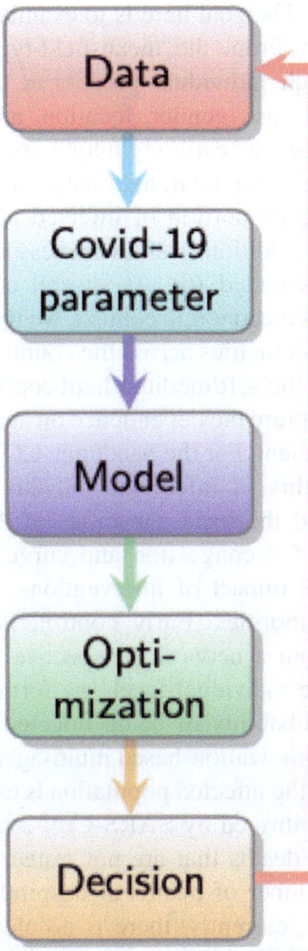

Fig. 10.1 Data-driven MFTG model: data, COVID-19 parameter, model, optimization, decision, and back to the data measurement. The verification–validation procedure used in this chapter

- Country-specific age demographics: The age distribution within a given country is a result of the data-driven model.
- Preexisting health conditions: Health conditions prevalent in specific cities or countries are embedded within the data-driven model.
- Probability estimations of exposure: This includes estimates of the likelihood of exposure through activities like coughing, sneezing, talking, and surface contact.
- Endogenous hospital capacity: Hospital capacity is determined internally within the model, aggregating data from various sources, including the number of hospital beds, intensive care facilities, ventilators, nurses, doctors, and point-of-care services.
- Data-driven testing capacity: The testing capacity, including the number of tests conducted per week and the availability of testing facilities in different areas, cities, and countries, is driven by the data within the model.
- Infection status of untested individuals: The infection status of individuals who have not undergone testing is an inherent component of the data-driven model.

The goal here is to examine the data-driven MFTG modeling aspect of COVID-19. From the mean-field-type game theory perspective, one examines carefully each individual process at the microscopic level. The individual characteristics are age, gender, location, and infection status. As each infected individual may affect its evolving neighbors and path, it creates an evolving mean-field term at the local, regional, national, and international level. Thus, a macroscopic description of the proportion of infected per location will emerge as a global mean-field term. In addition, risk-awareness at the individual level as well as the global level is accounted for. As several countries may not have the same season, culture, or socioeconomic context, we introduce temperature field, seasonal variation, and local specificities across the countries and couple them with the decisions on the duration of the soft/medium/hard confinement. The age-structured dynamics is introduced to capture the dependence on age and on the prior medical conditions of the COVID-19 patient. For the pandemic COVID-19, many local hospitals cannot sustain a massive influx of infection, and this will create a congestion and saturation. Saturation and the local capacities of the healthcare system are introduced into the model, and decongestion and curve flattening strategies are analyzed. We also examine the impact of interventions, control, and risk-aware mitigation policies from the authorities. Early control, late control, and no control scenarios are examined from a network perspective using a risk-aware mean-field-type game theory. At the individual level, we formulate a risk-aware decision-making problem that the probability of being infected and its possible consequences. We build upon prior optimization-based multi-agent models developed in Chap. 5. The number of deaths in the infected population is estimated in two ways: the number of deaths at hospitals confirmed by SARS-CoV-2 polymerase chain reaction (PCR) test and the number of deaths that are not tested (by PCR). While there is a reported data set for the number of deaths at hospitals that are directly and indirectly related to COVID-19, currently, there is no clear data set for the number of untested deaths due to COVID-19. We use a model to estimate it. The COVID-19 pandemic already has an enormous impact on economic, finance, agriculture, energy, and geopolitics. The decisions made by the authorities, such as confinement, lockdown, isolation, and curfew measures, may have severe side effects and consequences on other sectors such as employment, supply chain, agriculture, and food security. Therefore a cost–benefit analytics need to be conducted for each policy per area/city/country. The risk-aware mean-field-type game techniques presented here can be used to better understand the propagation of virus across geographical areas. As an integrated model, the impact of control strategies and switching between different locations in the propagation of the virus includes:

- international connectivity via flight, boat, or ground transportation,
- inter-city connectivity via flights, trains, buses, cars, and motorcycles,
- within-city connectivity with pedestrian mobility to pharmacies, local markets, open street air markets, supermarkets, workplaces, and offices.

Isolation/confinement strategies and testing strategy based on the spreading characteristics For finite networks, the mean-field term is stochastic and characterizes

Table 10.2 Heterogenous data sources

Data	Sources
International connectivity per country	Flightradar24
Age structure per country	World Bank Group
COVID-19 data set	CSSE
Local mobility	Google Community
	Mobility reports
Country-specific hospital capacity data	World Bank Group
Geographical map	OpenStreet map
Economic policy	IMF

the evolution of the total number of infected/recovered/tested/untested agents per location. By introducing control strategies at both atomic and non-atomic levels, the number of those simultaneously hospitalized and deaths can be reduced. However, the control is costly and needs to be carefully planned. Table 10.2 summarizes the heterogenous data sources.

Notation

Table 10.3 summarizes the notation used in the model introduced.

Table 10.3 Some notation used in the model introduced

Symbols	Meaning
l	Locality
$l = (l_1, l_2, l_3)$	Country, city, area
$T(t, l)$	Temperature field
η_1	Migration between countries
η_2	Migration between cities
η_3	Local mobility between areas of a city
z	Age variable
$\rho(l, z, t)$	Initial age distribution
$\bar{t}_l(t)$	Testing capacity per week
δ_l	Testing strategy at 1
$c = (c_1, c_2, c_3, c_4)$	Preexisting health condition, gender, height, family size, economic wealth
$x = (x_1, x_2, h)$	Position of an individual agent
v	Velocity of an individual agent
s	Individual epidemiological states
n_l	Subpopulation size at locality l
u	Meeting/activity strategy
m	Mean field
f	Drift
β	Interaction kernel
ε	Physical distancing of 2 m
e	Economic wealth

10.2 Basic Continuous Time Models

10.2.1 SIR

We start with a basic system of differential equations to estimate the propagation in a homogenous population with size $n = n(t_0) >> 1$ at time t_0 and three compartments (susceptible "S", infectious "I", removed "R"). There are three different variables (S, I, and R) representing the status of each compartment. β represents the disease transmission rate, and γ stands for the "removed" rate:

$$\begin{aligned} &\tfrac{dS}{dt} = -\lambda \cdot S, \\ &\tfrac{dI}{dt} = \lambda \cdot S - \gamma \cdot I, \\ &\tfrac{dR}{dt} = \gamma \cdot I, \\ &S(t_0) > 0, \\ &I(t_0) > 0 \\ &R(t_0) = 0, \\ &\lambda = \beta \cdot \tfrac{I}{n}. \end{aligned} \tag{10.1}$$

An important functional, called basic reproduction number, is given by

$$R_0 = \frac{\beta}{\gamma}.$$

10.2.2 Data-Driven SIR

Our goal now is to estimate the functions $I(t_0)$, β, γ, R_0 using the available dynamic data set:

- Given the data sets:
  - infected cases $\hat{I}(t_0),\ \hat{I}(t_0+1),\ \ldots,\ \hat{I}(t_0+k-1) and$
  - removed cases $\hat{R}(t_0),\ \hat{R}(t_0+1),\ \ldots,\ \hat{R}(t_0+k-1)$,

  obtain an estimate of $\beta_{data} = \hat{\beta}$ (for β) and $\gamma_{data} = \hat{\gamma}$ (for γ).
- Compute an estimate for R_0.

  We estimate the rate parameters by solving a dynamic optimization problem:
- Given the data sets, solve the basic optimization problem:

$$(\hat{\beta}, \hat{\gamma}, \hat{S}_0, \hat{I}_0) \in \arg \min_{(\beta,\gamma,S(t_0),I(t_0)) \in Spec} \int_{t_0}^{t_0+k} |I(t) - \hat{I}(t)|^2 + |R(t) - \hat{R}(t)|^2 \, dt.$$

- This provides $(\beta_{data,k} = \hat{\beta}, \gamma_{data,k} = \hat{\gamma})$.
-

$$R_{0,data,k} = \frac{\beta_{data,k}}{\gamma_{data,k}},$$

where $Spec$ means the specification parameters and $k \geq 1$.

This leads to a data-driven SIR:

$$\begin{array}{l} S(t_0) = S_0, I(t_0) = I_0 \neq 0, R(t_0) = 0, \\ \frac{dS}{dt} = -\hat{\beta} \cdot \frac{I}{N} \cdot S \\ \frac{dI}{dt} = \hat{\beta} \cdot \frac{I}{N} \cdot S - \hat{\gamma} \cdot I \\ \frac{dR}{dt} = \hat{\gamma} \cdot I. \end{array}$$

From the latter system, one can obtain the trajectories of (S, I, R) over time. Then, several questions arise:

- What if the data set changes regularly?
- Weekly estimates?
- Monthly estimates?
- Longer-term estimates?

It is important to notice that for some countries, the data set provided by the authorities need to be structured per week, 2 weeks, or month time windows or a proper moving average to learn a meaningful trend of the data:

- Pros:
  - We have a data-driven model of SIR model.
  - It is a simple model.
  - It is compartment based.
  - It is based on the data set $\{\hat{I}, \hat{R}\}$ during the time window $[t_0, t_0 + k - 1]$.
  - It is deterministic.
- Cons:
  - The number N is fixed and global.
  - Not localized.
  - It is homogenous.
  - Does the shape of I, R capture the observed data $\hat{I}, \hat{R}$?

Next, we relax the restrictive assumptions made in the previous model and include context-awareness into the data-driven model.

10.2.2.1 SIR with Location, Age, and Economic Wealth

We introduce the SIRD (susceptible–infected–recovered–deceased) dynamics with spatial location x over a map $\mathscr{D}_l$ at locality l, socioeconomic status e, and age structure z:

$$\begin{array}{ll} \text{Susceptible:} & S(x,z,e,t) \\ \text{Infected:} & I(x,z,e,t) \\ \text{Recovered:} & R(x,z,e,t), \\ \text{Deceased:} & D(x,z,e,t) \end{array}$$

$$\begin{array}{l} \text{Domain: } (x,z,e,t) \in \mathbb{R}^3 \times \mathbb{R} \times \mathbb{R} \times (t_0,t_1), \\ S_t(x,z,e,t) = -uS(x,z,e,t) \int \beta((x,z,e),(x',z',e'))I(dx',dz',de',t), \\ I_t(x,z,e,t) = uS(x,z,e,t) \int \beta((x,z,e),(x',z',e'))I(dx',dz',de',t) \\ -\gamma I(x,z,e,t) - \delta(x,z,e)I(x,z,e,t), \\ R_t(x,z,e,t) = \gamma I(x,z,e,t), \\ D_t(x,z,e,t) = \delta(x,z,e)I(x,z,e,t), \\ \quad \text{Initial condition} \\ S(dx,dz,de,t_0) = S_0(dx,dz,de), \\ I(dx,dz,de,t_0) = I_0(dx,dz,de), \\ R(dx,dz,de,t_0) = 0 = D(dx,dz,de,t_0), \\ x \in \mathscr{D}_l, \end{array} \tag{10.2}$$

where u is the meeting rate, γ is the recovery rate, and δ is the death rate. These are nonnegative real-valued. A solution to the above system, if any, will be denoted by $(S,I,R,D) = (S^{u,m_0}, I^{u,m_0}, R^{u,m_0}, D^{u,m_0})$ with the initial measure $m_0 = (S,I,R,D)(t_0)$.

From (10.2), we see that the function

$$t \mapsto \frac{d}{dt} \int (S+I+R+D)(dx,dz,de,t),$$

is identically zero. We set $\int (S+I+R+D)(dx,dz,de,t_0) = n_0$, at time $t = t_0$. The condition for I_t to be negative yields $uS(\int \beta I) < (\gamma+\delta)I$. This means that basic reproduction number R_0 depends not only on $\frac{u}{\gamma+\delta}$ but also on the contact rate kernel β, which captures physical distancing and economic wealth interaction in the society. We choose $I(x,z,e,t_0)$ to be higher for very high e to capture the initial infection from (international) travelers and people who are in touch with international businesses. Then the contact rate kernel β is inversely proportional with e. The spread is transferred to areas with low income; the dynamics exhibits a higher infection of COVID-19 in some places with low income per family size. However the spread transfer from higher income population to a population with lower income may be slower in some countries due to less interaction β with infectious people.

10.2.2.2 Control of SIRD with Spatial and Age Structure

We introduce the control of the SIRD (susceptible–infected–recovered–deceased) dynamics with spatial location x, socioeconomic status e, and age structure z. Let $t_0 < t_1$ and $(x, z, e, t) \in \mathbb{R}^3 \times \mathbb{R} \times \mathbb{R} \times (t_0, t_1)$:

$$
\begin{aligned}
&S_t + S_z - \varepsilon(s, z, e)\Delta_x S = -\tilde{d}(s, z, e)S - \mathrm{div}_x(vS)\\
&-uS(x, z, e, t) \int \beta((x, z, e), (x', z', e'))I(dx', dz', t),\\
&I_t + I_z - \varepsilon(i, z, e)\Delta_x I = -\tilde{d}(i, z, e)I - \mathrm{div}_x(vI)\\
&+uS(x, z, e, t) \int \beta((x, z, e), (x', z', e'))I(dx', dz', de', t)\\
&-\delta(x, z, e)I(x, z, e, t) - \gamma I(x, z, e, t),\\
&R_t + R_z - \varepsilon(r, z, e)\Delta_x R\\
&= -\tilde{d}(r, z, e)R + \gamma I(x, z, e, t) - \mathrm{div}_x(vR),\\
&D_t = \delta(x, z, e)I(x, z, e, t),
\end{aligned}
$$

Initial condition:

$$
\begin{aligned}
&S(dx, dz, de, t_0) = S_0(dx, dz, de),\\
&I(dx, dz, de, t_0) = I_0(dx, dz, de),\\
&R(dx, dz, de, t_0) = 0 = D(dx, dz, de, t_0),
\end{aligned}
$$

New borns (age 0):

$$
\begin{aligned}
&I(x, 0, e, t) = R(x, 0, e, t) = 0,\\
&S(x, 0, e, t) = \int \tilde{b}(x, z, e, s, t)S(x, dz, e, t)\\
&+ \int \tilde{b}(x, z, e, r, t)R(x, dz, e, t) + \int \tilde{b}(x, z, e, i, t)I(x, dz, e, t),
\end{aligned}
$$

where the number of infectious people inside the ball $\mathscr{B}_\varepsilon(x)$, within 2 meter radius ε is given by

$$
n_{I(t)}(\mathscr{B}_\varepsilon(x)) = \int_{x' \in \mathscr{B}_\varepsilon(x)} \int_{(z', e')} I(dx', dz', de', t);
$$

u is a control functional; $\varepsilon(.)$ is a diffusion coefficient; Δ_x denotes the Laplacian operator; div_x denotes the divergence operator; $\beta((x, z, e), (x', z', e'))$ is the interaction kernel by age and spatial position of the agents; $\tilde{b}$ is the fertility rate of the population of susceptible (S), infected (I), and recovered (R); $\tilde{d}(s, z, e)$ is the death rate by age and infection status; v is the velocity (three-dimensional); and e is the economic wealth. A solution to the augmented system is denoted by $(S, I, R, D) = (S^{u,v,m_0}, I^{u,v,m_0}, R^{u,v,m_0}, D^{u,v,m_0})$. Then, the optimal control of SIRD becomes a mean-field-type control problem and is given by

$$
\begin{aligned}
&\sup_{(u,v)} \int S(dx, dz, de, t_1) - D(dx, dz, de, t_1)\\
&- \int_{t_0}^{t_1} dt \int h_0(n_I(\mathscr{B}_\varepsilon(x))) \Big|v\Big|_2^2 (x, z, e, s, t) S(dx, dz, t)\\
&- \int_{t_0}^{t_1} dt \int \Big|v\Big|_2^2 (x, z, e, s, t) S(dx, dz, de, t)\\
&- \int_{t_0}^{t_1} dt \int \Big|v\Big|^2 (x, z, e, r, t) R(dx, dz, de, t)\\
&- \int_{t_0}^{t_1} dt \int \Big|v\Big|_2^2 (x, z, e, i, t) I(dx, dz, de, t)\\
&+ \int_{t_0}^{t_1} dt \int [\log(u) - u^2](x, z, e, s, t) S(dx, dz, de, t)\\
&+ \int_{t_0}^{t_1} dt \int [\log(u) - u^2](x, z, e, r, t) R(dx, dz, de, t),\\
&\text{subject to}\\
&\mu = (S, I, R, D) = (S^{u,v,\mu_0}, I^{u,v,\mu_0}, R^{u,v,\mu_0}, D^{u,v,\mu_0}),\\
&\mu(., t_0) = \mu_0.
\end{aligned}
\tag{10.3}
$$

Note that n_I depends on u and v, which are control actions.

10.2.2.3 A Basic SIRD Using Mean-Field-Type Game

We now introduce a basic mean-field-type game (MFTG, [106, 457, 458, 460, 463]) as follows:

$$
\begin{aligned}
&\sup_{(u,v)} \int S(dx, dz, de, t_1) - D(dx, dz, de, t_1)\\
&- \int_{t_0}^{t_1} [\varepsilon_1 \mathrm{var}_\mu(v) + \varepsilon_2 \mathrm{var}_\mu(u)] dt\\
&- \int_{t_0}^{t_1} dt \int h_0(n_{I^*}(\mathscr{B}_\varepsilon(x))) \Big|v\Big|^2 (x, z, e, s, t) S(dx, dz, de, t)\\
&- \int_{t_0}^{t_1} dt \int \Big|v\Big|^2 (x, z, e, s, t) S(dx, dz, de, t)\\
&- \int_{t_0}^{t_1} dt \int \Big|v\Big|^2 (x, z, r, t) R(dx, dz, de, t)\\
&- \int_{t_0}^{t_1} dt \int \Big|v\Big|^2 (x, z, e, i, t) I(dx, dz, de, t)\\
&+ \int_{t_0}^{t_1} dt \int [\log(u) - u^2](x, z, e, s, t) S(dx, dz, de, t)\\
&+ \int_{t_0}^{t_1} dt \int [\log(u) - u^2](x, z, e, r, t) R(dx, dz, de, t),\\
&\text{subject to}\\
&\mu = (S, I, R, D) = (S^{u,v,\mu_0}, I^{u,v,\mu_0}, R^{u,v,\mu_0}, D^{u,v,\mu_0}),\\
&\mu(., t_0) = \mu_0,\\
&m = (S^*, I^*, R^*, D^*) =\\
&(S^{u^*,v^*,m_0}, I^{u^*,v^*,m_0}, R^{u^*,v^*,m_0}, D^{u^*,v^*,m_0}),\\
&n_{I^*(t)}(\mathscr{B}_\varepsilon(x)) = \int_{x' \in \mathscr{B}_\varepsilon(x)} \int_{(z',e')} I^*(dx', dz', de', t).
\end{aligned}
\tag{10.4}
$$

The term n_{I^*} depends on u^* and v^*, which are equilibrium individual control actions. The parameters ε_1 and ε_2 are nonnegative real numbers. Due to the variance terms $\text{var}_\mu(v)$, $\text{var}_\mu(u)$, Problem (10.4) is non-linear with respect to the measure μ (and non-linear in m). Next we introduce the variance-aware data-driven MFTG.

10.3 Data-Driven MFTG

There are $K \geq 1$ countries distributed over a three-dimensional geographical area. The set of countries is represented by a set of locations $\{1, \ldots, K\}$ with geographical coordinates (latitude, longitude) on the map.

10.3.1 Authorities, Firms, and People

- Authorities: In each country/location, there is an authority who can decide on the testing policy (when, how many), migration, confinement policies of the country, and international connections via flights, boats, and ground transportations. The authority in l, has a testing capacity to be adjusted properly and the migration rate $\eta_{ll'}, l' \in \mathcal{N}_l$, where $\mathcal{N}_l$ is the set of connecting areas/cities/countries from l. The intra-city mobility in city l_3 is represented by the mobility rate η_3, the inter-city mobility from/to cities l_2 is represented by the mobility rate η_2, and the international connectivity from/to country l_1 is represented by the mobility rate η_1.

  The country-specific seasons and temperature field $T(t, l)$ are given. The temperature field will be taken into consideration when making a decision on the duration of a lockdown at a sector level, city level, or activity level. In each country in $\{1, \ldots, K\}$, there is a subpopulation of people, called agents. The distribution of age of all agents in country l is given by $\rho(l, dz, t)$ where the (chronological) age z ranges from 0 to 124 years. The function ρ is fitted from real data set making it heterogeneous from one country to another. The authorities play the role of atomic decision-makers as their decisions can influence significantly the internal and external migration from that country and the subpopulation at that location. The authority of country l aims to reduce infections, deaths, and economic losses. The authorities' strategic objectives are the following:

  - Reduce the number of deaths.
  - Interrupt agent-to-agent transmission including reducing secondary infections among close contacts and preventing further spread at a larger scale.
  - Identify, isolate, and care for patients early (testing–tracking–tracing).
  - Realizing a testing strategy at a given time window. The testing capacity $\bar{t}_l(t)$ can be improved by the authority in l over time t.
  - Minimize social and economic impact through multisectoral partnerships between countries, geographical locations, and airports hubs.

- Firms: Production goods are classified as essential, moderate-essential, and less-essential. Each firm decides its total working hours and aims to reduce the number of infected employees and economic losses.
- Agents per country: Each agent has an age z and a position located in one of the K countries. An agent of any age can be infected by a virus (SARS-CoV-2 for the COronaVIrus Disease 2019). An agent is parametrized by its physical location in area l_3 of city l_2 of province state/country l_1, age z, preexisting health condition (c), gender(g), height (h), family size (fs), and average revenue per family. We rename (c, g, h, fs) of the agent to be $c = (c_1, c_2, c_3, c_4)$. We denote $(l_1, l_2, l_3, z, c, g, h, fs$ by lac, which stands for location–age–condition. The agent has an infection status described as follows:
  - For untested individuals, there are six states. Non-infected agents are susceptible (lac-S). Non-infected agents including recovered agents can be reinfected by another mutated version (if any) of the virus. An agent can be exposed (lac-E), for example, by an infected agent who is coughing and sneezing, from a surface contact, or from a directional air conditioning (if any). Some of the exposed agents become passive (lac-P), which means dormant (asymptomatic infectious or mild symptom) and some others become active (lac-A). One absorbing state: deceased (lac-D). Some of the passive or active agents become recovered (lac-R).
  - Some of exposed people will be tested. There are ten additional states plus one specific state actioned by healthcare authorities to identify and isolate the potential exposed people, for example, by means of phone calls or contact-tracing and tracking. These states are testing (lac-T); lac-testing negative; lac-testing positive; lac-testing unclassified/unknown; lac-contact tracing of the last 2 weeks for each COVID-19 tested positive; lac-isolation at home, at a hotel, and at a point of care; lac-hospitalized (severely active case); lac-recovered; and lac-deceased.

  The 17 infection status are given by:

  - Lac-susceptible, lac-exposed.
  - Lac untested: lac-passive-untested, lac-active-untested, lac-recovered-untested, lac-deceased-untested.
  - Lac-testing, lac-testing negative, lac-testing positive, lac-testing unclassified/unknown.
  - Lac-contact-tracing-tracking: For each COVID-19 tested positive agent, a contact-tracing-tracking action is taken by the authorities to reach out and identify people who may have been potentially exposed by getting in the close contact of the COVID-19 positive agent in the last 2 weeks. This is not an individual but a collective action taken by the authorities. A person who has been identified from the list of contract-tracing person may be considered as potentially exposed, and a testing decision will be made after evaluation of the level of risk.
  - Lac-isolation at home, at a hotel, and at a point of care.
  - Lac-hospitalized (severely active case), lac-recovered, lac-deceased.

10.3.2 Individual State

The internal state of individual p is $(l_p(t), x_p(t), s_p(t), z_p(t), c_p(t))$ where:

- $l_p(t) = (l_{1,p}(t), l_{2,p}(t), l_{3,p}(t))$ is an index representing latitude–longitude of the location of person p at time t, $l_{1,p}$ for the country, $l_{2,p}$ for the city/province, and $l_{3,p}$ for the area of the city. $l_p(t) \in \mathcal{L}$ describes a discrete set of cities' map of the countries in $\{1, \ldots, K\}$.
- $x_p(t) = (x_{1,p}(t), x_{2,p}(t), h_p(t))$ represents the position of agent p at time t.
- $s_p(t) \in \{S, E, P, A, R, D, s_7, \ldots, s_{17}\}$ is the infection status of person p at time t.
- $z_p(t) \in [0, 124y]$ is the age of person p at time t, up to 124 years. The number y is a conversion from 1 year into a proper time units.
- $c_p(t) = (c_{1,p}(t), c_{2,p}(t), c_{3,p}(t), c_{4,p}(t))$ with $c_{1,p}(t) \in C_1$, which is the set of all subset of $C_1' = \{phc_1, phc_2, phc_3, \ldots\}$ a finite list of preexisting health conditions (phc) that are considered as relevant by health authorities. Note that C_1 contains the empty set $\varnothing$. If person p has no prior health issue, then $c_{1,p}(t)$ becomes the empty set $c_{1,p}(t) = \varnothing$. The set C_1 includes also some combinations of diabetes (type 2), cerebrovascular diseases, severe asthma, pulmonary hypertension, cardiovascular disease, high blood pressure, chronic obstructive pulmonary disease, and chronic kidney disease, among the others. Older people and people with preexisting medical conditions are assumed to be more vulnerable to becoming severely ill with the virus. This is reflected in λ_{hd} as a function of age z. The symbol $c_{2,p} \in C_2$ represents the gender of person p. C_2 is a finite set representing the spectrum of gender, and $c_{3,p}(t) = h_p(t) \in C_3$ is the height of person p at time t. C_3 represents the spectrum of height, and C_4 represents the set of family size or number of roommates at residential areas.

The set of states of an agent in location $l \in \mathcal{L}$ is $\mathcal{X} \times C \times$ Status, where the infection status of the agent can be $\{S, E, P, A, R, D, s_7, \ldots, s_{17}\}$ and $C = C_1 \times C_2 \times C_3 \times C_4$.

10.3.3 Global State

The population size is a large but finite number n. The global state of the system is the measure $(M(l, dx, s, dz, c, t),\ l \in \mathcal{L}, s \in \{s_1, \ldots, s_{17}\}, z \in [0, 124y], c \in \mathcal{C}$ where

$$M(l, \mathcal{V}_x, s, \mathcal{V}_z, c, t) = \sum_p \mathbb{I}_{\{(l_p(t),, s_p(t), c_p(t)) = (l,s,c)\}} \mathbb{I}_{x_p(t) \in \mathcal{V}_x, z_p(t) \in \mathcal{V}_z}, \tag{10.5}$$

where $\mathcal{V}_x$ is a Lebesgue measurable set of positions at l and $\mathcal{V}_z$ is a measurable set of ages at l. The intra-state status of the system in location l, age z, and characteristics (c_1, c_2, c_3, c_4) at time t is given by

$$M(l, dx, dz, c, t) \\ = (S, E, P, A, R, D, s_7, \ldots, s_{17})(l, dx, dz, c, t), \tag{10.6}$$

representing a measure of the total number of people located at l for each of the 17 status. Let $n(l, z, c, t)$ be the population size at location l with age z and characteristics (c_1, c_2, c_3, c_4) at time t. Let $n_l(t) = \sum_c \int n(l, dz, c, t)$ be the population size at location l at time t. The occupancy measure in area l_3 of city l_2 of country l_1 is

$$M^n(l, \mathcal{V}(x), z, c, t) = \frac{1}{n_l V} \int_{\hat{x} \in \mathcal{V}(x)} M(l, d\hat{x}, z, c, t),$$

where V is the volume of the neighborhood $\mathcal{V}(x)$.

10.3.4 Individual State Transitions

We describe below the transitions between infection status:

- Susceptible agents can be exposed to the virus in different ways. (i) become exposed from a surface contact that has been infected recently and (ii) exposed from a coughing or sneezing of an infectious person (with or without symptoms) with rate λ_{se}.
- Exposed agents: Recent observations have suggested that for the COVID-19, the risk of exposure was affected by surface contact, coughing, and sneezing (with a face mask vs without a face mask). We therefore examine the mathematics of coughing and sneezing in order to estimate the range and the risk of exposure of a person. A simple dynamics of coughing and sneezing is described below. The trajectory of the puff is governed by the evolution of its momentum I and buoyancy B. The buoyancy force is vertical, causing the cloud to follow a curvilinear trajectory denoted by (w, θ, h) or equivalently (x, y, h)

$$\begin{cases} V_t = 3\eta\alpha^3 w^2 w_t - v_p N_t, \\ \rho_{c,t} = \frac{1}{V}(v_p N_t(\rho_c - \rho_p) + 3\eta\alpha^3 w^2 w_t(\rho_a - \rho_c)), \\ B_t = g\rho_a V_t - g(V\rho_c)_t, \\ I_{y,t} = B, \\ w_t = \left(I_y^2 + \frac{I_0^2 \cos^2\theta_0}{V\rho_c}\right)^{\frac{1}{2}}, \\ N_t = -\frac{3NU_s}{2\alpha w}, \end{cases} \tag{10.7}$$

where V is the volume of the cloud; $\eta = \alpha = 0.1$; s is the distance from the source; v_p is the speed; N is the number of particles within the cloud, with particle concentration $\frac{N(t)}{V(t)}$ at time t; ρ_c is the density of the cloud; ρ_p, ρ_a is the density of the ambient air; B is the cloud buoyancy; g is the acceleration due to gravity; $I = (I_x, I_y)$ is the momentum; I_0 is the initial momentum; θ_0 is the initial angle; and $U_w = w_t$ is the speed of the suspended particles with a time scale of $\frac{h}{U_w}$.

The quantities of interest include the fallout time, position, and the horizontal range of deposition. These quantities affect the exposure to the range and possibly the transmission of the disease by emitted pathogens. We numerically compute the parameters $\beta, \lambda_{ep,l}$ by looking at the distance over which the cough/sneeze traveled around 0.7 m, with an initial v_p of 4 m/s. An area of 0.17m^2 is covered initially with an expansion rate equal to $1.3\text{m}^2/\text{s}$. We have run the system for 45 seconds. A susceptible person present in this area will be considered as a directly exposed person with a probability proportional to the intersection of their two path areas.

The intra-dynamics is driven by $D_{\text{intra}} :=$

$$\left(\begin{array}{ll}
s_1 = s: & -s_l\lambda_{se} + e_l\lambda_{\text{es}} + t_{-,l}\delta_l\lambda_{t_-s} + t_{u,l}\delta_l\lambda_{t_us} + \text{tr}_l\delta_l\lambda_{\text{tr},s} \\
s_2 := e: & s_l\lambda_{se} + p_l(1-\delta_l)\lambda_{pe} + a_l(1-\delta_l)\lambda_{ae} \\
 & +t_{-,l}\delta_l\lambda_{t_-e} + t_{u,l}\delta_l\lambda_{t_ue} - \delta_l\lambda_{et}(l)e_l - e_l\lambda_{es} - e_l(1-\delta_l)\lambda_{ea} \\
 & -e_l(u_l(o|s,z)(1-\delta_l)\lambda_{\text{Sm}} + u_l(m|p,z)(1-\delta_l)\lambda_{ep,l}p_l) + r_l(1-\delta_l)\lambda_{re} \\
s_3 := p: & -p_l(1-\delta_l)\lambda_{pr} - p_l(1-\delta_l)\lambda_{pe} - 2p_l(u_l(m|p,z)\bar{u}_l(m|p))(1-\delta_l)\lambda_{pa}\frac{np_l-1}{n_l} \\
 & +a_l(1-\delta_l)\lambda_{ap}\frac{p_l}{h_l+p_l} + e_l(u_l(o|e,z)(1-\delta_l)\lambda_{Sm} \\
 & +u_l(m|p,z)(1-\delta_l)\lambda_{ep,l}p_l) + r_l(1-\delta_l)\lambda_{rp} \\
s_4 := a: & 2p_l(u_l(m|p,z)\bar{u}_l(m|p))(1-\delta_l)\lambda_{pa}\frac{np_l-1}{n_l} - a_l(1-\delta_l)\lambda_{ae} - a_l(1-\delta_l)\lambda_{ap}\frac{p_l}{h_l+p_l} \\
 & +e_l(1-\delta_l)\lambda_{ea} - a_l(1-\delta_l)\lambda_{ad}(l,z,c) - a_l(1-\delta_l)\lambda_{ar}(l,z,c) \\
s_5 := r: & p_l(1-\delta_l)\lambda_{pr} + a_l(1-\delta_l)\lambda_{ar}(l,z,c) - r_l(1-\delta_l)\lambda_{rp} - r_l(1-\delta_l)\lambda_{rs} \\
s_6 := d: & (1-\delta_l)\lambda_{ad}(l,z,c)a_l \\
s_7 := t: & \delta_l\lambda_{tt,t}(l,c)\text{tt}_l + \delta_l\lambda_{et}(l)e_l - t_l\delta_l\lambda_{t,t_+} - t_l\delta_l\lambda_{t,t_-} - t_l\delta_l\lambda_{t,t_u} \\
s_8 := t_-: & t_l\delta_l\lambda_{t,t_-} - t_{-,l}\delta_l\lambda_{t_-s} - t_{-,l}\delta_l\lambda_{t_-e} \\
s_9 := t_u: & t_l\delta_l\lambda_{t,t_u} - t_{u,l}\delta_l\lambda_{t_us} - t_{u,l}\delta_l\lambda_{t_ue} \\
s_{10} := t_+: & t_l\delta_l\lambda_{t,t_+} - t_{+,l}\delta_l\lambda_{t_+,tt} - t_{+,l}\delta_l\lambda_{t_+,\text{th}_1} - t_{+,l}\delta_l\lambda_{t_+,\text{tpc}} - t_{+,l}\delta_l\lambda_{t_+,\text{th}_2} \\
s_{11} := \text{tt}: & \delta_l \text{ contacting, verifying, making a list, suggesting test} \\
s_{12} := \text{th}_1: & t_{+,l}\delta_l\lambda_{t_+,\text{th}_1} - \text{th}_{1,l}\delta_l\lambda_{\text{th}_1,\text{thos}}{}^{u_1\bar{u}_1(\bar{th}_{1,l}-\frac{1}{n_l})} - \text{th}_{1,l}\delta_l\lambda_{\text{th}_1,\text{tr}} - \text{th}_{1,l}\delta_l\lambda_{\text{th}_1,tpc} \\
s_{13} := \text{tpc}: & t_{+,l}\delta_l\lambda_{t_+,\text{tpc}} - \text{tpc}_l\delta_l\lambda_{\text{tpc},\text{thos}}{}^{u_3\bar{u}_3(\bar{\text{tpc}}_l-\frac{1}{n_l})} - \text{tpc}_l\delta_l\lambda_{\text{tpc},\text{tr}} \\
s_{14} := \text{th}_2: & t_{+,l}\delta_l\lambda_{t_+,\text{th}_2} - \text{th}_{2,l}\delta_l\lambda_{\text{th}_2,\text{thos}}{}^{u_2\bar{u}_2(\bar{th}_{2,l}-\frac{1}{n_l})} - \text{th}_{2,l}\delta_l\lambda_{\text{th}_2,tr} \\
s_{15} := \text{thos}: & \text{th}_{1,l}\delta_l\lambda_{\text{th}_1,\text{thos}}{}^{u_1\bar{u}_1(\bar{th}_{1,l}-\frac{1}{n_l})} + \text{tpc}_l\delta_l\lambda_{tpc,\text{thos}}{}^{u_3\bar{u}_3(\bar{\text{tpc}}_l-\frac{1}{n_l})} \\
 & +\text{th}_{2,l}\delta_l\lambda_{th_2,\text{thos}}{}^{u_2\bar{u}_2(\bar{th}_{2,l}-\frac{1}{n_l})} - \text{thos}_l\delta_l\lambda_{\text{thos},tr} - \text{thos}_l\delta_l\lambda_{\text{thos},\text{td}} \\
s_{16} := \text{tr}: & \text{thos}_l\delta_l\lambda_{\text{thos},\text{tr}} + \text{th}_{1,l}\delta_l\lambda_{\text{th}_1,\text{tr}} + tpc_l\lambda_{\text{tpc},\text{tr}} + \text{th}_{2,l}\delta_l\lambda_{\text{th}_2,\text{tr}} - \text{tr}_l\delta_l\lambda_{\text{tr},s} \\
s_{17} := \text{td}: & \text{thos}_l\delta_l\lambda_{\text{thos},\text{td}}.
\end{array}\right) \tag{10.8}$$

An exposed person:

- May become active with a probability $\lambda_{ea}(l)$.
- May become passive via two ways: (i) λ_{Sm} is the probability of getting infected by a surface of contact or in close contact with an infected agent. The exposed agent can decide to receive or not an item or a visit to a workplace, grocery store, pharmacy, $\{o, \bar{o}\}$. The contact probability is $u_l(o|e, z, c)$. In practice, the meeting or contact probability is estimated through the path trajectory of exposed people and the use of physical distancing to an exposed person or an exposed surface. (ii) $\lambda_{ep,l} P_l^n(t)$ models the probability of encounter with a passive agent where

$$P_l^n(t) = \int_{x',z'} \sum_c [P^n + t_+^n(l, dx', dz', c, t)].$$

An example of interaction between an exposed agent and an active or passive infectious agent would be

$$\begin{aligned} &m(l, x, s, z, c, t) \int_{x',z'} \sum_{c'} \beta((x', z'), (x, z)) \\ &\sum_{s' \in \{p, a, t_+, \text{tpc}, \text{th}_1, \text{th}_2, \text{thos}\}} m(l', dx', s', dz', c', t)], \end{aligned} \tag{10.9}$$

where β is the spatial exposure and cross-age kernel between positions x and x' as in (10.7), and $s \in$ {lac-S, lac-exposed}. For infection at homes and residential areas, we take the average family size per area/city/country. This is made endogenous to the model by setting c_4 to the "family" size (or number of roommates) of the individual.

- Untested passive agents: In practice, the confirmed cases (asymptomatic and symptomatic) are done through a two-step process via testing for COVID-19, which can be positive, negative, or underdetermined with some probabilities $(\lambda_{po}, \lambda_{ne})$. Since testing is still limited in many countries, we estimate the untested passive agents from a tracking of exposed agents obtained by phone calls, surveys, mobile tracking, and contact-tracing, among the others. A large number of the population have not been tested yet. A person in country l is untested with probability $1 - \delta_l$, where δ_l is the testing strategy adopted by the authority in l :

  - May become recovered with probability λ_{pr}.
  - May become susceptible with probability λ_{ps}.
  - May become active with a rate $\lambda_{pa,1}$.
  - May opportunistically encounter with another infectious agent or has comorbidities, and both become active. The rate is $\lambda_{pa,2}(z, c)(P_l^n(t) - \frac{1}{n_l})$. The passive agent can decide to be in physical contact or not with the other infected agent, so there are two possible actions: $\{m, \bar{m}\}$. The meeting probability is $u_l(m|p, z, c)$.
  - May cause death with age.

  - Newborns with passive status with age 0. This will be considered as a very small number as in the (publicly) reported data set.

- Untested active agents:

  - May become susceptible with probability $\lambda_{as}(l, z, c, \rho)$ and depends on age.
  - May become passive with probability $\lambda_{ap}(z, \rho, l)\frac{P_l^n(t)}{h_l+P_l^n(t)}$. This will be considered as a very small number as in the (publicly) reported data set.
  - May cause death with age. The death rate $\tilde{d}(z, c, \rho)$, due to the virus, increases with the age z of the agent. The distribution of age in location l solves

$$\rho_t + \tilde{v}\rho_z = -\tilde{d}(z, c, \rho, l)\rho(l, z, t),$$

    where $\tilde{v}$ is the aging speed and $\tilde{d}$ is the death rate. The initial distribution of age in location l is $\rho(l, dz, 0)$. The birth rate is $\rho(l, 0, t) = \int fert(l, z, \rho)\rho(l, dz, t)$, where fert$(l, z, \rho)$ is the fertility rate at age z in country l. Older people and people with preexisting medical conditions, such as severe asthma and pulmonary hypertension, high blood pressure, and diabetes, are assumed to be more vulnerable to becoming severely ill with the virus SARS-COV-2.
  - Newborns with active status with age 0. The newborn part of the population contributes as $\rho(l, 0, t)$.

- Untested recovered agents: Depending on the type of virus mutation and immunization of the agent, a recovered agent can be reinfected by a mutated version of the virus and hence:

  - May become susceptible with $\lambda_{rs}(z, c)$ but it is assumed to be small as suggested by the publicly reported data set.
  - May become passive with $\lambda_{rp}(z, c)$ but it is assumed to be small as suggested by the publicly reported data set.

- Untested death: This is an absorbing state. No transition to any other state once in. The death rate depends not only on the age of the patient but also on the temperature field or season at location l. It also depends on the capacity of the healthcare system at location l. The constraint is

$$\int \sum_{c'} \text{hospitalized}(l, dx', dz', c', t) \leq \bar{c}_l(t),$$

  where $\bar{c}_l(t)$ is the capacity of the healthcare system at location l at time t. When the capacity $\bar{c}_l(t)$ is exceeded, the death rate $\tilde{d}$ increases drastically. Authorities anticipate the saturation and find alternatives such as installations of equipment at stadiums and new hospitals. An aggregated hospital capacity data set $\bar{c}_l$ is available from the World Health Organization and World Bank Group. Depending on the country, there is also a certain rate λ_{co} of undetected COVID-19 patients (not reported at hospitals or from their homes) that lead to death.

- Lac-testing: During a period of asymptomatic infection, the testing can reveal infection depending on the outcome. The testing result can lead to different outcomes and the isolation can be at home, a hotel, or a point of care. Testing occurs in country l with probability δ_l.The result of the testing is often classified as positive, negative, or underdetermined (unknown). Quarantine at home, a hotel, or a point-of-care policy is case dependent. It can be dependent on whether a case is unknown, known positive, known negative, or recovered. Testing therefore makes possible the identification and quarantine of infected individuals and release of non-infected individuals. Each country has its own testing capacity $\bar{t}_l$ and testing rule δ_l (deciding by the authority at l), and the constraint is captured by

$$\sum_{c'} \int_{x',z'} \text{testing}(l, dx', dz', c', t) \leq \bar{t}_l(t).$$

- The testing state moves to one of the three states: lac-testing negative, lac-testing positive, and lac-testing unclassified/unknown.
- Lac-contact tracing of the last 2 weeks for each COVID-19 tested positive agent.
- Lac-isolation at home, at a hotel, and at a point of care:
  - May become hospitalized with probabilities

$$\begin{array}{l} \lambda_{\text{th}_1,\text{thos}}(z,c)u_1\bar{u}_1(\bar{\text{th}}^n_{1,l}(t) - \frac{1}{n_l}), \\ \lambda_{\text{th}_2,\text{thos}}(z,c)u_2\bar{u}_2(\bar{\text{th}}^n_{2,l}(t) - \frac{1}{n_l}), \textit{and} \\ \lambda_{\text{tpc},\text{thos}}(z,c)u_3\bar{u}_3(\bar{\text{tpc}}^n_l(t) - \frac{1}{n_l}). \end{array} \tag{10.10}$$

  - May become recovered. The rates are $\lambda_{\text{th}_1,\text{tr}}(z,c)$, $\lambda_{\text{th}_2,\text{tr}}(z,c)$, $\lambda_{\text{tpc},\text{tr}}(z,c)$,
  - There may be also a transition from home to one of the points of care with probability $\lambda_{\text{th}_1,\text{tpc}}(z,c)$.
- Lac-hospitalized (severely active case):
  - May become recovered with rate $\lambda_{\text{thos},\text{tr}}(z,c)(\text{thos}^n_l(t) - \frac{1}{n_l})$.
  - May become dead with rate $\lambda_{\text{thos},\text{td}}(z,c) + (\text{thos}_l - \bar{c}_l)_+$.
- Lac-recovered:
  - May become susceptible with rate $\lambda_{\text{tr},s}(z,c)$ but it is assumed to be very small.
  - May become exposed with rate $\lambda_{\text{tr},e}(z,c)$, but it is assumed to be very small.
- Lac-deceased: This is an absorbing state. There is no transition to any other state once in.

When $\delta_l = 0$, the model (without testing at locality l) reduces to six basic individual states.

10.3.5 Mobility per Area/City/Country

- The mobility pattern of the agents through ground transportation within the same street/area of the city, between areas of the cities, and between cities via ground and air transportation and international connectivity are given by a mobility map or graph with rate functions $\eta_{ll'}(T, C, c, \lambda, m, \rho, u, v, \delta, n)$ between locations l and l'. We decompose η into three layers: η_1 for connecting countries, η_2 for connecting cities, and η_3 for connecting areas of the same city.

The mobility pattern by area per city can be formulated as follows. Let $l = (l_1, l_2, l_3)$ be a location of a specific area l_3 of the city l_2 of region/country l_1 and $x = (x_1, x_2, h) \in \mathbb{R}^3$ a geographical location of an individual agent inside the area l_3. Let $m(l, x, s, z, c, t)$ be a measure of the agents satisfying the Kolmogorov equation (10.12). The limiting measure m can be written with the following components:

$$\begin{cases} \text{time-variation} \\ = \text{advection-jump-diffusion} \\ \text{local spread of the disease: coughing and sneezing} \\ \text{local spread of the disease via surface contact} \\ \text{intra-city mobility in } l_3 \\ \text{inter-city mobility from/to } l_2 \\ \text{international connectivity from/to } l_1 \\ \text{population aging and death} \\ \text{number of hospitalized patients vs hospital capacity in } l_3 \\ \text{number of tests vs testing capacity in } l_3 \\ \text{initial population} \\ \text{new borns with age 0,} \\ \text{city architecture constraint} \\ \text{map constraint} \\ \text{Allowed/feasible mobility areas/exits} \end{cases} . \qquad (10.11)$$

As in Chap. 5 of this volume as well as from Volume I, the evolution of fraction of agents per state is influenced by the following:

- The number of agents for which there is a transition in one time slot is always less than 2.
- Payoff contribution from switching: $\{0, \frac{1}{n_l}, -\frac{1}{n_l}, \frac{2}{n_l}, -\frac{2}{n_l}\}$.
- The intensity of interaction is in the order of $1/n_l$.
- The drift (the expected change of M^n in one time-step given the current state) of the system is $\frac{1}{n_l}[D_{\text{intra}} + D_{\text{inter}}]$, with D_{intra} given in (10.8) and

$$D_{inter} := \begin{pmatrix} \sum_{l'} s_{l'}\eta_{l'l} - s_l \sum_{l'} \eta_{ll'} - v s_{z,l} - \tilde{d} s_l \\ \sum_{l'} e_{l'}\eta_{l'l} - e_l \sum_{l'} \eta_{ll'} - v e_{z,l} - \tilde{d} e_l \\ \sum_{l'} p_{l'}\eta_{l'l} - p_l \sum_{l'} \eta_{ll'} - v p_{z,l} - \tilde{d} p_l \\ \sum_{l'} a_{l'}\eta_{l'l} - a_l \sum_{l'} \eta_{ll'} - v a_{z,l} - \tilde{d} a_l \\ \sum_{l'} r_{l'}\eta_{l'l} - r_l \sum_{l'} \eta_{ll'} - v r_{z,l} - \tilde{d} r_l \\ 0 \\ \sum_{l'} s_{l'j}\eta_{l'l} - s_{lj} \sum_{l'} \eta_{ll'} - v s_{lj,z} - \tilde{d} s_{lj} \\ 0 \end{pmatrix},$$

where $j \in \{7, \ldots, 16\}$, and m is the actual population mean-field profile as a function of (l, x, s, z, c, t) given in (10.12) where yq denotes a quarantined agent with status y. Agents with non-empty preexisting conditions $c \neq \varnothing$ are handled separately from the above dynamics with a certain quarantine rate at the early stage. Here $\tilde{b}_{00} = 0$ for $z > 0$, and the newborns are counted in the calculations of m from the functional $\tilde{b}$ with age $z = 0$. The transition rate $\tilde{q}(x, s, z, c; (l, x', s', z', c', t, m))$ is obtained from the functions λ following the diagram in (10.8).

$m = (m(l, x, s, z, c, t))_{l,s,c}$,	
$m_t = -\text{div}_x(vm) - div_v(u_3 m) + \frac{1}{2}\text{trace}[(\sigma^*\sigma m)_{xx}]$	movement and diffusion
$+\int_\Theta [m(l, x-\gamma, s, z, c, t) - m(l, x, s, z, c, t) + \langle m_x, \gamma\rangle]\nu(d\theta)$	jump
$+\int_{\mathcal{B}_v(x)} m(l, dx', s, z, t) q(x, s, z; (l, x', s, z, t, m, v))$	
$-m \int_{\mathcal{B}_v(x)} q(dx', s, z; (l, x, s, z, t, m, v))$	
$+\sum_{s' \neq s, c' \neq c} \int m(l, dx', s', dz', c', t) \tilde{q}(x, s, z, c; (l, x', s', z', c', t, m))$	
$-m(l, x, s, z, c, t) \sum_{s' \neq s, c' \neq c} \int \tilde{q}(dx', s', dz', c'; (l, x, s, z, c, t, m))$	local spread of the disease coughing, sneezing, surface contact
$+\sum_{l' \neq l_3} \int m(l', dx', s', z, c, t) \eta_3(l, x, s, z, c; (l', x', s', z, c, t, m))$	
$-m(l, x, s, z, c, t) \sum_{l' \neq l_3} \int \eta_3(l', dx', s', z, c; (l, x, s, z, c, t, m))$	intra-city mobility in l_3
$+\sum_{l' \neq l_2} \int m(l', dx', s', z, c, t) \eta_2(l, x, s, z, c; (l', x', s', z, c, t, m))$	
$-m(l, x, s, z, c, t) \sum_{l' \neq l_2} \int \eta_2(l', dx', s', z, c; (l, x, s, z, c, t, m))$	inter-city mobility from/to l_2
$+\sum_{l' \neq l_1} \int m(l', dx', s', z, c, t) \eta_1(l, x, s, z, c; (l', x', s', z, c, t, m))$	
$-m(l, x, s, z, c, t) \sum_{l' \neq l} \int \eta_1(l', dx', s', z, c; (l, x, s, z, c, t, m))$	international connectivity from/to l_1
$-\tilde{v} m_z - \tilde{d}(l, z, c, m) m$	aging and death
$\int m(l, dx, \text{hospitalized}, dz, t) \leq \bar{c}_l(t)$,	
$\int m(l, dx, \text{testing}, dz, t) \leq \bar{t}_l(t)$,	

$$
\begin{array}{ll}
m(l,x,s,z,c,0) = m_0(l,x,s,z,c) & \text{initial population} \\
m(l,x,s,0,c,t) = \int \tilde{b}(l,x,s,z,c,t,m)m(l,x,s,dz,c,t), & \\
x \in \mathscr{D}_l, & \text{City map constraint} \\
v_{|\partial \mathscr{D}_l} = 0, & \text{map constraint} \\
v_{Exits} = k(l,x,m), & \text{Allowed/feasible} \\
 & \text{ mobility areas/exits} \\
s' \in \{s, sq, p, pq, a, aq, r, d\}, & \text{infection status,} \\
s \in \{s, e, p, r, d, t_-, t_u, t_+, \text{home}, pc, \text{hotel}, tt, \text{hospitalized}, tr, td\}, & \text{infection status.}
\end{array}
\tag{10.12}
$$

Proposition 10.1 *As $n_l \longrightarrow +\infty$, the occupancy measure $M^n(t)$ converges in probability to the solution of the (controlled) differential equation given by*

$$
\dot{m}(l,x,s,z,c,t) = f(l,x,s,z,c,t,\lambda,u,v,\eta,\delta,m),
$$

starting from $m(.,t_0) = m_0(.)$.

The convergence is in distribution. It is, in general, weaker than the convergence in probability. However, here, the limiting object m is deterministic (parametrized), and hence the convergence is also in probability. The proof follows by using a martingale approach with an infinitesimal generator driven by the drift

$$
f := \lim_{n \to +\infty} [D_{\text{inter}} + D_{\text{intra}}],
$$

where the limit is taken in (10.8).

10.3.6 Data-Driven Mean-Field Trajectories

We build upon the works on mean-field-type filters to incorporate the economic data and the pandemic data together in the model. We tune the parameters of the model such that the solution of the dynamical system becomes as close as possible to the dynamically changing data set. We rewrite the model with the parameters λ to be tuned from the reported data set using the mean-field dynamics in Proposition 10.1:

- The geographical coordinates are taken from the world map data set.
- The sequences of reported data set $\hat{y}_l(t_k)$, $k \in \{0, 1, \ldots, \}$, $t_k \leq t$ are given. The reported data set is observed at discrete-time instants $t_0 < t_1 < t_2 < \ldots$, where t_k captures the time instant at which the k-th data point is officially reported by the corresponding state and healthcare authorities at location l. The reported data set is a sequence of aggregated data sets: $\hat{y}_l(s, t_k)$, where $s \in \{\text{tr,td,testing}, t_+\}$. The conditions of untested agents are not in the reported data set. The reported data set is assumed to be noisy and inaccurate.

- The reported data set is used to construct a filtered trajectory minimizing the global error within $[t_0, t_1]$ by choosing λ and h_l. Then the resulting dynamics is used to forecast the mean field within an uncertainty region. The sequence of optimization problems up to time t_k, is given by

$$\inf_{\lambda,h,m_0} \sum_s \int_{t_0}^{t_k} \left| \int_x \int_z \sum_c m(l, dx, s, dz, c, t) - \hat{m}_l(s,t) \right|_2^2 dt, \tag{10.13}$$

where $\hat{m}_l(t) = \hat{m}_l(t_{i-1}) + \frac{t-t_{i-1}}{(t_i-t_{i-1})} \mathbb{I}_{(t_{i-1},t_i)}(t)(\hat{y}_l(t_i) - \hat{m}_l(t_{i-1}))$ for $t \in (t_{i-1}, t_i)$. As a new data point $\hat{m}_l(t_{k+1})$ is observed, we need to re-update the estimation over the integral $[0, t_{k+1}]$ by solving the new minimization problem

$$\inf_{\lambda,h} \int_0^{t_{k+1}} \left| m(t) - \hat{m}(t) \right|_2^2 dt,$$

and so on. The continuation of the mean field starts from $m(t_k)$ at time t_k.

The above optimization provides $\lambda_{\text{data}}, h_{\text{data}}$ that are specific to COVID-19 country-specific spread. Moreover, λ_{data} and h_{data} are context aware as they depend on country-specific, age, gender, testing, and preexisting health conditions. The data-driven dynamics yields

$$\dot{m}(l, x, s, z, c, t) = f(l, x, s, z, c, t, \lambda_{\text{data}}, u, v, \eta, \delta, m),$$

with $m(., t_0) = m_0(.)$. Instead of $\hat{m}$ above, the estimation is replaced by the distribution of $m_{|\{\hat{y}(t_k),\ t_k \leq t\}}$ leading to a mean-field-type filtering problem.

10.3.7 Beyond Pairwise Interaction

We have detailed above the pairwise interaction case with a kernel β in (10.9). A 3-wise, 4-wise,... j-wise interaction case, depending on the local markets, supermarkets, pharmacies, family members at home, and the number of infected people at the neighborhood $\mathcal{V}(x)$ can be formulated as

$$\begin{aligned}
&\text{infectious} = \{p, a, t_+, \text{tpc}, \text{th}_1, \text{th}_2, \text{tpc}, \text{thos}\},\\
&m(l, x, s, z, c, t) \Big[\int_{x' \in \mathcal{V}(x)} \beta_1(x', x, t) \times \\
&\int \sum_{c'} \sum_{s' \in \text{infectious}} m(l', dx', s', dz', c', t) \\
&+ \sum_{j \geq 2} \lambda_j(l, x) \int_{(x'_1, \ldots x'_j) \in \mathcal{V}^j(x)} \\
&\beta_j(x'_1 \ldots x'_j, x; \eta) \left(1 - \prod_{\kappa=1}^{j} (1 - m_{++}(l, dx'_\kappa, t))\right) \Big],\\
&m_{++}(l, x'_\kappa, t) = \sum_{c'} \sum_{s \in \text{infectious}} \int_{z'} m(l, x'_\kappa, s, dz', c', t),
\end{aligned}$$

where $\lambda_j(l, x)$ denotes the probability of having j agents around x.

10.3.8 Risk-Aware MFTG

The risk-aware MFTG is given as follows:

- Decision-makers: authorities and individuals,
- Control actions: (η, δ) traveling restriction and testing policy for authorities, (u, v) movement, and meeting rate for individuals,
- Objective functionals: cost–benefit with variance awareness specified below.

Introduce $\bar{u} = \sum_k u_k \mu_k$ and $\mathrm{var}_\mu(u) = \sum_k (u_k - \bar{u})^2 \mu_k$. The benefit–cost functional is a risk-aware function as it depends explicitly on the variance.

Definition 10.2 (Best-Response) The best-response problem is one where, given (m, η) an individual agent chooses (u, v) to solve

$$
\begin{array}{l}
\sup_{u,v} \mathbb{E}[(s+r)(t_1) + \int_{t_0}^{t_1}((s+r) - q\mathrm{var}_\mu(u) - \bar{q}\bar{u}^2 - g_0(u, v, \mu, \eta, m))dt] \\
\text{subject to} \\
\dot{\mu}(l, x, s, z, c, t) = f(l, x, s, z, c, t, \lambda, u, v, \eta, \delta, m), \\
\mu(l, dx, s, dz, c, t_0) = m_0(l, dx, s, dz, c), \\
\dot{m}(l, x, s, z, c, t) = f(l, x, s, z, c, t, \lambda, u^*, v^*, \eta, \delta, m), \\
m(l, dx, s, dz, c, t_0) = m_0(l, dx, s, dz, c), \\
\int m(l, dx, \text{hospitalized}, dz, t) \le \bar{c}_l(t), \\
\int m(l, dx, \text{testing}, dz, t) \le \bar{t}_l(t), \\
x \in \mathcal{D}, \\
v_{|\partial\mathcal{D}} = 0, \\
v_{Exits} = k(l, x, m),
\end{array}
\tag{10.14}
$$

with $\partial\mathcal{D}$ being the border of the map $\mathcal{D}$:

$$
\begin{aligned}
& g_0(u, \mu, \eta, \delta, m) \\
&= \left|v\right|_2^2 [h_1(\int_{x' \in \mathcal{V}_x}(t_+ + p)(dx')) + h_2(x, \mathbb{I}_{\int_{x' \in \mathcal{V}_x}\int_z \sum_c (t_+ + p)(l, dx', dz, c, t) > 1})] \\
&+ \mathrm{var}(v) - b_1(v),
\end{aligned}
$$

and b_1 is the benefit of the agent, and h_1, h_2 are positive and nondecreasing functions (in the second variable for h_2).

The best-response problem given m of the l-th authority is to choose a finite-dimensional vector η_l, δ_l given the migration restrictions imposed by the other authorities:

$$
\begin{array}{l}
\inf_{\eta_l,\delta_l}(d_l + a_l + t_{+,l} + \text{th}_{1,l} + \text{th}_{2,l} + \text{tpc}_l + \text{thos}_l + td_l)(t_1) \\
+ \int_{t_0}^{t_1}((d_l + a_l + \text{thos}_l + \text{td}_l)dt \\
+ \int_{t_0}^{t_1} q_l \text{var}_m(\eta_l) + \bar{q}_l\bar{\eta}_l^2 + g_l(\eta,\delta,m))dt \\
\text{subject to} \\
\dot{m}(l,x,s,z,c,t) = f(l,x,s,z,c,t,\lambda,u,v,\eta,\delta,m), \\
m(l,dx,s,dz,c,t_0) = m_0(l,dx,s,dz,c), \\
\int m(l,dx,\text{hospitalized},dz,t) \leq \bar{c}_l(t), \\
\int m(l,dx,\text{testing},dz,t) \leq \bar{t}_l(t), \\
x \in \mathscr{D}_l, \\
v_{|\partial\mathscr{D}} = 0, \\
v_{\text{Exits}} = k(l,x,m),
\end{array}
\tag{10.15}
$$

where $\eta_l = (\eta_{l'l})_{l'}$, $\bar{\eta}_l = \sum_{l'} m_{l'}\eta_{l'l}$, $\text{var}_m(\eta_l) = \sum_{l'}(\eta_{l'l} - \bar{\eta}_l)^2 m_{l'}$, and $g_l(\eta,\delta,m)$ is the economic cost–benefit at locality l.

10.3.8.1 Mean-Field-Type Equilibrium

Definition 10.3 A mean-field-type Nash equilibrium is a strategy profile $(u^*, v^*, \eta^*, \delta^*)$, which constitutes a solution to the interdependent problems (10.14) and (10.15).

Proposition 10.4 *Let there exists a solution to the following Hamilton–Jacobi–Bellman system in the space of measure* $\mathscr{P}_2(\mathbb{R}^2 \times \mathscr{S} \times [0, 124y] \times C)$ *given by*

$$
\begin{array}{l}
W : [t_0, t_1] \times \mathscr{P}_2(\mathbb{R}^2 \times \mathscr{S} \times [0, 124y] \times C) \times \mathscr{P}_2(\mathbb{R}^2 \times \mathscr{S} \times [0, 124y] \times C) \\
\to \mathbb{R},\ V_l : [t_0, t_1] \times \mathscr{P}_2(\mathbb{R}^2 \times \mathscr{S} \times [0, 124y] \times C) \to \mathbb{R},
\end{array}
\tag{10.16}
$$

where $\mathscr{P}_2$ *is the set of square (in space and age) integrable (positive) measures and*

$$
\begin{array}{l}
W_t + H_0 = 0, \\
W(t_1, \mu, m) = (s + r + tr)(t_1) \\
V_{l,t} + H_l = 0, \\
V_l(t_1, m) = (d_l + a_l + t_{+,l} + th_{1,l} + th_{2,l} + tpc_l + thos_l + td_l)(t_1), \\
l \in \mathscr{L},
\end{array}
\tag{10.17}
$$

where the Hamiltonians are given by

$$
\begin{array}{l}
H_0 = \int \sup_{u,v}[b_0 + \langle f, W_\mu\rangle] + \langle f(., u^*, v^*), W_m\rangle, \\
H_l = \int \inf_{\eta_l,\delta_l}[b_l + \langle f, V_{l,\mu}\rangle], \\
b_0 = (s + r) - q var_\mu(u) - \bar{q}\bar{u}^2 - g_0(u, v, \mu, \eta, m), \\
b_l = d_l + a_l + thos_l + td_l + q_l var_m(\eta_l) + \bar{q}_l\bar{\eta}_l^2 + g_l(\eta, \delta, m), \\
l \in \mathscr{L},
\end{array}
\tag{10.18}
$$

with (u^*, v^*) optimizer in H_0 in (10.18) and (η^*, δ^*) optimizer in H_l in (10.18). Then, $(u^*, v^*, \eta^*, \delta^*)$ is a mean-field-type Nash equilibrium.

The proof of Proposition 10.4 follows from Bellman systems for measure-valued settings. From the formulation of the minimization problem, the number of infected people and the number of deaths are significantly reduced at the mean-field-type equilibrium. However, the cost–benefit analysis suggests that it depends on the functionals g_0, g_l, h_1, h_2.

The master adjoint system (MASS) is given by

$$\begin{array}{l} U_{0,t}(l, x, s, z, c, t, \mu, m) + H_{0,\mu} = 0, \\ U_0(l, x, s', z, c, t_1, \mu, m) = \mathbb{I}_{s' \in \{s, r, \mathrm{tr}\}} \\ U_{l,t} + H_l = 0, \\ U_l(t_1, m) = \\ (d_l + a_l + t_{+,l} + \mathrm{th}_{1,l} + \mathrm{th}_{2,l} + \mathrm{tpc}_l + \mathrm{thos}_l + \mathrm{td}_l)(t_1) \\ l \in \mathcal{L}. \end{array} \tag{10.19}$$

A solution to the MASS system above provides an equilibrium strategy via the best-response strategies from H_0, H_l but not necessarily the equilibrium costs of the decision-makers.

Other Relevant Payoffs

- Authorities: The authority in locality l decides on:
  - migration rules (lockdown, confinement, isolation, curfew) $(\eta_{1,l}, \eta_{2,l}, \eta_{3,l})$ and testing strategy δ_l,
  - budget allocation and incentives ls_l, tr_l.

  Multiple objectives for authority in locality l :
  - Reduce the number of deaths and infected.
  - Reduce economic losses.
  - Maximize the number of recovered–deaths.

$$\left\{ \begin{array}{l} \text{Optimize long-term objectives subject to} \\ \text{constraints} \\ \text{dynamics} \end{array} \right. .$$

- Consumption goods firms: Firms (producing essential, moderate-essential, and less-essential goods) in locality l decide on production, total hours worked $A_1 = \int a_1 dm$, and budget constraint.

  Multiple objectives for j : Reduce the number of infected employees and maximize profit:

$$\begin{cases} \text{Optimize long-term objectives subject to} \\ \text{constraints} \\ \text{dynamics.} \end{cases}$$

- Each individual's risk-awareness: An individual in locality l decides on the meeting rate a_2, local movement x, v, a_3, and consumption a_1. There are multiple objectives for an individual: reducing the risk of being infected, reducing the risk of exposing the others (if co-opetitive), and reducing economic losses:

$$\begin{cases} \text{Optimize long-term objectives subject to} \\ \text{constraints, dynamics} \\ \text{(locality, position, infection status, age, pre-existing health} \\ \text{gender, height, family size, income per family size).} \end{cases}$$

For susceptible:

- Economic:

$$\log\left(\frac{w\, a_1 + ls_g}{1 + tr_g}\right) + b_1(v, spot) - \frac{\hat{\lambda}_1}{2} a_1^2 - \hat{\lambda}_2\, var_\mu(a_1),$$

where w is the wealth and ls_g is the lump sum proposed by the government during COVID-19. $spot$ represents the hotspots of the locality and $b_1(v, spot)$ is the benefit to go the spot. Thus, utility captures some of the features observed in practice. If the individual decides to set its working hours a_1 to zero, then government supports (lump sum ls_g) can help to balance a bit, but this is costly to the government depending on the duration of the pandemic period.

- Pandemic:

$$s - q\, var_\mu(a_2) - \bar{q}\bar{a}_2^2 - h_0(n_{infectious}(\mathscr{B}_\varepsilon(x)))\left|v\right|^2 - \hat{\lambda}_3\, var_\mu(v),$$

where $q, \bar{q}$ are nonnegative numbers.

10.4 Model Outputs and Data Visualization

Data Used and Constraints

We present some illustrative examples with $K = 195$ countries and more detailed analysis for 66+ countries. The COVID-19 data set is taken from CDC, WHO, CSSE, World Bank, and IMF. The age structure data per country is provided in OurWorldinData. For the international connectivity graph, η_1, we have used Flightradar24, which is a global flight tracking service that provides information about flights from 1200 airlines and 4000 airports around the world. The data set does not capture all international connections but it provides a sample. As the initial propagation of COVID-19 started from flights, we have modulated the flight data set to capture it. In particular, we have Flightradar24 to adjust the number of COVID-19 who came via commercial flights. After a certain waiting time, the soft/partial/hard flight restriction decision has been made between some countries. Those countries we made $\eta_{1,l'l}$ to be zero or a very small number (0.001) if travels from l' to l are restricted. For the geographical mobility and co-location map of people in country l_1, we have used OpenStreetMap before and after confinement decisions (if any) per country. The Google Community Mobility Reports can be found online.

The country-specific hospital capacity data set is obtained from the World Bank Group and is used to capture

$$\int \sum_{c'} \text{hospitalized}(l, dx', dz', c', t) \leq \bar{c}_l(t).$$

However, the data set is limited to the country level and it is not per area/city/county. The country-specific testing capacity data set is not available but the number of tests per country is provided online. The set $\mathscr{D} = \cup_{l \in \mathscr{L}} \mathscr{D}_l$ where $\mathscr{D}_l$ is a constrained non-empty set. Given two countries l_1 and l_1', the dynamics $m(l, x, s, z, c, t)$ and $m(l', x, s, z, c, t)$ will be related to each other via the switching rates η_1.

We make the following observations from the model introduced:

- The model is flexible enough to capture some local context-awareness including local mobility and age structure per country. The provided model can be filtered to track the data set even for some of the countries with a few number of cases.
- The connectivity of the local graph η_3 plays a key role in the propagation of the virus across different areas of the city.
- The connectivity of the local graph η_2 plays a key role in the propagation of the virus across different cities.
- The connectivity of the local graph η_1 plays a key role in the propagation of the virus across states/countries.
- The case fatality rate (data-CFR) per country data set is the ratio between the number of (reported) confirmed cases and the total number of (reported) deaths from COVID-19 outbreak data set for that country. We see that there are some significant differences in terms of data-CFR per country even for those who have made similar decisions at similar infection times. Some of the differences comes from the implementation in practice and the response of the people to the measures. The MFTG model studied captures the data-CFR per country as illustrated in 66+ countries. However, different response strategies by individuals and authorities lead to different behaviors.

Visualization

The optimization algorithm is implemented in C++ using mean-field-type game theory tool for 3D risk-aware path choice. We combine several layers: the transportation layer (sea, air, and ground), economics layer, population mobility map after the first infection of COVID-19 in that city/country, and the epidemic data-driven model. This leads to a quadrilemma. We display below the outputs (tested positive, tested–recovered, tested–dead, tested–active) of the epidemic model in 66+ countries. We choose the constraints $\bar{c}_l, \bar{t}_l$ to be high enough so that the interior solution can be found. This means extending the capacity by means of new resources. The COVID-19 initial death rate is estimated for the countries where the death rate due to COVID-19 is unknown. We estimate the death rate from the model using the reported data. We choose $k = 0.7m(.)$ for the numerics at the exits on the grid over OpenStreet map. The rate λ_{ap} is set to 10^{-7} or zero depending on the country/locality. The other λ functions are obtained from the minimization with the smallest norm allowed to derive λ_{data}. η_l is set to 10^{-8} after the borders were closed at locality l. As new data comes, λ_{data} can be recalculated. The newborns are taken from the last year growth scale. From λ_{data}, we plot the behavior of the dynamics (10.12) and integrates it out. The initial t_0 is set to January 22, 2020, for most countries. We observe that the model captures the behavior in most countries. The data-driven MFTG model reveals that the number of active COVID-19-positive patients in countries such as Venezuela, Germany, Italy, Malaysia, Senegal, Zambia, and Togo have several local peaks and oscillating behaviors. Some of them cannot be fitted with a Gaussian or bi-Gaussian distribution. They cannot be fitted with a single exponential. By changing h_l from 0.1 to 0.9, we see a limit cycle of the dynamics. Also, non-Gaussianity of the reported active cases and hospitalized cases are observed in 15+ countries. This includes Albania, Azerbaijan, Bahamas, Bosnia, Costa Rica, Djibouti, Georgia, Iraq, Iran, Jordan, Lebanon, Lithuania, Malta, San Marino, Uzbekistan, etc. This context-aware data-driven MFTG model can be used in visualizing, comparing, planning, implementing, evaluating, and optimizing various detection, testing, prevention, and control programs (Fig. 10.5 and Table 10.4).

10.5 Data-Driven Discrete-Time Models

Even with all the control measures taken in place, the spread of COVID-19 is still dramatic. The efficiencies of the measures taken at different countries are still to be analyzed carefully. Pandemic modeling is an essential scientific tool to answer some questions by aiding people to understand the pandemic data, make predictions, and help the medical professionals and decision-makers allocate resources and design/evaluate intervention strategies to fight against COVID-19 and by looking at re-opening strategies to limit socioeconomic costs. Hence, the quadrilemma

Table 10.4 List of figures and illustrations

Illustration	Meaning
Figure 10.2	First wave of spread
Figure 10.3	Death rate distribution by age
Figure 10.4	Sample age distribution
Figure 10.6	Connection with SIR
Figure 10.7	Connection with SEIR
Figure 10.8	Connection with SIRD
Figure 10.9	Connection with SEIRD
Figure 10.10	Evolution of the epidemiological ODE with 17 states
Figure 10.11	Pandemic vs economic consequences of COVID-19
Figure 10.12	Spread worldwide
Figure 10.13	Propagation of the virus in 15 most reported active cases as of April 11, 2020
Figure 10.14	G7 countries
Figure 10.15	New York and Los Angeles testing results: data vs model
Figure 10.16	Brazil: sequential data-driven model
Figure 10.17	USA: sequential data-driven model
Figure 10.18	Iran: non-Gaussianity
Figure 10.19	Albania: non-Gaussianity
Figure 10.20	Illustrates the non-Gaussianity of the pandemic based on daily reported data and a mean-field-type filter of the SIRD that tracks it
Figure 10.21	COVID-19 samples in selected countries in Africa
Figure 10.22	Selected COMESA countries
Figure 10.23	Selected ECOWAS countries
Figure 10.24	Selected SADC countries

(pandemic–economic–supply chain security–mobility) modeling uses the reported data from 210 countries and 50 states. The table below illustrates some models that have been used to study the course of the COVID-19 pandemic.

Fig. 10.2 First flow of infection of COVID-19 from Asia to Europe to America and then to Africa

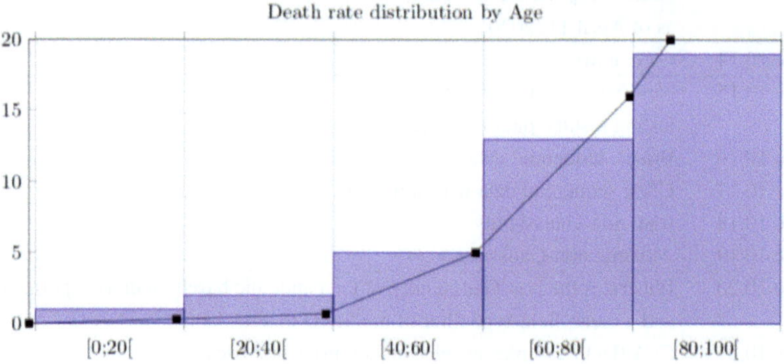

Fig. 10.3 Death rate distribution by age

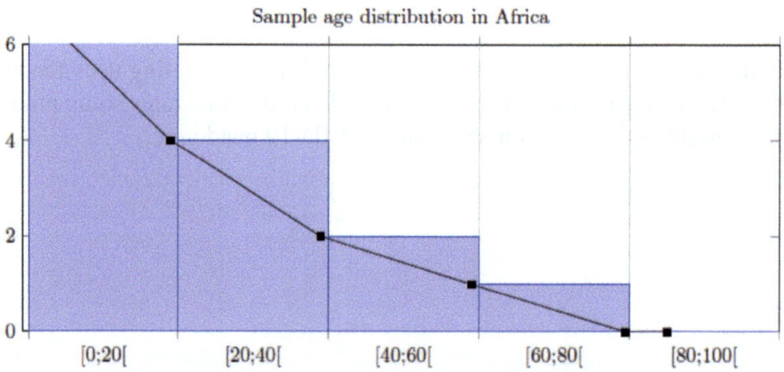

Fig. 10.4 A sample age distribution (unnormalized) for a specific sector

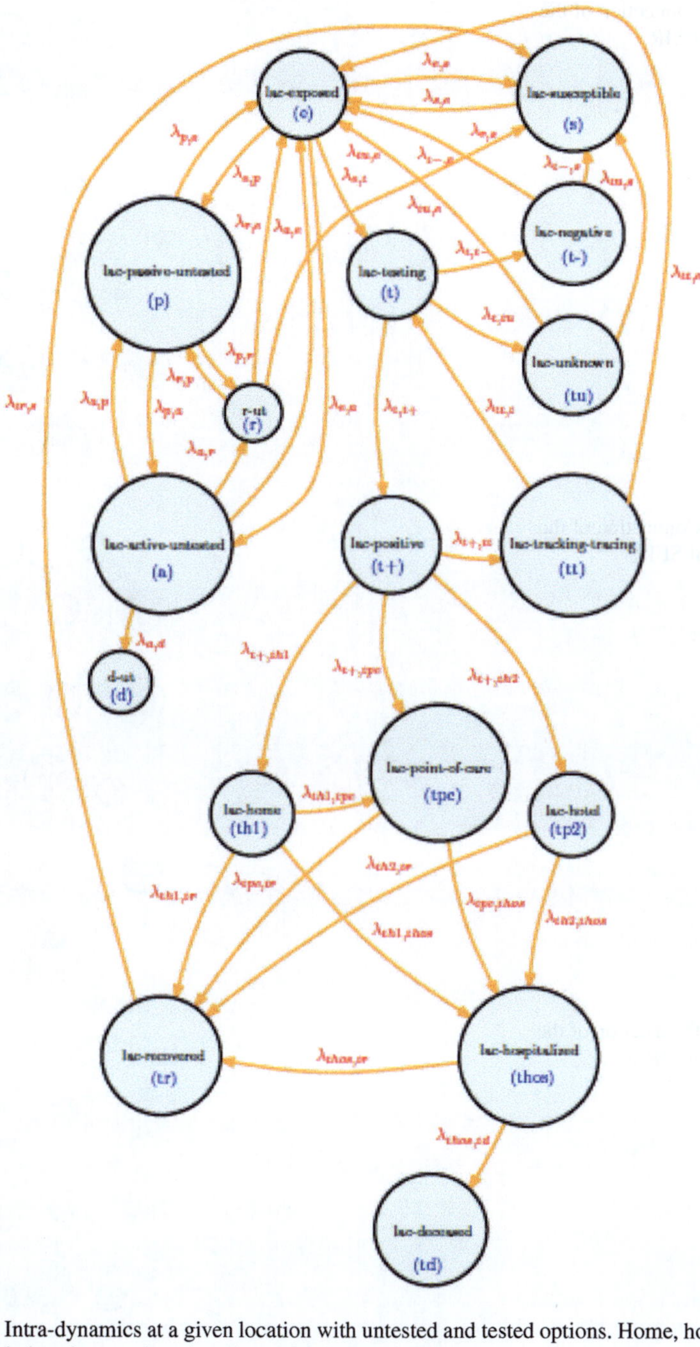

Fig. 10.5 Intra-dynamics at a given location with untested and tested options. Home, hotel, point-of-care isolation. l.a.c = location (l), age; (a), preexisting condition; (c), gender, height, family size, and economic status

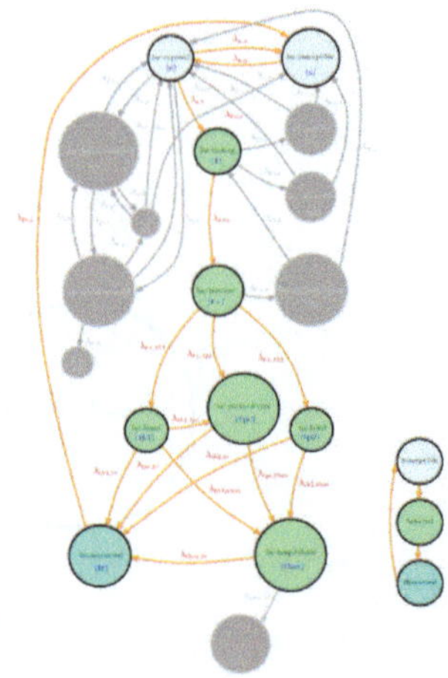

Fig. 10.6 Connection of the dynamics of SIR

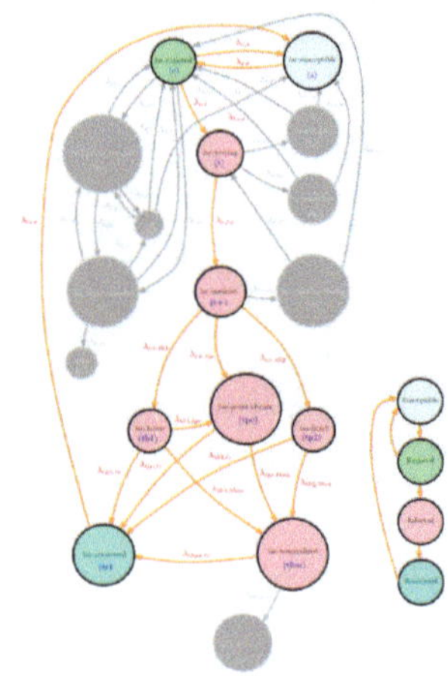

Fig. 10.7 Connection of the dynamics of SEIR

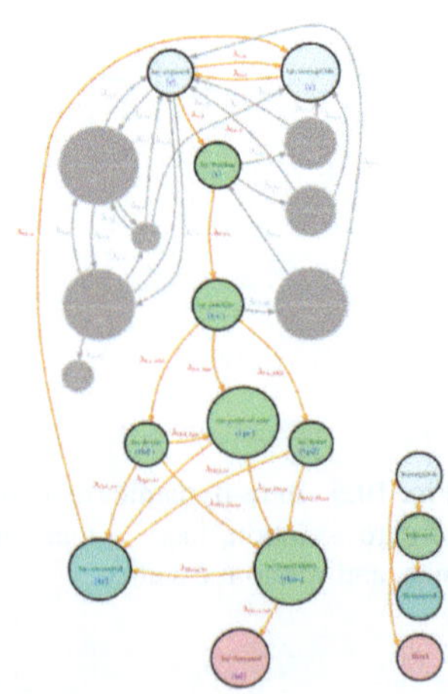

Fig. 10.8 Connection of the dynamics of SIRD

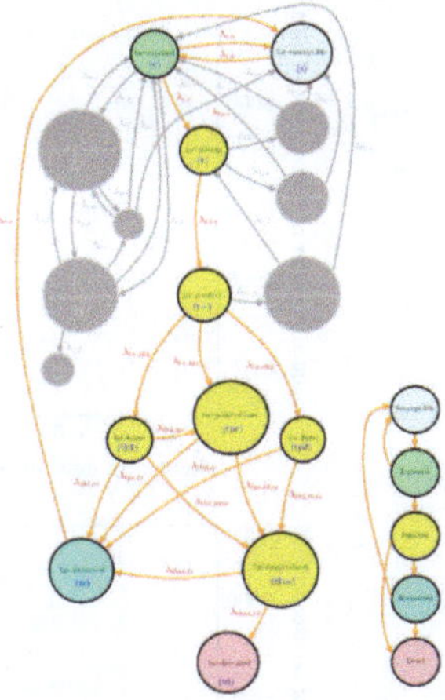

Fig. 10.9 Connection of the dynamics of SEIRD

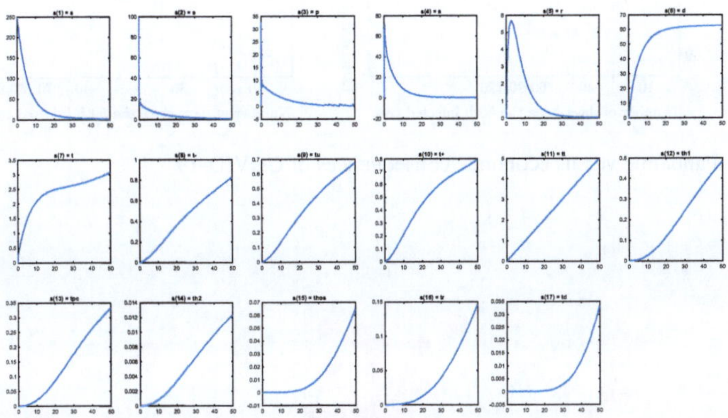

Fig. 10.10 Evolution of the ODE with 17 infection states

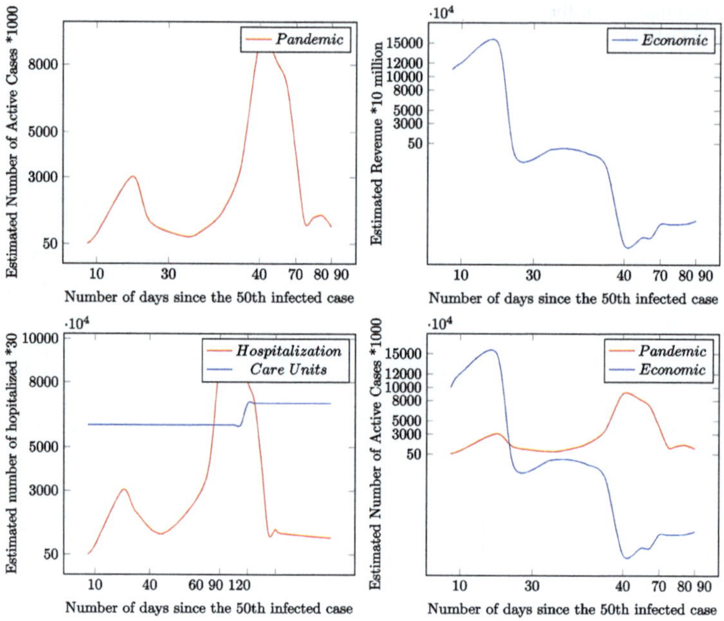

Fig. 10.11 Pandemic versus economic consequences of COVID-19

Fig. 10.12 World map: reported infectious as of March 30, 2020

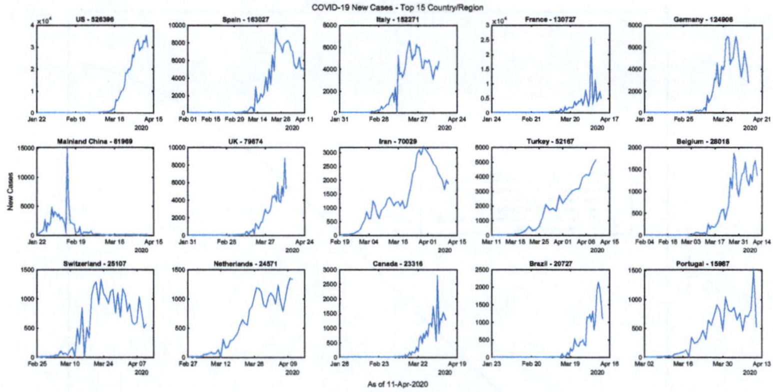

Fig. 10.13 Propagation of the virus in 15 most reported active cases as of April 11, 2020

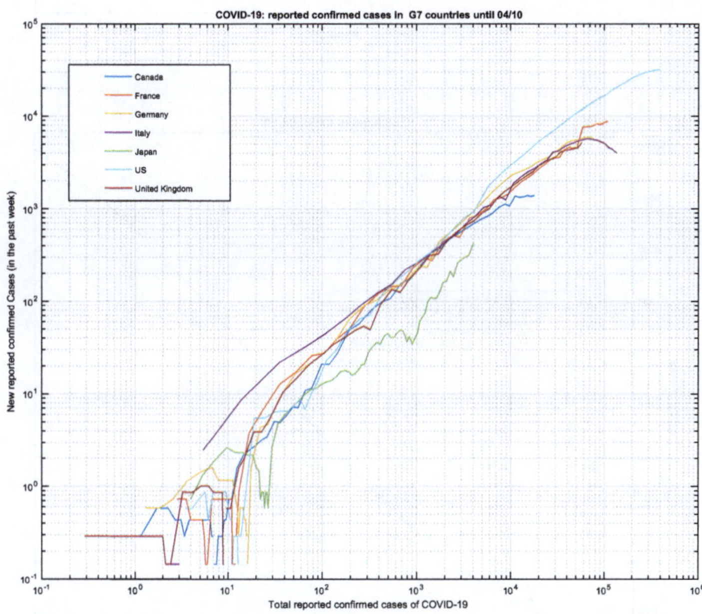

Fig. 10.14 COVID-19 samples in G7 countries

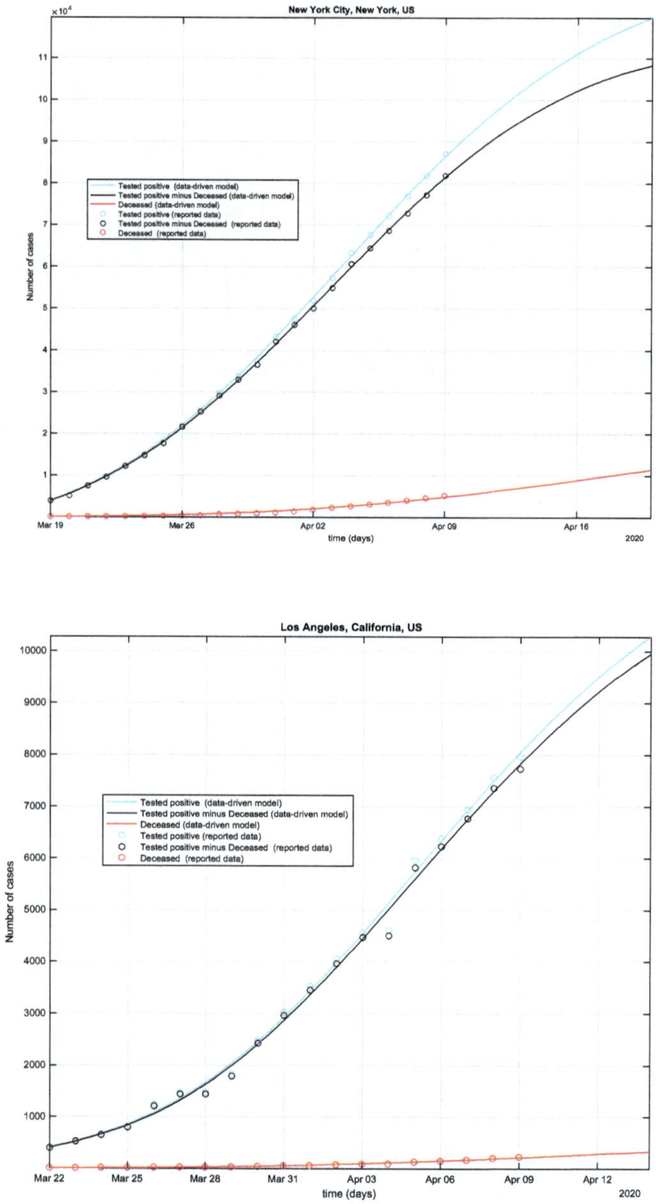

Fig. 10.15 New York and Los Angeles testing results: data versus model

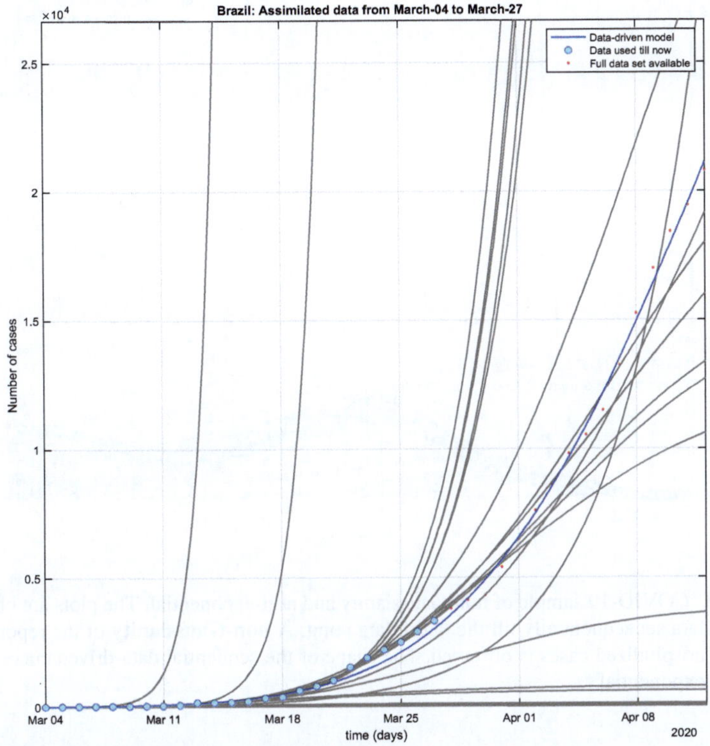

Fig. 10.16 Brazil: sequential data-driven model

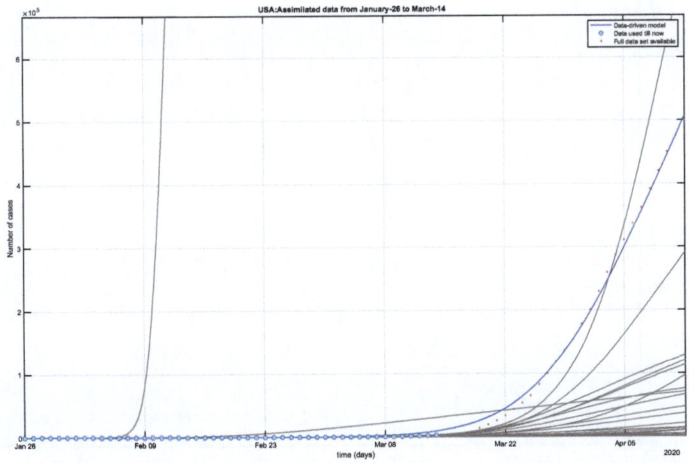

Fig. 10.17 USA: sequential data-driven model

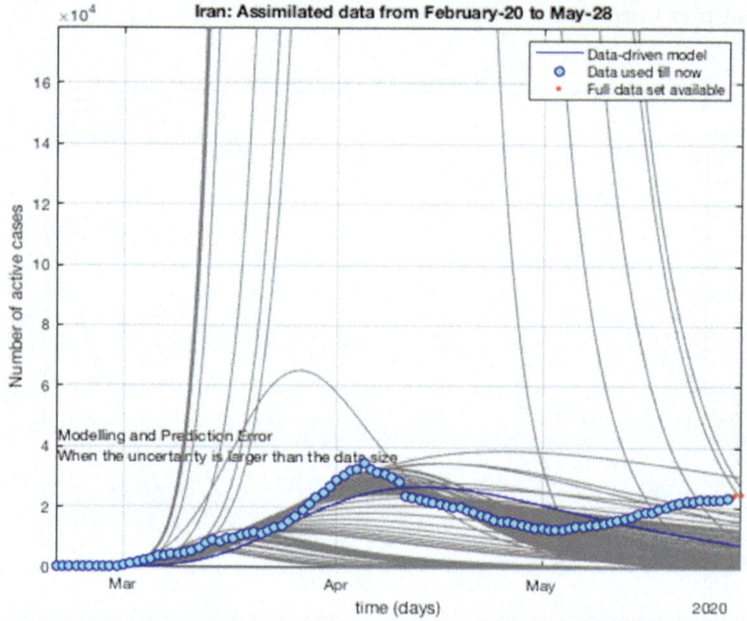

Fig. 10.18 COVID-19 sample of non-Gaussianity and non-exponential. The plots are obtained by using the data set sequentially till the latest data point. A **non-Gaussianity** of the reported active cases and hospitalized cases is observed. The shape of the sequential data-driven curve is clearly not single exponential

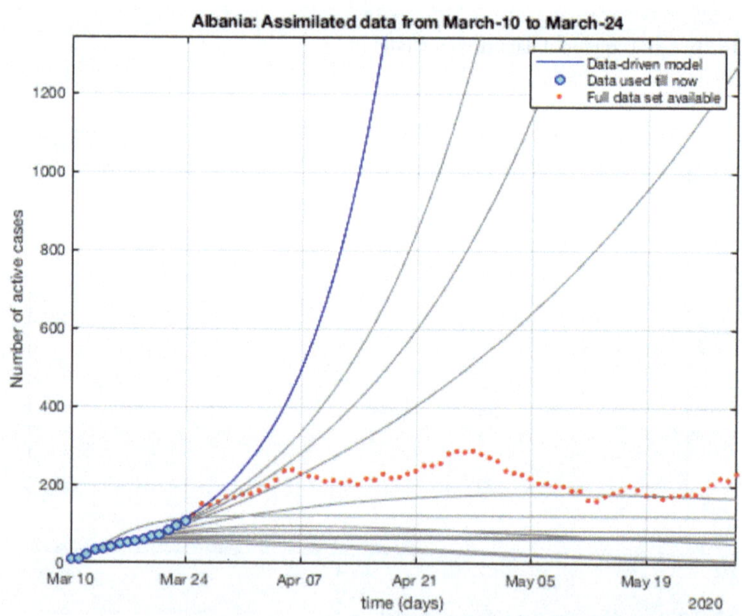

Fig. 10.19 Another COVID-19 sample of non-Gaussianity and non-exponential. The plots are obtained by using the data set sequentially till the latest data point. A **non-Gausisanity** of the reported active cases and hospitalized cases is observed. The shape of the sequential data-driven curve is clearly not single exponential

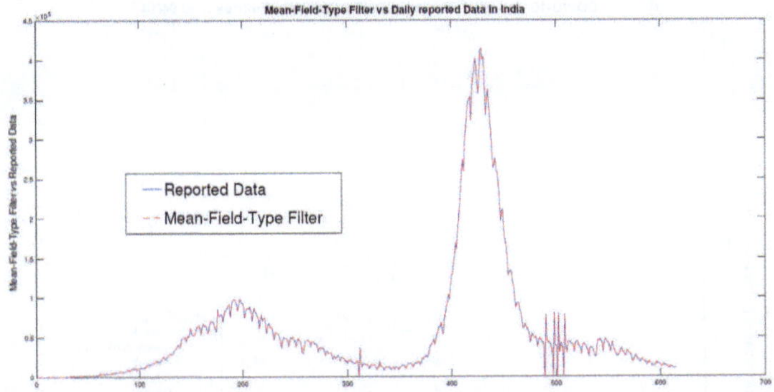

Fig. 10.20 Illustration of non-Gaussianity of the pandemic based on daily reported data and mean-field-type filter that tracks it

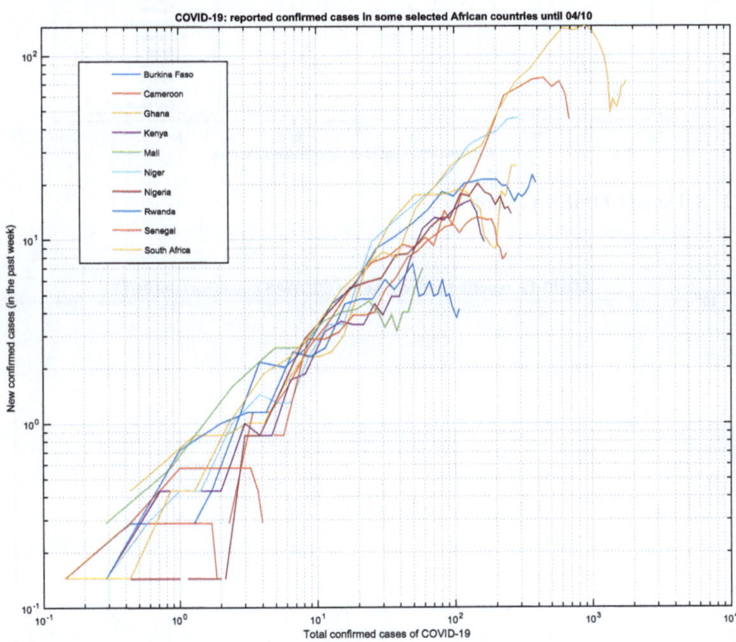

Fig. 10.21 COVID-19 samples in selected countries in Africa

10.5.1 Data Description

We consider the public data at time t as $(N^c, CC^c_k, R^c_k, D^c_k, V^c_k, T^c_k)_{c\in\mathscr{C},k\leq t}$ representing the population size of country c in 2019, reported confirmed COVID-19 cases, reported recovered cases, reported COVID-19 deaths, and reported vaccination status and reported number of PCR tests conducted where c belongs to the list of countries available in the database. PCR means polymerase chain reaction. It is a

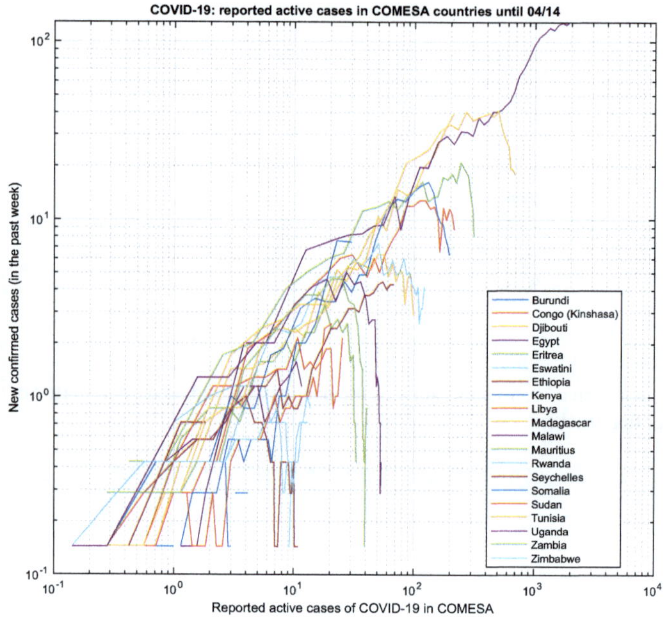

Fig. 10.22 Selected COMESA countries

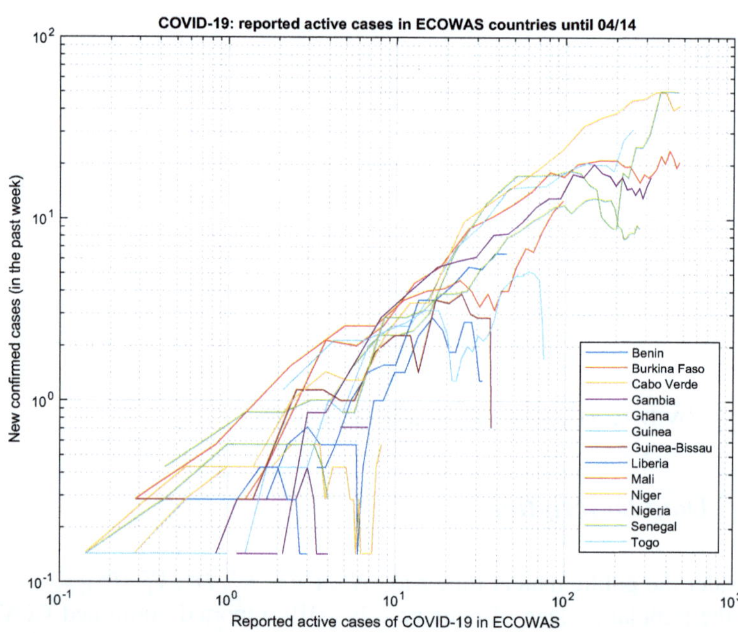

Fig. 10.23 Selected ECOWAS countries

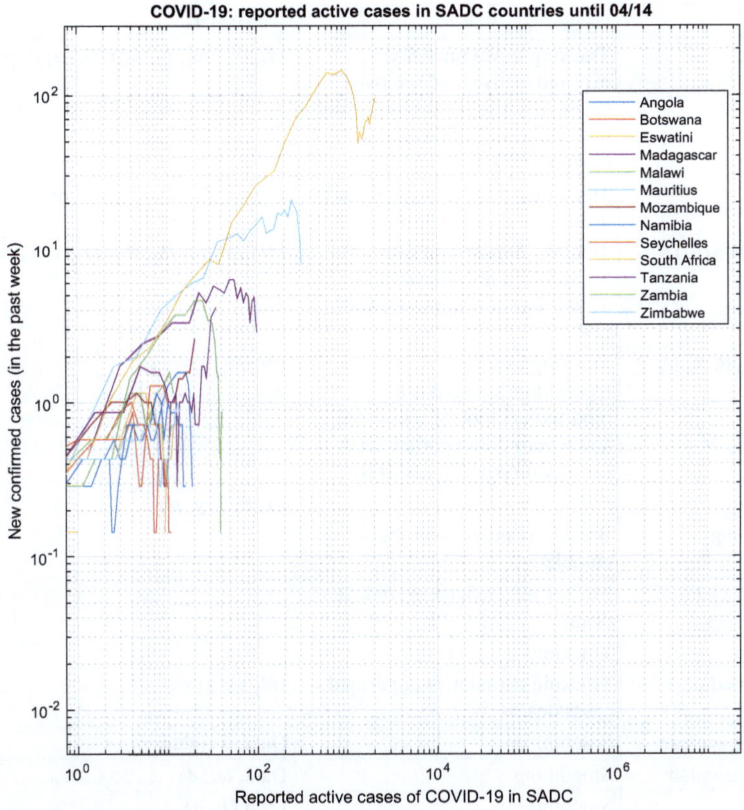

Fig. 10.24 Selected SADC countries

Model	Usage	Parameters
Logistic	Tracking cumulative confirmed cases/deaths	(K, x_0, r)
Beta logistic	Tracking cumulative confirmed cases/deaths	$(K, x_0, r, p, q, \alpha)$
SIR	Tracking active and removed cases. Compute R_0	(N, I_0, R_0), (β, γ)
SEIR	Tracking exposed, active, and removed cases. Compute $R0$	(N, I_0, E_0, R_0), (α, β, γ)
Fitted SEIR	Tracking exposed, active, and removed cases. Tracking confirmed cases and confirmed closed cases. Compute R_0	(N, I_0, E_0, R_0), (α, β, γ)

(continued)

Model	Usage	Parameters
STEIR	Tracking travel intensity, exposed, active, and removed cases. Tracking confirmed cases and confirmed closed cases. Compute $R0$	(N, T_0, E_0, I_0, R_0), $(\lambda, \alpha, \beta, \gamma)$
SIRD	Tracking active, recovered, and death cases. Compute R_0, IFR	(N, I_0, R_0, D_0), $(\beta, \gamma_r, \gamma_d)$
Fitted SEIHRD	Tracking exposed, active, hospitalized, recovered, and death cases. Compute R_0 and IFR	$(N, E_0, I_0, H_0, R_0, D_0)$ $(\alpha, \beta, \gamma, \mu)$
$SEI_rI_uR_rR_uD_rD_u$	Tracking exposed, reported/unreported active, reported/unreported recovered, and reported/unreported death cases. Compute R_0 and IFR	$(N, E_0, I_{r0}, I_{u0}, R_{r0}, R_{u0}, D_{r0}, D_{u0})$ $(\alpha, \beta, \delta, \gamma, \mu)$
Branching	Branching process for the cumulative	
Markov chain	MC for infected, recovered, and death	
Poisson	Tracking	
Euler method	Tracking cumulative confirmed cases/deaths	(N, x_0, x_1)
Moving average	Smoothing	Data, weights, x_0
Autoregression	Smoothing	Data, $(\mu, \phi_1, \ldots, \phi_p, x_{-p}, \ldots, x_{-1})$
ARMA	Smoothing	Data,(p, q), $(\mu, \phi_1, \ldots, \phi_p, \theta_1, \ldots, \theta_q)$ $(x_{-p}, \ldots, x_{-1}, \varepsilon_q, \ldots, \varepsilon_{-1})$
ARIMA	Smoothing	Data, (p, d, q) $(\mu, \phi_1, \ldots, \phi_p, \theta_1, \ldots, \theta_q)$ $(x_{-p-d}, \ldots, x_{-1}, \varepsilon_q, \ldots, \varepsilon_{-1})$
SARIMA	Smoothing	Data, $(p, d, q)(P, D, Q)s$ $(\mu, \phi_1, \ldots, \phi_p, \theta_1, \ldots, \theta_q)$ $(x_{-p-d}, \ldots, x_{-1}, \varepsilon_q, \ldots, \varepsilon_{-1})$ $(\Phi_1, \ldots, \Phi_P, \Theta_1, \ldots, \Theta_Q)$
SARIMAX	Smoothing	Data, SARIMA parameters $+(n, \ldots, \alpha_1, \ldots, \alpha_n)$
ETS	Smoothing	Data, $(l_0, b_0, s_{-L}, \ldots, s_{-1})$, $(\alpha, \beta, \gamma, h, L)$
TBATS	Tracking	Data, $(l_0, b_0, s_{-L}, \ldots, s_{-1})$, $(\alpha, \beta, \gamma, h, L)$
Holt	Smoothing	Data, $(l_0, b_0, \alpha, \beta, h)$

test to detect genetic material from a specific organism, such as a SARS-CoV-2. We describe below some basic models used to estimate the course of the pandemic.

Model	Usage	Parameters
Damped Holt	Smoothing	Data $(l_0, b_0, \alpha, \beta, \phi, h)$
Holt–Winters	Smoothing	Data, $(s_0, b_0, c_{-k}, \alpha, \beta, \gamma, h, L)$
Damped Holt–Winters	Tracking	Data, $(s_0, b_0, c_{-k}, \alpha, \beta, \gamma, h, L, \phi)$
NARNN	Tracking	Data, weights
LSTM	Tracking	Data, weights per layer
RNN	Tracking	Data, weights per layer

10.5.2 Some Time Series Models

We review some basic time series models for backcasting, nowcasting, and forecasting. The goal is to connect these time series to the epidemiology of COVID-19, i.e., the estimation of susceptible to infection, infected, actively infected, deceased, recovered cases, basic reproduction number per location, etc.

10.5.3 One-Step Ahead Multivariate Filters

Since the estimation target concerns multiple components, we look at multivariate filters. We start with one-step ahead multivariate filters. We denote the daily time-varying observation in country c at time t by $O_t^c = (CC_t^c, R_t^c, D_t^c, V_t^c, T_t^c)$, which is an s-dimensional vector with $s = 5$ and with nonnegative entries:

- At time t, we aim to estimate $\tilde{O}_{t+1}$ (or its distribution) based on the (past) observations $(O_k)_{k \leq t}$.

Below we present some basic examples of one-step ahead multivariate filters.

10.5.3.1 Multivariate Mann Iteration Method

A basic example is the Mann and Ishikawa iterates:

$$\begin{cases} \tilde{O}_0, \\ \Delta \tilde{O}_{t+1} = \Lambda_t (O_t - \tilde{O}_t), \end{cases} \tag{10.20}$$

where Λ_t is a (5×5)-matrix representing the learning rate. When Λ_t is the identity matrix $\Lambda = I_{5\times 5}$, the Mann iterate becomes the latest available observation. When Λ_t is zero, the Mann iterate remains equal to the previous estimated value. As O_t is a s-dimensional vector with $s = 5$, we work the norms of the error vector: $\left| O_t - \tilde{O}_t \right|_1$

and $\left|O_t - \tilde{O}_t\right|_2$. The 1-norm of $x = (x^1, x^2, x^3, x^4, x^5)$ is $\left|x\right|_1 = \sum_{i=1}^5 |x^i|$, and the 2-norm of x is $\left|x\right|_2 = \sqrt{\sum_{i=1}^5 |x^i|^2}$.

The application of every Mann iteration method and every simple exponential smoothing method requires the smoothing matrix parameters Λ_t and the initial values $\tilde{O}_0$ to be chosen.

10.5.3.2 Multivariate Hodrick–Prescott Filter

Another basic multivariate filter is the Hodrick–Prescott filter (HP filter). The HP filter output of the five-dimensional time series $(O_1, \ldots, O_T)$ of T instants is a minimizer $(\hat{O}_1, \ldots, \hat{O}_T)$ of the following function:

$$
\begin{aligned}
(\hat{O}_1, \ldots, \hat{O}_T) \in \mathbb{R}^{sT} \mapsto \sum_{t=1}^{T} \left|O_t - \hat{O}_t\right|_2^2 + \lambda \sum_{t=3}^{T} \Big|(\hat{O}_t - \hat{O}_{t-1}) \\
- (\hat{O}_{t-1} - \hat{O}_{t-2})\Big|_2^2,
\end{aligned}
\tag{10.21}
$$

where $\lambda \geq 0$ is a design parameter to be determined.

10.5.3.3 Evaluation Criteria

We will use a basic evaluation of the one-step ahead filter as

$$
ev_l(K, T) = \left(\frac{1}{K}\sum_{c=1}^{K} \frac{1}{T_c s}\sum_{t=1}^{T_c} \left|\hat{O}_t^c - O_t^c\right|_l^l\right)^{\frac{1}{l}}, \quad l \geq 1.
$$

Thus, for $l = 2$, one has

$$
ev_2(K, T) = \sqrt{\frac{1}{K}\sum_{c=1}^{K} \frac{1}{T_c s}\sum_{t=1}^{T_c} \left|\hat{O}_t^c - O_t^c\right|_2^2}.
$$

The evolution of $ev_2(K, T)$ for $T_c \in \{1, \ldots, 800\}$ reflects the global root mean square error (RMSE) for deterministic series as a function of the length of the observations. The local RMSE in country c is

$$
RMSE^c(T) = \sqrt{\frac{1}{T_c s}\sum_{t=1}^{T_c} \left|\hat{O}_t^c - O_t^c\right|_2^2}.
$$

These criteria are not suitable for stochastic processes. One way to modify $ev_l(K, T)$ could be through the expected value as

$$ev_l(K,T) = \left(\frac{1}{Ks}\sum_{c=1}^{K}\frac{1}{T_c}\sum_{t=1}^{T_c}\mathbb{E}\left|\hat{O}_t^c - O_t^c\right|_l^l\right)^{\frac{1}{l}}, \; l \geq 1,$$

where $\mathbb{E}(X)$ denotes the expectation operator.

10.5.4 Multistep Ahead Multivariate Filters

Let h be the length of the forecast horizon of the filter. The goal is to compute

$$\hat{O}_{t+h|t}, \hat{O}_{t+h-1|t}, \ldots, \hat{O}_{t+1|t}$$

based on the observations $O_t^c, \ldots, O_0^c$ up to t.

10.5.4.1 Multistep Mann Iterates

Using the Mann iteration method (10.20), we obtain

$$\begin{cases} \tilde{O}_0, \\ \Delta\tilde{O}_{t+1|t} = \Delta\tilde{O}_{t+1} = \Lambda_t(O_t - \tilde{O}_t), \\ \tilde{O}_{t+j|t} = \tilde{O}_{t+j-1|t} = \tilde{O}_{t+1}, \; j \in \{2, \ldots, h\}. \end{cases} \tag{10.22}$$

That is, all forecasts take the same value, equal to the last level component. A slightly modified version is

$$\begin{cases} \tilde{O}_0, \\ \Delta\tilde{O}_{t+1|t} = \Delta\tilde{O}_{t+1} = \Lambda_t(O_t - \tilde{O}_t), \\ \Delta\tilde{O}_{t+j|t} = \Lambda_{t+j-1}(O_t - \tilde{O}_{t+j-1|t}), \; j \in \{1, \ldots, h\}. \end{cases} \tag{10.23}$$

Another modified version is

$$\begin{cases} \tilde{O}_0, \\ \Delta\tilde{O}_{t+1|t} = \Delta\tilde{O}_{t+1} = \Lambda_t(O_t - \tilde{O}_t), \\ \tilde{O}_{t+j|t} = \tilde{O}_t + j(O_{t-1} - O_t), \;\; j \in \{2, \ldots, h\}. \end{cases} \tag{10.24}$$

10.5.4.2 Holt–Winters Additive Triple Exponential Smoothing

Here we use the Holt–Winters additive version of the method. Triple exponential smoothing with additive seasonality is given by

$$\begin{cases} (s_0, b_0, c_0), \\ \Delta s_t = b_{t-1} + \alpha(O_t - c_{t-L} - s_{t-1} - b_{t-1}), \\ \Delta b_t = \beta(s_t - s_{t-1} - b_{t-1}), \\ c_t = c_{t-L} + \gamma(O_t - s_{t-1} - b_{t-1} - c_{t-L}), \\ \Delta \hat{O}_{t+j} = -\hat{O}_{t+j-1} + s_t + jb_t + c_{t-L+j_m}, \\ j \in \{1, \ldots, h\}, \\ j_L = \lfloor (j-1) \, mod \, L \rfloor + 1, \end{cases} \tag{10.25}$$

where $s_0, b_0, c_0, \alpha, \beta, \gamma, h, L$, to be designed. s_t represents a smoothed value, b_t is the sequence of the linear trend with seasonal changes, and c_t is the sequence of seasonal correction factors.

The additive Holt–Winters model is a particular case of state space models. To see it, consider the error $\varepsilon_t = O_t - \mathbb{E}(O_t | O_k, \; k \leq t-1) = O_t - \hat{O}_t$. Thus, $O_t = \hat{O}_t + \varepsilon_t$, and $\hat{O}_t$ is obtained from

$$\begin{cases} \text{Given } (s_0, b_0, c_0), \\ \Delta s_t = b_{t-1} + \alpha \varepsilon_t, \\ \Delta b_t = \beta \varepsilon_t, \\ c_t = c_{t-L} + \gamma \varepsilon_t, \\ \Delta \hat{O}_{t+j} = -\hat{O}_{t+j-1} + s_t + jb_t + c_{t-L+j_L}, \\ j \in \{1, \ldots, h\}, \\ j_L = \lfloor (j-1) \, mod \, L \rfloor + 1. \end{cases} \tag{10.26}$$

10.5.4.3 Damped Holt–Winters

$$\begin{cases} s_0, b_0, c_k, k = -L, \ldots, 0, \\ \Delta s_t = \phi_t b_{t-1} + \alpha_t (O_t - c_{t-L} - s_{t-1} - \phi_t b_{t-1}), \\ \Delta b_t = (\phi_t - 1) b_{t-1} + \beta_t (s_t - s_{t-1} - \phi_t b_{t-1}), \\ c_t = c_{t-L} + \gamma_t (O_t - s_{t-1} - b_{t-1} - c_{t-L}), \\ \Delta \hat{O}_{t+j} = -\hat{O}_{t+j-1} + s_t + (\phi_t + \phi_t^2 + \ldots + \phi_t^j) b_t + c_{t-L+j_m}, \; j \in \{1, \ldots, h\}. \end{cases} \tag{10.27}$$

10.5.4.4 Singular Spectrum Analysis

The singular spectrum analysis (SSA) relates to the spectrum of eigenvalues in a singular value decomposition widely used in Kosambi–Karhunen–Loève decomposition. The base SSA model is as follows: $\hat{O}_t = s_t + \varepsilon_t$, and

$$s_t = \sum_{k=0}^{t} \theta_k s_{t-k}.$$

10.5.4.5 Multivariate ARIMA

A simple decomposition of a stochastic process O_t is

$$O_t = \Big(O_t - \mathbb{E}(O_t)\Big) + \mathbb{E}(O_t) = E_t + \eta_t,$$

where $E_t = O_t - \mathbb{E}(O_t)$ is a zero-mean process and $\eta_t = \mathbb{E}(O_t)$ is a deterministic sequence obtained from the expected values of O. It remains to decompose the zero-mean process E_t. Below we provide such a decomposition for a particular case of E_t.

The Wold representation theorem (given below), named after Herman Wold, states that every covariance–stationary time series can be written as the sum of two time series, one deterministic and one stochastic.

Theorem 10.5 *Let $\{O_t\}_{t\geq 0}$ be a covariance stationary process and then*

$$O_t = \eta_t + \sum_{j=1}^{+\infty} \theta_j \varepsilon_{t-j},$$

where:

- $\phi_0 = 1,\ \sum_{j=0}^{\infty} \phi_j^2 < +\infty,$
- $\varepsilon_t \sim N(0, \sigma^2)$ *with* $\sigma^2 < +\infty$, $cov(\varepsilon_s, \eta_t) = 0,\ \forall (t, s),$
- $\{\eta_t\}$ *is a deterministic time series.*

This representation can be seen as a moving average $MA(\infty)$ for a covariance–stationary time series. Note that this is not a finite representation/decomposition. Based on this theorem, one can build autoregressive integrated moving average, or ARIMA. Particular cases include multivariate AR, multivariate MA, and multivariate ARMA. A generic $ARIMA(p, d, q)$ has the following form:

$$\begin{cases} (O_{-p-d}, \ldots, O_{-1}, \varepsilon_{-q}, \ldots, \varepsilon_{-1}) \\ \phi(B)(1-B)^d O_t = \mu + \theta(B)\varepsilon_t,\ t \geq 0 \end{cases} \tag{10.28}$$

with $\phi(B) = 1 - \sum_{k=1}^{p} \phi_k B^k$, $\phi_p \neq 0$, $\theta(B) = 1 - \sum_{k=1}^{q} \theta_k B^k$, $\theta_q \neq 0$ with initial conditions $(O_{-p-d}, \ldots, O_{-1}, \varepsilon_{-q}, \ldots, \varepsilon_{-1})$. B denotes the backward shift (time lag) operator, i.e., $BO_t = O_{t-1}$, and so, $(1 - B)O_t = \Delta O_t$:

- $ARIMA(0, 0, 1) = MA(1)$.
- $ARIMA(0, 0, q) = MA(q)$.
- $ARIMA(0, 1, 1)$= Mann iteration = simple exponential smoothing.
- $ARIMA(0, 2, 2)$ =Holt method.
- $ARIMA(1, 1, 2)$ =Damped Holt method.
- $ARIMA(0, 1, 0) = I(1)$ means $\Delta \hat{O}_{t+1} = \mu + \varepsilon_t$, which is a random walk with drift if $\mu \neq 0$.
- $ARIMA(1, 0, 0) = AR(1)$.
- $ARIMA(p, 0, 0) = AR(p)$.
- $ARIMA(p, 0, q) = ARMA(p, q)$.

Some pros of ARIMA: ARIMA models are widely used in the literature. It requires the prior data of a time series to generalize the forecast iteratively. It performs relatively well for very short-term forecasts. The ARIMA model includes Holt, damped Holt, and simple exponential smoothing methods as special cases.

Some contras of ARIMA: ARIMA has several limitations when applied to COVID-19 data. COVID-19 has specific trends and shapes. It is difficult for ARIMA to predict turning points, peaks, waves, and wavelengths. There is a need of an extra method for choosing the (p, d, q) order of the ARIMA model using, for example, information criteria, in addition to selection of weights by minimizing a cost of error functional, which can be computationally expensive when the value of $p + d + q$ gets large. ARIMA seems to provide a poorer performance for long-term forecasts when tested on COVID-19 for 3-month horizon. Last but not the least, depending on the variant of SARS-CoV-2, the seasonality is not captured by ARIMA.

10.5.4.6 Fractional ARIMA

In an ARIMA model, the integrated part of the model includes the differencing operator $(1 - B)$ where B is the backshift operator raised to an integer power. In a fractional model, the power is allowed to be fractional. $\phi(B)(1 - B)^d O_t = \mu + \theta(B)\varepsilon_t$, where d is now fractional and $(1 - B)^d = \sum_{k=0}^{\infty} \left(\frac{\prod_{j=0}^{k-1}(d-j)}{k!} \right) (-B)^k$.

10.5.4.7 Multivariate SARIMA

Here we introduce the seasonal autoregressive integrated moving average (SARIMA) models. This is an extension to ARIMA that supports the direct modeling of the seasonal component of the series, which is called SARIMA. The basic setup is given by

$$\begin{cases} O_1, \ldots, O_t \\ \Delta \hat{O}_{t+1} = -\hat{O}_t + \mu \\ +\sum_{k=1}^{p} \phi_k O_{t-k} + \sum_{k=1}^{P} \Phi_k O_{t-ks} \\ -\sum_{k=1}^{q} \theta_k \varepsilon_{t-k} - \sum_{k=1}^{Q} \Theta_k \varepsilon_{t-ks} + \varepsilon_t, \end{cases} \tag{10.29}$$

with parameters $(\mu, (\phi_1, \ldots, \phi_p), (\Phi_1, \ldots, \Phi_P), \ (\theta_1, \ldots, \theta_q), (\Theta_1, \ldots, \Theta_Q))$ denoted by $(p, q, 0, P, Q)s$, $SARIMA(p, d, q), (P, D, Q)s$ is given by

$$\begin{cases} (O_{-p-d}, \ldots, O_{-1}, \varepsilon_{-q}, \ldots, \varepsilon_{-1}) \\ \phi_P(B^s)\phi_p(B)(1 - B^s)^D(1 - B)^d O_t = \mu + \theta_Q(B^s)\theta_q(B)\varepsilon_t, \ t \geq 0. \end{cases} \tag{10.30}$$

10.5.4.8 Multivariate SARIMAX

A seasonal autoregressive integrated moving average with eXogeneous parameters (SARIMAX) is given by

$$\begin{cases} (O_{-p-d}, \ldots, O_{-1}, \varepsilon_{-q}, \ldots, \varepsilon_{-1}) \\ \phi_P(B^s)\phi_p(B)(1 - B^s)^D(1 - B)^d O_t = \mu + \theta_Q(B^s)\theta_q(B)\varepsilon_t + \sum_{i=1}^{n} \beta_i x_t^i, \ t \geq 0, \end{cases} \tag{10.31}$$

where x_t^i are exogeneous variables with coefficient β.

10.5.4.9 BATS

BATS means "B" for Box–Cox transformation, "A" for ARIMA errors, "T" for Trend, and "S" for Seasonal components:

$$\begin{cases} (l_0, b_0, (s_0^{(i)})_i, \\ O_t^{(w)}(i) = \frac{(O_t(i))^w - 1}{w} \mathbb{I}_{\{w>0\}} + \log(O_t(i))\mathbb{I}_{\{w=0\}}, \\ O_t^{(w)} = l_{t-1} + \phi b_{t-1} + \sum_{i=1}^{M} s_{t-m_i}^{(i)} + d_t, \\ l_t = l_{t-1} + \phi b_{t-1} + \alpha d_t, \\ b_t = \phi b_{t-1} + \beta d_t, \\ s_t^{(i)} = s_{t-m_i}^{(i)} + \gamma d_t, \\ d_t = \sum_{i=1}^{p} \phi_i d_{t-i} + \sum_{j=1}^{q} \theta_j \varepsilon_{t-j} + \varepsilon_t. \end{cases} \tag{10.32}$$

The model is denoted as $BATS(w, p, q, \phi, m_i, i \in \{1, \ldots, M\})$, with parameters

$$w, \phi, \alpha, \beta, \gamma, \phi_1, \ldots, \phi_p, \theta_1, \ldots, \theta_q.$$

10.5.4.10 TBATS

We now add a trigonometric seasonality to BATS. This leads to TBATS in which "T" stands Trigonometric seasonality, "B" for Box–Cox transformation, "A" for ARMA errors, "T" for Trend, and "S" for Seasonal components:

$$\begin{cases} (l_0, b_0, (s_0^{(i)})_i), \\ O_t^{(w)}(i) = \frac{(O_t(i))^w - 1}{w}\mathbb{I}_{\{w>0\}} + \log(O_t(i))\mathbb{I}_{\{w=0\}}, \\ O_t^{(w)} = l_{t-1} + \phi b_{t-1} + \sum_{i=1}^{M} s_{t-m_i}^{(i)} + d_t, \\ l_t = l_{t-1} + \phi b_{t-1} + \alpha d_t, \\ b_t = \phi b_{t-1} + \beta d_t, \\ s_t^{(i)} = s_{t-m_i}^{(i)} + \gamma d_t, \\ d_t = \sum_{i=1}^{p} \phi_i d_{t-i} + \sum_{j=1}^{q} \theta_j \varepsilon_{t-j} + \varepsilon_t, \\ s_t^{(i)} = \sum_{j=1}^{k_i} s_{j,t}^{(i)}, \\ s_{j,t}^{(i)} = s_{j,t-1}^{(i)} \cos(\lambda_j^{(i)}) + s_{j,t-1}^{(i),*} \sin(\lambda_j^{(i)}) + \gamma_{1i} d_t, \\ s_{j,t}^{(i),*} = -s_{j,t-1}^{(i)} \sin(\lambda_j^{(i)}) + s_{j,t-1}^{(i),*} \cos(\lambda_j^{(i)}) + \gamma_{2i} d_t, \end{cases} \tag{10.33}$$

with $\lambda_j^{(i)} = 2\pi \frac{j}{m_i}$, $k_i = \lfloor m_i/2 \rfloor$. The model is denoted as $(w, p, q, \phi, (m_i, k_i), i \in \{1, \ldots, M)$:

- Strengths: The TBATS model has almost all the strengths that BATS has. One can do multi-seasonality.
- Weaknesses: The assumption on ε may not hold. The computation cost is high for a large data size.

10.5.4.11 $AR(p)$ as a $VAR(1)$

Consider an $AR(p)$ with

$$\begin{cases} (O_{-p}, \ldots, O_{-1}) \\ \phi_p(B) O_t = \mu + \varepsilon_t, \ t \geq 0, \end{cases} \tag{10.34}$$

which can be rewritten as

$$\begin{cases} (O_{-p}, \ldots, O_{-1}) \\ O_t = C + \varepsilon_t + \sum_{k=1}^{p} \phi_k O_{t-k}, \\ O_{t-1} = O_{t-1} \\ \vdots \\ \vdots \\ O_{t-p+1} = O_{t-p+1}, \end{cases} \tag{10.35}$$

which is

$$\begin{cases} M_0 = (O_{-p+1}, \ldots, O_0) \\ M_t = C + E_t + \Phi M_{t-1}, \end{cases} \tag{10.36}$$

with $M_t = (O_t, \ldots, O_{t-p+1})$, $C = (\mu, 0, \ldots, 0)$, Φ a matrix with entries $\Phi_{1j} = \phi_j$, and for $i > 1$, $\Phi_{ii} = 1$, and zero for all the other entries, and $E_t = (\varepsilon_t, 0, \ldots, 0)$.

The latter form is $VAR(1)$, and it proves $AR(p)$ can be seen as $VAR(1)$.

10.5.4.12 ARCH

The autoregressive conditional heteroskedasticity (ARCH) model is described by $\varepsilon_t = \sigma_t n_t$, where $n_t \sim N(0, 1)$ and

$$\sigma_t^2 = \theta_0 + \sum_{k=1}^{q} \theta_k \varepsilon_{t-k}^2,$$

with $\theta_i \geq 0$, $q \geq 1$, $\theta_0 > 0$, q is the length of ARCH lags.

10.5.4.13 GARCH Model

The generalized autoregressive conditional heteroskedasticity ($GARCH(p, q)$) model is given by $\varepsilon_t = \sigma_t n_t$, where $n_t \sim N(0, 1)$ and

$$\sigma_t^2 = \theta_0 + \sum_{k=1}^{q} \theta_k \varepsilon_{t-k}^2 + \sum_{k=1}^{p} \theta_k \sigma_{t-k}^2,$$

with $\theta_i \geq 0$, $q \geq 1$, $\theta_0 > 0$, q is the length of ARCH lags.

10.5.4.14 GJR GARCH Model

Glosten, Jagannathan, and Runkle models asymmetry in the ARCH process as $\varepsilon_t = \sigma_t n_t$, with

$$\sigma_t^2 = \mu + \theta \sigma_{t-1}^2 + \alpha \varepsilon_{t-1}^2 + \phi \varepsilon_{t-1}^2 \mathbb{I}_{\{\varepsilon_{t-1} < 0\}}.$$

10.5.5 Evaluation Criteria

The global RMSE with forecast horizon $h \geq 1$ and time horizon of length $T \geq 1$, is

$$ev_2(K, T, h) = \sqrt{\frac{1}{K}\sum_{c=1}^{K} \frac{1}{T_c hs} \sum_{t=1}^{T_c-h} \sum_{j=0}^{h-1} \mathbb{E}\left|\hat{O}^c_{t+j|t} - O^c_{t+j}\right|_2^2}.$$

Let

$$A_2 := \sqrt{\frac{1}{K}\sum_{c=1}^{K} \frac{1}{T_c h\ s} \sum_{t=1}^{T_c-h} \sum_{j=0}^{h-1} \mathbb{E}\left|\hat{O}^c_{t+j|t}\right|_2^2}.$$

$$B_2 = \sqrt{\frac{1}{K}\sum_{c=1}^{K} \frac{1}{T_c h\ s} \sum_{t=1}^{T_c-h} \sum_{j=0}^{h-1} \mathbb{E}\left| - O^c_{t+j}\right|_2^2}.$$

The global RMSE can be normalized with the symmetrized RMS terms of the data and estimate given by $\frac{ev_2(K,T,h)}{(A_2+B_2)}$.

The mean absolute error (MAE) is

$$ev_1(K, T, h) = \frac{1}{K}\sum_{c=1}^{K} \frac{1}{T_c hs} \sum_{t=1}^{T_c-h} \sum_{j=0}^{h-1} \mathbb{E}\left|\hat{O}^c_{t+j|t} - O^c_{t+j}\right|_1.$$

Let

$$A_1 = \frac{1}{K}\sum_{c=1}^{K} \frac{1}{T_c hs} \sum_{t=1}^{T_c-h} \sum_{j=0}^{h-1} \mathbb{E}\left|\hat{O}^c_{t+j|t}\right|_1$$

and

$$B_1 = \frac{1}{K}\sum_{c=1}^{K} \frac{1}{T_c hs} \sum_{t=1}^{T_c-h} \sum_{j=0}^{h-1} \mathbb{E}\left|O^c_{t+j}\right|_1.$$

The global MAE can be normalized with the symmetrized MAE terms of the data and estimates given by $\frac{ev_1(K,T,h)}{(A_1+B_1)}$. The goal is to study $(T, h) \mapsto ev_2(K, T, h)$ and $ev_1(K, T, h)$ for different multivariate multistep filters.

10.5.6 Some ML-Based Models

We briefly mention some machine learning models.

10.5.6.1 Regression-Based ML Method

There are several regression methods used in machine learning. These include, among the others, linear regression, ridge regression, neural network regression, Lasso regression, decision tree regression, random forest, K nearest neighbors (KNN) model, support vector machines (SVM), Gaussian regression, logistic regression, and polynomial regression.

10.5.6.2 Neural Network-Based ML Method

Examples of neural network-based machine learning methods include, among the others, backpropagation neural networks with genetic algorithms, backpropagation neural networks with particle swarm optimization, neural networks with principal component analysis, neural networks with multilayer perceptron–genetic algorithms, radial basis function neural networks with genetic algorithm, anatomically constrained neural networks, graph neural networks, artificial neural network-based ensemble approaches, convolutional neural network-based context-aware learning, recurrent neural network, long short-term memory, deep learning, generative adversarial networks, and strategic deep learning.

10.5.7 State Space Filtering Models

Time series models are particular cases of state space models. Consider the state $v^c = (S^c, I^c, R^c, D^c)$ generated by $v^c_{t+1} = \Phi^c(v^c_t) + \varepsilon^c_t,\ t \geq 0,\ v^c_0 \sim N(m^c_0, C^c_0)$, where ε^c_t is an i.i.d. sequence with $\varepsilon_0 \sim N(0, \Sigma^c_0)$. The observation sequence is given by $O^c_{t+1} = h^c(v^c_{t+1}) + \eta^c_{t+1},\ t \geq t^c_0$ with $\eta_1 \sim N(0, \Gamma)$. The objective of the data-driven analysis is to provide some information about the state v based the data and observation O.

Problem 10.6 (Filtering Problem) The evolution of $\mathbb{P}(v_t|O_1, \ldots, O_t)$. The filtering problem is the following: given observations $O_1, \ldots, O_t$, what is the best estimate $\hat{v}_t$ of the true state v_t of the system based on those observations?

Filtering is an operation that involves an extraction of information about a quantity of interest at time t, by using data measured up to and including t.

Problem 10.7 (Forecasting Problem) The evolution of $\mathbb{P}(v_{t+h}|O_1, \ldots, O_t)$, where $h \geq 1$ is the forecast horizon. Forecasting of an a priori form of estimation. It attempts to provide information about what the quantity of interest will be at some time $t + h$ in the future by using data measured up to and including $t - 1$.

10.5.7.1 Kalman Filter

$$\begin{cases} x_{t+1} = F_t x_t + G_t u_t + w_t, \\ O_t = H_t x_t + v_t, \\ var(w_t) = Q_t, \ var(v_t) = R_t, \\ x_0. \end{cases} \tag{10.37}$$

Given observations $O_1, \ldots, O_{t-1}$, we aim to determine $\hat{x}_t$ that minimizes the expectation of the squared-error loss function $\mathbb{E}\Big(\big|\hat{x}_t - x_t\big|_2^2\Big)$:

$$\begin{cases} \hat{x}_0 = E[x_0], \\ P_{0|0} = \mathbb{E}\Big((\hat{x}_0 - x_0)^\dagger (\hat{x}_0 - x_0) | O_1, \ldots, O_t\Big) \\ \hat{x}_{t+1|t} = F_t \hat{x}_{t|t} + G_t u_t \\ P_{t+1|t} = \mathbb{E}\Big((\hat{x}_{t+1|t} - x_{t+1})^\dagger (\hat{x}_{t+1|t} - x_{t+1}) | O_1, \ldots, O_t\Big) \\ = F_t P_{t|t} F_t^\dagger + Q_t, \\ \hat{x}_{t+1|t+1} = \hat{x}_{t+1|t} + \alpha_{t+1}(O_{t+1} - H_{t+1}\hat{x}_{k+1|t}), \\ P_{t+1|t+1} = \mathbb{E}\Big((\hat{x}_{t+1|t+1} - x_{t+1})^\dagger (\hat{x}_{t+1|t+1} - x_{t+1}) | O_1, \ldots, O_t\Big) \\ = (\mathbb{I} - \alpha_{t+1} H_{t+1}) P_{t+1|t} (\mathbb{I} - \alpha_{t+1} H_{t+1})^\dagger + \alpha_{t+1} R_{t+1} \alpha_{t+1}^\dagger, \\ \alpha_{t+1} = P_{t+1|t} H_{t+1}^\dagger \Big(H_{t+1} P_{t+1|t} H_{t+1}^\dagger + R_{t+1}\Big)^{-1}. \end{cases} \tag{10.38}$$

10.5.7.2 Extended Kalman Filter

Consider the non-linear system

$$\begin{cases} x_0, \\ x_{t+1} = \phi_t(x_t, u_t) + w_t, \\ O_t = h_t(x_t) + v_t, \\ var(w_t) = Q_t, \ var(v_t) = R_t. \end{cases} \tag{10.39}$$

Assuming ϕ_t, h_t are continuously differentiable, the linearized version at $\hat{x}$ of system (10.39) is generated by

$$\begin{cases} x_0, \\ x_{t+1} = \phi_{t,\hat{x}_t}(\hat{x}_t, u_t)x_t + [\phi_t(\hat{x}_t, u_t) - \phi_{t,\hat{x}_t}(\hat{x}_t, u_t)\hat{x}_t] + w_t, \\ O_t = h_{t,\hat{x}_t}(\hat{x}_t)x_t + [h_t(\hat{x}_t) - h_{t,\hat{x}_t}(\hat{x}_t)\hat{x}_t] + v_t. \end{cases} \quad (10.40)$$

We can then apply Kalman filter to the above linearized system.

Other important filtering methods include ensemble Kalman filter, particle filter, particle swarm filter, and mean-field-type filters.

10.5.7.3 Bayesian Updates

The Bayesian paradigm operates by first assigning to the parameters a prior distribution, which represents a prior belief about the unknown parameters, denoted by θ before seeing any data. Subsequently, this prior information is updated in the light of experimental data or observation (O) using Bayes theorem by multiplying it with the likelihood $p(O|\theta)$ and renormalizing, thus leading to the posterior distribution $p(\theta|O)$via:

$$p(\theta|O) = \frac{p(O|\theta)p(\theta)}{\int_y p(O|y)p(y)dy},$$

which is proportional to $p(O|\theta)p(\theta)$.

The stochastic processes are updated with new observation using Bayes rule. This includes Bayesian ETS, Bayesian ARMA, Bayesian ARIMA, and Bayesian SARIMAX.

10.5.8 Some Logistic Models

We present some logistic models.

10.5.8.1 Beta Logistic Models

The evolution of the number of deaths in the pandemic by means of beta logistic model is described by

$$C_O, \; \Delta C_{t+1} = rC_t^q\Big(1 - (\frac{C_t}{K})^\alpha\Big)^p,$$

where C is the cumulative deaths, with the parameters (r, q, K, α, p).

A two-wave logistic-like model is described by

$$D_O,\ \Delta D_{t+1} = \rho_1\Big(D_t - \xi_1)(1 - \frac{D_t}{\xi_2}\Big),$$

whose shape is similar to $D_t = \xi_1 + \frac{\xi_2-\xi_1}{2}\Big(1 + tanh(\frac{\rho_1(t-t_1)}{2})\Big)$, where t_1 is such that $D_{t_1} = \frac{\xi_2+\xi_1}{2}$.

A multiple wave extension is obtained with $D_t = \xi_1 + \sum_{i=1}^{w-1} \frac{\xi_{i+1}-\xi_i}{2}\Big(1 + tanh(\frac{\rho_i(t-t_i)}{2})\Big)$, where t_i is such that $D_{t_i} = \frac{\xi_{i+1}+\xi_i}{2}$, where $w \geq 2$ is the total number of waves.

10.5.9 Hawkes Models

The multivariate Hawkes with dimension s is an s-dimensional counting process $O_t = (O_t^j, j \in \{1, \ldots, 5\})$ with intensity $\lambda_i(t) = \mu_i(t) + \int_0^t \sum_{j=1}^s \phi_{ij}(t - t')O_{t'}^j(dt')$ where μ_i is nonnegative number and ϕ_{ij} are nonnegative functions.

10.5.10 Some Compartmental Models

In this subsection, we use a filtered compartmental method to study the basic reproduction numbers of SARS-CoV-2 and its variants. We evaluate the error of the proposed methods using variations of l-normed error such as a global (multivariate) root mean square error (RMSE), which is the root of mean norm squared of the difference between the forecasted and (ex post) observed values, and calibration with prediction interval coverage and the length of the observations in all countries available in the (public) database.

We describe the base compartmental model. During pandemics, there is a critical need to understand both the likely number of infections, mortality, and recovery and their time course to guide decision-makers for system responses. Approaches to backcasting, nowcasting, and forecasting the course of a pandemic vary and can include simulating the backward-forward dynamics of disease transmission and recovery and empirical fitting of data trends through risk quantification. Here we use a simple filtered and fitted pandemic compartmental model called data-driven susceptible-infected-recovered-deceased (SIRD) model, which extends the well-known SIR model to include parameter identification and estimation.

10.5.10.1 Why an SIRD Model?

The SIRD model aims to predict the number of individuals who are susceptible to infection, are actively infected, and have recovered or deceased from infection at any given time. This simplest version of the model was introduced in 1927 after the 1918 influenza pandemic, and its popularity may be due in part to its simplicity, which allows for approximation of disease behavior by estimating a small number of parameters as we will see below.

10.5.10.2 Description of the Multi-group SIRD Model

In multi-group compartmental models, individuals can be in one group at a time but can move depending on the migration policy in place. Individuals within a group are separated into mutually exclusive subgroups, or compartments of a group, based on their disease status. Each individual within a group is considered to be in one compartment at a given time, but can move from one compartment to another based on the parameters of the model. The multi-group SIRD model is one of the most basic compartmental models, named for its four compartments per group (susceptible, infected, recovered, and deceased). In this multi-group model, the assumed progression is for a susceptible individual in one of the groups to become infected through contact with another infected individual from any other group. Following a period as an infected individual, during which that person is assumed to be contagious, the individual advances to a noncontagious state, recovered, or death state. To mimic the empirical observations, the initial conditions of the multi-group SIRD must meet certain conditions: most individuals in each group of the population begins in the susceptible compartment. The infected compartment is defined as comprising individuals who have the ability to infect individuals in the susceptible compartment. As such, this compartment includes asymptomatic transmitters of the pathogen and untested infected as well as hospitalized patients who require intensive levels of care. The dynamics at which the susceptible individuals become infected is dependent on the number of individuals in each of the susceptible and infected compartments. At the start of an outbreak, when there are few infected individuals, the disease spreads slowly. As more individuals become infected, they contribute to the spread and increase the rate of infection. An additional factor in calculating the rate of spread is the effective contact rate (β). This matrix parameter accounts for the transmissibility of the disease as well as the mean number of contacts per individual. The transition from the infected to the recovered compartment is dependent on the amount of time that an individual is contagious, captured in the rate of recovery matrix (γ_r). The transition from the infected to the deceased compartment is captured in the death rate matrix (γ_d). Let $K = 210$ be the number of countries and $\mathscr{C} = \{1, 2, \ldots, K\}$ be the set of countries. The SIRD model ("S" for susceptible, "I" for infected, "R" for recovered, D for "deceased") is a particular mean-field model used in epidemiology. It represents the evolution of the (unnormalized) mass $(S^c, I^c, R^c, D^c)_c$ per country over time. Let $(S_0^c, I_0^c, R_0^c, D_0^c)$ be the initial distribution of the population in country c at time t_0^c. For two different countries $c \neq c'$, the initial starting times t_0^c and $t_0^{c'}$ of SARS-CoV-2 can be different:

$$\begin{cases} c \in \mathscr{C}, \\ (I_0^c, R_0^c, D_0^c) \text{ given at time } t_0^c, \\ \Delta I_{t+1}^c = \frac{\beta_t^c}{N^c} I_t^c S_t^c - (\gamma_r^c + \gamma_d^c) I_t^c, \\ \Delta R_{t+1}^c = \gamma_{r,t}^c I_t^c, \\ \Delta D_{t+1}^c = \gamma_{d,t}^c I_t^c, \\ S_t^c + I_t^c + R_t^c + D_t^c = N^c, \end{cases} \tag{10.41}$$

where $\Delta X_{i+1} := \frac{X_{i+1} - X_i}{t_{i+1} - t_i}$ denotes the difference operator, N^c is the population size of country c, and the vector $(I_0^c, R_0^c, D_0^c, \beta^c, \gamma^c)_{c \in \mathscr{C}}$ of nonnegative entries is to be determined, and N^c is the population size of country c, which can be corrected with the birth and the excess mortality in a yearly basis. The total number of COVID-19 cases by the SIRD model in country $c \in \mathscr{C}$ is $CC_t^c = I_t^c + R_t^c + D_t^c$. In words, the COVID-19 cases consist of actively infected, recovered, and deceased cases. *It is important to notice that here CC_t combines both reported and unreported cases.*

10.5.10.3 From Data to Multi-group SIRD Model

The SIRD model is determined by the starting time t_0, and the starting points, and the parameters (β, γ). It does require any data point to start running. However, in practice, we use few data points to estimate initial point and the starting time of the dynamics.

Can we obtain the shape in Fig. 10.25 as the daily estimated deaths from the SIRD model? To answer to this question, we plot $\gamma_r I_t$ over time t starting from the week 12 of year 2020.

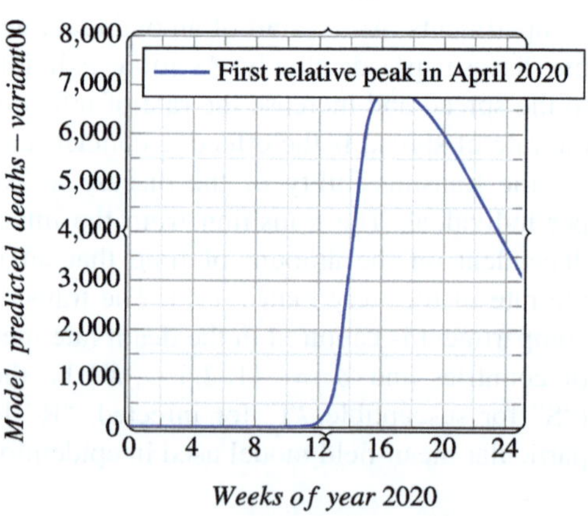

Fig. 10.25 Early stage of COVID-19: the shape of reported COVID-19 deaths worldwide

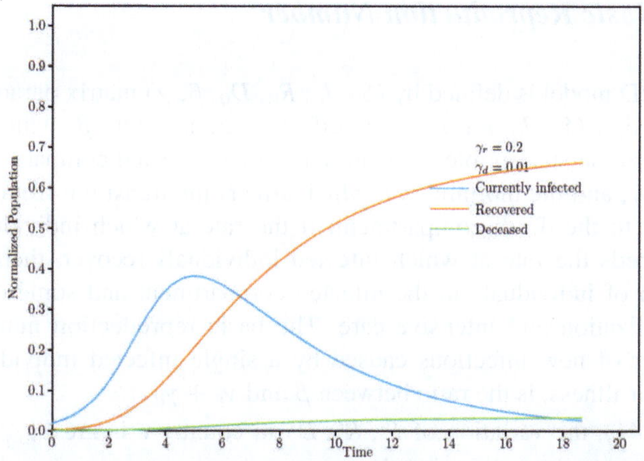

Fig. 10.26 Early stage of COVID-19: the shape of (I, R, D). Evolution of active infections, cumulative recovered, and cumulative deaths starting from $(I_0, R_0, D_0) = (0.03, 0, 0)$

As we can see in Fig. 10.26, $\gamma_r I_t$ has a similar shape as the one provided in Fig. 10.25. This means that the SIRD is a relevant candidate for describing the course of the COVID-19 pandemic for one single wave and variant.

10.5.11 Population Size Preserving Property

As the estimated population size per country is an available data, the estimated number of susceptible people in country c can be obtained as $S_t^c = N^c - CC_t^c$ and the actively infected people at time t as $I_t^c = CC_t^c - R_t^c - D_t^c$.

10.5.12 Analysis of the Long-Run Dynamics

We start our analysis by looking for a steady state, if any, of the dynamical system. We first observe that if $I_t^c = 0$ for some time t, then $I_{t'}^c = 0$ for any time $t' \geq t$. Thus, $(S^c, 0, R^c, D^c)$ with $S^c + R^c + D^c = N^c$ is a steady state of the dynamical system (10.41). If at some time instant t, $\frac{\beta_t^c}{N^c(\gamma_r^c+\gamma_d^c)} S_t^c < 1$, then $\Delta I_{t+1}^c < 0$, and the number of infected cases decreases. On the other hand, if $\frac{\beta_t^c}{N^c(\gamma_r^c+\gamma_d^c)} S_t^c > 1$, then $\Delta I_{t+1}^c > 0$, and the number of infected cases increases.

10.5.13 Basic Reproduction Number

The base SIRD model is defined by $(S_0, I_0, R_0, D_0, \beta, \gamma)$ matrix parameters: given the initial design (S_0, I_0, R_0, D_0), the effective contact rate β, which affects the transition from the susceptible compartment to the infected compartment, the rate of recovery γ_r, and the mortality γ_d, which affects the transition from the infected compartment to the death compartment. If the rate at which individuals become infected exceeds the rate at which infected individuals recover, there will be an accumulation of individuals in the infected compartment and some of them may need hospitalization and intensive care. The basic reproduction number R_0, the mean number of new infections caused by a single infected individual over the course of their illness, is the ratio between β and $\gamma_r + \gamma_d$.

From (10.41), the variation of I^c, R^c, D^c in country c is $\Delta CC_{t+1}^c = \frac{\beta_t^c}{N^c} I_t^c S_t^c$. From the latter equation, we compute

$$\beta_t^c = \frac{N^c \Delta CC_{t+1}^c}{I_t^c S_t^c}. \tag{10.42}$$

The time-varying basic reproduction number in country c is given by

$$R_{0,t}^c = \frac{\beta_t^c}{\gamma_{r,t}^c + \gamma_{d,t}^c}.$$

The vector $R_{0,t} = (R_{0,t}^c)_c$ can be interpreted as the average number of adequate contacts of a typical infectious person during the infectious period in all countries.

A decrease in the effective contact rate β through group control strategies decreases R_0, delaying and lowering the peak infection rate that occurs in the pandemic at hospitals. However, to maintain the decrease in total infections, the decrease in R_0 generally must be sustained. The goal is therefore to estimate $R_{0,t}^c$ per country using COVID-19 reported data.

10.5.14 Data-Driven SIRD Model

In this subsection, we combine data and filters for the parameters of the SIRD and build a data-driven SIRD model. We will use the reported available data $(N^c, \tilde{CC}_k^c, \tilde{R}_k^c, \tilde{D}_k^c)_{c,k\leq t}$ to estimate $(\beta_k^c, R_{0,k}^c)_{c,k\leq t}$ using the SIRD model. We can compute the data nowcasting version of β as

$$\tilde{\beta}_t^c = \frac{N^c \Delta \tilde{CC}_{t+1}^c}{(\tilde{CC}_t^c - \tilde{R}_t^c - \tilde{D}_t^c)(N^c - \tilde{CC}_t^c)}.$$

10.5.14.1 A Filtered SIRD

A filter-based SIRD in country c is obtained as follows:

where the filtering method is carefully chosen. Examples include ensemble Kalman filter, mean-field-type filter, HP filter, and HW filter.

10.5.14.2 A Fitted SIRD

A fitting-based SIRD in country c is obtained as follows:

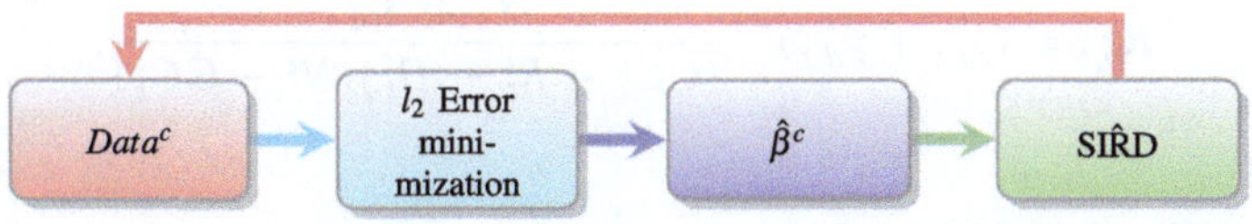

The time series $\hat{\beta}^c$ is obtained by minimizing $(\beta, \gamma, (I_0, R_0, D_0)) \mapsto ev_2(K, T, h)$. Notice that the optimization variables here include the initial conditions (I_0, R_0, D_0). The initial values of (I_0, R_0, D_0) do not need to coincide with the initial observations.

10.5.14.3 Bayesian SIRD

A Bayesian SIRD in country c is obtained as follows:

10.5.14.4 Data-Driven SIRD Error

From the reported data $(N^c, \tilde{CC}_k^c, \tilde{R}_k^c, \tilde{D}_k^c)_{c,k\leq t}$, we extract $(\tilde{S}, \tilde{I}, \tilde{R}, \tilde{D})_{data}$.

$$(S, I, R, D) = (\tilde{S}, \tilde{I}, \tilde{R}, \tilde{D})_{reported} + (\tilde{S}, \tilde{I}, \tilde{R}, \tilde{D})_{unreported} + e.$$

It remains to estimate the unreported components to compute the SIRD error evaluation:

$$\left|(S, I, R, D) - (\tilde{S}, \tilde{I}, \tilde{R}, \tilde{D})_{unreported} - (\tilde{S}, \tilde{I}, \tilde{R}, \tilde{D})_{reported}\right|_2.$$

For example, the actively infectious cases I is $I_{reported} + I_{unreported} + e_I$. The component $I_{unreported}$ (not confirmed by a PCR test) can be estimated through other sampling techniques such as antibody test and antigenic tests. We could also use the positivity rate (of PCR tests) to evaluate the recently infected percentage.

10.5.14.5 Basic Reproduction Number

The basic reproduction number in country c is obtained from the data-driven SIRD as

$$R_{0,t}^c = (\gamma_{r,t}^c + \gamma_{d,t}^c)^{-1} \frac{N^c \Delta CC_{t+1}^c}{(CC_t^c - R_t^c - D_t^c)(N^c - CC_t^c)}.$$

10.5.14.6 Empirical Measure

One can make the ratio of 2-week trends (previous versus the current one)

$$\frac{\sum_{k=t-27}^{t-14} I_k}{\sum_{i=t-13}^{t} I_k}$$

whenever the denominator is non-zero. This serves as an indicator of tendency based on the tested population.

10.5.15 Multi-group SIRD Models

The above SIRD is a list of independent system of SIRD. When one of the I_t^c is zero at time t and $I_t^{c'}$ is non-zero, then the infection status in c' will not affect the infections in country c and the epidemic would be over in that country. This is not the empirical observation of the propagation of COVID-19 since the epidemic has propagated to other countries via different channels and became a pandemic. SARS-CoV-2 started in one region before propagating worldwide and becoming a pandemic. We introduce inter-dependencies between the countries SIRD called multi-group SIRD.

$$\begin{cases} (I_0, R_0, D_0) = (I_0^c, R_0^c, D_0^c)_c \text{ given}, \\ \Delta I_{t+1} = \text{diag}((\frac{S_t^c}{N^c})_c)\beta_t I_t - (\gamma_r + \gamma_d) I_t \\ \Delta R_{t+1} = \gamma_{r,t} I_t \\ \Delta D_{t+1} = \gamma_{d,t} I_t \\ S_t + I_t + R_t + D_t = N, \end{cases} \tag{10.43}$$

where β is now a $(K \times K)$-matrix where K is the total number of countries, $N = (N^c)_c$, and the γ are diagonal matrices: $\gamma_{r,t} = \text{diag}((\gamma_{r,t}^c)_c)$ and $\gamma_{d,t} = \text{diag}((\gamma_{d,t}^c)_c)$. With this model, the effect of mobility of infectious people between country c and country c' can be captured via $\beta^{cc'}$ and $\beta^{c'c}$. To this model, we can also add a correction term, "birth-death" using the birth data and the excess mortality data (if any) in a yearly basis.

10.5.16 Multi-group Data-Driven SIRD

10.5.16.1 Estimation of Total Infected and Deaths Based on Reported Data

Let h be the forecast horizon of the SIRD model. The goal is to compute

$$(S_{t+h}^c, I_{t+h}^c, R_{t+h}^c, D_{t+h}^c)_c, \ldots, (S_{t+1}^c, I_{t+1}^c, R_{t+1}^c, D_{t+1}^c)_c$$

based on the observations $(O_t^c, \ldots, O_0^c)_c$ up to t.

10.5.16.2 Infection Fatality Rate

Confirmed case fatality rate is a measure of the proportion of confirmed COVID-19 cases that result in a patient's death. The key point here is that the total population that is being considered is only those who've been diagnosed with COVID-19 and had their disease status confirmed (e.g., by PCR test). This has the benefit that the total population is known and can be confirmed by counting all positive test results.

Infection fatality rate (IFR): IFR is a measure of the proportion of total infections that result in a patient's death. The total population here includes the confirmed cases above, plus everyone else that was infected and not diagnosed. Thus, unreported COVID-19 cases should be estimated in order to compute IFR.

Many early models assumed that confirmed cases were equal to infections or that the infection detection rate was constant worldwide over time. Early scarcity of PCR testing for COVID-19 was followed by rapid expansion of testing capacity in many high-income countries, while many low-resource settings continued to have low testing rates, all suggesting substantial variations in the detection rate over time and across locations.

Many data-driven models used deaths data, a metric that has been less affected by PCR testing availability in high-income countries, to estimate infections using empirical estimates of the infection–fatality ratio (IFR) and to estimate infection to hospitalization ratio.

10.5.16.3 Z-Transform and Fourier Analysis

Given the data $(O_t)_{t\leq T}$, the z-transform is $F(z) = \sum_{t=0}^{T} O_t z^t$. The Fourier series of the data is computed simply by using the Fourier formula, which is $F_k = F(e^{-2\pi ik/T})$.

10.5.16.4 Infection/Death Wave Duration per Country per Variant

Following Ransome's paper in 1868 [367], a wave of a variant is defined as follows: at each successive period of the duration of a variant, it advances at a diminishing speed until it reaches its highest point, followed by a relative plateau when it suddenly falls at a certain rate. A wave packet refers to the case where two (or more) waves exist simultaneously, may be due to two or more coexisting variants such as Delta and Omicron. A wave packet is often referred to as a wave group. Many variants of the SARS-CoV-2 coronavirus have arisen over the course of the pandemic. Some spread around the world, while others quickly faded away or were supplanted by other variants.

10.5.17 Filters with Exponentiated Cost

The risk-sensitive filter with an exponentiated $l-$normed cost with $l \geq 1$ is given by

$$\inf_{\{\hat{X}\in\mathscr{F}_O\}} \log \mathbb{E}\left(e^{\lambda\left|X-\hat{X}\right|_l^l}\right). \tag{10.44}$$

For a very small λ, this filter is related to the mean–variance filter described in the next subsection.

10.5.18 Mean-Field-Type Filters

Next we introduce a class of variance-aware filters.

10.5.18.1 Variance-Aware Filters

Consider the following mean–variance minimization problem with the l-norm:

$$\inf_{\{\hat{X}\in\mathscr{F}_O\}} \mathbb{E}\left(\left|X-\hat{X}\right|_l^l\right)+\frac{\hat{\lambda}}{\left|x_0\right|_l^l}\mathbb{E}\left(\left|X-\hat{X}\right|_l^l-\mathbb{E}[\left|X-\hat{X}\right|_l^l]]\right)^2, \tag{10.45}$$

with $x_0 \neq 0,\ l \geq 1, \hat{\lambda} \geq 0,\ \mathscr{F}_O$ is the filtration generated by the observation process O, and the conditional version of Problem (10.45) given by

$$\left\{ \inf_{\{\hat{X}\in\mathscr{F}_O\}} \mathbb{E}\left(\left|X-\hat{X}\right|_l^l\right)+\frac{\hat{\lambda}}{\left|x_0\right|_l^l}\mathbb{E}\left(\left|X-\hat{X}\right|_l^l-\mathbb{E}[\left|X-\hat{X}\right|_l^l|\mathscr{F}_O]\right)^2. \right. \tag{10.46}$$

Note that for $l = 2$, and $\hat{\lambda} = 0$, Problem (10.45) reduces to the expected (squared) error minimization, and the solution is the conditional expectation: $\hat{X}_{l=2,\lambda=0} = \mathbb{E}(X|\mathscr{F}_O)$.

Theorem 10.8 *For $l = 2$, the solution of Problem (10.46) is given by*

$$\begin{cases} \hat{X}_{l=2,\lambda} = [Id + 2\lambda\mathbb{E}\Big((X-\mathbb{E}(X|\mathscr{F}_O))'(X-\mathbb{E}(X|\mathscr{F}_O))|\mathscr{F}_O\Big)]^{-1}\times \\ \{\mathbb{E}(X|\mathscr{F}_O)+2\lambda\Big(\mathbb{E}(\left|X\right|_2^2 X|\mathscr{F}_O)-\mathbb{E}(\left|X\right|_2^2\mathbb{E}(X|\mathscr{F}_O)|\mathscr{F}_O)\Big)\}, \end{cases} \tag{10.47}$$

with $\lambda = \frac{\hat{\lambda}}{\left|x_0\right|_l^l}$.

Proof First, note that if the random process $\hat{X} \in \mathscr{F}_O$ and is integrable, then $\mathbb{E}(\hat{X}|\mathscr{F}_O) = \hat{X}$. Let the objective functional of mean-field type in (10.46) be

$$L = \mathbb{E}\left(\left|X-\hat{X}\right|_l^l\right)+\frac{\hat{\lambda}}{\left|x_0\right|_l^l}\mathbb{E}\left(\left|X-\hat{X}\right|_l^l-\mathbb{E}[\left|X-\hat{X}\right|_l^l|\mathscr{F}_O]\right)^2.$$

Let $l = 2$. Noting that if Z is a measurable and integrable random variable, then

$$\mathbb{E}[Z] = \mathbb{E}\left[\mathbb{E}\left(Z|\mathscr{F}_O\right)\right]$$

and can write L as

$$\begin{cases} L = \mathbb{E}(\hat{L}), \\ \hat{L} = \mathbb{E}\left(\left|X-\hat{X}\right|_2^2|\mathscr{F}_O\right)+\lambda\mathbb{E}[\left(\left|X-\hat{X}\right|_2^2-\mathbb{E}\left(\left|X-\hat{X}\right|_2^2|\mathscr{F}_O\right)\right)^2|\mathscr{F}_O]. \end{cases} \tag{10.48}$$

We expand the conditional variance term as follows:

$$\begin{cases} \mathbb{E}\left[\left(\left|X-\hat{X}\right|_2^2-\mathbb{E}\left(\left|X-\hat{X}\right|_2^2|\mathscr{F}_O\right)\right)^2|\mathscr{F}_O\right] \\ =\mathbb{E}\left[\left(\left|X-\hat{X}\right|_2^2\right)^2|\mathscr{F}_O\right]-\left[\mathbb{E}\left(\left|X-\hat{X}\right|_2^2|\mathscr{F}_O\right)\right]^2. \end{cases} \tag{10.49}$$

The first term in the expansion is

$$\begin{cases} \mathbb{E}\left[\left(\left|X-\hat{X}\right|_2^2\right)^2|\mathscr{F}_O\right] \\ =\mathbb{E}\left[\left(\left|X\right|_2^2-2\langle X,\hat{X}\rangle+\left|\hat{X}\right|_2^2\right)^2|\mathscr{F}_O\right] \\ =\mathbb{E}\left[\left|X\right|_2^4|\mathscr{F}_O\right]+4\mathbb{E}[\langle X,\hat{X}\rangle^2|\mathscr{F}_O]+\mathbb{E}[\left|\hat{X}\right|_2^4|\mathscr{F}_O] \\ +2\mathbb{E}[\left|X\right|_2^2\left|\hat{X}\right|_2^2|\mathscr{F}_O] \\ -4\mathbb{E}[\left|X\right|_2^2\langle X,\hat{X}\rangle|\mathscr{F}_O] \\ -4\mathbb{E}[\left|\hat{X}\right|_2^2\langle X,\hat{X}\rangle|\mathscr{F}_O], \end{cases} \tag{10.50}$$

and the second term is

$$\begin{cases} \left[\mathbb{E}\left(\left|X-\hat{X}\right|_2^2|\mathscr{F}_O\right)\right]^2 \\ =\left[\mathbb{E}\left(\left|X\right|_2^2-2\langle X,\hat{X}\rangle+\left|\hat{X}\right|_2^2|\mathscr{F}_O\right)\right]^2 \\ =\left[\mathbb{E}\left(\left|X\right|_2^2|\mathscr{F}_O\right)-2\langle\mathbb{E}(X|\mathscr{F}_O),\hat{X}\rangle+\left|\hat{X}\right|_2^2\right]^2 \\ =\left[\mathbb{E}\left(\left|X\right|_2^2|\mathscr{F}_O\right)\right]^2+4\langle\mathbb{E}(X|\mathscr{F}_O),\hat{X}\rangle^2+\left|\hat{X}\right|_2^4 \\ +2\mathbb{E}\left(\left|X\right|_2^2|\mathscr{F}_O\right)\left|\hat{X}\right|_2^2 \\ -4\mathbb{E}\left(\left|X\right|_2^2|\mathscr{F}_O\right)\langle\mathbb{E}(X|\mathscr{F}_O),\hat{X}\rangle \\ -4\left|\hat{X}\right|_2^2\langle\mathbb{E}(X|\mathscr{F}_O),\hat{X}\rangle. \end{cases} \tag{10.51}$$

Combining together, we arrive at

$$
\begin{cases}
\mathbb{E}\left[\left(\left|X-\hat{X}\right|_2^2\right)^2|\mathscr{F}_O\right]-\left[\mathbb{E}\left(\left|X-\hat{X}\right|_2^2|\mathscr{F}_O\right)\right]^2 \\
=\mathbb{E}\left[\left|X\right|_2^4|\mathscr{F}_O\right]+4\mathbb{E}\left[\langle X,\hat{X}\rangle^2|\mathscr{F}_O\right]+\mathbb{E}\left[\left|\hat{X}\right|_2^4|\mathscr{F}_O\right] \\
+2\mathbb{E}\left[\left|X\right|_2^2\left|\hat{X}\right|_2^2|\mathscr{F}_O\right] \\
-4\mathbb{E}\left[\left|X\right|_2^2\langle X,\hat{X}\rangle|\mathscr{F}_O\right] \\
-4\mathbb{E}\left[\left|\hat{X}\right|_2^2\langle X,\hat{X}\rangle|\mathscr{F}_O\right] \\
-\left[\mathbb{E}\left(\left|X\right|_2^2|\mathscr{F}_O\right)\right]^2-4\langle\mathbb{E}(X|\mathscr{F}_O),\hat{X}\rangle^2-\left|\hat{X}\right|_2^4 \\
-2\mathbb{E}\left(\left|X\right|_2^2|\mathscr{F}_O\right)\left|\hat{X}\right|_2^2 \\
+4\mathbb{E}\left(\left|X\right|_2^2|\mathscr{F}_O\right)\langle\mathbb{E}(X|\mathscr{F}_O),\hat{X}\rangle \\
+4\left|\hat{X}\right|_2^2\langle\mathbb{E}(X|\mathscr{F}_O),\hat{X}\rangle.
\end{cases}
\tag{10.52}
$$

Following the properties of the conditional expectation, we have

$$
\begin{cases}
+2\mathbb{E}\left[\left|X\right|_2^2\left|\hat{X}\right|_2^2|\mathscr{F}_O\right] \\
-2\mathbb{E}\left(\left|X\right|_2^2|\mathscr{F}_O\right)\left|\hat{X}\right|_2^2=0
\end{cases}
\tag{10.53}
$$

$$
\begin{cases}
-4\mathbb{E}\left[\left|\hat{X}\right|_2^2\langle X,\hat{X}\rangle|\mathscr{F}_O\right] \\
+4\left|\hat{X}\right|_2^2\langle\mathbb{E}(X|\mathscr{F}_O),\hat{X}\rangle=0.
\end{cases}
\tag{10.54}
$$

Hence, Eq. (10.52) becomes

$$\begin{cases} \mathbb{E}\left[\left(\left|X-\hat{X}\right|_2^2\right)^2|\mathscr{F}_O\right]-\left[\mathbb{E}\left(\left|X-\hat{X}\right|_2^2|\mathscr{F}_O\right)\right]^2 \\ =\mathbb{E}\left[\left|X\right|_2^4|\mathscr{F}_O\right]-\left[\mathbb{E}\left(\left|X\right|_2^2|\mathscr{F}_O\right)\right]^2+4\mathbb{E}\left[\langle X,\hat{X}\rangle^2|\mathscr{F}_O\right]-4\langle\mathbb{E}(X|\mathscr{F}_O),\hat{X}\rangle^2 \\ +4\mathbb{E}\left(\left|X\right|_2^2|\mathscr{F}_O\right)\langle\mathbb{E}(X|\mathscr{F}_O),\hat{X}\rangle-4\mathbb{E}[\left|X\right|_2^2\langle X,\hat{X}\rangle|\mathscr{F}_O] \\ =\mathbb{E}\left[\left|X\right|_2^4|\mathscr{F}_O\right]-\left[\mathbb{E}\left(\left|X\right|_2^2|\mathscr{F}_O\right)\right]^2-4\langle\mathbb{E}(X|\mathscr{F}_O),\hat{X}\rangle^2 \\ +4\mathbb{E}\left(\left|X\right|_2^2|\mathscr{F}_O\right)\langle\mathbb{E}(X|\mathscr{F}_O),\hat{X}\rangle-4\mathbb{E}\left[\left|X\right|_2^2\langle X,\hat{X}\rangle|\mathscr{F}_O\right] \\ +4\mathbb{E}\left[\langle X-\mathbb{E}(X|\mathscr{F}_O)+\mathbb{E}(X|\mathscr{F}_O),\hat{X}\rangle^2|\mathscr{F}_O\right] \\ =\mathbb{E}\left[\left|X\right|_2^4|\mathscr{F}_O\right]-\left[\mathbb{E}\left(\left|X\right|_2^2|\mathscr{F}_O\right)\right]^2 \\ +4\mathbb{E}\left[\langle X-\mathbb{E}(X|\mathscr{F}_O),\hat{X}\rangle^2|\mathscr{F}_O\right] \\ +4\mathbb{E}\left(\left|X\right|_2^2|\mathscr{F}_O\right)\langle\mathbb{E}(X|\mathscr{F}_O),\hat{X}\rangle-4\mathbb{E}\left[\left|X\right|_2^2\langle X,\hat{X}\rangle|\mathscr{F}_O\right]. \end{cases} \tag{10.55}$$

The last expression features a quadratic term in $\hat{X}$, which is suitable for convex/convex optimization. By combining (10.55) and (10.48), we obtain

$$\begin{cases} \hat{L}=\mathbb{E}\left(\left|X\right|_2^2|\mathscr{F}_O\right)-2\langle\mathbb{E}(X|\mathscr{F}_O),\hat{X}\rangle+\left|\hat{X}\right|_2^2 \\ +\lambda\left[\mathbb{E}\left[\left|X\right|_2^4|\mathscr{F}_O\right]-\left[\mathbb{E}\left(\left|X\right|_2^2|\mathscr{F}_O\right)\right]^2\right] \\ +4\lambda\mathbb{E}\left(\left[\langle X-\mathbb{E}(X|\mathscr{F}_O),\hat{X}\rangle\right]^2|\mathscr{F}_O\right) \\ -4\lambda\left(\mathbb{E}\left(\left|X\right|_2^2\langle X,\hat{X}\rangle|\mathscr{F}_O\right)-\mathbb{E}\left(\left|X\right|_2^2\langle\mathbb{E}(X|\mathscr{F}_O),\hat{X}\rangle|\mathscr{F}_O\right)\right), \end{cases} \tag{10.56}$$

where $\lambda=\frac{\hat{\lambda}}{\left|x_0\right|_l^l}$. Taking the variation of $\hat{L}$ over $\hat{X}$, we arrive at

$$\begin{cases} 2\left[Id+2\lambda\mathbb{E}\left(\langle X-\mathbb{E}(X|\mathscr{F}_O),\langle X-\mathbb{E}(X|\mathscr{F}_O)\rangle|\mathscr{F}_O\right)\right]\hat{X} \\ -2\mathbb{E}(X|\mathscr{F}_O)-4\lambda\left(\mathbb{E}(\left|X\right|_2^2X|\mathscr{F}_O)-\mathbb{E}\left(\left|X\right|_2^2\mathbb{E}\left(X|\mathscr{F}_O\right)|\mathscr{F}_O\right)\right)=0, \end{cases} \tag{10.57}$$

where Id denotes the identify mapping. Through simplification, we retrieve the announced expression in (10.47). This completes the proof. ∎

This theorem connects the mean–variance filters to risk-sensitive filter with exponentiated cost.

10.6 Discussion

10.6.1 Missing and Unreported Data

When some of the data points are missing (e.g., unreported deaths for some days/weeks) or when there is a limited number of data points, the filter performance is drastically affected. However, SIRD can still be used as in (10.41) for estimating the duration of single waves per country in Models 1, 2, and 3 given below (Tables 10.5, 10.6, and 10.7):

Table 10.5 Model 1

Model 1 is described by

$$\begin{cases} (N, I_0, R_0, D_0) \text{ given at time } t_0, \\ \Delta I_{t+1} = \frac{\beta_t}{N} I_t S_t - (\gamma_r + \gamma_d) I_t, \\ \Delta R_{t+1} = \gamma_r I_t, \\ \Delta D_{t+1} = \gamma_d I_t, \\ S_t + I_t + R_t + D_t = N, \end{cases} \tag{10.58}$$

with parameters (I_0, R_0, D_0), β, γ

Table 10.6 Model 2

Model 2 is described by

$$\begin{cases} s_0, b_0, c_0, O_1, \dots, O_t \\ s_t = \alpha(O_t - c_{t-L}) + (\mathbb{I} - \alpha)(s_{t-1} + b_{t-1}), \\ b_t = \beta(s_t - s_{t-1}) + (\mathbb{I} - \beta) b_{t-1}, \\ c_t = \gamma(O_t - s_{t-1} - b_{t-1}) + (\mathbb{I} - \gamma) c_{t-L}, \\ \Delta \hat{O}_{t+j} = -\hat{O}_{t+j-1} + s_t + j b_t + c_t, \\ j \in \{1, \dots, h\}, h \geq 1 \end{cases} \tag{10.59}$$

with parameters $\alpha, \beta, \gamma, h, L$,

Table 10.7 Model 3

Model 3 is described by

$$\begin{cases} O_1, \dots, O_t \\ \Delta \hat{O}_{t+1} = \mu + \sum_{k=1}^{p} \phi_k O_{t-k} \\ + \sum_{k=1}^{P} \Phi_k O_{t-ks} \\ - \sum_{k=1}^{q} \theta_k \varepsilon_{t-k} \\ - \sum_{k=1}^{Q} \Theta_k \varepsilon_{t-ks}, \end{cases} \tag{10.60}$$

with parameters $(\mu, (\phi_1, \dots, \phi_p), (\Phi_1, \dots, \Phi_P), (\theta_1, \dots, \theta_q), (\Theta_1, \dots, \Theta_Q))$ denoted by $(p, q, 1, P, Q)s$,

10.7 Notes

In this chapter, we have presented mean-field-type-based epidemiological models. High-quality data are crucial for tracking the course of the pandemics. Based on it, short-term projections can be useful for decision support. Both time series and mechanistic models are useful. Without accurate data or with only few data points, estimating the length of a wave of a specific COVID-19 variant is extremely challenging. By combining models, data, and the sequence of taken actions together, mean-field-type game theory has been used to examine the propagation of the pandemic. More details can be found in [8, 27, 155, 177, 178, 181, 183, 192, 292, 353, 359, 364, 376, 377, 497].

Part IV
Sociology, Psychology, Ecology, and Environment

Chapter 11
Mean-Field Games for Marriage

Marriage is a legally recognized union between two people, typically characterized by a commitment to share their lives, responsibilities, and emotional bonds. In the context of marriage, cohabitation, divorce, and remarriage, game dynamics model can be applied to understand how individuals within a society make choices related to forming and dissolving romantic partnerships. Cohabitation involves living together in an unmarried partnership, divorce signifies the legal dissolution of a marriage, and remarriage denotes entering into a new marriage after a previous one has ended. By applying the mean-field game approach, we can gain insights into the dynamics and decision-making processes that underlie these relationship-related phenomena on a broader societal scale. This chapter examines such issues, namely, mean-field games for marriage. The results support the argument that optimizing the long-term well-being through effort and social feeling state distribution (mean-field) will help to stabilize marriage. However, if the cost of effort is very high, the couple fluctuates in a bad feeling state or the marriage breaks down. We then examine the influence of society on a couple using mean-field sentimental games [460, 464]. We show that, in mean-field equilibrium, the optimal effort is always higher than the one-shot optimal effort. We illustrate numerically the influence of a couple's network on their feeling states and their well-being.

The chapter is organized as follows. Section 11.1 describes marital interaction modeling. Section 11.2 presents the model and some methods. Section 11.3 provides optimal open-loop effort control. Section 11.4 focuses on optimal feedback effort strategy for the couple. Mean-field sentimental games are analyzed in Sect. 11.5. Section 11.6 focuses on social welfare. Section 11.7 provides some illustrative examples. Section 11.8 provides some concluding remarks.

T. Başar et al., *Mean-Field-Type Game Theory II*, Static & Dynamic Game Theory: Foundations & Applications, https://doi.org/10.1007/978-3-032-07031-9_11

11.1 Marital Interaction Modeling

The myth of marriage has been and is still a fascinating historical societal phenomenon. Paradoxically, the empirical divorce rates are at an all-time high. In order to design and evaluate interventions, theoretical understanding of marital stability and dissolution is crucial. This chapter describes a unique paradigm for preserving relationships and marital stability using mean-field-type game theory [460, 464]. The approach is quite general for modeling social interactions and can be applied to empirical data generated over time. We look at marital interactions and relationships. Here relationship refers to the unit (couple), rather than to two persons. We focus on a very ephemeral thing: what happens to the couple as they interact over time. It is not either person; it is something that happens when they are together, i.e., as a couple. The couple will create something that we call a "feeling state" as they talk to each other, as they smile, and as they move. We model the feeling state of a couple as a noisy differential equation influenced by random events, parents, children, friends, and the social distribution of marital status (population mean field of states) of their networks, where the term mean field refers to the distribution of couple feeling states in the society [53, 460, 463, 478].

Why a Mathematical Model for Marital Interactions?
Putting the study of social relationships on a mathematical footing represents a major advance in our ability to understand and perhaps regulate these relationships for the betterment of all mankind. The unique feature of the study below is in the analysis of the social distribution of feeling states (mean field) based on a generic couple.

On the Influence of Couple-Feeling-States in the Society
Since the society, friends, and parents of a couple may influence them in an aggregative manner, a mean-field approach is suitable in capturing such an interaction. Here the distribution of states (mean field)—of all couples in the society—plays an important role in the payoff and in the effort used to preserve a marriage. We use population mean-field games to capture the influence of society distribution of states on a generic couple. We consider three main features:

- We introduce stochastic optimal control in marital interaction. We explicitly characterize the optimality equations under random fluctuations, which come from model uncertainty in the drift function of the feeling state.
- We examine the impact of the mean field and the effort in the feeling state of the couple.
- We introduce a population mean-field game model for marriage in a non-linear setup where generically two stable states (divorce and marriage) are observed. We study the impact of the network of couples on the feeling state. We show that, in mean-field equilibrium, the optimal effort of the mean-field sentimental game is always higher than the one-shot optimal effort if the social contribution is controlled.

Table 11.1 Summary of notations

Symbol	Meaning
$x(t)$	Feeling state of the couple at instant t
$e(t)$	Effort of the couple (control) at instant t
$s(\cdot)$	Instantaneous well-being function
$c(\cdot)$	Instantaneous cost function
$\tilde{c}(\cdot)$	Legendre–Fenchel transform of c
$g(\cdot)$	Terminal payoff function
B	Standard Brownian motion
E	Expectation operator
$P(\cdot, \cdot)$	Long-term payoff functional
v_t	Partial derivative of $v(t, x)$ with respect to t
v_x	Partial derivative of $v(t, x)$ with respect to x
v_{xx}	Second partial derivative
$m(t, \cdot)$	Distribution of states at instant t
$\mathscr{C}^k$	Set of k times continuously differentiable functions

While the study here focuses on marital interaction, the mathematical methodology can be applied to other situations as well, such as a society's ecological or smoking behavior.

We summarize some of the notations used in this chapter in Table 11.1.

11.2 Model and Methods

We first present a basic mathematical framework for feeling state and well-being of a couple. We will explain below how the model can be modified to include the (positive, negative, neutral) impact of the society (and the "social pressure") on a generic couple. We describe the sentimental dynamics, define the payoff functional for the marital interaction, and state the relevant optimal control problem.

Governing Feeling State Equation

The feeling state (feeling level) x of a generic couple is modeled as being generated an Itô stochastic differential equation:

$$dx = f(x, e)dt + \sigma dB(t), \quad \text{for } t \in [0, T], \tag{11.1}$$

where $f(x, e) = -h(x) + ae$, $a > 0$ represents the expected variation of the feeling in a short time window, $\sigma \geq 0$, h is smooth and goes to infinity with x, and $B(t)$ is a standard Brownian motion. The effort function $e(t)$ can be controlled by the couple. The sentimental dynamics is expected to start at a high feeling level x_0 greater than $\underline{x}$ where $\underline{x}$ is a certain threshold value below which the relationship of the couple is not viewed as satisfactory. The parameter a represents the efficiency of the effort.

Remark 11.1 The deterministic part of the governing feeling state is an experimental derivation from marital studies. However, here, the function h is not necessarily monotone in x. We introduce a stochastic term into the sentimental dynamics of the existing literature for multiple reasons: (1) shocks, (2) random events in the social network of the couple that may affect the couple, and (3) the drift function f that may not be perfectly known, and thus we add an uncertainty term to it.

Remark 11.2 If the marriage starts with a feeling level x_0 and is stopped at the first instant when $x(t)$ falls below a certain threshold level $\underline{x} - \epsilon$ or one member of the couple dies (life expectancy is set to $T_0 = 124$ years). Thus, the length of the horizon is expected to be finite, which is why we consider a finite horizon problem. Let $T_{\text{divorce}} = \inf\{t > 0 \mid x(t) \leq \underline{x} - \varepsilon\}$; one can take $T = \min(T_0, T_{\text{divorce}})$, which means that T is a random variable.

Problem Setup

Based on the feeling state and the effort, we define the payoff functional for a couple during $[0, T]$, $T > 0$, as

$$P(x_0, e) = E\left(g(x(T)) + \int_0^T [s(x(t)) - c(e(t))]dt\right), \tag{11.2}$$

where $s(x(t))$ represents the well-being of the couple at state $x(t)$ and $c(e(t))$ is the cost associated with the effort $e(t)$. The instantaneous payoff can capture the risk at time t but the analysis below is restricted to the long-term risk-neutral setup. We define a couple as a union of two adult persons having certain emotional and physical interactions and living together. We define family as the direct ascendants and the direct descendants of the conjoint. In this sense, we read the quantity $g(x(T))$ as the heritage that the couple bequeaths to the family in terms of well-being. This heritage will influence the next-generation sentimental game. It is reasonable to assume that the function g increases with the feeling state and $g'(x) \geq 0$. The function s is the well-being function of the couple. It depends on the feeling state x in the sense that, the higher the feeling state, the more joyful the couple is. The implicit purpose of the control problem is to increase well-being while taking into consideration the cost. This goal is achieved by pushing the feeling state above a certain satisfactory level. Contrastingly, couples with low feeling state are more or less happy and have the tendency to seize. This tendency is expressed by taking actions: caring for children, giving gifts, cooking special meals, making concessions, and in short, handling the couple according to his or her expectation. We define the effort as a constructive action from one conjoint to another one. It might be useful for the reader to notice that our definition of effort does not take into account counterproductive actions. For example, knowing that Juliette does not like football, it is a negative attitude for Romeo to force her to go to the Camp Nou Stadium for the FC Barcelona vs Milan AC game making her miss her preferred movies. The cost of effort can represent the psychological, financial, and/or emotional load provided by the couple when accomplishing this effort. Clearly, greater effort is more expensive as it imposes a higher cost. It is

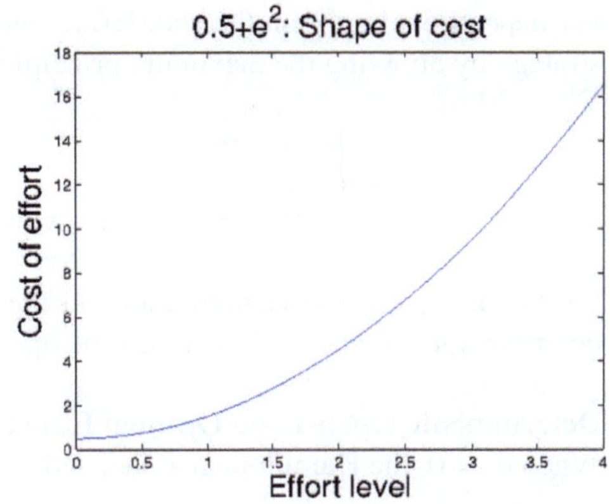

Fig. 11.1 Typical shape of the cost function

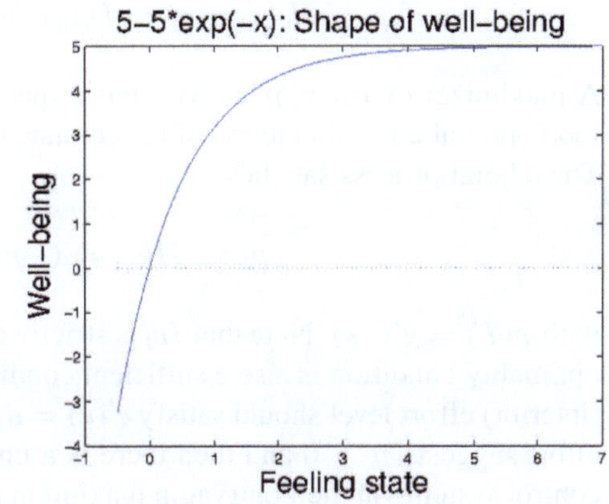

Fig. 11.2 Typical shape of well-being

appropriate then to say that the couple continuously provides efforts to achieve a greater well-being of the couple itself [460, 464].

We adopt the following assumptions:

- The drift function $f : \mathbb{R} \to \mathbb{R}$ is continuously differentiable ($\mathscr{C}^1$).
- The well-being function $s : \mathbb{R} \to \mathbb{R}$ is $\mathscr{C}^1$-differentiable, nondecreasing, concave, and saturated at $+\infty$.
- The cost function $c : \mathbb{R} \to \mathbb{R}$ is twice continuously differentiable ($\mathscr{C}^2$), nondecreasing, and strictly convex.

In Figs. 11.1, 11.2, we illustrate typical well-being and cost functions. In the next sections, we consider respectively an open-loop control problem, an optimal feedback strategy for the couple, and a mean-field sentimental strategy.

11.3 Optimal Open-Loop Effort Control

A control law of the form $e(t) = e(t; t_0, x_0)$ for $t \in [t_0, T]$, determined for a particular initial state value $x(t_0) = x_0$, is called an open-loop effort. This does not depend explicitly on the state $x(t)$. One seeks for an optimal open-loop effort strategy by applying the maximum principle to the following problem:

$$(\mathscr{P}_0) \begin{cases} \sup_{(e(t))_t} P(x_0, e), \\ \text{subject to } (11.1) \text{ with initial state } x_0 > \underline{x} + \varepsilon. \end{cases} \tag{11.3}$$

To do so, we use the Hamiltonian and the adjoint process. First, we analyze the deterministic case ($\sigma = 0$) and then the open-loop noisy case.

Deterministic Open-Loop Optimal Effort

When $\sigma = 0$, the Hamiltonian associated with the optimal control problem ($\mathscr{P}_0$) is

$$H_0(x, p, e) = f(x, e)p + s(x) - c(e). \tag{11.4}$$

A maximizer of $H_0(x, p, e)$ with the respect to the effort level provides an open-loop optimal control in terms of the co-state (adjoint) process p.
The adjoint process satisfies

$$p_t = -H_{0,x} = (-pf_x - s'(x)), \tag{11.5}$$

with $p(T) = g'(x_T)$. Note that H_0 is strictly concave in e, and hence, the first-order optimality condition is also a sufficient condition for an interior point solution. The (interior) effort level should satisfy $c'(e) = ap \geq 0$. The positivity constraint of the effort suggests $p \geq 0$ and then there is a unique solution. The open-loop optimal control system via the Pontryagin maximum principle yields

$$\begin{cases} e^*(t) = \max(0, (c')^{-1}[ap(t)]), \\ \dot{p}(t) = (p(t)h'(x(t)) - s'(x(t))), \\ \dot{x}(t) = -h(x(t)) + \max(0, a(c')^{-1}[ap(t)]), \\ x(0) = x_0, \quad p(T) = g'(x_T). \end{cases} \tag{11.6}$$

By differentiating $e^*(t)$ and combining with the above system, we arrive at

$$\dot{e} = \frac{1}{c''(e)} \left[c'(e)h'(x) - as'(x)\right] \mathbb{1}_{\{e(t) \geq 0\}}. \tag{11.7}$$

Hence, one has the following dynamical system between the optimal control and the optimal feeling:

$$\begin{cases} x(0) = x_0, \\ \dot{x} = -h(x) + ae, \\ \dot{e} = \dfrac{1}{c''(e)} \left[c'(e)h'(x) - as'(x)\right] \mathbb{1}_{\{e(t) \geq 0\}}, \\ e(T) = \max\left(0, (c')^{-1}[ag'(x(T))]\right). \end{cases} \tag{11.8}$$

We refer to Volume I, Chapter 4, of this book for more details on the maximum principle. Our first question is about the well-posedness of the above system (11.8). Our goal is to provide a sufficiency condition for existence of a solution. We use fixed-point theorem to establish existence under the above assumptions. Since the functions s and h are $\mathscr{C}^1$-differentiable and the strictly convex function c is $\mathscr{C}^2$-differentiable, (11.8) admits a local solution.

Next we use a stochastic maximum principle approach to analyze the case where σ is non-zero, i.e., the stochastic case.

Stochastic Optimal Open-Loop Effort
Utilizing stochastic maximum principle, the adjoint processes for constant variance coefficient case satisfy

$$\begin{aligned} &dp(t) = (-p(t) f_x - s'(x(t)))dt + q dB(t) \\ &\text{with } \; p(T) = g'(x(T)), \end{aligned} \tag{11.9}$$

where q is the adjoint variable associated with the diffusion term. Let $\tilde{c}$ be the Legendre–Fenchel transform of c :

$$\tilde{c}(ap(t)) = \sup_e \{ap(t) \cdot e - c(e)\}. \tag{11.10}$$

It is well-known that $\tilde{c}(ap(t))$ is convex in $p(t)$. Furthermore, the optimal control is given by

$$e^*(t) = \max(0, \tilde{c}'(ap(t))). \tag{11.11}$$

Using Ito's calculus, the Euler–Lagrange system is described by

$$\begin{cases} x(0) = x_0, \\ dx(t) = (-h(x(t)) + ae(t))dt + \sigma dB(t), \\ de(t) = \Big[\tilde{c}''(c'(e(t)))\left(c'(e(t))h'(x(t)) - as'(x(t))\right) \\ \qquad + \dfrac{q^2}{2} a^2 \tilde{c}'''(c(e(t)))\Big]\, dt + \tilde{c}''(c'(e(t))) aq dB(t), \\ e(T) = \max\left(0, (c')^{-1}[ag'(x(T))]\right). \end{cases} \tag{11.12}$$

A sufficiency condition for existence of a solution $(x(t), e(t))$ to the above is (locally) Lipschitz with respect to (x, e) of the coefficient functions and uniform Lipschitzness of $(c')^{-1}$ over the horizon $[0, T]$. We refer to Volume I of this book for more details on stochastic maximum principle.

11.4 Optimal Feedback Effort Strategy for the Couple

A feedback effort law is of the form $e(t, x(t)),\ t \in [t_0, T]$, i.e., the effort depends on time t and the feeling $x(t)$. The payoff functional is written as

$$P_{\text{closed}}(x, e) = E\left(g(x(T)) + \int_t^T \left[s(x(t')) - c(e(t'))\right] dt' \mid x(t) = x\right) \quad (11.13)$$

To find the optimal feedback control $e^*(t, x)$, we use dynamic programming. Let $v(t, x)$ be the value starting from x at time t, i.e.,

$$(\mathscr{P}_1) \begin{cases} v(t, x) = \sup_{e(t,.)\geq 0} P(x, e), \\ \text{subject to the feeling state dynamics (11.1).} \end{cases} \quad (11.14)$$

The Hamiltonian is given by

$$H(x, p) = -h(x)p + s(x) + \tilde{c}(ap). \quad (11.15)$$

Since $\tilde{c}(ap)$ is convex in p, we deduce that H is convex in p. It is well-known that if there exists a twice continuous differentiable function $v(t, x)$ as solution of

$$\begin{cases} v_t - h(x)v_x + s(x) + \tilde{c}(av_x) + \frac{\sigma^2}{2} v_{xx} = 0 \\ \text{with } \ v(T, x) = g(x), \end{cases} \quad (11.16)$$

then the stochastic optimal control effort is

$$e^*(t, x) = \max\left(0\,,\ \tilde{c}'(av_x)\right). \quad (11.17)$$

Under the assumptions on c, h, s, g and the strictly concavity of the payoff function in e implies that there is a unique optimal control. Thus, the existence of the value follows.

11.5 Mean-Field Sentimental Games

There is a significant consensus around the idea that the society state (particularly the feeling states of the friends, parents, friends' of friends, and social environment) may influence the status of the couple. For example, if many of the friends are in lower feeling states, their divorce can be contagious. Moreover, a couple's tendency to divorce not only depends on their friends' divorce status but also extends to their friend's friends and so on. Thus, we must consider the distribution of states within the entire society, i.e., the mean field of states [106]. We take into consideration the mean-field state of the society in the sentimental dynamics of the couple. This leads to a *mean-field sentimental game.*

Roughly speaking, we introduce a society's social influences (mean-field) and external shocks into the marital interaction model. To do this, we assume that the function h has the form $\hat{h}(x(t), m(t))$, where $m(t, .)$ is the distribution of states of all couples in the society at time t. Based on the model (11.1), we state that the feeling state (feeling level) $x(t)$ of the couple is modeled as a McKean–Vlasov Itô's stochastic differential equation:

$$dx(t) = \Big[-\hat{h}(x(t), m^*(t)) + ae(t)\Big]dt + \sigma dB(t), \tag{11.18}$$

where m^* is the equilibrium distribution of states.

To complete the formulation, the long-term payoff of a generic couple is written as

$$P_{\text{field}}(x_0, m^*, e) = E\left(g(x(T)) + \int_0^T \left[\hat{s}(x(t), m^*(t)) - c(e(t))\right]dt\right). \tag{11.19}$$

The objective of each generic couple is to maximize its long-term payoff through the mean-field fixed-point problem:

$$(\mathscr{P}_2)\begin{cases} \sup\limits_{(e(t))_t} P_{\text{field}}(x_0, m^*, e), \\ \text{subject to the feeling state dynamics (11.18).} \end{cases} \tag{11.20}$$

The happiness function $\hat{s}$ covers the well-being of the couple and the satisfaction of its network. It is natural that the happiness distribution over the network is concentrated in the family set. It is also known that the social network may contribute to enhance the well-being of the couple.

11.5.1 Mean-Field Sentimental Equilibria

We define a mean-field sentimental equilibrium as an (Nash/Wardrop) equilibrium of the mean-field sentimental game. A mean-field sentimental equilibrium in feedback strategies is a situation in which no couple has any incentive to move unilaterally from its effort feedback strategy. The mean-field equilibria are the solution of the following backward-forward system for $(t, x) \in [0, T] \times \mathbb{R}$ is

$$\begin{cases} v_t - \hat{h}(x, m)v_x + \hat{s}(x, m) + \tilde{c}(av_x) + \dfrac{\sigma^2}{2}v_{xx} = 0, \\ m_t + \partial_x\Big[-m\hat{h} + am\tilde{c}'(av_x))\Big] - \dfrac{\sigma^2}{2}m_{xx} = 0, \\ m(t=0, x) = m_0(x) \quad \text{for } x \in \mathbb{R}, \\ v(t=T, x) = v_T(x) = g(x) \quad \text{for } x \in \mathbb{R}. \end{cases} \tag{11.21}$$

Theorem 11.3 *Assume that the contribution of the mean field to the feeling state of a generic couple is small. Then, the optimal level of effort for a long-term viability of the couple (during their lifetime), which keeps a happy relationship going, is always greater than the effort level that would be chosen in a one shot, i.e.,* $e^*(t, x) \geq \underline{e}$, *where* $c'(\underline{e}) = 0$.

Proof By assumption, the functions c' and g are nondecreasing, and the term $h_2(m)$ has only small influence on the drift. By small influence of $h_2(m)$, we mean that the uncontrolled system leads to a divorce or fluctuate feeling state if no effort is injected. The optimality equation (11.21) and the strict convexity of the cost c give $c'(\underline{e}) = 0 \leq av_x(t, x) = c'(e^*(t, x))$. This means that $c'(\underline{e}) \leq c'(e^*(t, x))$. Using the inverse function $(c')^{-1}$, one gets $\underline{e} \leq e^*(t, x)$, and there are several time instants $t \in [0, T]$ for which $\underline{e} < e^*(t, x)$. ∎

11.5.2 How a High Divorce Rate Society Influences a Couple?

We analyze the influence of the society state (mean field) on the couple. Toward that end, we assume that the functionals $\hat{h}$, $\hat{s}$, and $\hat{c}$ take on the following forms:

$$\hat{h}(x, m) = h(x) - h_2(m), \quad \hat{s}(x, m) = s(x) + s_2(m),$$

where s_2 and h_2 are smooth functionals of the mean-field distribution. Next we present two important results on the contagion of divorce.

Theorem 11.4 *In a short horizon, breaking up is hard to do, unless everyone else is doing it too.*

Proof Consider a society where the majority of marriages are stable. This means that both $m(t)$ and $h_2(m(t)) \geq 0$ are high enough compared to $\underline{x}$. Thanks to the contribution of h_2 in the feeling state dynamics, the generic dynamics given in (11.18) has a higher value. The more the value of h_2 is, the more the feeling state will be. This means that $x(t, m(t))$ will go to a higher state than the case without the mean field. Thus, Breaking Up is Hard to Do in a short horizon even if there is no effort from the couple. ∎

Remark 11.5 (Societal benefit!) If $h_2(m(t)) \geq \epsilon > 0$, then the marriage remains preserved over time even if the couple effort is minimal. This is because the mean-field contribution plays the role of a positive effort. The network of the couple is having a big positive influence on their feeling state.

Theorem 11.6 *If the mean field has a tendency to the divorce states, then there is a contagious phenomenon for divorce, i.e., starting from* $x_0 \in (0, 2)$, *the couple state will degrade due to the influence of the mean field toward a negative feeling state. In particular, Breaking Up is not seen as a negative thing in that society because the majority is Doing it Too. Furthermore, stabilizing a marriage will require more effort, and hence it will be more expensive.*

Proof Let $h_2(m(t)) \leq -\epsilon < 0$. The mean field is concentrated to negative values below the divorce threshold range. The majority has a tendency to the divorce. Thus, the feeling state of a generic couple goes toward negative values if no effort is made. A high effort is required to bring back and maintain the feeling state to a satisfactory one. The threshold effort $e(t)$ to balance is $\frac{h(x(t))-h_2(m(t))}{a} > 0$, which requires some time and cost. In this configuration, a divorce is not seen as a negative thing by the society because the majority of the society is in a divorced state. The negative term from the society will required more effort (and hence more cost) for the stabilization of a marriage. ■

11.6 Social Welfare

In this section, we aim to maximize the social welfare. It entails solving the following mean-field control problem:

$$(\mathscr{P}_3)\begin{cases} \sup_{e(.)} P_{\text{social}}(m_0, m, e), \\ \text{subject to the feeling state dynamics (11.18)} and\ m(t) = \mathscr{L}(x(t)), \end{cases} \tag{11.22}$$

where $m(t) = \mathscr{L}(x(t))$ is the distribution of $x(t)$. Note that the social welfare problem $(\mathscr{P}_3)$ is different than $(\mathscr{P}_2)$ because now a control action e affects immediately the distribution m.

$$P_{\text{social}}(m_0, m, e) = E\left(\hat{g}(x(T), m(T)) + \int_0^T \left[\hat{s}(x(t), m(t)) - c(e(t))\right] dt\right). \tag{11.23}$$

Define the function $\hat{H}(x, m, p, q) = \hat{s}(x, m) - c(e) + fp + \sigma q$. The stochastic maximum principle of mean-field type leads to

$$dp = -[\hat{H}_x(x, m, p, q) + E\hat{H}_{x,m}(x, m, p, q)]dt + qdB$$

where e^* is the maximizer of $\hat{H}$. Note that without the impact of the society (m) on the functions $\hat{s}$, $\hat{h}$, and $\hat{g}$, one gets the same optimality equations as above. However, the presence of the mean-field term m changes drastically the behavior of the adjoint process p.

11.7 Illustrative Model of Marriage

In this section, we present a mathematical model of marriage for a specific functional h in (11.1). We take into consideration a control term ae and a noise term. The dynamical system in the case where

$$h(x) = rx - b \cdot \tanh(cx) \tag{11.24}$$

is written as

$$dx(t) = (-rx(t) + b \cdot \tanh(cx(t)) + ae(t))dt + \sigma dB(t), \tag{11.25}$$

with $r, a, b, c > 0$. The choice of the function $\tanh(x) = \frac{\exp(x)-\exp(-x)}{\exp(x)+\exp(-x)}$ in (11.24) is widely justified in the psychology literature as well as in evolutionary game theory as resulting from the imitative logit dynamics. The model is widely supported by many psychologists and sociologists. Note that the function h defined in (11.24) does not belong to the class of functions studied in the literature of sentimental games. We choose $b = c = 1$. Then, the sentimental dynamics with noise becomes

$$dx(t) = [-rx(t) + \tanh(x(t)) + ae(t)]dt + \sigma dB(t), \tag{11.26}$$

where the effort $e(t) \geq 0$ is a control variable. The parameter $r > 0$ refers to the type of the society where the couple lives. We also perform numerical examples to illustrate the effectiveness of the theoretical results. To do this, we consider the mathematical model (11.1) with h defined in (11.24). The cost and the well-being functionals are chosen as

$$c(e) = \frac{1}{2}e^2 \text{ and } s(x) = \bar{s} - 10\exp(-x) \text{ with } \bar{s} \geq 10. \tag{11.27}$$

It is worth noting that the above choices of c and s ensure the existence of solutions to problems (11.8), (11.12) (11.17), and (11.21).

11.7.1 Steady States of the Sentimental Dynamics

Depending on the type of the society, we analyze here the uncontrolled ($e = 0$) sentimental dynamics.

In a High Type Society ($r \geq 1$)
For $r \geq 1$, the uncontrolled system ($e = 0$) converges to zero independently of the starting point. This can be interpreted as if the couple feeling state being degraded over time if there is no effort. The uncontrolled feeling state has only one steady state, which is illustrated in Fig. 11.3.

In a Low Type Society ($r < 1$)
The degradation rate is less aggregative. We observe that in a society of low type, the uncontrolled system ($e = 0$) has three steady states. Two of them are stable and one is unstable. The unstable feeling state is the state $x = 0$ because the derivative at 0 is $1 - r > 0$. The other two steady states are symmetric with respect to the origin.

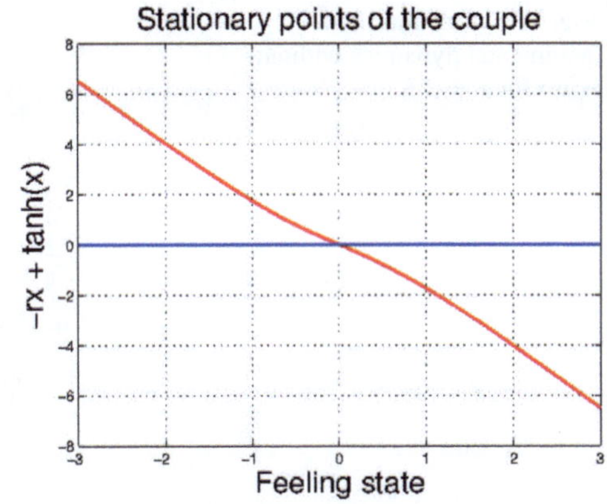

Fig. 11.3 Steady states of the uncontrolled ($e = 0$) dynamical system: high type society $r = 2.5$ (one steady state)

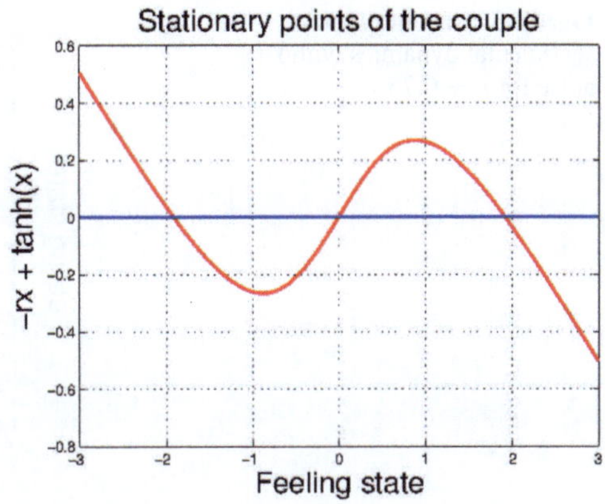

Fig. 11.4 Steady states of the uncontrolled ($e = 0$) dynamical system: low type society $r = 0.5$ (three steady states)

The positive one can be seen as if the marriage continues for life. A negative steady state characterizes a divorce situation. The two non-zero states are stable. Indeed, the derivative at the steady-state points can be written as $-r + (1 - \tanh^2(x^*)) = 1 - r - (rx^*)^2$, which is negative when rx is near -1 or 1. In Fig. 11.4, we illustrate the zeros of the drift function $-rx + \tanh(x)$ for different values of the societal parameter r.

11.7.2 Evolution of Uncontrolled Feeling States

Here, we demonstrate the societal influence on the sentimental dynamics of a couple without effort ($e = 0$). In this case, we observe the variation in time of the feeling state for the two types of society. Numerical experiments for the deterministic ($\sigma = 0$) evolution and stochastic ($\sigma > 0$) dynamics in high type society ($r \geq 1$) and in low type society ($r < 1$) have been performed.

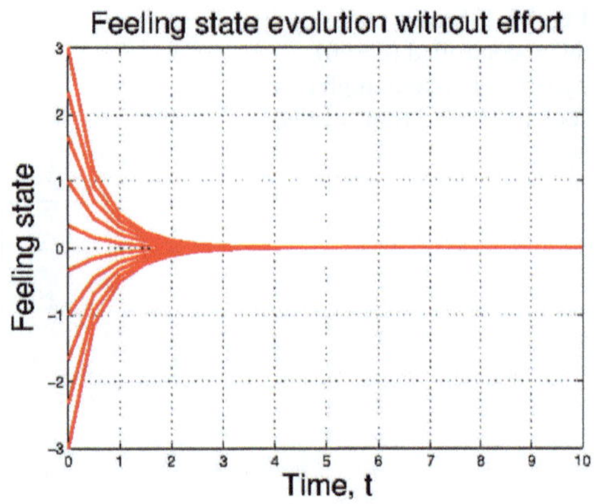

Fig. 11.5 Uncontrolled sentimental dynamics without noise for $r = 2.5$

Fig. 11.6 Uncontrolled sentimental dynamics without noise for $r = 0.75$

A sample of ten non-working couples with different initial feeling states is depicted in Fig. 11.5 and 11.6. We observe a fast convergence to the state 0 in the high type case ($r = 2.5$). In the low type case, $r = 0.5$, we observe three steady states. Two (one positive, one negative) of them are stable states.

11.7.3 Stochastic Evolution

Now we introduce a noise to better capture the fluctuations observed in real life. We observe that with a small variance σ, the feeling state of the couple leads to a fluctuating trajectory around the deterministic sentimental dynamics. However, a larger noise variance leads to two main branches and a non-zero probability to switch from one branch to another.

Figure 11.7 represents the stochastic evolution of sentimental dynamics without control for $r = 2$. Figure 11.8 represents the noisy evolution of feeling states for low type $r = 0.5$.

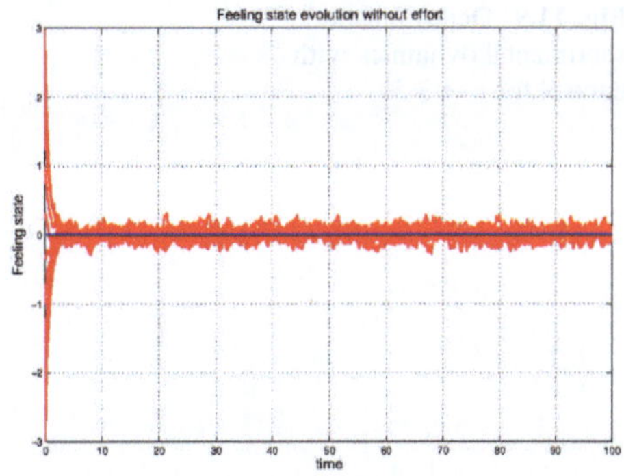

Fig. 11.7 Noisy sentimental dynamics without control for $r = 2$

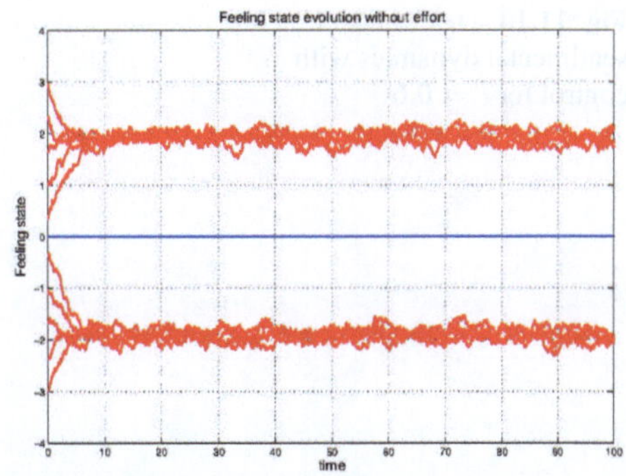

Fig. 11.8 Noisy sentimental dynamics without control for $r = 0.5$

11.7.4 Sentimental Dynamics with Effort

Interestingly, when we introduce the effort, a catastrophic situation (starting from $x_0 < 0$) can be turned to positive with effort and stabilize to a positive feeling state. These cases are illustrated in Figs. 11.9 and 11.10 for the deterministic dynamics and in Figs. 11.11, 11.12 for the stochastic dynamics. Thus, introduction of effort plays an important role in preserving the marriage or the cohabitation. On the other hand, a high effort level may be costly to the couple. Thus, it is crucial to find an acceptable trade-off.

Next we depict the dynamics driven by optimal control as a function of time. From the open-loop optimal control of the non-noisy dynamics (11.8), we observe two different situations. Figures 11.13 and 11.14 represent the trajectories of $(x(t), e(t))$ for different initial conditions for both high and low types for $a = 10$.

Figures 11.15 and 11.16 represent the vector field between feeling state and effort of the couple for high and low types. We observe that there is an invariant set that can maintain very high effort and high feeling state.

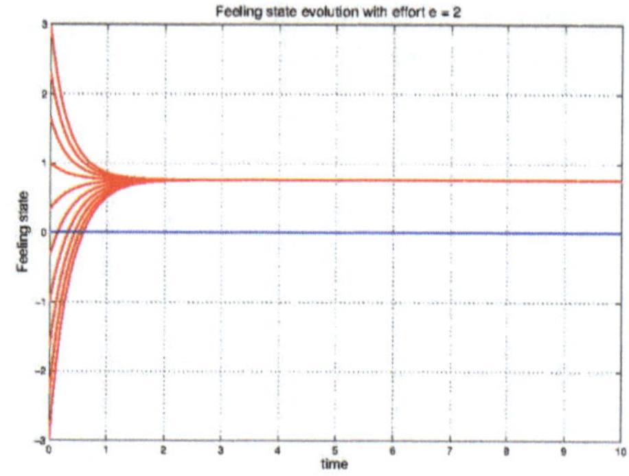

Fig. 11.9 Deterministic sentimental dynamics with control for $r = 3.5$

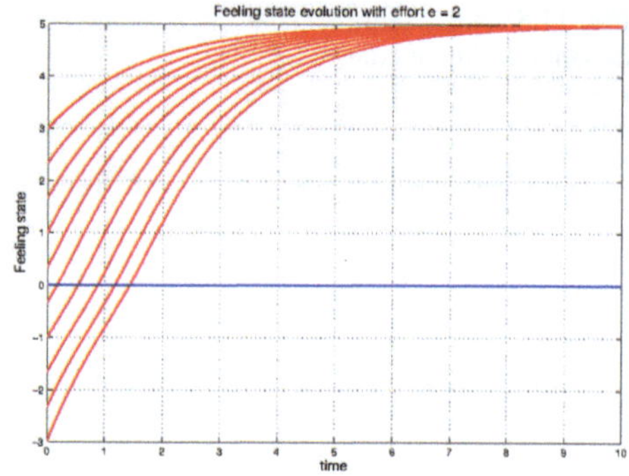

Fig. 11.10 Deterministic sentimental dynamics with control for $r = 0.6$

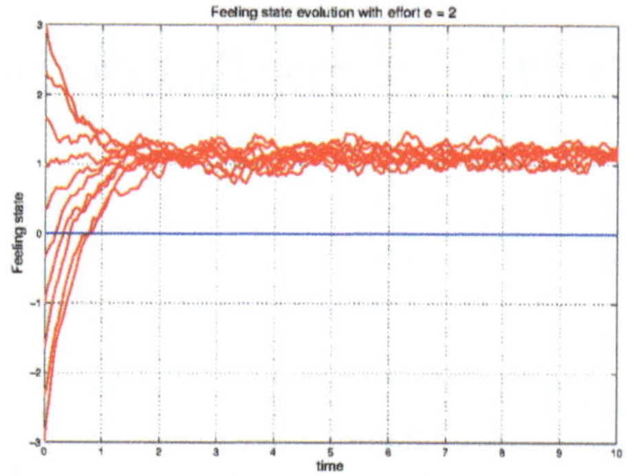

Fig. 11.11 Stochastic feeling state with control for $r = 2.5$

Stochastic Optimal Control

Figure 11.17 depicts the stochastic optimal control trajectories and optimal effort level. We observe that a small noise in the feeling state may have big consequences in the optimal effort (and the cost of effort [106]).

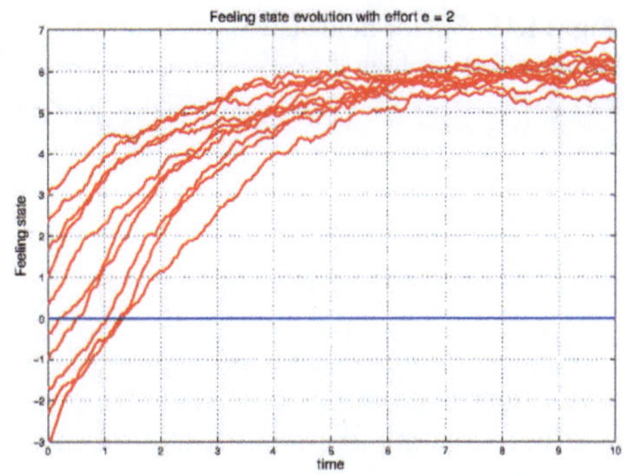

Fig. 11.12 Stochastic feeling state with control for $r = 0.5$

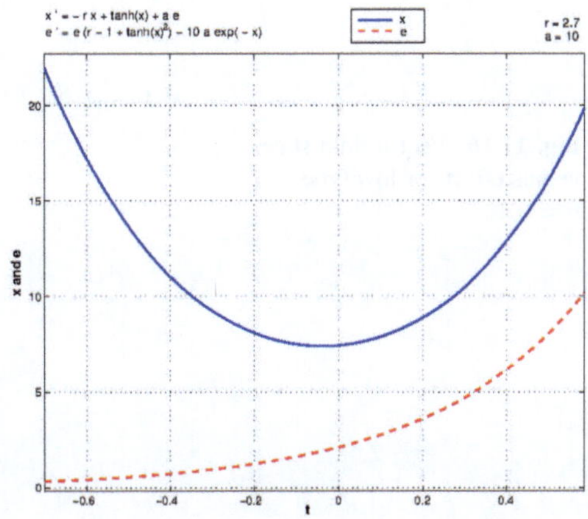

Fig. 11.13 Open-loop optimal control system with high type $r = 2.5$ as a function of time

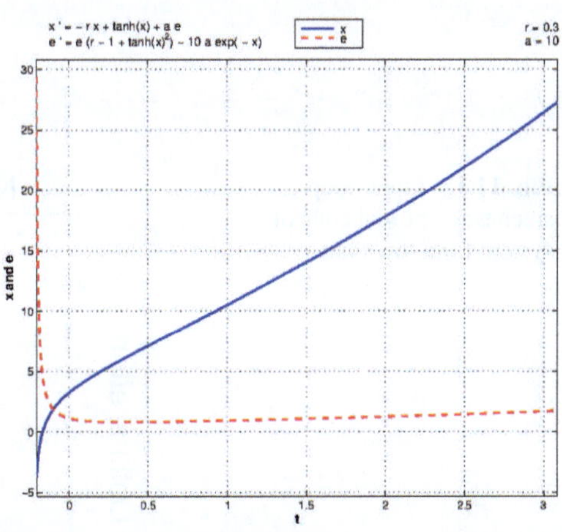

Fig. 11.14 Open-loop optimal control system with low type $r = 0.4$ as a function of time

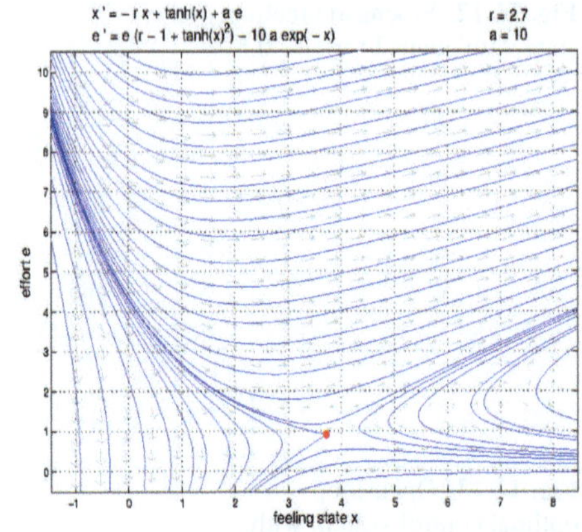

Fig. 11.15 Vector field state versus effort for high type $r = 2.5$

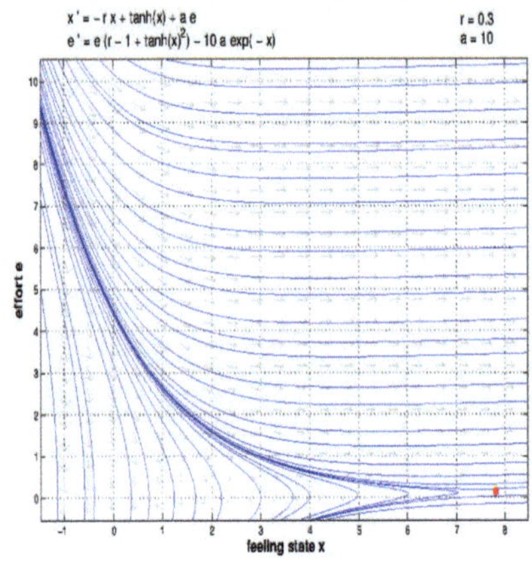

Fig. 11.16 Vector field state versus effort for low type $r = 0.1$

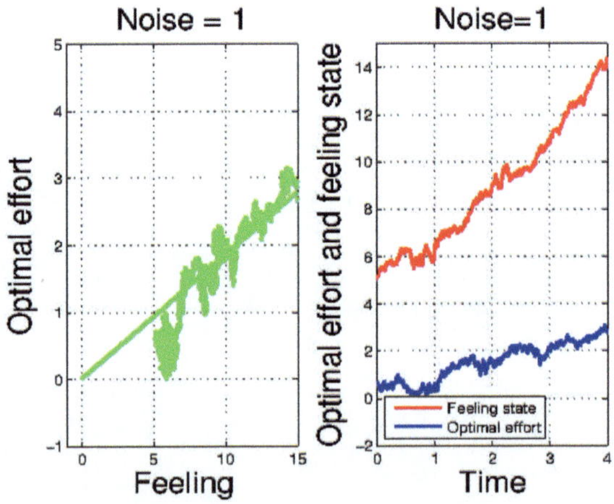

Fig. 11.17 Open-loop stochastic optimal control system for a low type

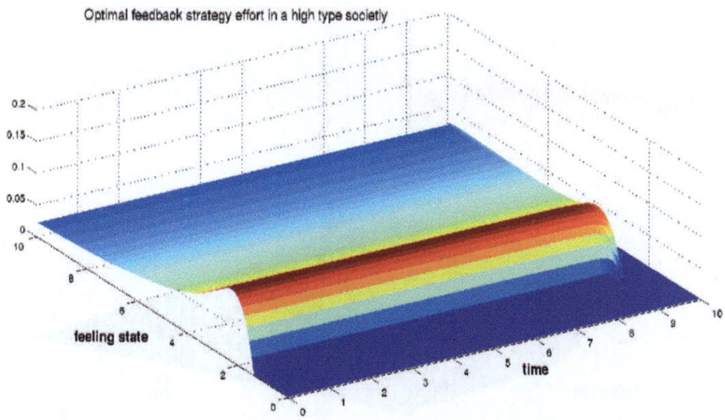

Fig. 11.18 The optimal strategy provided by a couple in a high type society $r = 2.4$

11.7.5 Dynamics of the Optimal Feedback Effort

The experiment involves in observing a couple over a finite horizon. We suppose that the couple reaches the end of horizon T with the same heritage of feeling state whatever the initial feeling state is. In other terms, the functional g is constant; therefore, the terminal optimal effort is zero. To describe the dynamics of the optimal feedback strategy of such a couple, we have applied an implicit (time) backward scheme to the backward differential equation (11.16). The Hamiltonian is approximated using the Lax–Friedrichs approach to preserve monotonicity, consistency, and continuity properties. Two numerical experiments are performed with respect to the type of the society.

The couple starting with a low feeling state is to provide no effort. For the conjoints of such a couple, it is hard to for them resist due to the influence of the high type society. From Fig. 11.18, one can see that for a couple starting with a fairly good feeling state ($\underline{x}+\epsilon \leq x \leq \underline{x}+K\epsilon$), the optimal strategy is to do actions. Indeed, the couple will break up because of the negative influence of society and will line up with a null optimal effort. Also, we observe the same phenomenon for a couple starting with a higher feeling state. The optimal effort is nevertheless important because starting at a higher feeling state can counterbalance the negative influence of the society.

The optimal strategy for a couple starting with low feeling state is to provide zero effort. This is not surprising since in a low type society living, a couple follows a societal phenomenon. We can see from Fig. 11.19 that the optimal strategy effort for a couple starting with high feeling state is a time-decreasing function. In that case, doing good action is motivated more by seeking ideal happiness than by upholding the couple. Overall, we observe that the optimal strategy is cheaper in a low type society than in a high type society.

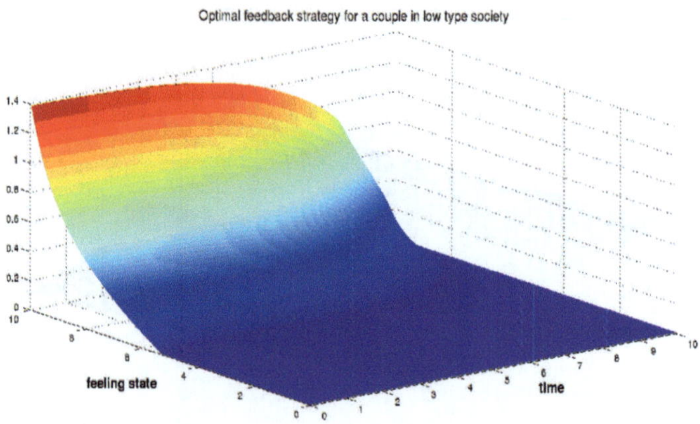

Fig. 11.19 The optimal effort strategy of a couple in a low type society $r = 0.5$

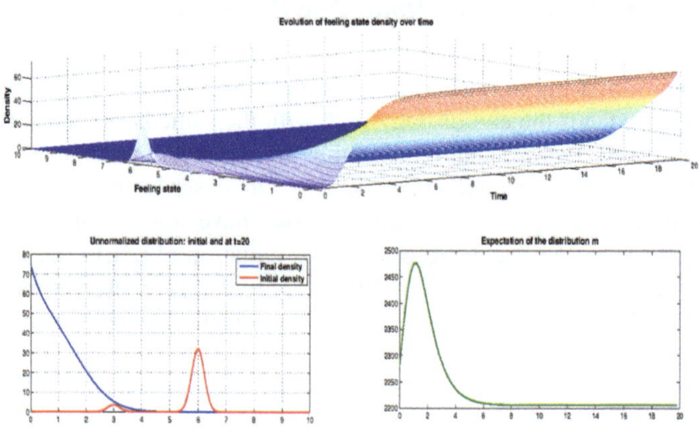

Fig. 11.20 Low type: evolution of mean field starting from Gaussian concentrated at $x = 6$ with standard deviation 1.5

11.7.6 Mean-Field Equilibrium Trajectories

We now address numerical simulations for the mean-field equilibrium (11.21). The functionals $\hat{h}$ and $\hat{s}$ are chosen as defined in ( ), the functional h is specified in (11.24), and the effort cost and the well-being intrinsic to the couple are given by (11.27). The numerical experiment here consists of observing the evolution of the distribution m representing the society in which a given couple lives. We start with a reference couple has an almost constant optimal effort in a particular window. This implies that the value v is linearized with respect to x and is constant in time.

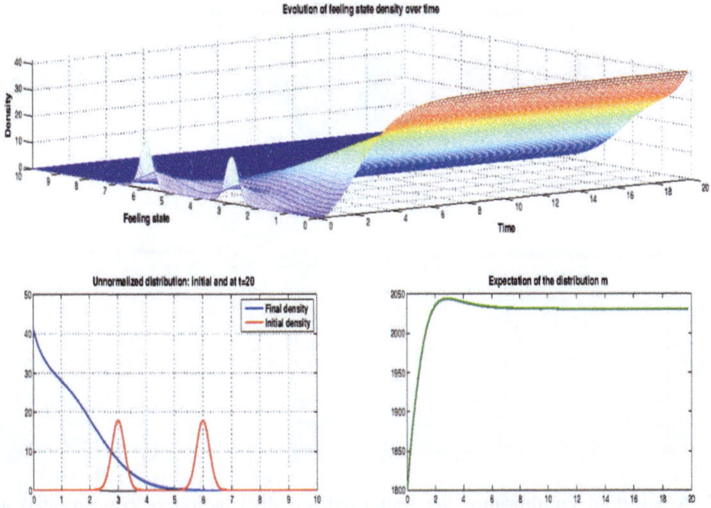

Fig. 11.21 Low type: evolution of mean field starting from mixture Gaussian concentrated at $x = 3$ and $x = 6$ with standard deviation 1.5

The numerical task is then to compute the Fokker–Planck–Kolmogorov equation in (11.21). The developed algorithm focuses on unnormalized distribution (positive measure). Figure 11.20 represents the evolution of the mean-field equilibrium with initial Gaussian distribution centered at $x = 6$ with standard deviation 1.5 for a low type. When the initial distribution is a mixture of Gaussian distributions centered at $x = 3$ and $x = 6$ with standard deviation 1.5, we observe in Fig. 11.21 that the state distribution will propagate rapidly to be concentrated at the extreme in a relatively short time. For the mean-field equilibrium, the symmetric steady states observed previously in the case where the control was a constant are not steady state anymore because the feedback control $e(t, x)$ is now dynamic in state as time proceeds. We observe that if the initial mean field is a mixture of Gaussians centered at -5 and at $+5$ with standard deviation 0.05, then the feeling state will propagate and will be concentrated at the two extremes at the final time (Fig. 11.22).

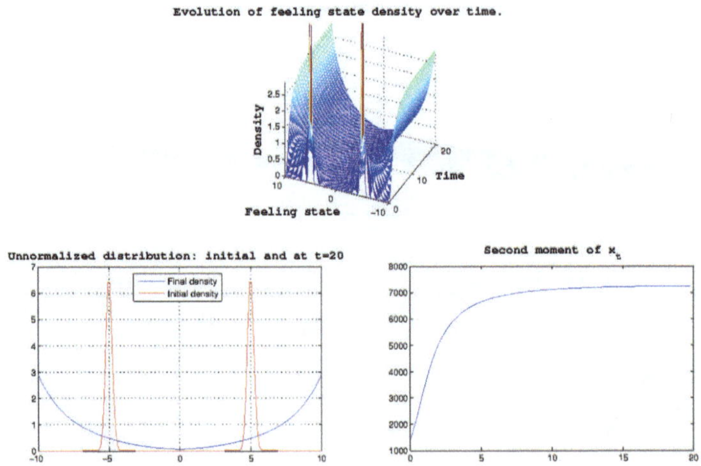

Fig. 11.22 Mean-field equilibrium: the unnormalized distribution is concentrated at the two extreme boundaries

11.8 Notes

In this chapter, we have introduced a mean-field game model for marriage, cohabitation, divorce, and remarriage. Our study suggests that the optimal effort may help in sustaining marriage and cohabitation if the cost of that effort is acceptable and the initial couple state is not too low. This helps us to understand at least theoretically the key processes related to marital dissolution, cohabitation separation, stability, and fluctuation. The study can be useful to design or evaluate an adequate intervention. It also suggests that knowing many divorced people may influence the status of a marriage. The more divorced people one knows, the riskier would be the own marriage. However, a marriage does not break down just because friends are divorcing. Marital breakdown depends on many factors including effort and mean field. Understanding marriage, cohabitation, divorce, and remarriage involves several research disciplines. Several dynamics have been proposed in the literature to capture how individuals within a society make choices related to forming and dissolving romantic partnerships. More details can be found in [2, 18, 75, 89, 90, 147, 153, 176, 196, 207, 227, 228, 237, 245, 337, 478].

Chapter 12
Pollution Control and Fishing Strategies in Smarter Cities

Pollution is the introduction of harmful or undesirable substances into the environment that can have detrimental effects on living organisms, ecosystems, and the overall quality of the environment. These substances, often referred to as pollutants, can be in various forms, including solid, liquid, or gas, and can come from natural sources or human activities. Pollution can occur in different environmental compartments, such as air, water, soil, and even noise and light. Pollution is a global concern because it can have significant impacts on the health of both the environment and human populations. Air pollution occurs when harmful substances are released into the atmosphere, primarily from activities like burning fossil fuels, industrial processes, and transportation. Air pollutants can include particulate matter, greenhouse gases (e.g., carbon dioxide), sulfur dioxide, nitrogen oxides, and volatile organic compounds. Air pollution can lead to respiratory problems, climate change, and damage to vegetation. Water pollution happens when contaminants are discharged into bodies of water such as rivers, lakes, oceans, and groundwater. Common water pollutants include chemicals, heavy metals, sewage, agricultural runoff, and oil spills. Water pollution can harm aquatic ecosystems, threaten human health, and affect drinking water quality. Soil pollution occurs when hazardous substances contaminate the soil. This can result from improper disposal of hazardous waste, agricultural chemicals, industrial runoff, and more. Soil pollution can harm plant growth, affect the food chain, and make land unsuitable for agriculture. Noise pollution also refers to excessive, annoying, or harmful noise levels in the environment, often caused by activities like transportation, construction, industrial operations, and urbanization. Prolonged exposure to high noise levels can have adverse effects on human health, including hearing loss, stress, and sleep disturbances.

Pollution control refers to strategies, technologies, regulations, and practices implemented to prevent, reduce, or mitigate pollution in various environmental media such as air, water, soil, and noise. The primary goal of pollution control is to minimize the adverse effects of pollutants on the environment and human

T. Başar et al., *Mean-Field-Type Game Theory II*, Static & Dynamic Game Theory: Foundations & Applications, https://doi.org/10.1007/978-3-032-07031-9_12

health. Pollution control measures aim to strike a balance between human activities and environmental protection by managing and limiting the release of harmful substances into the environment.

Fishing is the activity of capturing or harvesting fish and other aquatic organisms from bodies of water such as oceans, seas, rivers, lakes, and ponds. It is a practice that has been carried out by humans for thousands of years and is an important source of food, livelihood, and recreation for many people around the world. Commercial fishing involves the capture of fish and other aquatic species for sale in the marketplace. It is a major industry worldwide, supplying seafood for human consumption and a variety of other products such as fishmeal, fish oil, and fish byproducts used in various industries. Fishing plays a crucial role in global food security, as it is a primary source of protein for many people. However, sustainable fishing practices are essential to ensure the long-term health of aquatic ecosystems and to prevent overexploitation of fish stocks. Efforts to balance the economic, environmental, and social aspects of fishing are ongoing to promote responsible and sustainable fishing practices worldwide.

As cities continue to grow, accommodating attracted more than half of the world population, the issues pertaining to their efficiency and sustainability have become more and more relevant. In its simplest elements, a city can be divided into its hard infrastructure, its people, and the dynamic systems that connect the city within itself. Because of climate change, energy and resource supply, and economic drivers, the concept of smart cities has moved to its business development. The smart city is difficult to define because it can encompass such a large and extensive range of systems and infrastructure elements. But in this chapter, we use the term *smart city* to mean more intelligent, more efficient, and more sustainable. We will focus on key components including pollution control and economic fishing supply and how they interface with social behaviors of the city. In support of the environment, risk management and sustainable development, natural resources management including systems for reduction of pollutants, and managing human response to environmental stresses and to sustain biodiversity are key elements of smart cities.

This chapter focuses on pollution control and fishing strategies in smarter cities. The intersection of pollution control and fishing strategies takes on new importance as cities strive to create more environmentally friendly and resilient urban environments. We explore the multifaceted relationship between pollution control efforts and fishing strategies within the framework of smarter cities.

The chapter is organized as follows. Section 12.1 introduces water pollution management in smarter cities. Section 12.2 presents biodiversity and fishing industry. Section 12.3 presents a combined and balanced approach. The chapter ends with the concluding remarks of Sect. 12.4 and some historical notes.

12.1 Water Pollution in Smarter Cities

Water pollution management in smarter cities involves the use of advanced technologies, data-driven approaches, and innovative strategies to monitor, mitigate, and prevent water pollution within urban areas. Smarter cities aim to address water pollution comprehensively, recognizing that clean and accessible water is essential for public health, environmental sustainability, and overall urban well-being. A dynamic nonlinear model describes the economics of sea and shallow lakes (artificial or natural). Sea and lakes provide a variety of economic uses: they are sources of drinking water, fishing, recreational sports, and pleasant locations for smart homes. For agriculture, lakes are used for drainage. Runoff from the fields flows into the streams and rivers that feed the lakes. Much of the agricultural runoff includes phosphorus from fertilizers and animal wastes. Phosphorus is the primary nutrient of algae and weeds in water. When it grows excessively from an infusion of phosphorus, algal blooms reduce the oxygen content of the lakes and release toxins into them. Lakes then become unsafe for certain aquatic species and for recreational purposes by tourists and local residents. In addition, a large amount of untreated industrial discharges and municipal sewage enters the sea or lakes and pollute them with high concentrations of lead (Pb), zinc (Zn), copper (Cu), and cadmium (Cd) in their sediments. It is a common fact that inner city areas suffer from this agricultural and industrial pollution. There is a lot of work by the research community and industry to lower the emission of each individual farmer and each industry. This is a worthwhile effort but it misses two central aspects: normally, it is not the pollution caused by an individual farmer or one single small industry that harms people's health, but it is the aggregate effect of all polluting entities. That is the starting point of our study to introduce a mean-field approach. We develop a system that manages the aggregate pollution making use of longer-term cooperative mean-field-type game [106]. Using the stochastic maximum principle (see [125]), we show that the overall pollution could be managed and limited, leading to improved water quality and the quality of life (Qlife) in urban areas. We illustrate the economic and ecological feature of the smart city with the dynamics of pollution or eutrophication of sea and lakes and identify a basic class of players: farmers and mining industries. Industrial companies may need to reduce dumping of chemical waste into the sea. Farmers use artificial fertilizers to grow crops; these fertilizers contain chemical elements such as phosphorus, which in the end is washed into the sea (Fig. 12.1). The dynamics of the chemical waste (such as the amount of phosphorus in the sea) is given by

$$dx_1 = \left[\sum_{j=1}^{I} a_{1,j}(t) - bx_1 + r_1 \frac{x_1^{\beta}}{\varepsilon^{\beta} + x_1^{\beta}} \right] dt.$$

Here $a_{1,j}$ refers to the input of more chemical waste (due to industrial waste and farming) by player j; $b \geq 0$ is proportional to the rate of loss of chemical element due to sedimentation, outflow, and sequestration in other biomass. The term $f(x_1) = r_1 \frac{x_1^{\beta}}{\varepsilon^{\beta}+x_1^{\beta}}$, $\beta \geq 2,\ r_1 > 0$, models the biological production of chemical element in the sea (Fig. 12.2). We add uncertainty in parameter b and write the state

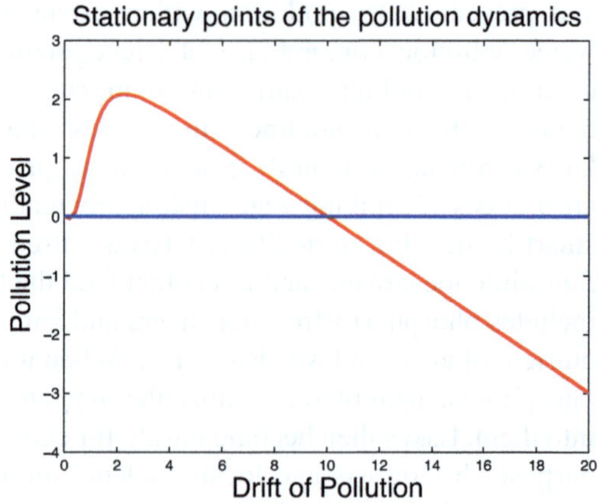

Fig. 12.1 Two steady states under null control action

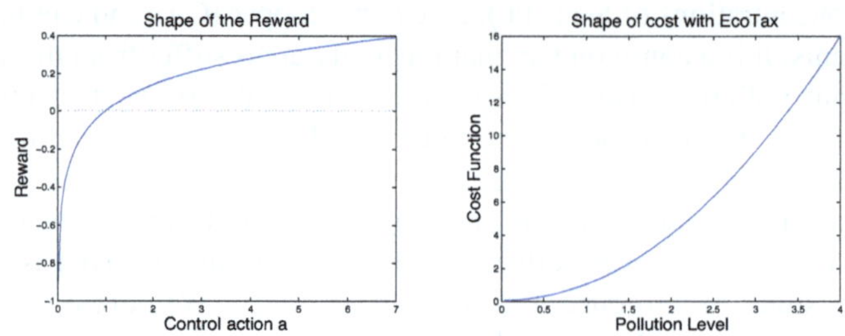

Fig. 12.2 Typical reward function and cost

as $dx_1 = [\sum_{j=1}^{I} a_{1,j}(t) + r_1 \frac{x_1^\beta}{\varepsilon^\beta + x_1^\beta}]dt - x_1 d\gamma_1,\ d\gamma_1 = bdt + \sigma_{11} dB_{11}$. Thus, the dynamics becomes

$$dx_1 = [\sum_{j=1}^{I} a_{1,j}(t) - bx_1 + r_1 \frac{x_1^\beta}{\varepsilon^\beta + x_1^\beta}]dt - \sigma_{11} x_1 dB_{11}. \qquad (12.1)$$

The instantaneous payoff function, which is captured by the benefits to be reaped from the use of the sea or lake, is modeled as $[\sum_{j=1}^{I} \log a_{1,j}(t)] - I\lambda_1 x_1^2$, and the terminal payoff is $-\ I\bar{\lambda}_1 var(x_1(T))$. Note that the presence of the variance term makes the problem of mean field type. The term $\sum_{j=1}^{I} \log a_{1,j}(t)$ models the benefits to farmers and industrial companies arising from using $[\sum_{j=1}^{I} a_{1,j}(t)]$ units of the chemical element. That water is also used by a second class of players such as fishermen, local residents, and tourists, who are interested in clean water. The term $\lambda_1 x_1^2$ represents the cost of pollution and the tax. The game interpretation from the industry and farming pollution should be clear: all actions add to the detriment of the public, and each player generates a negative externality for the other players. The non-optimized Hamiltonian is

$$H_1 = [\sum_{j=1}^{I} \log a_{1,j}(t)] - I\lambda_1 x_1^2 + p[\sum_{j=1}^{I} a_{1,j}(t) - bx_1 + r_1 \frac{x_1^\beta}{\varepsilon^\beta + x_1^\beta}] - \sigma_{11} x_1 q,$$

and maximization yields

$$a_{1,j} = -\frac{1}{p}, \quad j \in \mathscr{I}.$$

The stochastic maximum principle leads to

$$\begin{cases} dp = [-2I\lambda_1 x_1 + p(b - r_1 f') + \sigma_{11} q]dt + qdB, \\ p(T) = -2I\lambda_1[x_1(T) - \bar{x}_1(T)] \\ dx_1 = [\sum_{j=1}^{I} a_{1,j}(t) - bx_1 + r_1 \frac{x_1^\beta}{\varepsilon^\beta + x_1^\beta}]dt - \sigma_{11} x_1 dB_{11}, \\ x_1(0) = x_{10}, \end{cases}$$

providing the pair (x, p).

12.1.1 A Green Community Member Is Ready to Pay Higher Tax than a Classical Farmer

Within the smart city project, the community of farmers and company may be decomposed into two different subcategories. The first subcategory is composed of traditional farmers and industry. The second subcategory is the green community that attaches more importance to antipollution policy and is ready to pay higher tax (called EcoTax) than a classical farmer. We therefore modify the running payoff to be $\log a_{1,j}(t) - \lambda_{1,1} x_1^2$ and $\log a_{1,j}(t) - \lambda_{1,2} x_1^2$ for the first and second subcategories, respectively, $\lambda_{1,1} > \lambda_{1,2}$.

12.1.2 There Is a Simple EcoTax Scheme That Drives the Noncooperative Players to a Pareto Optimal Solution

If a single decision-maker were able to control all the farmers and industry, and can make a decision on tax, then a linear pricing function $EcoTax(t)a_{1,j}(t)$, $EcoTax > 0$ would achieve the Pareto optimal point. However, it is difficult to implement a tax schedule that is highly time varying. In that case, a stationary version, constant tax, would be preferred. The effect of the tax will be to modify each community's welfare function and induce each one to modify its chemical waste loading accordingly.

A second important issue is to identify such a social planner in the context of smart city. This is related to the politico-economic situation. If an election is implemented between these communities, then the social planner who has a political ambition will not necessarily implement the optimal tax if he or she is elected.

Another central point is that the smart city needs to take into consideration the fact that the pollution level affects the economics of the fish industry and tourism. In the next section, we describe a basic example of coupling between these systems.

12.2 Biodiversity and Fishing Industry

For centuries, our seas, oceans, lakes, and rivers have been viewed as a limitless bounty of food. But increasing fishing efforts over the last 50 years as well as unsustainable fishing practices are precipitating the collapse of many fish stocks and marine habitats all over the world. Never before has the need to shift from exploitation to good stewardship been more important. We face not only an environmental disaster, but a social one. Therefore, our fishery strategies need to be managed properly and in a sustainable way. According to the Food and Agriculture Organization of the United Nations (FAO), "85% of the world's fisheries are either fully exploited, over exploited, depleted, or recovering from depletion." As the number, size, and power of fishing boats have grown, an increasing number of commercial fisheries are being fished to the point of collapse. We need to move from overfishing to a smarter fishing strategy. Below we use a simple ecological dynamics for the fishing industry to find and promote smarter solutions that help to preserve the health of our fisheries and marine ecosystems.

We set up our model from the logistic equation, which is the most important and most widely used mathematical model to describe the growth of biological species. We start with the deterministic logistic equation $dx_2 = r_2 x_2(1 - \frac{x_2}{K(x_1)})dt$, where x_2 is the size of fish stock, $r_2 > 0$ is a growth rate, and K denotes the carrying capacity function. Note that $K(.) > 0$ depends on the pollution level x_1 so that the two systems are interacting. The function K is nonincreasing with the pollution level x_1. The more the pollution level x_1 is the smaller will be the size of fish stock x_2. For a fixed x_1, we can solve explicitly this equation. The steady states are 0 and $K(x_1)$. We know that 0 is unstable as the derivative at 0 of the drift is r_2, which is nonnegative. If we start from a strictly positive point, the logistic dynamics leads to $K(x_1)$ as time $t \longrightarrow +\infty$. However, in practice, we do not know the function K. To capture this, we introduce an uncertainty on the function $\frac{1}{K}$. Then, the logistic equation can be rewritten as

$$\begin{cases} dx_2 = r_2 x_2 d\gamma_{21} - r_2 x_2^2 d\gamma_{22}, \\ d\gamma_{21} = dt + \sigma_{21} dB_{21}(t), \\ d\gamma_{22} = \frac{1}{K(x_1)} dt + \sigma_{22} dB_{22}(t), \end{cases}$$

where B_{21} and B_{22} are independent Brownian motions, and σ_{21} and σ_{22} represent the diffusion coefficients. Thus, the stochastic logistic equation becomes

$$dx_2 = r_2x_2(1 - \frac{x_2}{K(x_1)})dt + r_2x_2\sigma_{21}dB_{21}(t) - r_2\sigma_{22}x_2^2dB_{22}(t).$$

In this equation, we include a negative term relating to the harvesting activity $h = kx_2a_2$ associated with the action a_2 :

$$dx_2 = r_2[x_2(1 - \frac{x_2}{K(x_1)}) - kx_2a_2]dt + r_2x_2\sigma_{21}dB_{21}(t) - r_2\sigma_{22}x_2^2dB_{22}(t).$$

Parameter k measures the catchability of the fish stock. Function h is the harvesting activity. Harvest size is increasing in both the size of the stock x_2 and the amount of effort a_2. The size of the fish stock evolves according to

$$dx_2 = [g(x_1, x_2) - h(x_2, a_2)]dt + r_2x_2\sigma_{21}dB_{21}(t) - r_2\sigma_{22}x_2^2dB_{22}(t),$$

where the first term $g = r_2x_2(1 - \frac{x_2}{K(x_1)})$ is the inflow and the second term $h = r_2kx_2a_2$ is the outflow due to harvesting activities. Note that for $k = 0$, the term "h" is zero, and the stationary solutions can be found through the stationary solution of the Fokker–Planck–Kolmogorov equation given by

$$\partial_t m_2(t,x) + \partial_x[m_2(t,x)g] - \frac{1}{2}r_2^2\partial^2_{xx}[\left(x_2^2\sigma_{21}^2 + \sigma_{22}^2x_2^4\right)m_2(t,x)] = 0.$$

Generically, the stationary solutions are explicitly given by $x_2 = 0$ and the probability density

$$m_2^*(x) = \mu x^{2(\frac{1}{\sigma_{21}^2}-1)}\left(1 + \frac{\sigma_{22}^2}{\sigma_{21}^2}x^2\right)^{-\frac{1}{\sigma_{21}^2}-1} e^{-\frac{2}{K\sigma_{21}\sigma_{22}}\arctan(x\frac{\sigma_{22}}{\sigma_{21}})},$$

where $x > 0$, $\sigma_{21}^2 < 2$, and μ is a normalizing constant, i.e.,

$$\frac{1}{\mu} = \int_0^{+\infty} x^{2(\frac{1}{\sigma_{21}^2}-1)}\left(1 + \frac{\sigma_{22}^2}{\sigma_{21}^2}x^2\right)^{-\frac{1}{\sigma_{21}^2}-1} e^{-\frac{2}{K\sigma_{21}\sigma_{22}}\arctan(x\frac{\sigma_{22}}{\sigma_{21}})}\,dx.$$

For $\sigma_{22} = 0$, the Fokker–Planck equation reduces to

$$\partial_t m_2(t,x) + \partial_x[m_2(t,x)g] - \frac{1}{2}r_2^2\partial^2_{xx}[x_2^2\sigma_{21}^2m_2(t,x)] = 0,$$

whose nontrivial steady-state distribution is the Gamma distribution.

12.3 Combined and Balanced Approach

Smart cities start with smart systems, working for the benefit of residents, industry, and the environment. Traditionally, industrial pollution via waste and biodiversity in water quality of a city are separate and operate in silos independent of each other. A truly efficient city requires not only that the performance of each system is optimized, but also that these systems are managed in an integrated way to better prioritize investment and maximize value of the city. On the one hand, we have farmers and industry that drop contaminants indirectly benefiting them but polluting the lake. On the other hand, the fishing industry and the tourists suffer from a polluted lake. If the lake is polluted, the number of fish stock drops to a low level with mean $K(x_1)$, which is less than the demand for fish in the population. The fish industry will lose substantial profit because the cost is higher and the harvest is smaller. Thus, a smarter way would be to design a fish industry that may get better outcome by cooperating with the farmers to control the pollution level. On the other hand, the farmers may be compensated by tax reduction. Putting together, we get

$$\begin{cases} dx_1 = & [\sum_{j=1}^{I} a_{1,j}(t) - bx_1 + r_1 \frac{x_1^\beta}{\varepsilon^\beta + x_1^\beta}]dt - \sigma_{11} x_1 dB_{11}, \\ dx_2 = r_2[x_2(1 - \frac{x_2}{K(x_1)}) - kx_2 a_2]dt + r_2 x_2 \sigma_{21} dB_{21}(t) - r_2 \sigma_{22} x_2^2 dB_{22}(t). \end{cases}$$

We look at numerically the interaction by choosing $K(p) = 1/(1+p)$, and $K(p) = exp(p)$. Figure 12.3 illustrate the dynamics of both states when there is no control action. As we can see, these two particular cases, the pollution grows and the fish stock decreases.

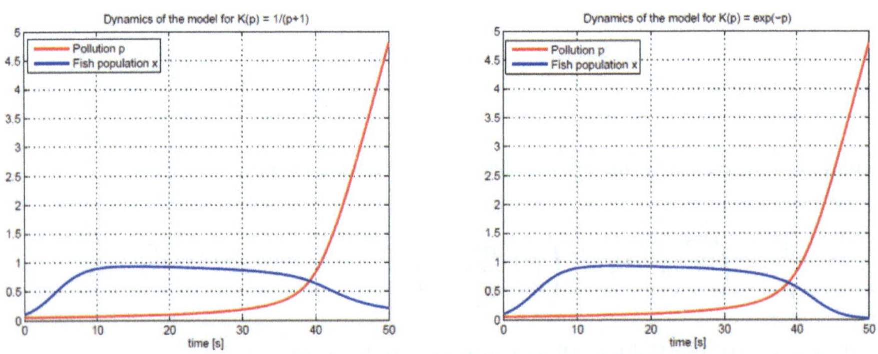

Fig. 12.3 The state dynamics for null control for $K(p) = 1/(p+1)$ and $K(p) = exp(-p)$

12.4 Notes

This chapter has focused on pollution control and fishing strategies in smarter cities using mean-field-type game theory. The intersection of pollution control and fishing strategies takes on new importance as cities strive to create more environmentally friendly and resilient urban environments. We have introduced game dynamics that captures the interaction between the two ecosystems. More details on pollution control and fishing strategies can be found in [36, 128, 147, 169, 199, 218, 338, 344, 410, 464].

Part V
Network Economics, Finance, and Insurance

Chapter 13
Economics of Networks

This chapter examines some economics of the smart grid, charging electric vehicles in smarter cities, cloud networks, and distribution power networks.

13.1 Economics of the Smart Grid

The energy market system is moving toward an important transformation. The success of the new paradigm created by the smart grid vision will require not only the creation and integration of new technologies into the grid but also the redesign of the market structures coupled with it. In order to design the market structures for the grid of the future, network economic models able to capture the new physical reality are one of the first requirements. The focus here is the market side of the smart grid by contributing to the understanding of the impact of dynamics, constraints, and uncertainty, which will become more acute with the increased deployment of renewable resources. The energy market is a coupling of different constrained and highly complex dynamical systems including physical and economical systems. The physical system is a complex network consisting of power, owing through transmission lines, modulated by distributed generation units, flow constraints, and operational and security constraints. Loads and generation are each subject to uncertainty. The economic system typically consists of coupled markets such as day-ahead and real-time auctions. These auctions are dynamic, prices vary, and they are subject to uncertainty because of the users driving the economic side by reacting to the prices and their objectives during their sojourn time. The presence of such complicating factors in these coupled systems, along with the sometimes orthogonal nature of the physical objectives with respect to the economic goals of the market players, makes energy market design a challenging task.

T. Başar et al., *Mean-Field-Type Game Theory II*, Static & Dynamic Game Theory: Foundations & Applications, https://doi.org/10.1007/978-3-032-07031-9_13

How Can the Smart Grid Drive Down Energy Costs and Peak Demand?
Demand response, or consumer side adjustment to energy demand based on supply conditions, is a key element of the smart grid. In the classical approach, consumers have been passive, not clearly understanding the costs associated with their energy demand on a real-time basis, and therefore unable to take significant action with regard to their usage patterns. In fact, most consumers today pay a fixed price per unit of electricity independent of the cost of producing the electricity at the time of use. In economic terms, this type of consumption of electricity is inelastic because consumers do not see or reap any benefit from the differences in the cost of energy production during peak and nonpeak times. If consumers were able to know and benefit from lower energy costs during off-peak periods, they could increase and decrease their use of electricity in reaction to real-time prices. The smarter the grid, the better consumers are able to flatten their own demand during peaks. The smart grid will enable consumers to interact with the electrical grid and make informed choices about when they use energy, the rates they pay, and the source of that energy. Intuitively, this choice will drive down cost because during off-peak periods, energy will cost less and consumers will be able to use that to their advantage.

Peaks and High Variance
Since electricity must be consumed the moment it is generated, we cannot produce energy and store it for when we need it most—say on a hot summer day when we all run our air-conditioners. These times when energy is most in demand are referred to as *peak periods*. Electric grids must be built to provide enough capacity to meet the needs during peak periods, although to do so creates inefficiencies during off peak periods. One of the primary goals of the smart grid is to help flatten the peak and smooth out the demand. By means of strategic game design, one can propose a payoff function that respects the preferences and flattens the peak as much as possible.

We summarize some of the notation used in the chapter in Table 13.1.

Table 13.1 Summary of notation

Symbol	Meaning
p^d	Price demanded
p^s	Price supplied
q^d	Quantity demanded
q^s	Quantity supplied
$p_t^I(.)$	Price of the market at time t
D(.)	Demand function
S(.)	Supply function
$\bar{M}_t^I$	Aggregative term
$r_t(.)$	Instantaneous payoff function
$q_{j,t}^I$	Control of user j

13.1.1 Energy Market on the Network

We assume that all users face the same demand price p^d. In a static setting, we aim to work with the following equalities:

$$q_*^d = D(p_*^d), \ q_*^s = S(p_*^s); \ q_*^d = q_*^s, p_*^d = p_*^s. \tag{13.1}$$

This says that in the energy market, the demand equals the supply.

13.1.1.1 Market Price

Since the network is stochastic and the market is dynamically changing, there is no reason for the price supplied to be an inverse demand at each time. Therefore, we choose the price to be a certain function of the aggregate quantity $\bar{M}_t^I = \frac{1}{I}\sum_{j=1}^{I} x_{j,t}^I$. We choose the price as a monotone function of the demand:

- **Price Model 1:** The price can be chosen as

$$(*) \ p_t^I(x_t^I, q_t^I) = \alpha_t(e^{\frac{\beta_t}{I}\sum_{j=1} x_{j,t}^I} - 1), \ (\alpha_t, \beta_t) \in \mathbb{R}^2,$$

where $x_{j,t}^I$ is the state of user j at time t when the network size is I. If the demand is zero, the price is zero (this is why we have the term -1 in $(*)$). The total sum $\sum_{j=1}^{I} x_{j,t}^I$ will be subject to a positivity constraint. The coefficients α_t and β_t capture the seasonal aspect of the price. For example, α_t, can be chosen as periodical function of time to illustrate the price variation over a year (α_t differs from winter to summer).
- **Price model 2** Another price function could be

$$(**) \ p_t^I(x_t^I, q_t^I) = \alpha_t e^{\frac{\beta_t}{I}\sum_{j=1} q_{j,t}^I}, \ (\alpha_t, \beta_t) \in \mathbb{R}^2.$$

The difference with price model 1 is that now we have a function of the average quantity (average control) instead of average state. This price model is related to the Cournot oligopoly model where the exponential is replaced by the identity function.

13.1.1.2 State Dynamics

We assume that the state of user j evolves according to controlled stochastic differential equation (SDE). A state may represent an energy storage level, reserve, or battery state. The state evolves stochastically depending on the investment, production, and the amount of quantity of the user sell/buy on the market (which plays the role of the control). The control input is given by $q_{j,t}^I$. A negative $q_{j,t}^I$

is interpreted as the quantity bought and a positive $q^I_{j,t}$ as the quantity sold. As an example, we consider the well-known geometric state evolution given by

$$dx^I_{j,t} = \delta_t x^I_{j,t}\left(\bar{x}^I_t - q_{j,t}\right)dt + \sigma_t x^I_{j,t} dB_{j,t},\ x^I_{j,0} \in \mathbb{R},\ (\delta_t, \sigma_t) \in \mathbb{R}^2, \quad (13.2)$$

where $f_j(x^I_{j,t}, q^I_{j,t}) := \delta_t x^I_{j,t}\left(\bar{x}^I_t - q_{j,t}\right)$ represents a drift term and $\sigma_t x^I_{j,t}$ is the noise coefficient term, and $B_{j,t}$ is a standard Brownian motion. The state $x^I_{j,t}$ is subject to a positivity constraint. This leads the state dynamics as **reflected** stochastic differential equation with control. The variable $\bar{x}^I_t$ is a referent state (benchmark). The geometric state model (13.2) is well justified empirically in stock market and in **electricity market** over the past years. The coefficients are bounded deterministic functions of time and $q^I_{j,t}$ is a control j, which is an element of $\mathcal{L}^2(\mathcal{F}_{j,t})$ where $\mathcal{F}_{j,t}$ is the natural filtration generated by $B_{j,t}$. This price model is inspired from the empirical evolution of price in the electricity market. In many cases, the state dynamics is non-linear.

13.1.1.3 Price Dynamics

In order to derive the price dynamics for price model 1, we use the Itô formula. Next we recall Itô's formula for Itô SDE: let $dx_t = \tilde{f}(t, x_t)dt + \tilde{\sigma}(t, x_t)dB_t$, and $l(t, x)$ be a real-valued function, defined for $t \in [0, T]$ and $x \in \mathbb{R}^I$, with continuous partial derivatives $\partial_t l(t, x)$, $\partial_x l(t, x)$, $\partial^2_{xx} l(t, x)$. If $y_t = l(t, x_t)$, then

$$dy_t = \left[\frac{\partial}{\partial t} l(t, x_t) + \langle \tilde{f}(t, x_t), \partial_x l(t, x_t)\rangle + \frac{1}{2} tr[\tilde{\sigma}\partial^2_{xx} l(t, x_t)\tilde{\sigma}']\right] dt + (\partial_x l)'\tilde{\sigma} dB_t,$$

where $\partial_x l(t, x_t)$ is the gradient with the respect to the vector x, $\partial^2_{xx} l$ is the Hessian matrix, and σ'_t is the transposition of σ_t. Hence, the SDE satisfied by the price $p^I_t(x^I_t, q^I_t) = \alpha_t(e^{\frac{\beta_t}{I}\sum_{j=1}^I x^I_{j,t}} - 1)$ (nonautonomous) is

$$\begin{aligned} dp^I_t &= \partial_t[\alpha_t(e^{\frac{\beta_t}{I}\sum_{j=1}^I x^I_{j,t}} - 1)] \\ &+ (p^I_t + \alpha_t)\left([\delta_t \bar{x}^I_t \log(\frac{p^I_t + \alpha_t}{\alpha_t}) + (\frac{\sigma_t^2 \beta_t^2}{2I} - \delta_t \beta_t)\frac{\sum_{i=1}^I (x^I_i)^2}{I}]dt \right. \quad (13.3) \\ &\left. + \frac{\beta\sigma}{I}\sum_{i=1}^I x^I_{i,t} d\mathbb{B}_{i,t}\right), \end{aligned}$$

where we have used the fact that

$$\bar{M}^I_t = \frac{1}{\beta_t}\log(\frac{p^I_t + \alpha_t}{\alpha_t}),\ p^I_t + \alpha_t = \alpha_t e^{\beta_t \bar{M}^I_t}.$$

13.1.1.4 Payoff Function

Each user in the market has its own benefit and own cost. User $j's$ payoff is given by

$$R^I_{j,T}(q_1,\ldots,q_I) = \mathbb{E}\left(g(x^I_{j,T}) + \int_0^T [q^I_{j,t} p^I_t(x^I_t, q^I_t) - c_t(x^I_{j,t}, q^I_{j,t})]\, dt \right),$$

where $T < +\infty$ is the length of the horizon of the interaction and g is a terminal payoff. As we can see, each user's control $q^I_{j,t}$ influences its state, which influences the price and the price influences the payoff of each user in the energy market. Thus, there is an interdependency in the user outcome. In most of our analysis, we will choose the cost in the form $\underline{c}_t(q^I_{j,t}) + \bar{c}_t(x^I_{j,t})$.

13.1.2 Energy Market with Intervention of a Controller

We introduce below designer tax and purchase.

13.1.2.1 Designer Tax

A network controller or market controller (such as government) imposes a quantity tax $\bar{t}$ on the quantity q, so that equation (13.1) is changed to $D(p_* + \bar{t}) = S(p_*)$, $p^d_* = p^s_* + \bar{t} = p_* + \bar{t}$ (a subsidy can be seen as a negative tax). Combining all of the equations now yields

$$q^d_* = D(p^d_* + \bar{t}),\; q^s_* = S(p^s_*);\; q^d_* = q^s_*,\; p^d_* = p^s_* + \bar{t}. \tag{13.4}$$

13.1.2.2 Designer Purchase

Now, the controller/designer (government) purchases $\bar{q}$ in the energy market (which we assume does not go to consumers/users), changing the system (13.1) to $q^d + \bar{q} = q^s$ (similarly a government sale can be expressed as a negative $\bar{q}$). Combining all of the equations now yields $D(p_*) + \bar{q} = S(p_*)$, $q^s = S(p_*)$, $q^d = S(p_*) - \bar{d}$. This means that there are two quantities. For I large, we will consider the designer as a "big player" or atomic player (in the sense that the designer can have a big influence in the market) and the users as "small players" or non-atomic (in the sense that the effect of one user is in order of $\frac{\beta_t}{I}$). This leads to a stochastic differential game with an atomic player and many non-atomic players.

The tax and/or intervention of the designer influences the equilibrium outcome as well as the users' density.

13.1.3 Dynamic Equilibrium Approach to Smart Grid

The problem formulated above is a stochastic differential game described by the following:

- $\mathcal{I} = \{1, 2, \ldots, I\}$ is the set of users.
- Let $0 < T < +\infty$. For $t \in [0, T] =: \mathcal{T}, q^I_{j,t} \in \mathbb{R}$, is the control of user $j \in \mathcal{I}$ at time t.
- The instantaneous payoff of user j is defined by $r_t : \mathbb{R}^3 \longrightarrow \mathbb{R}$,

$$r_t(x^I_{j,t}, q^I_{j,t}, \bar{M}^I_t) = q^I_{j,t} p^I_t(x^I_t, q^I_t) - c_t(x^I_{j,t}, q^I_{j,t}).$$

- Each individual state evolves according to the SDE (13.2).

Let $\mathcal{F}^I_t = \sigma(x^I_{t'}, B_{t'}, t' \leq t), t \in \mathcal{T})$ be a sigma-algebra generated by state processes and the standard Brownian motion. The filtration $\mathcal{F}^I_t$ plays an important role in the construction of a strategy at time t. Next, we define key concepts for the control of a user:

- An open-loop control for user j at time t is a function of time that belongs to the space of controls.
- A state-feedback control for user j at time t is a function of $x^I_t = (x^I_{1,t}, \ldots, x^I_{I,t})$ and time.
- An own-state-feedback control for user j at time t is a function of $x^I_{j,t}$ and time.
- An aggregate-based and own-state feedback control at time t is a function of $x^I_{j,t}$, the current aggregative term $\bar{M}^I_t$, and time.

A strategy in the differential game is a collection of mapping that depends on the past observations, the information patterns, and the probabilistic state space. Due to the fact that the network is complex and limited in terms of state observations, we restrict our attention to a very simple classes of strategies. This is a strategic decision-making problem, so the state dynamics cannot be assumed to be independent processes.

Where does the interdependence between the individual states come from? Because the users will be reacting to that they get as revenue via their payoffs, which depend on the other state, their controls may change and result in implicit dependence of the users' decisions and states. This is one of the main reasons why the law of large numbers (LLN) cannot be directly applied to the above interaction problem when I is large. We will clarify how to deal with large scale via *asymptotic indistinguishability per class* argument.

We say that a collection of mappings $\{q^I_{j,t}(.)\}_{j,t}$ constitutes an *own-state-and-aggregate-based feedback (Nash) equilibrium* if for every user j, $q^I_{j,t}$ maximizes $R^I_{j,T}(., q^I_{-j})$, where $\{q^I_{i,t}(.)\}_{i\neq j, t\leq T}$. Similarly, we say that $\{q^I_{j,t}(.)\}_{j,t}$ is own-state feedback Nash equilibrium if it is a Nash equilibrium and the control is of own-state feedback type.

Clearly, there is no reason for own-state feedback strategies to be optimal. However, due to the limitations in terms of observations, this type of control is easily implementable. One can use own-state-and-aggregate-based feedback strategy if the environment allows the user to observe the statistics of the aggregative term $\bar{M}^I_t$ at time t or if these can be pre-computed as the interaction moves. In the large population game setting, when $I \longrightarrow +\infty$, we will discuss how to compute $\bar{M}^I_t$, using the mean of the Fokker–Planck–Kolmogorov (FPK) equation. Below we state a general methodology for finding own-state-and-aggregate-based feedback strategy, which is a response to the measure-valued random process measure $\bar{M}^I$.

Let η_t be the information pattern at time t. Then, a feedback equilibrium is defined as follows.

Definition 13.1 A strategy profile in the form $q^I_{j,t} = \gamma^*_{j,t}(t, \eta_t)$ constitutes a feedback Nash equilibrium solution if there exist functions $v^I_j(t, x^I_1, \ldots, x^I_I)$ defined on $\mathcal{T} \times \mathbb{R}^I$, and satisfying the following relations for each $j \in \mathcal{I}$:

$$\begin{cases} v^I_j(T, x^I) = g(x^I), \\ v^I_j(t, x^I) = g(x^*_{j,T}) + \int_{t'\in[t,T]} r_{j,t'}(x^*_{j,t}, \gamma^*_{i,t'}(\eta_{t'}), \frac{1}{I}\sum_{i\in\mathcal{I}} x^*_{i,t'}(\eta_{t'}))\, dt' \\ \geq g(x^{*,(j)}_T) + \int_{t'\in[t,T]} r_{j,t'}(x^{*,(j)}_{j,t'}, a_{j,t'}(\eta_{t'}), \frac{1}{I} x^{*,(j)}_{j,t'}(\eta_{t'}) + \frac{1}{I}\sum_{i\neq j} x^*_{i,t'}(\eta_{t'}))\, dt', \end{cases}$$

for all $a_{j,.}(.), x^I \in \mathbb{R}^I$, where on the interval $[t, T]$,

$$\begin{cases} dx^*_{j,t'} = \delta_t x^*_{j,t'}(\bar{x}^I_{t'} - \gamma^*_{j,t'}(\eta_{t'}))dt + \sigma x^*_{j,t'} dB_{j,t}, \\ x^*_{j,t} = x^I_j,\ j \in \mathcal{I}, \\ dx^{*,(j)}_{t'} = \delta_t x^{*,(j)}_{t'}(\bar{x}^I_{t'} - a_{j,t'}(\eta_{t'}))dt + \sigma x^{*,(j)}_{t'} dB_{j,t}, \\ x^{*,(j)}_{j,t} = x^I_j,\ j \in \mathcal{I}. \end{cases}$$

Lemma 13.2 *A strategy profile* $q_{j,t} = \gamma_{j,t}(x^I) \in \mathbb{R},\ j \in \mathcal{I}, t \in \mathcal{T}$ *provides a feedback Nash equilibrium if there exist continuously differentiable functions* v^I_j : $\mathcal{T} \times \mathbb{R}^I \longrightarrow \mathbb{R},\ j \in \mathcal{I}$ *satisfying the following set of partial differential equations called Hamilton–Jacobi–Bellman equations:*

$$
\begin{cases}
v_j^I(T, x^I) = g(x^I), \\
0 = \frac{\partial}{\partial t} v_j(t, x^I) + \sup_{a'_j(.)} \Big\{ r_t(x_j^I, a'_{j,t}(x^I), \frac{1}{I} x'_{j,t}(x^I) + \frac{1}{I} \sum_{i \neq j} x_{i,t}(x^I)) \\
\quad + \langle \partial_{x^I} v_j^I(t, x^I), f(x^I, a'_{j,t}(x^I)) \rangle \Big\} \\
\quad + \frac{1}{2} \sigma_t^2 (x_j^I)^2 \partial^2_{x^I x^I} v_j^I \\
= r_t(x_j^I, \gamma_{j,t}, \bar{M}_t^I) + \langle \partial_{x^I} v_j^I(t, x^I), f(x^I, \gamma_{j,t}(x^I)) \rangle \\
\quad + \frac{1}{2} \sigma_t^2 \sum_k (x_k^I)^2 \partial^2_{x_k^I x_k^I} v_j^I.
\end{cases}
$$

Note that here the vector $(x_{j,t}^I)_j$ is considered. As I gets larger, the cross terms $\partial^2_{x_k^I x_k^I} v_j^I$ take an aggregate form. A necessary condition for an interior feedback strategy is easily obtained for model 1 by solving

$$
-\delta_t \langle x^I, \partial_{x^I} v_j^I(t, x^I) \rangle + p_t^I(x_t^I) - \partial_q c_t(x_{j,t}^I, q_{j,t}^I) = 0. \tag{13.5}
$$

Lemma 13.3 *If every x, $\partial_q c_t(x, .)$ is a one-to-one map, then the above equation (13.5) provides a unique control given by*

$$
q_{j,t}^I = (\partial_q c_t)^{-1} [p_t^I(x_t^I) - \delta_t \langle x^I, \partial_{x^I} v_j^I(t, x^I) \rangle].
$$

Since this requires solving I partial differential equations (PDE) in $I-$dimensional space where each PDE depends on the knowledge of the state of all the users (I SDEs), it is clear that the computation of the value $v_j^I(.)$ becomes very complex when I is large.

Remark 13.4 For price model 2, finding the equilibrium strategy requires solving the following fixed point equation:

$$
q_{j,t}^I = (\partial_q c_t)^{-1} [p_t^I(q_t^I) - \delta_t \langle x^I, \partial_{x^I} v_j^I(t, x^I) \rangle].
$$

Below we show how to use large population mean-field stochastic game in order to overcome the curse of dimensionality problem [460, 464].

13.1.4 Micro to Macro: Homogeneous Case

13.1.4.1 Convergence of the Prices

The convergence of the equilibrium prices depends on the behavior of the aggregative term $\bar{M}_t^I$. To show the convergence of $\bar{M}_t^I$ to $\bar{m}_t$ when $I \longrightarrow +\infty$, we examine the mean-field object $M_t^I = \frac{1}{I} \sum_{i=1}^I \delta_{x_{i,t}^I}$, where $\delta_{x_{j,t}^I}$ is the Dirac measure. For price model 1, one has

$$\bar{M}_t^I = \frac{1}{I}\sum_{i=1}^{I} x_{i,t} = \int_{\mathbb{R}} y\left(\frac{1}{I}\sum_{i=1}^{I}\delta_{x_{i,t}^I}\right)(dy) = \int_{\mathbb{R}} yM_t^I(dy), \quad (13.6)$$

where $M_t^I = \frac{1}{I}\sum_{i=1}^{I}\delta_{x_{i,t}^I}$.

To prove the (weak) convergence of $\bar{M}_t^I$, it suffices to prove that the **mean-field object** M_t^I converges weakly and checks integrability. Suppose that $q_{j,t}^I(.,.) \longrightarrow q_t(.,)$ as I goes to infinity. The collection of the system states $(x_1^I, \ldots, x_I^I)$ satisfies the asymptotic indistinguishability per class property when I goes to infinity. In particular, the system preserves the indistinguishability property if a homogeneous control law is employed. If M_0^I converges in probability to $\rho_0(.)$ when I goes to infinity and $q_{j,t}^I$ preserves the asymptotic indistinguishability per class property, then M_t^I converges in distribution to m_t. Moreover, $\bar{m}_t$, the limit of $\bar{M}_t^I$, is the expectation of the random variable m_t with probability distribution (renamed $m_t(x)$) where $m_t(.)$ solves the Fokker–Planck–Kolmogorov equation:

$$\partial_t m_t(x) + \partial_x(\delta_t x(\bar{x} - q_t^*(.))m_t(x)) = \frac{\sigma_t^2}{2}\partial_{xx}^2 m_t(x)$$
$$m_0(y) = \rho_0(y), \ \forall y \in \mathbb{R}, \quad (13.7)$$

where $q_t^*(.)$ will be the optimal control, which is a function of $t, x, \bar{m}_t$ and the value function v_t for price model 1.

13.1.5 Hamilton–Jacobi–Bellman at the Asymptotics

We now study the Hamilton–Jacobi–Bellman equation in the large population limit regime. In the homogeneous case, each generic user is facing an evolving mean quantity $\bar{m}_t$ and optimizes its long-term payoff during the time window $\mathcal{T}$. The Hamilton–Jacobi–Bellman equation associated with a generic user is given by

$$\partial_t v(t,x) + \sup_q\,[qp_t(\bar{m}_t) - c_t(q) + \delta_t x(\bar{x} - q)\partial_x v(t,x)] + \frac{\sigma_t^2}{2}x^2\partial_{xx}^2 v(t,x) = 0, \quad (13.8)$$

with terminal condition $v(T, x) = g(x)$. The optimal response (model 1) to $\bar{m}_t$ is

$$q_t^* = (\partial_q c_t)^{-1}[p_t(\bar{m}_t) - \delta_t x \partial_x v(t,x)].$$

Next, we need to find the mean of the optimal trajectory, $\mathbb{E}x_t^*$.

13.1.6 *Continuum of Decision-Makers*

We now consider the game with a continuum of users in a homogeneous population.

13.1.6.1 Analysis of the Static Game

Consider a one-shot game in which each decision-maker chooses its action in a non-empty and compact set $\mathbb{S} \subset \mathbb{R}$. Let $\Delta(\mathbb{S})$ be the set of probability distributions on $\mathbb{S}$. We say that $m \in \Delta(\mathbb{S})$ is a static equilibrium if

$$\text{supp}(m) \subseteq \arg \max_{q \in \Delta(\mathbb{S})} [qp(\bar{m}) - c(q)].$$

Such a mean field of actions is a static mean-field equilibrium [460, 464].

The next result connects subsequence of static equilibria of finite population games to the static mean-field equilibrium.

Theorem 13.5 *Assume that each game has at least one symmetric pure equilibrium. Consider a converging subsequence* $q_j^{\phi(I)}$ *of symmetric pure equilibria* $(q_j^{\phi(I)})_{j \in \mathcal{I}}$ *of the game with* I *decision-makers. Then, any limiting subsequence of* $\frac{1}{I}\sum_{i=1}^{I} \delta_{q_i^I}$ *provides a static equilibrium.*

The proof is immediate. It uses the continuity of functions p and c. ∎

13.1.6.2 Continuum of Buyers/Sellers

We formulate the following mean-field response problem:

Mean-field response problem:
Let B be a one-dimensional Brownian motion defined on $(\Omega; \mathcal{F}; \mathbb{P})$, where $\mathcal{F}$ is the natural filtration generated by B and x_0 is any random variable independent of B having density $m_0(x)$. Maximize the functional

$$\sup_{\{q_t\}_{t \in \mathcal{T}}} \mathbb{E}\left(g(x_T) + \int_0^T [q_t p_t(\bar{m}_t^*) - c_t(x_t, q_t)]\, dt \right),$$

with the dynamics of x_t given by

$$dx_t = \delta_t x_t\, (\bar{x}_t - q_t)\, dt + \sigma_t x_t dB_t,\ t \in (0, T], \tag{13.9}$$

(continued)

$$x_0 \in \mathbb{R}, (\delta_t, \sigma_t) \in \mathbb{R}^2, t \in (0, T],$$

where $\bar{m}_t^*$ is the mean of the infinite population mean field of states.

The mean-field equilibrium is a solution of the backward/forward system:

$$\begin{aligned}
&v_T(x) = g(x),\\
&\partial_t v_t(x) + q_t^* p(\bar{m}_t) - c_t(q_t^*) + \delta_t x(\bar{x} - q_t^*)\partial_x v_t(x) + \tfrac{\sigma_t^2}{2} x^2 \partial_{xx}^2 v_t(x) = 0,\\
&q_t^* = (c_t')^{-1}[p(\bar{m}_t) - \delta_t x \partial_x v_t(t, x)],\\
&\partial_t m_t(x) + \delta_t \partial_x (x(\bar{x} - q_t^*) m_t(x)) = \tfrac{\sigma_t^2}{2} \partial_{xx}^2 (x^2 m_t(x)),\\
&m_0(.)\ fixed,\\
&\bar{m}_t = \textstyle\int_{\mathbb{R}} x m_t(dx).
\end{aligned} \tag{13.10}$$

Lemma 13.6 ([438]) *If $F(.)$ is the cumulative distribution function of the mean-field limit, then $F(.)$ is a solution of the PDE:*

$$\partial_t F_t(x) + \delta_t x(\bar{x} - q_t^*)\partial_x F_t(x) - \frac{\sigma_t^2}{2}\partial_x (x^2 \partial_x F_t(x)) = 0 \tag{13.11}$$

$$\partial_t F_t(x) + \delta_t x(\bar{x} - (c_t')^{-1}[p\left(\int (1 - F_t(x')dx')\right) - \delta_t x \partial_x v_t(t, x)])\partial_x F_t(x)$$
$$-\frac{\sigma_t^2}{2}\partial_x (x^2 \partial_x F_t(x)) = 0.$$

13.1.7 Deterministic Mean-Field Game: Infinite of Number Users

Deterministic mean-field response problem:
Maximize the payoff functional

$$\sup_{\{q_t\}_{t \in \mathcal{T}}} g(x_T) + \int_0^T [q_t p_t(\bar{m}_t^*) - c_t(x_t, q_t)]\, dt,$$

with the dynamics of x_t given by

(continued)

$$\dot{x}_t = \delta_t x_t\left(\bar{x}_t - q_t\right),\ t \in (0, T],$$
$$x_0 \in \mathbb{R}, \tag{13.12}$$

where $\bar{m}_t^*$ is the mean of the mean-field trajectory.

Note that for a deterministic coefficient q_t, the solution of (13.12) is

$$x_t = x_0 e^{\int_0^t \delta_s(\bar{x}_s - q_s)\,ds}.$$

The mean-field equilibrium (for the deterministic case) is a solution of the backward/forward system:

$$\begin{array}{l} v_T(x) = g(x), \\ \partial_t v_t(x) + q_t^* p(\bar{m}_t) - c_t(q_t^*) + \delta_t x(\bar{x} - q_t^*)\partial_x v_t(x) = 0, \\ q_t^* = (c_t')^{-1}[p(\bar{m}_t) - \delta_t x \partial_x v_t(t, x)], \\ \partial_t m_t(x) + \delta_t \partial_x(x(\bar{x} - q_t^*)m_t(x)) = 0, \\ m_0(.)\ fixed, \\ \bar{m}_t = \int_{\mathbb{R}} x m_t(dx). \end{array} \tag{13.13}$$

13.1.8 Nonhomogeneous System

We now study the localized energy market. The price depends on the zone (geographical localization). Denote by Θ the set of locations of the smart grid network. For $\theta \in \Theta$, we introduce $c_{\theta,t}(.)$, $q_{\theta,t}(.)$, $g_{\theta,t}(.)$, $\delta_{\theta,t}$, $\sigma_{\theta,t}$ as the parameters for users in zone θ :

$$\begin{array}{l} v_{\theta,T}(x) = g_\theta(x),\ (\theta, x) \in \Theta \times \mathbb{R} \\ \partial_t v_{\theta,t} + q_{\theta,t}^* p(\bar{m}_t) - c_{\theta,t}(q_t^*) + \delta_{\theta,t} x(\bar{x} - q_{\theta,t}^*)\partial_x v_{\theta,t} + \frac{\sigma_{\theta,t}^2}{2} x^2 \partial_{xx}^2 v_{\theta,t} = 0 \\ q_{\theta,t}^* = (c_{\theta,t}')^{-1}[p(\bar{m}_t) - \delta_{\theta,t} x \partial_x v_{\theta,t}(t, x)], \\ \partial_t m_{\theta,t}(x) + \delta_{\theta,t}\partial_x(x(\bar{x} - q_{\theta,t}^*)m_{\theta,t}(x)) = \frac{\sigma_{\theta,t}^2}{2}\partial_{xx}^2(x^2 m_{\theta,t}(x)), \\ m_{\theta,0}(.)\ fixed, \\ \bar{m}_t = \int_{\Theta\times\mathbb{R}} x m_t(d\theta, dx). \end{array} \tag{13.14}$$

Here there is an interaction between the different areas (zones, classes) and inside the same class. The resulting mean-field equilibrium is a multi-class one.

13.2 Charging Electric Vehicles in Smarter Cities

A plug-in hybrid electric vehicle (PHEV) is a vehicle powered by a combination of an internal combustion engine and an electric motor with a battery pack. The battery pack can be charged by plugging the vehicle into the electric grid and from using excess engine power. A PHEV allows for all electric operation for limited distances. Electric vehicles (EVs) are seen as a midterm solution to reduce the transportation sector's dependency on oil. PHEVs enjoy special attention since they decrease a fleet's average gasoline consumption and shift it to the electricity sector without violating consumer preference. Typically, the drive train of such a vehicle includes an electrical motor, an internal combustion engine, a battery, and a device to connect utility grid and vehicle. During most of the utilization time, the vehicle operates in an all-electric drive mode. This is most efficient with respect to gasoline consumption and usually ranges around 60 miles. If the battery's charge is insufficient, the internal combustion engine can be engaged to extend the range. As an asset, EV can provide valuable service to power systems, more than its transportation function. Firstly, the battery of the EV can be considered as a controllable load. With a smarter charging strategy for the battery, the peak load can be shaved (demand response can be addressed), and by doing this, EV owners could maximize their profits by purchasing energy at the lower electricity price. Secondly, the battery of the EV can also be seen as an energy storage equipment, which has possibility to provide vehicle-to-grid (V2G) and grid-to-vehicle service, and thereby earning profit by saving the cost of storage. Therefore, a smarter charging strategy is required to coordinate the needs from the power system and maximize the EV owner's profit. For a power system operator, a smarter strategy can be interpreted as enabling the large penetration of renewable and distributed energy resources, like wind power and photovoltaic (solar), and also accommodate the new loads, like EV and heat pump, both reliably. From an EV user's perspective, a smarter strategy means minimizing their charging cost without interfering their daily drive profile. However, if implemented in a large scale without control, peak load increases significantly and the grid may be overloaded. The goal here is to develop a methodology for optimizing a PHEV charging design and determine the optimum designs for charging and reducing market demand peak periods. A non-linear battery model is characterized and presented, and different future electricity price formats are examined. The charging game is a mean-field-type game if the price is a nontrivial function of the demand. Mean-field-type algorithms to address this problem are introduced and analyzed in this chapter. The methodology is based on a forecast of future electricity prices and uses the maximum principle to find the economically optimal solution for the vehicle owner. This optimizes the charging time and energy flows. It reduces daily electricity cost substantially without increasing battery degradation.

Model We consider plug-in electric vehicle with battery packs, which are represented by electric circuits. The circuit is described by a voltage source V in series with an internal resistance. These parameters are functions of the state of charge

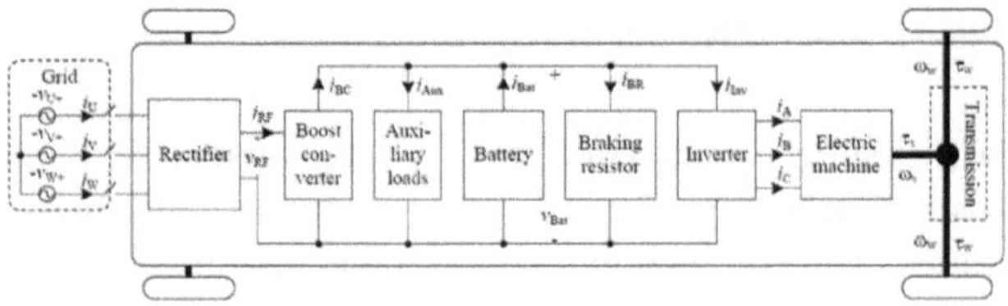

Fig. 13.1 EV architecture composed of rectifier, boost converter, auxiliary loads, battery, braking resistor, inverter, and electric machine

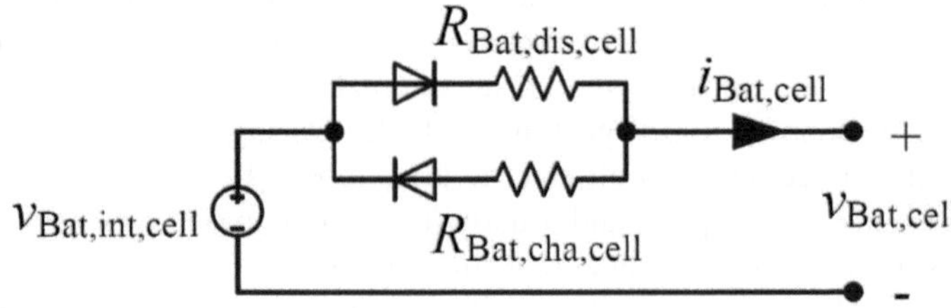

Fig. 13.2 Electric equivalent circuit for battery cell. $R_{Bat,dis,cell}$ (resp. $R_{Bat,cha,cell}$) is discharge (resp. charge) resistance of the cell. $V_{Bat,int,cell}$ is the internal battery cell voltage

(SoC) of the battery. The battery pack is the heart of an electric vehicle. Many different battery types exist, e.g., lead-acid, nickel-metal hydride, and lithium ion. However, today, the lithium ion is the preferred choice due to its relatively high specific energy and power. We use Ohm's law to derive the important structure of the evolution of the state of charge of the battery. Ohm's law deals with the relationship between voltage and current in an ideal conductor. This relationship states that the potential difference (voltage) across an ideal conductor is proportional to the current through it. The constant of proportionality is called the "resistance."

Considering the circuits in Figs. 13.1 and 13.2, we have the following relationships:

$$P_{RF} = V_{RF} I_{RF} = P_{BC} + P_{loss,BC},$$

where P_{RF}[in Watts] is the input power of boost converter (BC), P_BC [W] is the output power of the boost converter, $P_{loss,BC}$ [W] is the power loss of the boost converter, V_{RF} [Volts] is the input voltage of the boost converter, I_{RF} [Amps] is the input current of the boost converter, I_{BC} [A] is the output current of the boost converter, $P_{loss,BC}$ is the power loss, and RF stands for radio frequency. From Fig. 13.2, the battery cell voltage is

$$\begin{aligned} V_{Bat,cell} &= [V_{Bat,int,cell} - R_{Bat,dis,cell} I_{Bat,cell}] \mathbb{1}_{\{I_{Bat,cell}>0\}} \\ &\quad + [V_{Bat,int,cell} - R_{Bat,cha,cell} I_{Bat,cell}] \mathbb{1}_{\{I_{Bat,cell}<0\}}. \end{aligned}$$

The battery pack consists of $N_{Bat,s}$ series connected cells and $N_{Bat,p}$ parallel connected strings. The total voltage and current of the battery pack can therefore be calculated as

$$V_{Bat} = N_{Bat,s} V_{Bat,cell}, \quad I_{Bat} = N_{Bat,p} I_{Bat,cell}.$$

During the charging of the battery, the battery cell voltage $V_{Bat,cell}$ should not exceed a certain maximum (e.g., 4.2 V), and the maximum cell charging current should not be higher than a certain maximum current (e.g., 7 A). The output power of the boost converter is $P_{BC} = V_{Bat} I^*_{BC} = -\eta(t)u(t)P_{max-plug}\mathbb{1}_{\{t\in plug-in\}} + P_{driving}\mathbb{1}_{\{t\in driving\}}$, where $P_{driving}$ is the power requirement for driving cycling, $P_{max-plug}$ is the maximum power (which is around 4KW), and $u(t)$ is a dimensionless control variable. u represents the charging rate of the plug-in, and η represents the efficiency parameter, which is around 95% in practice. Note that if the electric vehicle does not have an internal combustion engine, the dimensionless control variable u is set to zero in the driving intervals. The power P_{BC} is negative when the battery is charging. In general, the power will be a negative number if energy enters the circuit by way of the element. Energy is being converted from a nonelectrical form to electrical current.

13.2.1 Derivation of the Electric Current

From the above relations, we deduce the system

$$\begin{cases} V_{RF} I_{RF} = P_{BC} + P_{loss,BC} \\ P_{loss,BC} = R_{BC} I_{RF}^2 + I_{RF} V_{th,BC}, \end{cases} \tag{13.15}$$

where $V_{th,BC}$ is the threshold voltage and R_{BC} is the resistance of the switch and diode (in the order of $25.10^{-4} Ohm$). Thus,

$$R_{BC} I_{RF}^2 + (V_{th,BC} - V_{RF}) I_{RF} + P_{BC} = 0.$$

This relation can be obtained directly by considering the difference voltage as $V_{th} = V_s - R_{BC} I_{RF}$, where V_s is the voltage source. Thus, the value of the input current I_{RF} is given by

$$I_{RF}^{\pm} = \frac{-(V_{th,BC} - V_{RF}) \pm ((V_{th,BC} - V_{RF})^2 - 4R_{BC}P_{BC})^{\frac{1}{2}}}{2R_{BC}},$$

provided that $(V_{th,BC} - V_{RF})^2 - 4R_{BC}P_{BC} \geq 0$.

In order to determine which of the two electric currents I_{RF}^{-}, I_{RF}^{+} corresponds to I_{RF}, we use the open-circuit argument using the standard conventions. For the purpose of the optimization, it is sufficient to choose $I(x, u) = I_{RF}^{+}$.

13.2.2 State of Charge of the Battery

The state of charge (SoC) is modeled by $x = \frac{Q}{Q_{max}}$ where Q is the electric charge (in Coulomb or the Ampere-hour (Ah)) and Q_{max} is the total capacity of charge (in order of 50 Ah). Since the time derivative of the charge is the current I, one has $I = \frac{dQ}{dt}$. Considering that the equivalent open-circuit voltage and internal resistance are functions of SoC, the time derivative of x is a function of SoC and the control u :

$$\dot{x} = \frac{d}{dt}x = \frac{d}{dt}\frac{Q}{Q_{max}} = \frac{\dot{Q}}{Q_{max}} = I(x, u)/Q_{max}. \tag{13.16}$$

Remark 13.7 In contrast to the prior works that considered linear models, we observe that the SoC of the battery in Equation (13.16) is not linear in the control variable. Approximation by its linear counterpart may not capture the battery state. Therefore, a careful error estimate needs to be conducted.

Since batteries have a finite energy capacity, PHEV must be recharged on a regular basis, typically by connecting to the power grid in some form. According to Texas Instruments, for most users, level 1 charging (120 volt alternating current (VAC) at 15 A–20 A) will be the most available power supply that all on-board chargers should be capable of handling as it will be easily accessible to all users. In its current state, most users would prefer to take advantage of the faster charging level 2 (240 VAC at 40 A) that will allow for faster charging times compared to Level 1, but will require a larger power source to supply the appropriate current and voltage. By having the capacity to utilize both types of charging, you provide the user with more flexibility in their charging options and a larger number of locations where a charge can be delivered. Levels 1 and 2 (slow or fast) charging systems for these vehicles consist of an alternating/direct current (AC/DC) converter to generate a DC voltage from the AC line. The fast charge (supercharger) can charge up to 80% in around 20 mn. There is also DC fast charging (DC level 2), which can go to around 80 miles of range in 20 minutes of charging. Smart charging is an alternative to these options when controlling the rate. Figure 13.3 shows that the state model considered here captures the observed time-scale (Fig. 13.4).

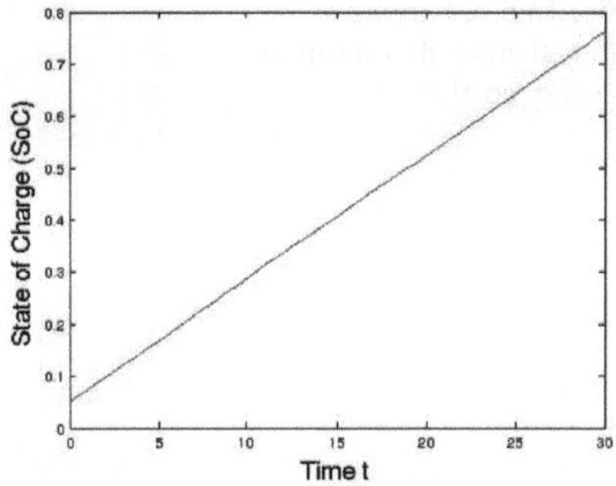

Fig. 13.3 State of charge of the battery: 75% of the battery is charged in 30 minutes in open-loop control

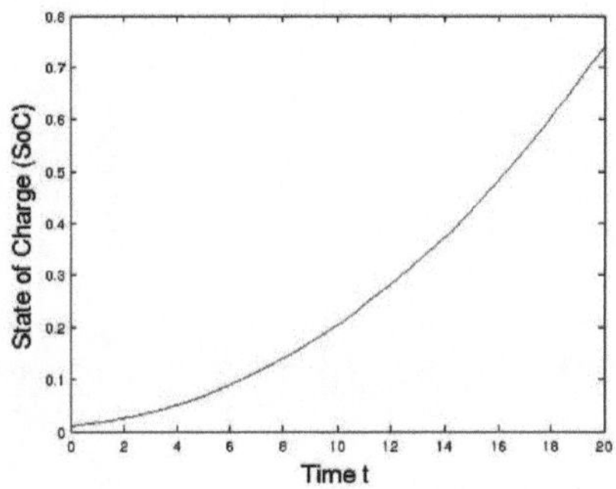

Fig. 13.4 State of charge of the battery: 75% of the battery is charged in 20 minutes in feedback control. A highly non-linear battery behavior is observed

13.2.2.1 Economic Consideration

When the control u is used, the quantity $\eta(t)u(t)P_{max-plug}$ is consumed, where $\eta(t)$ is an efficiency parameter and $P_{max-plug}$ is the maximum available charge power. A reasonable charging cost function is therefore

$$C(t, x, u, m, l) = \eta(t)u(t)P_{max-plug}.price(t, m, l),$$

where l denotes the location of the charging station. m is the total demand of the electricity in the market, which can be decomposed as $m = \varepsilon m_{ev} + (1-\varepsilon)m_{non-ev}$, where m_{ev} (resp. m_{non-ev}) is the total demand of all the EVs (resp. non-EVs). $\varepsilon \in (0, 1)$ represents the proportion of EVs in the entire population of users. The price is dependent on the geographical location l. The EV owners can visualize the most recent price at different locations. We can distinguish between three major different functions for $price(t, m, l)$: (a) constant price per location, (b) piecewise constant price (in time), and (c) price as a function of total demand of the entire energy market and time. As illustrated in Fig. 13.5, charging immediately every time the vehicle completes a trip without considering the price may be very costly. By taking into consideration the electricity price, a lower cost can be achieved by estimating the peak hours of the next day. In particular, if there is an off-period during the phase, where the vehicle is not needed by its owner, then that slot should be considered for

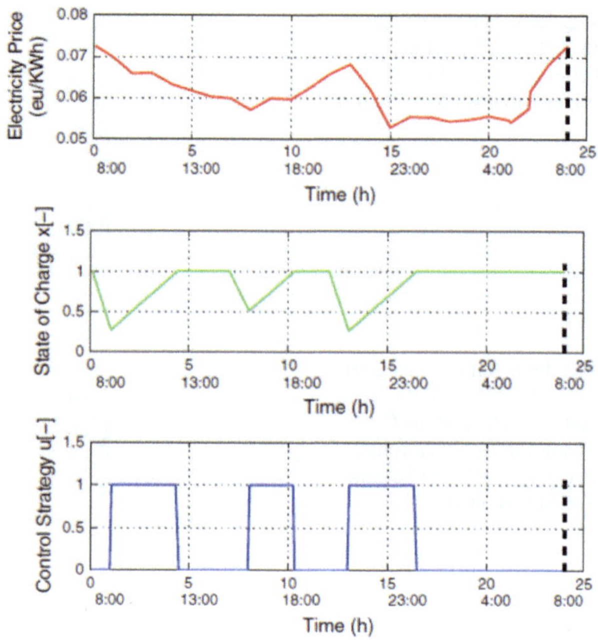

Fig. 13.5 Charging immediately after a trip may be very costly

a smarter charging strategy and the charging should be done at the lowest electricity price of the corresponding period (see Fig. 13.6).

13.2.3 Discretize-and-Optimize Approach

The discretize-and-then-optimize approach consists of transforming the continuous variables (e.g. time) into discrete ones and then optimize the cumulative sum of the instantaneous objectives. In this setup, the discrete time optimization problem for each EV yields

$$\min_{u^h=(u_h,u_{2h},\ldots,u_{Nh})} \sum_{k=1}^{I} hC(kh, x_{kh}, u_{kh}, m_{kh}, l) \text{ subject to}$$

$$x_{(k+1)h} = x_{kh} + hI(x_{kh}, u_{kh})/Q_{max}, \tag{13.17}$$

where $h > 0$ is a small stepsize and $N \geq 2$ is set to $N = \lfloor \frac{T}{h} \rfloor$, $\lfloor x \rfloor$ being the floor of x. The next step of discretization is to have the control variable to be piecewise constant (e.g.,$u_{kh} \in \{0, 1\}$).

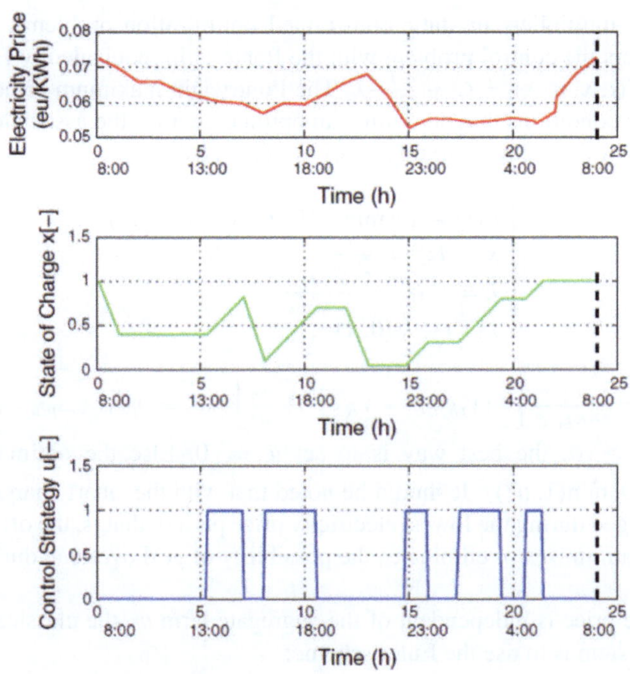

Fig. 13.6 A smarter charging strategy: exploit off-peak period opportunities when possible

13.2.4 Optimize-and-Discretize Approach

The second way of dealing with continuous time is to optimize first in continuous time and then solve for the resulting optimality equations by discretizing them:

$$\min_{u(.)} \int_0^T C(t, x(t), u(t), m(t), l)\, dt \;\text{ subject to}$$

$$dx = I(x(t), u(t))dt/Q_{max}. \tag{13.18}$$

13.2.5 EV Charging as a Mean-Field-Type Game

As we have discussed extensively, mean-field-type games are games where the payoffs are functions not only of the states and actions profiles but also on the distributions of states and the distribution of actions.

Pontryagin Maximum Principle (See Volume I) We can now turn our attention to characterizing optimal controls. To do so, we first introduce another variable, $\lambda(t)$, called a costate variable (or adjoint variable). This variable will serve a role similar to Lagrange multipliers in static constrained optimization problems. The adjoint variable enters the control problem with the Pontryagin (equivalently Hamiltonian) function $H^0(t, x, u, \lambda) = C + \frac{I}{Q_{\max}}\lambda$. The Pontryagin maximum principle is a set of necessary conditions that determine an optimal control, the associated response, and the path of the adjoint variable. They are given by

$$\begin{cases} u(t) \in \arg\min_w H^0(t, x(t), w, \lambda(t)) \\ \dot{x} = H^0_\lambda = \frac{I}{Q_{\max}} \\ \dot{\lambda} = -H^0_x = -\frac{I_x}{Q_{\max}} \\ \lambda(T) \in -[0, 1]^{\perp}. \end{cases} \tag{13.19}$$

Let $u^0 = \frac{1}{4R_{BC}\bar{P}}\left[-(V_{th,BC} - V_{RF})^2 + \frac{b^2}{a^2}\right]$. $a = \eta P_{\max - plug}\ price$, $b = \frac{\lambda\bar{P}}{Q_{\max}}$. If $\lambda \geq 0$, the best way is to set $u = 0$. Else the optimal control is $u_* = max(0, \min(1, u^0))$. It should be noted that with the smart charging strategy, EV is recharged during the lowest electricity price period, that is, the off-peak hours. It naturally minimizes or eliminates the possibility of grid overload during the peak load hours.

When the price is independent of the aggregate term m, the classical scheme to solve this system is to use the Euler scheme:

$$\begin{cases} x_{(k+1)h} = x_{kh} + hH^0_\lambda(kh, x_{kh}, u_*, \lambda_{(k+1)h}) \\ \lambda_{(k+1)h} = \lambda_{kh} - hH^0_x(kh, x_{kh}, u_*, \lambda_{(k+1)h}), \end{cases} \tag{13.20}$$

which characterizes the optimal control with error in order of $O(h)$.

Bellman Equation We introduce the value function

$$v(t, x) = \inf_{u(.)} \int_t^T C(t', x(t'), u(t'), m(t'), l)\, dt' \text{ subject to} \tag{13.21}$$

$$dx(t') = I(x(t'), u(t'))dt'/Q_{max}. \tag{13.22}$$

$$x(t') := x. \tag{13.23}$$

Let $H^1(t, x, v_x) = \min_w H^0(t, x, w, v_x)$. Then, the Hamilton–Jacobi–Bellman (HJB) equation associated with system (13.18) is given by

$$\begin{cases} u(t, .) \in \arg\min_w H^0(t, x, w, v_x) \\ v_t + H^1(t, x, v_x) = 0 \\ v(T, .) = 0. \end{cases} \tag{13.24}$$

Solving the HJB equation along the characteristics yields to the system (13.19), which is solved numerically using the scheme (13.20). In addition, one needs a fixed-point equation, when the price is dependent on m. This is because the optimal control will be a function $m_{ev}(t)$ via the price and hence

$$m_{ev}(t) = \int_0^1 u_*(t, y)\, \mu(t, dy),$$

where μ the EVs population of state measure.

13.2.6 Cooperative Mean-Field-Type Game Between EV End-Users

Smart cities invest in infrastructure like e-station for recharging electric vehicles. Collaborative vehicle renting, compact rented electric vehicles, or peer-to-peer car-sharing is therefore a great opportunity for users to reduce transportation cost and provide an environmentally friendly city. The associated problem is similar to the system in Equation (13.18), but now the optimization is over a global control and global aggregate term, which is the mean of control actions. This feature is not well-known in mean-field games. However, it comes naturally in this context where the aggregate demand of energy plays an important role. We derive optimality equations for the global optimization for charging the EVs.

The maximum principle yields

$$\begin{cases} u(t) \in \arg\min_w H^0(t, x(t), w, \bar{w}, \lambda(t)) \\ \dot{x} = H^0_\lambda = \frac{I}{Q_{\max}} \\ \dot{\lambda} = -H^0_x - \bar{H}^0_m \\ m(t) = \bar{u}(t) = \int_0^1 u_*(t, y)\, \mu(t, dy), \\ \lambda(T) \in -[0, 1]^\perp, \end{cases} \tag{13.25}$$

where $\bar{H}^0_m$ is the average (over states) of the derivative of H^0 with the respect to m.

13.2.7 Charging Under Stochastic Environment

EVs can be considered as environmentally friendly, provided that the energy used for their charging is obtained from renewable energy sources. The energy generated from renewable sources such as sunlight, wind, and waves is, however, dependent on weather conditions. As a consequence, the electricity production from these sources is inherently uncertain in time and quantity. Furthermore, electricity has to be produced and consumed at the same time, as the large-scale storage of the energy generated is, still today, very limited. Demand for electricity is stochastic and varies over the day and so does the electricity generated from renewable sources. This leads to a stochastic environment and a stochastic optimization problem. For more details, see [131, 150, 151, 265, 379, 384, 400, 461, 471, 552].

Price Competition Between Charging Stations Based on the above, we note that if the end-user preference for driving is compatible, the most straightforward strategy for charging electric vehicle is to charge it when the price of electricity is low, e.g., at off-peak hours. Considering that conventional internal combustion engine vehicles refuel at gas stations, electric vehicles might also be charged at other facilities that provide charging service, that is, EV charging stations. A station can buy electricity from the power grid at a cheaper rate compared to the rate that an EV owner has to pay to charge at home such that EV charging at the station is cheaper than charging at home, and thus an EV might run out of electricity while traveling and immediate charging becomes unavoidable. Since charging an EV is more time consuming than refueling a conventional vehicle, some stations can be located at public parking lots and shopping centers, where the EVs are parked for a long time. As a market, the electricity price will be rationalized, and EV stations will start generating revenue from selling electricity to electric vehicles. Therefore, when multiple stations with different ownerships coexist in the neighboring area, the competition between the stations to attract EVs comes into play. With the heterogeneity of the gas price at different gas stations, the price of electricity can be different for different stations in the smart city. However, the similarity with gas station ends here because unlike gas stations, EV stations can generate electricity without incurring additional generation cost by using their own renewable generation such as photovoltaic (PV). Electric vehicles are just one piece of the puzzle, but they could be a driver for other systems. When a city wants to support lots of EVs, they typically have to invest in the electric grid as well. Upgrading the grid infrastructure then enables other smart services, such as demand response programs, and can be a catalyst for creating new distributed energy generation offerings. Such offerings may include localized renewable energy and waste-to-energy plants, for example. After all, a city is a network of systems, and an investment in one area should include analysis of opportunities in the connecting systems. Thus, EVs are seen in a positive light for the environment and for the concept of smart cities if the battery problem is solved properly.

In this section, we have addressed a real-world problem arising from the increasing demand for charging plug-in electric vehicles from the electrical energy distribution grid. By means of strategic decision-making and dynamic optimization, we have analyzed the effect of charging requests on the electrical grid and examined how smart battery charging methods may mitigate load and save charging cost.

13.3 Network Economics for Cloud Computing

A cloud computing infrastructure typically consists of collections of computing servers deployed at multiple sites distributed over the Internet. At a basic level of abstraction, the infrastructure is required to provide a specified computing capacity aggregated among the currently available servers. A computational task may be executed on these servers, typically at locations unknown to users. The total computing power of all servers that are up and connected to the Internet is the available capacity and is a primary performance measure. This infrastructure critically depends on the continued functioning of the cyber components, including computers and routers, as well as physical components, including fiber routes, cooling, and power systems. These components may be degraded by deliberate cyberattacks on servers and routers as well as physical attacks on fiber and power routes. To reduce the energy costs, research efforts are being made in two main directions:

- Combine traditional electricity grid power with renewable energy sources such as solar, wave, water, and wind, motivated by long-term cost reductions and growing unreliability of the electric grid.
- Exploit the temporal and geographical variation of electricity prices.

Cost

Energy costs are becoming an important contributor in smart grid operation costs from one sector to another. One basic approach to reduce these costs is to exploit the spatiotemporal variation in electricity prices and population density by moving production to sectors in which energy is available at a cheaper price. However, injudicious storage migration between zones/sectors might increase the overall operation cost due to the transport cost, routing cost, maintenance of migration, and transferring storage state and technology over the network. To address this challenge, we propose here a novel mean-field stochastic game formulation with migration costs, which handles the fundamental trade-offs between benefit, energy, and migration costs. A distinctive feature of our approach is that it considers not only the current availability and cost of (possibly multiple) energy nodes but also the future variability and uncertainty thereof.

13.3.1 Resource-Sharing in Cloud Networking

Data traffic in wireline and wireless networks continues to grow rapidly. The challenge that operators face is to support more subscribers with higher bandwidth requirements. To meet the bandwidth demand, operators need to seek new technologies to efficiently utilize available network resources, in particular efficient resource-sharing, resource allocation, and flow management. In wireless networks, the problem is particularly relevant due to the inherently limited resources. The success of powerful mobile devices like smart phones or tablets to access different kinds of Internet service is not only promoting the growth in traffic volume but also changing the traffic pattern. Certain applications automatically pull or push large amounts of data without the user necessarily being aware. Cloud-based applications transfer more and more data (sometimes even with real-time requirements). This strong growth and continual changes in the traffic characteristics and usage behavior raises questions on how to share limited capacity resources fairly and more efficiently. The challenge is to enable these different kinds of applications and to apply traffic management that maintains high degrees of resource utilization, a reasonable distribution of capacity without affecting transport protocol performance and impeding innovation. Such traffic management has to be able to adapt to innovation and be scalable with respect to growing traffic, growing numbers of flows, different application types, etc. One of the questions that we address in this chapter is whether and how cloud resource-sharing can support demand management evolving in this direction. Let there be $I \geq 2$ users all seeking resources from a cloud network with total capacity C_I. The reward for user j is given by $r_j(a)$ where $a_j \geq 0$ is the action/demand of user j. p_j is the price per unit of action/demand of user j. A generic expression for $r_j(a)$ is given below, where it should be clear from the expression that the reward of user j depends on the total demand of all the users. So this is an interactive strategic decision-making problem where all users compete for the same resources and try to maximize their respective rewards in the process.

$$\underbrace{r_j(a)}_{\text{reward}} = \underbrace{C_I}_{\text{available capacity}} \overbrace{\frac{a_j}{\underbrace{D}_{\text{Total demand}}}}^{\text{\% of total capacity}} - \overbrace{\underbrace{a_j}_{\text{demand}} \underbrace{p_j}_{\text{price}}}^{\text{Cost for user j}}. \tag{13.26}$$

Let $D \triangleq \varepsilon_I + \sum_{k=1}^{I} a_k$ where $\varepsilon_I > 0$ is a reservation price parameter necessary to keep the denominator greater than zero. The reward r_j of each user depends on the percentage of total capacity C_I it is demanding minus the cost that it has to pay for those resources. An alternate interpretation of a_j/D could be the probability of getting access to C_I, if only one user can access all the resources at one time. It is important to mention here that the set of active users (i.e., a set of users $\{j \mid a_j > 0\}$) need to be at least two.

The interactive strategic decision-making problem between these users can be formulated as follows:

$$\sup_{a_j \geq 0} r_j(a), \ \forall\, j \in \mathcal{I}. \tag{13.27}$$

Each user wants to maximize its respective reward; the demand profile is $a = (a_1, \ldots, a_I) \in [0, +\infty)^I$. Hence, the problem has multiple objectives (each user has its own objective). It is important to note here that the demand from each user is nonnegative, $a_j \geq 0$.

Then, the (simultaneous) best-response algorithm is given by

$$x_{t+1} = \left[\sqrt{\frac{c_I}{p}(\varepsilon_I + (I-1)x_{t-1})} - (\varepsilon_I + (I-1)x_{t-1})\right]_+ .$$

Let $f(x) := \left[\sqrt{\frac{c_I}{p}(\varepsilon_I + (I-1)x)} - (\varepsilon_I + (I-1)x)\right]_+$, which is the best response function. A direct computation of the derivative (at the interior) of f gives $f'(x) = -(I-1) + \frac{c_I(I-1)}{p}\left(\frac{c_I}{2p}(\varepsilon_I + (I-1)x_{t-1})\right)^{-1/2}$. Clearly, f is not a contraction. This means that the standard Banach–Picard iterates should not be applied directly. Chapter 6 provides a detailed analysis of alternative algorithms associated with f.

13.4 Bayesian Mean-Field Games for the Power Grid

In this section, we study a power grid network with finitely many producers (players), each having a certain capacity to produce and an individual production cost per energy unit, which represents the type of that producer. The price of the energy market is affected by the total quantity produced and the demand of the users. Initially, a producer cannot know the production cost of the others and cannot conjecture the resulting decision of the others. We model this scenario using the framework of Bayesian aggregative games where the aggregative term is in the energy market price. When the number of players (producers) is sufficiently large, we analyze the interaction as a Bayesian mean-field game. We address the question of ex post resilience by unilateral deviation, i.e., no player wants to switch after learning the types and actions of the others. We identify a large class of mean-field games with incomplete information that does not satisfy the indistinguishability property and establishes an ex post resilience property. Then, we focus on dynamic games with individual state driven by a standard Brownian motion and disturbance parameter. We establish a mean-field system for such Bayesian robust games. We apply the methodology to power grid where ex post equilibria are very important since they ensure a directional stability of the market after acquisition of the information. Finally, we show that the dimensionality of the mean-field system can be significantly reduced by considering a functional of the first moments of the mean-field process.

13.4.1 Incomplete Information in Power Grid

In this section, we focus on basic aggregative game with incomplete information and examine the robustness of equilibria in such games when the number of players is sufficiently large. Aggregative game models, such as Cournot oligopoly and Bertrand competition, are widely used in studies of imperfectly competitive industries. Its standard version, which is concerned with the case of firms producing a good with complete information about demand and production costs, has been extensively studied. However, in the last few decades, a large amount of research has been dedicated to questions that arise when the information is incomplete, i.e., when there is uncertainty about the market demand and/or the product cost functions, and the decision-makers have asymmetric information about the others. In aggregative games with incomplete information, some of the questions that had been addressed concern the *value of information* to a player (i.e., whether and by how much a player can benefit from receiving additional information), as well as players incentives to *share a specific information* such as their individual characteristics. Naturally, treating these questions requires comparisons of the pure Bayesian Cournot equilibrium outcomes (when they exists) that differ with respect to the information endowments of the players.

Ex post resilience is an important property of equilibria of one-shot simultaneous-act Bayesian games. An equilibrium of a Bayesian game is *ex post resilient* if no player has an incentive to change her action after she learns the realized types and the realized actions of all the players. Equilibria that are not ex post resilient may not provide useful predictions of outcomes in games in which players have opportunities to change their choices based on recent information.

In aggregative Bayesian games, it is well-known that a pure Cournot equilibrium may not exist. But even if a pure Bayesian equilibrium exists in the game with finite number of players, the equilibrium may not be ex post resilient. The reason is that, each player makes its ex ante decision based on distribution, beliefs, and expectations, but after seeing the realized actions and types (the produced quantities of energy and unit costs) of all the other players, she would most likely regret her choice.

13.4.2 Energy Market

Consider a power grid network consisting of I producers (players). The set of players is denoted by $\mathscr{I} = \{1, 2 \ldots, I\}$. Each player j has her own-type θ_j, which represents her production cost per unit of power. Each player knows her type, which is randomly drawn by a probability distribution of type, μ_j, over some compact set $I_j \subset \mathbb{R}_+$. With knowledge of his own type, each player decides on a production level $a_j \in [0, K]$ where K is a positive parameter. This results in the profile of types $(\theta_j)_{j \in \mathscr{I}}$ and the action profile $(a_j)_{j \in \mathscr{I}} \in [0, K]^I$.

Assumptions:

- player's no knowledge of other's action, $a_i, i \neq j$,
- price formation dependence on demand/geography,
- game played static/dynamic/simultaneous.

13.4.2.1 Price Formation

We will choose the price to a certain function of the aggregate quantity of the market, i.e., a decreasing function of $\frac{1}{I}\sum_{j\in\mathcal{I}} a_j$. An example of relation between the market price and the energy produced is described by the function:

$$p(a_1, \ldots, a_I) = K - \frac{1}{I}\sum_{j\in\mathcal{I}} a_j. \tag{13.28}$$

The price, as an inverse demand function, has the property that a higher level of production results in a lower energy trading price.

13.4.2.2 Payoff Function

The payoff function of player j is given by $r_j : \left(\prod_{j\in\mathcal{I}} I_j\right) \times [0, K]^I \longrightarrow \mathbb{R}$

$$r_j(\theta, a) = a_j p(a_1, \ldots, a_I) - \theta_j a_j. \tag{13.29}$$

The above payoff function is a Bayesian potential game if the price function is twice continuously differentiable and satisfies $\partial^2_{a_i a_j} p(a) = 0$, and $\partial_{a_i} p(a) = \partial_{a_j} p(a)$. These conditions are satisfied by the expression in (13.28). Then, the one-shot game between the energy producers will be a Bayesian potential game with a Bayesian potential function $V^*(\theta, a)$. Moreover, $V^*(a, \theta) = \sum_{j\in\mathcal{I}} \int_0^1 r_j(\theta, a_{\gamma(s)})[\frac{d}{ds} a_{\gamma(s)}]ds$ where $a_{\gamma(s)}$ is a piecewise continuously differentiable (in s) path in $[O, K]^I$ that connects an arbitrary action profile $a_{\gamma(0)}$ to the action profile $a_{\gamma(1)} = a$.

The explicit expression of $V^*(\theta, a)$ is given by

$$V^*(\theta, a) = K\sum_{j\in\mathcal{I}} a_j - \frac{1}{I}\sum_{j\in\mathcal{I}} a_j^2 - \frac{1}{2I}\sum_{(i,j)\in\mathcal{I},\, i\neq j} a_i a_j - \sum_{j\in\mathcal{I}} \theta_j a_j.$$

13.4.2.3 Equilibrium Under Perfect Information of the Types

Given a type profile, the optimal strategy for each player can be calculated by setting to zero the derivative of the payoff function with respect to player's strategy. It can be then derived as the solution to the following simultaneous I equations:

$$2a_j + \sum_{i\neq j} a_i = I(K - \theta_j),$$

which can be written as $Sa = b$ where $S \in \mathbb{Z}_+^{I\times I}$ is the matrix

$$S = \begin{pmatrix} 2 & 1 & \dots & 1 & 1 \\ 1 & 2 & & \dots & 1 \\ \dots & & \dots & & \dots \\ 1 & & & 2 & 1 \\ 1 & \dots & & 1 & 2 \end{pmatrix},$$

and $b := (I(K - \theta_j))_{j \in \mathscr{I}}$.

Notice that the best response action of player j is decreasing in the output of each other players. Hence, when a player is expected to increase its own production, any other player has an incentive to choose a lower production level.

The optimal strategy is explicitly given by

$$a_j^* = \frac{I}{I+1}(K - \theta_j) + \frac{I}{I+1}\left(-I\theta_j + \sum_{j \in \mathscr{I}} \theta_j\right).$$

The strategy of each player in this case would depend on the sum of type of the other players (aggregate type). The total quantity produced at the perfect information equilibrium is

$$\sum = \frac{I^2 K - I \sum_{j \in \mathscr{I}} \theta_j}{I+1}.$$

Hence, the mean production $\frac{\sum}{I}$ converges to $K - \int_{\bar{\theta}} \bar{\theta} \mu(d\bar{\theta})$. This implies that the equilibrium price converges to the mean of production cost per unit of the population of producers, i.e., $p(a_1^*, \dots, a_I^*) \longrightarrow \int_{\bar{\theta}} \bar{\theta} \mu(d\bar{\theta})$.

13.4.3 Equilibrium of the Bayesian Game

Definition 13.8 (Strategy) A strategy for player j is a map $s_j : I_j \to [0, K]$ prescribing an action for each possible type of player j.

Given μ_j, player j can compute the conditional distribution $\mu_j(\theta_{-j}|\theta_j)$ using Bayes rule, where $\theta_{-j} = (\theta_1, \dots, \theta_{j-1}, \theta_{j+1}, \dots, \theta_I)$. Player j can then evaluate her expected payoff based on the expected types of other players.

Definition 13.9 (Bayesian Nash Equilibrium) The strategy $s = (s_1, s_2, \dots, s_I)$ is a Bayesian Nash equilibrium if for all $j \in \mathscr{I}$ and θ_j, we have

$$s_j \in \arg \max_{s_j' \in [0,K]} \int r_j(\theta, s_j', s_{-j}) \mu(d\theta_{-j}|\theta_j).$$

Definition 13.10 We say that a strategy is an ex post resilient if for each player $j \in \mathcal{I}$,

$$\arg \max_{s'_j \in [0,K]} \int r_j(\theta, s'_j, s_{-j}) \mu(d\theta_{-j} | \theta_j) = \arg \max_{s'_j \in [0,K]} r_j(\theta, s'_j, s_{-j}).$$

Here the ex ante equilibrium depends on the realized average type and it is therefore not an ex post equilibrium. However, as I goes to infinity, at the infinite population regime, one obtains an ex post equilibrium. It does not depend on the realized types but the expected value of the types. The proof follows from the convergence of $p(a_1^*, \ldots, a_I^*)$ above.

13.5 Demand–Supply Management Under Operational Constraints

Consider a demand–supply management under operational constraints and outage scheduling. We first develop a theoretical framework for finding cooperative production strategies between electricity producers. We formulate the problem as an optimal control problem under constraints. If several producers need to match the demand, a cooperative strategy is proposed in order to reduce the cost of production and hence reduce electricity price for the consumers. Thanks to Pontryagin maximum principle, we have closed-form expressions for the optimal supply and the optimal production strategy. We show that the method provides production strategies that are time-consistent with the respect to possible renegotiation between the producers. Smart power generation is a concept of matching electricity production (supply) with demand using multiple smart generators that can start, stop, and operate efficiently at a chosen load. These generators can coordinate autonomously, making them suitable for smart grid and smart cities. Matching supply and demand, sometimes referred to as "load balancing," is essential for a stable and reliable supply of electricity. Short-term deviations between supply and demand lead to frequency variations, and a prolonged mismatch may result in blackouts if the other alternative sources fail to compensate. Power transmission producers are charged with the balancing task, matching the supply of all the generators to the demand load of their electrical grid. The load balancing task has become much more challenging as increasingly intermittent and variable generators such as wind turbines and solar cells and costly renewable renewables such as geothermal are being added to the grid, forcing other producers to adapt their output much more frequently than has been required in the past.

We formulate a dynamic cooperative game between the producers and between the generators that minimize the mismatch between supply and demand under production, maintenance, and refueling constraints. We use the Pontryagin maximum principle technique to solve the constrained optimal control problem. Explicit structural results are derived. We then address the question of credible cooperation in the smart grid. We provide a closed-form expression of the optimal supply given the operational constraints of the power plants. We obtain explicitly the optimal production strategies as a function of time, demand, and the capacity of the power plants. We investigate the time consistency of the solution, which allows sustainable cooperative behavior between the producers in the smart grid.

We first present the energy production model under scheduling and maintenance constraint. Then, we formulate the problem as a cooperative game between electricity producers. Explicit structural results are provided. Finally, we examine the credibility of the cooperative production strategies in the smart grid. We summarize in the table below some of the notation used in this section.

Symbol	Meaning
$\mathcal{T} = [0, T]$	Time horizon of the energy production
t	Time $t \in \mathcal{T}$
$\mathcal{I}$	$\{1, 2, \ldots, I\}$: set of the energy producers
j	$j \in \mathcal{I}$
K_j	Number of power plants of producer j
$e_{jk}(t)$	Energy stock of power plant k of j at time t
T_{jk}	Inactivity periods of power plant k of j
$s_{jk}(t)$	Production rate of power plant k of j at time t
$\bar{s}_{jk}$	Maximum production of power plant k of j
$c_{jk}(t)$	Refueling rate of power plant k of j at time t
$D(t)$	Demand of electricity supply
$S(t)$	Energy to be produced at t
$l(t, .)$	Instant cost function at time t
$l_T(.)$	Terminal cost function

13.5.1 Maintenance and Refueling of Power Plant

Power plants must sometimes be switched off in order to reduce unused energy in off-peak hours or to perform refueling and maintenance operations. The scheduling of the outages is one of the major problems to be solved in energy production management. In order to get structural results, we assume that the scheduling parts have already been decided by the producer (who knows the functionalities). Given various constraints, we would like to understand how the production can fulfill the demand. Most power stations in the world burn fossil fuels such as coal, oil, natural gas, or nuclear power to generate electricity, but there is an increasing use of cleaner renewable sources such as solar, wind, wave, geothermal, and hydroelectric. The competitive environment for electricity generation has significant implications for power plant operations, including, among others, the need for efficient use of resources and effective management of plant activities such as maintenance and outages. Power plant outage management is a key factor for good, safe, and economic power plant performance before and during the outage. During a refueling outage (to supply again with fuel), plant workers have greater access to many areas of the plant that are sometimes difficult to access during normal operations, such as areas close to the reactor core. This lends itself to being a great opportunity to conduct routine inspections and maintenance and perform any necessary repairs. The type of maintenance that is conducted varies by plant, but an example includes inspecting or testing pumps and valves to ensure that they are working properly. The producer will place many of these routine maintenance activities on a schedule so that certain equipment and instrumentation is evaluated at each refueling outage, whereas others may only need to be inspected every other outage or less frequently depending on their functions. Although the producer could potentially perform refueling on several power plants at a time, it is generally not done for many reasons. First, a refueling outage requires a large amount of resources. Typically, about thousands of contract workers are brought on-site to support the outage, allowing the producer to work continuously until the outage is complete. Second, nuclear energy facilities provide baseload electricity—meaning they pump out electricity around-the-clock—and when they are not operating, it is difficult and costly for the producer to purchase all the electricity from other sources, sometimes in other states or other nations. This is another reason why utilities conduct refueling in the specific frame when electricity demand is generally lower, so that it minimizes the impact to the electricity grid. The new fuel is ordered from the fuel vendor or from the market before the reactor goes into its outage. Once it arrives at the facility, it is stored safely on-site in new fuel vaults or in pools.

Mismatch Between Demand and Supply of Electricity

The producer aims to produce energy in order to satisfy the demand. However, due to unknown demand and outages, the energy produced may not match the demand. This creates a cost because the producer needs other alternative (more costly) sources of energy or to buy from the market or have an extra energy that may create a storage cost. We denote the instant loss by $l(t, D(t) - S(t))$, where

$$S(t) = \sum_{j=1}^{I} s_j(t) = \sum_{j=1}^{I} \sum_{k=1}^{K_j} s_{j,k}(t)\, \mathbb{1}_{\{t \in T_{jk}\}},$$

where $s_{j,k}(t)$ is the production rate of plant/generator k of j at time t. The stock of energy is given by the following classical motion:

$$\frac{d}{dt} e_{jk} = \text{incoming flux}_{jk}(t) - \text{outgoing flux}_{jk}(t). \tag{13.30}$$

The incoming flux happens only when the power station is in the maintenance phase. In that case, the arrival rate is $c_{jk}(t)\mathbb{1}_{\{k \in A_j^c(t)\}}$ where $c_{jk}(t) \geq 0$, and the set of active power plants of producer j is defined by

$$A_j(t) = \{k \mid t \notin T_{jk}\}, \tag{13.31}$$

where T_{jk} is the set of scheduled outage periods of power plant k of producer j. The set of all active power plants is $A(t) = \cup_j A_j(t)$. The outgoing flux is the production term: $s_{jk}(.)$. Thus, the dynamics is

$$\frac{d}{dt} e_{jk} = c_{jk}(t)\mathbb{1}_{\{k \in A_j^c(t)\}} - s_{jk}(t), \tag{13.32}$$

where $A_j^c(t)$ is the set of non-active power plants of producer j at time t. The production rate s_{jk} is a control variable when the power plant is active. It is nonnegative and must be bounded by the maximum production rate $\bar{s}_{jk}$. The initial stock of energy in plant k is $e_{jk}(0)$. $e_{jk}(t) = 0$ if t is at the beginning of a maintenance period of the plant k. $e_{jk}(T) \geq 0$. Lack of electricity causes not only inconvenience but also economic loss due to reduced industrial production. It is indispensable to individual users, energy efficient buildings, factories, commercial establishments, homes, and even most recreational facilities, creating a certain $D(t)$ at time t.

Uncertainties in Energy Demand and Outage Period

Past performance and historical data may affect the current decision of a producer. However, past outcomes are not future outcomes. Past outcomes can be very misleading if there are reasons to believe that future network (both economic and engineering) conditions are likely to be significantly different from those that shaped past outcomes. This seems to be true in particular for demand profile that may depend on weather conditions (which is stochastic) and the growing number of devices per user. Thus, the demand $D(t)$ becomes a random process, and hence the optimal outage scheduling periods become stochastic intervals. Since future demands are not known in advance, we work with estimated demand function $\hat{D}(t)$ for which we solve the scheduling problem by optimizing an objective $L_0(\hat{T}, \hat{D})$. Thus, we look at the case where the outage scheduling is fixed to $(T_{jk})_{jk}$.

Formulation of the Cooperative Game

$$\begin{cases} (*) \ \inf_{\hat{T}=(\hat{T}_{jk})_{jk}} L_0(\hat{T}, \hat{D}) \\ (**) \ \inf_{s,e} L(s, e, T). \end{cases}$$

We solve the scheduling problem and find the outage period T based on an estimated demand $\hat{D}$ (because the future demands are unknown and highly stochastic):

$$\{ (*) \ \inf_{\hat{T}} L_0(\hat{T}, \hat{D}).$$

The outage periods of the very costly plants are then planned in these periods where the estimated demand is lower. Once the scheduling is fixed, we aim to find the optimal production strategies:

$$
\begin{cases}
(**)\ \inf_{s,e} L(s,e,T) \\
L(s,e) = \mathbb{E}\left[l_T(e(T)) + \int_0^T l(t, D(t) - S(t))\, dt \right] \\
\frac{d}{dt} e_{jk}(t) = c_{jk}(t) \mathbb{1}_{\{k \in A_j^c(t)\}} - s_{jk}(t) \\
c_{jk}(t) \geq 0. \\
s_{jk}(t) \in [0, \bar{s}_{jk}],\ \forall j,k,t \\
s_{jk}(w) = 0 \text{ ifis a starting time of a maintenance period.} \\
e_{j,k}(0) \text{ given.}
\end{cases}
$$

Structure of the Optimal Supply Strategy
Our objective is to find or to characterize (via the Pontryagin maximum principle) an optimal cooperative strategy $s^* \in \arg\min_s L(s,e)$.

Solving the Constrained Optimal Control
We introduce the Pontryagin function

$$H(t,s,p) = l(t, D - S) + \sum_{jk} p_{jk}[c_{jk} \mathbb{1}_{\{k \in A_j^c(t)\}} - s_{jk}].$$

Note that this function that does not directly depend on the energy stock e. The minimized version of H corresponds to the Hamiltonian. We assume that the loss functions are continuously differentiable and strictly convex. The Pontryagin maximum (minimum) principle yields

$$\frac{d}{dt} p_{jk} = H_{e_{jk}}(t, s(t), p(t)) = 0.$$

The first-order adjoint variable (co-state) $p_{jk}(t)$ can take at most two values in $\{p_{jk}(0), p_{jk}(T)\}$. Moreover,

$$p_{jk}(T) \leq l_{T,e_{jk}}(e(T)),$$

$$(p_{jk}(T) - l_{T,e_{jk}}(e(T))) e_{jk}(T) = 0.$$

The optimal control minimizes the Pontryagin function that can be decomposed as

$$
\begin{aligned}
& H(t,s,p) \\
&= \left[l(t, D(t) - \textstyle\sum_{jk} s_{jk}(t) \mathbb{1}_{\{k \in A_j(t)\}}) + \sum_{jk} (-p_{jk}) s_{jk}(t) \right] \\
&\quad + \textstyle\sum_{jk} p_{jk} c_{jk}(t) \mathbb{1}_{\{k \in A_j^c(t)\}}.
\end{aligned}
$$

The minimization problem can be easily transformed from $(\sum_{j=1}^{I} K_j)-$ optimization variables into a one-dimensional optimization problem. This allows us to reduce significantly the curse of dimensionality:

$$\begin{aligned}&\inf_{s\in\mathbb{R}^{\sum_j K_j}} H(t,s,p) \\ &= \inf_{S\in\mathbb{R}} [l(t,D(t)-S(t))+f(S(t),p(t))]+Cst(t),\end{aligned} \tag{13.33}$$

where

$$f(S,p)=\inf_{s\in\Delta_S}\sum_{jk}(-p_{jk})s_{jk},$$

$$\Delta_S=\{s_{jk}\geq 0 \mid \sum_{jk} s_{jk}=S\}.$$

and $Cst(t)=\sum_{jk} p_{jk}c_{jk}\mathbb{1}_{\{k\in A_j^c(t)\}}$ is independent of the supply S. Note that the numbers $(-p_{jk})$ are positive and are interpreted as prices.

This inf-convolution form of the problem helps us to understand the structure of the optimal control. For t greater than the ending outage period of k, $p_{jk}(t)=p_{jk}(T)=l_{T,e_{jk}}(e(T))<0$. For $S\leq\sum_{jk}\bar{s}_{jk}\mathbb{1}_{\{k\in A_j\}}$, the linear optimization can be easily solved as follows:

$$f(S,p)=\inf_{s\in\Delta_S}\sum_{jk}(-p_{jk})s_{jk}.$$

The solution is to start saturating the cheapest power stations, i.e., the ones with the lowest price (adjoint variable) $\bar{p}_1=\min_{jk}(-p_{jk})>0$, for any of these cheapest stations, the maximum control $\bar{s}_{jk}$ is used. If

$$\sum_{\substack{j_1k_1\\ \bar{p}_1=-p_{j_1k_1}=\min_{jk}(-p_{jk})}} \bar{s}_{jk}<S,$$

then we start saturating the power station with second lowest prices, i.e.,

$$P_2=\{jk \mid -p_{jk}=\min_{jk\notin P_1} -p_{jk}>\bar{p}_1\}$$

where $P_1=\{j_1k_1 \mid \bar{p}_1=\min_{jk}-p_{jk}\}$. We reiterate the process until the entire budget S is satisfied.

No Cohabitation Between Maintenance and Production Phase

Even if it was possible from system perspective, we show that it is suboptimal to do maintenance and production at the same time. This corroborates the decision-makers' intuition that the power stations will not be able to produce while refueling and maintenance: $t \in T_{jk}$ implies that $s_{jk}(t) = 0$. Since $(-p_{jk}) > 0$ and $S(t) = \sum_{jk} s_{jk}(t)\mathbb{1}_{\{k\in A_j(t)\}}$, the extra term in Hamiltonian will bring an extra cost $\sum_{jk}(-p_{jk})s_{jk}(t)\mathbb{1}_{\{k\in A_j^c(t)\}}$, which can be reduced to zero by setting $s_{jk}(t) = 0$ if $k \in A_j^c(t)$. Hence, for an optimal control,

$$s^*_{jk}(t) = 0 \text{ if } k \in A_j^c(t).$$

Explicit Solution of the Minimization f

The function $S \mapsto f(S, p)$ is convex and block-wise affine in S. We provide the explicit expression of the solution depending on maximum control bounds $\bar{s}_{jk}$. Denote by $W_1 = \sum_{jk\in P_1} \bar{s}_{jk}$ the total production capacity of the cheapest power stations (the one in P_1). Similarly, we have P_2 as the set of second cheapest power stations and W_2 the associated total capacity and so on. We then have

$$f(S, p) =$$

$$\begin{cases} \bar{p}_1 S \text{ if } S \in [0, W_1] \\ f(W_1, p) + \bar{p}_2(S - W_1) \text{ if } S \in [W_1, W_1 + W_2] \\ f(W_2, p) + \bar{p}_3(S - W_2 - W_1) \text{ if } S \in [W_1 + W_2, W_1 + W_2 + W_3] \\ \dots \end{cases}$$

The Optimal Control May Not Be Unique

We now show that the optimal control may not be unique and explain where the nonuniqueness comes from. Let the demand satisfy

$$\begin{aligned} & (l')^{-1}[\bar{p}_m(t)] + W_1 + \ldots + W_{m-1} \\ & < D(t) \\ & < (l')^{-1}[\bar{p}_m(t)] + W_1 + \ldots + W_m. \end{aligned} \tag{13.34}$$

Then, it may happen that there are at least two power plants jk and $j'k'$ for which $p_{jk} = p_{j'k'}$ on a subinterval of $[0, T]$, corresponding to the $m-$th cheapest prices $(\bar{p}_m(T))$. In that case, we have infinite number of optimal s at the level m such that $W_1 + \ldots + W_{m-1} + \sum_{jk\in P_m} s_{jk}(t) = D(t) - (l')^{-1}[\bar{p}_m(t)] = S^*(t)$.

The Optimal Supply Is Unique

Thanks to the strict convexity conditions, the one-dimensional problem (which focuses on the supply S) has at most one solution. The problem also has at least one solution because of the piecewise continuity of the l and f, compactness (and non-emptiness) of the control space. Thus, there is exactly one optimal supply $S^*(t)$ solving Equation (13.33).

If $D(t) \in \left(\sum_{k=1}^{m-1} W_k, \sum_{k=1}^{m} W_k\right)$, $m > 1$, then there is a unique supply S^*, which is an interior solution given by

$$S^*(t) = D(t) + (l')^{-1}[t, \bar{p}_m] \tag{13.35}$$

$$\bar{p}_1 = \min_{jk} \; p_{jk} \text{ and } P_1 = \arg\min_{jk} p_{jk} \tag{13.36}$$

$$\bar{p}_2 = \min_{jk \notin P_1^c} \; p_{jk} \text{ and } P_2 = \arg \min_{jk \notin P_1^c} \; p_{jk} \tag{13.37}$$

$$\bar{p}_3 = \min_{jk \notin (P_1 \cup P_2)^c} \; p_{jk} \text{ and } P_3 = \arg \min_{jk \notin (P_1 \cup P_2)^c} \; p_{jk} \tag{13.38}$$

$$\ldots$$

The Structure of Optimal Controller

We are now ready for a complete solution to the problem. To do so, we distinguish between different demands $D(t)$. If $D(t) \in \left(\sum_{k=1}^{m-1} W_k, \sum_{k=1}^{m} W_k\right)$, for some $m \geq 2$, then

$$\begin{cases} S^*(t) = D(t) + (l')^{-1}[t, \bar{p}_m] \\ s_{jk}^*(t) = \bar{s}_{jk} \text{ for } (jk) \in P_1 \cup \ldots \cup P_{m-1} \\ \sum_{jk \in P_m} s_{jk}^*(t) = D(t) - (l')^{-1}[\bar{p}_m(t)] - \sum_{k=1}^{m-1} W_k \\ s_{jk}^* = 0 \text{ for } (jk) \notin P_1 \cup \ldots \cup P_m. \end{cases}$$

If $S^*(t) = \sum_{k=1}^{m} W_k$, then

$$\begin{cases} s_{jk}^*(t) = \bar{s}_{jk} & \text{for } (jk) \in P_1 \cup \ldots \cup P_m \\ s_{jk}^* = 0 & \text{for } (jk) \notin P_1 \cup \ldots \cup P_m \; . \end{cases}$$

Remark 13.11 We now provide a sufficient condition for uniqueness of the optimal control. If all the prices p_{jk} (i.e., production cost of power plant k of producer j) are all different, then all the sets P_m are reduced to singletons and $m = \sum_{j=1}^{I} K_j$. The prices are ordered by their membership to P_m in a strict increasing way and therefore the optimal control is unique.

Toward a Credible Cooperation Between Electricity Producers

Noncooperative equilibrium plays the role of a benchmark in a cooperative game, i.e., gives what producers could secure for themselves if there is no agreement. One needs to find a better way to share the cost $L(\mathscr{I}, \mathscr{T}) + c(\mathscr{I}, \mathscr{T})$. This leads to the introduction of the notion of imputation, i.e., a vector profile $\gamma_1, \ldots, \gamma_I$ such that

$$\sum_{i \in \mathscr{I}'} \gamma_i \leq L(\mathscr{I}', \mathscr{T}) + c(\mathscr{I}', \mathscr{T}),$$

for any $\mathscr{I}' \subset \mathscr{I}$, and

$$\sum_{i \in \mathscr{I}} \gamma_i = L(\mathscr{I}, \mathscr{T}) + c(\mathscr{I}, \mathscr{T}).$$

The set of all such profiles is called core of the dynamic cooperative game.

Definition 13.12 A cooperative solution is time consistent if, at any position $t \in [0, T]$, the cooperative solution cost-to-go is less than the deviating cost:

$$\gamma(\{i\}, [t, T]) \leq L(\{i\}, [t, T]),$$

where the deviating cost $L(\{i\}, [t, T])$ is computed along the state trajectory of the cooperative state trajectory $e_c^*(t)$.

This notion of time consistency and its implementation is essential in cooperative differential games. A stringer notion of time consistency is that the cooperative cost-to-go is lower than the noncooperative cost-to-go for any energy state $e(t),\ t \in \mathcal{T}$. This is called subgame consistent solution.

Dynamical Proportional Allocation

Even though one of the most commonly used allocation principles is the dynamical Shapley value, in the case when players may be asymmetric in their powers and sizes of payoffs, equal imputation of cooperative gains may not be totally agreeable to asymmetric producers. To overcome this, one can suggest the allocation principle in which the producers' shares of the gain from cooperation (cost savings) are proportional to the relative sizes of their expected deviating costs. Thus, a proportional time-consistent solution is given by

$$\begin{aligned}
\gamma_i(t) &= \frac{L(\{i\}, [t, T])}{\sum_j L(\{j\}, [t, T])} \sum_k \gamma(\{k\}, [t, T]) \\
&= \frac{L(\{i\}, [t, T])}{\sum_j L(\{j\}, [t, T])} [L(\mathcal{I}, [t, T]) + c(\mathcal{I}, [t, T])] \\
&\leq L(\{i\}, [t, T]).
\end{aligned} \tag{13.39}$$

Since these quantities are positive, one gets

$$\gamma(\{i\}, [t, T]) \leq L(\{i\}, [t, T]), \forall t,$$

(individual rationality) and

$$\sum_i \gamma(\{i\}, [t, T]) = L(\mathcal{I}, [t, T]) + c(\mathcal{I}, [t, T]).$$

(efficiency at any time).

Then one needs to find the appropriate instantaneous distribution of the production cost γ_i, i.e., find function $\hat{\gamma}_i$ such that

$$\gamma_i(0) = \hat{\gamma}_i(T) + \int_0^T \hat{\gamma}_i(t)\, dt.$$

For dynamic games, an additional and stringent condition on the solutions is required: the specific optimality principle must remain optimal at any instant of time throughout the game duration along the optimal state trajectory. This condition is known as dynamic stability or time consistency.

The notion of time consistency is crucial since the initial distribution of states and starting time influences naturally the energy stock. A cooperative solution is subgame consistent if an extension of the cooperative strategy to a situation with a later starting time and to any possible state brought about by the prior optimal behavior of the players remains optimal. Subgame consistency is a stronger notion than time consistency. In the presence of stochastic elements, subgame consistency is required in a credible cooperative solution.

Fix $(t', e) \in [0, T] \times \mathbb{R}_+$ and define the control law s_ε as the spike variation of $\hat{s}$ over the set $[t', t' + \varepsilon],\ \varepsilon > 0$, i.e.,

$$s_\varepsilon(t) = s(t)\mathbb{1}_{[t',t'+\varepsilon]}(t) + \hat{s}(t)\mathbb{1}_{\mathcal{T}\setminus[t',t'+\varepsilon]}(t),$$

where s is an arbitrary admissible control and $\mathbb{1}_{[t',t'+\varepsilon]}$ is the indicator function over the set $[t', t' + \varepsilon]$.

Definition 13.13 Let L be the objective function to be minimized in s under production and maintenance constraints. We say that $\hat{s}$ is a subgame perfect cooperative strategy under spike variation if for any t_0, e_0, s,

$$\lim_{\varepsilon\to 0} \frac{1}{\varepsilon} \left[L([t_0, T], e_0, \hat{s}) - L([t_0, T], e_0, s_\varepsilon) \right] \leq 0.$$

Note that a subgame perfect cooperative strategy under spike variation is in particular a time-consistent solution.

The key difference here is that the solution that we are looking for should not depend on the initial data (when and where we started). Let H^{t_0,e_0} be the Pontryagin function associated with the state trajectory x that starts from e_0 at $t_0 \in \mathcal{T}$. $H^{t_0,e_0}(t, e, s, p) = fp^{t_0,e_0} + l^{t_0,e_0}$, where the notation l^{t_0,e_0} is obtained from the cost function l and $f = c_{jk}(t)\mathbb{1}_{\{k\in A_j^c(t)\}} - s_{jk}(t)$. It follows that $H_e^{t_0,e_0}(t, e, s, p) = 0$ and the first-order adjoint variable p^{t_0,e_0} is piecewise constant. The first-order adjoint equation is given by

$$\begin{aligned} p^{t_0,e_0} &\in \{p^{t_0,e_0}(t_0), p^{t_0,e_0}(T)\}, \\ p^{t_0,e_0}(T) &\leq l_{T,e}^{t_0,e_0}(e(T)) \\ p^{t_0,e_0}(T) &= l_{T,e}^{t_0,e_0}(e(T)) \text{ if } e_{jk}(T) > 0. \end{aligned}$$

Examining Failure Consequences
If one of the power plants (jk) fails during the production process in $[0, T]$, there are several consequences:

- If $s_{jk}(t) = 0$, i.e., that power plant was not used in the production. Then, the electricity production is not affected directly.
- $s_{jk}(t) > 0$. If the production price $-\ p_{jk}$ was among the cheapest to match the demand $D(t)$, its failure leads to a cost of taking the next cheapest and available power plant. This leads to an extra cost of electricity production. It should be noted that the process may be delayed in restarting producing from another power plant.

The structural results obtained here can be useful in many other scenarios highlighted below.

13.5.2 Energy Harvesting in Cognitive Wireless Communications

Energy harvesting sensors are increasingly being deployed in place of their traditional, battery-operated counterparts, when factors such as the sheer number of nodes or inaccessibility render battery replacement difficult and cost prohibitive. The applications of energy-harvesting sensors are diverse, ranging from industrial and environmental monitoring to smart buildings and grid asset monitoring. In contrast to battery-operated sensors, where minimizing energy consumption is crucial to prolong lifetime, in energy-harvesting sensors, the objective is the intelligent management of the harvested energy to ensure long-term, uninterrupted operation:

$$\frac{d}{dt}e_j(t) = -s_j(t) + c_j(t),$$

where $s_j(t)$ is the transmission power of the sensor node j at time t. These nodes are communicating through wireless channels that may be random. Here $c_j(t)$ is the power arriving at the battery of node j from the renewable energy sources (solar, wind). The variable $\bar{s}_j$ is now interpreted as the maximum power of node j. The wireless communication from the secondary users generates a certain interference level at a base station of a primary user (PU), which is a function of $I(t) = \sum_j s_j(t)$. The goal is to exploit the resources left by primary users (spectrum holes) under energy-harvesting constraint. $D(t) = I^{ref}(t)$ is a reference interference level to satisfy the secondary users. Here the loss function $l(I^{ref} - I)$ can be chosen as the relative loss due to unexploited resource left by the primary users. The methodology discussed earlier in this chapter provides explicit structural solutions to this opportunistic spectrum access problem.

13.5.3 Wireless Power Charging of Smart Cars

Here a player is a taxi operator or a transportation company. The operator deploys several wireless power charging stations in hybrid smart cars (and cabs). Usually, it involves huge charging stations, electric motor for charging electric vehicles. However, recently, wireless charging technology has been proposed. At each 40 seconds stop (during the pickup of new clients), it picks up a charge strong enough to drive for a further few miles. The goal is to reduce the gap between the charged energy and the energy needed to take from the hybrid system. This can be formulated as the dynamic optimization problem presented above. At the same time, there are some constraints: the bus (or the taxi) cannot wait more than the time required at the stop. At the first step, the operator estimates the trip cycle and the duration based on traffic congestion and the distance. Based on these estimates, every car j can be charged wirelessly when it is near a wireless power station k. $c_{jk}(t)$ is the wireless charging rate. s_{jk} is the consumption rate when driving. Thus, the fuel stock evolves according to the dynamics:

$$\frac{d}{dt}e_{jk}(t) = c_{jk}(t)\mathbb{1}_{\{k\in A_j^c(t)\}} - s_{jk}(t),$$

where $k \in A_j^c(t)$ means that the car j is in the range of the station (bus/taxi stop) k at time t. The fuel needed for car j is $\sum_k s_{jk}$, and the one from wireless charging is $\sum_k \sum_t \mathbb{1}_{k\in A_j^c(t)} c_{jk}(t)$. The extra energy needs to be compensated from other sources, hence creating a certain cost. The methodology developed above can be applied to this situation.

13.5.4 Energy Efficiency in Next-Generation Networks

In mobile networks, more than 90% of budget is used to power the radio sites. The energy consumption of base stations (macro, micro, femto, pico, small) needs to be managed by satisfying users' demand. Base stations (of all kinds) are deployed all around. The state of energy stock can be modeled as $\frac{d}{dt}e_{jk} = -s_{jk} + c_{jk}\mathbb{1}_{\{k\in A_j^c\}}$ where $s_{jk}(t)$ now denotes the power consumed by the base station k of the operator j. Each station has a maximum power budget constraint $\bar{s}_{jk}$. Power plants are used to generate energy, at say the rate of $c_{jk}(t)$. Then, one gets exactly the state dynamics we have considered earlier, and the operator needs to find a trade-off between the energy consumption and user satisfaction (quality of experience and quality of service).

13.6 Notes

We have used both aggregative games with finite number of decision-makers and static Bayesian mean-field games to illustrate the economics of networks. More details on energy-efficiency analysis using aggregative games, mean-field games, and mean-field-type games can be found [33, 34, 98, 118, 147, 186, 293, 300, 303, 387, 388, 414, 478, 541].

Chapter 14
Electricity Price Dynamics in the Smart Grid

In this chapter, an interactive decision-making process between electricity producers is formulated. The interaction is through the market price, the demand (of consumers and prosumers), and the supply (of producers and prosumers). The price dynamics is described by a linear jump-diffusion system of conditional mean-field type where the conditioning is with respect to a common noise representing a global market uncertainty. The common noise is perfectly observed by all decision-makers during the interaction phase. The production cost functional of each decision-maker is quadratic and involves the second moment, the square of the conditional expectation of the control actions of the producers. The gain is the one obtained by selling the supplied quantity. The direct method developed in Volume I is revisited to provide a semi-explicit solution of the corresponding mean-field-type game with common noise. The equilibrium strategies are in state-and-conditional mean-field-type feedback form, where the mean-field term is the conditional expectation of the price given the realization of the global uncertainty [131, 147, 150, 151, 265, 373, 374, 379, 384, 400, 460, 461, 463, 471, 552].

The rest of the chapter is organized as follows. The next section introduces the model and the key interaction terms that affect the price. Section 14.2 examines longer-term horizon. Section 14.3 briefly presents the computation of the Cournot equilibrium in the case of a static price. We then discuss the sequential decision-making in Sect. 14.4. The chapter ends with the concluding remarks and historical notes of Sect. 14.5.

14.1 Log-Price Dynamics Under Jump-Diffusion Process

Consider a mean-field-type game[460, 464] described by the following settings. Let $\mathscr{T} := [t_0, T]$ be the time horizon with $t_0 = 0 < T$, and $I \geq 2$ potential interacting energy producers over the horizon $\mathscr{T}$. At time $t \in \mathscr{T}$, producer i's

T. Başar et al., *Mean-Field-Type Game Theory II*, Static & Dynamic Game Theory: Foundations & Applications, https://doi.org/10.1007/978-3-032-07031-9_14

output is $u_i(t) \geq 0$. The log-price dynamics is described by the following stochastic differential equation with the initial condition $p(0) = p_0$:

$$dp = s[a - D - p]dt + \left(\sigma dB + \int_{\theta\in\Theta} \mu(\theta)\tilde{N}(dt, d\theta)\right) + \sigma_o dB_o, \quad (14.1)$$

where

$$D(t) := \sum_{i=1}^{I} u_i(t)$$

is the supply at time t and B_o is a standard Brownian motion representing a global uncertainty observed by all participants in the market. The processes B and N are local uncertainty or noise. B is a standard Brownian motion, and N is a jump process with Lévy measure $\nu(d\theta)$ defined over Θ. It is assumed that ν is a Radon measure over Θ (the jump space). The process

$$\tilde{N}(dt, d\theta) = N(dt, d\theta) - \nu(d\theta)dt$$

is a compensated martingale. *We assume that all these processes are mutually independent.* We denote by $\mathscr{F}_t^{B_o}$ the filtration generated by the observed common noise up to t, $\{B_o(t'),\ t' \leq t\}$. The parameter s is positive, with larger values of s the market price adjusts quicker along the inverse demand. a, σ, σ_o are fixed parameters. The jump rate size $\mu(.)$ is in $L^2_\nu(\Theta, \mathbb{R})$, i.e.,

$$\int_\Theta \mu^2(\theta)\nu(d\theta) < +\infty.$$

The initial distribution of p_0 is square integrable: $Ep_0^2 < \infty$. At time $t \in \mathscr{T}$, producer i receives $\hat{p}(t)u_i - C_i(u_i)$, where $C_i : \mathbb{R} \to \mathbb{R}$, given by

$$C_i(u_i) = c_i u_i + \frac{r_i u_i^2}{2} + \frac{\hat{r}_i \hat{u}_i^2}{2},$$

is the instant cost function of i. The term $\hat{u}_i = E(u_i \mid \mathscr{F}_t^{B_o})$ is the conditional expectation of producer i's output given the global uncertainty B_o observed in the market. The last term in the expression of the instant cost C_i, i.e., $\frac{\hat{r}_i \hat{u}_i^2}{2}$, aims to capture the risk sensitivity of the producer. The conditional expectation of the price given the global uncertainty B_o up to time t is $\hat{p}(t) = E(p(t) \mid \mathscr{F}_t^{B_o})$. At the terminal time T, the payoff is $-\frac{1}{2}e^{-\lambda_i T}\left(p(T) - \hat{p}(T)\right)^2$. The long-term revenue of producer i is

$$R_{i,\mathscr{T}}(p_0, u) = -\frac{q}{2}e^{-\lambda_i T}\left(p(T) - \hat{p}(T)\right)^2 + \int_0^T e^{-\lambda_i t}\left[\hat{p}u_i - C_i(u_i)\right]dt,$$

where λ_i is a discount factor of producer i. Finally, each producer optimizes its long-term expected revenue.

A number of remarks are in order here:

- The price dynamics model above can be reinterpreted as an error to the standard inverse demand model.
- The flexibility in the choice of s allows us to navigate between several regimes.
- The jump term $\tilde{N}$ captures some of the big changes in the market that may happen randomly.
- The global uncertainty B_o captures the common noise in the market.
- Based on the common noise that is observed by all decision-makers, a conditional price is calculated. The revenue is computed from the conditional price.

Why Is This a Conditional Mean-Field-Type Game?

First, the setup considered here is a strategic game because the producers are coupled through the price functional. Second, two conditional mean-field terms are involved:

- The conditional price $\hat{p}(t)$ based on the observations of the common noise B_o up to t.
- The square of the conditional control action

$$\hat{u}_i^2(t) = \left[E(u_i \mid \mathcal{F}_t^{B_o})\right]^2$$

based on the observations of the common noise B_o up to t.

These mean-field-type terms make the problem a game of conditional mean-field type [106]. The resulting price dynamics is of conditional mean-field type. A strategy for each producer is a mapping that is progressively measurable to respect the information available to the producer. A pure strategy at time t maps an information set of the producer to its set of control actions $\mathbb{R}_+$. Let $\mathcal{U}_i$ be the set of strategies of producer i.

Definition 14.1 (BR_i **: Best response of producer** i) Any strategy $u_i^*(\cdot) \in \mathcal{U}_i$ satisfying the supremum in (14.2)

$$\begin{cases} \sup_{u_i \in \mathcal{U}_i} E\left[R_{i,\mathcal{T}}(p_0, u)\right], \\ dp = s\left[a - D - p\right]dt + \left(\sigma dB + \int_\Theta \mu(\theta)\tilde{N}(dt, d\theta)\right) + \sigma_o dB_o, \\ p_0, \end{cases} \tag{14.2}$$

is a best-response strategy of producer i to the other producers' strategy $u_{-i} \in \prod_{j \neq i} \mathcal{U}_j$. The set of best-response strategies of i is denoted by $BR_i : \prod_{j \neq i} \mathcal{U}_j \to 2^{\mathcal{U}_i}$, where $2^{\mathcal{U}_i}$ denotes the set of subsets of $\mathcal{U}_i$.

Finding State-and-Conditional Mean-Field Feedback Nash Solution
The study of open-loop equilibrium in (deterministic) differential games goes back at least to Roos (1925 [373], pages 163–164). In the open-loop adapted setting, the information structure of the producer is restricted to the common noise B_o, and producers are allowed to employ production output functions of time and that are $\mathcal{F}^{p_0,B_o}$–measurable. This means that the adapted control action law is in the form $\phi_i(t, p_0, B_o)$. It does not explicitly depend on the price $p(t)$. For the integrability of the cost functional, we impose an integrability condition on u_i^2. The set of square integrable, progressively $\mathcal{F}^{p_0,B_o}$–measurable open-loop strategies is

$$\mathcal{U}_i^{ol} = L^2_{\mathcal{F}^{p_0,B_o}}\left([0, T], \mathbb{R}\right).$$

In our case, however, the information structure of each producer is the price process in addition to the common noise. Therefore, we look for state-and-mean-field-type feedback strategies [460, 464]. The producer is allowed to employ production output functions of state and time, which are $\mathcal{F}^{p_0,p,B_o}$–measurable. This means that the control action law is in the form $\phi_i\left(t, p_0, p, B_o\right)$. At each time t, the realized log-price $p(t)$ will be observed by the agent. The set of square integrable, progressively $\mathcal{F}^{p_0,p,B_o}$–measurable feedback strategies is

$$\mathcal{U}_i^{fb} = L^2_{\mathcal{F}^{p_0,p,B_o}}\left([0, T] \times \mathbb{R}, \mathbb{R}\right).$$

A state-and-mean-field-type feedback Nash equilibrium is a strategy profile $u_i \in \mathcal{U}_i^{fb}$ such that $u_i \in BR_i((u_j)_{j\neq i})$ and $u_i(t)$ can be expressed as a function of $(t, p(t), \hat{p}(t))$.

Dynamics for the Conditional Price
Since the conditional price appears in the cost function, we would like to derive a simple equation for $\hat{p}$ as well. From the price dynamics, we deduce that the conditional price solves the following stochastic differential equation:

$$d\hat{p} = s\left[a - \hat{D} - \hat{p}\right]dt + \sigma_o dB_o,\ \hat{p}(0) = \hat{p}_0, \tag{14.3}$$

where $\hat{D} = \sum_i \hat{u}_i$ is the conditional expectation of the supply. Thus, the mean-field-type term $\hat{p}$ is a stochastic process driven by the common noise B_o. For a given $\hat{D}$, this is an OU (Ornstein–Uhlenbeck) process, stabilizing around the trajectory of $a - D$ as time T gets larger.

Proposition 14.2 *Generically, the problem formulated above has the following interior solution (if any):*
Equilibrium strategy in state-and-conditional mean-field-type feedback form:

$$u_i^* = -\frac{s\tilde{\alpha}_i}{r_i}(p - \hat{p}) + \frac{\hat{p}(1 - s\tilde{\beta}_i) - (c_i + s\tilde{\gamma}_i)}{r_i + \hat{r}_i}. \tag{14.4}$$

Conditional equilibrium price:

$$\begin{cases} \hat{p} = s\left\{a + \sum_{j=1}^{I} \frac{c_j + s\tilde{\gamma}_i}{r_j + \hat{r}_j} - \hat{p}\left(1 + \sum_{j=1}^{I} \frac{1 - s\tilde{\beta}_j}{r_j + \hat{r}_j}\right)\right\} dt + \sigma_o dB_o, \\ \hat{p}(0) = \hat{p}_0. \end{cases} \tag{14.5}$$

Riccati system:

$$\begin{cases} d\tilde{\alpha}_i = \left\{(\lambda_i + 2s)\tilde{\alpha}_i - \frac{s^2}{r_i}\tilde{\alpha}_i^2 - 2s^2\tilde{\alpha}_i \sum_{j\neq i} \frac{\tilde{\alpha}_j}{r_j}\right\} dt, \\ \tilde{\alpha}_i(T) = -q, \\ d\tilde{\beta}_i = \left\{(\lambda_i + 2s)\tilde{\beta}_i - \frac{(1 - s\tilde{\beta}_i)^2}{r_i + \hat{r}_i} + 2s\tilde{\beta}_i \sum_{j\neq i} \frac{1 - s\tilde{\beta}_j}{r_j + \hat{r}_j}\right\} dt, \\ \tilde{\beta}_i(T) = 0, \\ d\tilde{\gamma}_i = \left\{(\lambda_i + s)\tilde{\gamma}_i - s\tilde{\beta}_i a - \tilde{\beta}_{i,o}\sigma_o \right. \\ \qquad + \dfrac{(1 - s\tilde{\beta}_i)(c_i + s\tilde{\gamma}_i)}{r_i + \hat{r}_i} + s\tilde{\gamma}_i \displaystyle\sum_{j\neq i} \frac{1 - s\tilde{\beta}_j}{r_j + \hat{r}_j} \\ \qquad \left. - s\tilde{\beta}_i \sum_{j\neq i} \frac{c_j + s\tilde{\gamma}_i}{r_j + \hat{r}_j}\right\} dt - \tilde{\beta}_i\sigma_o dB_o, \\ \tilde{\gamma}_i(0) = 0, \\ d\tilde{\delta}_i = -\left\{-\lambda_i\tilde{\delta}_i + \frac{1}{2}\sigma_o^2\tilde{\beta}_i + \frac{1}{2}\tilde{\alpha}_i\left(\sigma^2 + \int_\Theta \mu^2(\theta)\nu(d\theta)\right) \right. \\ \qquad \left. + s\tilde{\gamma}_i a + \tilde{\gamma}_{i,o}\sigma_o + \dfrac{1}{2}\dfrac{(c_i + s\tilde{\gamma}_i)^2}{r_i + \hat{r}_i} + s\tilde{\gamma}_i \sum_{j\neq i} \frac{c_j + s\tilde{\gamma}_i}{r_j + \hat{r}_j}\right\} dt - \sigma_o\tilde{\gamma}_i dB_o, \\ \tilde{\delta}_i(T) = 0. \end{cases} \tag{14.6}$$

Equilibrium conditional revenue of producer i:

$$E\frac{1}{2}\alpha_i(0)(p(0) - \hat{p}_0)^2 + \frac{1}{2}\beta_i(0)\hat{p}_0^2 + \gamma_i(0)\hat{p}_0 + \delta_i(0). \tag{14.7}$$

Note that, by taking the expected value of the conditional equilibrium revenue, we obtain

$$\frac{1}{2}\alpha_i(0)\mathrm{var}_{\mathcal{F}_t^{B_o}}(p(0)) + \frac{1}{2}E\beta_i(0)\hat{p}_0^2 + E\gamma_i(0)\hat{p}_0 + E\delta_i(0),$$

which is a combination of the weighted conditional variance and the second moment.

Proof of Proposition 14.2 We provide a proof of the proposition; we first write down a generic structure of the discounted revenue functional, with unknown deterministic functions to be identified. Inspired from the structure of the terminal revenue function, we try a guess functional in a quadratic form. Let

$$f_i(t,p) = \frac{1}{2}\alpha_i(t)(p-\hat{p})^2 + \frac{1}{2}\beta_i(t)\hat{p}^2 + \gamma_i(t)\hat{p} + \delta_i(t),$$

where $\alpha,\ \beta,\ \gamma,\ \delta$ are random functions of time t, which are $\mathscr{F}_t^{B_o}$−measurable such that

$$f_i(T,p(T)) = -\frac{q}{2}e^{-\lambda_i T}(p(T)-\hat{p}(T))^2,$$

i.e., $\alpha_i(T) = -qe^{-\lambda_i T},\ \beta_i(T) = \gamma_i(T) = \delta_i(T) = 0$. Using Itô's formula for jump-diffusion processes, one obtains

$$d\hat{p}^2 = \left(2s\hat{p}(a-\hat{D}-\hat{p}) + \sigma_o^2 s\right)dt + 2\hat{p}\sigma_o dB_o, \tag{14.8}$$

$$d\left[\frac{\beta_i\hat{p}^2}{2}\right] = \frac{\hat{p}^2 d\beta_i}{2} + \frac{\beta_i d[\hat{p}^2]}{2} + \beta_{i,o}\hat{p}\sigma_o dt. \tag{14.9}$$

Using (14.8) in (14.9) yields

$$d\left[\frac{\beta_i\hat{p}^2}{2}\right] = \tfrac{1}{2}\hat{p}^2 d\beta_i + \tfrac{1}{2}\beta_i\left(2s\hat{p}\left(a-\hat{D}-\hat{p}\right) + \sigma_o^2\right)dt + \beta_i\hat{p}\sigma_o dB_o + \beta_{i,o}\hat{p}\sigma_o dt.$$

We compute the difference between p and $\hat{p}$:

$$d[p-\hat{p}] = -s\left(D-\hat{D}+p-\hat{p}\right)dt + \left(\sigma dB + \int_\Theta \mu(\theta)\tilde{N}(dt,d\theta)\right), \tag{14.10}$$

$$\begin{aligned} d[p-\hat{p}]^2 &= 2(p-\hat{p})\left(\sigma dB + \int_\Theta \mu(\theta)\tilde{N}(dt,d\theta)\right) \\ &\quad -2s(p-\hat{p})(D-\hat{D}+p-\hat{p})dt + \left(\sigma^2 + \int_\Theta \mu^2(\theta)\nu(d\theta)\right)dt. \end{aligned} \tag{14.11}$$

Moreover,

$$d\left[\frac{\alpha_i(p-\hat{p})^2}{2}\right] = \frac{(p-\hat{p})^2}{2}d\alpha_i + \tfrac{1}{2}\alpha_i d\left[(p-\hat{p})^2\right] + 0, \tag{14.12}$$

and using (14.11) in (14.12) yields

$$d\left[\frac{\alpha_i(p-\hat{p})^2}{2}\right] = \tfrac{(p-\hat{p})^2}{2}d\alpha_i + \tfrac{1}{2}\alpha_i\left(\sigma^2 + \int_\Theta \mu^2(\theta)\nu(d\theta)\right)dt$$
$$-s(p-\hat{p})\alpha_i(D-\hat{D}+p-\hat{p})dt + (p-\hat{p})\alpha_i\sigma dB$$
$$+(p-\hat{p})\alpha_i\int_\Theta \mu(\theta)\tilde{N}(dt,d\theta). \quad (14.13)$$

Finally,

$$d[\gamma_i\hat{p}] = \hat{p}d\gamma_i + \left(s\gamma_i\left(a-\hat{D}-\hat{p}\right)+\gamma_{i,o}\sigma_o\right)dt + \sigma_o\gamma_i dB_o. \quad (14.14)$$

Taking the conditional expectation with respect to the filtration $\mathscr{F}_T^{B_o}$, we arrive at

$$E\left[f_i(T,p(T)) - f_i(0,p(0))|\,\mathscr{F}_T^{B_o}\right]$$
$$= \frac{1}{2}\int_0^T (p-\hat{p})^2(d\alpha_i - 2s\alpha_i dt) - 2s\alpha_i(p-\hat{p})(D-\hat{D})dt$$
$$+\frac{1}{2}\int_0^T \alpha_i\left(\sigma^2 + \int\mu^2\nu(d\theta)\right)dt + \frac{1}{2}\int_0^T \hat{p}^2(d\beta_i - 2s\beta_i dt)$$
$$+\frac{1}{2}\int_0^T 2s\beta_i\hat{p}(a-\hat{D})dt + (\sigma_o^2\beta_i + 2\beta_{i,o}\hat{p}\sigma_o)dt \quad (14.15)$$
$$+\frac{1}{2}\int_0^T 2\beta_i\hat{p}\sigma_o dB_o + \int_0^T \hat{p}(d\gamma_i - s\gamma_i dt)$$
$$+\int_0^T \left(s\gamma_i(a-\hat{D}) + \gamma_{i,o}\sigma_o\right)dt + \int_0^T \sigma_o\gamma_i dB_o + d\delta_i.$$

The expected revenue can be expressed as

$$e^{-\lambda_i t}E\left[\hat{p}u_i - C_i(u_i)|\,\mathscr{F}_T^{B_o}\right]$$
$$= e^{-\lambda_i t}E\left[(\hat{p}-c_i)\hat{u}_i|\,\mathscr{F}_T^{B_o}\right] - e^{-\lambda_i t}E\left[\tfrac{1}{2}r_i(u_i-\hat{u}_i)^2 + \tfrac{1}{2}(r_i+\hat{r}_i)(\hat{u}_i)^2|\,\mathscr{F}_T^{B_o}\right]. \quad (14.16)$$

We now express the difference between the long-term revenue and the guess functional:

$$E[R_{i,\mathscr{T}} - f_i(0, p(0))|\ \mathscr{F}_T^{B_o}] = -\frac{q}{2}e^{-\lambda_i T} E\left[(p(T) - \hat{p}(T))^2|\ \mathscr{F}_T^{B_o}\right]$$
$$+\frac{1}{2}E\int_0^T e^{-\lambda_i t}\Big(2(\hat{p} - c_i)\hat{u}_i - r_i(u_i - \hat{u}_i)^2$$
$$-(r_i + \hat{r}_i)(\hat{u}_i)^2|\ \mathscr{F}_T^{B_o}\Big)dt + E\frac{1}{2}\int_0^T (p - \hat{p})^2(d\alpha_i - 2s\alpha_i dt)$$
$$+\frac{1}{2}\int_0^T -2s\alpha_i(p - \hat{p})(D - \hat{D})dt + \alpha_i\left(\sigma^2 + \int_\Theta \mu^2(\theta)\nu(d\theta)\right)dt$$
$$+\frac{1}{2}\int_0^T \hat{p}^2\,(d\beta_i - 2s\beta_i dt) + 2s\beta_i\hat{p}(a - \hat{D})dt \tag{14.17}$$
$$+\frac{1}{2}\int_0^T \left(\sigma_o^2\beta_i + 2\beta_{i,o}\hat{p}\sigma_o\right)dt + 2\beta_i\hat{p}\sigma_o dB_o$$
$$+\int_0^T \hat{p}\,(d\gamma_i - s\gamma_i dt) + [s\gamma_i(a - \hat{D}) + \gamma_{i,o}\sigma_o]dt$$
$$+\int_0^T \sigma_o\gamma_i dB_o + d\delta_i.$$

Therefore,

$$E\left[R_{i,\mathscr{T}} - f_i(0, p(0))|\ \mathscr{F}_T^{B_o}\right] = \frac{(\alpha_i(T) - qe^{-\lambda_i T})}{2}E\left[(p(T) - \hat{p}(T))^2|\ \mathscr{F}_T^{B_o}\right]$$
$$+E\int_0^T \frac{e^{-\lambda_i t}}{2}\Big(2(\hat{p} - c_i)\hat{u}_i - r_i(u_i - \hat{u}_i)^2 - (r_i + \hat{r}_i)(\hat{u}_i)^2\Big)$$
$$-2s\alpha_i(p - \hat{p})(D - \hat{D}) - 2s(\beta_i\hat{p} + \gamma_i)\hat{D}dt$$
$$+E\int_0^T \frac{(p - \hat{p})^2}{2}(d\alpha_i - 2s\alpha_i dt) + \frac{\alpha_i}{2}\left(\sigma^2 + \int_\Theta \mu^2(\theta)\nu(d\theta)\right)dt$$
$$+\frac{1}{2}\int_0^T \hat{p}^2\,(d\beta_i - 2s\beta_i dt) + 2s\beta_i\hat{p}a\,dt$$
$$+\frac{1}{2}\int_0^T (\sigma_o^2\beta_i + 2\beta_{i,o}\hat{p}\sigma_o)dt + 2\beta_i\hat{p}\sigma_o dB_o$$
$$+\int_0^T \hat{p}\,(d\gamma_i - s\gamma_i dt) + [s\gamma_i a + \gamma_{i,o}\sigma_o]dt$$
$$+\int_0^T \sigma_o\gamma_i dB_o + d\delta_i. \tag{14.18}$$

We now use a completion of squares technique. The term

$$e^{-\lambda_i t}\left(2(\hat{p}-c_i)\hat{u}_i - r_i(u_i-\hat{u}_i)^2 - (r_i+\hat{r}_i)(\hat{u}_i)^2\right)$$
$$-2s\alpha_i(p-\hat{p})(D-\hat{D}) - 2s(\beta_i\hat{p}+\gamma_i)\hat{D}$$

is equal to

$$-\left(r_i+\hat{r}_i\right)e^{-\lambda_i t}(\hat{u}_i)^2 + 2\left[(\hat{p}-c_i)e^{-\lambda_i t} - s(\beta_i\hat{p}+\gamma_i)\right]\hat{u}_i$$
$$-r_i e^{-\lambda_i t}(u_i-\hat{u}_i)^2 - 2s\alpha_i(p-\hat{p})(u_i-\hat{u}_i)$$
$$-2s\alpha_i(p-\hat{p})(D_{-i}-\hat{D}_{-i}) - 2s(\beta_i\hat{p}+\gamma_i)\hat{D}_{-i}$$

$$= -(r_i+\hat{r}_i)e^{-\lambda_i t}\left(u_i - \tfrac{\hat{p}(e^{-\lambda_i t}-s\beta_i)-(c_i e^{-\lambda_i t}+s\gamma_i)}{(r_i+\hat{r}_i)e^{-\lambda_i t}}\right)^2$$
$$-r_i e^{-\lambda_i t}\left(u_i-\hat{u}_i+\tfrac{s\alpha_i(p-\hat{p})}{r_i e^{-\lambda_i t}}\right)^2 + (p-\hat{p})^2\left(\tfrac{s^2\alpha_i^2}{r_i e^{-\lambda_i t}}\right.$$
$$\left.+2s^2\alpha_i\textstyle\sum_{j\neq i}\tfrac{\alpha_j}{r_j e^{-\lambda_j t}}\right) + \hat{p}^2\left(\tfrac{(e^{-\lambda_i t}-s\beta_i)^2}{(r_i+\hat{r}_i)e^{-\lambda_i t}}\right.$$
$$\left.-2s\beta_i\textstyle\sum_{j\neq i}\tfrac{e^{-\lambda_j t}-s\beta_j}{(r_j+\hat{r}_j)e^{-\lambda_j t}}\right) - 2\hat{p}\left(\tfrac{(e^{-\lambda_i t}-s\beta_i)(c_i e^{-\lambda_i t}+s\gamma_i)}{(r_i+\hat{r}_i)e^{-\lambda_i t}}\right.$$
$$\left.+s\gamma_i\textstyle\sum_{j\neq i}\tfrac{(e^{-\lambda_j t}-s\beta_j)}{(r_j+\hat{r}_j)e^{-\lambda_j t}} - s\beta_i\textstyle\sum_{j\neq i}\tfrac{(c_j e^{-\lambda_j t}+s\gamma_i)}{(r_j+\hat{r}_j)e^{-\lambda_j t}}\right)$$
$$+\tfrac{(c_i e^{-\lambda_i t}+s\gamma_i)^2}{(r_i+\hat{r}_i)e^{-\lambda_i t}} + 2s\gamma_i\textstyle\sum_{j\neq i}\tfrac{(c_j e^{-\lambda_j t}+s\gamma_i)}{(r_j+\hat{r}_j)e^{-\lambda_j t}},$$

where we have used

$$D_{-i}-\hat{D}_{-i} = -s(p-\hat{p})\sum_{j\neq i}\frac{\alpha_j}{r_j e^{-\lambda_j t}},$$

and

$$\hat{D}_{-i} = \hat{p}\sum_{j\neq i}\frac{e^{-\lambda_j t}-s\beta_j}{(r_j+\hat{r}_j)e^{-\lambda_j t}} - \sum_{j\neq i}\frac{c_j e^{-\lambda_j t}+s\gamma_i}{(r_j+\hat{r}_j)e^{-\lambda_j t}}.$$

It follows that

$$E\left[R_{i,\mathscr{T}}-f_i(0,p(0))|\,\mathscr{F}_T^{B_o}\right]=\frac{\alpha_i(T)-qe^{-\lambda_i T}}{2}E\left[(p(T)-\hat{p}(T))^2|\,\mathscr{F}_T^{B_o}\right]$$

$$-E\int_0^T\frac{r_i+\hat{r}_i}{2}e^{-\lambda_i t}\left(\hat{u}_i-\frac{\hat{p}(e^{-\lambda_i t}-s\beta_i)-(c_ie^{-\lambda_i t}+s\gamma_i)}{(r_i+\hat{r}_i)e^{-\lambda_i t}}\right)^2dt$$

$$-E\int_0^T\frac{r_i}{2}e^{-\lambda_i t}\left(u_i-\hat{u}_i+\frac{s\alpha_i(p-\hat{p})}{r_ie^{-\lambda_i t}}\right)^2dt$$

$$+E\int_0^T\frac{(p-\hat{p})^2}{2}\left\{d\alpha_i+\left(-2s\alpha_i+\frac{s^2\alpha_i^2}{r_ie^{-\lambda_i t}}+2s^2\alpha_i\sum_{j\neq i}\frac{\alpha_j}{r_je^{-\lambda_j t}}\right)dt\right\}\tag{14.19}$$

$$+\frac{1}{2}\int_0^T\hat{p}^2\left\{d\beta_i+(-2s\beta_i+\frac{(e^{-\lambda_i t}-s\beta_i)^2}{(r_i+\hat{r}_i)e^{-\lambda_i t}}-2s\beta_i\sum_{j\neq i}\frac{(e^{-\lambda_j t}-s\beta_j)}{(r_j+\hat{r}_j)e^{-\lambda_j t}})dt\right\}$$

$$+\int_0^T\hat{p}\left\{d\gamma_i-s\gamma_i dt+s\beta_i a dt+\beta_{i,o}\sigma_o dt+\beta_i\sigma_o dB_o-\frac{(e^{-\lambda_i t}-s\beta_i)(c_ie^{-\lambda_i t}+s\gamma_i)}{(r_i+\hat{r}_i)e^{-\lambda_i t}}-s\gamma_i\sum_{j\neq i}\frac{(e^{-\lambda_j t}-s\beta_j)}{(r_j+\hat{r}_j)e^{-\lambda_j t}}+s\beta_i\sum_{j\neq i}\frac{(c_je^{-\lambda_j t}+s\gamma_i)}{(r_j+\hat{r}_j)e^{-\lambda_j t}}\right\}\tag{14.20}$$

$$+\int_0^T d\delta_i+\left[\frac{1}{2}\sigma_o^2\beta_i+\frac{1}{2}\alpha_i\left(\sigma^2+\int_\Theta\mu^2(\theta)\nu(d\theta)\right)+s\gamma_i a+\gamma_{i,o}\sigma_o\right]dt+\sigma_o\gamma_i dB_o+\frac{1}{2}\frac{(c_ie^{-\lambda_i t}+s\gamma_i)^2}{(r_i+\hat{r}_i)e^{-\lambda_i t}}+s\gamma_i\sum_{j\neq i}\frac{(c_je^{-\lambda_j t}+s\gamma_i)}{(r_j+\hat{r}_j)e^{-\lambda_j t}}.$$

We deduce that $E\left[R_{i,\mathscr{T}}-f_i(0,p(0))|\,\mathscr{F}_T^{B_o}\right]\leq 0$ and equality holds when the random coefficient solves the following system:

$$\hat{u}_i^*=\frac{\hat{p}(e^{-\lambda_i t}-s\beta_i)-(c_ie^{-\lambda_i t}+s\gamma_i)}{(r_i+\hat{r}_i)e^{-\lambda_i t}}.\tag{14.21}$$

Equilibrium strategy in state-and-mean-field type feedback form is

$$\begin{aligned} u_i^* &= -\frac{s\alpha_i(p-\hat{p})}{r_i e^{-\lambda_i t}} + \hat{u}_i^*, \\ &= -\frac{s\alpha_i}{r_i e^{-\lambda_i t}}(p-\hat{p}) + \frac{\hat{p}(e^{-\lambda_i t} - s\beta_i) - (c_i e^{-\lambda_i t} + s\gamma_i)}{(r_i+\hat{r}_i)e^{-\lambda_i t}}. \end{aligned} \tag{14.22}$$

Conditional equilibrium price is

$$\begin{cases} \hat{p} = s\Big[a + \sum_{j=1}^{I} \frac{c_j e^{-\lambda_j t} + s\gamma_i}{(r_j+\hat{r}_j)e^{-\lambda_j t}} \\ \qquad -\hat{p}\left(1 + \sum_{j=1}^{I} \frac{e^{-\lambda_j t} - s\beta_j}{(r_j+\hat{r}_j)e^{-\lambda_j t}}\right)\Big] dt + \sigma_o dB_o, \\ \hat{p}(0) = \hat{p}_0. \end{cases} \tag{14.23}$$

Riccati system is

$$\begin{cases} d\alpha_i = -\left(-2s\alpha_i + \frac{s^2\alpha_i^2}{r_i e^{-\lambda_i t}} + 2s^2\alpha_i \sum_{j\neq i} \frac{\alpha_j}{r_j e^{-\lambda_j t}}\right) dt, \\ \alpha_i(T) = -qe^{-\lambda_i T}, \\ d\beta_i = -\Big(-2s\beta_i + \frac{(e^{-\lambda_i t} - s\beta_i)^2}{(r_i+\hat{r}_i)e^{-\lambda_i t}} \\ -2s\beta_i \sum_{j\neq i} \frac{(e^{-\lambda_j t} - s\beta_j)}{(r_j+\hat{r}_j)e^{-\lambda_j t}}\Big) dt, \\ \beta_i(T) = 0, \\ \gamma_i = \Big\{s\gamma_i - s\beta_i a - \beta_{i,o}\sigma_o \\ + \frac{(e^{-\lambda_i t} - s\beta_i)(c_i e^{-\lambda_i t} + s\gamma_i)}{(r_i+\hat{r}_i)e^{-\lambda_i t}} + s\gamma_i \sum_{j\neq i} \frac{(e^{-\lambda_j t} - s\beta_j)}{(r_j+\hat{r}_j)e^{-\lambda_j t}} \\ -s\beta_i \sum_{j\neq i} \frac{(c_j e^{-\lambda_j t} + s\gamma_i)}{(r_j+\hat{r}_j)e^{-\lambda_j t}}\Big\} dt, \\ \gamma_i(0) = 0, \\ d\delta_i = -\Big\{\frac{1}{2}\sigma_o^2\beta_i + \frac{1}{2}\alpha_i(\sigma^2 + \int_\Theta \mu^2(\theta)\nu(d\theta)) \\ \qquad + s\gamma_i a + \gamma_{i,o}\sigma_o + \frac{1}{2}\frac{(c_i e^{-\lambda_i t} + s\gamma_i)^2}{(r_i+\hat{r}_i)e^{-\lambda_i t}} \\ + s\gamma_i \sum_{j\neq i} \frac{(c_j e^{-\lambda_j t} + s\gamma_i)}{(r_j+\hat{r}_j)e^{-\lambda_j t}}\Big\} dt \\ -\sigma_o\gamma_i dB_o, \\ \delta_i(T) = 0. \end{cases} \tag{14.24}$$

Conditional equilibrium revenue is

$$\frac{1}{2}\alpha_i(0)(p(0)-\hat{p}_0)^2+\frac{1}{2}\beta_i(0)\hat{p}_0^2+\gamma_i(0)\hat{p}_0+\delta_i(0). \tag{14.25}$$

Set $\tilde{\alpha}_i=\alpha_i e^{\lambda_i t},\ \tilde{\beta}_i=\beta_i e^{\lambda_i t}$. Then:
Equilibrium strategy in state-and-mean-field type feedback form is:

$$u_i^*=-\frac{s\tilde{\alpha}_i}{r_i}(p-\hat{p})+\frac{\hat{p}(1-s\tilde{\beta}_i)-(c_i+s\tilde{\gamma}_i)}{r_i+\hat{r}_i}. \tag{14.26}$$

Conditional equilibrium price is

$$\begin{cases} \hat{p}=s\left\{a+\sum_{j=1}^{I}\frac{c_j+s\tilde{\gamma}_i}{r_j+\hat{r}_j}-\hat{p}\left(1+\sum_{j=1}^{I}\frac{1-s\tilde{\beta}_j}{r_j+\hat{r}_j}\right)\right\}dt \\ \qquad +\sigma_o dB_o, \\ \hat{p}(0)=\hat{p}_0. \end{cases} \tag{14.27}$$

Riccati system is

$$\begin{cases} d\tilde{\alpha}_i=\left\{(\lambda_i+2s)\tilde{\alpha}_i-\frac{s^2}{r_i}\tilde{\alpha}_i^2-2s^2\tilde{\alpha}_i\sum_{j\neq i}\frac{\tilde{\alpha}_j}{r_j}\right\}dt, \\ \tilde{\alpha}_i(T)=-q, \\ d\tilde{\beta}_i=\left\{(\lambda_i+2s)\tilde{\beta}_i-\frac{(1-s\tilde{\beta}_i)^2}{r_i+\hat{r}_i}+2s\tilde{\beta}_i\sum_{j\neq i}\frac{1-s\tilde{\beta}_j}{r_j+\hat{r}_j}\right\}dt, \\ \tilde{\beta}_i(T)=0, \\ d\tilde{\gamma}_i=\left\{(\lambda_i+s)\tilde{\gamma}_i-s\tilde{\beta}_i a-\tilde{\beta}_{i,o}\sigma_o \right. \\ \qquad +\frac{(1-s\tilde{\beta}_i)(c_i+s\tilde{\gamma}_i)}{r_i+\hat{r}_i}+s\tilde{\gamma}_i\sum_{j\neq i}\frac{1-s\tilde{\beta}_j}{r_j+\hat{r}_j} \\ \left. -s\tilde{\beta}_i\sum_{j\neq i}\frac{c_j+s\tilde{\gamma}_i}{r_j+\hat{r}_j}\right\}dt-\tilde{\beta}_i\sigma_o dB_o, \\ \tilde{\gamma}_i(0)=0, \\ d\tilde{\delta}_i=-\left\{-\lambda_i\tilde{\delta}_i+\frac{1}{2}\sigma_o^2\tilde{\beta}_i+\frac{1}{2}\tilde{\alpha}_i\left(\sigma^2+\int_{\Theta}\mu^2(\theta)\nu(d\theta)\right)\right. \\ \left. +s\tilde{\gamma}_i a+\tilde{\gamma}_{i,o}\sigma_o+\frac{1}{2}\frac{(c_i+s\tilde{\gamma}_i)^2}{(r_i+\hat{r}_i)}+s\tilde{\gamma}_i\sum_{j\neq i}\frac{(c_j+s\tilde{\gamma}_i)}{(r_j+\hat{r}_j)}\right\}dt-\sigma_o\tilde{\gamma}_i dB_o, \\ \tilde{\delta}_i(T)=0, \end{cases} \tag{14.28}$$

thus completing the proof. ■

14.2 Larger Horizon and Stationary Solution

We consider in this section the maximization of the performance criterion $\lim_{T\to+\infty} ER_{\mathcal{T}}(p_0, u)$. We are interested in studying Nash equilibria within the class of linear time-invariant strategies.

Limiting Feedback Strategies

Denoting a set of feedback Nash equilibrium strategies for the finite horizon game by $u^{fb}_{i,\mathcal{T}}$, and those for the infinite horizon game by $u^{fb}_{i,+\infty}$, we will study the relationship between $u^{fb}_{i,\mathcal{T}}$ and $u^{fb}_{i,+\infty}$. As $T \to +\infty$, the coefficient α_i vanishes and the (unconstrained) equilibrium strategies become

$$\begin{cases} u^{fb}_{i,+\infty} = \dfrac{\hat{p}(1-s\tilde{\beta}_i)-(c_i+s\tilde{\gamma}_i)}{r_i+\hat{r}_i}, \\ (-\lambda_i-2s)\tilde{\beta}_i + \dfrac{(1-s\tilde{\beta}_i)^2}{r_i+\hat{r}_i} - 2s\tilde{\beta}_i \displaystyle\sum_{j\neq i} \dfrac{1-s\tilde{\beta}_j}{r_j+\hat{r}_j} = 0, \\ \tilde{\beta}_{i,o} = 0. \end{cases} \tag{14.29}$$

Producer i will participate if the log-price is high enough to receive some reasonable revenue. Let $\underline{p}_i := \frac{c_i+s\tilde{\gamma}_i}{1-s\tilde{\beta}_i}$. If, in addition, σ_o vanishes, then the following structure

$$u_i^* = \begin{cases} \dfrac{\hat{p}(1-s\tilde{\beta}_i^*)-(c_i+s\tilde{\gamma}_i^*)}{r_i+\hat{r}_i} & \text{if } \hat{p} \geq \underline{p}_i, \\ 0 & \text{otherwise} \end{cases} \tag{14.30}$$

provides a stationary equilibrium. When the price increases, the equilibrium output of the producer i increases because $1 - s\tilde{\beta}_i^* > 0$. There is therefore a stationary equilibrium price and it is given by

$$\hat{p}^* = \frac{a + \sum_{j=1}^{I} \frac{c_j+s\tilde{\gamma}_i}{r_j+\hat{r}_j}}{1 + \sum_{j=1}^{I} \frac{1-s\tilde{\beta}_j}{r_j+\hat{r}_j}}. \tag{14.31}$$

Moreover, if the game starts at $\hat{p}_0 \neq \hat{p}^*$, the closed-loop equilibrium price converges exponentially to the stationary equilibrium price. The learning dynamics stabilizes quickly to the steady-state price. Note that when the speed s goes to infinity, $s\beta_i$ has a limit β_i^*, and the limiting electricity price would then be

$$p(s \to \infty) = \frac{a + \sum_{j=1}^{I} \frac{c_j + \tilde{\gamma}_i^*}{r_j + \hat{r}_j}}{1 + \sum_{j=1}^{I} \frac{1 - \tilde{\beta}_j^*}{r_j + \hat{r}_j}}. \tag{14.32}$$

Note that the results presented in this section can be extended to the case of linear coefficients σ, $\mu(\theta, .)$, and σ_0.

14.3 Static Cournot Competition Equilibrium Price

In this section, we are interested in investigating the implications of considering the dynamic learning price model as in (14.1) as compared with the static price scenario, i.e., considering $p^{static} = a - D$. We compute the static revenue of producer i as

$$R_{i,\mathcal{T}}(u) = \{[a - D]u_i - C_i(u_i)\}\Lambda_i,$$

where $\Lambda_i = \int_0^T e^{-\lambda_i t} dt$ and

$$C_i(u_i) = c_i u_i + \frac{1}{2}\left(r_i + \hat{r}_i\right) u_i^2.$$

Hence, the interior best response (if any) of the static Cournot game is given by

$$u_i^* \in \arg\max_{u_i \in \mathbb{R}}\{(a - D)u_i - C_i(u_i)\} = \frac{a - \sum_{j \neq i} u_j - c_i}{2 + r_i + \hat{r}_i}.$$

It follows that the static Cournot equilibrium price is given by

$$p^{C} = \frac{a + \sum_{j=1}^{I} \frac{c_j}{1 + r_j + \hat{r}_j}}{1 + \sum_{j=1}^{I} \frac{1}{1 + r_j + \hat{r}_j}}. \tag{14.33}$$

Finally, notice that the presented analysis allows us to determine whether the consideration of the price dynamics (14.1) in the market would be beneficial for the consumers. This is evaluated by comparing the price of the previous section in (14.32) and the price of this section (14.33).

14.4 Sequential Decision-Making

In this section, we consider a hierarchical mean-field-type control problem described as follows. Let $\mathcal{T} := [0, T]$ be the time horizon with $T > 0$. The energy market in our model is described by $I \geq 2$ agents, one producer and $I - 1$ prosumers following a hierarchical structure (explained below). Each agent $i \in \{1, \ldots, I\}$, at time $t \in \mathcal{T}$, has an output $u_i(t) \geq 0$. The price dynamics is given by (14.1).

Sequential Mean-Field-Type Game

The hierarchical control problem we consider consists of the following leader–follower order: given $I-1$ prosumers and one producer, the ith prosumer is a leader of the $(i-1)$th prosumer and is a follower of the $(i+1)$th prosumer, for $i < I$. The Ith agent (the producer) is the leader at the top level and the 1st prosumer is the follower at the bottom level. More precisely, let

$$\mathcal{U}_i = L^2_{\mathcal{F}^{p_0,p,B_o}}([0,T] \times \mathbb{R}, \mathbb{R})$$

the set of square integrable, progressively measurable $\mathcal{F}^{p_0,p,B_o}$ feedback controls. All the agents aim to maximize their revenues as follows:

For given $u_i \in \mathcal{U}_i$, $i \in \{2, \ldots, I\}$, Prosumer 1 aims to find her optimal control:

$$u_1^* = u_1^*(u_2, \ldots, u_I) \in \mathcal{U}_1,$$

such that

$$E[\mathcal{R}_1(p_0, u_1^*, \ldots, u_I)] = \sup_{u_1 \in \mathcal{U}_1} E[\mathcal{R}_1(p_0, u_1, \ldots, u_I)],$$

under the dynamics (14.1). Given the optimal control of the first prosumer, Prosumer 2 finds her optimal control:

$$u_2^* = u_2^*(u_3, \ldots, u_I) \in \mathcal{U}_2,$$

such that

$$E[\mathcal{R}_2(p_0, u_1^*, u_2^*, \ldots, u_I)] = \sup_{u_2 \in \mathcal{U}_2} E[\mathcal{R}_2(p_0, u_1^*(u_2, \ldots, u_I), u_2, \ldots, u_I)],$$

under the price dynamics

$$dp = s[a - D_1^* - p]dt + \left(\sigma dB + \int_{\theta \in \Theta} \mu(\theta)\tilde{N}(dt, d\theta)\right) + \sigma_o dB_o, \tag{14.34}$$

where

$$D_1^*(t) = u_1^*(t) + \sum_{i=2}^{I} u_i(t).$$

Thereby, we can define $u_1^*, u_2^*, \ldots, u_I^*$ inductively, and the producer's optimal control u_I^* will not depend on the controls of $I-1$ prosumers.

Next we present the structure of the equilibrium strategies.

Sequential Mean-Field-Type Equilibrium

Proposition 14.3 *The optimal control strategies for the agents are given by*

$$u_i^* = -\frac{s((p-\bar{p})\alpha_i + \xi_i)}{e^{-\lambda_i t} r_i} + \frac{e^{-\lambda_i t}(\bar{p}-c_i) - \beta_i s \bar{p} - s\gamma_i}{e^{-\lambda_i t}(r_i + \bar{r}_i)}.$$

The conditional optimal price:

$$\begin{array}{l} d\bar{p} = s\left[a - \sum_{i=1}^{I}\left(\frac{e^{-\lambda_i t}(\bar{p}-c_i)-\beta_i s\bar{p}-s\gamma_i}{e^{-\lambda_i t}(r_i+\bar{r}_i)}\right) - \bar{p}\right]dt + \sigma_0 dB_o, \\ \bar{p}(0) = \bar{p}_0. \end{array} \tag{14.35}$$

The stochastic functions $\alpha_i, \beta_i, \gamma_i, \delta_i$ and ξ_i are $\mathscr{F}_t^{B_o}$—measurable and solve the following Riccati system: for $2 \le i \le I-1$

$$\left\{\begin{array}{l}
d\alpha_i = \left(2s\alpha_i - \frac{s^2\alpha_i^2}{e^{-\lambda_i t}r_i} - 2s^2\alpha_i \sum_{j=1}^{i-1}\frac{\alpha_j}{e^{-\lambda_j t}r_j}\right)dt + \alpha_{i,0}dB_o,\\
\alpha_i(T) = -qe^{-\lambda_i T},\\
d\beta_i = \left(2s\beta_i - \frac{(e^{-\lambda_i t}-\beta_i s)^2}{e^{-\lambda_i t}(r_i+\bar{r}_i)} + 2s\beta_i\sum_{j=1}^{i-1}\frac{e^{-\lambda_j t}-s\beta_j}{e^{-\lambda_j t}(r_j+\bar{r}_j)}\right)dt + \beta_{i,0}dB_o,\\
\beta_i(T) = 0,\\
d\gamma_i = -\left(\beta_i s\left(a - \sum_{j=i+1}^{I}\bar{u}_j + \sum_{j=1}^{i-1}\frac{e^{-\lambda_j t}c_j+s\gamma_j}{e^{-\lambda_j t}(r_j+\bar{r}_j)}\right) - s\gamma_i\sum_{j=1}^{i-1}\frac{e^{-\lambda_j t}-s\beta_j}{e^{-\lambda_j t}(r_j+\bar{r}_j)}\right.\\
\qquad \left. + \sigma_0\beta_{i,0} - s\gamma_i - \frac{(e^{-\lambda_i t}-\beta_i s)(s\gamma_i+e^{-\lambda_i t}c_i)}{(r_i+\bar{r}_i)e^{-\lambda_i t}}\right)dt - \beta_i\sigma_0 dB_o,\\
\gamma_i(0) = 0,\\
d\xi_i = s\left(\xi_i + \alpha_i\sum_{j=i+1}^{I}(u_j-\bar{u}_j) - s\alpha_i\sum_{j=1}^{i-1}\frac{\xi_j}{e^{-\lambda_j t}r_j} - s\xi_i\sum_{j=1}^{i-1}\frac{\alpha_j}{e^{-\lambda_j t}r_j} - \frac{s\alpha_i\xi_i}{e^{-\lambda_i t}r_i}\right)dt\\
\qquad +\xi_{i,0}dB_o,\\
\xi_i(T) = 0,\\
d\delta_i = -\left(\frac{1}{2}\frac{(s\gamma_i+e^{-\lambda_i}c_i)^2}{e^{-\lambda_i}(r_i+\bar{r}_i)} + s^2\xi_i\sum_{j=1}^{i-1}\frac{\xi_j}{e^{-\lambda_j t}r_j} + \frac{1}{2}\frac{s^2\xi_i^2}{e^{-\lambda_i}r_i} + \frac{1}{2}\beta_i\sigma_0^2\right.\\
\qquad +\frac{\alpha_i}{2}(\sigma^2 + \int_\Theta \mu^2(\theta)\nu(d\theta)) + s\gamma_i(a - \sum_{j=i+1}^{I}\bar{u}_j) + s\gamma_i\sum_{j=1}^{i-1}\frac{(e^{-\lambda_j t}c_j+s\gamma_j)}{e^{-\lambda_j t}(r_j+\bar{r}_j)}\\
\qquad \left.+\gamma_{i,0}\sigma_0 - s\xi_i\sum_{j=i+1}^{I}(u_j-\bar{u}_j)\right)dt - \sigma_0\gamma_i dB_o,\\
\delta_i(T) = 0.
\end{array}\right.$$

For $i = 1$

$$
\begin{cases}
d\alpha_1 = \left(2s\alpha_1 - \frac{s^2\alpha_1^2}{e^{-\lambda_1 t}r_1}\right)dt + \alpha_{1,0}dB_o,\\
\alpha_1(T) = -qe^{-\lambda_1 T},\\
d\beta_1 = \left(2\beta_1 s - \frac{(e^{-\lambda_1 t}-\beta_1 s)^2}{e^{-\lambda_1 t}(r_1+\bar{r}_1)}\right)dt + \beta_{1,0}dB_o,\\
\beta_1(T) = 0,\\
d\gamma_1 = -\Big(\beta_1 s(a - \sum_{j=2}^{I}\bar{u}_j) + \sigma_0\beta_{1,0} - s\gamma_1\\
\quad -\frac{(e^{-\lambda_1 t}-\beta_1 s)(s\gamma_1+e^{-\lambda_1 t}c_1)}{(r_1+\bar{r}_1)e^{-\lambda_1 t}}\Big)dt - \beta_1\sigma_0 dB_o,\\
\gamma_1(0) = 0,\\
d\xi_1 = s\left(\xi_1 + \alpha_1\sum_{j=2}^{I}(u_j - \bar{u}_j) - \frac{s\alpha_1\xi_1}{e^{-\lambda_1 t}r_1}\right)dt + \xi_{1,0}dB_o,\\
\xi_1(T) = 0,\\
d\delta_1 = -\Big(\frac{1}{2}\frac{(s\gamma_1+e^{-\lambda_1}c_1)^2}{e^{-\lambda_1}(r_1+\bar{r}_1)} + \frac{1}{2}\frac{\xi_1^2 s^2}{e^{-\lambda_1 t}r_1} - s\xi_1\sum_{j=2}^{I}(u_j - \bar{u}_j)\\
\quad +\frac{\beta_1}{2}\sigma_0^2 + \frac{\alpha_1}{2}(\sigma^2 + \int_\Theta \mu^2(\theta)\nu(d\theta)) + s\gamma_1(a - \sum_{j=2}^{I}\bar{u}_j)\\
\quad +\gamma_{1,0}\sigma_0\Big)dt - \sigma_0\gamma_1 dB_o,\\
\alpha(T) = 0.
\end{cases}
$$

For $i = I$

$$
\begin{cases}
d\alpha_I = \left(2s\alpha_I - \frac{s^2\alpha_I^2}{e^{-\lambda_I t}r_I} - 2s^2\alpha_I\sum_{i=1}^{I-1}\frac{\alpha_i}{e^{-\lambda_i t}r_i}\right)dt + \alpha_{I,0}dB_o,\\
\alpha_I(T) = -qe^{-\lambda_2 T},\\
d\beta_I = \left(2\beta_I s - \frac{(e^{-\lambda_I t}-\beta_I s)^2}{e^{-\lambda_I t}(r_I+\bar{r}_I)} + 2s\beta_I\sum_{i=1}^{I-1}\frac{(e^{-\lambda_i t}-\beta_i s)}{e^{-\lambda_i t}(r_i+\bar{r}_i)}\right)dt + \beta_{I,0}dB_o,\\
\beta_I(T) = 0,\\
d\gamma_I = -\Big(\beta_I s(a + \sum_{i=1}^{I-1}\frac{e^{-\lambda_i t}c_i+s\gamma_i}{e^{-\lambda_i t}(r_i+\bar{r}_i)}) - s\gamma_I\sum_{i=1}^{I-1}\frac{(e^{-\lambda_i t}-s\beta_i)}{e^{-\lambda_i t}(r_i+\bar{r}_i)}\\
\quad + \sigma_0\beta_{I,0} - s\gamma_I - \frac{(e^{-\lambda_I t}-\beta_I s)(s\gamma_I+e^{-\lambda_I t}c_I)}{(r_I+\bar{r}_I)e^{-\lambda_I t}}\Big)dt - \beta_I\sigma_0 dB_o,\\
\gamma_I(0) = 0,\\
d\xi_I = s\left(\xi_I - s\alpha_I\sum_{i=1}^{I-1}\frac{\xi_i}{e^{-\lambda_i t}r_i} - s\xi_I\sum_{i=1}^{I-1}\frac{\alpha_i}{e^{-\lambda_i t}r_i} - \frac{s\alpha_I\xi_I}{e^{-\lambda_I t}r_I}\right)dt + \xi_{I,0}dB_o,\\
\xi_I(T) = 0,\\
d\delta_I = -\Big(\frac{1}{2}\frac{(s\gamma_I+e^{-\lambda_I}c_2)^2}{e^{-\lambda_I}(r_I+\bar{r}_I)} + s^2\xi_I\sum_{i=1}^{I-1}\frac{\xi_i}{e^{-\lambda_i t}r_i} + \frac{1}{2}\frac{s^2\xi_I^2}{e^{-\lambda_I}r_I} + \frac{1}{2}\beta_I\sigma_0^2\\
\quad +\frac{\alpha_I}{2}(\sigma^2 + \int_\Theta \mu^2(\theta)\nu(d\theta)) + s\gamma_I a + s\gamma_I\sum_{i=1}^{I-1}\frac{e^{-\lambda_i t}c_i+s\gamma_i}{e^{-\lambda_i t}(r_i+\bar{r}_i)}\\
\quad +\gamma_{I,0}\sigma_0\Big)dt - \sigma_0\gamma_I dB_o,\\
\delta_I(T) = 0.
\end{cases}
$$

Proof of Proposition 14.3 We provide a proof of Proposition 14.3 by an inductive argument. We start with Prosumer 1 and use a direct method consisting of the following guess functional for the quadratic discounted revenue functional:

$$f_1(t,p) = \frac{1}{2}\alpha_1(t)(p-\bar{p})^2 + \frac{1}{2}\beta_1(t)\bar{p}^2 + \xi_1(t)(p-\bar{p}) + \gamma_1(t)\bar{p} + \delta_1(t), \quad (14.36)$$

where $\alpha_1, \beta_1, \gamma_1, \delta_1$ and ξ_1 are random functions of time t, which are $\mathcal{F}_t^{B_o}$−measurable such that

$$f_1(T, p(T)) = \frac{q}{2}e^{-\lambda_1 T}\alpha_1(t)(p(T) - \bar{p}(T))^2.$$

Using Itô's formula, we obtain

$$\begin{aligned}
&df_1(t,p) = \tfrac{(p-\bar{p})^2 d\alpha_1}{2} + \tfrac{\alpha_1}{2}\left(\sigma^2 + \int_\Theta \mu^2(\theta)\nu(d\theta)\right)dt \\
&-(s(p-\bar{p})\alpha_1 + s\xi_1)(D - \bar{D} + p - \bar{p})dt + \tfrac{1}{2}\bar{p}^2 d\beta_1 \\
&+\tfrac{1}{2}\beta_1(2s\bar{p}(a - \bar{D} - \bar{p}) + \sigma_0^2)dt + \beta_1\bar{p}\sigma_0 dB_o + \beta_{1,0}\bar{p}\sigma_0 dt \\
&+\bar{p}d\gamma_1 + \left(s\alpha_1(a - \bar{D} - \bar{p}) + \gamma_{1,0}\sigma_0\right)dt + \xi_1\sigma dB + \sigma_0\gamma_1 dB_o + d\delta_1 \\
&+((p-\bar{p})\alpha_1 + \xi_1)(\sigma dB + \int_\Theta \mu(\theta)\tilde{N}(dt,d\theta)) + (p-\bar{p})d\xi_1.
\end{aligned} \quad (14.37)$$

We have

$$\begin{aligned}
&E[\mathcal{R}_1 - f_1(0,p(0))|\mathcal{F}_T^{B_o}] \\
&= \int_0^T e^{-\lambda_1 t}E[(\bar{p} - c_1)\bar{u}_1 - \tfrac{1}{2}r_1(u_1 - \bar{u}_1)^2 - \tfrac{1}{2}(r_1 + \bar{r}_1)\bar{u}_1^2|\mathcal{F}_T^{B_o}]dt \\
&+E\int_0^T \tfrac{(p-\bar{p})^2 d\alpha_1}{2} + \int_0^T \tfrac{\alpha_1}{2}\left(\sigma^2 + \int_\Theta \mu^2(\theta)\nu(d\theta)\right)dt \\
&-E\int_0^T (s(p-\bar{p})\alpha_1 + s\xi_1)(D - \bar{D} + p - \bar{p})dt + \tfrac{1}{2}\int_0^T \bar{p}^2 d\beta_1 \\
&+\tfrac{1}{2}\int_0^T \beta_1(2s\bar{p}(a - \bar{D} - \bar{p}) + \sigma_0^2)dt + \int_0^T \beta_1\bar{p}\sigma_0 dB_o \\
&+\int_0^T \beta_{1,0}\bar{p}\sigma_0 dt + \int_0^T \bar{p}d\gamma_1 + \int_0^T \left(s\gamma_1(a - \bar{D} - \bar{p}) + \gamma_{1,0}\sigma_0\right)dt \\
&+\int_0^T \sigma_0\gamma_1 dB_o + \int_0^T d\delta_1 + \int_0^T (p-\bar{p})d\xi_1.
\end{aligned} \quad (14.38)$$

We rearrange and complete the square for the terms

$$\begin{aligned}
&e^{-\lambda_1 t}\left(2(\bar{p}-c_1)\bar{u}_1 - r_1(u_1-\bar{u}_1)^2 - (r_1+\bar{r}_1)\bar{u}_1^2\right) - 2s((p-\bar{p})\alpha_1 + \xi_1)(D - \bar{D}) \\
&-2\beta_1 s\bar{p}\bar{D} - 2s\gamma_1\bar{D},
\end{aligned}$$

where

$$\bar{D} = \sum_{j=1}^{I}\bar{u_j}, \quad D = \sum_{j=1}^{I} u_j.$$

We have

$$\begin{aligned}
&e^{-\lambda_1 t}\left(2(\bar{p}-c_1)\bar{u}_1 - r_1(u_1-\bar{u}_1)^2 - (r_1+\bar{r}_1)\bar{u}_1^2\right) - 2s((p-\bar{p})\alpha_1 + \xi_1)(u_1 - \bar{u}_1) \\
&-2s((p-\bar{p})\alpha_1 + \xi_1)\textstyle\sum_{j=2}^{I}(u_j - \bar{u}_j) - 2\beta_1 s\bar{p}\bar{u}_1 \\
&-2s\gamma_1\bar{u}_1 - 2s\beta_1\bar{p}\textstyle\sum_{j=2}^{I}\bar{u}_j - 2s\gamma_1\textstyle\sum_{j=2}^{I}\bar{u}_j,
\end{aligned} \quad (14.39)$$

which is equal to

$$
\begin{aligned}
&-(r_1+\bar{r}_1)e^{-\lambda_1 t}\left(\bar{u}_1-\tfrac{e^{-\lambda_1 t}(\bar{p}-c_1)-\beta_1 s\bar{p}-s\gamma_1}{e^{-\lambda_1 t}(r_1+\bar{r}_1)}\right)^2-r_1 e^{-\lambda_1 t}\left(u_1-\bar{u}_1+\tfrac{s((p-\bar{p})\alpha_1+\xi_1)}{e^{-\lambda_1 t}r_1}\right)^2\\
&+\tfrac{s^2(p-\bar{p})^2\alpha_1^2+2s^2(p-\bar{p})\alpha_1\xi_1+s^2\xi_1^2}{e^{-\lambda_1 t}r_1}+\tfrac{(e^{-\lambda_1 t}(\bar{p}-c_1)-\beta_1 s\bar{p}-s\gamma_1)^2}{e^{-\lambda_1 t}(r_1+\bar{r}_1)}\\
&-2s((p-\bar{p})\alpha_1+\xi_1)\textstyle\sum_{j=2}^{I}(u_j-\bar{u}_j)-2s\beta_1\bar{p}\sum_{j=2}^{I}\bar{u}_j-2s\gamma_1\sum_{j=2}^{I}\bar{u}_j.
\end{aligned}
\tag{14.40}
$$

Then,

$$
\begin{aligned}
&E[\mathscr{R}_1-f_1(0,p(0))|\mathscr{F}_T^{B_o}]\\
&=-E\left[\int_0^T\tfrac{r_1+\bar{r}_1}{2}e^{-\lambda_1 t}\left(\bar{u}_1-\tfrac{e^{-\lambda_1 t}(\bar{p}-c_1)-\beta_1 s\bar{p}-s\gamma_1}{e^{-\lambda_1 t}(r_1+\bar{r}_1)}\right)^2 dt|\mathscr{F}_T^{B_o}\right]\\
&-E\left[\int_0^T\tfrac{r_1}{2}e^{-\lambda_1 t}\left(u_1-\bar{u}_1+\tfrac{s((p-\bar{p})\alpha_1+\xi_1)}{e^{-\lambda_1 t}r_1}\right)^2 dt|\mathscr{F}_T^{B_o}\right]\\
&+E\left[\int_0^T\tfrac{(p-\bar{p})^2}{2}\left\{d\alpha_1-(2s\alpha_1-\tfrac{s^2\alpha_1^2}{e^{-\lambda_1 t}r_1})dt\right\}\right]\\
&+\int_0^T\tfrac{(p-\bar{p})}{2}\left\{2d\xi_1-2s\left(\xi_1+\alpha_1\textstyle\sum_{j=2}^{I}(u_j-\bar{u}_j)-\tfrac{s\alpha_1\xi_1}{e^{-\lambda_1 t}r_1}\right)dt\right\}\\
&+\tfrac{1}{2}\int_0^T\bar{p}\left\{2d\gamma_1+(2\beta_1 s(a-\textstyle\sum_{j=2}^{I}\bar{u}_j)+2\sigma_0\beta_{1,0}\right.\\
&-2s\gamma_1-\tfrac{2(e^{-\lambda_1 T}-\beta_1 s)(s\gamma_1+e^{-\lambda_1 T}c_1)}{(r_1+\bar{r}_1)e^{-\lambda_1 T}})dt\\
&+2\beta_0\sigma_0 dB_o\}+\int_0^T\tfrac{\bar{p}^2}{2}\left\{d\beta_1-(2\beta_1 s-\tfrac{(e^{-\lambda_1 t}-\beta_1 s)^2}{e^{-\lambda_1 t}(r_1+\bar{r}_1)})dt\right\}\\
&+\tfrac{1}{2}\int_0^T\left\{2d\delta_1+\tfrac{(s\gamma_1+e^{-\lambda_1}c_1)^2}{e^{-\lambda_1}(r_1+\bar{r}_1)}dt\right.\\
&+\tfrac{\xi_1^2 s^2}{e^{-\lambda_1 t}r_1}dt-2s\xi_1\textstyle\sum_{j=2}^{I}(u_j-\bar{u}_j)dt+\beta_1\sigma_0^2 dt\\
&+\alpha_1(\sigma^2+\int_\Theta\mu^2(\theta)\nu(d\theta))dt+2sa\gamma_1 dt\\
&\left.-2s\gamma_1\textstyle\sum_{j=2}^{I}\bar{u}_j dt+2\gamma_{1,0}\sigma_0 dt+2\sigma_0\gamma_1 dB_o\right\}.
\end{aligned}
$$

We deduce that $E[\mathscr{R}_1-f_1(0,p(0))|\mathscr{F}_T^{B_o}]\leq 0$, with equality holding when

$$
\bar{u}_1^*=\frac{e^{-\lambda_1 t}(\bar{p}-c_1)-\beta_1 s\bar{p}-s\gamma_1}{e^{-\lambda_1 t}(r_1+\bar{r}_1)},
$$

and

$$
u_1^*=-\frac{s((p-\bar{p})\alpha_1+\xi_1)}{e^{-\lambda_1 t}r_1}+\bar{u}_1^*.
$$

Moreover,

$$\begin{cases} d\alpha_1 = \left(2s\alpha_1 - \frac{s^2\alpha_1^2}{e^{-\lambda_1 t}r_1}\right)dt + \alpha_{1,0}dB_o, \\ \alpha_1(T) = -qe^{-\lambda_1 T}, \\ d\beta_1 = \left(2\beta_1 s - \frac{(e^{-\lambda_1 t}-\beta_1 s)^2}{e^{-\lambda_1 t}(r_1+\bar{r}_1)}\right)dt + \beta_{1,0}dB_o, \\ \beta_1(T) = 0, \\ d\gamma_1 = -\left(\beta_1 s(a - \sum_{j=2}^I \bar{u}_j) + \sigma_0\beta_{1,0} - s\gamma_1 \right. \\ \quad \left. - \frac{(e^{-\lambda_1 t}-\beta_1 s)(s\gamma_1 + e^{-\lambda_1 t}c_1)}{(r_1+\bar{r}_1)e^{-\lambda_1 t}} \right)dt - \beta_1\sigma_0 dB_o, \\ \gamma_1(0) = 0, \\ d\xi_1 = s\left(\xi_1 + \alpha_1\sum_{j=2}^I(u_j - \bar{u}_j) - \frac{s\alpha_1\xi_1}{e^{-\lambda_1 t}r_1}\right)dt + \xi_{1,0}dB_o, \\ \xi_1(T) = 0, \\ d\delta_1 = -\left(\frac{1}{2}\frac{(s\gamma_1+e^{-\lambda_1}c_1)^2}{e^{-\lambda_1}(r_1+\bar{r}_1)} + \frac{1}{2}\frac{\xi_1^2 s^2}{e^{-\lambda_1 t}r_1} - s\xi_1\sum_{j=2}^I(u_j - \bar{u}_j) \right. \\ \quad +\frac{\beta_1}{2}\sigma_0^2 + \frac{\alpha_1}{2}(\sigma^2 + \int_\Theta \mu^2(\theta)\nu(d\theta)) + s\gamma_1(a - \sum_{j=2}^I \bar{u}_j) \\ \quad \left. +\gamma_{1,0}\sigma_0 \right)dt - \sigma_0\gamma_1 dB_o, \\ \alpha(T) = 0. \end{cases}$$

Similarly, we can find $u_2^*, \ldots, u_{I-1}^*$. Now we find the optimal control for u_I^*. Given $\bar{u}_2^*, \ldots, \bar{u}_{I-1}^*$ the conditional price dynamics becomes

$$d\bar{p} = s[a - \sum_{j=1}^{I-1} \frac{e^{-\lambda_j t}(\bar{p} - c_j) - \beta_j s\bar{p} - s\gamma_j}{e^{-\lambda_j t}(r_j + \bar{r}_j)} - \bar{u}_I - \bar{p}]dt + \sigma_0 dB_o, \quad \bar{p}(0) = \bar{p}_0, \tag{14.41}$$

use the guess functional

$$f_I(t, p) = \frac{1}{2}\alpha_I(t)(p-\bar{p})^2 + \frac{1}{2}\beta_I(t)\bar{p}^2 + \xi_I(t)(p-\bar{p}) + \gamma_I(t)\bar{p} + \delta_I(t), \tag{14.42}$$

where $\alpha_I, \beta_I, \gamma_I$ and δ_I and ξ_I are random functions of time t, which are $\mathscr{F}_t^{B_o}$–measurable such that

$$f_I(T, p(T)) = \frac{q}{2}e^{-\lambda_I T}\alpha_I(t)(p(T) - \bar{p}(T))^2.$$

Using Itô's formula, we obtain

$$\begin{aligned} &E[\mathscr{R}_I - f_I(0, p(0))|\mathscr{F}_T^{B_o}] = \int_0^T e^{-\lambda_I t}E[(\bar{p} - c_I)\bar{u}_I - \tfrac{1}{2}r_I(u_I - \bar{u}_I)^2 - \tfrac{1}{2}(r_I + \bar{r}_I)\bar{u}_I^2|\mathscr{F}_T^{B_o}]dt \\ &+E\int_0^T \tfrac{(p-\bar{p})^2 d\alpha_I}{2} + \int_0^T \tfrac{\alpha_I}{2}\left(\sigma^2 + \int_\Theta \mu^2(\theta)\nu(d\theta)\right)dt \\ &-E\int_0^T (s(p - \bar{p})\alpha_I + s\xi_I)(D^* - \bar{D}^* + p - \bar{p})dt + \tfrac{1}{2}\int_0^T \bar{p}^2 d\beta_I \\ &+\tfrac{1}{2}\int_0^T \beta_I(2s\bar{p}(a - \bar{D}^* - \bar{p}) + \sigma_0^2)dt + \int_0^T \beta_I\bar{p}\sigma_0 dB_o \\ &+\int_0^T \beta_{I,0}\bar{p}\sigma_0 dt + \int_0^T \bar{p}d\gamma_I + \int_0^T \left(s\gamma_I(a - \bar{D}^* - \bar{p}) + \gamma_{I,0}\sigma_0\right)dt \\ &+\int_0^T \sigma_0\gamma_I dB_o + \int_0^T d\delta_I + \int_0^T (p - \bar{p})d\xi_I, \end{aligned} \tag{14.43}$$

where

$$D^* = \sum_{j=1}^{I-1} u_j^* + u_I \ , \quad \bar{D}^* = \sum_{j=1}^{I-1} \bar{u}_j^* + \bar{u}_I.$$

We have

$$\begin{aligned}
&e^{-\lambda_I t}\ (\ 2(\bar{p}-c_I)\bar{u}_I - r_I(u_I-\bar{u}_I)^2-(r_I+\bar{r}_I)\bar{u}_I^2\ )-2s((p-\bar{p})\alpha_I+\xi_I)(u_I-\bar{u}_I)\\
&-2s((p-\bar{p})\alpha_I+\xi_I)\textstyle\sum_{j=1}^{I-1}(u_j^*-\bar{u}_j^*)-2s\beta_I\bar{p}\sum_{j=1}^{I-1}\bar{u}_j^*\\
&-2s\gamma_I\textstyle\sum_{j=1}^{I-1}\bar{u}_j^*-2s\beta_I\bar{p}\bar{u}_I-2s\gamma_I\bar{u}_I,
\end{aligned} \tag{14.44}$$

which is equal to

$$\begin{aligned}
&-(r_I+\bar{r}_I)e^{-\lambda_I t}\left(\bar{u}_I-\tfrac{e^{-\lambda_I t}(\bar{p}-c_I)-\beta_I s\bar{p}-s\gamma_I}{e^{-\lambda_I t}(r_I+\bar{r}_I)}\right)^2-r_I e^{-\lambda_I t}\left(u_I-\bar{u}_I+\tfrac{s((p-\bar{p})\alpha_I+\xi_I)}{e^{-\lambda_I t}r_I}\right)^2\\
&+\tfrac{s^2(p-\bar{p})^2\alpha_I^2+2s^2(p-\bar{p})\alpha_I\xi_I+s^2\xi_I^2}{e^{-\lambda_I t}r_I}+\tfrac{(e^{-\lambda_I t}(\bar{p}-c_I)-\beta_I s\bar{p}-s\gamma_I)^2}{e^{-\lambda_I t}(r_I+\bar{r}_I)}\\
&+2s((p-\bar{p})\alpha_I+\xi_I)\textstyle\sum_{j=1}^{I-1}\tfrac{s((p-\bar{p})\alpha_j+\xi_j)}{e^{-\lambda_j t}r_j}-2s(\beta_I\bar{p}+\gamma_I)\sum_{j=1}^{I-1}\tfrac{e^{-\lambda_j t}(\bar{p}-c_j)-\beta_j s\bar{p}-s\gamma_j}{e^{-\lambda_j t}(r_j+\bar{r}_j)}.
\end{aligned} \tag{14.45}$$

Then,

$$\begin{aligned}
&E[\mathscr{R}_I-f_I(0,p(0))|\mathscr{F}_T^{B_o}]\\
&=-E\left[\int_0^T \tfrac{r_I+\bar{r}_I}{2}e^{-\lambda_I t}\left(\bar{u}_I-\tfrac{e^{-\lambda_I t}(\bar{p}-c_I)-\beta_I s\bar{p}-s\gamma_I}{e^{-\lambda_I t}(r_I+\bar{r}_I)}\right)^2 dt|\mathscr{F}_T^{B_o}\right]\\
&-E\left[\int_0^T \tfrac{r_I}{2}e^{-\lambda_I t}\left(u_I-\bar{u}_I+\tfrac{s((p-\bar{p})\alpha_I+\xi_I)}{e^{-\lambda_I t}r_I}\right)^2 dt|\mathscr{F}_T^{B_o}\right]\\
&+E\left[\int_0^T \tfrac{(p-\bar{p})^2}{2}\left\{d\alpha_I-(2s\alpha_I-\tfrac{s^2\alpha_I^2}{e^{-\lambda_I t}r_I}-2s^2\alpha_I\textstyle\sum_{j=1}^{I-1}\tfrac{\alpha_j}{e^{-\lambda_j t}r_j})dt\right\}\right]\\
&+\int_0^T(p-\bar{p})\left\{d\xi_I-s\left(\xi_I-s\alpha_I\textstyle\sum_{j=1}^{I-1}\tfrac{\xi_j}{e^{-\lambda_j t}r_j}-s\xi_I\sum_{j=1}^{I-1}\tfrac{\alpha_j}{e^{-\lambda_j t}r_j}-\tfrac{s\alpha_I\xi_I}{e^{-\lambda_I t}r_I}\right)dt\right\}\\
&+E\left[\int_0^T\bar{p}\left\{d\gamma_I+s\beta_I a+s\beta_I\textstyle\sum_{j=1}^{I-1}\tfrac{e^{-\lambda_j t}c_j+s\gamma_j}{e^{-\lambda_j t}(r_j+\bar{r}_j)}-s\gamma_I\sum_{j=1}^{I-1}\tfrac{e^{-\lambda_j t}-s\beta_j}{e^{-\lambda_j t}(r_j+\bar{r}_j)}+\sigma_0\beta_{I,0}\right.\right.\\
&\left.\left.-s\gamma_I-\tfrac{(e^{-\lambda_I t}-\beta_I s)(s\gamma_I+e^{-\lambda_I t}c_I)}{(r_I+\bar{r}_I)e^{-\lambda_I t}})dt+\beta_I\sigma_0 dB_o\right\}\right]\\
&+E\left[\int_0^T\tfrac{\bar{p}^2}{2}\left\{d\beta_I-(2\beta_I s-\tfrac{(e^{-\lambda_I t}-\beta_I s)^2}{e^{-\lambda_I t}(r_I+\bar{r}_I)}\right.\right.\\
&\left.\left.+2s\beta_I\textstyle\sum_{j=1}^{I-1}\tfrac{(e^{-\lambda_j t}-\beta_j s)}{e^{-\lambda_j t}(r_j+\bar{r}_j)})dt\right\}\right]\\
&+\int_0^T\left\{d\delta_I+\tfrac{1}{2}\tfrac{(s\gamma_I+e^{-\lambda_I}c_2)^2}{e^{-\lambda_I}(r_I+\bar{r}_I)}dt+s^2\xi_I\textstyle\sum_{j=1}^{I-1}\tfrac{\xi_j}{e^{-\lambda_j t}r_j}dt+\tfrac{1}{2}\tfrac{s^2\xi_I^2}{e^{-\lambda_I}r_I}dt\right.\\
&+\tfrac{1}{2}\beta_I\sigma_0^2dt+\tfrac{\alpha_I}{2}(\sigma^2+\textstyle\int_\Theta\mu^2(\theta)\nu(d\theta))dt+s\gamma_I a+s\gamma_I\sum_{j=1}^{I-1}\tfrac{e^{-\lambda_j t}c_j+s\gamma_j}{e^{-\lambda_j t}(r_j+\bar{r}_i)}dt+\gamma_{I,0}\sigma_0 dt\\
&\left.+\sigma_0\gamma_I dB_o\ \right\}
\end{aligned}$$

We deduce that $E[\mathscr{R}_I-f_I(0,p(0))|\mathscr{F}_T^{B_o}]\leq 0$, with equality holding when

$$\bar{u}_I^* = \frac{e^{-\lambda_I t}(\bar{p} - c_I) - \beta_I s \bar{p} - s\gamma_I}{e^{-\lambda_I t}(r_I + \bar{r}_I)},$$

and

$$u_I^* = -\frac{s((p - \bar{p})\alpha_I + \xi_I)}{e^{-\lambda_I t} r_I} + \bar{u}_I^*.$$

Moreover,

$$\begin{cases} d\alpha_I = \left(2s\alpha_I - \frac{s^2\alpha_I^2}{e^{-\lambda_I t} r_I} - 2s^2\alpha_I \sum_{j=1}^{I-1} \frac{\alpha_j}{e^{-\lambda_j t} r_j}\right) dt + \alpha_{I,0} dB_o, \\ \alpha_I(T) = -qe^{-\lambda_2 T}, \\ d\beta_I = \left(2\beta_I s - \frac{(e^{-\lambda_I t} - \beta_I s)^2}{e^{-\lambda_I t}(r_I + \bar{r}_I)} + 2s\beta_I \sum_{j=1}^{I-1} \frac{(e^{-\lambda_j t} - \beta_j s)}{e^{-\lambda_j t}(r_j + \bar{r}_j)}\right) dt + \beta_{I,0} dB_o, \\ \beta_I(T) = 0, \\ d\gamma_I = -\left(\beta_I s(a + \sum_{j=1}^{I-1} \frac{e^{-\lambda_j t} c_i + s\gamma_j}{e^{-\lambda_j t}(r_j + \bar{r}_j)}) - s\gamma_I \sum_{j=1}^{I-1} \frac{(e^{-\lambda_j t} - s\beta_j)}{e^{-\lambda_j t}(r_j + \bar{r}_j)}\right. \\ \qquad \left. + \sigma_0 \beta_{I,0} - s\gamma_I - \frac{(e^{-\lambda_I t} - \beta_I s)(s\gamma_I + e^{-\lambda_I t} c_I)}{(r_I + \bar{r}_I) e^{-\lambda_I t}}\right) dt - \beta_I \sigma_0 dB_o, \\ \gamma_I(0) = 0, \\ d\xi_I = s\left(\xi_I - s\alpha_I \sum_{j=1}^{I-1} \frac{\xi_j}{e^{-\lambda_j t} r_j} - s\xi_I \sum_{j=1}^{I-1} \frac{\alpha_j}{e^{-\lambda_j t} r_j} - \frac{s\alpha_I \xi_I}{e^{-\lambda_I t} r_I}\right) dt + \xi_{I,0} dB_o, \\ \xi_I(T) = 0, \\ d\delta_I = -\left(\frac{1}{2} \frac{(s\gamma_I + e^{-\lambda_I} c_2)^2}{e^{-\lambda_I}(r_I + \bar{r}_I)} + s^2 \xi_I \sum_{j=1}^{I-1} \frac{\xi_j}{e^{-\lambda_j t} r_i} + \frac{1}{2} \frac{s^2 \xi_I^2}{e^{-\lambda_I} r_I} + \frac{1}{2} \beta_I \sigma_0^2\right. \\ \qquad + \frac{\alpha_I}{2}(\sigma^2 + \int_\Theta \mu^2(\theta) \nu(d\theta)) + s\gamma_I a + s\gamma_I \sum_{j=1}^{I-1} \frac{e^{-\lambda_j t} c_j + s\gamma_j}{e^{-\lambda_j t}(r_j + \bar{r}_j)} \\ \qquad \left. + \gamma_{I,0}\sigma_0 \right) dt - \sigma_0 \gamma_I dB_o, \\ \delta_I(T) = 0. \end{cases}$$

This completes the proof. ■

Sequential Mean-Field-Type Equilibrium in Feedback Form

Proposition 14.4 *The optimal control strategies for the agents in feedback form are given by*

$$u_i^* = -\frac{s((p - \bar{p})\alpha_i + \xi_i)}{e^{-\lambda_i t} r_i} + \frac{e^{-\lambda_i t}(\bar{p} - c_i) - \beta_i s \bar{p} - s\gamma_i}{e^{-\lambda_i t}(r_i + \bar{r}_i)}.$$

The conditional optimal price:

$$\begin{aligned} d\bar{p} &= s\left[a - \sum_{i=1}^{I} \left(\frac{e^{-\lambda_i t}(\bar{p} - c_i) - \beta_i s \bar{p} - s\gamma_i}{e^{-\lambda_i t}(r_i + \bar{r}_i)}\right) - \bar{p}\right] dt + \sigma_0 dB_o, \\ \bar{p}(0) &= \bar{p}_0. \end{aligned} \tag{14.46}$$

The stochastic functions $\alpha_i, \beta_i, \gamma_i, \delta_i$ and ξ_i are $\mathscr{F}_t^{B_o}$−measurable and solve the following Riccati system: for $2 \leq i \leq I-1$

$$\begin{cases} d\alpha_i = \left(2s\alpha_i - \frac{s^2\alpha_i^2}{e^{-\lambda_i t}r_i} - 2s^2\alpha_i \sum_{j\neq i} \frac{\alpha_j}{e^{-\lambda_j t}r_j}\right) dt + \alpha_{i,0} dB_o, \\ \alpha_i(T) = -qe^{-\lambda_i T}, \\ d\beta_i = \left(2\beta_i s - \frac{(e^{-\lambda_i t}-\beta_i s)^2}{e^{-\lambda_i t}(r_i+\bar{r}_i)} + 2s\beta_i \sum_{j\neq i} \frac{(e^{-\lambda_j t}-\beta_j s)}{e^{-\lambda_j t}(r_j+\bar{r}_j)}\right) dt + \beta_{i,0} dB_o, \\ \beta_i(T) = 0, \\ d\gamma_i = -\left(\beta_i s(a + \sum_{j\neq i} \frac{e^{-\lambda_j t}c_i+s\gamma_j}{e^{-\lambda_j t}(r_j+\bar{r}_j)}) - s\gamma_i \sum_{j\neq i} \frac{(e^{-\lambda_j t}-s\beta_j)}{e^{-\lambda_j t}(r_j+\bar{r}_j)} \right. \\ \qquad \left. + \sigma_0\beta_{i,0} - s\gamma_i - \frac{(e^{-\lambda_i t}-\beta_i s)(s\gamma_i+e^{-\lambda_i t}c_i)}{(r_i+\bar{r}_i)e^{-\lambda_i t}}\right) dt - \beta_i\sigma_0 dB_o, \\ \gamma_i(0) = 0, \\ d\xi_i = s\left(\xi_i - s\alpha_i \sum_{j\neq i} \frac{\xi_j}{e^{-\lambda_j t}r_j} - s\xi_i \sum_{j\neq i} \frac{\alpha_j}{e^{-\lambda_j t}r_j} - \frac{s\alpha_i\xi_i}{e^{-\lambda_i t}r_i}\right) dt + \xi_{i,0} dB_o, \\ \xi_i(T) = 0, \\ d\delta_i = -\left(\frac{1}{2}\frac{(s\gamma_i+e^{-\lambda_i}c_2)^2}{e^{-\lambda_i}(r_i+\bar{r}_i)} + s^2\xi_i \sum_{j\neq i} \frac{\xi_j}{e^{-\lambda_j t}r_i} + \frac{1}{2}\frac{s^2\xi_i^2}{e^{-\lambda_i}r_i} + \frac{1}{2}\beta_i\sigma_0^2 \right. \\ \qquad + \frac{\alpha_i}{2}(\sigma^2 + \int_\Theta \mu^2(\theta)\nu(d\theta)) + s\gamma_i a + s\gamma_i \sum_{j\neq i} \frac{e^{-\lambda_j t}c_j+s\gamma_j}{e^{-\lambda_j t}(r_j+\bar{r}_j)} \\ \qquad \left. + \gamma_{i,0}\sigma_0 \right) dt - \sigma_0\gamma_i dB_o, \\ \delta_i(T) = 0. \end{cases}$$

Note that the process ξ is identical to the zero process in this case. The process α, β, γ can be made deterministic because their terminal value is deterministic (and hence independent of the process B_o). Hence, one can replace the δ system by its deterministic counterpart by taking the expected values.

14.5 Notes

In this chapter, we have examined a price formation in smart energy systems using a price dynamics model introduced by Roos in 1925. We have introduced a common noise in the environment. Since the common noise is observed, producers can condition on it and exploit that information. The new conditioned price affects the revenues of the producers. We have provided explicit solution to the master system of the corresponding mean-field-type game with jump-diffusion and common noise. In practice, it is often not desirable to change continuously the price. Thus, the "virtual" price adjustment model will be used as a learning feature to help the decision-maker

(continued)

before acting on the “real" price for the next time-block. For example, if the price is decided on a 3-month cycle, then this base model can be used to determine the electricity market trends and tendencies. If the price is almost in real time, this simple adjustment technique can also be used with higher speed s. More details on state-and-mean-field-type feedback equilibria can be found in [131, 147, 171, 460, 463, 478].

Chapter 15
Blockchain-Based Distributed Power Networks

This chapter examines mean-field-type games in blockchain-based distributed power networks with several different entities: investors, consumers, prosumers, producers, and miners. Under a simple model of jump diffusion and regime-switching processes, we identify risk-aware mean-field-type optimal strategies for the decision-makers.

The cryptocurrency system consists in a peer-to-peer electronic payment platform in which the transactions are made without the need for a centralized entity in charge of authorizing them. Therefore, the aforementioned transactions are validated/verified by means of a coded scheme called *blockchain*. In addition, the blockchain is maintained by its participants, which are called *miners*.

Blockchain or distributed ledger technology is an emerging technology for peer-to-peer transaction platforms that use decentralized storage to record all transaction data. One of the first blockchain applications was developed in the e-commerce sector to serve as the basis for the cryptocurrency "bitcoin." Since then, several other altcoins and cryptocurrencies including Ethereum, Litecoin, Dash, Ripple, SolarCoin, BitShares, etc. have been widely adopted and are all based on blockchain. More and more new applications have recently been emerging that add to the technology's core functionality—decentralized storage of transaction data—by integrating mechanisms that allow for the actual transactions to be implemented on a decentralized basis. The lack of a centralized entity, which could have control over the security of transactions, requires the development of a sophisticated verification procedure to validate transactions. Such task is known as proof-of-work, which brings new technological and algorithmic challenges.

The possibility of implementing distributed power networks on the blockchain and its pros and contras are presented. The core model (Fig. 15.1) uses a Bayesian mean-field-type game theory on the blockchain. The base interaction model considers producers, consumers, and a new important element of distributed power networks called prosumers. A prosumer (producer–consumer) is a user that not only consumes electricity but can also produce and store electricity. We identify and

T. Başar et al., *Mean-Field-Type Game Theory II*, Static & Dynamic Game Theory: Foundations & Applications, https://doi.org/10.1007/978-3-032-07031-9_15

formulate the key interactions between consumers, prosumers, and producers on the blockchain. Based on forecasted demand generated from the blockchain, each producer determines its production quantity and its mismatch cost and engages an auction mechanism to the prosumer market on the blockchain. The resulting supply is completed by the prosumers' auction market. This determines a market price, and the consumers react to the offers and the price and generate a certain demand. The consistency relationship between demand and supply provides a fixed-point system, whose solution is a mean-field-type equilibrium.

The rest of the chapter is organized as follows. The next subsection presents the emergence of decentralized platform. Section 15.2 focuses on the game model. Section 15.3 examines mean-field-type game with complete information. Incomplete information scenarios are presented in Sect. 15.4. Section 15.5 presents risk-awareness and price stability analysis. Section 15.6 focuses on consumption–insurance and investment trade-offs.

15.1 Toward a Decentralized Platform

What Is a Ledger?
A ledger in computer science is a software that stores transactions. A database is different from a ledger such that in a database, we can add, remove, and modify records, whereas in a ledger, we can only append but not delete or modify.

What Is a Blockchain?
A blockchain is a chain of blocks connected to each other. Every block contains a list of transactions and certain other meta-data, such as when it was created, which is its previous block, the block number, who is the creator of the block, and so on. Every block maintains a hash of the previous block, therefore creating a chain of blocks linked with each other. Every node in the network should hold the complete copy of the blockchain and, when a new node comes in, it will request and download the blockchain from other nodes. A blockchain contains a data structure that implements a decentralized ledger. A distributed ledger technology is a peer-to-peer transaction platform that integrates mechanisms that allow decentralized transactions or decentralized and distributed exchange system. These mechanisms, called "smart contracts," operate on the basis of individually defined rules (e.g., specifications as to quantity, quality, price, and location) that enable an autonomous matching of distributed producers and their prospective customers.

What Is Forking in Blockchain?
A fork is said to have happened when there is a conflict among the nodes regarding the validity of the blockchain, that is, more than one blockchain happens to be in the network. There are three kinds of fork: regular, soft, and hard.

A regular fork is said to be happening when there are two or more blocks at the same height. It is a temporary conflict and is resolved automatically. This is resolved by nodes by selecting the most accurate blockchain. For example, in proof-of-work, if two miners mine a block at the same time, then it creates a regular fork. And this is resolved by selecting the blockchain with the highest difficulty as the most accurate one.

A soft fork, by contrast, is any change to the blockchain protocol that's backward compatible. Say, instead of 10 MB blocks, a new rule might only allow 3 MB blocks. Non-upgraded nodes will still see the new transactions as valid (3 MB is less than 10 MB in this example). However, if non-upgraded nodes continue to create blocks, the blocks they create will be rejected by the upgraded nodes. So, if the minority of a nodes in the network are upgraded, then the chain they will form will become less accurate and overridden by the blockchain created by the non-upgraded nodes. Soft forks are resolved when the majority of the nodes in the network upgrade their node software.

A hard is a software upgrade that introduces a new rule to the network that isn't with the older software. You can think of a hard fork as an expansion of the rules. For example, a new rule that allows the block size to be 5 MB instead of 2 MB would require a hard fork. Nodes that continue running the old version of the software will see new transactions as invalid. So, the fork can only be resolved when all the nodes in the network upgrade their node software. Until then, there will be two different blockchains in the network.

> The synchronization and group-consensus problem can be seen as a majority game between nodes.

Blockchain Pros

Some of the advantages of using a blockchain are the following:

- It enables the facilitation of transactions without a central trusted party.
- Data (inside the blockchain) is secured using cryptography.
- Data is immutable, as blockchain removes friction and reduces risk, so settlements happen in real time.

Basically, it automates auditing, makes the application transparent, and provides a single source of verifiable information.

Blockchain Contras

Some of the challenges of blockchains are the following:

- Humans-in-the-loop security: Blockchain has currently humans-in-the-loop. Endpoints or connecting points or last mile are the points where humans and the blockchain meet. Tablets, smartphones, and computers are the devices that users use to access blockchain networks and services. There is an input from these computers and it goes to other computers as an output. The links for accessing the blockchain are among the weakest links of the network and therefore they are vulnerable. This is a sort of limitation in user security issue, which affects the entire blockchain security.

- Complexity, scalability, robustness, fraud, and security: Public and private keys are essential to access the blockchain. Guessing the keys is difficult. However, hackers spent time in dealing with keys from the personal computer of the users. There is a need of adequate protection of these devices. Currently, there is no standard protocol. Without protocol, developers cannot easily benefit from the mistakes of others as the network size grows in the decentralized autonomous organization.
- Transaction cost and speed.
- Consensus, fork, and majority rule: Forking a blockchain, a process that involves updating the blockchain protocol when a majority of a blockchain users have agreed to it. To reach a majority, the members need to repeat a task by burning electricity and time on the way. This process could be costly. The computation is far slower and more expensive than on a traditional single computer. There are many initiatives that seek to reduce this cost focusing on alternative means of maintaining consensus, such as proof-of-stake.

Technically, the evolution of the number of blocks in the blockchain can be modeled using a multidimensional Markov chain. Recently, the energy sector is also moving toward a distributed and semi-decentralized platform with the integration of prosumers' market and aggregators to the power grid. Distributed power is a power generated at or near the point of use. This includes technologies that supply both electric power and mechanical power. In electrical applications, distributed power systems stand in contrast to central power stations that supply electricity from a centralized location, often far from users. The rise of distributed power is being driven by broader decentralization movement of smarter cities. With blockchain transaction, every participant in a network can transact directly with every other network participant without involving a third-party intermediary (aggregator, operator). In other words, aggregators and the third parties are replaced by the blockchain. All transaction data is stored on a blockchain, with all relevant information being stored identically on the computers of all participants, and all transactions are made on the basis of smart contracts, i.e., based on predefined individual rules concerning quality, price, quantity, location, feasibility, etc.

15.1.1 A Blockchain for Underserved Areas

One of the first questions that rises in blockchain is the service to the society. An authentication service offering to make environment-friendly (solar/wind/hydro) energy certificates available via a blockchain. The new service works by connecting solar panels and wind farms to an Internet of Things (IoT)-enabled device that measures the quality (of the infrastructure), quantity, and location of the power produced and fed into the grid. Certificates supporting PV growth and wind power can be bought and sold anonymously via a blockchain platform. Then, solar and wind energy produced by prosumers in underserved areas can be transmitted to end-users. SolarCoin was developed following that idea, with blockchain technology to generate an additional reward for solar electricity producers. Solar installation owners registering to the SolarCoin network receive one SolarCoin for each MWh of solar electricity that they produce. This digital asset will allow solar electricity producers to receive an additional reward for their contribution to the energy transition, which will develop itself through network effect. SolarCoin is freely distributed to any owner of a solar installation owner. Participating in the SolarCoin program can be done online, directly on the SolarCoin website. As of October 2017, more than 2,134,893 MWh of solar energy has been incentivized through SolarCoin across 44 countries. The ElectriCChain aims to provide the bulk of blockchain recording for the solar installation owners in order to micro-finance the solar installation, incentivize it (through the SolarCoin tool), and monitor the install production. The idea of Wattcoin is to build this scheme for other renewable energies such as wind, thermo, and hydropower plants to incentivize global electricity generation from several renewable energy sources. The incentive scheme influences the prosumers' decision because they will be rewarded in Wattcoins as an additional incentive to initiate the energy transition and possibly to compensate a fraction of the peak hours energy demand.

15.1.2 Security, Energy Theft, and Regulation Issues

If fully adopted, blockchain-based distributed power network (b-DIPONET) is not without challenge. One of the challenges is security. This includes not only network security but also robustness, double spending, and false/fake accounts. Stokens are regulated securities tokens built on the blockchain using smart contracts. They provide a way for accredited investors to interact with regulated companies through a digital ecosystem. Currently, the cryptoeconomics industry has enormous potential—but it needs to be accompanied properly.

The blockchain technology can be used to reduce energy theft and unpaid bills by means of the automation of the prosumers who are connected to the power grid, and their produced energy data is monitored in the network.

15.2 Mean-Field-Type Game

This section presents the base mean-field-type game model. We identify and formulate the key interactions between consumers, prosumers, and producers (see Fig. 15.1). Based on the forecasted demand from the blockchain-based history

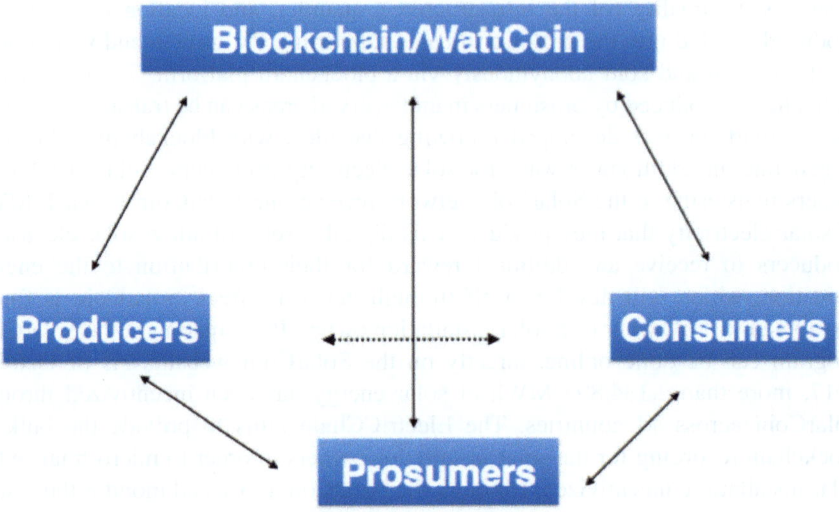

Fig. 15.1 Interaction blocks for blockchain-based distributed power networks

matching, each prosumer determines its production quantity and its mismatch cost and uses the blockchain to respond directly to consumers. All the energy producers together are engaged in a competitive energy market share. The resulting supply is completed by the prosumers' energy market. This determines a market price, and the consumers react to the price and generate a demand. The consistency relationship between demand and supply of the three components provides a fixed-point system, whose solution is a mean-field equilibrium [106, 457, 458, 460, 463].

The Game Setup

Consumer i can decide to install a solar panel on her roof or a wind power station. Depending on sunlight or wind, speed consumer i may produce surplus energy. She is no longer just an energy consumer but a prosumer. *A prosumer can decide to participate or not to the blockchain.* If the prosumer decides to participate to the blockchain to sell her surplus energy, the energy produced by this prosumer is measured by a dedicated meter, which is connected and linked to the blockchain. The measurement and the validation are done ex post from the quality-of-experience of the consumers of prosumer i. The characteristics and the bidding price of the energy produced by the prosumer are registered in the blockchain. This allows to give a certain score or Wattcoin to that prosumer for incentivization and participation level. This data is public if in the public blockchain's distributed register. All the transactions are verified and validated by the users of the blockchain ex post. If the energy transaction does not happen in the blockchain platform, the proof-of-validation is simply an ex post quality-experience measurement, and therefore it does not need to use the heavy proof-of-work used by some cryptocurrencies. The adoption of energy transactions to be blockchain requires a significant reduction of the energy consumption of the proof-of-work itself. If the proof-of-work is energy consuming (and costly), then the energy transactions are kept to the traditional channel and only proof-of-validation is used as a recommendation system to monitor and to incentivize the prosumers. The blockchain technology makes it public and more transparent. If j and k are neighbors of the location of where i produced the energy, j and k can buy electricity off him and the consumption needs recorded in the blockchain ex post. The transactions need to be technically secure and automated. Once prosumer i reaches a total of 1 MWh of energy sold to its neighbors, consumer i gets an equivalent of a certain unit of blockchain crypto-tokens such as Wattcoin, WindCoin, and SolarCoin. It is an extra reward to the revenue of the prosumer. This scheme incentivizes prosumers to participate and promotes environment-friendly energy. Instead of a digitally mined product (transaction), the WattCoin proof-of-validity happens in the physical world, and those who have wind/thermo/photovoltaic arrays can earn Wattcoin just for generating electricity and serving it successfully. It is essentially a global rewarding/loyalty program and is designed to help and incentivize more renewable electricity production while also serving as a lower-carbon cryptocurrency than bitcoin and similar alternative currencies.

Each entity can decide on the following:

- Purchase and supply energy and have automated and verifiable proof of the amounts of green energy purchased/supplied via the information stored on the blockchain.
- Ensure that local generation (and feasibility) is supported, as it becomes possible to track the exact geographical origin of each energy MWh produced. For example, it becomes possible to pay additional premiums for green energy if it is generated locally, to promote further local energy generation capacity. Since the incentive reward is received only ex post by the prosumer after checking the quality-of-experience, the proof-of-validity will improve the feasibility status of the energy supply and demand.
- Spatial energy price (price field) is publicly available to the consumers and prosumers who would like to purchase. This includes production cost and migration/distribution fee for moving energy from its point of production to its point of use.
- Each producer can supply energy on the platform and make smart contract for the delivery.
- Miners can decide to mine environment-friendly energy blocks. Honest miners are entities or people who validate the proof-of-work or proof-of-stakes (or other schemes). This can be an individual, a pool, or a coalition. There should be an incentive for them to mine. Selfish miners are those who may aim to pool their effort to maximize their own interest. This can be an individual, a pool, or a coalition. Deviators or malicious miners are entities or people who buy tokens for market and vote to impose their version of blockchain (different assigns at different blocks).

The game is described by the following four key elements:

- Platform: A blockchain.
- Players: Investors, consumers, prosumers, producers, and miners.
- Decisions: Each player can decide and act via the blockchain.
- Outcomes: The outcome is given by gain minus loss for each participant.

Note that in this model, there is no energy trading option on the blockchain. However, the model can be modified to include trading at some part of the private blockchain. The electricity price dynamics regulation and stability will be discussed below.

How Can Blockchain Improve the Penetration Rate of Renewable Energy?
Thanks to the blockchain-based incentive, a non-negligible portion of prosumers will participate in the program. This will give an opportunity to increase the produced renewable energy volumes. A basic rewarding scheme that is simple and easy to implement is a Tullock-like scheme, where probabilities to win a winner-take-all contest are considered, defining some *contest success functions*.

The incentive consists of taking a spatial rewarding scheme for the prosumers and contributors if a certain number of criteria are satisfied. As a concrete example of incentive, a prosumer producing energy from location x will be rewarded ex post $R(x)$ with probability $\frac{h_j(x,a_j)}{\sum_{i=1}^{I} h_i(x,a_i)}$ if $\sum_{i=1}^{I} h_i(x, a_i) > R(x) > 0$, where h_i is nondecreasing in its second component.

Clearly, with this incentive scheme, a non-negligible portion of producers can reinvest more funds in the renewable energy production.

Implementation Cost
We identify basic costs for the blockchain-based energy system that need to be implemented properly with the largest coverage. As the next-generation wireless communication and Internet-of-everything is moving toward advanced devices with high speed, better connectivity, and improved security and reliability than the previous version, blockchain technology should take advantage of it to enable decentralized operation. The wireless communication devices can be used as hotspots to connect to the blockchain as mobile calls are using wireless access points and hotspots as relays. Thus, a large coverage of the technology is related to the wireless coverage and connectivity of the location. Thus, the cost is reflected to the consumers and to the producers from their Internet subscription fees. In addition to that cost, miner operations consume energy and powers. The cost of supercomputers (CPUs, GPUs) and operating machines should be added too.

Demand–Supply Mismatch Cost

Let $\mathcal{T} := [t_0, t_1]$ be the time horizon with $t_0 < t_1$. In the presence of blockchain, prosumers aim to anticipate their production strategies by solving the following problem:

$$\begin{cases} L(s, e, \mathcal{T}) = l_{t_1}(e(t_1)) + \int_{t_0}^{t_1} l(t, D(t) - S(t))\, dt, \\ \inf_s EL(s, e, \mathcal{T}) \\ \frac{d}{dt} e_{jk}(t) = x_{jk}(t)\mathbb{1}_{\{k\in A_j(t)\}} - s_{jk}(t), \\ I \geq 1,\ j \in \{1, \ldots, I\},\ k \in \{1, \ldots, K_j\},\ K_j \geq 1, \\ x_{jk}(t) \geq 0,\ s_{jk}(t) \in [0, \hat{S}_{jk}],\ \forall j, k, t,\ \hat{S}_{jk} \geq 0, \\ e_{jk}(t_0) \text{ given,} \end{cases} \tag{15.1}$$

where:

- The instant loss is $l(t, D(t) - S(t)) and\ l_{t_1}$ is the terminal loss function.
- The energy supply at time t is

$$S(t) = \sum_{j=1}^{I}\sum_{k=1}^{K_j} s_{jk}(t),$$

  $s_{jk}(t)$ is the production rate of power plant/generator k of prosumer j at time $t, and\ \hat{S}_{jk}$ is an upper bound for s_{jk}, which will be used as a control action.
- The stock of energy $e_{jk}(t)$ of prosumer j at power plant k at time t is given by the following classical motion dynamics:

$$\frac{d}{dt} e_{jk}(t) = \text{incoming flow}_{jk}(t) - \text{outgoing flow}_{jk}(t). \tag{15.2}$$

  The incoming flow happens only when the power station is active. In that case, the arrival rate is $x_{jk}(t)\mathbb{1}_{\{k\in A_j(t)\}}$ where $x_{jk}(t) \geq 0$, and the set of active power plant of j is defined by $A_j(t), and$ the set of all active power plants is $A(t) = \cup_j A_j(t)$. $D(t)$ is the demand on the blockchain at time t.

In general, the demand needs to be anticipated/estimated/predicted so that the produced quantity is enough to serve the consumers. If the supply S is below D, some of the consumers will not be served; hence, it is costly for the operator. If the supply S is above D, then the operator needs to store the excess amount of energy. It will be lost if the storage is enough. Thus, it is costly in both cases, and the cost is represented by $l(\cdot, D - S)$. The demand–supply mismatch cost is determined by solving (15.1).

15.3 Oligopoly with Complete Information

There are $I \geq 2$ potential interacting energy producers over the horizon $\mathcal{T}$. At time $t \in \mathcal{T}$, producer i's output is $u_i(t) \geq 0$. The log-price dynamics is given by $p(t_0) = p_0$ and

$$dp(t) = \eta[a - S(t) - p(t)]dt + \left(\sigma dB(t) + \int_{\theta\in\Theta} \mu(\theta)\tilde{N}(dt, d\theta)\right) + \sigma_o dB_o(t), \tag{15.3}$$

where

$$S(t) := \sum_{i=1}^{I} u_i(t),$$

is the supply at time $t \in \mathcal{T}$ and B_o is a standard Brownian motion representing a global uncertainty observed by all participant to the market. The processes B and N describe local uncertainties or noises. B is a standard Brownian motion, and N is a jump process with Lévy measure $\nu(d\theta)$ defined over Θ. It is assumed that ν is a Radon measure over Θ (the jump space), which is subset of $\mathbb{R}^m$. The process

$$\tilde{N}(dt, d\theta) = N(dt, d\theta) - \nu(d\theta)dt$$

is the compensated martingale. We assume that all these processes are mutually independent. Denote by $\mathcal{F}_t^{B,N,B_o}$ the natural filtration generated by the union of events $\{B, N, B_o\}$ up to time t, and by $\mathcal{F}_t^{B_o}$ the natural filtration generated by the observed common noise up to time t. The number η is positive. For larger values of the real number η, the market price adjusts quicker along the inverse demand, all in the logarithmic scale. The terms a, σ, σ_o are fixed constant parameters. The jump rate size $\mu(\cdot)$ is in $L^2_\nu(\Theta, \mathbb{R})$, i.e.,

$$\int_\Theta \mu^2(\theta)\nu(d\theta) < +\infty.$$

The initial distribution of $p(0)$ is square integrable: $E[p_0^2] < \infty$.

At time $t \in \mathcal{T}$, producer i receives $\hat{p}(t)u_i - C_i(u_i)$ where $C_i : \mathbb{R} \to \mathbb{R}$, given by

$$C_i(u_i) = c_i u_i + \frac{1}{2} r_i u_i^2 + \frac{1}{2}\bar{r}_i \hat{u}_i^2,$$

is the instant cost function of i. The term $\hat{u}_i = E[u_i \mid \mathcal{F}_t^{B_o}]$ is the conditional expectation of producer i's output given the global uncertainty B_o observed in the market. The last term $\frac{1}{2}\bar{r}_i\hat{u}_i^2$, in the expression of the instant cost C_i, aims to capture the risk sensitivity of producer i. The conditional expectation of the price given the global uncertainty B_o up to time t is $\hat{p}(t) = E[p(t) \mid \mathscr{F}_t^{B_o}]$. At the terminal time t_1, the revenue is $-\frac{q}{2}e^{-\lambda_i t_1}\left(p(t_1) - \hat{p}(t_1)\right)^2$. The long-term revenue of producer i is

$$R_{i,\mathscr{T}}(p_0, u) = -\frac{q}{2}e^{-\lambda_i t_1}\left(p(t_1) - \hat{p}(t_1)\right)^2 + \int_{t_0}^{t_1} e^{-\lambda_i t}\left[\hat{p}u_i - C_i(u_i)\right]dt, \tag{15.4}$$

where λ_i is a discount factor of producer i. Finally, each producer optimizes her long-term expected revenue:

- This price dynamics model is interesting because it can be reinterpreted as an error to the standard inverse demand model.
- The changing η allow us to navigate between several regimes.
- The jump term $\tilde{N}$ captures some of the big changes in the market that may happen randomly.
- The global uncertainty B_o captures common noise in the market, for example, weather conditions and temperature field in specific season.
- Based on the common noise that is observed, a conditional price is calculated. The revenue is computed from the conditional price.
- Each of the I firms can influence the price and cannot be neglected.

Since the conditional price appears in the cost function, we would like to derive a simple equation for $\hat{p}$ as well. From the price dynamics, we deduce that the conditional log-price solves the following stochastic differential equation:

$$\begin{aligned} d\hat{p} &= \eta\left[a - \hat{S} - \hat{p}\right]dt + \sigma_o dB_o, \\ \hat{p}(t_0) &= \hat{p}_0, \end{aligned} \tag{15.5}$$

where $\hat{S} = \sum_i \hat{u}_i$ is the conditional expectation of the supply. Thus, the mean-field term $\hat{p}$ is a stochastic process [106] driven by the common noise B_o. For given $\bar{D}$, this is an OU (Ornstein–Uhlenbeck) process, stabilizing around the trajectory of $a - S$ as time t_1 gets larger.

Proposition 15.1 *Generically, the problem has the following interior solution:*

For each i, the equilibrium strategy is in state-and-conditional mean-field feedback form:

$$u_i^* = -\frac{\eta\tilde{\alpha}_i}{r_i}(p - \hat{p}) + \frac{\hat{p}(1 - \eta\tilde{\beta}_i) - (c_i + \eta\tilde{\gamma}_i)}{r_i + \bar{r}_i}, \tag{15.6}$$

where the conditional equilibrium log-price:

$$\begin{cases} d\hat{p} = \eta\left\{a + \sum_{j=1}^{I}\dfrac{c_j + \eta\tilde{\gamma}_j}{r_j + \bar{r}_j} - \hat{p}\left(1 + \sum_{j=1}^{I}\dfrac{1 - \eta\tilde{\beta}_j}{r_j + \bar{r}_j}\right)\right\}dt + \sigma_o dB_o, \\ \hat{p}(t_0) = \hat{p}_0, \end{cases} \tag{15.7}$$

and the functions $\tilde{\alpha}, \tilde{\beta}, \tilde{\gamma}, \tilde{\delta}_i$ *solve the Riccati system:*

$$\begin{cases} d\tilde{\alpha}_i = \left\{ (\lambda_i + 2\eta)\tilde{\alpha}_i - \frac{\eta^2}{r_i}\tilde{\alpha}_i^2 - 2\eta^2\tilde{\alpha}_i \sum_{j\neq i} \frac{\tilde{\alpha}_j}{r_j} \right\} dt + \tilde{\alpha}_{i,o} dB_o, \\ \tilde{\alpha}_i(t_1) = -q, \\ d\tilde{\beta}_i = \left\{ (\lambda_i + 2\eta)\tilde{\beta}_i - \frac{(1-\eta\tilde{\beta}_i)^2}{r_i + \bar{r}_i} + 2\eta\tilde{\beta}_i \sum_{j\neq i} \frac{1-\eta\tilde{\beta}_j}{r_j + \bar{r}_j} \right\} dt + \tilde{\beta}_{i,o} dB_o, \\ \tilde{\beta}_i(t_1) = 0, \\ d\tilde{\gamma}_i = \left\{ (\lambda_i + \eta)\tilde{\gamma}_i - \eta\tilde{\beta}_i a - \tilde{\beta}_{i,o}\sigma_o + \frac{(1-\eta\tilde{\beta}_i)(c_i + \eta\tilde{\gamma}_i)}{r_i + \bar{r}_i} + \eta\tilde{\gamma}_i \sum_{j\neq i} \frac{1-\eta\tilde{\beta}_j}{r_j + \bar{r}_j} \right. \\ \qquad \left. -\eta\tilde{\beta}_i \sum_{j\neq i} \frac{c_j + \eta\tilde{\gamma}_j}{r_j + \bar{r}_j} \right\} dt + \tilde{\gamma}_{i,o} dB_o, \\ \tilde{\gamma}_i(0) = 0, \\ d\tilde{\delta}_i = -E\left\{ -\lambda_i\tilde{\delta}_i + \frac{1}{2}\sigma_o^2\tilde{\beta}_i + \frac{1}{2}\tilde{\alpha}_i \left(\sigma^2 + \int_{\Theta} \mu^2(\theta)\nu(d\theta) \right) + \eta\tilde{\gamma}_i a \right. \\ \qquad \left. + \tilde{\gamma}_{i,o}\sigma_o + \frac{1}{2}\frac{(c_i + \eta\tilde{\gamma}_i)^2}{r_i + \bar{r}_i} + \eta\tilde{\gamma}_i \sum_{j\neq i} \frac{c_j + \eta\tilde{\gamma}_j}{r_j + \bar{r}_j} \right\} dt, \\ \tilde{\delta}_i(t_1) = 0. \end{cases} \tag{15.8}$$

The equilibrium revenue of producer i is

$$E\frac{1}{2}\alpha_i(t_0)(p(t_0) - \hat{p}_0)^2 + \frac{1}{2}\beta_i(t_0)\hat{p}_0^2 + \gamma_i(t_0)\hat{p}_0 + \delta_i(t_0).$$

Proof The proof is based on the direct method used in Volume I of this book. To do so, we first write a generic structure of the discounted revenue functional, with unknown deterministic functions to be identified. Inspired from the structure of the terminal revenue function, we try a guess functional in a quadratic form. Let

$$f_i(t, p) = \frac{1}{2}\alpha_i(t)(p - \hat{p})^2 + \frac{1}{2}\beta_i(t)\hat{p}^2 + \gamma_i(t)\hat{p} + \delta_i(t),$$

where $\alpha, \beta, \gamma, \delta$ are random functions of time t, which are $\mathscr{F}_t^{B_o}-$measurable such that

$$f_i(t_1, p(t_1)) = -\frac{q}{2}e^{-\lambda_i t_1}(p(t_1) - \hat{p}(t_1))^2,$$

i.e., $\alpha_i(t_1) = -qe^{-\lambda_i t_1}$, $\beta_i(t_1) = \gamma_i(t_1) = \delta_i(t_1) = 0$. Using Itô's formula for jump-diffusion process, one obtains

$$d(\tfrac{\beta_i \hat{p}^2}{2}) = \tfrac{\hat{p}^2 d\beta_i}{2} + \tfrac{\beta_i d[\hat{p}^2]}{2} + \beta_{i,o}\hat{p}\sigma_o dt. \tag{15.9}$$

Using (15.9) yields

$$d(\frac{\beta_i \hat{p}^2}{2}) = \frac{1}{2}\hat{p}^2 d\beta_i + \frac{1}{2}\beta_i \left(2\eta\hat{p}\left(a - \bar{S} - \hat{p}\right) + \sigma_o^2\right) dt + \beta_i \hat{p}\sigma_o dB_o + \beta_{i,o}\hat{p}\sigma_o dt. \tag{15.10}$$

We compute the difference between p and $\hat{p}$.

$$\begin{aligned} d\left[p-\hat{p}\right]^2 =& 2(p-\hat{p})\left(\sigma dB + \textstyle\int_\Theta \mu(\theta)\tilde{N}(dt, d\theta)\right) - 2\eta(p-\hat{p})(S-\hat{S}+p-\hat{p})dt \\ &+ \left(\sigma^2 + \textstyle\int_\Theta \mu^2(\theta)\nu(d\theta)\right) dt. \end{aligned} \tag{15.11}$$

Moreover,

$$d\left(\frac{\alpha_i (p-\hat{p})^2}{2}\right) = \frac{(p-\hat{p})^2}{2} d\alpha_i + \frac{1}{2}\alpha_i d\left[(p-\hat{p})^2\right] + 0, \tag{15.12}$$

and replacing (15.11) in (15.12) yields

$$\begin{aligned} & d\left(\tfrac{\alpha_i (p-\hat{p})^2}{2}\right) \\ &= \tfrac{(p-\hat{p})^2}{2} d\alpha_i + \tfrac{1}{2}\alpha_i \left(\sigma^2 + \textstyle\int_\Theta \mu^2(\theta)\nu(d\theta)\right) dt \\ &\quad -\eta(p-\hat{p})\alpha_i (S - \hat{S} + p - \hat{p})dt + (p-\hat{p})\alpha_i \sigma dB \\ &\quad +(p-\hat{p})\alpha_i \textstyle\int_\Theta \mu(\theta)\tilde{N}(dt, d\theta). \end{aligned} \tag{15.13}$$

Similarly, we have

$$d(\gamma_i \hat{p}) = \hat{p} d\gamma_i + \left(\eta\gamma_i \left(a - \hat{S} - \hat{p}\right) + \gamma_{i,o}\sigma_o\right) dt + \sigma_o \gamma_i dB_o. \tag{15.14}$$

Taking the conditional expectation with respect to the filtration $\mathscr{F}_{t_1}^{B_o}$, we arrive at

$$\begin{aligned} & E\left[f_i(t_1, p(t_1)) - f_i(t_0, p(t_0)) | \mathscr{F}_{t_1}^{B_o}\right] \\ &= \frac{1}{2}\int_{t_0}^{t_1} (p-\hat{p})^2 (d\alpha_i - 2\eta\alpha_i dt) - 2s\alpha_i (p-\hat{p})(S-\hat{S})dt \\ &+ \frac{1}{2}\int_{t_0}^{t_1} \alpha_i \left(\sigma^2 + \int \mu^2 \nu(d\theta)\right) dt + \frac{1}{2}\int_{t_0}^{t_1} \hat{p}^2 (d\beta_i - 2\eta\beta_i dt) \\ &+ \frac{1}{2}\int_{t_0}^{t_1} 2\eta\beta_i \hat{p}(a - \bar{D})dt + (\sigma_o^2 \beta_i + 2\beta_{i,o}\hat{p}\sigma_o)dt \\ &+ \frac{1}{2}\int_{t_0}^{t_1} 2\beta_i \hat{p}\sigma_o dB_o + \int_{t_0}^{t_1} \hat{p}(d\gamma_i - s\gamma_i dt) \\ &+ \int_{t_0}^{t_1} \left(s\gamma_i (a - \hat{S}) + \gamma_{i,o}\sigma_o\right) dt + \int_{t_0}^{t_1} \sigma_o \gamma_i dB_o + d\delta_i. \end{aligned} \tag{15.15}$$

The expected revenue can be expressed as

$$e^{-\lambda_i t} E\left[\hat{p}u_i - C_i(u_i)|\ \mathscr{F}_{t_1}^{B_o}\right]$$
$$= e^{-\lambda_i t} E\left[(\hat{p} - c_i)\hat{u}_i|\ \mathscr{F}_{t_1}^{B_o}\right] - e^{-\lambda_i t} E\left[\tfrac{1}{2}r_i(u_i - \hat{u}_i)^2 + \tfrac{1}{2}(r_i + \bar{r}_i)(\hat{u}_i)^2|\ \mathscr{F}_{t_1}^{B_o}\right]. \tag{15.16}$$

We now express the difference between the long-term revenue and the guess functional:

$$\begin{aligned}
E[R_{i,\mathscr{T}} - f_i(t_0, p(t_0))|\ \mathscr{F}_{t_1}^{B_o}] &= -\tfrac{q}{2}e^{-\lambda_i t_1} E\left[(p(t_1) - \hat{p}(t_1))^2|\ \mathscr{F}_{t_1}^{B_o}\right]\\
&+\tfrac{1}{2}E\int_{t_0}^{t_1} e^{-\lambda_i t}\left(2(\hat{p} - c_i)\hat{u}_i - r_i(u_i - \hat{u}_i)^2 - (r_i + \bar{r}_i)(\hat{u}_i)^2|\ \mathscr{F}_{t_1}^{B_o}\right)dt\\
&+E\tfrac{1}{2}\int_{t_0}^{t_1}(p - \hat{p})^2\,(d\alpha_i - 2\eta\alpha_i dt) + \tfrac{1}{2}\int_{t_0}^{t_1} -2\eta\alpha_i(p - \hat{p})(S - \hat{S})dt\\
&+\alpha_i\left(\sigma^2 + \int_\Theta \mu^2(\theta)\nu(d\theta)\right)dt + \tfrac{1}{2}\int_{t_0}^{t_1}\hat{p}^2\,(d\beta_i - 2\eta\beta_i dt) + 2\eta\beta_i\hat{p}(a - \hat{S})dt\\
&+\tfrac{1}{2}\int_{t_0}^{t_1}\left(\sigma_o^2\beta_i + 2\beta_{i,o}\hat{p}\sigma_o\right)dt + 2\beta_i\hat{p}\sigma_o dB_o\\
&+\int_{t_0}^{t_1}\hat{p}\,(d\gamma_i - \eta\gamma_i dt) + [\eta\gamma_i(a - \hat{S}) + \gamma_{i,o}\sigma_o]dt\\
&+\int_{t_0}^{t_1}\sigma_o\gamma_i dB_o + d\delta_i.
\end{aligned} \tag{15.17}$$

Therefore,

$$\begin{aligned}
E\left[R_{i,\mathscr{T}} - f_i(t_0, p(t_0))|\ \mathscr{F}_{t_1}^{B_o}\right] &= \tfrac{(\alpha_i(t_1) - qe^{-\lambda_i t_1})}{2} E\left[(p(t_1) - \hat{p}(t_1))^2|\ \mathscr{F}_{t_1}^{B_o}\right]\\
&+E\int_{t_0}^{t_1}\left\{\tfrac{e^{-\lambda_i t}}{2}\left(2(\hat{p} - c_i)\hat{u}_i - r_i(u_i - \hat{u}_i)^2 - (r_i + \bar{r}_i)(\hat{u}_i)^2\right)\right.\\
&\left.-2s\alpha_i(p - \hat{p})(S - \bar{S}) - 2\eta(\beta_i\hat{p} + \gamma_i)\bar{S}\right\}dt\\
&+E\int_{t_0}^{t_1}\tfrac{(p-\hat{p})^2}{2}(d\alpha_i - 2s\alpha_i dt) + \tfrac{\alpha_i}{2}\left(\sigma^2 + \int_\Theta \mu^2(\theta)\nu(d\theta)\right)dt\\
&+\tfrac{1}{2}\int_{t_0}^{t_1}\hat{p}^2\,(d\beta_i - 2\eta\beta_i dt) + 2\eta\beta_i\hat{p}a\,dt\\
&+\tfrac{1}{2}\int_{t_0}^{t_1}(\sigma_o^2\beta_i + 2\beta_{i,o}\hat{p}\sigma_o)dt + 2\beta_i\hat{p}\sigma_o dB_o\\
&+\int_{t_0}^{t_1}\hat{p}\,(d\gamma_i - \eta\gamma_i dt) + [\eta\gamma_i a + \gamma_{i,o}\sigma_o]dt\\
&+\int_{t_0}^{t_1}\sigma_o\gamma_i dB_o + d\delta_i.
\end{aligned} \tag{15.18}$$

We have

$$\begin{aligned}
&e^{-\lambda_i t}\left(2(\hat{p} - c_i)\hat{u}_i - r_i(u_i - \hat{u}_i)^2 - (r_i + \bar{r}_i)(\hat{u}_i)^2\right)\\
&\quad -2s\alpha_i(p - \hat{p})(S - \hat{S}) - 2s(\beta_i\hat{p} + \gamma_i)\hat{S}\\
&= -(r_i + \bar{r}_i)\,e^{-\lambda_i t}(\hat{u}_i)^2 + 2\left[(\hat{p} - c_i)e^{-\lambda_i t} - s(\beta_i\hat{p} + \gamma_i)\right]\hat{u}_i - r_i e^{-\lambda_i t}(u_i - \hat{u}_i)^2\\
&\quad -2\eta\alpha_i(p - \hat{p})(u_i - \hat{u}_i) - 2\eta\alpha_i(p - \hat{p})(S_{-i} - \hat{S}_{-i}) - 2\eta(\beta_i\hat{p} + \gamma_i)\hat{S}_{-i}\\
&= -(r_i + \bar{r}_i)e^{-\lambda_i t}\left(u_i - \tfrac{\hat{p}(e^{-\lambda_i t} - \eta\beta_i) - (c_i e^{-\lambda_i t} + \eta\gamma_i)}{(r_i + \bar{r}_i)e^{-\lambda_i t}}\right)^2\\
&\quad -r_i e^{-\lambda_i t}\left(u_i - \hat{u}_i + \tfrac{\eta\alpha_i(p - \hat{p})}{r_i e^{-\lambda_i t}}\right)^2 + (p - \hat{p})^2\left(\tfrac{\eta^2\alpha_i^2}{r_i e^{-\lambda_i t}} + 2\eta^2\alpha_i\sum_{j\neq i}\tfrac{\alpha_j}{r_j e^{-\lambda_j t}}\right)\\
&\quad +\hat{p}^2\left(\tfrac{(e^{-\lambda_i t} - \eta\beta_i)^2}{(r_i + \bar{r}_i)e^{-\lambda_i t}} - 2\eta\beta_i\sum_{j\neq i}\tfrac{e^{-\lambda_j t} - \eta\beta_j}{(r_j + \bar{r}_j)e^{-\lambda_j t}}\right)\\
&\quad -2\hat{p}\left(\tfrac{(e^{-\lambda_i t} - \eta\beta_i)(c_i e^{-\lambda_i t} + \eta\gamma_i)}{(r_i + \bar{r}_i)e^{-\lambda_i t}} + s\gamma_i\sum_{j\neq i}\tfrac{(e^{-\lambda_j t} - \eta\beta_j)}{(r_j + \bar{r}_j)e^{-\lambda_j t}}\right)\\
&\quad -2\hat{p}\left(-\eta\beta_i\sum_{j\neq i}\tfrac{(c_j e^{-\lambda_j t} + \eta\gamma_j)}{(r_j + \bar{r}_j)e^{-\lambda_j t}}\right) + \tfrac{(c_i e^{-\lambda_i t} + s\gamma_i)^2}{(r_i + \bar{r}_i)e^{-\lambda_i t}} + 2\eta\gamma_i\sum_{j\neq i}\tfrac{(c_j e^{-\lambda_j t} + \eta\gamma_j)}{(r_j + \bar{r}_j)e^{-\lambda_j t}},
\end{aligned} \tag{15.19}$$

where we used

$$S_{-i} - \hat{S}_{-i} = -\eta(p-\hat{p})\sum_{j\neq i}\frac{\alpha_j}{r_j e^{-\lambda_j t}},$$

and

$$\hat{S}_{-i} = \hat{p}\sum_{j\neq i}\frac{e^{-\lambda_j t} - \eta\beta_j}{(r_j+\bar{r}_j)e^{-\lambda_j t}} - \sum_{j\neq i}\frac{c_j e^{-\lambda_j t} + \eta\gamma_j}{(r_j+\bar{r}_j)e^{-\lambda_j t}}.$$

It follows that

$$\begin{aligned}
E\left[R_{i,\mathscr{T}} - f_i(t_0, p(t_0))|\, \mathscr{F}_{t_1}^{B_o}\right] &= \tfrac{\alpha_i(t_1) - qe^{-\lambda_i t_1}}{2} E\left[(p(t_1)-\hat{p}(t_1))^2|\, \mathscr{F}_{t_1}^{B_o}\right] \\
&-E\int_{t_0}^{t_1}\tfrac{r_i+\bar{r}_i}{2}e^{-\lambda_i t}\left(\hat{u}_i - \tfrac{\hat{p}(e^{-\lambda_i t}-\eta\beta_i)-(c_i e^{-\lambda_i t}+s\gamma_i)}{(r_i+\bar{r}_i)e^{-\lambda_i t}}\right)^2 dt \\
&-E\int_{t_0}^{t_1}\tfrac{r_i}{2}e^{-\lambda_i t}\left(u_i - \hat{u}_i + \tfrac{\eta\alpha_i(p-\hat{p})}{r_i e^{-\lambda_i t}}\right)^2 dt \\
&+E\int_{t_0}^{t_1}\tfrac{(p-\hat{p})^2}{2}\left\{d\alpha_i + \left(-2\eta\alpha_i + \tfrac{\eta^2\alpha_i^2}{r_i e^{-\lambda_i t}} + 2\eta^2\alpha_i\sum_{j\neq i}\tfrac{\alpha_j}{r_j e^{-\lambda_j t}}\right)dt\right\} \\
&+\tfrac{1}{2}\int_{t_0}^{t_1}\hat{p}^2\left\{d\beta_i + (-2\eta\beta_i + \tfrac{(e^{-\lambda_i t}-\eta\beta_i)^2}{(r_i+\bar{r}_i)e^{-\lambda_i t}} - 2\eta\beta_i\sum_{j\neq i}\tfrac{(e^{-\lambda_j t}-\eta\beta_j)}{(r_j+\bar{r}_j)e^{-\lambda_j t}})dt\right\} \\
&+\int_{t_0}^{t_1}\hat{p}\Big\{d\gamma_i - \eta\gamma_i dt + \eta\beta_i a dt + \beta_{i,o}\sigma_o dt + \beta_i\sigma_o dB_o - \tfrac{(e^{-\lambda_i t}-\eta\beta_i)(c_i e^{-\lambda_i t}+\eta\gamma_i)}{(r_i+\bar{r}_i)e^{-\lambda_i t}}dt \\
&+\left[-\eta\gamma_i\sum_{j\neq i}\tfrac{(e^{-\lambda_j t}-\eta\beta_j)}{(r_j+\bar{r}_j)e^{-\lambda_j t}} + \eta\beta_i\sum_{j\neq i}\tfrac{(c_j e^{-\lambda_j t}+s\gamma_j)}{(r_j+\bar{r}_j)e^{-\lambda_j t}}\right]dt\Big\} \\
&+\int_{t_0}^{t_1} d\delta_i + \left[\tfrac{1}{2}\sigma_o^2\beta_i + \tfrac{1}{2}\alpha_i\left(\sigma^2 + \int_\Theta \mu^2(\theta)\nu(d\theta)\right) + \eta\gamma_i a + \gamma_{i,o}\sigma_o\right]dt \\
&+\int_{t_0}^{t_1}\sigma_o\gamma_i dB_o + \left[\tfrac{1}{2}\tfrac{(c_i e^{-\lambda_i t}+\eta\gamma_i)^2}{(r_i+\bar{r}_i)e^{-\lambda_i t}} + \eta\gamma_i\sum_{j\neq i}\tfrac{(c_j e^{-\lambda_j t}+\eta\gamma_j)}{(r_j+\bar{r}_j)e^{-\lambda_j t}}\right]dt.
\end{aligned} \tag{15.20}$$

We deduce that $E\left[R_{i,\mathscr{T}} - f_i(t_0, p(t_0))|\, \mathscr{F}_{t_1}^{B_o}\right] \leq 0$, and the equality occurs when the coefficient solves the following system:

$$\hat{u}_i^* = \frac{\hat{p}(e^{-\lambda_i t} - \eta\beta_i) - (c_i e^{-\lambda_i t} + \eta\gamma_i)}{(r_i+\bar{r}_i)e^{-\lambda_i t}}.$$

Equilibrium strategy in state-and-conditional mean-field-type feedback form:

$$\begin{aligned}
u_i^* &= -\tfrac{\eta\alpha_i(p-\hat{p})}{r_i e^{-\lambda_i t}} + \hat{u}_i^*, \\
&= -\tfrac{\eta\alpha_i}{r_i e^{-\lambda_i t}}(p-\hat{p}) + \tfrac{\hat{p}(e^{-\lambda_i t}-\eta\beta_i)-(c_i e^{-\lambda_i t}+\eta\gamma_i)}{(r_i+\bar{r}_i)e^{-\lambda_i t}}.
\end{aligned} \tag{15.21}$$

Conditional equilibrium log-price:

$$\begin{cases} d\hat{p}=\eta\left[a+\sum_{j=1}^{I}\frac{c_je^{-\lambda_jt}+s\gamma_j}{(r_j+\bar{r}_j)e^{-\lambda_jt}}-\hat{p}\left(1+\sum_{j=1}^{I}\frac{e^{-\lambda_jt}-\eta\beta_j}{(r_j+\bar{r}_j)e^{-\lambda_jt}}\right)\right]dt+\sigma_o dB_o,\\ \hat{p}(t_0)=\hat{p}_0. \end{cases}$$

Riccati system:

$$\begin{cases} d\alpha_i=-\left(-2\eta\alpha_i+\frac{\eta^2\alpha_i^2}{r_ie^{-\lambda_it}}+2\eta^2\alpha_i\sum_{j\neq i}\frac{\alpha_j}{r_je^{-\lambda_jt}}\right)dt+\alpha_{i,o}dB_o,\\ \alpha_i(t_1)=-qe^{-\lambda_it_1},\\ d\beta_i=-\left(-2\eta\beta_i+\frac{(e^{-\lambda_it}-\eta\beta_i)^2}{(r_i+\bar{r}_i)e^{-\lambda_it}}-2\eta\beta_i\sum_{j\neq i}\frac{(e^{-\lambda_jt}-\eta\beta_j)}{(r_j+\bar{r}_j)e^{-\lambda_jt}}\right)dt+\beta_{i,o}dB_o,\\ \beta_i(t_1)=0,\\ d\gamma_i=\left\{\eta\gamma_i-\eta\beta_ia-\beta_{i,o}\sigma_o+\frac{(e^{-\lambda_it}-\eta\beta_i)(c_ie^{-\lambda_it}+\eta\gamma_i)}{(r_i+\bar{r}_i)e^{-\lambda_it}}\right.\\ \qquad\left.+\eta\gamma_i\sum_{j\neq i}\frac{(e^{-\lambda_jt}-\eta\beta_j)}{(r_j+\bar{r}_j)e^{-\lambda_jt}}-\eta\beta_i\sum_{j\neq i}\frac{(c_je^{-\lambda_jt}+\eta\gamma_j)}{(r_j+\bar{r}_j)e^{-\lambda_jt}}\right\}dt-\beta_i\sigma_o dB_o,\\ \gamma_i(0)=0,\\ d\delta_i=-\left\{\frac{1}{2}\sigma_o^2\beta_i+\frac{1}{2}\alpha_i(\sigma^2+\int_\Theta\mu^2(\theta)\nu(d\theta))+\eta\gamma_ia+\gamma_{i,o}\sigma_o\right.\\ \qquad\left.+\frac{1}{2}\frac{(c_ie^{-\lambda_it}+\eta\gamma_i)^2}{(r_i+\bar{r}_i)e^{-\lambda_it}}+\eta\gamma_i\sum_{j\neq i}\frac{(c_je^{-\lambda_jt}+\eta\gamma_j)}{(r_j+\bar{r}_j)e^{-\lambda_jt}}\right\}dt-\sigma_o\gamma_idB_o,\\ \delta_i(t_1)=0. \end{cases}$$

Conditional equilibrium revenue:

$$\frac{1}{2}\alpha_i(t_0)(p(t_0)-\hat{p}_0)^2+\frac{1}{2}\beta_i(t_0)\hat{p}_0^2+\gamma_i(t_0)\hat{p}_0+\delta_i(t_0).$$

Set $\tilde{\alpha}_i=\alpha_ie^{\lambda_it}$, $\tilde{\beta}_i=\beta_ie^{\lambda_it}$. Then:
Equilibrium strategy in state-and-mean-field feedback form:

$$u_i^*=-\frac{\eta\tilde{\alpha}_i}{r_i}(p-\hat{p})+\frac{\hat{p}(1-\eta\tilde{\beta}_i)-(c_i+\eta\tilde{\gamma}_i)}{r_i+\bar{r}_i}.$$

Conditional equilibrium log-price:

$$\begin{cases} d\hat{p} = \eta \left\{ a + \sum_{j=1}^{I} \frac{c_j + \eta\tilde{\gamma}_j}{r_j + \bar{r}_j} - \hat{p}\left(1 + \sum_{j=1}^{I} \frac{1 - \eta\tilde{\beta}_j}{r_j + \bar{r}_j}\right)\right\} dt + \sigma_o dB_o, \\ \hat{p}(t_0) = \hat{p}_0. \end{cases}$$

Riccati system:

$$\begin{cases} d\tilde{\alpha}_i = \left\{ (\lambda_i + 2\eta)\tilde{\alpha}_i - \frac{\eta^2}{r_i}\tilde{\alpha}_i^2 - 2\eta^2\tilde{\alpha}_i \sum_{j\neq i} \frac{\tilde{\alpha}_j}{r_j} \right\} dt + \tilde{\alpha}_{i,o} dB_o, \\ \tilde{\alpha}_i(t_1) = -q, \\ d\tilde{\beta}_i = \left\{ (\lambda_i + 2\eta)\tilde{\beta}_i - \frac{(1-\eta\tilde{\beta}_i)^2}{r_i + \bar{r}_i} + 2\eta\tilde{\beta}_i \sum_{j\neq i} \frac{1-\eta\tilde{\beta}_j}{r_j + \bar{r}_j} \right\} dt + \tilde{\beta}_{i,o} dB_o, \\ \tilde{\beta}_i(t_1) = 0, \\ d\tilde{\gamma}_i = \left\{ (\lambda_i + \eta)\tilde{\gamma}_i - \eta\tilde{\beta}_i a - \tilde{\beta}_{i,o}\sigma_o + \frac{(1-\eta\tilde{\beta}_i)(c_i + \eta\tilde{\gamma}_i)}{r_i + \bar{r}_i} \right. \\ \quad \left. + \eta\tilde{\gamma}_i \sum_{j\neq i} \frac{1-\eta\tilde{\beta}_j}{r_j + \bar{r}_j} - \eta\tilde{\beta}_i \sum_{j\neq i} \frac{c_j + \eta\tilde{\gamma}_j}{r_j + \bar{r}_j} \right\} dt - \tilde{\beta}_i \sigma_o dB_o, \\ \tilde{\gamma}_i(0) = 0, \\ d\tilde{\delta}_i = -\left\{ -\lambda_i\tilde{\delta}_i + \frac{1}{2}\sigma_o^2\tilde{\beta}_i + \frac{1}{2}\tilde{\alpha}_i\left(\sigma^2 + \int_\Theta \mu^2(\theta)\nu(d\theta)\right) + \eta\tilde{\gamma}_i a \right. \\ \quad \left. + \tilde{\gamma}_{i,o}\sigma_o + \frac{1}{2}\frac{(c_i + \eta\tilde{\gamma}_i)^2}{(r_i + \bar{r}_i)} + \eta\tilde{\gamma}_i \sum_{j\neq i} \frac{(c_j + \eta\tilde{\gamma}_j)}{(r_j + \bar{r}_j)} \right\} dt - \sigma_o\tilde{\gamma}_i dB_o, \\ \tilde{\delta}_i(t_1) = 0, \end{cases} \tag{15.22}$$

completing the proof.

Taking the expected value of the conditional equilibrium revenue, we obtain

$$\frac{1}{2}\alpha_i(t_0)\text{var}(p(t_0)) + \frac{1}{2}E\beta_i(t_0)\hat{p}_0^2 + E\gamma_i(t_0)\hat{p}_0 + E\delta_i(t_0).$$

15.3.1 Larger Horizon and Stationary Solution

We consider in this section the maximization of the performance criterion:

$$\lim_{t_1\to+\infty} ER_{\mathscr{T}}(p_0, u).$$

We are interested in studying Nash equilibria within the class of linear time-invariant strategies.

Limiting Feedback Strategies

Denote a set of feedback Nash equilibrium strategies for the finite horizon game $u^{fb}_{i,\mathcal{T}}$ and those for the infinite horizon game by $u^{fb}_{i,+\infty}$. We will study the relationship between $u^{fb}_{i,\mathcal{T}}$ and $u^{fb}_{i,+\infty}$. As $t_1 \to +\infty$, the coefficient α_i vanishes and the (unconstrained) equilibrium strategies become

$$\begin{cases} u^{fb}_{i,+\infty} = \dfrac{\hat{p}(1-\eta\tilde{\beta}_i) - (c_i + \eta\tilde{\gamma}_i)}{r_i + \bar{r}_i}, \\ (-\lambda_i - 2\eta)\tilde{\beta}_i + \dfrac{(1-\eta\tilde{\beta}_i)^2}{r_i + \bar{r}_i} - 2\eta\tilde{\beta}_i \sum_{j\neq i} \dfrac{1-\eta\tilde{\beta}_j}{r_j + \bar{r}_j} = 0, \\ \tilde{\beta}_{i,o} = 0. \end{cases} \tag{15.23}$$

Producer i will participate if the price is interesting enough to get some revenue. Let

$$\underline{p}_i := \frac{c_i + \eta\tilde{\gamma}_i}{1-\eta\tilde{\beta}_i}.$$

If, in addition, σ_o vanishes, then the following structure

$$u_i^* = \begin{cases} \dfrac{\hat{p}(1-\eta\tilde{\beta}_i^*) - (c_i + \eta\tilde{\gamma}_i^*)}{r_i + \bar{r}_i}, & \text{if } \hat{p} \geq \underline{p}_i, \\ 0, & \text{otherwise}, \end{cases} \tag{15.24}$$

provides a stationary equilibrium. When the price increases, the equilibrium output of producer i increases because $1-\eta\tilde{\beta}_i^* > 0$. There is a stationary equilibrium price and it is given by

$$\hat{p}^* = \frac{a + \sum_{j=1}^{I} \dfrac{c_j + \eta\tilde{\gamma}_j}{r_j + \bar{r}_j}}{1 + \sum_{j=1}^{I} \dfrac{1-\eta\tilde{\beta}_j}{r_j + \bar{r}_j}}. \tag{15.25}$$

Moreover, if the game starts at $\hat{p}_0 \neq \hat{p}^*$, the closed-loop equilibrium price converges exponentially to stationary equilibrium price. The learning dynamics stabilizes quickly to the steady-state price. Note that when the speed η goes to infinity, $\eta\beta_i$ has a limit β_i^* and the limiting electricity price is

$$p(\eta \to \infty) = \frac{a + \sum_{j=1}^{I} \frac{c_j + \tilde{\gamma}_j^*}{r_j + \bar{r}_j}}{1 + \sum_{j=1}^{I} \frac{1 - \tilde{\beta}_j^*}{r_j + \bar{r}_j}}. \tag{15.26}$$

Extensions of the result presented in this section to linear coefficients (state, control, conditional mean-state, conditional mean-control) for σ, μ, and σ_o can be found in Volume I.

15.3.2 Static Cournot Competition Equilibrium Price

We are interested in investigating the implications of considering the dynamic learning price model as in (15.3) with respect to the static price scenario, i.e., considering

$$p = a - S.$$

We compute the static revenue of producer i being

$$R_{i,\mathcal{T}}(u) = \{[a - S]u_i - C_i(u_i)\}\Lambda_i, \tag{15.27}$$

where $\Lambda_i = \int_{t_0}^{t_1} e^{-\lambda_i t} dt$,

$$C_i(u_i) = c_i u_i + \frac{1}{2}(r_i + \bar{r}_i)\, u_i^2. \tag{15.28}$$

Hence, the interior best response (if any) of the static Cournot game is given by

$$u_i^* \in \arg\max_{u_i \in \mathbb{R}} \{(a - S)u_i - C_i(u_i)\} = \{\tfrac{a - \sum_{j \neq i} u_j - c_i}{2 + r_i + \bar{r}_i}\}. \tag{15.29}$$

It follows that the static Cournot equilibrium price is given by

$$p^{\mathrm{C}} = \frac{a + \sum_{j=1}^{I} \frac{c_j}{1 + r_j + \bar{r}_j}}{1 + \sum_{j=1}^{I} \frac{1}{1 + r_j + \bar{r}_j}}. \tag{15.30}$$

Finally, notice that the presented analysis allows us to determine if the consideration of the price dynamics (15.3) in the market is beneficial for the consumers. This is evaluated by comparing the price in (15.26) and (15.30).

15.3.2.1 Decision Support

In practice, it is often not desirable to change continuously the price. Thus, the above virtual price adjustment model will be used as a learning feature to help the decision-maker before acting on the real price for the next time-block. For example, if the price is decided every 3-month cycle, then this base model can be used to determine the electricity market trends and recent features. If the price is almost in real time, this simple adjustment technique can also be used with higher speed η. Figure 15.2 plots a real data illustrating the evolution of electricity price versus the numerical parameters displayed in Table 15.1.

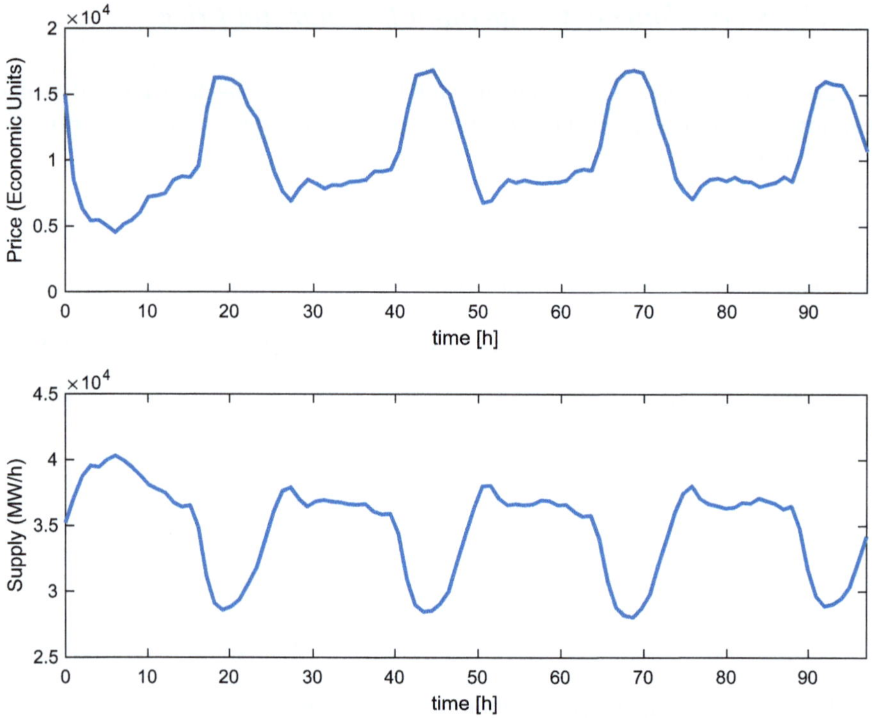

Fig. 15.2 Evolution of the price for real data regional supply from the *US Electric System Operating Data* for 4 days during October 2018

Table 15.1 Parameters mimicking the price observed in real data

Parameter	Value
t_1	97
p_0	1.5e4
s	100
a	45,000
S	Real data

Table 15.2 Parameters of Scenario 1

Parameter	Value
t_1	1
p_0	170
s	100
a	30
S	$30 + 20\sin(10\pi t)$

Table 15.3 Parameters of Scenarios 2 and 3

Parameter	Value
t_1	1.5
p_0	50
$\hat{p}_0$	50
c_1	1
c_2	5
c_3	10
$\lambda_1, \ldots, \lambda_3$	0.1
s	0.5
a	1
$r_1, \bar{r}_1$	1
$r_2, \bar{r}_2$	2
$r_3, \bar{r}_3$	3
q	1
S	$u_1 + u_2 + u_3$

In order to illustrate both the price dynamics behavior and the performance of the mean-field-type game [106], we present some illustrative examples:

- Scenario 1: First, consider the case with $\sigma = 0, \sigma_0 = 0$, without jump terms, and the arbitrary parameters presented in Table 15.2.
- Scenario 2: Now, we consider the case with $\sigma, \sigma_0 = 0$, without jump terms, and the arbitrary parameters presented in Table 15.3.
- Scenario 3: Now, we consider the case with $\sigma_0 = 0$, with $\sigma = 2$ and jump terms (see Fig. 15.4), and the arbitrary parameters presented in Table 15.3.

Figure 15.3 shows that the price dynamics behaves in a way such that the price tracks the term $a - S$. Furthermore, it is shown that this track performs faster as the parameter s increases. Thus, the price behaves depending on the evolution of the demand (Fig. 15.4).

Figures 15.5, 15.6, and 15.7 present the behavior of the mean-field-type games corresponding to Scenarios 2 and 3, respectively. Notice that Scenario 2 is deterministic, for which the evolutions of $p(t)$ and $\hat{p}(t)$ are equal. On the contrary, Fig. 15.6 shows different evolutions for $p(t)$ and $\hat{p}(t)$, and for $u_i(t)$ and $E[u_i(t)]$, for all i. Besides, it can be seen that the Riccati equations satisfy the boundary conditions

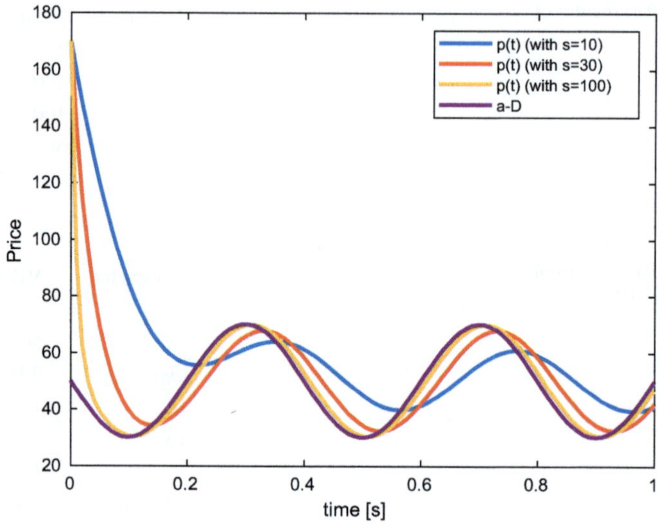

Fig. 15.3 Evolution of the price given some demand profile for different values of η. This shows a tracking behavior

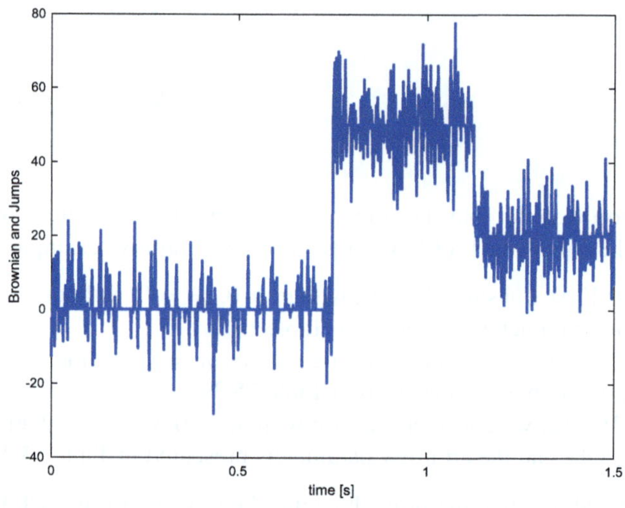

Fig. 15.4 Brownian motion + two jumps

presented in Proposition 15.1, e.g., $\alpha_i(t_1) = -q$. Moreover, Table 15.4 summarizes the costs corresponding to Scenarios 2 and 3, where it can be seen that the costs are higher when the Brownian motion with jumps is considered.

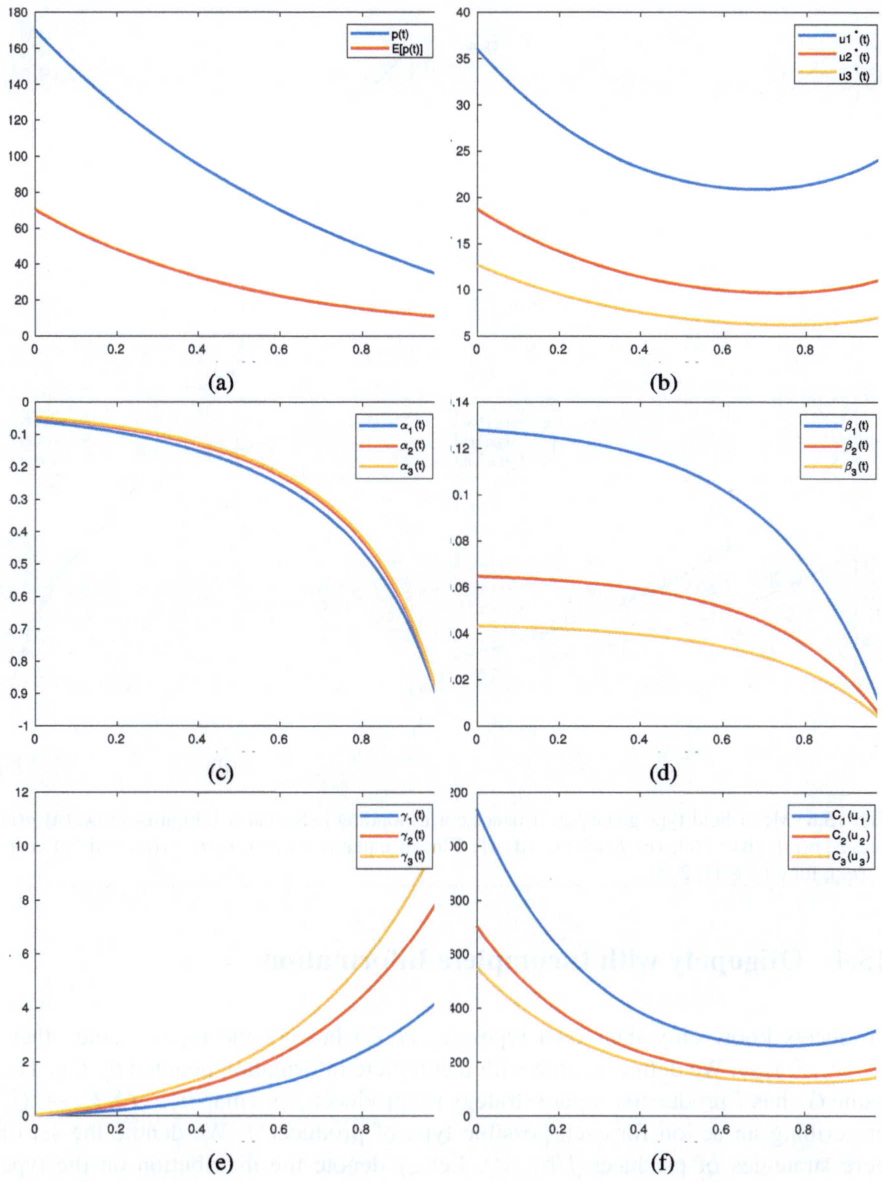

Fig. 15.5 Mean-field-type game performance corresponding to Scenario 2. Figures show: **(a)** $p(t)$ and $E[p(t)]$, **(b)** $u_i^*(t)$, **(c)–(e)** Riccati equations $\alpha_i(t)$, $\beta_i(t)$, $\gamma_i(t)$, and **(f)** costs $C_i(u_i)$, for all $i = \{1, \ldots, 3\}$

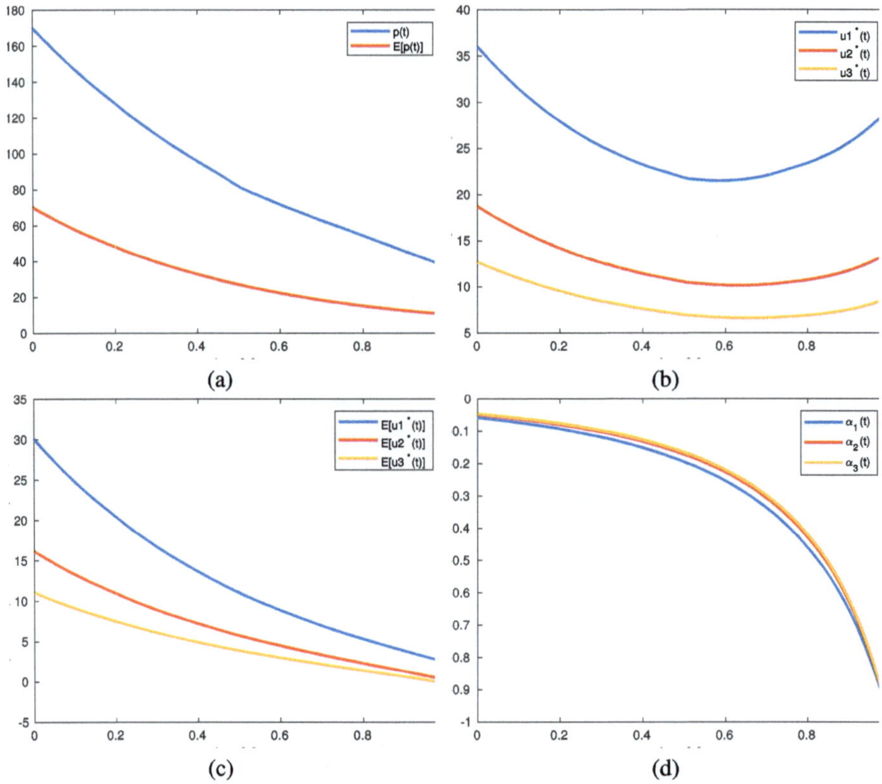

Fig. 15.6 Mean-field-type game performance corresponding to Scenario 3. Figures show: (**a**) $p(t)$ and $E[p(t)]$, (**b**) $u_i^*(t)$, (**c**) $E[u_i^*(t)]$, (**d**)–(**f**) Riccati equations $\alpha_i(t)$, $\beta_i(t)$, $\gamma_i(t)$, and (**g**) costs $C_i(u_i)$, for all $i \in \{1, 2, 3\}$

15.4 Oligopoly with Incomplete Information

Producers know only their own types $(c_i, r_i, \bar{r}_i)$ but not the types of the others $(c_j, r_j, \bar{r}_j)_{j \neq i}$. We define a game with incomplete information denoted by G_ξ. The game G_ξ has I producers. A pure strategy for producer j is a map $\tilde{u}_j : INF_j \to U_j$ prescribing an action for each possible type of producer j. We denote the set of pure strategies of producer j by $\tilde{\mathscr{U}}_j$. Let ξ_j denote the distribution on the type vector $(c_j, r_j, \bar{r}_j)$ from the perspective of the jth producer. Given ξ_j, producer j can compute the conditional distribution $\xi_{-j}(c_{-j}, r_{-j}, \bar{r}_{-j} | c_j, r_j, \bar{r}_j)$, where

$$c_{-j} = (c_1, \ldots, c_{j-1}, c_{j+1}, \ldots, c_I) \in \mathbb{R}^{I-1}.$$

Producer j can evaluate her expected payoff based on the expected types of other producers. An equilibrium of G_ξ is a Bayesian equilibrium.

Below, we investigate the equilibrium solution under incomplete information.

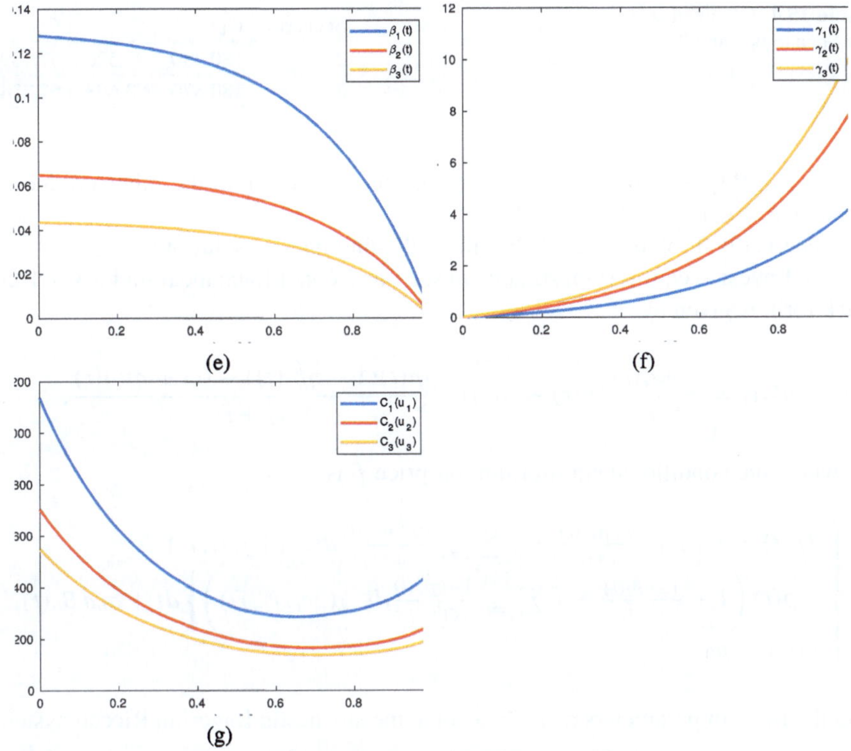

Fig. 15.7 Mean-field-type game performance corresponding to Scenario 3. Figures show: **(a)** $p(t)$ and $E[p(t)]$, **(b)** $u_i^*(t)$, **(c)** $E[u_i^*(t)]$, **(d)–(f)** Riccati equations $\alpha_i(t)$, $\beta_i(t)$, $\gamma_i(t)$, and **(g)** costs $C_i(u_i)$, for all $i \in \{1, 2, 3\}$

15.4.1 Bayesian Mean-Field-Type Equilibria

A Bayesian–Nash mean-field-type equilibrium is a strategy profile and beliefs specified for each producer about the types of the other producers that minimize the expected performance functional for each producer given their beliefs about the other producers' types and given the strategies played by the other producers. We compute the generic expression of the Bayesian mean-field-type equilibria [106]. Any strategy $u_i^* \in \tilde{\mathscr{U}_i}$ satisfying the maximum in

$$\begin{cases} \max_{u_i \in \tilde{\mathscr{U}_i}} E\left[R_{i,\mathscr{T}}(p_0, u) \,|c_i, r_i, \bar{r}_i, \xi\right], \\ dp(t) = \eta\left[a - S(t) - p(t)\right]dt + \left(\sigma dB(t) + \int_\Theta \mu(\theta)\tilde{N}(dt, d\theta)\right) \\ \qquad\quad +\sigma_o dB_o(t), \\ p(t_0) = p_0, \end{cases} \tag{15.31}$$

Table 15.4 Obtained costs for Scenarios 1 and 2

Scenario	C_1	C_2	C_3
2	180,430	93,526	57,159
3	180,570	93,608	57,224

is called a Bayesian best-response strategy of producer i to the other producers' strategy $u_{-i} \in \prod_{j\neq i} \tilde{\mathscr{U}}_j$.

Generically, Problem (15.31) has the following interior solution:

The Bayesian equilibrium strategy in state-and-conditional mean-field-type feedback form is given by

$$\tilde{u}_i^*(t) = -\frac{\eta\hat{\alpha}_i(t)}{r_i}(p(t)-\hat{p}(t)) + \frac{\hat{p}(t)(1-\eta\hat{\beta}_i(t)) - (c_i+\eta\hat{\gamma}_i)(t)}{r_i+\bar{r}_i},$$

where the conditional equilibrium log-price $\hat{p}$ is

$$\begin{cases} d\hat{p}(t) = \eta\left\{a + \frac{c_i+\eta\hat{\gamma}_i(t)}{r_i+\bar{r}_i} + \int\sum_{j\neq i}\frac{c_j+\eta\hat{\gamma}_j(t)}{r_j+\bar{r}_j}d\xi_{-i}(.|c_i,r_i,\bar{r}_i) \right. \\ \left. -\hat{p}(t)\left(1+\frac{1-\eta\hat{\beta}_i(t)}{r_i+\bar{r}_i} + \int\sum_{j\neq i}\frac{1-\eta\hat{\beta}_j(t)}{r_j+\bar{r}_j}d\xi_{-i}(.|c_i,r_i,\bar{r}_i)\right)\right\}dt + \sigma_o dB_o(t), \\ \hat{p}(t_0) = \hat{p}_0, \end{cases}$$

and the random parameters $\hat{\alpha}, \hat{\beta}, \hat{\gamma}, \hat{\delta}$ solve the stochastic Bayesian Riccati system:

$$\begin{cases} d\hat{\alpha}_i(t) = \left\{(\lambda_i+2\eta)\hat{\alpha}_i(t) - \frac{\eta^2}{r_i}\hat{\alpha}_i^2(t) - 2\eta^2\hat{\alpha}_i(t)\int\sum_{j\neq i}\frac{\hat{\alpha}_j(t)}{r_j}d\xi_{-i}(.|c_i,r_i,\bar{r}_i)\right\}dt \\ \quad +\hat{\alpha}_{i,o}(t)dB_o(t), \\ \hat{\alpha}_i(t_1) = -q, \\ \\ d\hat{\beta}_i(t) = \left\{(\lambda_i+2\eta)\hat{\beta}_i(t) - \frac{(1-\eta\hat{\beta}_i(t))^2}{r_i+\bar{r}_i} + 2\eta\hat{\beta}_i(t)\int\sum_{j\neq i}\frac{1-\eta\hat{\beta}_j(t)}{r_j+\bar{r}_j}d\xi_{-i}(.|c_i,r_i,\bar{r}_i)\right\}dt \\ \quad +\hat{\beta}_{i,o}(t)dB_o(t), \\ \hat{\beta}_i(t_1) = 0, \\ \\ d\hat{\gamma}_i(t) = \left\{(\lambda_i+\eta)\hat{\gamma}_i(t) - \eta a\hat{\beta}_i(t) - \hat{\beta}_{i,o}(t)\sigma_o + \frac{(1-\eta\hat{\beta}_i(t))(c_i+s\hat{\gamma}_i(t))}{r_i+\bar{r}_i}\right. \\ \left. +\eta\hat{\gamma}_i(t)\int\sum_{j\neq i}\frac{1-\eta\hat{\beta}_j(t)}{r_j+\bar{r}_j}d\xi_{-i}(.|c_i,r_i,\bar{r}_i) - \eta\hat{\beta}_i(t)\int\sum_{j\neq i}\frac{c_j+\eta\hat{\gamma}_j(t)}{r_j+\bar{r}_j}d\xi_{-i}(.|c_i,r_i,\bar{r}_i)\right\}dt \\ +\hat{\gamma}_{i,o}dB_o(t), \\ \hat{\gamma}_i(0) = 0, \\ \\ d\hat{\delta}_i(t) = -E\left\{-\lambda_i\hat{\delta}_i(t) + \frac{1}{2}\sigma_o^2\hat{\beta}_i(t) + \frac{1}{2}\hat{\alpha}_i(t)\left(\sigma^2 + \int_\Theta \mu^2(\theta)\nu(d\theta)\right) + \eta a\hat{\gamma}_i(t)\right. \\ \quad \left. +\hat{\gamma}_{i,o}(t)\sigma_o + \frac{1}{2}\frac{(c_i+\eta\hat{\gamma}_i(t))^2}{r_i+\bar{r}_i} + \eta\hat{\gamma}_i(t)\int\sum_{j\neq i}\frac{c_j+s\hat{\gamma}_j(t)}{r_j+\bar{r}_j}d\xi_{-i}(.|c_i,r_i,\bar{r}_i)\right\}dt, \\ \hat{\delta}_i(t_1) = 0, \end{cases}$$

and the equilibrium revenue of producer i is

$$E\left[\frac{1}{2}\hat{\alpha}_i(t_0)(p(t_0)-\hat{p}_0)^2+\frac{1}{2}\hat{\beta}_i(t_0)\hat{p}_0^2+\hat{\gamma}_i(t_0)\hat{p}_0+\hat{\delta}_i(t_0)\right].$$

The proof of the Bayesian Riccati system follows from a direct method by conditioning on the type $(c_i, r_i, \bar{r}_i, \xi)$.

Note that the Riccati system of the Bayesian mean-field-type game is different from the Riccati system of mean-field-type game; it follows that the Bayesian equilibrium costs are different. They become equal when $\xi_{-j} = \delta_{(c_{-j}, r_{-j}, \bar{r}_{-j})}$. This also shows that there is a value of information in this game. Note that the equilibrium supply is

$$\sum_i \tilde{u}_i^*(t) = -\eta(p(t)-\hat{p}(t))\sum_i \frac{\hat{\alpha}_i(t)}{r_i} + \sum_i \frac{\hat{p}(t)(1-\eta\hat{\beta}_i(t))-(c_i+s\hat{\gamma}_i(t))}{r_i+\bar{r}_i}.$$

Ex Post Resilience

Definition 15.2 A strategy profile $\tilde{u}$ is ex post resilient if for every type profile $(c_j, r_j, \bar{r}_j)_j$, and for each producer i,

$$\begin{aligned}\arg\max_{\tilde{u}_i\in\tilde{\mathscr{U}}_i} E\int R_{i,\mathscr{T}}(p_0, c_i, , r_i, \bar{r}_i, \tilde{u}_i, \tilde{u}_{-i})\xi_{-i}(dc_{-i}dr_{-i}d\bar{r}_{-i} \mid c_i, r_i, \bar{r}_i)\\ = \arg\max_{\tilde{u}_i\in\tilde{\mathscr{U}}_i} E R_{i,\mathscr{T}}(p_0, \tilde{u}_i, \tilde{u}_{-i}).\end{aligned}$$

We observe that generically, the Bayesian equilibrium is not ex post resilient. An $I-$tuple of strategies is said to be ex post resilient if each producer's strategy is a best response to the other producers' strategies, under all possible realizations of the others' types. An ex post-resilient strategy must be an equilibrium of every game with the realized type profile $(c, r, \bar{r})$. Thus, any ex post-resilient strategy is a robust strategy of the game in which all the parameters $(c, r, \bar{r})$ are taken. Here, each producer makes her ex ante decision based on ex ante information, that is, distribution and expectation, which is not necessarily identical to her ex post information, that is, the realized actions and types of other producers. Thus, ex post, or after the producer observes the quantities of energy actually produced of all the other producers, she may prefer to alter her ex ante optimal production decision.

15.5 Price Stability and Risk-Awareness

This section examines the price stability of a stylized blockchain-based market under regulation designs. As a first step, we design a target price dynamics that allows a high volume of transactions while fulfilling the regulation requirement. However, the target price is not the market price. In a second step, we propose and examine a simple price market dynamics under jump-diffusion process. The market

Fig. 15.8 Coindesk database: the price of bitcoin went from 10K USD to 20 K USD and back to below 7 K USD within 2–3 months in 2017–2018

price model builds on the market demand, supply, and token quantity. We use three different token supply strategies to evaluate the proposed market price motion. The first strategy designs a supply of tokens to the market more frequently balancing the mismatch between market supply and market demand. The second strategy is a mean-field control strategy. The third strategy is a mean-field-type control strategy that incorporates the risk of deviating from the regulation bounds.

15.5.1 Unstable and High Variance Market

As an illustration of high variance price, we take the fluctuations of bitcoin price between December 2017 and February 2018. The data is from coindesk (https://www.coindesk.com/price/). The price went from 10 K USD to 20 K USD and back to 7 K USD within 3 months. The variance was extremely high within that period, which implied very high risks in the market (Fig. 15.8). This extremely high variance and unstable market is far beyond the risk-sensitivity index distributions of users and investors. Therefore, the market needs to be re-designed to fit investors and users' risk-sensitivity distributions.

15.5.2 Fully Stable and Zero Variance

We have seen that the above example is too risky given the short time window and is beyond the risk-sensitivity index of many users. Thus, it is important to have a more stable market price in the blockchain.

A fully stable situation is the case of constant price. For that case, the variance is zero and there is no risk on that market. However, this case may not be interesting for producers and investors: if they know that the price will not vary, they will not buy. Thus, the volume of transactions will be significantly reduced, which is not convenient for the blockchain technology, which aims to be a place of innovations and investments.

Electricity market price cannot be constant because demand is variable on a daily basis or from one season to another within the same year. Peak hours price may be different from off-peak hours price as it is already the case in most countries.

Below we propose a price dynamics that is somehow in between the two scenarios: it is of relatively low variance and it allows several transaction opportunities.

15.5.3 What Is a More Stable Price Dynamics?

An example of a more stable cryptocurrency within similar time frame as the bitcoin is the tether USD (USDT), which oscillates between 0.99 and 1.01 but with an important volume of transactions (see Fig. 15.9). The maximum magnitude variation of the price remains very small, while the number oscillations in between is large, allowing several investment and buying/selling opportunities.

Is Token Supply Possible in the Blockchain?
Tokens in blockchain-based cryptocurrencies are generated by blockchain algorithms. Token supply is a decision process that can be incorporated in the algorithm. Thus, token supply can be used to influence the market price. In our model below, we will use it as a control action variable.

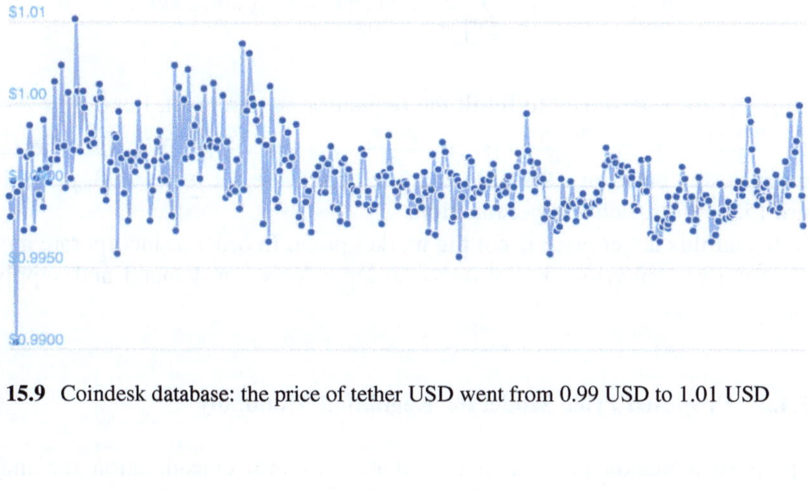

Fig. 15.9 Coindesk database: the price of tether USD went from 0.99 USD to 1.01 USD

15.5.4 A More Stable and Regulated Market Price

Let $\mathscr{T} := [t_0, t_1]$ be the time horizon with $t_0 < t_1$. There are I potential interacting regulated blockchain-based technologies over the horizon $\mathscr{T}$. The regulation authority of each blockchain-based technology has to choose the regulation bounds: the price of cryptocurrency i should be between $[\underline{p}_i, \bar{p}_i]$, $\underline{p}_i < \bar{p}_i$. We construct a target price $p_{tp,i}$ from an historical data-driven price dynamics of i. The target price should stay within the interval $[\underline{p}_i, \bar{p}_i]$ target range. The market price $p_{mp,i}$ depends on the quantity of token supplied and demanded and is given by a simple price adjustment dynamics obtained from Charles Roos (1925); see [373, 374]. The idea of the Roos's model is very simple: Suppose that the cryptocurrency authority supplies a very small number of token in total, it will result in high prices, and if the authorities expect these high price conditions not to continue in the following period, they will raise the number of tokens and, as a result, the market price will decrease a bit. If low prices are expected to continue, the authorities will decrease the number of token, resulting again in higher prices. Thus, oscillating between periods of low number of tokens with high prices and high number of tokens with low prices, the set price-quantity traces out an oscillatory phenomenon, which will allow large volume of transactions.

15.5.4.1 Designing a Regulated Price Dynamics

For any given $\underline{p}_i < \bar{p}_i$, one can choose the coefficients $c, \hat{c}$ such that the target price $p_{tp,i}(t) \in [\underline{p}_i, \bar{p}_i]$ for all time t. An example of such an oscillatory function is as follows:

$$p_{tp,i}(t) = c_{i0} + \sum_{k=1}^{2} c_{ik} \cos(2\pi kt) + \hat{c}_{ik} \sin(2\pi kt),$$

with $c_{ik}, \hat{c}_{ik}$ to be designed to fulfill the regulation requirement. Let $c_{i0} := \frac{\underline{p}_i+\bar{p}_i}{2}$, $c_1 := \frac{\bar{p}_i-\underline{p}_i}{100}$, $\hat{c}_{i1} := \frac{\bar{p}_i-\underline{p}_i}{150}$, $c_{12} := \frac{\bar{p}_i-\underline{p}_i}{200}$, $\hat{c}_{12} := \frac{\bar{p}_i-\underline{p}_i}{250}$. We want the target function to stay between 0.98 USD and 1.02 USD. We set $\underline{p}_i = 0.98$, $\bar{p}_i = 1.02$. Figure 15.10 plots such a target function.

Note that this target price is not the market price. In order to incorporate a more realistic market behavior, we introduce a dependence on demand and supply of tokens.

15.5.4.2 Proposed Price Model for Regulated Monopoly

We propose a market price dynamics that takes into consideration the market demand and the market supply. The blockchain-based market log-price dynamics is given by $p_i(t_0) = p_0$ and

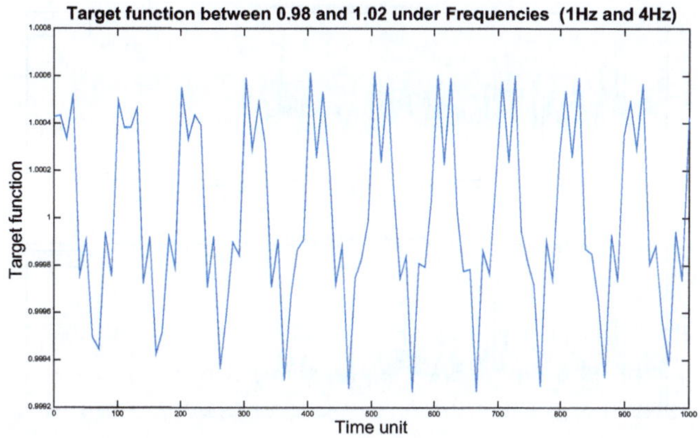

Fig. 15.10 Target price function $p_{tp,i}(t)$ between 0.98 and 1.02 under frequencies (1Hz and 4Hz)

$$
\begin{aligned}
dp_i(t) &= \eta_i[D_i(t) - p_i(t) - (S_i(t) + u_i(t))]dt \\
&\quad + \left(\sigma_i dB_i(t) + \int_{\theta\in\Theta} \mu_i(\theta)\tilde{N}_i(dt, d\theta)\right) + \sigma_o dB_o(t), \qquad (15.32)
\end{aligned}
$$

where $u_i(t)$ is the total token injected to the market at time t and B_o is a standard Brownian motion representing a global uncertainty observed by all participant to the market. As above, processes B and N are local uncertainty or noise. B is a standard Brownian motion, while N is a jump process with Lévy measure $\nu(d\theta)$ defined over Θ. It is assumed that ν is a Radon measure over Θ (the jump space). The process

$$\tilde{N}(dt, d\theta) = N(dt, d\theta) - \nu(d\theta)dt,$$

is the compensated martingale. We assume that all these processes are mutually independent. Denote by $\mathcal{F}_t^{B_o}$ the filtration generated by the observed common noise up to t, $\{B_o(t'),\ t' \leq t\}$. The number η_i is positive. For larger values of η_i, the market price adjusts quicker along the inverse demand. a, σ, σ_o are fixed constant parameters. The jump rate size $\mu(.)$ is in $L^2_\nu(\Theta, \mathbb{R})$, i.e., $\int_\Theta \mu^2(\theta)\nu(d\theta) < +\infty$. The initial distribution p_0 is square integrable: $E[p_0^2] < \infty$.

15.5.4.3 A Control Design that Tracks the Past Price

We formulate a basic control design that tracks the past price and the trend. A typical example is to choose the control action $u_{ol,i}(t) = -p_{tp,i}(t) + D_i(t) - S_i(t)$. This is an open-loop control strategy if D_i and S_i are explicit functions of time. Then the log-price dynamics becomes

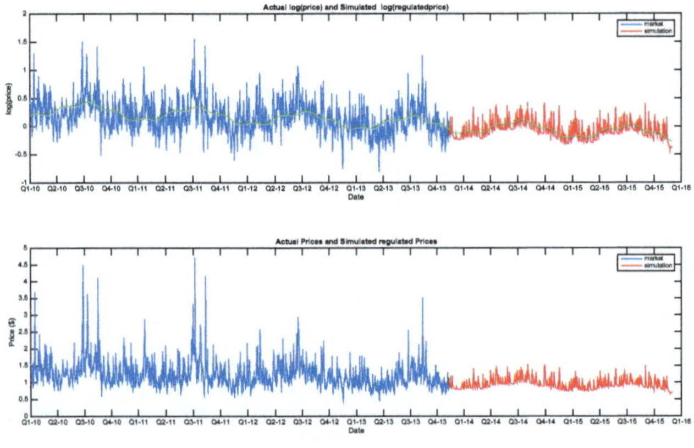

Fig. 15.11 Real market price and simulation of the regulated price dynamics as a continuation price under open-loop strategy

$$dp_i(t) = \eta_i[p_{tp,i}(t) - p_i(t)]dt + \sigma_i dB_i(t) + \int_{\theta\in\Theta} \mu_i(\theta)\tilde{N}_i(dt, d\theta) + \sigma_o dB_o(t). \tag{15.33}$$

Figure 15.11 illustrates an example of real price evolution from prosumer electricity markets in which we have incorporated a simulation of a regulated price dynamics as a continuation of real market. We observe that the open-loop control action $u_{ol,i}(t)$ decreases the magnitude of the fluctuations under similar circumstances.

15.5.4.4 An LQR Control Design

We formulate a basic LQR problem to a control strategy. Choose the control action that minimizes $E\{(p_i(t_1)-p_{tp,i}(t_1))^2+\int_{t_0}^{t_1}(p_i(t)-p_{tp,i}(t))^2dt\}$. Then the log-price dynamics becomes

$$\begin{aligned} dp_i(t) &= \eta_i[D_i(t) - p_i(t) - (S_i(t) + u_i(t))]dt \\ &\quad +\sigma_i dB_i(t) + \int_{\theta\in\Theta} \mu_i(\theta)\tilde{N}_i(dt, d\theta) + \sigma_o dB_o(t). \end{aligned} \tag{15.34}$$

The cost function here is degenerate in u but there is a best-response sequence.

15.5.4.5 Mean-Field Game Strategy

The mean-field game strategy is obtained by freezing the mean-field term $Ep_i(t) := m(t)$ resulting from other cryptocurrencies and choosing the control action that minimizes

$$\begin{array}{l} Eq(t_1)(p_i(t_1) - f(t_1))^2 + \bar{q}(t_1)[m(t_1) - f(t_1)]^2 \\ +E\int_{t_0}^{t_1} q(t)(p_i(t) - f(t))^2 + \bar{q}(t)[m(t) - f(t)]^2 dt. \end{array} \quad (15.35)$$

The mean-field term $Ep_i(t) := m(t)$ is a frozen quantity and does not depend on the individual control action $u_{mfg,i}$. Then, the price dynamics becomes

$$\begin{array}{rl} dp_i(t) = & \eta[D_i(t) - p_i(t) - (S_i(t) + u_{mfg,i}(t))]dt \\ & +\sigma_i dB_i(t) + \int_{\theta\in\Theta} \mu_i(\theta)\tilde{N}_i(dt, d\theta) + \sigma_o dB_o(t). \end{array} \quad (15.36)$$

Again the cost function here is degenerate in u but there is a best-response sequence.

15.5.4.6 A Mean-Field-Type Game Strategy

A mean-field-type game strategy consists of a choice of a control action $u_{mftg,i}$ that minimizes

$$\begin{array}{rl} L_{mftg} = & Eq_i(t_1)(p_i(t_1) - p_{tp,i}(t_1))^2 + \bar{q}_i(t_1)[E(p_i(t_1) - p_{tp,i}(t_1))]^2 \\ & +E\int_{t_0}^{t_1} q_i(t)(p_i(t) - p_{tp,i}(t))^2 + \bar{q}_i(t)[Ep_i(t) - p_{tp,i}(t)]^2\, dt. \end{array} \quad (15.37)$$

Note that here the mean-field-type term $Ep_i(t)$ is not a frozen quantity. It depends significantly on the control action $u_{mftg,i}$. The performance index can be rewritten in terms of variance as

$$\begin{array}{l} L_{mftg} = Eq_i(t_1)var(p_i(t_1) - p_{tp,i}(t_1)) + [q_i(t_1) + \bar{q}_i(t_1)][Ep_i(t_1) - p_{tp,i}(t_1)]^2 \\ +\int_{t_0}^{t_1} q_i(t)\text{Var}(p_i(t) - p_{tp,i}(t))dt \\ +E\int_{t_0}^{t_1} [q_i(t) + \bar{q}_i(t)][Ep_i(t) - p_{tp,i}(t)]^2\, dt. \end{array} \quad (15.38)$$

Then the log-price dynamics becomes

$$\begin{array}{l} dp_i(t) = \eta_i[D_i(t) - p_i(t) - (S_i(t) + u_{mftg,i})(t)]dt \\ +\left(\sigma_i dB_i(t) + \int_{\theta\in\Theta} \mu_i(\theta)\tilde{N}_i(dt, d\theta)\right) + \sigma_o dB_o(t). \end{array} \quad (15.39)$$

The cost to be paid to the regulation authority if the price does not stay within $[\underline{p}_i, \hat{p}_i]$ is $\bar{c}_i(1 - \mathbb{1}_{[\underline{p}_i, \hat{p}_i]}(p_i(t)))$, $\bar{c}_i > 0$. Since the market price is stochastic due to demand, exchange, and random events, there is still a probability to be out of the regulation range $[\underline{p}_i, \hat{p}_i]$. The outage probabilities under the three strategies $u_{ol,i}, u_{mfg,i}, u_{mftg,i}$ can be computed and used as a decision support with respect to the regulation bounds. However, these continuous time strategies may not be convenient.

Very often, the supply of token decision is made in fixed times τ_i and not continuously. We look for a simpler strategy that is piecewise constant and takes a finite number of values within the horizon $\mathcal{T}$. Since the price may fluctuate very quickly due the jump terms, we propose an adjustment based on the recent moving average called the trend: $y(t) = \int_{t-\tau_i}^{t} x(t')\phi(t,t')\lambda(dt')$, implemented at different discrete time block units.

Different regulated blockchain technologies may choose different ranges $[\underline{p}_i, \hat{p}_i]$, so that investors and users can diversify their portfolios depending on their risksensitivity index distribution across the assets. This means that there will be an interaction between the cryptocurrencies and the altcoins. The demand $D = \sum_{i=1}^{I} D_i$ will be shared between them. Users may exchange between coins and switch into another altcoins. The payoff of the blockchain-based technology i is $R_i = \hat{p}_i D_i - \bar{c}_i(1 - \mathbb{1}_{[\underline{p}_i,\hat{p}_i]}(\hat{p}_i(t)))$, where $\hat{p}_i(t) = E[p_i(t) \mid \mathcal{F}_t^{B_o}]$ is the conditional expectation of the market price with respect to $\mathcal{F}_t^{B_o}$.

15.5.5 Handling Positive Constraints

The price of the energy asset under cryptocurrency k is $x_k = e^{p_k} \geq 0$. The wealth of decision-maker i is $x = \sum_{k=0}^{d} \kappa_k x_k$. Set $u_k^I = \kappa_k x_k$ to get the state dynamics. The sum of all the u_k is x. The variation is

$$\begin{aligned} dx = [\kappa_0(r_0 + \hat{\mu}_0)x + \textstyle\sum_{k=1}^{d}[\hat{\mu}_k - (r_0 + \hat{\mu}_0)\kappa_0]u_k^I]dt \\ + \textstyle\sum_{k=1}^{d} u_k^I\{Drift_k dt + Diffusion_k + Jump_k\}, \end{aligned} \tag{15.40}$$

where

$$\begin{aligned} Drift_k &= \eta_k[D_k - p_k - (S_k + u_{mftg,k})] + \tfrac{1}{2}(\sigma_i^2 + \sigma_o^2) \\ &\quad + \textstyle\int_{\Theta}[e^{\gamma_k} - 1 - \gamma_k]\nu(d\theta), \\ Diffusion_k &= (\sigma_k dB_k + \sigma_o dB_o), \\ Jump_k &= \textstyle\int_{\Theta}[e^{\gamma_k} - 1]\tilde{N}_k(dt, d\theta). \end{aligned}$$

15.6 Consumption–Investment–Insurance

A generic agent wants to decide between consumption, investment, and insurance when the blockchain market constituted of a bond with price p_0 and several stocks with prices $p_k,\ k > 0$ and is under different switching regimes defined over a complete probability space $(\Omega, \mathcal{F}, \mathbb{P})$ in which a standard Brownian motion B, a jump process N, an observable Brownian motion B_o, and an observable continuous-time finite-state Markov chain $\tilde{s}(t)$ representing a regime switching, with $\tilde{S}$ being the set of regimes, and $\tilde{q}_{\tilde{s}\tilde{s}'}$ a generator (intensity matrix) of $\tilde{s}(t)$. The log-price processes are the ones given above. The total wealth of the generic agent follows the dynamics:

$$\begin{aligned} dx &= \kappa_0(r_0(\tilde{s}) + \hat{\mu}_0(\tilde{s}))x dt \\ &\quad + \textstyle\sum_{k=1}^d [\hat{\mu}_k - (r_0(\tilde{s}) + \hat{\mu}_0(\tilde{s}))\kappa_0 + \text{Drift}_k(\tilde{s})] u_k^I dt - u^c dt \\ &\quad - \bar{\lambda}(\tilde{s})(1 + \bar{\theta}(\tilde{s})) E[u^{ins}] dt + \textstyle\sum_{k=1}^d u_k^I \text{Diffusion}_k(\tilde{s}) \\ &\quad + \textstyle\sum_{k=1}^d u_k^I \text{Jump}_k(\tilde{s}) - (L - u^{ins}) dN, \end{aligned} \tag{15.41}$$

where $L = l(\tilde{s})x$.

In the dynamics (15.41), we have considered per-claim insurance of u^{ins}. That is, if the agent suffers a loss L at time t, the indemnity pays $u^{ins}(L)$. Such indemnity arrangements are common in private insurance at the individual level, among others. Motivated by new blockchain-based insurance products, we allow not only the cryptocurrency market but also the insurable loss to depend on the regime of the cryptocurrency economy and mean-field terms.

The payoff functional of the generic agent is

$$R = q e^{-\lambda t_1}\{\hat{x}(t_1) - [x(t_1) - \hat{x}(t_1)]^2\} + \int_{t_0}^{t_1} e^{-\lambda t} \log u^c(t)\, dt,$$

where the process $\hat{x}$ denotes $\hat{x}(t) = E[x(t) \mid \mathscr{F}_t^{\tilde{s}_0, B_o}]$. The generic agent seeks for a strategy $u = (u^c, u^I, u^{ins})$ that optimizes the expected value of R given $x(t_0), \tilde{s}(t_0)$ and the filtration generated by the common noise B_o.

For $q = 0$, an explicit solution can be found. To prove it, we choose a guess functional of the form

$$f = \alpha_1(t, \tilde{s}(t)) \log x(t) + \alpha_2(t, \tilde{s}(t)).$$

Applying Itô's formula for jump-diffusion-regime switching yields

$$\begin{aligned} & f(t, x, \tilde{s}) = f(t_0, x_0, \tilde{s}_0) + \textstyle\int_{t_0}^t \big\{\dot{\alpha}_1 \log x + \dot{\alpha}_2 + \frac{\alpha_1}{x}\kappa_0(r_0(\tilde{s}) + \hat{\mu}_0(\tilde{s}))x \\ & + \frac{\alpha_1}{x} \textstyle\sum_{k=1}^d [\hat{\mu}_k - (r_0(\tilde{s}) + \hat{\mu}_0(\tilde{s}))\kappa_0 + \text{Drift}_k(\tilde{s})] u_k^I \\ & - \frac{\alpha_1}{x} u^c - \frac{\alpha_1}{x} \bar{\lambda}(\tilde{s})(1 + \bar{\theta}(\tilde{s})) E[u^{ins}] - \frac{\alpha_1}{x^2}\frac{1}{2} \textstyle\sum_{k=1}^d \{(u_k^I \sigma_k)^2 + (u_k^I \sigma_o)^2\} \\ & + \textstyle\sum_{k=1}^d \int_\Theta \alpha_1 \log\{x + u_k^I (e^{\gamma_k} - 1)\} - \alpha_1 \log x - \frac{\alpha_1}{x} u_k^I (e^{\gamma_k} - 1)\nu(d\theta) \\ & + \bar{\lambda}[\alpha_1 \log(x - (L - u^{ins})) - \alpha_1 \log x + \frac{\alpha_1}{x}(L - u^{ins})] \\ & \textstyle\sum_{\tilde{s}'} [\alpha_1(t, \tilde{s}') - \alpha_1(t, \tilde{s})] \log x + \sum_{\tilde{s}'} \alpha_2(t, \tilde{s}') - \alpha_2(t, \tilde{s}) \big\} dt + \int_{t_0}^t d\tilde{\varepsilon}, \end{aligned} \tag{15.42}$$

where $\tilde{\varepsilon}$ is a martingale. The term $\bar{\theta}(\tilde{s})$ represents $\frac{\theta(\tilde{s})}{1+\bar{m}(t)}$ where $\bar{m}(t)$ is the average amount invested by other agents for insurance:

$$
\begin{aligned}
&R - f(t_0, x_0, \tilde{s}_0) = -f(t_1, x(t_1), \tilde{s}(t_1)) - qe^{-\lambda t_1}[x(t_1) - \hat{x}(t_1)]^2 \\
&+ \int_{t_0}^{t_1} \{\dot{\alpha}_1 \log x + \dot{\alpha}_2 + \tfrac{\alpha_1}{x}\kappa_0(r_0(\tilde{s}) + \hat{\mu}_0(\tilde{s}))x + e^{-\lambda t} \log u^c - \tfrac{\alpha_1}{x} u^c \\
&+ \tfrac{\alpha_1}{x} \textstyle\sum_{k=1}^d [\hat{\mu}_k - (r_0(\tilde{s}) + \hat{\mu}_0(\tilde{s}))\kappa_0 + \text{Drift}_k(\tilde{s})] u_k^I \\
&- \tfrac{\alpha_1}{x^2} \tfrac{1}{2} \textstyle\sum_{k=1}^d \{(u_k^I \sigma_k)^2 + (u_k^I \sigma_o)^2\} \\
&+ \textstyle\sum_{k=1}^d \int_\Theta \alpha_1 \log\{x + u_k^I (e^{\gamma_k} - 1)\} - \alpha_1 \log x - \tfrac{\alpha_1}{x} u_k^I (e^{\gamma_k} - 1) \nu(d\theta) \\
&- \tfrac{\alpha_1}{x} \bar{\lambda}(\tilde{s})(1 + \bar{\theta}(\tilde{s})) E[u^{ins}] \\
&+ \bar{\lambda}[\alpha_1 \log(x - (L - u^{ins})) - \alpha_1 \log x + \tfrac{\alpha_1}{x}(L - u^{ins})] \\
&+ \textstyle\sum_{\tilde{s}'} [\alpha_1(t, \tilde{s}') - \alpha_1(t, \tilde{s})] \log x + \sum_{\tilde{s}'} \alpha_2(t, \tilde{s}') - \alpha_2(t, \tilde{s}) \,\} \, dt + \int_{t_0}^{t_1} d\tilde{\varepsilon}.
\end{aligned} \tag{15.43}
$$

The optimal u^c is obtained by direct optimization of $e^{-\lambda t} \log u^c - \frac{\alpha_1}{x} u^c$. This is a strictly concave function, and its maximum is achieved at $u^c = \frac{e^{-\lambda t}}{\alpha_1} x$, provided that $\alpha_1(t, \cdot) > 0$ and $x(\cdot) > 0$.

This latter result can be interpreted as follows. The optimal consumption strategy process is proportional to the wealth process, i.e., the ratio $\frac{u^c(t)}{x^*(t)} > 0$. This means that the blockchain-based cryptocurrency investors will consume proportionally more when they become wealthier in the market. Similarly, the insurance strategy u^{ins} can be obtained by optimizing

$$
-\frac{1}{x}(1 + \bar{\theta}(\tilde{s})) E[u^{ins}(\tilde{s})] + \log(x - (L(\tilde{s}) - u^{ins}(\tilde{s}))) + \frac{1}{x}(L(\tilde{s}) - u^{ins}(\tilde{s}))],
$$

which yields that

$$
\frac{1}{x - L + u^{ins}} = \frac{1}{x}(2 + \bar{\theta}).
$$

Thus, noting that we have set $L(\tilde{s}) = l(\tilde{s})x$, we obtain

$$
u^{ins}(\tilde{s}) = \left[l(\tilde{s}) - \frac{1 + \bar{\theta}(\tilde{s})}{2 + \bar{\theta}(\tilde{s})} \right]^+ x = \max \left\{ 0, l(\tilde{s}) - \frac{1 + \bar{\theta}(\tilde{s})}{2 + \bar{\theta}(\tilde{s})} \right\} x.
$$

We observe that, for each fixed regime $\tilde{s}$, the optimal insurance is proportional to the blockchain investor's wealth x. We note that it is optimal to buy insurance only if $l(\tilde{s}) > \frac{1+\bar{\theta}(\tilde{s})}{2+\bar{\theta}(\tilde{s})}$. When this condition is satisfied, the insurance strategy is $u^{ins}(\tilde{s}) := \left[l(\tilde{s}) - \frac{1+\bar{\theta}(\tilde{s})}{2+\bar{\theta}(\tilde{s})} \right] x$, which is a decreasing and convex function of $\bar{\theta}$. This monotonicity property means that, as the premium loading $\bar{\theta}$ increases, it is optimal to reduce the purchase of insurance.

The optimal investment strategy u_k^I can be found explicitly by mean-field-type optimization. Incorporating all together, a system of backward ordinary differential equations can be found for the coefficient functions $\{\alpha(t, \tilde{s})\}_{\tilde{s} \in \tilde{S}}$. Lastly, a fixed-point problem is solved by computing the total wealth invested in insurance to match with $\bar{m}$.

15.7 How Different Is MFTG from Multi-population Coalitional MFG?

In classical game theory, the structure of payoff functions influences the equilibrium system. Payoff functions that are linear in joint mixed strategies are known as von Neumann expected utility. However, many applications consider non-von Neumann utility functions, which are not necessarily linear in the measure of states or actions. Such games are particularly relevant in risk quantification, where key risk measures are non-linear in the underlying probability measure. A similar issue arises in mean-field-type game (MFTG) theory, a class of games where the payoff function depends not only on state–action pairs but also on their distribution. MFTG can be a finite number of decision-makers, infinite number of decision-makers, or a combination of both within several subpopulations and atoms. MFTG equilibrium systems are nonstandard and involve the so-called master adjoint systems (MASS). Mean-field-type games differ from mean-field games, multi-population mean-field games, and multi-population coalitional mean-field games. To illustrate these distinctions, we examine a power distribution network (DIPONET) with a finite number of renewable energy producers competing in a Cournot market. Given the high volatility of renewable energy sources, variance awareness is introduced, with stochastic fluctuations modeled using a Rosenblatt process. This work elucidates fundamental differences between these models.

A class of MFTG is designed to analyze the interactions among a finite number of renewable energy producers. This class of MFTG incorporates the dynamics of price formation, demand, and supply, alongside the impact of individual production decisions. The price dynamics in this MFTG is mean-reverting around demand–supply and is influenced by the Rosenblatt process, and the payoff for each producer is evaluated over a long-term horizon. The key feature of MFTG is its ability to capture non-linearities in the payoff functions along with a variance-awareness of the supply, making it highly relevant for scenarios where individual actions significantly affect the overall system.

In contrast, multi-population coalitional MFG mimics of the same scenario consider multiple coalitions or subpopulations, each of infinite size. These models are particularly virtual agents, not true decision-makers, as individual actions have a negligible impact on the overall mean field on that coalition. The best-response problem for each decision-maker within a coalition is formulated in a linear-and-mean-field feedback form. Unlike MFTG, the payoff functions in multi-population coalitional MFG are typically linear in the individual mean field, and the equilibrium strategies are derived based on the collective behavior of the coalitions.

Key differences are established. In this particular scenario, MFTG involves a finite number of true decision-makers, whereas multi-population coalitional MFG assumes an infinite number of virtual agents within each coalition. In MFTG, individual actions significantly influence the mean field, capturing higher-order payoffs and real-world scenarios more accurately. In contrast, multi-population coalitional MFG assumes that individual actions have a negligible impact on the coalition mean field and on the other coalitions' mean field. MFTG models are non-linear in the measure of the state and control actions, making them more suitable for measure-non-linear payoff functions. Multi-population coalitional MFG, on the other hand, assumes measure-linear payoffs, which may not capture the complexities of real-world scenarios as effectively. MFTG models are better suited for scenarios where the production cost and storage are high, and the supply needs to be accurately modeled. Multi-population coalitional MFG may not adequately represent such scenarios due to its simplifying assumptions.

There are two types of mean-field terms: individual mean field and population mean field:

- Individual mean field: This can refer to the mean field of an individual state, the mean field of an individual control action, or the joint distribution of the individual state–action pair.
- Population mean field: This includes the mean field of the population state, the mean field of the control action of all decision-makers, or the joint distribution of the state–action pair of all decision-makers.

In multi-population coalitional MFG, each member is non-atomic, meaning their individual actions do not significantly impact the overall coalition mean field. The MFG framework can of course be extended to include both atomic and non-atomic agents but still with linear payoff function with respect to own state mean field. Conversely, in MFTG, a decision-maker can be atomic, non-atomic, or a combination of both, and the payoff is not necessarily linear in the individual mean field.

For interactions among renewable energy producers, we show that multi-population coalitional MFG is not suitable. This is because the simplifying assumptions of non-atomic members and linear payoffs do not capture the variance-awareness and significant impacts of individual production decisions on the overall system. MFTG, with its ability to model non-linearities and individual impacts, provides a more accurate and effective framework for optimizing renewable energy production. Table 15.5 displays some comparison of MFTG and multi-population MFG in the same scenario. By scaling the parameters $(p, c_i, r_i, \bar{r}_i)$ be $\lambda(p, c_i, r_i, \bar{r}_i)$ with $\lambda > 0$ the price of simplicity (which is the profit gap obtained after simplifying or freezing the mean-field terms) is unbounded. It turns out multi-population MFG, which consists of freezing the mean-field terms, is irrelevant, inefficient, and suboptimal for this simple variance-awareness problem. We show that the result extends to a dynamic setting with state driven by the Rosenblatt process.

In this particular scenario, there is no dynamic programming principle, no Pontryagin maximum principle, and no Wiener chaos expansion available in the literature to solve games with state dynamics driven by the Rosenblatt noise. We therefore proceed with a direct method. The direct method offers a semi-explicit representation of the solution under the restriction of the class of strategies.

Table 15.5 Comparison of MFTG and multi-population MFG. By scaling the parameters $(p, c_i, r_i, \bar{r}_i)$ to be $\lambda(p, c_i, r_i, \bar{r}_i)$ with $\lambda > 0$, the price of simplicity is unbounded. It turns out multi-population MFG, which consists of freezing the mean-field terms, is irrelevant, inefficient, and suboptimal for this simple variance-awareness problem

Features	MFTG	Multi-population MFG
I	True decision-makers	$K = I$ subpopulations of infinite size
p	Market price (random variable)	
Best response	$\sup_{u_i\in U_i}\mathbb{E}(pu_i - c_iu_i - \frac{r_i}{2}u_i^2 - \frac{\bar{r}_i}{2}(\int \mu(dy)u_i(y,\mu))^2)$	$\sup_{u_k\in U_k}\mathbb{E}(pu_k - c_ku_k - \frac{r_k}{2}u_k^2 - \frac{\bar{r}_k}{2}(\int m^*(dy)u_k^*(y,m^*))^2)$
Linearity	Non-linear in μ	linear in μ
Decomposition	$\sup_{u_i\in U_i} cov(p,u_i) - \frac{r_i}{2}var(u_i) + (\int y\mu(dy) - c_i)(\int \mu(dy)u_i(y,\mu)) - \frac{r_i+\bar{r}_i}{2}(\int \mu(dy)u_i(y,\mu))^2$	
Completion	$\sup_{u_i\in U_i} -\frac{r_i}{2}\mathbb{E}(u_i - \int \mu(dy)u_i(y,\mu) - \frac{p-\int y\mu(dy)}{r_i})^2 + \frac{var(p)}{2r_i} - \frac{r_i+\bar{r}_i}{2}(\int \mu(dy)u_i(y,\mu) - \frac{\int y\mu(dy)-c_i}{r_i+\bar{r}_i})^2 + \frac{(\int y\mu(dy)-c_i)^2}{2(r_i+\bar{r}_i)}$	$\sup_{u_k\in U_k}\mathbb{E}(-\frac{r_k}{2}(u_k - \frac{(p-c_k)}{r_k})^2 + \frac{(p-c_k)^2}{2r_k}) - \frac{\bar{r}_k}{2}(\int m^*(dy)u_k^*(y,m^*))^2)$
Mean field	$(\int \mu(dy)u_i(y,\mu)), (\int \mu(dy)u_i(y,\mu))^2,$ $cov(p,u_i), var(u_i)$	$(\int m^*(dy)u_k^*(y,m^*))^2$
Equilibrium feedback strategies	$\frac{p-\int y\mu(dy)}{r_i} + \frac{\int y\mu(dy)-c_i}{r_i+\bar{r}_i}$	$\frac{(p-c_k)}{r_k}$
Equilibrium payoff	$\frac{var(p)}{2r_i} + \frac{(\int y\mu(dy)-c_i)^2}{2(r_i+\bar{r}_i)}$	$\int \mu(dy)\frac{(y-c_k)^2}{2r_k} - \frac{\bar{r}_k}{2}(\int m^*(dy)u_k^*(y,m^*))^2)$
Gap	$\frac{var(p)}{2r_i} + \frac{(\int y\mu(dy)-c_i)^2}{2(r_i+\bar{r}_i)}$	$\frac{var(p)}{2r_k} + \frac{1}{2}\frac{r_k-\bar{r}_k}{r_k^2}(\int y\mu(dy) - c_k)^2$
Price of simplicity	0	$\frac{1}{2}[\frac{1}{(r_i+\bar{r}_i)} - \frac{r_i-\bar{r}_i}{r_i^2}](\int y\mu(dy) - c_i)^2$
Price of simplicity	0	Unbounded (by scaling)

15.7.1 Non-zero-Sum MFTG for Renewable Energy Producers

The transition to renewable energy sources has introduced new challenges and opportunities for energy producers, particularly for prosumers. In competitive markets, understanding the dynamics of price formation and the impact of individual production decisions is crucial for maximizing profitability. This section provides a basic model that integrates demand, supply, and Rosenblatt price fluctuations to evaluate the long-term average payoff for renewable energy producers. Let $I \geq 2$ be the number of producers, and denote by $\mathcal{I} = \{1, \ldots, I\}$ the set of producers. We consider the price dynamics given by

$$p(t) = p_0 + \int_0^t \frac{1}{\varepsilon}\left[a + D - S(t') - p(t')\right] dt' + R^H(t),$$

where $p(t)$ is the price at time t; p_0 is the initial price; a is a constant; D represents the demand; $S(t) = \sum_{i \in \mathcal{I}} u_i(t)$ is the total supply, with $u_i(t)$ being the production of agent i; $R^H(t)$ is the Rosenblatt process; $\varepsilon > 0$ is a speed adjustment parameter; and $r_i, \bar{r}_i > 0,\ i \in \mathcal{I}, H \in (\frac{1}{2}, 1)$. The instant payoff for producer i is given by

$$\pi_i(t, p, u) = p(t)u_i(t) - c_i u_i(t) - \frac{r_i}{2}u_i^2(t) - \frac{\bar{r}_i}{2}\bar{u}_i^2(t),$$

where $\bar{u}_i = \mathbb{E}[u_i]$ is the expected production of agent i. To evaluate the long-run average payoff as the horizon goes to infinity, we consider the time-averaged payoff:

$$\bar{\Pi}_i(p, u) = \liminf_{T\to\infty}\frac{1}{T}\mathbb{E}\int_0^T \pi_i(t, p, u)\, dt.$$

Due to the memory dependence and non-Gaussianity, we consider a very specific class of strategies as an illustrative example. Given $(u_j,\ j \in \mathcal{I}\backslash\{i\})$ in linear-and-mean-field-type feedback form, the best-response problem of the decision-maker i is

$$\begin{array}{l} \sup_{u_i \in \mathcal{U}_i} \liminf_{T\to\infty}\frac{1}{T}\mathbb{E}[\int_0^T \pi_i(t, p, u)\, dt\ |p(0) = p_0], \\ \text{subject to} \\ p(t) = p_0 + \int_0^t \frac{1}{\varepsilon}[a + D - \sum_{j\in\mathcal{I}} u_j(t') - p(t')]dt' + R^H(t), \end{array} \tag{15.44}$$

with the restricted control action set given by

$$\mathcal{U}_i := \{(u_i(t))_{t\geq 0} : u_i(t) = \eta_i(p(t) - \bar{p}(t)) + \bar{\eta}_i\bar{p}(t) + \rho_i \text{ with } (\eta_i, \bar{\eta}_i, \rho_i) \in \mathbb{R}^3, (1 + \sum_{j\in\mathcal{I}} \eta_j) > 0, (1 + \sum_{j\in\mathcal{I}} \bar{\eta}_j) > 0\}.$$

In practice, we are not in favor of considering such infinite horizon problems. The horizon length T for such interaction between producers can be relatively large but not infinite. Here we use this structure in order to get explicit representation of semi-explicit solution. In this MFTG, there is finite number of true decision-makers. The mean-field-type terms are $\bar{p}, \bar{u}_i, var(u_i)$, and $\bar{u}_i^2$. The quantities-of-interest in the payoff functional $\mathbb{E}[\pi_i(t, x, u)]$ is clearly non-linear in the measure of the state and in the measure of control action.

15.7.1.1 Two Subproblems

In order to solve the problem in (15.44), we formulate two subproblems within this MFTG framework. The best-response problem of a decision-maker is analyzed, considering scenarios with linear-and-mean-field-type feedback. The first subproblem is as follows. Given $(\tilde{u}_j,\ j \in \mathscr{I}\backslash\{i\})$ in linear-and-mean-field-type feedback form, the best-response problem of the decision-maker i is

$$\begin{array}{l} \sup_{\tilde{u}_i\in\tilde{\mathscr{U}_i}} \liminf_{T\to\infty}\frac{1}{T}\mathbb{E}[\int_0^T \tilde{\pi}_i(t, \tilde{p}(t), \tilde{u}(t))\, dt\ |\tilde{p}(0) = \tilde{p}_0], \\ \text{subject to} \\ \tilde{p}(t) = \tilde{p}_0 + \int_0^t \frac{1}{\varepsilon}[-\sum_{j\in\mathscr{I}} \tilde{u}_j - \tilde{p}]dt' + R^H(t), \end{array} \tag{15.45}$$

with $\tilde{p}_0 = p_0 - \bar{p}_0$, $\tilde{\pi}_i(t, \tilde{p}(t), \tilde{u}(t)) = \tilde{p}(t)\tilde{u}_i(t) - \frac{r_i}{2}\tilde{u}_i^2(t)$, and

$$\begin{array}{l} \tilde{\mathscr{U}_i} := \{(u_i(t))_{t\geq 0} : \tilde{u}_i(t) = \eta_i\tilde{p}(t) \text{ with } \eta_i \in \mathbb{R}, \\ \mathbb{E}[\tilde{p}(t)] = 0,\ (1 + \sum_{j\in\mathscr{I}} \eta_j) > 0\}. \end{array}$$

Subproblem (15.45) requires a stochastic calculus from the Rosenblatt process to evaluate the long-run payoff.

The second subproblem is as follows. Given $(\bar{u}_j,\ j \in \mathscr{I}\backslash\{i\})$ in linear-and-mean-field-type feedback form, the best-response problem of the decision-maker i is

$$\begin{array}{l} \sup_{\bar{u}_i\in\bar{\mathscr{U}_i}} \liminf_{T\to\infty}\frac{1}{T}[\int_0^T \bar{\pi}_i(t, \bar{p}(t), \bar{u}(t))\, dt\ |\bar{p}(0) = \bar{p}_0], \\ \text{subject to} \\ \bar{p}(t) = \bar{p}_0 + \int_0^t \frac{1}{\varepsilon}[a + D - \sum_{j\in\mathscr{I}} \bar{u}_j(t') - \bar{p}(t')]dt', \end{array} \tag{15.46}$$

with $\bar{\pi}_i(t, \bar{p}(t), \bar{u}(t)) = \bar{p}(t)\bar{u}_i(t) - c_i\bar{u}_i(t) - \frac{r_i+\bar{r}_i}{2}\bar{u}_i^2(t)$,

$$\begin{array}{l} \bar{\mathscr{U}_i} := \{(\bar{u}_i(t))_{t\geq 0} : \bar{u}_i(t) = \bar{\eta}_i\bar{p}(t) + \rho_i \text{ with } (\bar{\eta}_i, \rho_i) \in \mathbb{R}^2, \\ (1 + \sum_{j\in\mathscr{I}} \bar{\eta}_j) > 0\}. \end{array}$$

Subproblem (15.46) is a standard deterministic differential game.

15.7.1.2 Combining the Solutions of These Two Subproblems

We observe that the state is orthogonally decomposed as $x(t) = \tilde{x}(t) + \bar{x}(t)$, the control action is decomposed as $u_i(t) = \tilde{u}_i(t) + \bar{u}_i(t)$, and the payoff functional is decomposed as $\mathbb{E}[\pi_i(t, p(t), u(t))] = \mathbb{E}[\tilde{\pi}_i(t, \tilde{p}, \tilde{u}) + \bar{\pi}_i(t, \bar{p}, \bar{u})]$.

15.7.1.3 Explicit Solution of the MFTG

The MFTG (15.45) is reduced to

$$\begin{array}{l} \sup_{\eta_i \in \tilde{\mathscr{G}}_i} \liminf_{T\to\infty} \frac{1}{T}\mathbb{E}[\int_0^T (\eta_i - \frac{r_i}{2}\eta_i^2)\tilde{p}^2(t)\, dt \mid \tilde{p}(0) = \tilde{p}_0], \\ \text{subject to} \\ \tilde{p}(t) = \tilde{p}_0 + \int_0^t \frac{1}{\varepsilon}[-(1 + \sum_{j\in\mathscr{I}} \eta_j)\tilde{p}]dt' + R^H(t). \end{array} \tag{15.47}$$

Then the MFTG (15.45) is semi-explicitly solved as the following constrained game:

Theorem 15.3 *The payoff of producer i is explicitly computed as* $\tilde{\Pi}_{i,\infty}(\eta) = \frac{\Gamma(2H+1)}{2[\frac{1}{\varepsilon}(1+\sum_{j\in\mathscr{I}} \eta_j)]^{2H}} \left(\eta_i - \frac{r_i}{2}\eta_i^2\right)$ *and the gain in the strategies spaces become the coupled space* $\tilde{\mathscr{G}}_i = \{\eta_i \mid 1 + \sum_{j\in\mathscr{I}} \eta_j > 0\}$.

The first-order derivative of $\tilde{\Pi}_{i,\infty}(\eta)$ at η_i yields

$$\begin{array}{l} i \in \mathscr{I}, \\ \partial_{\eta_i}\tilde{\Pi}_{i,\infty}(\eta) = \frac{\Gamma(2H+1)}{2[\frac{1}{\varepsilon}(1+\sum_{j\in\mathscr{I}} \eta_j)]^{2H+1}} \left(1 - r_i\eta_i)(1 + \sum_{j\in\mathscr{I}} \eta_j) - 2H\eta_i + Hr_i\eta_i^2\right) \\ = \frac{\Gamma(2H+1)}{2[\frac{1}{\varepsilon}(a+\eta_i)]^{2H+1}} \left((1 - r_i\eta_i)(a + \eta_i) - 2H\eta_i + Hr_i\eta_i^2\right) \\ = \frac{\Gamma(2H+1)}{2[\frac{1}{\varepsilon}(a+\eta_i)]^{2H+1}} \left((H - 1)r_i\eta_i^2 - (r_i a + 2H - 1)\eta_i + a\right), \\ a = (1 + \sum_{j\in\mathscr{I}\setminus\{i\}} \eta_j) \end{array}$$

at the positive critical point $\partial_{\eta_i^2}\tilde{\Pi}_{i,\infty}(\eta) < 0$ making a concavity in η_i. Therefore, we have the following best-response gain:

Theorem 15.4 *Given the admissible strategies of the other agents* ($u_j \in \mathscr{U}_j,\ j \neq i$), *the best-response gain of producer i in the family of admissible feedback gains is given by* $\eta_i^* = \frac{-(r_i a+2H-1)+\sqrt{(r_i a+2H-1)^2+4(1-H)r_i a}}{2(1-H)r_i}$, *and the best-response cost is given by* $\tilde{\Pi}_{i,\infty}(\eta^*)$.

Theorem 15.5 *If there exists a solution to the following system:*

$i \in \mathcal{I}$,

$$\eta_i = \frac{-(r_i(1+\sum_{j\in\mathcal{I}\backslash\{i\}}\eta_j)+2H-1)+\sqrt{(r_i(1+\sum_{j\in\mathcal{I}\backslash\{i\}}\eta_j)+2H-1)^2+4(1-H)r_i(1+\sum_{j\in\mathcal{I}\backslash\{i\}}\eta_j)}}{2(1-H)r_i}$$

$1+\sum_{j\in\mathcal{I}}\eta_j > 0$

then the non-zero-sum stochastic differential game under the Rosenblatt noise (15.45) has an equilibrium in stationary linear state-feedback strategies given by $\eta_i^*(p(t)-\bar{p}(t))$.

The MFTG (15.46) is reduced to

$$\begin{aligned}
&\sup_{\bar{\eta}_i\in\bar{\mathcal{G}}_i} \liminf_{T\to\infty}\tfrac{1}{T}\mathbb{E}[\textstyle\int_0^T -\frac{r_i+\bar{r}_i}{2}(\bar{u}_i-\frac{(\bar{p}(t)-c_i)}{(r_i+\bar{r}_i)})+\frac{(\bar{p}(t)-c_i)^2}{2(r_i+\bar{r}_i)}\,dt \text{ |subject to}\\
&\bar{p}(t)=\bar{p}_0+\textstyle\int_0^t \frac{1}{\varepsilon}[a+D-\sum_{i\in\mathcal{I}}\rho_i-(1+\sum_{j\in\mathcal{I}}\eta_j)\bar{p}(t')]dt',
\end{aligned} \tag{15.48}$$

$\bar{\mathcal{G}}_i := \{(\bar{\eta}_i,\rho_i)\in\mathbb{R}^2 : (1+\sum_{j\in\mathcal{I}}\bar{\eta}_j) > 0\}$.

By direct computation, we obtain the following:

Theorem 15.6 *The non-zero-sum deterministic differential game with long-run payoff (15.46) has an equilibrium in linear state feedback strategies given by* $\bar{\eta}_i^*\bar{p}(t)+\rho_i$ *where*

$$\begin{aligned}
&\lim_{t\to+\infty}\bar{p}(t)=\bar{p}^*=\frac{a+D+\sum_{i\in\mathcal{I}}\frac{c_i}{r_i+\bar{r}_i}}{1+\sum_{j\in\mathcal{I}}\frac{1}{r_i+\bar{r}_i}},\\
&\bar{\eta}_i^*=\frac{1}{r_i+\bar{r}_i}\\
&\rho_i^*=\frac{-c_i}{r_i+\bar{r}_i}\\
&\bar{u}_i^*(t)=\frac{(\bar{p}(t)-c_i)}{(r_i+\bar{r}_i)},\\
&\bar{\Pi}_{i,\infty}^*(\bar{\eta},\rho)=\frac{(\bar{p}^*-c_i)^2}{2(r_i+\bar{r}_i)}.
\end{aligned} \tag{15.49}$$

We now combine the two subproblems to solve the original problem in (15.44):

Theorem 15.7 *Consider (15.44). The price-taking Cournot–Nash equilibrium is given by*

$$\begin{aligned}
&u_i^*(t)=\eta_i^*(p^*(t)-\bar{p}^*(t))+\bar{\eta}_i\bar{p}^*(t)+\rho_i^*,\\
&\eta_i^*=\frac{-(r_i(1+\sum_{j\in\mathcal{I}\backslash\{i\}}\eta_j^*)+2H-1)+\sqrt{(r_i(1+\sum_{j\in\mathcal{I}\backslash\{i\}}\eta_j^*)+2H-1)^2+4(1-H)r_i(1+\sum_{j\in\mathcal{I}\backslash\{i\}}\eta_j^*)}}{2(1-H)r_i}\\
&\lim_{t\to+\infty}\bar{p}(t)=\bar{p}^*=\frac{a+D+\sum_{i\in\mathcal{I}}\frac{c_i}{r_i+\bar{r}_i}}{1+\sum_{j\in\mathcal{I}}\frac{1}{r_i+\bar{r}_i}},\\
&\bar{\eta}_i^*=\frac{1}{r_i+\bar{r}_i}\\
&\rho_i^*=\frac{-c_i}{r_i+\bar{r}_i}\\
&\bar{u}_i^*(t)=\frac{(\bar{p}^*(t)-c_i)}{(r_i+\bar{r}_i)},\\
&\Pi_{i,\infty}^*(\eta,\bar{\eta},\rho)=\frac{\Gamma(2H+1)}{2[\frac{1}{\varepsilon}(1+\sum_{j\in\mathcal{I}}\eta_j^*)]^{2H}}\left(\eta_i^*-\frac{r_i}{2}(\eta_i^*)^2\right)+\frac{(\bar{p}^*-c_i)^2}{2(r_i+\bar{r}_i)}.
\end{aligned} \tag{15.50}$$

We now compute explicitly the limiting price as a function of associated control strategies.

Theorem 15.8 *The non-zero-sum deterministic differential game with long-run payoff (15.46) has an equilibrium in stationary strategies given by* $\bar{\eta}_i^* \bar{p}^* + \rho_i^*$ *where*

$$\begin{array}{l} \bar{p} = \frac{a+D-\sum_{i\in\mathcal{I}}\rho_i}{1+\sum_{j\in\mathcal{I}}\bar{\eta}_j}, \\ \bar{\Pi}_{i,\infty}^*(\bar{\eta},\rho)=(\bar{\eta}_i-\frac{r_i+\bar{r}_i}{2}\eta_i^2)(\bar{p}^*)^2+(\rho_i-c_i\bar{\eta}_i-(r_i+\bar{r}_i)\bar{\eta}_i\rho_i)(\bar{p}^*)-(c_i\rho_i+\frac{r_i+\bar{r}_i}{2}\rho_i^2) \\ = (\bar{\eta}_i - \frac{r_i+\bar{r}_i}{2}\bar{\eta}_i^2)(\frac{a+D-\sum_{i\in\mathcal{I}}\rho_i}{1+\sum_{j\in\mathcal{I}}\bar{\eta}_j})^2 + (\rho_i - c_i\bar{\eta}_i - (r_i+\bar{r}_i)\bar{\eta}_i\rho_i)(\frac{a+D-\sum_{i\in\mathcal{I}}\rho_i}{1+\sum_{j\in\mathcal{I}}\bar{\eta}_j}) \\ \quad -(c_i\rho_i + \frac{r_i+\bar{r}_i}{2}\rho_i^2) \\ (\bar{\eta}_i^*, \rho_i^*) \in \arg\max_{\bar{\eta}_i,\rho_i} \bar{\Pi}_{i,\infty}^*(\bar{\eta}_i, \rho_i, \bar{\eta}_{-i}^*, \rho_{-i}^*) \\ \bar{u}_i^* = \bar{\eta}_i^*\bar{p}^* + \rho_i^* \\ \bar{p}^* = \frac{a+D-\sum_{i\in\mathcal{I}}\rho_i^*}{1+\sum_{j\in\mathcal{I}}\bar{\eta}_j^*}. \end{array} \tag{15.51}$$

Theorem 15.9 *Consider (15.44). The stationary closed-loop Cournot–Nash equilibrium is given by*

$$\begin{array}{l} u_i^* = \eta_i^*(p^* - \bar{p}^*) + \bar{\eta}_i^*\bar{p}^* + \rho_i^*, \\ \eta_i^* = \frac{-(r_i(1+\sum_{j\in\mathcal{I}\backslash\{i\}}\eta_j^*)+2H-1)+\sqrt{(r_i(1+\sum_{j\in\mathcal{I}\backslash\{i\}}\eta_j^*)+2H-1)^2+4(1-H)r_i(1+\sum_{j\in\mathcal{I}\backslash\{i\}}\eta_j^*)}}{2(1-H)r_i} \\ (\bar{\eta}_i^*, \rho_i^*) \in \arg\max_{\bar{\eta}_i,\rho_i} \bar{\Pi}_{i,\infty}^*(\bar{\eta}_i, \rho_i, \bar{\eta}_{-i}^*, \rho_{-i}^*) \\ \bar{u}_i^* = \bar{\eta}_i^*\bar{p}^* + \rho_i^* \\ \bar{p}^* = \frac{a+D-\sum_{i\in\mathcal{I}}\rho_i^*}{1+\sum_{j\in\mathcal{I}}\bar{\eta}_j^*}. \end{array} \tag{15.52}$$

15.7.2 Multi-population Coalitional MFG

Consider $K \geq 2$ number of coalitions (or subpopulation) and denote by $\mathcal{K} = \{1, 2, \ldots, K\}$ the set of all coalitions. Each coalition is assumed to be of infinite size with a proper scaling. The multi-population coalitional MFG driven by the Rosenblatt noise can be defined. Consider a generic decision-maker from coalition k. Given u_k^* of all other members of coalition k and $(u_j^*,\ j \in \mathcal{K}\backslash\{k\})$ of the other coalitions in linear-and-mean-field feedback form, the best-response problem of the generic decision-maker in coalition k in linear-and-mean-field feedback form is

$$
\begin{array}{l}
\sup_{u_k\in\mathcal{U}_k} \liminf_{T\to\infty}\frac{1}{T}\mathbb{E}[\int_0^T p(t)u_k(t) - c_k u_k(t) - \frac{r_k}{2}u_k^2(t) \\
-\frac{\bar{r}_k}{2}\int m^*(t,dy)u_k^*(t,y,m^*)\,dt\ |p(0)=p_0], \\
\text{subject to} \\
p(t) = p_0 + \int_0^t \frac{1}{\varepsilon}[a + D - \sum_{j\in\mathcal{K}} u_j(t') - p(t')]dt' + R^H(t), \\
\bar{m}^*(t) = \bar{m}_0 + \int_0^t \frac{1}{\varepsilon}[a + D - \sum_{j\in\mathcal{K}} \int m^*(t',dy)u_j^*(t,y,m^*) - \bar{m}^*(t')]dt' \\
m^*(t,dy) = \mathbb{P}_{p^*(t)}(dy)
\end{array}
\tag{15.53}
$$

with the restricted control action set given by

$$
\begin{array}{l}
\mathcal{U}_k := \{(u_k(t))_{t\geq 0} : u_k(t) = \eta_k p(t) + \rho_k \text{ with } (\eta_k,\rho_k)\in\mathbb{R}^2, \\
(1+\sum_{j\in\mathcal{K}}\eta_j) > 0\}.
\end{array}
$$

By definition of multi-infinite population MFG, the control action u_k of a generic decision-maker in coalition c has no impact on mean of the population mean-field $\bar{m}^*(t)$. It has also no effect on the population mean-field $m^*(t,dy)$, the subpopulation mean of mean field of actions $\int m^*(t,dy)u_k^*(t,y,m^*)$, and any other subpopulation mean of mean field of actions $\int m^*(t,dy)u_j^*(t,y,m^*)$, $j\in\mathcal{K}\backslash\{k\}$. Next we rewrite the functionals involved.

The best response is decoupled of m^*, and it is given by

$$
\begin{array}{l}
\sup_{(\eta_k,\rho_k)\in\mathcal{G}_{k,mfg}} (\eta_k - \frac{r_k}{2}\eta_k^2)[\frac{\Gamma(2H+1)}{2[\frac{1}{\varepsilon}(1+\sum_{j\in\mathcal{J}}\eta_j)]^{2H}} + \frac{(a+D)^2}{\varepsilon^2} + 2\frac{a+D}{\varepsilon}(\frac{a+D-\sum_{i\in\mathcal{J}}\rho_i}{1+\sum_{j\in\mathcal{J}}\eta_j})] \\
+(\rho_k - c_k\eta_k - r_k\eta_k\rho_k)(\frac{a+D-\sum_{i\in\mathcal{J}}\rho_i}{1+\sum_{j\in\mathcal{J}}\eta_j}) - (c_k\rho_k + \frac{r_k}{2}\rho_k^2) - \frac{\bar{r}_k}{2}(\eta_k^*\frac{a+D-\sum_{i\in\mathcal{J}}\rho_i^*}{1+\sum_{j\in\mathcal{J}}\eta_j^*} + \rho_k^*)^2, \\
\text{subject to} \\
\bar{m}^*(t) = \bar{m}_0 + \int_0^t \frac{1}{\varepsilon}[a + D - \sum_{k\in\mathcal{K}}\rho_k^* - (1+\sum_{j\in\mathcal{K}}\eta_j^*)\bar{m}^*(t')]dt' \\
m^*(t,dy) = \mathbb{P}_{p^*(t)}(dy).
\end{array}
\tag{15.54}
$$

Theorem 15.10 *The mean-field equilibrium of virtual multi-population coalitional MFG does not depend on $\bar{r}_k$. In particular, for very large r_k, the equilibrium stays the same.*

Note that the term $-\frac{\bar{r}_k}{2}(\eta_k^*\frac{a+D-\sum_{i\in\mathcal{J}}\rho_i^*}{1+\sum_{j\in\mathcal{J}}\eta_j^*} + \rho_k^*)^2$ is from frozen as it is coming from the infinite population mean field. The consequence of this observation is that the mean-field equilibrium does not capture the energy producer market as when the production cost and storage are extremely high, the supply should be nearly zero as observed in the true MFTG above. Thus, the virtual game created to represent the MFTG into a multi-population MFG does not do the job. In this particular case, it does not even simplify the analysis. This multi-population MFG does not have the orthogonal decomposition advantage used in MFTG. This proves also that multi-population coalitional MFG differs from MFTG.

15.7.3 *Difference Between MFTG and Multi-population Coalitional MFG*

True Number of Decision-Makers
In this particular scenario, MFTG involves a finite number of true decision-makers, whereas multi-population coalitional MFG assumes an infinite number of virtual agents within each coalition to represent the same scenario. As we can see, these virtual agents are not true decision-makers.

No Asymptotic Indistinguishability Assumption
In this MFTG, each decision-maker has a strong influence on its own mean-field as well as in the other mean-field terms. The asymptotic indistinguishability assumption does not hold in MFTG. The permutation of index of the decision-maker changes the payoff. However, in the multi-population coalitional MFG off the same scenario, it is the asymptotic indistinguishability that is assumed within each coalition.

Effect of a Generic Decision-Maker on Its Individual Mean Field
In MFTG, individual actions significantly influence the mean field, capturing higher-order payoffs and real-world scenarios more accurately. In contrast, this multi-population coalitional MFG assumes that individual actions have a negligible impact on the coalition mean field and on the other coalitions' mean field.

The MFTG Payoff Is Non-linear in the Individual State Mean Field
MFTG models are non-linear in the measure of the state and control actions, making them more suitable for measure-non-linear payoff functions. Multi-population coalitional MFG, on the other hand, assumes measure-linear payoffs, which may not capture the risk-awareness of real-world scenarios as effectively.

The MFTG Payoff Captures Higher-Order Payoffs
The MFTG payoff captures higher-order payoffs. Multi-population coalitional MFG is limited to measure-linear payoff. The equilibrium equation of MFGs is nonvalid when it comes to non-linear payoff such as variance, quantile, expected-value-at-risk, expectile value-at-risk, extremile value-at-risk, and entropic value-at-risk.

MFTGs Capture More Real-World Scenarios
MFTG models are better suited for scenarios where the production cost and storage are high and the supply needs to be accurately modeled. Multipopulation coalitional MFG may not adequately represent such scenarios due to its simplifying assumptions. Both MFTG and multi-population coalitional MFG offer valuable insights in the demand side. When it comes to the dynamics of renewable energy production, MFTG stands out as the more suitable framework for measure-non-linear payoff functions. MFTG's ability to capture the variance-awareness of individual actions

and their impact on the overall system makes it a powerful tool for optimizing renewable energy markets. Multi-population coalitional MFG approach freezes all the mean-field terms and uses a matching argument. This leads to sub-optimality in important scenarios such as risk-awareness renewable energy production.

15.8 Notes

Mean-field-type games have been used to study blockchain-based distributed power networks with several different entities: investors, consumers, prosumers, producers, miners, verifiers, and validators; see [47, 48, 129, 244, 464, 465]. The blockchain-enabled energy platforms are online, but the execution is sometimes influenced by the geographical location of their servers and customers. In this context, the entire market has something similar to *too domestic to fail* [161] where here the energy producers are national champions by their production homes.

Chapter 16
Blockchain Token Economics

This chapter studies blockchain cryptographic tokens by means of mean-field-type game theory. It introduces variance-aware utility function to capture the risk of cryptographic tokens associated with the uncertainties of adoption, network security, regulatory legislation, and market volatility. A relationship between network characteristics, price, number of token holders, and token supply is established. Both in-chain and cross-chain diversifications between tokens are examined using a mean–variance approach. The results suggest that the number of tokens in circulation need to be adjusted in order to capture risk-awareness and self-regulatory behavior in blockchain token economics. Sharpe and Modigliani ratios for crypto-tokens are revisited.

Blockchain Interactions
In basic game theory, a utility function of an agent should capture its preferences and its satisfaction for using an action. The satisfaction drives the demand of the agent, which in turn drives the price together with the supply. Network economics applies a similar principle by including network characteristics. A blockchain is a peer-to-peer network. A blockchain is a distributed cryptographic ledger shared among all participating nodes, over which every data or transaction is recorded. The transactions are collected in blocks, which are found in a random process from the protocol. Some blockchains do have cryptographic tokens and some others not. In the context of cryptographic assets, a utility function of an agent would be dependent on the number of tokens in circulation in the blockchain, the demand which is the aggregated drives of agents or entities, the network characteristics (security, vulnerability, delay), and the incentives to verifiers (miners, validators, stakeholders, delegates, enough-token holders, enough-reputation holders, etc.). The game theoretic incentives are key elements for the verifiers to engage in a desired set of behaviors [47, 129, 460, 463].

The rest of the chapter is structured as follows: Sect. 16.1 introduces blockchain token economics. Section 16.2 presents some blockchain models. Section 16.3

T. Başar et al., *Mean-Field-Type Game Theory II*, Static & Dynamic Game Theory: Foundations & Applications, https://doi.org/10.1007/978-3-032-07031-9_16

analyzes the token outcomes: adoption, demand, and supply. Section 16.4 focuses on token diversification and exchange cross chain and within the chain. Section 16.5 discusses the use of generative machine intelligence for decision support in blockchain token economics using variance and squared bias reduction algorithms which are of mean-field type.

16.1 What Is Blockchain Token Economics?

A blockchain is a decentralized, distributed ledger that records transactions across many computers in a secure, transparent, and immutable manner. The core structure of a blockchain consists of a chain of blocks, where each block contains a list of transactions. Each block includes a cryptographic hash of the previous block, a timestamp, and the transaction data, linking all blocks together in a chronological order, forming a "chain." At a high level, the process starts with a user, often referred to as a node, initiating a transaction, which is broadcasted to the entire network of participating nodes. These nodes can either be full nodes, which store and verify the entire blockchain, or lightweight nodes, which rely on full nodes to access data. The transaction is then validated through consensus algorithms, the most common being proof of work and proof of stake. In PoW, miners compete to solve complex mathematical puzzles to create the next block, using computational power, and the first to solve the puzzle gets to add the block to the chain and is rewarded with cryptocurrency. In proof of stake, validators are chosen to propose and validate blocks based on the amount of cryptocurrency they hold and are willing to "stake" as collateral. Once a consensus is reached, the block containing the transaction is added to the blockchain, making it tamper-resistant, as altering any block would require altering all subsequent blocks, which is computationally impractical. Cryptographic keys play a significant role in ensuring security; users have a private key, which they use to sign transactions, and a public key, which acts as an address for receiving transactions. Smart contracts, which are self-executing contracts with the terms of the agreement written into code, can also be deployed on blockchain platforms like Ethereum, enabling automatic and trustless execution of predefined actions. Network nodes continuously propagate the updated blockchain to ensure all participants have a synchronized copy. Mechanisms such as trees and directed graphs are used to efficiently and securely verify the integrity of data without needing to access the entire chain. As a decentralized system, blockchains eliminate the need for a central authority, ensuring transparency, security, and trust in environments where multiple, often unknown, entities interact. However, this decentralization comes with trade-offs in terms of scalability, energy consumption (especially in PoW-based systems), and transaction throughput, which ongoing innovations like sharding, layer-2 solutions and new consensus mechanisms aim to address. In essence, a blockchain ecosystem consists of miners or validators, nodes, users, smart contracts, consensus algorithms, and cryptographic protocols, all working together to provide a secure, transparent, and decentralized ledger of transactions.

16.1.1 Low-Energy Consensus Algorithms

Alternative consensus mechanisms, such as mobile-based verification and validation, aim to address the inefficiencies and energy-intensive nature of traditional proof-of-work consensus mechanisms. These alternative approaches focus on leveraging lightweight, accessible, and less resource-intensive devices like mobile phones to contribute to the validation of transactions and maintain blockchain security. In traditional proof of work, miners use vast computational power to solve complex cryptographic puzzles, leading to enormous energy consumption, particularly as difficulty levels rise. This has prompted the search for more sustainable solutions. Mobile-based consensus mechanisms could potentially replace or supplement traditional methods by utilizing the ubiquity and processing capabilities of mobile devices, which are far less energy-demanding than industrial-grade mining rigs. One such method might involve utilizing a proof-of-stake-like approach, where users' mobile devices are employed to validate transactions based on their stake or participation in the network. In this model, users with a mobile device holding a certain amount of cryptocurrency could act as validators or witnesses, signing off on transactions without the need for intensive computations. A variant of this could be delegated proof of stake, where mobile users can vote for delegates to validate transactions on their behalf, thereby reducing the number of active validators and streamlining the process. Mobile-based consensus can also incorporate proof of activity, where users' devices contribute to the network based on their activity levels, such as the number of transactions they initiate or verify. This model may also incentivize users to remain online and keep their devices active, contributing to the blockchain's security while maintaining energy efficiency. Additionally, mobile phones are well-suited for consensus mechanisms like proof of burn, where users burn (or render unusable) tokens to participate in transaction validation, providing a non-computational mechanism to secure the network. Mobile phones could also be integral to proof of capacity, where users allocate unused storage space on their devices to support the blockchain. Another innovative mobile-based consensus approach could leverage the existing mobile infrastructure, such as geographical position signals or sensor data, to validate user location or behavior as part of a decentralized consensus mechanism, ensuring a more energy-efficient network without relying on high-power mining. The underlying idea is to replace computationally expensive tasks with tasks that mobile devices can perform more efficiently, using smaller amounts of energy, while still maintaining robust security and decentralized control. By leveraging mobile devices for consensus, blockchain networks could become more accessible to the general public, enabling more people to participate without needing specialized hardware, thus democratizing the validation process. These mobile-based consensus mechanisms hold promise for reducing the environmental footprint of blockchain technology while ensuring scalability, security, and decentralization in a more sustainable manner. However, challenges remain in ensuring the robustness, fairness, and resilience of these systems, particularly in maintaining network security without the substantial computational guarantees provided by proof of work, and in preventing issues like attacks or centralization of power among a few high-stake mobile users. Nonetheless, the shift toward mobile-based consensus represents an exciting evolution in blockchain technology that aligns more closely with principles of sustainability and accessibility.

16.1.2 Who Are the Decision-Makers in a Blockchain?

In a blockchain ecosystem, decision-making involves several entities, each playing a critical role in ensuring the network's functionality, security, and decentralization. These entities range from technical participants, like miners or validators, to governance actors such as developers, node operators, users, and even external stakeholders. At the core, miners, verifies, or validators are key decision-makers, as they are responsible for validating transactions, adding new blocks to the chain, and ensuring the integrity of the blockchain. Their role is fundamental to maintaining consensus and upholding the security of the network, as they either solve complex cryptographic puzzles, vote, or stake their cryptocurrency as collateral to ensure honest behavior. Another critical decision-making entity is the node operator, which can be a full node or a light node. Full nodes store a complete copy of the blockchain and validate the authenticity of transactions and blocks independently. They ensure that the consensus rules are strictly followed, rejecting invalid blocks or transactions that do not adhere to the network's protocol. Full nodes are decentralized arbiters that prevent any single party from unilaterally altering the blockchain's rules. Developers are another key group of decision-makers, particularly when it comes to technical upgrades, protocol changes, and governance proposals. Developers maintain the blockchain's software, propose updates, fix bugs, and often introduce new features or optimizations that affect the entire network. Their proposals often require approval or consensus from other stakeholders like miners, validators, and node operators. Governance within a blockchain can also involve decision-making by token holders, who often participate in decentralized governance systems. In platforms like Ethereum and others with governance tokens, token holders can vote on key issues like protocol upgrades, fee structures, and other parameters that affect the operation of the blockchain. Their voting power is usually proportional to the amount of tokens they hold, and they directly influence the future direction of the blockchain. This form of decentralized governance, sometimes referred to as on-chain governance, allows users to have a say in how the network evolves. In decentralized autonomous organizations, token holders collectively make decisions regarding the allocation of resources, the execution of smart contracts, and other governance-related tasks, further decentralizing power. Consensus algorithms themselves are also decision-making entities. Algorithms like proof of work, proof of stake, delegated proof of stake, or models like proof of authority define how the blockchain reaches agreement on the state of the ledger and how conflicting versions of the blockchain are resolved. These algorithms, coded into the network, act as arbiters for validating blocks and preventing double-spending or malicious behavior. In addition to internal stakeholders, external actors can also play a role in decision-making on blockchain networks, especially when considering interoperability with other blockchains, regulatory compliance, or community-driven initiatives, for example, exchanges that list blockchain tokens or businesses that use blockchain technology might have indirect influence over network decisions, particularly if they hold significant amounts of the token or if their integration depends on specific network functionalities. Oracle providers, who supply off-chain data to smart contracts, also participate in decision-making by ensuring the accuracy and reliability of real-world data that influences on-chain outcomes. Furthermore, in more advanced blockchain governance models, formal councils, committees, or governance groups are created with specific roles in making or ratifying decisions. These specialized governance bodies are tasked with executing key decisions, such as protocol upgrades, and ensuring the network remains decentralized, transparent, and secure.

16.1.3 What Is Blockchain Token Economics?

Blockchain token economics involves the study of incentives and strategies that govern the behavior of participants within a decentralized network, where tokens serve as the primary unit of value exchange. Game theory is used to analyze how rational actors—miners, verifiers, validators, users, developers, and other stakeholders—make decisions based on the rules and rewards embedded in the system. In token economics, these participants aim to maximize their utility (profits, rewards, or influence), while the network ensures that their individual incentives align with the broader goal of securing and maintaining the blockchain. The verification, validation, and incentive mechanisms are structured to create a mean-field-type equilibrium where it is in every participant's best interest to act honestly, for example, verifiers are incentivized to follow the protocol because deviating from the rules—such as attempting a double-spend—would lead to loss of mining rewards or reputation or even devaluation of the currency, which they hold. Similarly, proof of stake ensures that validators stake their tokens as collateral, aligning their economic interests with the security of the network; any dishonest behavior risks losing their staked tokens. The entire ecosystem functions as a risk-aware and incentive-driven mean-field type game [460, 464] where participants are rewarded for contributing to network security, while penalties for malicious actions create disincentives. This equilibrium is maintained through the careful design of rewards (e.g., block rewards, transaction fees) and punishments (e.g., slashing), ensuring that the optimal strategy for each participant is to participate honestly, cooperate, thus maintaining data integrity, information integrity, and network integrity. Additionally, in decentralized governance models, token holders vote on protocol changes, creating a dynamic where stakeholders must balance their short-term gains with long-term network health.

16.1.4 MFTG-Based Incentive Loop

An incentive loop design between verifiers, developers, investors, regulators, and users in blockchained token economics is essential because it creates a self-sustaining ecosystem where the interests of all parties are aligned, ensuring the long-term stability, security, and growth of the blockchain network. In such a decentralized system, each of these actors plays a crucial role, and their motivations need to be harmonized to prevent any misalignment that could threaten the network's health. Verifiers (miners or validators) are responsible for maintaining the integrity of the blockchain by validating transactions and securing the network. They need incentives, typically in the form of block rewards or transaction fees, to continue performing these tasks honestly and efficiently. Developers, on the other hand, design the protocol, upgrade the network, and build applications. Their incentives are tied to the long-term usability and growth of the platform, often via token rewards or governance rights, making them essential in maintaining the blockchain's functionality and introducing innovations that keep the network competitive. Investors bring capital into the system by purchasing tokens or funding projects, and they are incentivized by the potential appreciation of token value. If the blockchain is well-designed, investor confidence grows, leading to greater network participation, funding for further development, and an increase in token value. Users, who interact with the blockchain through transactions or decentralized applications, are incentivized through the utility provided by the network, whether it be financial gains, data security, or decentralized services. Their demand drives network activity, which in turn generates more transaction fees and token circulation, benefiting verifiers and investors. Regulators, though often external actors, play a critical role in ensuring that the system remains compliant with legal frameworks. Proper regulation can help prevent fraudulent behavior, protect users, and foster broader adoption, ensuring the blockchain's legitimacy and sustainability. Without proper incentives, any of these actors could undermine the network—verifiers might not secure it, developers may stop innovating, users may abandon the platform, investors could lose interest, and regulators might clamp down with restrictive policies. An incentive loop ensures that each actor benefits from the success of the network and has a vested interest in its continued operation, for instance, verifiers are rewarded for maintaining security, which ensures users trust the platform, driving more usage and token demand, benefiting investors, whose financial backing supports further development, which in turn attracts more users and verifiers, creating a virtuous cycle. Moreover, a well-designed incentive loop helps create network effects, where the more participants join, the more valuable the system becomes for everyone. This mutual interdependence ensures that the network remains decentralized and resilient, even as it scales. A breakdown in this incentive loop could lead to centralization, reduced security, or abandonment of the network, which is why designing a balanced, self-reinforcing incentive structure between verifiers, developers, investors, regulators, and users is crucial for the success of any blockchain-based economy.

16.1.5 Mean–Variance in Blockchain Token Economics

Mean–variance is an important utility function in blockchain token economics because it helps participants—whether they are investors, validators, or users—balance the trade-off between risk and return when making decisions in a volatile and often uncertain environment. Blockchain networks, especially those that use cryptocurrencies or tokens as part of their economic model, are highly volatile due to factors like market speculation, technological developments, regulatory shifts, and network usage fluctuations. In this context, mean–variance utility functions, derived from modern portfolio theory, provide a framework for decision-makers to optimize their portfolios by maximizing expected returns (mean) while minimizing exposure to risk (variance). For investors, the application of mean–variance allows them to select an optimal portfolio of blockchain tokens, balancing the expected gains from holding a volatile asset with the risk of potential losses due to price swings or network failures. Since blockchain projects often involve speculative investment, the mean–variance utility function provides a way to quantify and manage the uncertainty surrounding token prices, helping investors to diversify their token holdings and hedge against risks associated with a single token's volatility.

For validators and miners who are often compensated in tokens for securing the network, mean–variance is also relevant because it helps them evaluate the risk of future token price fluctuations against the expected rewards for participating in consensus mechanisms. If the token price is highly volatile, validators may face significant uncertainty in the value of their future rewards, which can influence their decision to participate in the network or to sell their holdings immediately. Mean–variance utility helps validators assess the balance between the expected value of their future rewards and the risk of devaluation, ensuring that they are compensated fairly for the work they perform to secure the network. This is especially critical in PoS systems, where validators lock up tokens as collateral, effectively tying their capital to the future success and stability of the network.

Moreover, blockchain token economics often involves incentives for users to hold and use tokens, whether for transaction fees, governance participation, or staking. Mean–variance utility is important here because it allows users to make informed decisions about when to hold tokens, when to transact, and when to exit their positions, based on their risk tolerance and expected returns, for example, a user may want to weigh the benefits of staking tokens in a proof-of-stake system, where they earn rewards, against the potential risk that the value of the staked tokens might decrease due to market volatility. Using a mean–variance approach, the user can estimate the optimal amount of tokens to stake that balances the potential rewards from staking with the risks of price fluctuation.

In a broader sense, the use of mean–variance as a utility function in blockchain token economics supports the creation of more stable and resilient ecosystems. It encourages participants to make rational, data-driven decisions about their involvement in the network, which can reduce speculative cycles, panic selling, and other destabilizing behaviors. By understanding the relationship between expected returns and risks, blockchain stakeholders can better navigate the volatility inherent in token markets, contributing to the long-term stability and success of blockchain networks.

16.1.6 Token Price Correlation Is Empirically Observed

Token prices in blockchain ecosystems are correlated not only with other internal and external blockchain tokens but also with the entire cryptocurrency market and increasingly with central bank digital currencies (CBDCs) of major countries and organizations, due to a combination of interdependencies, market psychology, macroeconomic factors, and evolving regulatory environments. At the core of this correlation is the fact that most cryptocurrencies, tokens, and even CBDCs operate within the same broad financial ecosystem, where their value is often perceived through the lens of overall market sentiment. Cryptocurrencies tend to be influenced by the same set of investors and traders, many of whom allocate capital across multiple blockchain projects and tokens, driving correlations between prices, for example, a significant price movement in Bitcoin (BTC), which acts as a bellwether for the cryptocurrency market, often triggers similar movements across a wide range of tokens, including both internal tokens (native to a specific blockchain) and external tokens (from other blockchain networks). This is because Bitcoin's dominance in market capitalization leads to it being viewed as a proxy for the health of the broader crypto ecosystem. When BTC experiences price volatility, it reflects shifting market sentiment, which can cause ripple effects on the prices of altcoins and tokens as investors adjust their portfolios in response to the changing risk environment.

Internal blockchain tokens, such as those used in decentralized finance (DeFi) platforms or governance tokens in specific ecosystems, often display correlations due to shared dependencies on the same network infrastructure or underlying assets, for instance, on the Ethereum network, tokens that represent different DeFi projects or NFTs might experience price fluctuations in tandem because they share the same underlying network dynamics, such as gas fees, network congestion, or upgrades. External tokens, operating on different blockchains but within the same crypto ecosystem, also correlate due to cross-chain liquidity pools, token bridges, and interoperability solutions that allow capital to flow more freely between networks. These mechanisms create price interdependencies, as the value of one token can affect liquidity in another token's pool, influencing its price and vice versa. This is further amplified by the rise of stablecoins like Tether (USDT) or USD Coin (USDC), which provide liquidity and act as a medium of exchange across multiple blockchains, tying together different token economies through their peg to fiat currencies.

The broader cryptocurrency market, too, influences token prices due to macroeconomic factors like inflation rates, interest rates, and global financial conditions. Many investors treat cryptocurrencies as a hedge against traditional financial markets, so when macroeconomic uncertainty rises, capital often flows into cryptocurrencies as an alternative asset class. However, this can lead to synchronized movements within the cryptocurrency space, as major cryptocurrencies like Bitcoin or Ethereum rise in value, smaller tokens tend to follow suit, driven by the same market participants. Conversely, in times of market fear or regulatory crackdowns, investors may exit the entire crypto market, causing token prices to drop simultaneously across the board. The increasing involvement of institutional investors, hedge funds, and large corporations in the cryptocurrency space further ties token prices to global market trends, as these institutions often make allocation decisions across a range of cryptocurrencies rather than focusing on individual tokens, reinforcing correlations across the entire crypto market.

The rise of CBDCs is also beginning to affect token prices. As CBDCs from major countries and organizations, like the European Central Bank's digital euro or China's digital yuan, become integrated into the global financial system, they introduce new dynamics between cryptocurrencies and traditional financial markets. CBDCs, while centralized, represent digital forms of fiat currency that may eventually compete with or complement decentralized cryptocurrencies. Investors and traders might view CBDCs as less volatile alternatives to stablecoins, shifting capital flows and liquidity in the process, thereby influencing token prices across the cryptocurrency market. Moreover, the introduction of CBDCs could impact regulatory frameworks for cryptocurrencies, as governments may impose stricter rules on crypto usage or tax regimes in response to their CBDC initiatives. This interplay between traditional fiat systems and decentralized tokens further correlates token prices with macroeconomic factors, including CBDC policies and adoption rates, as they can directly impact liquidity, market access, and investor sentiment in the broader cryptocurrency space.

Additionally, global economic events, such as geopolitical tensions, fiscal policies, and inflation, which affect major fiat currencies and CBDCs, often influence investor behavior in the cryptocurrency markets as well. When fiat currencies or CBDCs of major economies experience fluctuations due to inflation or monetary policy changes, investors might seek refuge in cryptocurrencies, leading to price increases across multiple tokens. Conversely, confidence in CBDCs or traditional markets could lead to an outflow of capital from cryptocurrencies, depressing token prices. Thus, the interconnectedness of blockchain tokens with other digital currencies, macroeconomic conditions, and investor psychology makes token prices highly correlated not just within the crypto market but also with the broader financial landscape, including the emergence and development of CBDCs.

16.1.7 Log-Price in Blockchain Token Economics

Log-price modeling in blockchain token markets cannot be reduced to simple Gaussian noise because the dynamics of token prices are far too complex and exhibit behaviors that deviate from the assumptions of normality, constant volatility, and independence that underlie Gaussian noise models. Gaussian noise assumes that price changes are random, memoryless, and follow a standard bell curve distribution, where extreme events or volatility clusters are highly unlikely. However, the token market is characterized by nonlinear behaviors, memory effects, sudden regime shifts (like halving events), jumps, and long-range dependencies that cannot be adequately captured by Gaussian models. To more accurately model token price dynamics, a variety of alternative noise processes must be considered to account for these irregularities.

For instance, fractional Brownian motion (fBm) is crucial for including memory in the model, which reflects the persistence or anti-persistence in price movements over time. Unlike standard Brownian motion, where price increments are independent, fBm allows for long-range dependence, where past price movements influence future dynamics. This is particularly important in token markets, where trends can persist over time due to investor sentiment, market momentum, or feedback loops driven by network effects or staking mechanisms. Additionally, fractional Brownian motion can capture the phenomenon of "volatility clustering," where high-volatility periods are followed by similar periods, a common feature in the token market due to speculative behavior and external shocks.

Regime-switching models are essential for capturing sudden shifts in market conditions, such as cryptocurrency halving events, forks, or regulatory announcements, which can dramatically alter the price trajectory. These events create different regimes with distinct volatility and return characteristics, causing abrupt changes in price behavior. A simple Gaussian noise model, which assumes a single, constant regime, would fail to capture these shifts. Regime-switching models allow for the inclusion of multiple states, such as high-volatility and low-volatility periods, and the transitions between these states, reflecting how the market reacts to major events.

Poisson noise is also important for modeling the occurrence of discrete, random events in token markets, such as large buy or sell orders, airdrops, or exchange listings. These events can cause sudden jumps or drops in token prices that occur unpredictably and cannot be described by continuous Gaussian processes. Poisson processes introduce a jump component that reflects these discrete movements, capturing the irregular and often sharp changes in prices driven by such events. This aligns with real-world trading activity, where price movements can be significantly influenced by a few large trades rather than a continuous flow of small transactions.

The inclusion of Gaussian–Volterra noise is essential to capture non-Markovian behaviors that arise in token markets, where the future price evolution depends not just on the current state but on a range of past states. This type of noise helps model the complex, nonlinear relationships between price changes and past market conditions, which can be influenced by factors such as the network's transaction volume, staking rewards, or even external economic indicators. By incorporating Gaussian–Volterra noise, we can model how token prices evolve under the influence of past conditions that are not adequately captured by memoryless processes like Brownian motion.

Rosenblatt noise is necessary to capture non-Gaussian scenarios with memory effects, where extreme price movements or heavy tails in the distribution are common. Unlike Gaussian noise, which underestimates the likelihood of extreme events, Rosenblatt noise accounts for more frequent large deviations from the mean, which are characteristic of token markets due to factors like speculative bubbles, sudden crashes, or unexpected regulatory changes. Rosenblatt processes also incorporate memory, meaning that large movements in the past can increase the likelihood of similar large movements in the future, reflecting how market shocks can propagate and influence future prices.

These types of noises—fractional Brownian motion, regime-switching, Poisson noise, Gaussian–Volterra noise, and Rosenblatt noise—create a more comprehensive framework for modeling the log-price dynamics of token markets. Token prices are influenced by a wide array of factors, including market sentiment, investor behavior, network events (like halving or hard forks), and external economic conditions, all of which exhibit complex behaviors that cannot be reduced to Gaussian randomness. By incorporating different types of noise, we can better model the reality of token price dynamics, which involve memory effects, jumps, volatility clustering, regime changes, and heavy-tailed distributions, thus providing a more realistic representation of the highly intricate and multifaceted blockchain token market. This approach allows us to more accurately predict token price movements, manage risk, and design trading strategies that account for the full spectrum of market behaviors, rather than oversimplifying them into Gaussian processes that ignore the rich structure of the token economy.

16.1.8 CDI: Immediate Consumption, Diversification/Investment, and Insurance

Under the more realistic and empirically observed combined noise models—such as fractional Brownian motion, regime-switching processes, Poisson noise, Gaussian–Volterra noise, and Rosenblatt noise—employing a consumption, diversification, and insurance (CDI) strategy becomes critical for effective risk quantification and management in a multi-token portfolio. These noise processes reflect the complex, nonlinear, and often unpredictable nature of token price movements, characterized by memory effects, sudden jumps, regime shifts, and extreme events. As a result, traditional risk management strategies that rely on simpler assumptions, such as Gaussian models with independent and identically distributed (i.i.d.) returns, are insufficient. The CDI strategy, in this context, provides a holistic approach to navigating the volatile and interdependent nature of token markets, particularly under the influence of such realistic noise models.

Consumption refers to how investors or users take away for instant consumption or allocate their capital to tokens to derive utility or benefits over time. In a market with fractional Brownian motion, where price movements exhibit long-range dependence and memory, consumption strategies must account for the fact that past market conditions can influence future returns. By using a CDI strategy, an investor can manage the timing and size of their investments, recognizing that token prices are not uniform but may follow trends or cycles that persist over time, for instance, in staking or yield farming, where users "consume" tokens by locking them for rewards, understanding the memory effects and long-term dependencies of price movements becomes critical in deciding when to stake or redeem tokens. The CDI strategy would help users optimize their consumption behavior to maximize returns while minimizing the risk of large losses due to long-term market trends that fractional Brownian motion captures.

Diversification is essential because these noise models reveal that token prices are subject to various types of shocks and market dynamics that cannot be adequately hedged with a single asset or token. In a multi-token portfolio, diversification helps spread risk across tokens with different behaviors under these noise scenarios, for example, Poisson noise introduces jump risks where certain tokens may experience sudden price movements due to discrete events (such as an airdrop or large institutional buy), while other tokens might exhibit smoother price paths influenced by Gaussian–Volterra noise with memory effects. A CDI strategy recognizes that by holding a diversified portfolio of tokens, investors can reduce the impact of a sudden jump in one token while benefiting from the steady growth or hedging characteristics of others. Diversification across tokens also mitigates the regime-switching risks where certain tokens may thrive in a particular economic or market regime (e.g., during a bull market or after a halving event), while others may underperform, allowing the portfolio to remain resilient across different market conditions.

Insurance under the CDI strategy refers to hedging against extreme risks and ensuring protection from adverse market events, which are more accurately modeled by non-Gaussian noise such as Rosenblatt noise. Rosenblatt noise accounts for the heavy-tailed nature of price distributions, where extreme events (e.g., token crashes, flash crashes, or regulatory shocks) are more common than predicted by Gaussian models. In a CDI framework, investors can use various financial instruments—such as derivatives, options, or stablecoins—as insurance against these tail risks. Insurance strategies may also include rebalancing the portfolio frequently in response to regime switches or ensuring liquidity through more stable assets that are less sensitive to sudden Poisson jumps. By integrating insurance into the CDI strategy, investors can safeguard their portfolios against the extreme volatility, and rare but significant events that these more sophisticated noise models reveal are possible in token markets.

The CDI strategy in a multi-token portfolio helps investors navigate the complexities introduced by realistic noise models in blockchain token economics. These models reflect the true behavior of token prices, which are subject to persistent memory, sudden jumps, regime shifts, and heavy-tailed risks. By focusing on optimized consumption, strategic diversification across tokens, and effective insurance against extreme events, the CDI strategy enables investors to better quantify and manage risks. This holistic approach ensures that they are not overly exposed to any one type of market behavior and instead can adapt to the full spectrum of risks and opportunities presented by the token market's inherently unpredictable and noisy dynamics. This is particularly important in blockchain markets, where the stakes are high due to the high volatility, speculation, and rapid pace of technological and regulatory change.

16.2 Modeling a Blocktree and a Blockchain

We introduce a blocktree and a blockchain. A blocktree is a 2-tuple $(\mathscr{B}, \mathscr{E})$, where $\mathscr{B} \subset \mathbb{B}$, is a non-empty set of valid blocks, including at least one element b_0 (the genesis block). $\mathscr{E} \subset \mathscr{B}^2$ such as it exists a unique path from the initial block b_0 to any other block. A total order $\succ$ over $\mathscr{B}$, preserving $\mathscr{E}$, i.e.

$$(b, b') \in \mathscr{E} \implies b' \succ b.$$

The blockchain is the unique path from the genesis block b_0 to the last block $\max_{\succ} \mathscr{B}$. A blockchain is built by a network of processes applying sequentially the expansion operation starting from the initial blocktree $\{b_0, \{\}\}$. In the distributed case, every processing node refers to its own view of the blocktree and strives to build a consensus with each other on the shared blockchain while increasing its depth. Regular nodes try to build valid blocks. When such valid block is found by a node, it expands its local blocktree from its last block (i.e., the last block of its own view of the blockchain) and broadcasts it to the others. When a regular node receives a new block from another one, it updates its local blocktree expanding it with the new block. During this updating phase, forks may actually be observed due to blocks broadcasting delay. Building a new valid block is an operation that may take many forms depending on the blockchain protocols. Several types of nodes are considered depending on the behaviors: empathy-altruism, regular, spiteful, and malicious behaviors. There are several consensus algorithms implemented in the blockchain literature (see Table 16.1). As of today, there is no chain-based protocol that outperforms all the others. Each protocol has the pros and contras (network performance, end-to-end delay, security, cost), and the economic and game theoretic analysis (utility, gain, cost, adoption, medium of exchange) should be taken into consideration. The evolution of the protocols may suggest coexistence and randomization among the better protocols. However, the randomized better protocols have a different cost and different outcome. Thus, the trade-offs need to be designed and analyzed (see Fig. 16.1)

Table 16.1 Some consensus algorithms implemented in blockchain

Consensus algorithms	Key observations
Proof of work	Blockchain platform: Bitcoin
	Programmed in C++
	No smart contract
	Pros: less opportunity for 51% attack
	Contras: greater energy consumption
	Miners centralization
Proof of stake	Blockchain platform: Peercoin, Nxt, BlackCoin
	Programmed in Java
	With smart contract
	Pros: less energy consumption
	Pros: more decentralized
	Contras: nothing-at-stake problem
	Double-spend attacks
	Miners centralization
Delegated proof of stake	Blockchain platform: Lisk, Steem, EOS, BitShares
	Programmed in JavaScript
	No smart contract
	Pros: less energy consumption
	Contras: required participation, double-spend attacks
	Partial centralization of miners
Leased proof of stake	Blockchain platform: Waves, Lunes
	Programmed in Scala
	With smart contract
	Pros: fair usage, lease coins
	Contras: decentralization problem
Pure proof of stake	Blockchain platform: Vault, Algorand
	Programmed in Testnet
	With smart contract
	Pros: scalability
	Contras: incentive issue
Proof of elapsed time	Blockchain platform: Hyperledger
	Programmed in Python, C++, Java
	With smart contract
	Pros: cheap participation
	Contras: specialized hardware
Practical Byzantine Fault Tolerance	Blockchain platform: Hyperledger
	Programmed in Java
	With smart contract
	Pros: no need for confirmation
	Contras: communication issue
	Sybil attack

(continued)

Table 16.1 (continued)

Consensus algorithms	Key observations
Delegated Byzantine Fault Tolerance	Blockchain platform: NEO
	Programmed in Python, Java, C++
	With smart contract
	Pros: scalable, fast
	Contras: greater energy consumption
	Miners centralization
Federated Byzantine Fault Tolerance	Blockchain platform: Stellar, Ripple
	Programmed in Java
	With smart contract
	Pros: low latency, scalable
	Contras: quorum slice choice
Directed Acyclic Graphs	Blockchain platform: IOTA
	Programmed in JavaScript, C++
	Pros: scalability
	Contras: double-spending issues
Proof of activity	Blockchain platform: Decred
	Programmed in Go
	With smart contract
	Pros: lower probability of the 51% attack
	Contras: double signature
Proof of importance	Blockchain platform: NEM
Proof of devotion (Nebulas)	Programmed in C++, Java
	With smart contract
	Pros: partnership
	Contras: decentralization issue
Proof of capacity	Blockchain platform: Burstcoin
	Programmed in Java
	With smart contract
	Pros: cheap
	Contras: decentralization problem
Proof of burn	Blockchain platform: Slimcoin
	Programmed in Java
	No smart contract
	Pros: cheap
	Contras: long-term investors
	Miners centralization
Proof of weight	Blockchain platform: Filecoin, Algorand
	Programmed in Snark
	With smart contract
	Pros: scalability
	Contras: incentives issue

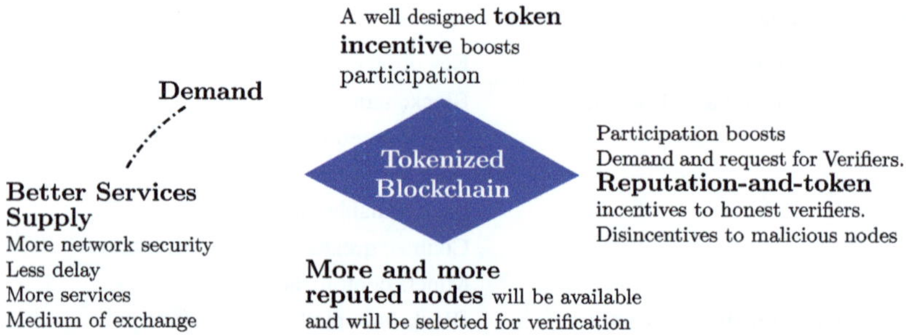

Fig. 16.1 Blockchain-based cryptographic token economics examined in this chapter

16.3 Tokenless and Tokenized Blockchains

We now introduce two types of blockchains.

16.3.1 Blockchain Without Tokens

To proceed, let us consider a basic setup where risk-aware users can conduct businesses and enjoy the exchange surplus on the blockchain platform without holding tokens but use a base currency (dollar, euro, yuan, etc.), the numéraire, or central bank digital currency (CBDC) as the medium of exchange.

Tokenless: Decision-Maker i's Problem

$$\begin{aligned}
&\sup_{x_i} \mathbb{E}\{w_i(T) - \tfrac{\eta_i}{2w_i(0)} var[w_i(T)]\},\\
&dw_i = [\{(x_i)^{\rho}(h(n)A\lambda_i)^{1-\rho}\}\mathbb{1}_{\{a=0\}} - \xi - x_i r]dt,\\
&w_i(0) > 0,\\
&dA = A(d_A dt + \sigma_A dB),\\
&A(0) > 0,
\end{aligned} \tag{16.1}$$

where $\rho \in (0, 1)$, w_i is the flow of good achieved on the blockchain by transacting in its native crypto-token by decision-maker i, x_i is the amount in dollar used in decision-maker i, n is the total number of decision-makers that decide to join the blockchain network (i.e., $x_i > 0,$), r is a marginal unit cost, h is a positive one-to-one mapping, a is a binary random variable representing the state of the attack, $a = 1$ when the attack is successful, $i \in \mathcal{I} = \{1, 2, \ldots, I\}$ where I is the total number of decision-makers including those who are not participating, and $A\lambda_i$ is the productivity of i. The initial flow of good is positive $w_i(t_0) > 0$. Here, the interaction with other decision-makers is through $n(t) = \sum_{i\in\mathcal{I}} \mathbb{1}_{\{x_i(t)>0\}}$. The parameter η_i is a risk awareness index of decision-maker i. The term ξ is the instant access cost; it can be expressed as a function of waiting cost. An example of such a function is given by

$$\xi(d_i) = \xi_0 + \left(\frac{\hat{\lambda}}{\hat{\mu}K}\right) \int_0^{d_i} F'(x)x \frac{D'_K(\frac{\hat{\lambda}(1-F(x))}{\hat{\mu}K})}{\hat{\mu}} dx.$$

Delays are costly to the users, and the delay cost per unit time is distributed randomly with cumulative distribution F where D_K is the expected waiting time measured in blocks of size K, $\hat{\lambda}$ is the arrival rate of request, and $\hat{\mu}$ is the service rate. Note that Problem (16.1) is not a standard stochastic differential game problem because of the variance term. This is a variance-aware mean-field-type game problem [106, 457, 458, 460, 463].

Proposition 16.1 (No Tokens) *The equilibrium strategy of i in the token-less blockchain technology is*

$$x_i^* = \left(\frac{\rho}{r}\right)^{\frac{1}{1-\rho}} h(n)\tilde{A}\lambda_i,$$

where $\tilde{A} = A(1-\nu)^{\frac{1}{(1-\rho)}}$ and ν is the probability that an attack (by malicious node or a group of malicious nodes) succeeds under the blockchain protocol.

The number of participants in the token-free blockchain technology $n_{nto} := \kappa_{nto}\bar{n}$ solves the fixed-point equation

$$\begin{aligned} \kappa_{nto} &:= \mathbb{P}[\lambda_i \geq \tfrac{\xi}{h(\kappa_{nto}\bar{n})A} \tfrac{\rho}{r(1-\rho)} (\tfrac{r}{\rho})^{\frac{1}{1-\rho}}] \\ &= 1 - F_{\lambda_i}\left[\tfrac{\xi}{h(\kappa_{nto}\bar{n})A} \tfrac{\rho}{r(1-\rho)} (\tfrac{r}{\rho})^{\frac{1}{1-\rho}}\right], \end{aligned} \tag{16.2}$$

where F_{λ_i} is the cumulative function of the productivity random variable.

To prove this proposition, we can use the dynamic programming principle in infinite dimensions for mean-field-type games (see Volume I of this book). As the drift does not contain the state variable w_i and the reformulation does not have a running payoff, the equilibrium strategies are obtained by one-shot direct optimization for each decision-maker. ■

Notice that the optimal investment of decision-maker i in the token-free blockchain shares the following interesting features:

- increases with the total number of active participants to the token-free blockchain,
- increases with the productivity $A\lambda_i$ of the token-free blockchain,
- decreases with the probability of success ν of an attack of the blockchain by malicious nodes.

16.3.2 Reduction of the Cost of Verification

Settlement is the process of transferring an asset or information to another party. Reconciliation ensures that internal records pertaining to a particular transaction are consistent across the relevant parties. Transfers of crypto-assets through a distributed digital ledger require verification because electronic object files, unlike physical objects, can be easily duplicated or changed. For this, a blockchain network needs verifiers. In a centralized network, a specific node (a firm or an institution) is entrusted with the responsibility of verification. In exchange, it charges users with fees. In a distributed network, verification tasks are not delegated to a single node but to different members of the network. Trust does not rely on a node but on the behavior of the network and its protocols. The verifiers contribute resources that deliver a certain level of trust. A consensus protocol allows participants to agree on a common output that aggregates private inputs when some dishonest (who do not follow the process) participants may attack the process. This question is known as the byzantine agreement. The cost of verification of the blockchain will be affected by the protocol used and the economic incentives to the verifiers. If there is a significant reduction of cost of verification of the blockchain compared with the traditional operation, part of this saving can be used to fund token at the beginning of the blockchain. Below, we examine the effect of tokens on the blockchain economy.

16.3.3 Blockchain with a Native Token

In the current blockchain technology, token economics and legislation are important challenges to be addressed. Before issuing a token, it is important to define the goal, context, and objectives of that token. Does it add value to the blockchain project? Does the token price has an economic demand–supply values fit ?

Next, we evaluate a native token introduced in the blockchain from a nonzero price $p(0)$ and nonzero production $A(0)$.

Tokenized Blockchain: Decision-Maker i's Problem

$$\begin{aligned}
&\sup_{u_i} \mathbb{E} w_i(T) - \frac{\eta_i}{2w_i(0)} var[w_i(T)],\\
&dw_i = \{(pu_i)^{\rho}(h(n)A\lambda_i)^{1-\rho}dt + u_i p\mathbb{E}[\tfrac{dp}{p}]\}\mathbb{1}_{\{a=0\}}\\
&-\xi dt - u_i prdt,\\
&w_i(0) > 0,\\
&dA = A(d_A dt + \sigma_A dB_A),\\
&A(0) > 0,\\
&dp = p(d_p dt + \sigma_p dB_p),\\
&p(0) > 0.
\end{aligned} \tag{16.3}$$

Now, n is the total number of decision-makers who decide to join the blockchain network (i.e., $u_i > 0$ to hold some tokens). $u_i p\mathbb{E}[\frac{dp}{p}]$ is the token appreciation and ξ is the access cost. The number d_p is changed to $\tilde{d}_p := d_p(1-\nu)^{\frac{1}{(1-\rho)}}$.

We provide the changes compared with the results of Proposition 16.1.

Proposition 16.2 (With Tokens) *The equilibrium strategy of i is*

$$u_i^* = \frac{1}{p}(\frac{\rho}{r-d_p})^{\frac{1}{1-\rho}}[h(n)\tilde{A}\lambda_i].$$

The number of participants in the blockchain with tokens is

$$n_{to} := \kappa_{to}\bar{n},$$

solving the following fixed-point equation:

$$\begin{aligned}\kappa_{to} &:= \mathbb{P}[\lambda_i \geq \frac{\xi}{h(\kappa_{to}\bar{n})\tilde{A}}\frac{\rho}{(1-\rho)[r-d_p]}(\frac{r-d_p}{\rho})^{\frac{1}{1-\rho}}] \\ &= 1 - F_{\lambda_i}\left[\frac{\xi}{h(\kappa_{to}\bar{n})\tilde{A}}\frac{\rho}{(1-\rho)[r-d_p]}(\frac{r-d_p}{\rho})^{\frac{1}{1-\rho}}\right].\end{aligned} \tag{16.4}$$

To prove this proposition, we have used again dynamic programming principle in infinite dimensions for mean-field-type games [106]. Only the drift is modified, and the new drift does not contain the state variable w_i. There is no running payoff. Thus, the equilibrium strategies are obtained by one-shot direct optimization for each decision-maker. ■

The optimal number of tokens of decision-maker i has the following interesting features:

- increases with the total number of active participants to the blockchain,
- decreases with the probability of success ν of an attack,
- decreases with the token price p,
- increases with the productivity $A\lambda$,
- increases with interest rate d_p of the token price.

As observed in practice, the network security affects the price, the adoption, and the sustainability of blockchain technologies. This simple model captures that behavior.

Proposition 16.3 *The network security is identical $\nu_{to} = \nu_{nto} = \nu$,, and then the following inequality holds:*

$$n_{to} > n_{nto}.$$

This means that the introduction of token can boost the adoption and participation to the blockchain technology.

Proof The proof follows from the monotonicity of F and the fact that $\frac{\rho}{[r-d_p]}(\frac{r-d_p}{\rho})^{\frac{1}{1-\rho}}$ increases with d_p. ■

Demand–Supply per Token

The demand is $D = \frac{1}{p}(\frac{\rho}{r-d_p})^{\frac{1}{1-\rho}}[h(\kappa_{to}\bar{n})\tilde{A}\kappa_{to}\bar{n}]$ and the supply of tokens is S_{to}. Hence, the following formula holds in equilibrium:

$$S_{to} = \frac{1}{p}(\frac{\rho}{r-d_p})^{\frac{1}{1-\rho}}[h(\kappa_{to}\bar{n})\tilde{A}\kappa_{to}\bar{n}]. \tag{16.5}$$

This determines a token price formula p_{to} of the tokenized blockchain technology as a function of the total population size $\bar{n}$, token user base κ_{to}, and the supply of tokens S_{to} :

$$p_{to} = \frac{1}{S_{to}}(\frac{\rho}{r-d_p})^{\frac{1}{1-\rho}}[h(\kappa_{to}\bar{n})\tilde{A}\kappa_{to}\bar{n}]. \tag{16.6}$$

The velocity of a token is the number of times (frequency) token is exchanged from one transaction to another over the period of time $[0, T]$ or, in other words, how often token is turned over. The token velocity is a crucial parameter in influencing the value of the token. Based on (16.6), if the target is a big market size n with a scarce supply S, when the token has high velocity, then there will only be limited value appreciation potential to the token. Token velocity should be taken into consideration in the design of d_p.

Note that if we modify the utility function by inserting $u_i dp$ instead of $u_i p\mathbb{E}[\frac{dp}{p}]$ the result would completely different in terms of the nice properties mentioned above. This is because dp will bring the controlled diffusion term $u_i p\sigma_p dB$ which affects the optimization part with a term $\frac{1}{2}[u_i p\sigma_p]^2$. If we use $u_i\mathbb{E}[dp]$ instead of $u_i p\mathbb{E}[\frac{dp}{p}]$, one obtains a function of the ratio $d_p\frac{\mathbb{E}[p]}{p}$ instead of d_p.

When the probability of successful attack ν_{to} of the tokenized blockchain is high compared with the probability of successful attack ν_{nto} of the tokenless blockchain, the inequality can be reverse, i.e., a not well-designed token can have a negative effect.

As an example, Algorand is based on the cryptographic sortition mechanism, i.e., for every round, efficiently select only few verifiers, at random from the set of all users, and in a way that is both unpredictable until the very last moment. The original protocol of Algorand does not specify an incentive to verifiers. As the temporary verifiers are among the token holders, they may have want to protect their own tokens. This makes a lower-cost validation procedure if there is enough participation and if the number of regular nodes is high enough.

How to get enough participation of honest nodes to be connected online at any random time?

- voluntary contributors from token holders and/or those with good reputations (above a certain threshold),
- community or federation-based contributors who are token holders and with good reputation states,
- incentives for truly decentralized nodes who are token holders and good reputation states.

In order to encourage participation from more than a majority of randomly selected committee members, the nodes who honestly participate gets an award (in tokens). If a malicious node is detected, it will not be awarded. Such an incentive supports the main assumption of Algorand on the proportion of honest nodes.

16.3.3.1 Incentives for Verifiers

In order to consolidate the incentives to the verifiers, we propose to add a relatively small award (in token) to the randomly selected verifiers for their service. They consume an energy to send and propagate messages whenever active. Thus, the production (supply) is not at free cost. One should consider the cost of delivering (end-to-end delay and timely) such a task. Each verifier will have a specific state depending on their reputation score. The verifier will be its reputation increased after a successful completion of a transaction. A detected malicious behavior will have its reputation state to be decreased if detected. After a certain timestep in the evolution of protocol, the number of nodes with high reputation state is big enough. The probability of selection can now combine token and reputation or between two types of tokens as illustrated in Fig. 16.2.

A typical example of committee selection is to choose a proportional probability to

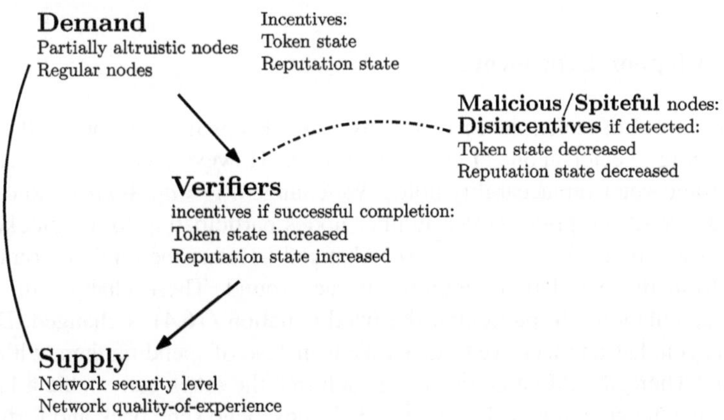

Fig. 16.2 Incentives to users and verifiers with token and reputation

$$[e^{(to_i - \bar{to})}]^{\lambda}[e^{(r_i - \bar{r})}]^{(1-\lambda)} \text{ or}$$

$$\lambda[e^{(to_i - \bar{to})}] + (1 - \lambda)e^{(r_i - \bar{r})}$$

where to_i is the number of the tokens of i and r_i is the current reputation of i in the blockchain. If a malicious node is detected, that user's reputation is to be set down to zero as well as its token state. Other disincentive schemes apply.

In a single interaction between a set of verifiers and a user, the verifiers have no incentive to provide services to the user. The mechanism we study for creating incentives to provide service involves the exchange of cryptographic tokens. Each agent can hold an arbitrary nonnegative number of cryptographic tokens but cannot hold a negative number of cryptographic tokens and cannot borrow.

The blockchain creates incentives for the verifiers and nodes to provide or share resources by providing a supply of tokens and recommending strategies on their reputation. The best response strategies depend on current token holdings and reputation.

For node i, sending an information to node k, a fully distributed and random number of nodes are selected to form a committee J. J is selected based on specific random process. A strategy profile is a collection $(s_i, s_k, \{s_j\}_{j \in J})$. s_j output maps to acceptance or rejection for j to participate to the vote as an honest node or to attack or not if a malicious node. If the requested service is provided, a portion of token is transferred to the committee members and to be shared equally. So the client's holding of tokens decreases by to_s, and each verifier's holding of tokens increases by $\frac{1}{|J|} to_s$. Let $\bar{u}$ is the per capita supply of tokens. The distribution of token holdings between verifiers should satisfy the following equalities:

$$\begin{aligned} &\textstyle\sum_{to} \mu(to) = 1, \\ &\textstyle\sum_{to} to\, \mu(to) = \bar{u}. \end{aligned} \qquad (16.7)$$

16.3.3.2 Adoption-Dependent Price

With a pure proof of stake, the more tokens a user keeps in its accounts, the more chance of being selected at one of the voting rounds. However, its vote counts as one single vote and voter replaceability holds. We change $\tilde{d_p} = d_p + \tilde{\lambda} h(n)$ to capture the dependence of the price on the number nodes participating to the blockchain. Also, the productivity $\tilde{d_A} = d_A + \tilde{\lambda} h(n)$ changed because the random committee is drawn from the population sampling at each round. These changes do affect the above calculations. In particular, the fixed equation (16.4) is changed. Does a token-rich agent has an incentive to save tokens instead of spending them within the blockchain? There should be an incentive such that the token can enable a kind of medium of exchange while avoiding rewarding those who send back and forth.

16.4 Managing Tokens: Diversification and Exchange

Token diversification is a risk engineering strategy that creates a mix of various investments within a token portfolio. A diversified token portfolio contains a mix of distinct token asset types and investment strategies in an attempt to limit exposure to risk.

Once the native token is adopted in the system, one can diversify with several other projects and tokens. We consider two different tokens with price dynamics given by

$$\begin{aligned} dp_1 &= p_1(d_1dt + \sigma_1 dB_1), \\ p_1(0) &> 0, \\ dp_2 &= p_2(d_2dt + \sigma_2 dB_2), \\ p_2(0) &> 0, \end{aligned} \tag{16.8}$$

where B_1 and B_2 are two Brownian motions (possibly correlated such that $\mathbb{E}[dB_i dB_j] = \beta_{ij}dt$ with $\beta_{ii} = 1$). Empirical observation from CoinMarketCap shows a nonnegligible positive correlation of some crypto-tokens with the major tokens, and some others have negative correlation. Tables 16.2 and 16.3 show the correlation between some major cryptographic tokens in 2018 and 2017. Daily prices of Bitcoin and ten other largest cryptocurrencies by market capitalization are used (Ethereum, XRP, Bitcoin Cash, EOS, Stellar, Litecoin, Tron, Cardano, IOTA, and Monero). The entries are obtained using

$$cov(X, Y) := \frac{\mathbb{E}[(X - \mathbb{E}(X))(Y - \mathbb{E}(Y))]}{\sqrt{\mathbb{E}[(X - \mathbb{E}(X))^2]\mathbb{E}[(Y - \mathbb{E}(Y))^2]}}.$$

So correlation between the random variables seems to be a reasonable setup in the context of crypto-token networks.

By stochastic integration formula, one gets

$$\begin{aligned} &d[(p_1 - \mathbb{E}p_1)(p_2 - \mathbb{E}p_2)] \\ &= (p_1 - \mathbb{E}p_1)[(p_2 - \mathbb{E}p_2)d_2dt + p_2\sigma_2 dB_2] \\ &+(p_2 - \mathbb{E}p_2)[(p_1 - \mathbb{E}p_1)d_1dt + p_1\sigma_1 dB_1] \\ &+p_2\sigma_2 p_1\sigma_1\beta_{12}dt. \end{aligned} \tag{16.9}$$

Taking the expectation yields

$$\begin{aligned} &d[cov(p_1, p_2)] = d\mathbb{E}[(p_1 - \mathbb{E}p_1)(p_2 - \mathbb{E}p_2)] \\ &= [d_1 + d_2 + \sigma_1\sigma_2\beta_{12}]cov(p_1, p_2) - \mathbb{E}[p_1]\mathbb{E}[p_2]\sigma_1\sigma_2\beta_{12}. \end{aligned} \tag{16.10}$$

By direct computation, one has

$$\mathbb{E}[p_1]\mathbb{E}[p_2] = \mathbb{E}[p_1(0)]\mathbb{E}[p_2(0)]e^{(d_1+d_2)t}.$$

Table 16.2 Empirical evidence on correlation between some cryptographic tokens in 2018. Symmetric matrix (except the average column). The numbers are computed from the formula $cov(X, Y)$. The source data is available at https://coinmarketcap.com/historical/

2018	Bitcoin	Ethereum	Ripple	BTC Cash	EOS	Stellar	Litecoin	Tron	Cardano	IOTA	Monero
BTC	1	0.815	0.668	0.759	0.687	0.667	0.885	0.556	0.736	0.748	0.829
ETH		1	0.738	0.753	0.698	0.697	0.827	0.531	0.784	0.780	0.790
Ripple			1	0.612	0.653	0.728	0.695	0.545	0.772	0.703	0.676
BTC Cash				1	0.632	0.583	0.746	0.451	0.653	0.720	0.733
EOS					1	0.610	0.686	0.501	0.711	0.637	0.656
Stellar						1	0.632	0.414	0.836	0.645	0.675
LTC							1	0.481	0.714	0.758	0.806
TRON								1	0.528	0.487	0.465
Cardano									1	0.738	0.727
IOTA										1	0.739
Monera											1

Table 16.3 Empirical evidence on correlation between some cryptographic tokens in 2018. The numbers are computed from the covariance formula $cov(X, Y)$. The source data is available at https://coinmarketcap.com/historical/

2017	Bitcoin	Ethereum	Ripple	Stellar	Litecoin	Monero	Average
BTC	1	0.355	0.107	0.22	0.369	0.412	0.293
ETH		1	0.111	0.194	0.343	0.461	0.292
Ripple			1	0.474	0.213	0.156	0.212
Stellar				1	0.273	0.356	0.304
Litecoin					1	0.348	0.309
Monero						1	

Hence, the covariance is

$$\begin{aligned} & cov(p_1, p_2) = \\ & e^{[d_1+d_2+\sigma_1\sigma_2\beta_{12}]t} \left\{cov(p_1(0), p_2(0)) \right. \\ & \left. -\mathbb{E}[p_1(0)]\mathbb{E}[p_2(0)]\sigma_1\sigma_2\beta_{12} \int_0^t e^{-[\sigma_1\sigma_2\beta_{12}]t'} dt' \right\} \\ & = e^{[d_1+d_2+\sigma_1\sigma_2\beta_{12}]t} cov(p_1(0), p_2(0)) \\ & + e^{[d_1+d_2]t}\mathbb{E}[p_1(0)]\mathbb{E}[p_2(0)](e^{\sigma_1\sigma_2\beta_{12}t} - 1). \end{aligned} \tag{16.11}$$

This means that even if B_1 and B_2 were uncorrelated, an initial correlation of the token prices will propagate and affect both tokens prices law later on. The total token value is $m = p_1u_{i1} + p_2u_{i2}$, where u_{ik} is the number of tokens of type $k \in \{1, 2\}$ held by decision-maker i. The token diversification problem for decision-maker i is the following risk-aware mean-field-type control problem (16.12):

$$\begin{aligned} & \sup_{u_{i1}} \mathbb{E}m(T) - \frac{\eta_i}{2m(0)} var[m(T)], \\ & dm = [d_2 m + p_1u_{i1}(d_1 - d_2)]dt + \sigma_1 p_1 u_{i1} dB_1 + \sigma_2(m - p_1u_{i1})dB_2, \\ & m(0) > 0, \\ & 0 \le p_1u_{i1} \le m. \end{aligned} \tag{16.12}$$

Note that the control action u_{i1} is constrained. The correlation between the noises and constraint makes the problem more involved.

Proposition 16.4 (Diversification Between Two Tokens) *The interior (if any) optimal allocation strategy of i is*

$$\begin{aligned} u_{i1}^* &= \frac{1}{p_1}[\tau_i(m - \mathbb{E}m) + \bar{\tau}_i\mathbb{E}m + e_i], \\ u_{i2}^* &= \frac{1}{p_2}[m - \tau_i(m - \mathbb{E}m) - \bar{\tau}_i\mathbb{E}m - e_i], \end{aligned} \tag{16.13}$$

with

$$\begin{array}{l} d[\mathbb{E}m] = [d_2 + \bar{\tau}_i(d_1 - d_2)]\mathbb{E}[m] + e_i(d_1 - d_2)]dt, \\ \mathbb{E}m(0) > 0, \\ c\tau_i = (d_1 - d_2) + \sigma_2^2 - \frac{1}{2}\beta_{12}\sigma_1\sigma_2, \\ c\bar{\tau}_i = [\frac{\bar{\alpha}_i}{\alpha_i}(d_1 - d_2) + \sigma_2^2 - \frac{1}{2}\beta_{12}\sigma_1\sigma_2], \\ c = \sigma_1^2 + \sigma_2^2 - \beta_{12}\sigma_1\sigma_2, \\ ce_i = m_0(d_1 - d_2)\frac{\gamma_i}{\alpha_i}. \end{array} \tag{16.14}$$

Proof Consider the following problem:

$$\begin{array}{l} \sup_{x_{i1}} \mathbb{E}m(T) - \frac{\eta_i}{2m(0)} var[m(T)], \\ dm = [d_2 m + x_{i1}(d_1 - d_2)]dt + \sigma_1 x_{i1} dB_1 + \sigma_2(m - x_{i1})dB_2, \\ m(0) > 0, \\ 0 \leq x_{i1} \leq m. \end{array} \tag{16.15}$$

Using the decompositions $X = (X - \mathbb{E}[X]) + \mathbb{E}[X]$, we rewrite the problem as

$$\begin{array}{l} \sup_{x_{i1}} \mathbb{E}m(T) - \frac{\eta_i}{2m(0)} var[m(T)], \\ d[\mathbb{E}m] = [d_2\mathbb{E}[m] + \mathbb{E}[x_{i1}](d_1 - d_2)]dt, \\ \mathbb{E}m(0) > 0, \\ \\ d(m - \mathbb{E}m) = [d_2(m - \mathbb{E}m) + (x_{i1} - \mathbb{E}x_{i1})(d_1 - d_2)]dt \\ +[\sigma_1(x_{i1} - \mathbb{E}x_{i1}) + \sigma_1\mathbb{E}x_{i1}]dB_1 \\ +\sigma_2(m - \mathbb{E}m + \mathbb{E}m - (x_{i1} - \mathbb{E}x_{i1}) - \mathbb{E}x_{i1})dB_2, \\ m(0) - \mathbb{E}m(0), \\ \\ 0 \leq (x_{i1} - \mathbb{E}x_{i1}) \leq (m - \mathbb{E}m), \\ 0 \leq \mathbb{E}x_{i1} \leq \mathbb{E}m. \end{array} \tag{16.16}$$

Based on the structure of the terminal mean-variance utility, we consider the following guess functional

$$f_i = \gamma_i \mathbb{E}m - \frac{\alpha_i}{2m_0}(m - \mathbb{E}m)^2 + \frac{\bar{\alpha}_i}{2m_0}(\mathbb{E}m)^2,$$

where $(\alpha_i, \bar{\alpha}_i, \gamma_i)$ are deterministic functions (of time) to be determined below:

$$\begin{array}{l} \mathbb{E}[R_i - f_i(0)] = \mathbb{E}[(1 - \gamma_i(T))\mathbb{E}m(T) + \frac{\alpha_i(T)-1}{2m_0}(m - \mathbb{E}m)^2 - \frac{\bar{\alpha}_i(T)}{2m_0}(\mathbb{E}m)^2] \\ +\mathbb{E}\int_0^T (\dot{\gamma}_i + d_2\gamma_i)\mathbb{E}[m] - [\dot{\alpha}_i + 2\alpha_i d_2 + \alpha_i\sigma_2^2]\frac{(m-\mathbb{E}m)^2}{2m_0}dt \\ -\frac{\alpha_i}{m_0}(d_1 - d_2)(m - \mathbb{E}m)(x_{i1} - \mathbb{E}x_{i1})dt + 2\frac{\alpha_i}{2m_0}\sigma_2^2(m - \mathbb{E}m)(x_{i1} - \mathbb{E}x_{i1})dt \\ -\frac{\alpha_i}{2m_0}\beta_{12}\sigma_1\sigma_2(x_{i1} - \mathbb{E}x_{i1})(m - \mathbb{E}m) - (\sigma_1^2 + \sigma_2^2 - \beta_{12}\sigma_1\sigma_2)\frac{\alpha_i}{2m_0}(x_{i1} - \mathbb{E}x_{i1})^2 \\ +\mathbb{E}[x_{i1}](d_1 - d_2)\gamma_i dt - \frac{\alpha_i}{2m_0}\beta_{12}\sigma_1\sigma_2[\mathbb{E}x_{i1}]\mathbb{E}m dt + 2\frac{\alpha_i}{2m_0}\sigma_2^2[\mathbb{E}m]\mathbb{E}x_{i1}dt \\ +\frac{\bar{\alpha}_i}{m_0}(d_1 - d_2)(\mathbb{E}m)\mathbb{E}[x_{i1}] - (\sigma_1^2 + \sigma_2^2 - \beta_{12}\sigma_1\sigma_2)\frac{\alpha_i}{2m_0}[\mathbb{E}x_{i1}]^2 dt \\ +[\dot{\bar{\alpha}}_i + 2\bar{\alpha}_i d_2 - \alpha_i\sigma_2^2]\frac{(\mathbb{E}m)^2}{2m_0}dt. \end{array} \tag{16.17}$$

Assume that

$$m(0) := m_0 > 0,\ \sigma_1^2 + \sigma_2^2 - \beta_{12}\sigma_1\sigma_2 > 0. \tag{16.18}$$

It follows that

$$\begin{aligned}
x_{i1}(t) &= \tau_i(m(t) - \mathbb{E}m) + \bar{\tau}_i\mathbb{E}m + e_i,\\
c\tau_i &= (d_1 - d_2) + \sigma_2^2 - \tfrac{1}{2}\beta_{12}\sigma_1\sigma_2,\\
c\bar{\tau}_i &= [\tfrac{\bar{\alpha}_i}{\alpha_i}(d_1 - d_2) + \sigma_2^2 - \tfrac{1}{2}\beta_{12}\sigma_1\sigma_2],\\
c &= \sigma_1^2 + \sigma_2^2 - \beta_{12}\sigma_1\sigma_2,\\
ce_i &= m_0(d_1 - d_2)\tfrac{\gamma_i}{\alpha_i}.
\end{aligned} \tag{16.19}$$

■

As the risk awareness index η_i of decision-maker i plays an important role in the optimal allocation strategy, from the terminal condition, $\alpha_i(T) = \eta_i > 0$. This means that η_i affects significantly the value of $(\bar{\tau}_i, e_i)$ which in turn affects the value of $\mathbb{E}m$.

16.4.1 Sharpe and Modigliani Ratios for Crypto-tokens

The Sharpe ratio was developed by Nobel laureate William F. Sharpe and is used to help investors understand the return of an investment compared to its risk (Table 16.4 shows a sample data of the Sharpe ratio from https://coinmarketcap.com/historical/). The classical Sharpe ratio is the average return earned in excess of the risk-free rate per unit of volatility or total risk. For cryptographic tokens, we do not recommend to use the risk-free model. We instead recommend a risk-reference model because of empirical volatility. We consider two cryptographic token assets with parameters (d_k, σ_k). The parameters $\tau_i, \bar{\tau}_i, and e_i$ can be seen as functions of modified Sharpe ratios:

$$\tau_i = \frac{(d_1 - d_2) + \sigma_2^2 - \frac{1}{2}\beta_{12}\sigma_1\sigma_2}{\sigma_1^2 + \sigma_2^2 - \beta_{12}\sigma_1\sigma_2},$$

$$\bar{\tau}_i = \frac{\frac{\bar{\alpha}_i}{\alpha_i}(d_1 - d_2) + \sigma_2^2 - \frac{1}{2}\beta_{12}\sigma_1\sigma_2}{\sigma_1^2 + \sigma_2^2 - \beta_{12}\sigma_1\sigma_2},$$

$$e_i = m_0\frac{\gamma_i}{\alpha_i}\frac{(d_1 - d_2)}{\sigma_1^2 + \sigma_2^2 - \beta_{12}\sigma_1\sigma_2}.$$

As both cryptographic token prices are volatile, we modify the evaluation criteria. One can consider the ratio

Table 16.4 Sharpe ratio of some cryptographic tokens

2018	Ethereum	Ripple	Litecoin	EOS	Bitcoin Cash	Bifinance coin	Stellar	Tether
Sharpe	0.0513	−0.024	0.209	0.1440	0.0725	0.325	0.258	−0.507
2018	IOTA	Dash	Maker	NEO	Ontology	ETH classic	NEM	Chain
Sharpe	0.0717	0.0552	0.145	0.0787	0.277	0.00593	0.109	0.342
2018	Tron	Cardano	Bitcoin Sv	Monero	Zcash	Vechain	Waves	Tezos
Sharpe	−0.132	0.124	0.00921	0.00671	−0.0322	0.1841	−0.289	0.0117

$$M^2 := \mathbb{E}[r_k - r_{ref}]\frac{\sigma_{ref}}{\sigma_{r_k}} + \mathbb{E}[r_{ref}],$$

with r_k return on cryptographic token k, σ_{r_k} is the variance on cryptographic token k, r_{ref} return on the market reference, and σ_{ref} variance on the market reference. It is in units of percentage return, which is instantly interpretable by the crypto users and investors. Note that e_i can be rewritten as

$$e_i = \left(\frac{m_0 \frac{\gamma_i}{\sigma_2^2 \alpha_i}}{\frac{\sigma_1}{\sigma_2} + \frac{\sigma_2}{\sigma_1} - \beta_{12}}\right)\left((d_1 - d_2)\frac{\sigma_2}{\sigma_1}\right).$$

The last factor $\left((d_1 - d_2)\frac{\sigma_2}{\sigma_1}\right)$ can be seen as $M^2 - \mathbb{E}[r_{ref}]$ and is therefore related to the Modigliani ratio.

16.4.2 In-Chain Diversification

Consider two tokens within the same blockchain technology. The network characteristics are similar. In terms of utility function, the token diversification problem between decision-makers is the following risk-aware mean-field-type game problem:

$$\begin{array}{l}
\sup_{u_{i1}} \mathbb{E}w_i(T) - \frac{\eta_i}{2w_i(0)} var[w_i(T)], \\
dw_i = \{m^{\rho}(h(n)A\lambda_i)^{1-\rho}dt + [md_2 + u_{i1}p_1(d_1 - d_2)]dt \\
+[u_{i1}^{\rho} + \mathbb{E}u_{i1}^{\rho}]^{\frac{1}{2}} p_1^{\frac{\rho}{2}} \sigma_1 dB_1 \\
+\sigma_2[(m - p_1u_{i1})^{\rho} + \mathbb{E}(m - p_1u_{i1})^{\rho}]^{\frac{1}{2}} dB_2\}\mathbb{1}_{\{a=0\}} \\
-\xi dt - [mr_2 + u_{i1}p_1(r_1 - r_2)]dt, \\
dm = [d_2m + p_1u_{i1}(d_1 - d_2)]dt \\
+\sigma_1 p_1 u_{i1} dB_1 + \sigma_2(m - p_1u_{i1})dB_2, \\
m(0) > 0, \\
dA = A(d_A dt + \sigma_A dB), \\
0 \leq p_1u_{i1} \leq m, \\
n = \sum_i \mathbb{1}_{\{u_{i1}>0\}} + \mathbb{1}_{\{p_1u_{i1}<m\}}.
\end{array} \tag{16.20}$$

In the in-chain procedure, there is a need of token circulation at a later stage. To so do, one can incentivize user for spending tokens. However, sending back and forth between two accounts may be penalized.

16.4.3 Cross-Chain Diversification

We now consider cross-chain diversification between two tokens from two different blockchains. The network characteristics, productivity and security level A_k, a_k, ν_k, are different. The utility characteristics, $h_k, \lambda_{ik}, \rho_k, \xi_{kk'}, r_k$, are also different. The number of token adopters n_{to1} and n_{to2} are also different:

$$\begin{aligned} dw_i &= \{(p_1 u_{i1})^{\rho_1}(h_1(n_{to1})A_1\lambda_{i1})^{1-\rho_1}dt \\ &+u_{i1}p_1\mathbb{E}[\tfrac{dp_1}{p_1}]\}\mathbb{1}_{\{a_1=0\}} - \xi_{12}dt - u_{i1}p_1 r_1 dt \\ &+\{(m-p_1u_{i1})^{\rho_2}(h_2(n_{to2})A_1\lambda_{i2})^{1-\rho_2}dt \\ &+(m-u_{i1}p_1)\mathbb{E}[\tfrac{dp_2}{p_2}]\}\mathbb{1}_{\{a_2=0\}}dt \\ &-\xi_{21}dt-(m-u_{i1}p_1)r_2dt. \end{aligned} \tag{16.21}$$

16.4.4 Arbitrary Number of Cryptographic Tokens

Variance-aware games offer valuable applications in the context of blockchain autoregulation, where decentralized and self-governing mechanisms are employed to maintain the integrity and efficiency of blockchain networks. Here are some real-world applications of variance-aware games in blockchain autoregulation: These MFTGs can optimize the consensus protocol used in blockchain networks. By considering variance-related factors such as network latency, node performance, and transaction verification time, these games enable blockchain networks to dynamically adjust their consensus mechanisms for improved efficiency and security. They facilitate the dynamic management of block sizes in blockchain networks. These games account for variance-related aspects such as transaction volume, network congestion, and validation time, allowing the block size to adapt in real time to ensure optimal throughput and minimize latency.

MFTGs can assist in determining gas fees for blockchain transactions. These games consider variance-related factors like network activity, transaction complexity, and miner incentives to dynamically adjust gas fees, ensuring timely processing of transactions while optimizing user costs. They aid in handling network partitions and forks in blockchain networks. These games consider variance-related uncertainties in network connectivity and consensus divergence, helping nodes make decisions that maintain network integrity and recover from disruptions.

Variance-aware MFTGs support decentralized governance mechanisms in blockchain networks. These games consider variance-related factors such as token holder preferences, voting behavior, and decision outcomes to ensure fair and effective decision-making processes for network upgrades and changes. They optimize the execution of smart contracts on blockchain platforms. These games consider variance-related aspects such as contract complexity, execution time, and resource consumption, leading to more efficient and cost-effective contract deployments.

They guide the allocation of resources in proof-of-work and proof-of-stake blockchain networks. These games consider variance-related factors like mining power, staking amounts, and reward distribution, helping participants make strategic decisions that enhance network security and participation. They assist in designing economic incentives for participants in blockchain networks. These games consider variance-related aspects such as token distribution, inflation rates, and reward mechanisms, ensuring that economic incentives align with desired network behaviors. They contribute to the adaptation of security measures in blockchain networks. These games consider variance-related threats, attack probabilities, and network vulnerabilities to dynamically adjust security protocols and countermeasures.

They enhance the resilience of blockchain networks to external shocks and adversarial actions. These games consider variance-related factors such as network attacks, node failures, and economic incentives, guiding the implementation of measures that strengthen network robustness and sustainability.

By incorporating variance-aware MFTGs into blockchain autoregulation, blockchain networks can achieve greater adaptability, efficiency, and self-governance, ultimately leading to more reliable and secure decentralized systems.

16.4.4.1 Token Price Dynamics

Consider a market with $L \geq 2$ tokens and a time horizon length $T > 0$. The price dynamics of token $l \in \{1, \ldots, L\}$ is given by the following:

$$\begin{aligned}
&\text{On the interval: } (0, T),\\
&dp_l = p_l \left(d_l dt + \textstyle\sum_{k=1}^{K_l} \sigma_{lk} dB_{lk} + \int \gamma_l \tilde{N}_l(dt, d\theta) + \sigma_{gv,l} dB_{gv,l} + \sigma_{r,l} dR_l^H \right),\\
&p_l(0) > 0.
\end{aligned} \tag{16.22}$$

The coefficient functions $d_l, \sigma_{lk}, \gamma_l, \sigma_{gv,l}, and \sigma_{r,l}$ will be designed to capture each token dynamics. They may also be function of time, number of active decision-makers in the system, and global event such as halving.

Under square integrability of the coefficients over $(0, T)$, there is a unique solution to the uncontrolled SDE 16.22, and the unique solution process is recursively constructed by the stochastic integral

$$\begin{aligned}
&p_l(t) = p_l(0)\\
&+ \int_0^t p_l \left(d_l dt' + \textstyle\sum_{k=1}^{K_l} \sigma_{lk} dB_{lk} + \int \gamma_l \tilde{N}_l(dt', d\theta) + \sigma_{gv,l} dB_{gv,l} + \sigma_{r,l} dR_l^H \right).
\end{aligned} \tag{16.23}$$

Furthermore, the dynamics preserves almost surely the positive orthant: it is initially $p_l(0) > 0$ and then $p_l(t) > 0$ a.s. for any $(0, T)$ thanks to the multiplicative factor. This exponentiated structure allows us to interpret random process p_l as a token price as it should be positive. In this scenario, negative prices (where you pay users to hold a token) are not allowed. The presence of regime-switching process inside the coefficient d_l is particularly relevant during a halving period.

16.4.4.2 Evolution of Return of Decision-Maker *i*

The return of decision-maker i on that market at time t is $m_i(t) = \sum_{l=1}^{L} p_l(t)u_{il}(t)$ where $u_{il}(t) \geq 0$ is the share of i on token l. The evolution of the return of i is given by

$$\begin{aligned}
dm_i &= [d_1 m_i + \textstyle\sum_{l=2}^{L}(d_l - d_1)p_l u_{il}]dt \\
&+(\textstyle\sum_{k=1}^{K_1} \sigma_{1k} dB_{1k} + \int \gamma_1 \tilde{N}_1(dt, d\theta) + \sigma_{gv,1} dB_{gv,1} + \sigma_{r,1} dR_1^H)(m_i - \textstyle\sum_{l=2}^{L} p_l u_{il}) \\
&+ \textstyle\sum_{l=2}^{L}[\textstyle\sum_{k=1}^{K_l} \sigma_{lk} dB_{lk} + \int \gamma_l \tilde{N}_l(dt, d\theta) + \sigma_{gv,l} dB_{gv,l} + \sigma_{r,l} dR_l^H)] p_l u_{il}, \\
& m_i(0) = m_{i0} > 0, \\
& 0 \leq \textstyle\sum_{l=2}^{L} p_l u_{il} \leq m_i, \\
& u_{il} \geq 0, \quad \forall (i, l).
\end{aligned} \tag{16.24}$$

We consider an investment problem over multiple correlated tokens. The considered payoff is the mean-variance payoff with a risk-sensitivity index associated with decision-maker i. As there are L tokens, the number of variables of i is L, but the dimension can be reduced to $L - 1$ thanks to the weighted simplex constraint:

$$\sum_{l=1}^{L} p_l u_{il} = m_i, \ u_{il} \geq 0, \ \forall (i, l).$$

Diversification in Tokenized Blockchain

We examine a market characterized by a collection of L tokens, where $L \geq 2$, operating within a time span denoted as $(0, T)$, with $T > 0$. Each token l is associated with a price p_l, and the market includes multiple decision-makers $\mathcal{I}$. Decision-maker $i \in \mathcal{I}$ possesses a total amount denoted as m_i invested in the market and holds a proportion represented by u_{il} of the l-th token. The set $\mathcal{A}_i(t)$ defines the range of feasible actions available to decision-maker i at time t and is characterized by conditions such as

$$u_{il}(t) \geq 0 \text{ and } \sum_{l=1}^{L} p_l(t)u_{il}(t) = m_i(t), \ \ \forall (i, l).$$

(continued)

These decision-makers are interconnected through the dynamics of the market price process, encompassing the variations in prices of tokens $(p_1, \ldots, p_L)$. Each token's price evolution is subject to both internal drift and interrelated sources of stochastic disturbance, thereby creating a complex system. The sources of disturbance encompass several stochastic processes, namely, Brownian motion, Poisson jump processes, Gauss–Volterra processes which include fractional Brownian motions, and Rosenthal processes. It's noteworthy that Rosenthal processes deviate from conventional Gaussian, Poisson, and Markovian characteristics, contributing to diverse price patterns of the overall market dynamics. In this market, decision-maker i aims to look at a mean-variance trade-off with risk-sensivity index $\eta_i \geq 0$ as

$$\begin{aligned}
&\sup_{u_i\in\mathscr{A}_i} mv(m_i) = \sup_{u_i\in\mathscr{A}_i} \mathbb{E}m_i(T) - \tfrac{\eta_i}{2m_i(0)} var[m_i(T)],\\
dm_i &= [d_1 m_i + \textstyle\sum_{l=2}^{L}(d_l - d_1)p_l u_{il}]dt\\
&+(\textstyle\sum_{k=1}^{K_1}\sigma_{1k}dB_{1k} + \int \gamma_1\tilde{N}_1(dt,d\theta) + \sigma_{gv,1}dB_{gv,1} + \sigma_{r,1}dR_1^H)(m_i - \sum_{l=2}^{L} p_l u_{il})\\
&+\textstyle\sum_{l=2}^{L}[\sum_{k=1}^{K_l}\sigma_{lk}dB_{lk} + \int \gamma_l\tilde{N}_l(dt,d\theta) + \sigma_{gv,l}dB_{gv,l} + \sigma_{r,l}dR_l^H]p_l u_{il},\\
&m_i(0) = m_{i0} > 0,\\
&0 \leq \textstyle\sum_{l=2}^{L} p_l u_{il} \leq m_i, \quad u_{il} \geq 0, \forall l
\end{aligned} \tag{16.25}$$

with the correlation $\mathbb{E}[dB_{ij}dB_{kl}] = \beta_{ijkl}\mathbb{1}_{\{(ij)\neq(kl)\}}dt$ and $\beta_{ijij} = 1$ for any (i, j).

The unconstrained problem can be solved using mean-variance MFTG. However, the problem is constrained by the state m. Thus, the vector u_i should belong to the weighted simplex at time $t \in (0, T)$. For token $l \in \{1, \ldots, L\}$, the term d_l can be modified to capture token demand-token supply. As d_l depends on the market cap and number of the decision-maker who joined the system, there is an interaction between them via the prices.

16.4.4.3 Insurance in Blockchain Token Economics

There are several decision-makers in the token economic insurance sector. Insured are the blockchainized platform users who seek protection against a set of L tokens. Insureds are compensated when the hazard they are insured against occurs. In our model, this happens with a Poisson jump process with intensity rate $(L_i - u_{i,ins})$ with the loss $L_i = \hat{l}(\theta)m_i$ with $\hat{l} > 0$. There are insurers whose role is played by the smart contracts of the blockchain. The insurer redistribute funds to insureds with blockchain-approved claims. There are also underwriters who assess the risk of all the L tokens. Reinsurers provide extra protection by increasing an insurance of blockchain protocol solvency. The reinsurers share part of the blockchain's profit. Claim assessors are part of smart contract and the blockchain protocol. They either approve or object a claim from the insured. The insured pays $\bar{\lambda}_i(1 + \bar{\theta})E[u_{i,ins}]$

which is nonrandom. This also makes the dynamics of McKean–Vlasov type as the expected value of the insurance control action $E[u_{i,ins}]$ will enter into the drift of user i. Below, we explicitly include the impact of the insurance in the token dynamics of each user.

16.4.4.4 Consumption, Diversification, and Insurance in Blockchain Token Economics

Decider-maker i wants to decide between consumption, diversification, and Insurance (CDI) when the tokenized blockchain market is constituted of central bank digital currency bond and several crypto-token stocks. The total wealth of decision-maker i follows the dynamics

$$\begin{array}{l} dm_i = [d_1 m_i + \sum_{l=2}^{L}(d_l - d_1)p_l u_{il}]dt \\ \quad +(\sum_{k=1}^{K_1}\sigma_{1k}dB_{1k} + \int \gamma_1 \tilde{N}_1(dt, d\theta) + \sigma_{gv,1}dB_{gv,1} + \sigma_{r,1}dR_1^H)(m_i - \sum_{l=2}^{L} p_l u_{il}) \\ \quad +\sum_{l=2}^{L}[\sum_{k=1}^{K_l}\sigma_{lk}dB_{lk} + \int \gamma_l \tilde{N}_l(dt, d\theta) + \sigma_{gv,l}dB_{gv,l} + \sigma_{r,l}dR_l^H]p_l u_{il} \\ \quad -u_{i,c}dt - \bar{\lambda}_i(1+\bar{\theta})E[u_{i,ins}]dt - \int(L_i - u_{i,ins})d\tilde{N}_{ins}, \\ \quad m_i(0) = m_{i0} > 0, \\ \quad 0 \leq \sum_{l=2}^{L} p_l u_{il} \leq m_i, \\ \quad u_{il} \geq 0, \forall l, \end{array} \tag{16.26}$$

where $L_i = \hat{l}(\theta)m_i$ with $\hat{l} > 0$; the allocation (diversification) strategy is now denoted by $u_{i,a} = (u_{il})_{l\in\{1,\ldots,L\}}$. In the dynamics (16.26), we have considered per-claim insurance of $u_{i,ins}$, that is, if the decision-maker i suffers a loss L_i at time t, the indemnity pays $u_{i,ins}(L_i)$. Such indemnity arrangements are common in private insurance at the individual level, among others. The term $\bar{\theta}$ represents a premium loading. The term $\bar{\lambda}_i$ is $\tilde{\lambda}_i E[m_i]$. Motivated by new blockchain-based insurance products, we allow not only the cryptocurrency market but also the insurable loss to depend on the regime of the cryptocurrency economy and mean-field terms. One may think of insurance of the trend before halving and after halving events. The consumption strategy of the decision-maker i is denoted by $u_{i,c}(t)$. This consumption makes a utility of $\log E[u_{i,c}(t)]$.

Here, the set of control actions $\mathscr{A}_i(t)$ of decision-maker i becomes

$$u_{il}(t) \geq 0, \sum_{l=1}^{L} p_l(t)u_{il}(t) = m_i(t), \ \ \forall (i,l), \ u_{i,c}(t) \geq 0, \ u_{i,ins}(t) \geq 0.$$

The payoff functional of the decision-maker i is modified to be

$$R_i = e^{-\lambda_i T}(\log(E[m_i(T)]) - \frac{\eta_i}{2m_i(0)} var[m_i(T)]) + \int_0^T e^{-\lambda_i t}(\log(E[u_{i,c}(t)]) - r_i var[u_{i,c}(t)])dt.$$

Decision-maker i seeks for a strategy $u_i = (u_{i,a}, u_{i,c}, u_{i,ins})$ that optimizes the expected value of R_i.

The choice of log-utility is common in blockchain token economics and can be obtained from the power law or α-fairness utility $\frac{x^\alpha - 1}{\alpha} = \frac{e^{\alpha \log x} - 1}{\alpha} \rightarrow \log x$ as α approaches zero.

Consumption–Diversification–Insurance Strategy in Blockchain Token Economics

The best response of decision-maker i given the parameters in d_l is given by

$$\begin{array}{l} \sup_{(u_{i,a}, u_{i,c}, u_{i,ins}) \in \mathscr{A}_i} \mathbb{E}[R_i], \\ \text{subject to (16.26).} \end{array} \tag{16.27}$$

16.4.4.5 Variation of the Wealth of the Decision-Maker

Let the state dynamics m_i of decision-maker i be real-valued, driven by different types of processes: Brownian motion, Gauss–Volterra, two Poisson jump processes and Rosenblatt process, and a regime-switching process. Let the wealth of i be

$$\begin{array}{l} m_i(t) = m_{i0} + \int_0^t \hat{d}_1\, dt' + \int_0^t \sum_{l=1}^{L} \sum_{k=1}^{K_l} \hat{d}_{2lk} d B_{lk}(t') + \int_0^t \sum_{l=1}^{L} \hat{d}_{3,l} d B_{gv,l}(t') \\ + \int_0^t \sum_{l=1}^{L} \hat{d}_{4,l} \tilde{N}_l(dt', d\theta) + 2\tilde{c} \int_0^t \sum_{l=1}^{L} \hat{d}_{5,l} d B_l^{\frac{H+1}{2}}(t') + \int_0^t \sum_{l=1}^{L} \hat{d}_{6,l} d R_l^H(t') \\ + \int_0^t \hat{d}_7 N_{ins}(dt', d\theta), \\ \\ \mathbb{P}(s(t+\delta) = s' | s) = \int_t^{t+\delta} \tilde{q}_{ss'} dt' + o(\delta), \ s' \neq s, \end{array} \tag{16.28}$$

for coefficient functions $(\hat{d}_1, \hat{d}_2, \hat{d}_3, \hat{d}_4, \hat{d}_5, \hat{d}_6, \hat{d}_7, \tilde{q})$ given by

$$\begin{array}{l} \hat{d}_1 = [d_1 m_i + \sum_{l=2}^{L} (d_l - d_1) p_l u_{il}] - u_{i,c} - \bar{\lambda}_i (1+\theta) E[u_{i,ins}], \\ \hat{d}_{2,lk} = \sigma_{1k}(m_i - \sum_{l=2}^{L} p_l u_{il}) \mathbb{I}_{\{l=1\}} + \sigma_{lk} p_l u_{il} \mathbb{I}_{\{l \neq 1\}}, \\ \\ \hat{d}_{3,l} = \sigma_{gv,1}(m_i - \sum_{l=2}^{L} p_l u_{il}) \mathbb{I}_{\{l=1\}} + \sigma_{gv,l} p_l u_{il} \mathbb{I}_{\{l \neq 1\}}, \\ \\ \hat{d}_{4,l} = \gamma_1 (m_i - \sum_{l=2}^{L} p_l u_{il}) \mathbb{I}_{\{l=1\}} + \gamma_l p_l u_{il} \mathbb{I}_{\{l \neq 1\}}, \\ \\ \hat{d}_{5,l}(.) = 0 \\ \\ \hat{d}_{6,l} = \sigma_{r,1}(m_i - \sum_{l=2}^{L} p_l u_{il}) \mathbb{I}_{\{l=1\}} + \sigma_{r,l} p_l u_{il} \mathbb{I}_{\{l \neq 1\}}, \\ \\ \hat{d}_7 = -(L_i - u_{i,ins}), \end{array} \tag{16.29}$$

where:

- Drift coefficients: $\hat{d}_1(t, s, n, m_i, p, u_i) : [0, T] \times \mathscr{S} \times \mathbb{R} \times \mathbb{R} \times \mathbb{R}^L \times \mathbb{R}^{L+2} \to \mathbb{R}$.
- Diffusion coefficient: $\hat{d}_{2,lk}(t, s, n, m_i, p, u_i)$, Brownian motion B_{lk}.
- Gauss–Volterra coefficient: $\hat{d}_{3,l}(t, s, n, m_i, p, u_i) : [0, T] \times \mathscr{S} \times \mathbb{R} \times \mathbb{R} \times \mathbb{R}^L \times \mathbb{R}^{L+2} \to \mathbb{R}$, Gauss–Volterra process $B_{gv,l}$.

  The variance of the process $\int_0^t \hat{d}_{3,l}(t', s, n, m_i, p, u_i) dB_{gv}(t')$ is

$$\hat{d}^2_{3,l,cogv,} = \frac{d}{dt}\left[\int_0^t \left\{K(t'_+, t')\hat{d}_{3,l}(t') + \int_{t'}^t \hat{d}_{3,l}(t'')K(t', t'')dt''\right\}^2 dt'\right]$$

  for a certain kernel K.
- Set of jump size Θ, jump rate $\hat{d}_{4,l}(t, s, n, m_i, p, u_i, \theta)$, $[0, T] \times \mathscr{S} \times \mathbb{R} \times \mathbb{R} \times \mathbb{R}^L \times \mathbb{R}^{L+2} \times \Theta \to \mathbb{R}$; jump process, $N_l(dt, d\theta)$; compensated jump, $\tilde{N}_l(dt, d\theta) = N_l(dt, d\theta) - \nu_l(d\theta)dt$,.
- The fractional Brownian motion coefficient, $\hat{d}_{5,l}(t, s, n, m_i, p, u_i)$, $[0, T] \times \mathscr{S} \times \mathbb{R} \times \mathbb{R} \times \mathbb{R}^L \times \mathbb{R}^{L+2} \to \mathbb{R}$. The fractional Brownian motion is $B^H(t) = \tilde{c}_2 \int_{\mathbb{R}} (\int_0^t (t' - x)_+^{H-\frac{3}{2}} dt') dW(x)$, $H \in (\frac{1}{2}, 1)$, $\tilde{c}_2 = \sqrt{\frac{H(2H-1)}{Beta(2-2H, H-\frac{1}{2})}}$.
- Let the Wiener–Itô integral be

$$Z^{\delta_1,\ldots,\delta_k}(t) = c(\delta) \int_{\mathbb{R}^k \setminus D_k} [\int_0^t \prod_{m'=1}^k (t' - x_{m'})_+^{\delta_{m'}} dt'] \prod_{l=1}^k dB(x_l),$$

  with

$$D_k = \{(x_1, \ldots, x_k),\ x_1 = \ldots = x_k\} \subset \mathbb{R}^k,$$

$$c(\delta) \neq 0,\ \sum_{m'=1}^k \delta_{m'} \in (-\frac{k}{2} - \frac{1}{2}, -\frac{k}{2}).$$

  Rosenblatt coefficient: $\hat{d}_{6,l}(t, s, n, m_i, p, u_i) : [0, T] \times \mathscr{S} \times \mathbb{R} \times \mathbb{R} \times \mathbb{R}^L \times \mathbb{R}^{L+2} \to \mathbb{R}$.

  The Rosenblatt process is $r_l^H(t) := Z^{\delta,\delta}(t)$, with $\delta = \frac{H}{2} - 1$, $c(\frac{H}{2} - 1, \frac{H}{2} - 1) = \frac{\sqrt{2H(2H-1)}}{2Beta(1-H, \frac{H}{2})} \Gamma^2(\frac{H}{2})$.

  Beta function: $Beta(z_1, z_2) = \int_0^1 t^{z_1-1}(1 - t)^{z_2-1} dt$, (z_1, z_2) : $\mathrm{Re}\,(z_1), \mathrm{Re}\,(z_2) > 0$.

  Gamma function: $\Gamma(z) = \int_0^\infty t^{z-1} e^{-t} dt$, $\mathrm{Re}\,(z) > 0$.
- The insurance jump coefficient is $\hat{d}_7$, and insurance jump process is $N_{ins}(dt, d\theta)$.
- The regime-switching process $s(t) \in \mathscr{S}$ plays a crucial role in modeling the different halving phase of crypto-tokens or regulations. Its switching rate is

$\tilde{q}_{ss'} > 0, s \neq s', \tilde{q}_{ss} = -\sum_{s' \neq s} \tilde{q}_{ss'}, \ s \in \mathcal{S}$, where $\mathcal{S}$ denotes a finite set of switching regimes.

We assume that s, $x(0)$, B, N, R^H, and N_{ins}, are mutually independent.

Let $f(., s, .) \in C^{1,3}$ in (t, m_i) for any regime-switching component s. Then, the process y defined by $y(t) = f(t, m_i(t), s(t))$ satisfies

$$
\begin{aligned}
&y(t) = y_0 + \int_0^t \tilde{d}_1 \, dt' + \int_0^t \sum_{l=1}^L \sum_{k=1}^{K_l} \tilde{d}_{2lk} dB_{lk}(t') + \int_0^t \sum_{l=1}^L \tilde{d}_{3,l} dB_{gv,l}(t') \\
&+ \int_0^t \sum_{l=1}^L \tilde{d}_{4,l} \tilde{N}_l(dt', d\theta) + 2\tilde{c} \int_0^t \sum_{l=1}^L \tilde{d}_{5,l} dB_l^{\frac{H+1}{2}}(t') + \int_0^t \sum_{l=1}^L \tilde{d}_{6,l} dR_l^H(t') \\
&+ \int_0^t \tilde{d}_7 N_{ins}(dt', d\theta),
\end{aligned}
\tag{16.30}
$$

where

$$
\begin{aligned}
&\tilde{d}_1(t') = f_{t'}(t', m_i(t'), s(t')) + f_{m_i}(t', m_i(t'), s(t'))\hat{d}_1 \\
&+ \sum_{l=1}^L \sum_{k=1}^{K_l} f_{m_i m_i}(t', m_i(t'), s(t')) \tfrac{1}{2} \hat{d}_{2lk}^2 \\
&+ \sum_{l,k} \sum_{l',k'} f_{m_i m_i}(t', m_i(t'), s(t')) \tfrac{1}{2} \hat{d}_{2lk} \hat{d}_{2l'k'} \beta_{lk,l'k'} \mathbb{I}_{\{(lk) \neq (l'k')\}} \\
&+ \sum_{l=1}^L f_{m_i m_i}(t', m_i(t'), s(t')) \tfrac{1}{2} \hat{d}_{3,l,cogv}^2 \\
&+ \int [f(t', m_i(t') + \hat{d}_{4,l}, s(t')) - f(t', m_i(t'), s(t')) - \hat{d}_{4,l} f_{m_i}(t', m_i(t'), s(t'))] \nu(d\theta) \\
&+ \sum_{s'} [f(t', m_i(t'), s') - f(t', m_i(t'), s)] \tilde{q}_{s's} \\
&+ 2c f_{m_i m_i}(t', m_i(t'))(\nabla^{\frac{H}{2}} m_i(t'))(t') \hat{d}_{5,l} \\
&+ c f_{m_i m_i}(t', m_i(t'), s(t'))(\nabla^{\frac{H}{2}, \frac{H}{2}} m_i)(t', t') \hat{d}_{6,l} \\
&+ c f_{m_i m_i m_i}(t', m_i(t'), s(t'))[(\nabla^{\frac{H}{2}} m_i(t'))(t')]^2 \hat{d}_{6,l} \\
&+ \int [f(t', m_i(t') + \hat{d}_7, s(t')) - f(t', m_i(t'), s(t')) - \hat{d}_7 f_{m_i}(t', m_i(t'), s(t'))] \nu_{ins}(d\theta), \\[1em]
&\tilde{d}_{2,l}(t') = f_{m_i}(t', m_i(t'), s(t')) \hat{d}_{2,l}, \\[1em]
&\tilde{d}_{3,l}(t') = f_{m_i}(t', m_i(t'), s(t')) \hat{d}_{3,l}, \\[1em]
&\tilde{d}_{4,l}(t') = [f(t', m_i(t'), s(t')) + \hat{d}_{4,l}, s(t')) - f(t', m_i(t'), s(t'))], \\[1em]
&\tilde{d}_{5,l}(t') = f_{m_i}(t', m_i(t'), s(t')) \hat{d}_{5,l} + f_{m_i m_i}(t', m_i(t'))(\nabla^{\frac{H}{2}} m_i)(t') \hat{d}_{6,l}, \\[1em]
&\tilde{d}_{6,l}(t') = f_{m_i}(t', m_i(t'), s(t')) \hat{d}_{6,l}, \\[1em]
&\tilde{d}_7(t') = [f(t', m_i(t'), s(t')) + \hat{d}_7, s(t')) - f(t', m_i(t'), s(t'))],
\end{aligned}
\tag{16.31}
$$

with

$$c = c(\delta)\Gamma^2(\tfrac{H}{2}),\ \tilde{c} = \sqrt{\frac{(2H-1)\Gamma(1-\frac{H}{2})\Gamma(\frac{H}{2})}{(H+1)\Gamma(1-H)}},$$
$$\nabla^{\frac{H}{2}} = I_+^{\frac{H}{2}} \circ D,\ \ \nabla^{\frac{H}{2},\frac{H}{2}} = I_{+,+}^{\frac{H}{2},\frac{H}{2}} \circ D^2,$$
$$f\ : \mathbb{R} \to \mathbb{R},$$
$$(I_+^{\alpha} f)(t) = \frac{1}{\Gamma(\alpha)} \int_{-\infty}^{t} f(u)(t-u)^{\alpha-1} du,$$
$$(\alpha_1, \alpha_2) \in (0,1)^2,\ f\ : \mathbb{R}^2 \to \mathbb{R},$$
$$(I_{+,+}^{\alpha_1,\alpha_2} f)(t_1,t_2) = \frac{1}{\Gamma(\alpha_1)\Gamma(\alpha_2)} \int_{-\infty}^{t_1}\int_{-\infty}^{t_2} f(u,v)(t_1-u)^{\alpha_1-1}(t_2-v)^{\alpha_2-1} du dv. \tag{16.32}$$

16.4.4.6 Variance of the Wealth of Decision-Maker i

The dynamics of the expected value of wealth of i is given by

$$\begin{aligned}\bar{m}_i(t) &= \bar{m}_{i0} + \int_0^t E[\hat{d}_1]\, dt' \\ &= \bar{m}_{i0} + \int_0^t d_1 \bar{m}_i + \sum_{l=2}^{L} (d_l - d_1) E[p_l u_{il}] - \bar{u}_{i,c} - \bar{\lambda}_i (1+\theta)\bar{u}_{i,ins}\, dt'.\end{aligned} \tag{16.33}$$

Note that the term $E[p_l u_{il}]$ is not decoupled. However, one can directly work with the product process $\tilde{u}_{il} = p_l u_{il}$.

The dynamics of $m_i - \bar{m}_i$ of i is given by

$$\begin{aligned}& m_i(t) - \bar{m}_i(t) = \bar{m}_{i0} - \bar{m}_{i0} \\ & + \int_0^t d_1(m_i - \bar{m}_i) + \sum_{l=2}^{L}(d_l - d_1)(\tilde{u}_{il} - E[\tilde{u}_{il}]) - (u_{i,c} - \bar{u}_{i,c})\, dt' \\ & + \int_0^t \sum_{l=1}^{L}\sum_{k=1}^{K_l} \hat{d}_{2lk} dB_{lk}(t') + \int_0^t \sum_{l=1}^{L} \hat{d}_{3,l} dB_{gv,l}(t') \\ & + \int_0^t \sum_{l=1}^{L} \hat{d}_{4,l} \tilde{N}_l(dt', d\theta) + 2\tilde{c} \int_0^t \sum_{l=1}^{L} \hat{d}_{5,l} dB_l^{\frac{H+1}{2}}(t') + \int_0^t \sum_{l=1}^{L} \hat{d}_{6,l} dR_l^H(t') \\ & + \int_0^t \hat{d}_7 N_{ins}(dt', d\theta),\end{aligned}$$

$$\mathbb{P}(s(t+\delta) = s'|s) = \int_t^{t+\delta} \tilde{q}_{ss'} dt' + o(\delta),\ s' \neq s. \tag{16.34}$$

In order to obtain the dynamics of the variance, we apply (16.30) with $f(y) = y^2$ to the process $m_i - \bar{m}_i$ and then take the expectation.

16.4.4.7 Performance of the Uniform Allocation Strategy

The uniform diversification is a simple strategy and it is feasible. It is therefore a candidate strategy to initialize many learning algorithms. The time-varying uniform strategy given by $\tilde{u}_{il}(t) = \frac{m_i(t)}{L},\ 1 \leq l \leq L$ is distributed among the L tokens. For this diversification strategy at token l time t, the allocation is $u_{il}(t) = \frac{m_i(t)}{Lp_l},\ 1 \leq l \leq L$. Then, the expected return is given by

$$\bar{m}_i(T) = \bar{m}_{i0} e^{\frac{1}{L}\int_0^T \sum_{l=1}^{L} d_l(t')dt'}. \tag{16.35}$$

As expected, the cumulative uniform average demand–supply across all the tokenized blockchains $\frac{1}{L}\int_0^T \sum_{l=1}^L d_l(t')dt'$ plays a crucial role as well as the initial amount invested.

16.4.4.8 Solving the Diversification Problem

We now solve Problem (16.25). We consider a candidate functional of decision-maker i in the following form:

$$f_i(t, m_i, s) = \alpha_i m_i + \frac{\bar{\beta}_i}{2m_{i0}}(\bar{m}_i)^2 - \frac{\beta_i}{2m_{i0}}(m_i - \bar{m}_i)^2 + \delta_i$$

We apply (16.30) with $f_i(x(t), s(t)) = \frac{\beta_i(t,s(t))}{2m_{i0}}x(t)^2$ with the process $x = m_i - \bar{m}_i$ and then take the expectation. The expected value of the difference between the mean-variance payoff and the candidate functional yields

$$\begin{aligned}
&E[R_i - \alpha_i(0)m_i(0) - \tfrac{\bar{\beta}_i(0)}{2m_{i0}}\bar{m}_i(0)^2 + \tfrac{\beta_i(0)}{2m_{i0}}(m_i - \bar{m}_i)^2 - \delta_i(0)]\\
&= E[(1-\alpha_i(T))m_i(T) - \tfrac{\bar{\beta}_i(T)}{2m_{i0}}\bar{m}_i(T)^2 - \tfrac{\eta_i - \beta_i(T)}{2m_{i0}}(m_i(T) - \bar{m}_i(T))^2 - \delta_i(T)]\\
&+ \textstyle\int_0^t [\dot{\alpha}_i(t,s) + \sum_{s'}(\alpha_i(t,s') - \alpha_i(t,s))\tilde{q}_{s's}]\bar{m}_i + \alpha_i d_1 \bar{m}_i\\
&+ \alpha_i \textstyle\sum_{l=2}^L (d_l - d_1)E[\tilde{u}_{il}]\\
&+ \textstyle\int_0^t [\frac{d}{dt}\bar{\beta}_i(t,s) + \sum_{s'}(\bar{\beta}_i(t,s') - \bar{\beta}_i(t,s))\tilde{q}_{s's}]\frac{1}{2m_{i0}}(\bar{m}_i)^2 dt'\\
&+ \textstyle\int_0^t \frac{\bar{\beta}_i}{m_{i0}}\bar{m}_i[d_1\bar{m}_i + \sum_{l=2}^L (d_l - d_1)E[\tilde{u}_{il}]]dt'\\
&+ \textstyle\int_0^t [\dot{\delta}_i(t,s) + \sum_{s'}(\delta_i(t,s') - \delta_i(t,s))\tilde{q}_{s's}]dt'\\
&- [\dot{\beta}_i + \textstyle\sum_{s'}(\beta_i(t,s') - \beta_i(t,s))\tilde{q}_{s's}]\frac{1}{2m_{i0}}(m_i - \bar{m}_i)^2\\
&- \tfrac{\beta_i}{m_{i0}}(m_i - \bar{m}_i)[d_1(m_i - \bar{m}_i) + \textstyle\sum_{l=2}^L (d_l - d_1)(\tilde{u}_{il} - E[\tilde{u}_{il}])]\\
&- \textstyle\sum_{l=1}^L \sum_{k=1}^{K_l} \frac{\beta_i}{m_{i0}}\frac{1}{2}\hat{d}_{2lk}^2 - \sum_{l,k}\sum_{l',k'} \frac{\beta_i}{m_{i0}}\frac{1}{2}\hat{d}_{2lk}\hat{d}_{2l'k'}\rho_{lk,l'k'}\mathbb{1}_{\{(lk)\neq(l'k')\}}\\
&- \textstyle\sum_{l=1}^L \frac{\beta_i}{m_{i0}}\frac{1}{2}\hat{d}_{3,l,cogv}^2 - \sum_{l=1}^L \int \frac{\beta_i}{m_{i0}}\frac{1}{2}\hat{d}_{4,l}^2 \nu(d\theta)\\
&- c\tfrac{\beta_i}{m_{i0}}(\nabla^{\frac{H}{2},\frac{H}{2}}(m_i - \bar{m}_i))(t',t')\hat{d}_{6,l}.
\end{aligned} \tag{16.36}$$

Allocation Action Decomposition

The allocation action space of decision-maker $i \in \{1, \ldots, I\}$ at time t is decomposed as follows:

$$\tilde{A}_i(t) = \left\{ \begin{array}{l}
((\tilde{u}_{il}(t) - E[\tilde{u}_{il}(t)]), E[\tilde{u}_{il}(t)])_{l\in\{1,\ldots,L\}} :\\
E[\tilde{u}_{il}(t)] \geq 0,\ \sum_{l=1}^L E[\tilde{u}_{il}(t)] = \bar{m}_i(t),\\
\sum_{l=1}^L (\tilde{u}_{il}(t) - E[\tilde{u}_{il}(t)]) = m_i(t) - \bar{m}_i(t),\\
(u_{il}(t) - E[\tilde{u}_{il}(t)]) + E[\tilde{u}_{il}(t)] \geq 0,\ l \in \{1, \ldots, L\}.
\end{array} \right\} \tag{16.37}$$

Note that the two new variables $(u_{il} - E[\tilde{u}_{il}])$ and $E[\tilde{u}_{il}]$ remain coupled via the negativity constraints $(u_{il} - E[\tilde{u}_{il}]) + E[\tilde{u}_{il}] \geq 0,\ l \in \{1, \ldots, L\}$.

Squared Mean–Variance–Covariance Decomposition

We use the following orthogonal decomposition:

$$X^2 = (X - \bar{X} + \bar{X})^2 = (X - \bar{X})^2 + \bar{X}^2 + 2(X - \bar{X})\bar{X}$$

As the last term is of zero expected value, one gets $E[X^2] = var(X) + \bar{X}^2$. We apply this rule to the following terms $\hat{d}^2_{2lk}$, $\hat{d}^2_{3,l,cogv}$, and $\hat{d}^2_{4,l}$ and to the cross term $\hat{d}_{2lk}\hat{d}_{2l'k'}$ to use

$$XY = (X-\bar{X}+\bar{X})(Y-\bar{Y}+\bar{Y}) = (X-\bar{X})(Y-\bar{Y})+(X-\bar{X})\bar{Y}+(Y-\bar{Y})\bar{X}+\bar{X}\bar{Y},$$

which means that $E[XY] = cov(X, Y) + \bar{X}\bar{Y}$.

By applying the above with $X = \chi(m_i - \sum_{l=2}^{L} \tilde{u}_{il})$, we get

$$E[X^2] = \chi^2 E[m_i - \bar{m}_i - \sum_{l=2}^{L}(\tilde{u}_{il} - E\tilde{u}_{il})]^2 + \chi^2[\bar{m}_i - \sum_{l=2}^{L} E\tilde{u}_{il}]^2,$$

and for $X = \chi\tilde{u}_{il}$, , we get

$$E[X^2] = \chi^2 E[\tilde{u}_{il} - E\tilde{u}_{il}]^2 + \chi^2[E\tilde{u}_{il}]^2.$$

Here, the coefficient functions are

$$\begin{aligned}
\hat{d}_{2,lk} &= \sigma_{1k}(m_i - \textstyle\sum_{l=2}^{L} \tilde{u}_{il})\mathbb{I}_{\{l=1\}} + \sigma_{lk}\tilde{u}_{il}\mathbb{I}_{\{l\neq 1\}}, \\
\hat{d}_{3,l} &= \sigma_{gv,1}(m_i - \textstyle\sum_{l=2}^{L} \tilde{u}_{il})\mathbb{I}_{\{l=1\}} + \sigma_{gv,l}\tilde{u}_{il}\mathbb{I}_{\{l\neq 1\}}, \\
\hat{d}_{4,l} &= \gamma_1(m_i - \textstyle\sum_{l=2}^{L} \tilde{u}_{il})\mathbb{I}_{\{l=1\}} + \gamma_l\tilde{u}_{il}\mathbb{I}_{\{l\neq 1\}}, \\
\hat{d}_{5,l}(.) &= 0 \\
\hat{d}_{6,l} &= \sigma_{r,1}(m_i - \textstyle\sum_{l=2}^{L} \tilde{u}_{il})\mathbb{I}_{\{l=1\}} + \sigma_{r,l}\tilde{u}_{il}\mathbb{I}_{\{l\neq 1\}}, \\
\hat{d}_7 &= -(L_i - u_{i,ins}).
\end{aligned} \tag{16.38}$$

Applying the squared mean-variance-covariance decomposition yields

$$E\hat{d}^2_{2,lk} = \left[\sigma^2_{1k}(\bar{m}_i - \textstyle\sum_{o=2}^L E\tilde{u}_{io})^2 + \sigma^2_{1k}E(m_i - \bar{m}_i - \sum_{o=2}^L \tilde{u}_{io} - E\tilde{u}_{io})^2\right]\mathbb{I}_{\{l=1\}}$$
$$+\left[\sigma^2_{lk}E(u_{il} - E\tilde{u}_{il})^2 + \sigma^2_{lk}(E\tilde{u}_{il})^2\right]\mathbb{I}_{\{l\neq 1\}},$$

$$E\hat{d}^2_{3,l,cogv} = \left[\sigma^2_{1,cogv}(\bar{m}_i - \textstyle\sum_{o=2}^L E\tilde{u}_{io})^2 + \sigma^2_{1,cogv}E(m_i - \bar{m}_i - \sum_{o=2}^L \tilde{u}_{io} - E\tilde{u}_{io})^2\right]\mathbb{I}_{\{l=1\}}$$
$$+\left[\sigma^2_{l,cogv}E(u_{il} - E\tilde{u}_{il})^2 + \sigma^2_{l,cogv}(E\tilde{u}_{il})^2\right]\mathbb{I}_{\{l\neq 1\}},$$

$$E\hat{d}^2_{4,l} = \left[\gamma_1^2(\bar{m}_i - \textstyle\sum_{o=2}^L E\tilde{u}_{io})^2 + \gamma_1^2 E(m_i - \bar{m}_i - \sum_{o=2}^L \tilde{u}_{io} - E\tilde{u}_{io})^2\right]\mathbb{I}_{\{l=1\}}$$
$$+\left[\gamma_l^2 E(u_{il} - E\tilde{u}_{il})^2 + \gamma_l^2(E\tilde{u}_{il})^2\right]\mathbb{I}_{\{l\neq 1\}},$$

$$E\hat{d}_{2,lk}\hat{d}_{2,l'k'}\mathbb{I}_{\{(lk)\neq(l'k')\}} = \sigma_{lk}\sigma_{l'k'}\times$$
$$\begin{cases}(\bar{m}_i - \sum_{o=2}^L E\tilde{u}_{io})^2 + E(m_i - \bar{m}_i - \sum_{o=2}^L \tilde{u}_{io} - E\tilde{u}_{io})^2, & l=1, l'=1, k\neq k',\\ (\bar{m}_i - \sum_{o=2}^L E\tilde{u}_{io})(E\tilde{u}_{il'}) + E[(m_i - \bar{m}_i - \sum_{o=2}^L \tilde{u}_{io} - E\tilde{u}_{io})(\tilde{u}_{il'} - E\tilde{u}_{il'})], & l=1, l'\neq 1,\\ (E\tilde{u}_{il})(\bar{m}_i - \sum_{o=2}^L E\tilde{u}_{io}) + E[(\tilde{u}_{il} - E\tilde{u}_{il})(m_i - \bar{m}_i - \sum_{o=2}^L \tilde{u}_{io} - E\tilde{u}_{io})] & l\neq 1, l'=1\\ (E\tilde{u}_{il})(E\tilde{u}_{il'}) + E[(\tilde{u}_{il} - E\tilde{u}_{il})(\tilde{u}_{il'} - E\tilde{u}_{il'})] & l\neq 1, l'\neq 1, (lk)\neq(l'k').\end{cases} \tag{16.39}$$

We evaluate the last term $E[(\nabla^{\frac{H}{2},\frac{H}{2}}(m_i - \bar{m}_i))(t', t')\hat{d}_{6,l}]$ with $(\nabla^{\frac{H}{2},\frac{H}{2}}(m_i - \bar{m}_i))(t, t) = I_{+,+}^{\frac{H}{2},\frac{H}{2}}(D^2(m_i - \bar{m}_i))(t, t) = o(t)$.

Recall that

$$\begin{aligned} m_i(t) - \bar{m}_i(t) &= \bar{m}_{i0} - \bar{m}_{i0}\\ &+ \textstyle\int_0^t d_1(m_i - \bar{m}_i) + \sum_{l=2}^L (d_l - d_1)(\tilde{u}_{il} - E[\tilde{u}_{il}])dt'\\ &+ \textstyle\int_0^t \sum_{l=1}^L \sum_{k=1}^{K_l} \hat{d}_{2lk} dB_{lk}(t') + \int_0^t \sum_{l=1}^L \hat{d}_{3,l} dB_{gv,l}(t')\\ &+ \textstyle\int_0^t \sum_{l=1}^L \hat{d}_{4,l}\tilde{N}_l(dt', d\theta) + 2\tilde{c}\int_0^t \sum_{l=1}^L \hat{d}_{5,l} dB_l^{\frac{H+1}{2}}(t') + \int_0^t \sum_{l=1}^L \hat{d}_{6,l} dR_l^H(t')\\ &+ \textstyle\int_0^t \hat{d}_7 N_{ins}(dt', d\theta). \end{aligned} \tag{16.40}$$

We compute the following terms:

$$\begin{aligned} &E[(\nabla^{\frac{H}{2},\frac{H}{2}}(\textstyle\int_0^t \chi dB))(t', t')\hat{d}_{6,l}],\\ &E[(\nabla^{\frac{H}{2},\frac{H}{2}}(\textstyle\int_0^t \chi dB_{gv,l}))(t', t')\hat{d}_{6,l}],\\ &E[(\nabla^{\frac{H}{2},\frac{H}{2}}(\textstyle\int_0^t \chi d\tilde{N}_l))(t', t')\hat{d}_{6,l}],\\ &E[(\nabla^{\frac{H}{2},\frac{H}{2}}(\textstyle\int_0^t \chi dR_l^H))(t', t')\hat{d}_{6,l}], \end{aligned} \tag{16.41}$$

with χ having the form of $a(m_i - \bar{m}_i) + b\bar{m}_i + e$, which is also the form of $\hat{d}_{6,l}$.

We use the following properties:

$$E[(\nabla^{\frac{H}{2},\frac{H}{2}}(\int_0^t \chi dR_l^H))(t', t')\hat{d}_{6,l}(t)] = c\int_0^t E[(\nabla^{\frac{H}{2},\frac{H}{2}}\hat{d}_{6,l})(t', t')\chi(t')]dt'.$$

In particular,

$$
\begin{aligned}
&E[\chi(t)\int_0^t (\nabla^{\frac{H}{2},\frac{H}{2}}\chi(t'))(t,t)dR_l^H(t')] \\
&= c\int_0^t E[(\nabla^{\frac{H}{2},\frac{H}{2}}\chi(t))(t',t')(\nabla^{\frac{H}{2},\frac{H}{2}}\chi(t'))(t,t)]dt'.
\end{aligned}
$$

We deduce the following expressions:

$$
\begin{aligned}
&E[(\nabla^{\frac{H}{2},\frac{H}{2}}(\int_0^t (m_i-\bar{m}_i)dR_l^H))(t',t')(m_i-\bar{m}_i)(t)], \\
&E[(\nabla^{\frac{H}{2},\frac{H}{2}}(\int_0^t \bar{m}_i dR_l^H))(t',t')(m_i-\bar{m}_i)(t)], \\
&E[(\nabla^{\frac{H}{2},\frac{H}{2}}(\int_0^t (m_i-\bar{m}_i)dR_l^H))(t',t')\bar{m}_i(t)], \\
&E[(\nabla^{\frac{H}{2},\frac{H}{2}}(\int_0^t \bar{m}_i dR_l^H))(t',t')\bar{m}_i(t)].
\end{aligned}
\tag{16.42}
$$

16.4.4.9 Solving the Consumption–Diversification–Insurance Problem

We now solve Problem (16.27). We consider a candidate functional of decision-maker i in the following form:

$$
f_i(t,m_i,s) = -\frac{\beta_i}{2m_{i0}}(m_i-\bar{m}_i)^2 + \gamma_i\log(\bar{m}_i) + \delta_i.
$$

$$
\begin{aligned}
&E[R_i + \tfrac{\beta_i(0)}{2m_{i0}}(m_i-\bar{m}_i)^2 - \gamma_i(0)\log(\bar{m}_i(0)) - \delta_i(0)] \\
&= E[-\tfrac{\eta_i e^{-\lambda_i T}-\beta_i(T)}{2m_{i0}}(m_i(T)-\bar{m}_i(T))^2 + (e^{-\lambda_i T}-\gamma_i(T))\log(\bar{m}_i(T)) - \delta_i(T)] \\
&+E\int_0^t[\dot{\delta}_i(t,s) + \textstyle\sum_{s'}(\delta_i(t,s')-\delta_i(t,s))\tilde{q}_{s's}] \\
&-\tfrac{\dot{\beta}_i}{2m_{i0}}(m_i-\bar{m}_i)^2 - \tfrac{\beta_i}{m_{i0}}(m_i-\bar{m}_i)[d_1(m_i-\bar{m}_i) \\
&+\textstyle\sum_{l=2}^L(d_l-d_1)(\tilde{u}_{il}-E[\tilde{u}_{il}]) - (u_{i,c}-\bar{u}_{i,c})] \\
&-\textstyle\sum_{l=1}^L\sum_{k=1}^{K_l}\tfrac{\beta_i}{m_{i0}}\tfrac{1}{2}\hat{d}_{2lk}^2 - \sum_{l,k}\sum_{l',k'}\tfrac{\beta_i}{m_{i0}}\tfrac{1}{2}\hat{d}_{2lk}\hat{d}_{2l'k'}\beta_{lk,l'k'}\mathbb{1}_{\{(lk)\neq(l'k')\}} \\
&-\textstyle\sum_{l=1}^L\tfrac{\beta_i}{m_{i0}}\tfrac{1}{2}\hat{d}_{3,l,cogv}^2 - \sum_{l=1}^L\int\tfrac{\beta_i}{m_{i0}}\tfrac{1}{2}\hat{d}_{4,l}^2\nu(d\theta) \\
&-\textstyle\sum_{s'}[(\beta_i(t,s')-\beta_i(t,s))]\tilde{q}_{s's}\tfrac{1}{2m_{i0}}(m_i-\bar{m}_i)^2 \\
&-2c\tfrac{\beta_i}{m_{i0}}(\nabla^{\frac{H}{2}}m_i-\bar{m}_i(t'))(t')\hat{d}_{5,l} - c\tfrac{\beta_i}{m_{i0}}(\nabla^{\frac{H}{2},\frac{H}{2}}m_i-\bar{m}_i)(t',t')\hat{d}_{6,l} \\
&-\textstyle\int\tfrac{\beta_i}{m_{i0}}\tfrac{1}{2}\hat{d}_7^2\nu_{ins}(d\theta)dt' \\
&+\textstyle\int_0^t[\dot{\gamma}_i(t,s) + \sum_{s'}(\gamma_i(t,s')-\gamma_i(t,s))\tilde{q}_{s's}]\log(\bar{m}_i)dt' \\
&+\textstyle\int_0^t\gamma_i\tfrac{1}{\bar{m}_i}[d_1\bar{m}_i + \sum_{l=2}^L(d_l-d_1)E[\tilde{u}_{il}] - \bar{u}_{i,c} - \bar{\lambda}_i(1+\bar{\theta})\bar{u}_{i,ins}]dt' \\
&+\textstyle\int_0^t e^{-\lambda_i t}(\log(\bar{u}_{i,c}) - r_i var[u_{i,c}(t)])dt'.
\end{aligned}
\tag{16.43}
$$

We next determine the CDI strategy of mean-field type [106].

Optimal Consumption Strategy
The payoff function

$$\bar{u}_{i,c} \mapsto -\gamma_i \frac{1}{\bar{m}_i}\bar{u}_{i,c} + e^{-\lambda_i t}\log(\bar{u}_{i,c}) + \frac{\beta_i}{m_{i0}}(m_i - \bar{m}_i)(u_{i,c} - \bar{u}_{i,c}) - e^{-\lambda_i t} r_i var[u_{i,c}(t)]$$

is concave, and its first-order optimality condition yields

$$\begin{aligned} \bar{u}_{i,c} &= \frac{e^{-\lambda_i t}}{\gamma_i}\bar{m}_i, \\ (u_{i,c} - \bar{u}_{i,c}) &= \frac{\beta_i}{2r_i e^{-\lambda_i t} m_{i0}}(m_i - \bar{m}_i). \end{aligned} \tag{16.44}$$

This is also a sufficient condition for optimality as by maximization of strictly concave function with $r_i > 0$, $\gamma_i > 0$. Therefore, the optimal consumption strategy of decision-maker i is

$$u^*_{i,c} = \frac{1}{\gamma_i e^{\lambda_i t}}\bar{m}_i + \frac{\beta_i e^{\lambda_i t}}{2r_i m_{i0}}(m_i - \bar{m}_i). \tag{16.45}$$

Optimal Diversification Allocation Strategy
The allocation is potentially across all L tokens. For $1 \le l \le L$, the share at token l is $\tilde{u}_{il}$ to be determined. It is uniquely determined by the maximization of the following strictly concave function

$$\begin{aligned} &-\tfrac{\beta_i}{m_{i0}}(m_i - \bar{m}_i)[\textstyle\sum_{l=2}^{L}(d_l - d_1)(\tilde{u}_{il} - E[\tilde{u}_{il}])] \\ &-\textstyle\sum_{l=1}^{L}\sum_{k=1}^{K_l}\tfrac{\beta_i}{m_{i0}}\tfrac{1}{2}\hat{d}^2_{2lk} - \sum_{l,k}\sum_{l',k'}\tfrac{\beta_i}{m_{i0}}\tfrac{1}{2}\hat{d}_{2lk}\hat{d}_{2l'k'}\beta_{lk,l'k'}\mathbb{I}_{\{(lk)\neq(l'k')\}} \\ &-\textstyle\sum_{l=1}^{L}\tfrac{\beta_i}{m_{i0}}\tfrac{1}{2}\hat{d}^2_{3,l,cogv} - \sum_{l=1}^{L}\int\tfrac{\beta_i}{m_{i0}}\tfrac{1}{2}\hat{d}^2_{4,l}\nu(d\theta) \\ &-c\tfrac{\beta_i}{m_{i0}}(\nabla^{\frac{H}{2},\frac{H}{2}}m_i - \bar{m}_i)(t',t')\hat{d}_{6,l} \\ &+\gamma_i\tfrac{1}{\bar{m}_i}\textstyle\sum_{l=2}^{L}(d_l - d_1)E[\tilde{u}_{il}] \end{aligned} \tag{16.46}$$

over the convex set.

From the term $\tilde{u}_{il}$, we deduce the optimal allocation strategies $u^*_{il} = \frac{\tilde{u}_{il}}{p_l}$.

Optimal Insurance Strategy
The terms that involve $u_{i,ins}$ are

$$-\int \frac{\beta_i}{m_{i0}}\frac{1}{2}\hat{d}^2_7\nu_{ins}(d\theta) - \gamma_i\frac{1}{\bar{m}_i}\bar{\lambda}_i(1+\bar{\theta})u_{i,ins}$$

$$= -\int \frac{\beta_i}{m_{i0}}\frac{1}{2}(L_i - u_{i,ins})^2\nu_{ins}(d\theta) - \gamma_i\frac{1}{\bar{m}_i}\bar{\lambda}_i(1+\bar{\theta})u_{i,ins}.$$

The latter term $u_{i,ins}$ is strictly concave. The maximization yields

$$u_{i,ins} = \frac{1}{\int\frac{\beta_i}{m_{i0}}\nu_{ins}(d\theta)}\{-\gamma_i\bar{\lambda}_i(1+\bar{\theta}) + \int\frac{\beta_i}{m_{i0}}L_i\nu_{ins}(d\theta)\}$$

which is linear (affine) in m_i as L_i is linear in m_i.

16.5 There Is No Best Generative Machine Intelligence

Transformers, diffusions, and other large learning models can be used to analyze blockchain tokens, cryptocurrencies, and central bank digital currencies. These generative foundational machine intelligence models can look at past data (backcasting) and current data (nowcasting) and estimate future trends (forecasting) of the financial technology (FinTech) market. By doing this, they can help spot investment opportunities, track market movements, and provide valuable insights. The results can be delivered through an easy-to-use dashboard, making it accessible for investors, traders, and FinTech professionals to make informed decisions. This dashboard could highlight key trends, risks, and opportunities in almost real time, helping users stay ahead in the market. The precision of the output from generative machine intelligence displayed in a dashboard is crucial because it directly affects the quality of the decisions users make. Inaccurate or imprecise insights can lead to poor investment choices, misinterpreting market trends, or overlooking key threads and opportunities. For investors, traders, and FinTech professionals, even small errors in AI-generated forecasts or recommendations can result in significant financial losses. Therefore, ensuring high precision not only builds trust in the machine intelligence system but also helps users confidently rely on it for accurate, actionable insights that support smarter decision-making.

Let us start with the case of backcasting. FinTech has generated vast amounts of online data, particularly in the blockchain space. The dashboard should be tested on this historical data. If the model does not perform well on past data generated by FinTech, it is unlikely that it will be trusted for identifying future opportunities. That is why backcasting is a critical first step. It is also worth noting that this is easily verifiable because the data already exists and there is no need for a prediction window to assess its performance ex post.

Reviewing the literature on transformer-based deep neural networks, such as masked multi-head Boltzmann–Gibbs self-attention with two-layer feedforward networks using ReLU, SiLU, GeLU, weighted Kolmogorov–Arnold networks, or diffusion transformers (transfusion), all tends to introduce statistical bias after optimizing trillions of parameters. Transformers, with their large number of parameters, are trained on massive data sets that may contain imbalances or patterns that lead the model to favor certain predictions. Additionally, the training process involves minimizing a loss function or a variational inequality across many parameters, which can lead to overfitting or underfitting, especially when the data distribution is skewed or lacks diversity. Moreover, techniques like masked self-attention, which transformers rely on, can unintentionally emphasize certain parts of the input data, further contributing to (statistically) biased outputs. Having such a statistical bias is not a good indicator of learning accuracy. In the business world, sometimes, being able to foresee a trend in advance, even without full market details, can make a big difference. However, in some areas, errors in trend prediction can lead to significant financial losses in the FinTech market, which can have global consequences. The same explanation holds for nowcasting, forecasting, forward-looking, estimation, approximation, etc.

This means that after the post-design, post-training, post-optimization, post-fine-tuning, and post-self-consistency check of these class of learning algorithms, most of these would lead to

$$E[\hat{\mathcal{O}}_L \circ \hat{\mathcal{O}}_{L-1} \circ \ldots \circ \hat{\mathcal{O}}_1(x_{t-1}) - (a_{t+0}, \ldots, a_{t+h-1})] \neq 0$$

where $x_{t-1} = (a_1, \ldots, a_{t-1})$ is the historical data up to $t-1$.

The connection with the mean-field-type game theory [460, 464] here is the variance reduction component of the algorithms.

16.5.1 Point Forecasting Setup

We consider $A \geq 2$ interactive assets distributed across a platform of platform blockchain companies, crypto-tokens, and central bank digital currencies. In practice, the number of assets of top investors is around $A = 30$. For each asset i, we display a dashboard on trends and indicators. An example of price dynamics for asset o is given by

$$\begin{cases} p_{o,0} > 0, \ o \in \{1, 2, \ldots, A\}, \\ p_{o,t+1} = p_{o,t}(1 + f_{o,t}), \ o \in \{1, 2, \ldots, A\}, \\ f_{o,t} = (d_{o,t}(r'_t, n_{o,t}) - s_{o,t}(r'_t, n_{o,t}))\delta + \eta_{o,t+1}, \end{cases} \tag{16.47}$$

where d_o is the demand (usage) of o in the market, s_o is the supply, δ is a timescale, and $\eta_{o,t+1}$ is a combination of difference noise times $\sqrt{\delta}$. The interaction between A assets is through the demand–supply which depends on the amount invested by the active decision-makers. We choose d_o, s_o to be functions of the total number of active decision-makers $n_{o,t} := \sum_{j=1}^{I} \mathbb{I}_{a_{j,o,t}>0}$, where r'_t is a regime-switching process over a finite set S. The action $a_{j,o,t}$ is the number of unit tokens o held by decision-maker j at time t. $a_{j,o,t} > 0$ means that j is active on token o at time t.

The evolution of return is given by

$$\begin{cases} w_{i,0} > 0, \ i \in \{1, 2, \ldots, I\}, \\ w_{i,t+1} = w_{i,t} \\ +(d_{1,t} - s_{1,t})\delta(w_{i,t} - \sum_{o=2}^{A} p_{o,t}a_o) + (w_{i,t} - \sum_{o=2}^{A} p_{o,t}a_o)\eta_{1,t+1} \\ +\sum_{o=2}^{A} p_{o,t}(d_{o,t} - s_{o,t})\delta a_o + \sum_{o=2}^{A} p_{o,t}a_o\eta_{o,t+1}. \end{cases} \tag{16.48}$$

The returns are correlated via the prices which in turn are correlated via demand–supply.

Consider the system (16.48) and associate with it a vector of quantities of interest which are, for example:

- the wealth $w_{i,T}$, the variance $var(w_{i,T})$, the trends such as the (weekly, monthly, quarterly, yearly) moving averages

$$\frac{1}{M}\sum_{k=1}^{M} w_{i,T-k}, \quad M \in \{7, 30, 90, 365\},$$

  and the (de-)growth rates $\frac{w_{i,T}}{w_{i,T-1}}, \frac{w_{i,T}}{w_{i,T-730}}, \frac{w_{i,T}}{w_{i,T-365}}, \frac{w_{i,T-365}}{w_{i,T-730}}$ for each decision-maker i,
- the o-th asset wealth $w_{i,o,T}$, the variance $var(w_{i,o,T})$, the trends such as the (weekly, monthly, quarterly, yearly) moving averages

$$\frac{1}{M}\sum_{k=1}^{M} w_{i,o,T-k}, \quad M \in \{7, 30, 90, 365\},$$

  and the (de)growth rates $\frac{w_{i,o,T}}{w_{i,o,T-1}}, \frac{w_{i,o,T}}{w_{i,o,T-730}}, \frac{w_{i,o,T}}{w_{i,o,T-365}}, \frac{w_{i,o,T-365}}{w_{i,o,T-730}}$ per asset o for decision-maker i,
- the market price $p_{o,T}$, the variance price $var(p_{o,T})$, the (weekly, monthly, quarterly, yearly) moving averages

$$\frac{1}{M}\sum_{k=1}^{M} p_{o,T-k}, \quad M \in \{7, 30, 90, 365\},$$

  and the (de)growth rates $\log(\frac{p_{o,T}}{p_{o,T-1}}), \log(\frac{p_{o,T}}{p_{o,T-730}}), \log(\frac{p_{o,T}}{p_{o,T-365}}), \log(\frac{p_{o,T-365}}{p_{o,T-730}})$ for each asset o.

We collect all these quantities of interest and obtain the element of $\mathbb{R}^D$ where D is the total number of these quantities of interest to be displayed in the dashboard of the decision-makers. These quantities of interest are quantized into a set V which is a non-empty and finite set. We then consider A historical sequences $(r_{1,k}, \ldots, r_{A,k})_{1\leq k\leq t-1} \in \mathbb{R}^{(t-1)AD}$ for $t \geq 2$. The realized result $r_{o,t}$ of token o at time t takes values in a non-empty and finite set V of $\mathbb{R}^D$. Example of a component of V could be as follows:

- increase by 10+ %,
- decrease by 2 %, or
- no change from the previous time.

A well-trained financial technology generative artificial intelligence is used to forecast the next $h \geq 1$ trends of the market: h-ahead forecast of the vector $(r_{1,t+j}, \ldots, r_{A,t+j})_{0\leq j\leq h-1}$ given $(r_{1,k}, \ldots, r_{A,k})_{k\leq t-1}$ the past realization up to $t-1$. The generative machine intelligence can be a top deep learning, transformer-based, diffusion-based, generative adversarial network, or any other considered as state-of-the-art algorithm for processing sequences, time series, tabular, sheets, etc.

The output of the h-ahead forecast by generative machine intelligence is denoted by

$$(f_{1,t+j}, \ldots, f_{A,t+j})_{0 \le j \le h-1}.$$

The transformer-based generated output is

$$\hat{\mathcal{O}}_L \circ \hat{\mathcal{O}}_{L-1} \circ \ldots \circ \hat{\mathcal{O}}_1(h_{t-1})$$

with the operator at layer l given by

$$\hat{\mathcal{O}}_l = (Id + \mathcal{O}_{l,ff} \circ \mathcal{O}_{l,nn}) \circ (Id + \mathcal{O}_{l,att} \circ \mathcal{O}_{l,nn}),$$

with:

- L is the total number of layers. $1 \le l \le L$ denotes the l-th layer of the neural network.
- Input space: $\mathcal{H}_0 := \mathbb{R}^{(t-1)AD}$, where $(t-1)$ is the size of the historical data window length. The input is three-entry tensor with time index, asset index, and trend index.
- $\mathcal{O}_{l,nn}$ is the normalization operator followed by a linear operator. It could be for normalization part $\frac{z}{1+|z|}$ or $proj_{C_2} \circ proj_{C_1}(z)$ where $C_1 = \{w \in \mathbb{R}^k,\ \langle w, \mathbb{1} \rangle = 0\}$ and $C_2 = \{w \in \mathbb{R}^k,\ \langle w, w \rangle \le \sqrt{D-1}\}$.
- $\mathcal{O}_{l,att}$ is the masked self-attention mechanism. It is the part where there occurs an interaction between the historical data of the token. Examples of masked self-attentions include Boltzmann–Gibbs (applied to each row of each same token), sigmoid (entry-wise), etc.
- $\mathcal{O}_{l,ff}$ is a (two-layered) feedforward neural network such as $W_{2l}r_l(W_{1l}x + b_{1l}) + b_{2l}$ with r_l being $ReLU, GeLU, SiLU$, etc., and $(W_{2l}, b_{2l}, W_{1l}, b_{1l})$ are bounded linear weights and bias. One can also replace it by a properly designed weighted Kolmogorov–Arnold network with rational polynomial basis.
- The output at layer L is extracted to be in the form

  $$(f_{1,t+j'}, \ldots, f_{A,t+j'})_{0 \le j' \le h-1} \in \mathbb{R}^{hAD} =: \mathcal{H}_L,$$

  where h is the forecasting window. When $h = 0$, it corresponds to nowcasting. The change of dimension can be obtained from a linear bounded operator $W_* : \mathbb{R}^{(t-1)AD} \to \mathbb{R}^{hAD}$; it could also be a final masked softmax operator applied to the vector.

Definition 16.5 The generative machine intelligence is given by

$$(L, t, A, D, \hat{\mathcal{O}}_L \circ \hat{\mathcal{O}}_{L-1} \circ \ldots \circ \hat{\mathcal{O}}_1, \mathbb{R}^{(t-1)AD}, \mathbb{R}^{hAD}).$$

We define the squared error of the generative AI-based backcasting/nowcasting/forecasting evaluated ex post as

$$Err_{h,t}(f|r) = \frac{1}{A}\sum_{1\le o\le A}\frac{1}{t_o-h}\sum_{1\le t'\le t_o-h}\frac{1}{hD}\sum_{0\le h'\le h-1}\mathbb{E}d^2(f_{o,t'+h'}, r_{o,t'+h'}),$$

where $d = d_2$ is the distance induced by the 2-norm of $\mathbb{R}^D$.

This h-ahead averaged forecasting error is not scaled by data magnitude but normalized by the time horizon length, forecasting horizon length, number of assets, and number of forecasting trend indicators. $Err_{h,t}(f|r)$ will be used as a performance evaluation criteria of the algorithm. The forecasting f is considered as good when r is not too small but $Err_{h,t}(f|r)$ is small enough. We have chosen different updating time. Each asset has its own clock t_o which are updated at time of the day/night. We work with the available data (the number of times it has been updated but not the number of seconds since the beginning). To build f, we are allowed to convexify the domain V which arises naturally when the outcomes are seen and one uses averaging or frequency of occurrences.

Problem 16.6 Consider a well-trained generative machine intelligence which takes a historical data as input $(r_{1,k},\ldots,r_{A,k})_{1\le k\le t-1} \in \mathbb{R}^{(t-1)AD}$, operates over L layers, and provides an output $(f_{1,t+j'},\ldots,f_{A,t+j'})_{0\le j'\le h-1} \in \mathbb{R}^{hAD}$. Define the nowcasting/forecasting error as

$$Err_{h,t}((f_{1,t+j'},\ldots,f_{A,t+j'})_{0\le j'\le h-1}|(r_{1,k},\ldots,r_{A,k})_{1\le k\le t-1}).$$

Can we design a new learning algorithm based on the input and the output and that does better than f in terms of error?

We show that the answer to this question is positive. This is an important result as it shows the limitation of current generative machine intelligence architectures in pointwise backcasting/nowcasting/forecasting. The forecasting here is point-wise-related. And it is clear that point-wise forecasting is much asking to an approximation method such as generative machine intelligence foundational learning models. This is one of the reasons why one has attempted to focus on weaker version such as probability-based, cumulative function-based, interval-based, and localized or aggregated quantities of interest. Typically, given the historical data, forecasting a mean (expected value) is more easier than forecasting a (point) value that will be realized. The limitation is also coming from the fact that the approximation tool used during the training yields an optimal weight and bias that are dependent on input which is not a constant. A constant weight and bias are obtained at the end of the training process. This constant weight and bias is therefore unrelated to the specific historical data in hand. The conditional expectation however depends strongly on the nature of the historical data.

Next, we explain in an explicit way how we can do better than the best generative AI.

16.5.2 Doing Better than the Existing Best Machine Intelligence in Pointwise Forecasting

As $\mathbb{E}X^2 = var(X) + (\mathbb{E}X)^2$, one can decompose $Err_{h,t}(f|r)$ into two parts. To do so, we introduce the frequency of realization at entry z of asset i. $\nu_{r,t_o}(z) = \frac{1}{t_o}\sum_{1\leq t\leq t_o} \mathbb{I}_{r_{o,t}=z}$. One can write the time-average realization up to t_o as $\bar{r}_{o,t_o} = \frac{1}{t_o}\sum_{1\leq t\leq t_o} r_{o,t} = \sum_{z\in V} z(\frac{1}{t_o}\sum_{1\leq t\leq t_o} \mathbb{I}_{r_{o,t}=z}) = \mathbb{E}_{\nu_{r,to}} r_o$. We do this operation for all the A assets available on the platform.

The following real-valued random variables $d_2(f_{o,t}, \bar{f}_{o,t}), d_2(\bar{f}_{o,t}, \bar{r}_{o,t}), d_2(\bar{r}_{o,t}, r_{o,t})$ are all within two times the diameter of V :

$$
\begin{aligned}
&Err_{h,t}(f|r)\\
&= \tfrac{1}{D}\mathbb{E}\sum_{(o,t',h')} \lambda_{o,t',h'} d^2(f_{o,t'+h'}, r_{o,t'+h'})\\
&= \tfrac{1}{D}\mathbb{E}\sum_{(o,t',h')} \sum_{z\in V} \lambda_{o,t',h'} d^2(z, r_{o,t'+h'})\mathbb{I}_{f_{o,t'+h'}=z},
\end{aligned}
\tag{16.49}
$$

where $\lambda_{o,t',h'} = \frac{1}{A(t_o-h)h} > 0$.

The idea here is to write this expression in terms of the second moment of a conditional random variable. We write

$$
\mathbb{E}(X^2) = \mathbb{E}(\mathbb{E}(X^2|f)) = \mathbb{E}(\text{var}(X|f)) + \mathbb{E}\{(\mathbb{E}(X|f))^2\}.
$$

The idea now is to build another mechanism g that does better than f for the performance $Err_{h,t}(f|r)$. Let $\mathscr{F}_t$ be the filtration generated by the union of events $\{r_{t'}, t' \leq t\}$. The function $g \mapsto \mathbb{E}\sum_{1\leq o\leq A, 1\leq t'\leq t} d^2(g_o, r_{o,t'})$ is minimized at the conditional expectation $g_{o,t} = \mathbb{E}(r_{o,t}| \mathscr{F}_{t-1})$. The next h step ahead forecast is $\mathbb{E}((r_{o,t}, \ldots, r_{o,t+h-1})| \mathscr{F}_{t-1})$. This estimation is unbiased. However, the output of the generative machine intelligence is an estimated version of $(g_{o,t}, \ldots, g_{o,t+h-1})$ and not the exact conditional expectation. This leads to an estimation error or approximation for the well-trained generative machine intelligence f.

Let us replace the conditional expectation by a very simple mechanism which averages the previous actions:

$$
g_{o,t}(z) = \frac{\sum_{1\leq t'\leq t} \lambda_{o,t'} r_{o,t'} \mathbb{I}_{f_{o,t'}=z}}{\sum_{1\leq t''\leq t} \lambda_{o,t''} \mathbb{I}_{f_{o,t''}=z}}.
$$

Proposition 16.7 *We have*

$$
Err_{h,t}(g|r) \leq Err_{h,t}(f|r) - e_t^2,
$$

where e_t goes zero as t goes to infinity.

This means in particular that for every generative machine intelligence output, there is another simple algorithm one that performs strictly better in terms of accuracy as long as the term $\mathbb{E}(X|f) \neq 0$. This answers positively to the question of

Problem 16.6. The strategy g_t uses $\{(r_{t'}, f_{t'}), t' \le t-h\}$ but does not the past $\{g_{t'}, t' \le t-2\}$. One can also build a sequence that does not use f and that does better than f. It suffices to replace $\{f_{t'}, t' \le t-h\}$ by $\{r_{t'}, t' \le t-h\}$ in the definition of g. This also says that generative machine intelligence combined with reinforcement learning can do better than generative machine intelligence alone.

We prove Proposition 16.7 step by step. The algorithm induced by g tracks the entire average. We write it in an iterative version so that the next average is obtained by a simple convex combination between the new data and the previous average. Indeed, the weighted average

$$m_t = \frac{\sum_{1\le t'\le t} \lambda_{t'} r_{t'} \mathbb{I}_{f_{t'}=z}}{\sum_{1\le t'\le t} \lambda_{t'} \mathbb{I}_{f_{t'}=z}},$$

with $\sum_{1\le t'\le t} \lambda_{t'} \mathbb{I}_{f_{t'}=z} > 0$ and $\lambda_{t'} \ge 0$ for every t', can be rewritten as

$$m_t = m_{t-1} + \frac{\lambda_t \mathbb{I}_{f_t=z}}{\sum_{1\le t'\le t} \lambda_{t'} \mathbb{I}_{f_{t'}=z}} (r_t - m_{t-1}).$$

This recursive equation is crucial in online update. As we can see, the latter $m_t(z)$ is $g_t(z)$.

The recursive equation for m is made as an online update as the next average is obtained from the previous update and the new data. We ask whether a similar online exists for the error induced by g. Let

$$\begin{aligned}
\tilde{m}_{2,t} &= \textstyle\sum_{1\le t'\le t} \lambda_{t'} \mathbb{I}_{f_{t'}=z} d^2(r_{t'}, m_t(z)) \\
&= \textstyle\sum_{1\le t'\le t} \lambda_{t'} \mathbb{I}_{f_{t'}=z} d^2(r_{t'}, m_{t-1} + \frac{\lambda_t \mathbb{I}_{f_t=z}}{\sum_{1\le t'\le t} \lambda_{t'} \mathbb{I}_{f_{t'}=z}} (r_t - m_{t-1})) \\
&= \textstyle\sum_{1\le t'\le t} \lambda_{t'} \mathbb{I}_{f_{t'}=z} d^2(r_{t'}, m_{t-1}) \\
&\quad + \textstyle\sum_{1\le t'\le t} \lambda_{t'} \mathbb{I}_{f_{t'}=z} d^2(0, \frac{\lambda_t \mathbb{I}_{f_t=z}}{\sum_{1\le t'\le t} \lambda_{t'} \mathbb{I}_{f_{t'}=z}} (r_t - m_{t-1})) \\
&\quad -2\langle \textstyle\sum_{1\le t'\le t} \lambda_{t'} \mathbb{I}_{f_{t'}=z} r_{t'}, \frac{\lambda_t \mathbb{I}_{f_t=z}}{\sum_{1\le t''\le t} \lambda_{t''} \mathbb{I}_{f_{t''}=z}} (r_t - m_{t-1})\rangle) \\
&= \tilde{m}_{2,t-1} + \lambda_t \mathbb{I}_{f_t=z} d^2(r_t, m_{t-1}) \\
&\quad + \textstyle\sum_{1\le t'\le t} \lambda_{t'} \mathbb{I}_{f_{t'}=z} d^2(0, \frac{\lambda_t \mathbb{I}_{f_t=z}}{\sum_{1\le t''\le t} \lambda_{t''} \mathbb{I}_{f_{t''}=z}} (r_t - m_{t-1})) \\
&\quad -2\langle m_{t-1} + \frac{\lambda_t \mathbb{I}_{f_t=z}}{\sum_{1\le t'\le t} \lambda_{t'} \mathbb{I}_{f_{t'}=z}} (r_t - m_{t-1}), \lambda_t \mathbb{I}_{f_t=z} (r_t - m_{t-1})\rangle \\
&= \tilde{m}_{2,t-1} + \lambda_t \mathbb{I}_{f_t=z} (1 - \frac{\lambda_t \mathbb{I}_{f_t=z}}{\sum_{1\le t'\le t} \lambda_{t'} \mathbb{I}_{f_{t'}=z}}) d^2(r_t, m_{t-1}) \\
&\quad -2\lambda_t \mathbb{I}_{f_t=z} \langle m_{t-1}, (r_t - m_{t-1}).\rangle
\end{aligned} \tag{16.50}$$

Hence,

$$\begin{aligned}
\tilde{m}_{2,t} &= \textstyle\sum_{1\le t'\le t} \lambda_{t'} \mathbb{I}_{f_{t'}=z} (1 - \frac{\lambda_{t'} \mathbb{I}_{f_{t'}=z}}{\sum_{1\le t''\le t'} \lambda_{t''} \mathbb{I}_{f_{t''}=z}}) d^2(r_{t'}, m_{t'-1}) \\
&\quad -2\textstyle\sum_{1\le t'\le t} \lambda_{t'} \mathbb{I}_{f_{t'}=z} \langle m_{t'-1}, r_{t'}\rangle + 2\sum_{1\le t'\le t} \lambda_{t'} \mathbb{I}_{f_{t'}=z} d^2(m_{t'-1}, 0).
\end{aligned} \tag{16.51}$$

Thus,

$$
\begin{aligned}
m_{2,t} &= \frac{\tilde{m}_{2,t}}{\sum_{1\le t''\le t}\lambda_{t''}\mathbb{I}_{f_{t''}=z}} \\
&= \frac{\sum_{1\le t'\le t}\lambda_{t'}\mathbb{I}_{f_{t'}=z}d^2(r_{t'},m_{t'-1})}{\sum_{1\le t''\le t}\lambda_{t''}\mathbb{I}_{f_{t''}=z}} \\
&= -\frac{\sum_{1\le t'\le t}\frac{(\lambda_{t'}\mathbb{I}_{f_{t'}=z})^2}{\sum_{1\le t''\le t'}\lambda_{t''}\mathbb{I}_{f_{t''}=z}}d^2(r_{t'},m_{t'-1})}{\sum_{1\le t''\le t}\lambda_{t''}\mathbb{I}_{f_{t''}=z}} \\
&+\frac{(-2\sum_{1\le t'\le t}\lambda_{t'}\mathbb{I}_{f_{t'}=z}\langle m_{t'-1},r_{t'}\rangle+2\sum_{1\le t'\le t}\lambda_{t'}\mathbb{I}_{f_{t'}=z}d^2(m_{t'-1},0))}{\sum_{1\le t''\le t}\lambda_{t''}\mathbb{I}_{f_{t''}=z}}.
\end{aligned}
\tag{16.52}
$$

The extra error made by working with online computation and the error is given by

$$
\begin{aligned}
&\frac{\sum_{1\le t'\le t}\lambda_{t'}\mathbb{I}_{f_{t'}=z}d^2(r_{t'},m_{t'-1})}{\sum_{1\le t''\le t}\lambda_{t''}\mathbb{I}_{f_{t''}=z}} - \frac{\sum_{1\le t'\le t}\lambda_{t'}\mathbb{I}_{f_{t'}=z}d^2(r_{t'},m_{t})}{\sum_{1\le t''\le t}\lambda_{t''}\mathbb{I}_{f_{t''}=z}} \\
&= \frac{\sum_{1\le t'\le t}\frac{(\lambda_{t'}\mathbb{I}_{f_{t'}=z})^2}{\sum_{1\le t''\le t'}\lambda_{t''}\mathbb{I}_{f_{t''}=z}}d^2(r_{t'},m_{t'-1})}{\sum_{1\le t''\le t}\lambda_{t''}\mathbb{I}_{f_{t''}=z}} \\
&+\frac{(2\sum_{1\le t'\le t}\lambda_{t'}\mathbb{I}_{f_{t'}=z}\langle m_{t'-1},r_{t'}\rangle-2\sum_{1\le t'\le t}\lambda_{t'}\mathbb{I}_{f_{t'}=z}d^2(m_{t'-1},0))}{\sum_{1\le t''\le t}\lambda_{t''}\mathbb{I}_{f_{t''}=z}}.
\end{aligned}
\tag{16.53}
$$

Using the diameter and the convex combination in V, we arrive at

$$
\begin{aligned}
&\frac{\sum_{1\le t'\le t}\lambda_{t'}\mathbb{I}_{f_{t'}=z}d^2(r_{t'},m_{t'-1})}{\sum_{1\le t''\le t}\lambda_{t''}\mathbb{I}_{f_{t''}=z}} - \frac{\sum_{1\le t'\le t}\lambda_{t'}\mathbb{I}_{f_{t'}=z}d^2(r_{t'},m_{t})}{\sum_{1\le t''\le t}\lambda_{t''}\mathbb{I}_{f_{t''}=z}} \\
&\le \frac{1+\log(\sum_{1\le t'\le t}\lambda_{t'}\mathbb{I}_{f_{t'}=z})}{\sum_{1\le t''\le t}\lambda_{t''}\mathbb{I}_{f_{t''}=z}} 4\, diameter^2(V) \\
&+\frac{(2\sum_{1\le t'\le t}\lambda_{t'}\mathbb{I}_{f_{t'}=z}\langle m_{t'-1},r_{t'}\rangle-2\sum_{1\le t'\le t}\lambda_{t'}\mathbb{I}_{f_{t'}=z}d^2(m_{t'-1},0))}{\sum_{1\le t''\le t}\lambda_{t''}\mathbb{I}_{f_{t''}=z}}.
\end{aligned}
\tag{16.54}
$$

Based on the above calculations, it follows that

$$
\begin{aligned}
&0 \le Err_{h,t}(f|r) - Err_{h,t}(g|r) \\
&= \frac{1}{D}\mathbb{E}\sum_{(o,t',h')}\lambda_{o,t',h'}(d^2(f_{o,t'+h'},r_{i,t'+h'}) - d^2(g_{o,t'+h'},r_{o,t'+h'})) \\
&\le \mathbb{E}\frac{1+\log(\sum_{1\le t'\le t}\lambda_{t'}\mathbb{I}_{f_{t'}=z})}{\sum_{1\le t''\le t}\lambda_{t''}\mathbb{I}_{f_{t''}=z}}4\text{diameter}^2(V).
\end{aligned}
\tag{16.55}
$$

■

Why Is Generative Machine Intelligence Alone Very Limited for Pointwise Forecasting?

In order to understand the limitation of generative AI, let us consider a very simple and well-known example which is the linear regression of $y_j = Wx_j + b + n_j$, where (W, b) are fixed and $\mathbb{E}n_j = 0$. We sample $(t-1)$ times and obtain a sequence (x_j, y_j) from that equation for $1 \le j \le t-1$. This sequence is sent to a top generative AI. The goal is to reduce the global error. The coefficient b is easily obtained as $b = \mathbb{E}y - W\mathbb{E}x$ which clearly depends on the distribution via the expected values. From the relation $\mathbb{E}(y-\mathbb{E}y)(x-\mathbb{E}x)^\dagger = W\mathbb{E}(x-\mathbb{E}x)(x-\mathbb{E}x)^\dagger$, we obtain that W is also a function of the distribution of (x, y) via the covariance and the variance term. This means the optimal weight/bias that minimizes the error is dependent on the moments. They are not distribution-independent. The fixed weight and bias in the architecture of generative machine intelligence can approximate but are not optimal even for a basic linear regression problem:

- Pointwise matching: Based on this reasoning, we can see that the equation $W_L x_L + b_L = y_L + f_L(y_L)$ cannot be satisfied with all sequences (x_L, y_L) with a constant (W_L, b_L), where f_L is such that the activation function at L is $r_L = (Id+\partial f_L)^{-1}$, where ∂f_L is the sub-differential operator of f_L. The optimal choice (W_L, b_L) to match $W_L x_L + b_L = y_L + f_L(y_L)$ is possible only when we allow (W_L, b_L) to be dependent on (x_L, y_L).
- Data set-based average matching: If we work with a given sequence, we get

$$b_L = \bar{y}_L + \bar{f}_L(y_L) - W_L \bar{x}_L$$

  and

$$W_L(x_L - \bar{x}_L)(x_L - \bar{x}_L)^\dagger$$
$$= (y_L - \bar{y}_L)(x_L - \bar{x}_L)^\dagger + (f_L(y_L) - \bar{f}_L(y_L))(x_L - \bar{x}_L)^\dagger,$$

  where $\bar{a} = \frac{1}{t}\sum_{t'=1}^{t} a_{t'}$. If x_L is zero, there is no W_L needed and b_L should match $y_L + f_L(y_L)$. Clearly, the optimal weight/bias are sequence-dependent via the empirical average and the empirical covariance and variance.
- And in the stochastic case, given the distribution of (y_L, x_L), the optimal bias is

$$b_L = \mathbb{E}y_L + \mathbb{E}f_L(y_L) - W_L \mathbb{E}x_L$$

  and

$$W_L \mathbb{E}(x_L - \mathbb{E}x_L)(x_L - \mathbb{E}x_L)^\dagger$$
$$= \mathbb{E}(y_L - \mathbb{E}y_L)(x_L - \mathbb{E}x_L)^\dagger + \mathbb{E}(f_L(y_L) - \mathbb{E}f_L(y_L))(x_L - \mathbb{E}x_L)^\dagger.$$

  By taking the inverse of the matrix $\mathbb{E}(x_L - \mathbb{E}x_L)(x_L - \mathbb{E}x_L)^\dagger$, we obtain the optimal weight which involves again the second moment of x_L, the covariance (y_L, x_L), and the covariance $(f_L(y_L), x_L)$. We conclude the optimal training weight/bias are data-dependent in the finite regime and moment-/covariance-dependent in the long run. Restricting the design to constant (data-independent, distribution-independent) coefficients leads therefore to suboptimality. In practice, however, we may not need the optimal weight/bias to perform relatively well in some specific tasks.

Observe that, here, the time average is performing better than any other approximation tool such as the current golden standard generative machine intelligence because the average $\frac{1}{t}\sum_{t'=1}^{t} a_{t'}$ is the argmin of the cumulative pointwise squared error function $x \mapsto \sum_{t'=1}^{t} d_2^2(x, a_{t'})$.

The fact that the weighted time average performs better than the current golden standard generative machine intelligence in a pointwise forecasting is because the average $\frac{\sum_{t'=1}^{t} \lambda_{t'} a_{t'}}{\sum_{t'=1}^{t} \lambda_{t'}}$ is the argmin of the cumulative pointwise squared error $x \mapsto \sum_{t'=1}^{t} \lambda_{t'} d_2^2(x, a_{t'})$.

Note that if we change d_2 or the 2-norm to be another one such as p-norm, $p \geq 1, p \neq 2$, the argmin is changed; the average is not the argmin anymore.

In this chapter, we have studied basic crypto-token on blockchain using variance-aware mean-field-type games [460, 464]. We have seen that incentives, network security, and network delay (via the productivity) affect the behavior of the agents. The incentive proposed here exhibits the following features:

- It is shown that a well-designed token can incentivize the agents and boost participation.
- The participation boosts demand and requests for verifiers. Then reputation and token-based incentive for verifiers and disincentives for malicious nodes are designed. The reputation allows to select verifiers who are really known for their good work in the blockchain. They do not necessarily have large amount of native tokens.
- Then, more and more reputed nodes will be available and will be selected for verification.
- This enables better network security, less delay, better supply, and a medium of exchange, which in turn enables for demand.

We have also examined the use of generative machine intelligence in pointwise backcasting/nowcasting/forecasting algorithms in blockchain token economics. We have shown that a transformer-based generative machine intelligence is limited in terms of statistical bias and approximation bias. These biases create in turn a strong limitation for reducing the global accuracy. As a consequence, one can always find another simple averaging algorithm that does better than generative machine intelligence in terms of accuracy. Of course, accuracy is not the only key performance metric in generative machine intelligence, and pointwise forecasting is not the only golden standard approach. However, in the FinTech area where important decisions are made by central banks, regulators, investors, insurers, traders, and users based on generative machine intelligence output displayed on dashboards, the accuracy of these algorithms need to be checked and improved with knowledge such as the one proposed here.

16.6 Notes

The structure of the network and protocol used play important roles in the verification and validation of new information, data, or transaction to be added to the network [47, 130, 244, 478]. Direct acyclic graph has been proposed as an alternative network and extension *beyond blockchains*. It removes blocks and works directly with Markov chain Monte Carlo or Poisson point process. This involves also mean-field-type games with McKean–Vlasov Markov chains [103].

Chapter 17
Hub-Based Airline Networks

This chapter studies the performance of hub-based airline networks using mean-field-type game theory. We identify three types of interactions within the game: interaction between passengers, interaction between passengers and airlines, and interaction between airlines. The key mean-field terms [460, 464] are the traffic flow (or frequency of flights) and the number of people at the same slot. In the absence of congestion, there is a dominating strategy for a designer to adopt the hub network leading a negation of Braess paradox [70]. However, when the frequency of flights increases due to demand, the hub network is no longer superior, and all the links will be used, leading to an enhanced hub network. At an enhanced hub network, higher prices (could be congestion-dependent) are charged to passengers taking the longer direct flights compared to the ones who transit via the hub. We show that the resulting Stackelberg MFTG with multiple leaders and multiple followers has an equilibrium and the equilibrium payoffs are compared across both types of networks. Distributed strategic learning algorithms are designed to compute the best airlines for passengers, the equilibrium flows, and the optimal pricing schemes. The analogy with communication networks is established with hybrid small base stations (femto, pico) and macro base stations. The small cell base stations help to work with low-power regime and save some energy. However, the femtocell network alone does not cover the entire area and does not dominate the market. It is shared with the enhanced macro cell.

Congestion in Airline Transportation

Growing congestion in the airline transportation network poses a substantial threat to the economy and to the quality of life of millions of people. At its fundamental level, airline congestion is caused by the lack of a mechanism to efficiently manage the use of capacity, scheduling, and flight frequencies. When searching for a solution to the congestion problem, most people immediately think of adding a new flight to an overburdened airlines and to create hubs. The performance of such redesign of the network has been widely studied, and several counterintuitive results have

T. Başar et al., *Mean-Field-Type Game Theory II*, Static & Dynamic Game Theory: Foundations & Applications, https://doi.org/10.1007/978-3-032-07031-9_17

Fig. 17.1 Hub-based network

been observed including Braess paradox as well as its negation. In the context of airline networks, a hub is a node of the network that concentrates traffic [planes and passengers] from several origins and distributes it to the final destinations (see Fig. 17.1). In multi-hub networks, the traffic is concentrated at a hub and sent from there to a second hub and possibly a third hub and so on, which distributes it to the final destinations.

In an airline network with a lot of traffic at the hubs, adding a new flight with a transit at the hub to distribute the traffic may seem like a sensible idea. But according to the Braess paradox, just the opposite occurs: a new flight added to an airline network results in increase in the cost to all airlines.

Here, we identify several regimes where Braess and its negation may occur. We find that paradox disappears as the frequencies of flights and the congestion level increase. In the low-congestion regime, hub networks may have some benefits over the non-hub networks. The reason is that when congestion is low, the transportation between hubs is less expensive per unit flow than the transportation between a hub and a non-hub. However, as traffic levels increase, hub airports become more congested than non-hub airports, because they receive higher traffic levels. Braess paradox disappears if the congestion at hub is priced in a smarter way, and some passengers will be willing to pay a higher price for direct flights. We know from Pigou's approach that there is a global welfare maximizing pricing for such a network based on congestion.

In order to efficiently design and schedule flights, several analyses of resulting effects, costs, delays, and congestions should be conducted. We develop a game theoretic tool to analyze both users' behavior and airline's reactions. The analysis help us to prevent high levels of congestion at particular hubs.

The rest of the chapter is structured as follows: Sect. 17.2 describes the network interaction model and the topology design policies. We then develop a Stackelberg MFTG to model the demand–supply interactions. Section 17.3 analyzes different network configurations. In Sect. 17.4, we explain how the approach developed here can be applied to hybrid networks with small cell hotspots and macro-cell cellular networks.

17.1 Braess Paradox, Its Negation, and Crowd Effect

In this section, we examine Braess paradox, its negation, and crowd effect.

Consider a normalized traffic flow from source (S) to destination (D) (see Fig. 17.2). We examine how closing the road H1–H2 in Fig. 17.2 can improve the travel time for everyone. The following observations are in order:

- When segments H2–H1 and H1–H2 are open: For example, for a flow traffic less than half, everyone will go through the route S–H2–H1–D which minimizes the individual travel time. Moreover, the travel time is $flow + 0.01 + flow$. This is illustrated in Fig. 17.3.
- Now, if the authority closes the road H2–H1, then the new network does not have the edge H1–H2 and H2–H1. The traffic will be split equally between the two routes: S–H1–D and S–H2–D. The individual travel time will be $1 + \frac{flow}{2}$. This is illustrated in Fig. 17.4.
- We compare $0.01+2*flow$ to $1+\frac{flow}{2}$. This means that $3\frac{flow}{2}$ will be compared to 0.99.

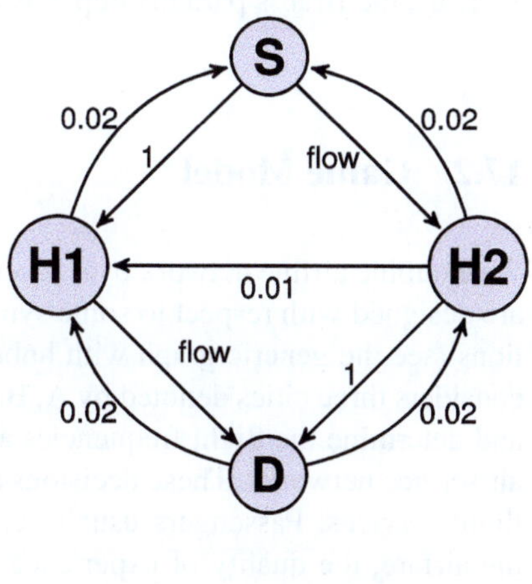

Fig. 17.2 Typical network: the closure of the middle road may lead to better outcome for all travelers

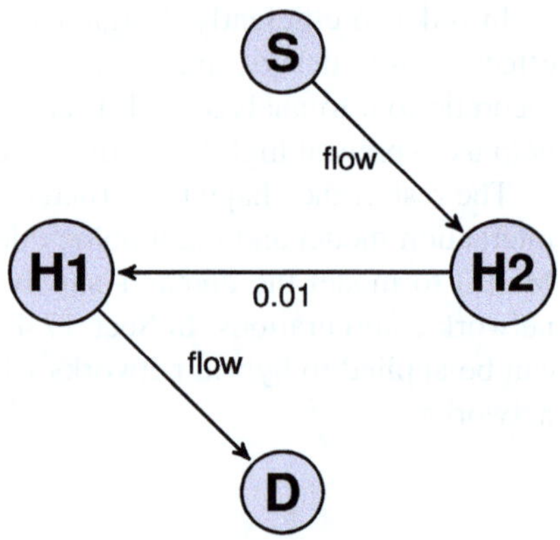

Fig. 17.3 Selfish behavior passing the flow

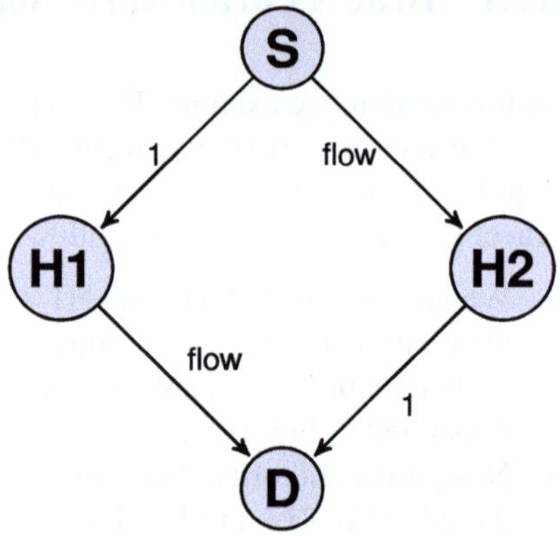

Fig. 17.4 Half traffic flow split between the two routes. The closure of the road H1–H2 leads to a longer travel time for everyone from S to D

- Thus, if the normalized flow is greater than $\frac{0.99*2}{3}$, then closing the road decreases the travel time. However, if the normalized flow is less than $\frac{0.99*2}{3}$, the opposite effect occurs.
- Thus, the Braess paradox depends on the total flow/initial mass.

17.2 Game Model

We examine airline network over several cities and possible hubs. The possible hubs are designed with respect to some symmetry to some cities and topology configurations (see the generic graph with hubs displayed above). Our typical configuration considers three cities denoted by A, B, and H where the airlines can choose the fares and determine the flight frequencies and aircraft types on individual routes of their air service networks. These decisions affect the cost and quality of airline passenger flight services. Passengers usually choose airlines based on their service quality, the airfare, the quality of experience, flight frequencies, etc., so airlines decisions on frequencies and type of aircraft further affect airlines passenger demand, cost, and revenue and airline reputation on congestion. Demand for travel exists between each pair of cities, yielding three city pair flight connections: AH, BH, and AB. This leads to a bi-level interaction. For a given specific time window, passengers select flights offered by different airlines based on airfares, flight frequencies, the availability of nonstop flights, travel and layover times, and the airlines reputations and frequent-flyer (FFlyer) programs. Airlines may use high flight frequencies or nonstop flights to attract passengers, but excess frequencies or nonstop flights may entail diseconomies of scale. Conversely, airlines may consolidate passenger flows from several city pair routes, combining these individual routes into hub-and-spoke networks, using larger aircraft and offering fewer nonstop flights. However, the choice of these strategies may also increase passenger schedule delays and thus lose some time-sensitive passengers. This interaction between demand and supply can be analyzed using the so-called mean-field-type games [460, 464]. The schedule delay time T_{as} arising from the difference between the time a passenger desires to travel and the time he or she can travel, due to the existing flight frequencies offered by airline a, is inversely proportional to the flight frequency of airline a on its route r between origin and destination pair. So the schedule time should be added to the line-haul travel time (block time) T_{ar} spent flying route r on airline a aircraft between origin and destination pair. Thus, the individual passenger's total perceived cost of time on a trip via airline a route r is $\alpha_s T_{a,s} + \alpha_r T_{a,r}$, where $\alpha_s, \alpha_r > 0$, and $T_{a,r} = \frac{1}{f_{a,r}}$ where $f_{a,r}$ is the flight frequency and $T_{a,s} = \omega_r \sum_a f_{a,r}$ is a congestion-based queueing time. Considering a pair of cities, we distinguish two types of routes:

- The first one is a fully connected (FC) route (nonstop) which corresponds to a direct link between the origin and the destination:

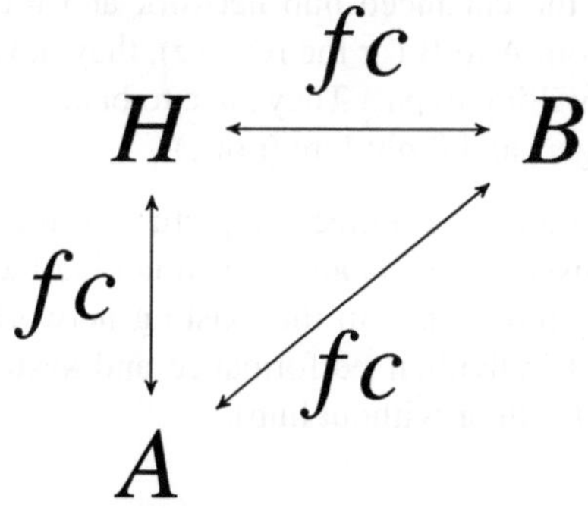

  In an FC network, each airline operates fights between each pair of cities, so that nonstop travel occurs in each city pair market.
- The second type of routing is the hub-and-spoke routing in which the airline operates flights on only two routes, those connecting cities A and B to the hub H. Although passengers in AH and BH still enjoy nonstop service, AB passengers must make a connecting trip, changing planes at the hub H:

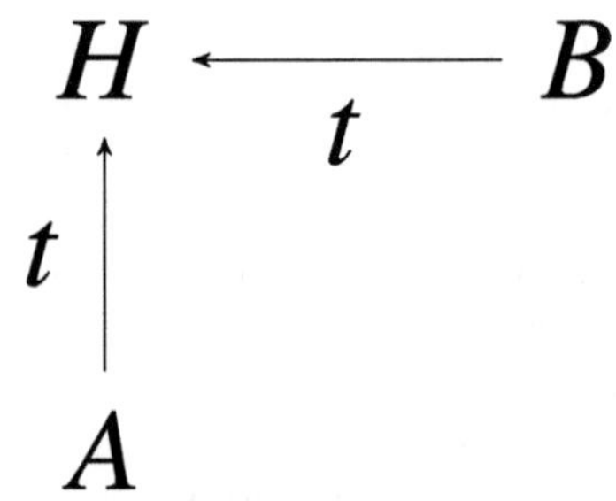

The connecting flights HA and HB are not represented in the graph.

- Enhanced Hub: The third network is a mixture between the hub network and some direct flight between A and B for VIPs and business purposes. The price of the direct flight will be differentiated from the price of those who will transit via the hub:

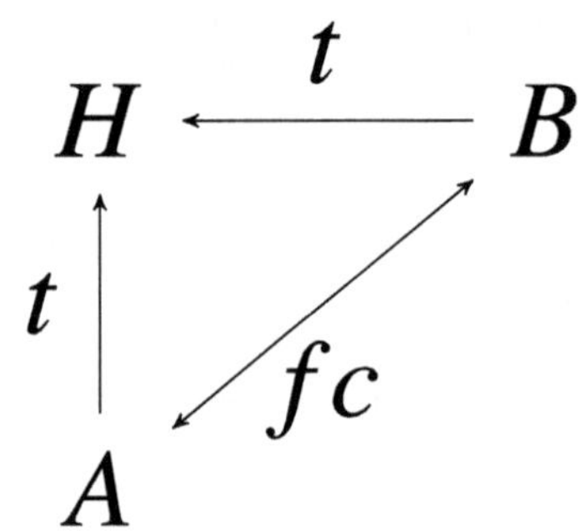

The connecting flights HA and HB are not represented in the graph. The difference between the enhanced hub network and the FC network is that for passengers flying from A to B (or the reverse), they have two options depending how much they are willing to pay. They have to balance between delay [transit at the hub may be longer] and flight fare (price).

It is therefore important to examine the picture of the traffic, cost, and benefit when adding a hub between cities A and B. Mimicking the idea of Braess [70], the situation is like adding more links to the existing network between A and B. We address the question of individual performance and system performance in these two types of networks (with or without hub).

17.2.1 *Payoff of a Passenger*

Payoff to a passenger i when he or she chooses airline a is given by

$$P_{i,a} = W_i + s_i + b_{i,a} - \text{time cost}_a - \text{fare}_a,$$

where s_i is consumption satisfaction and $b_{i,a}$ is travel benefit, which varies across consumers and across airline according a certain distribution in $[\underline{b}, \bar{b}]$. Time cost consists of two components explained above and equals

$$\text{time cost}_a = \alpha_r \frac{1}{f_{a,r}} + \omega_r \sum_{a'} f_{a',r}.$$

$\text{fare}_a = p_a$ is the price (fare) fixed by airline a. This is a congestion-dependent cost due the fact that there may be a high concentration of passengers at the hub and flight departure might delayed when several airlines are scheduled within the same window.

17.2.1.1 Passenger Choice

Given the price vector $(p_a)_a$ and the time cost per airline, passengers have to decide with which airline to travel with. Passenger i would prefer a to a' if $P_{i,a}$ is greater than $P_{i,a'}$ which leads to

$$b_{i,a} - b_{i,a'} \geq p_a - p_{a'} + \frac{\alpha_r}{f_{a,r}} - \frac{\alpha_r}{f_{a',r}}.$$

17.2.1.2 Formulation of Mean-Field and Market Share

The choice of the passengers determine a repartition of the users' population creating mean-field terms. One can determine the market size of each airline via the mean-field terms [106].

Since $b_{i,a} - b_{i,a'}$ follows the random variable, one can compute the expected fraction of number of the passengers with airline a as the mean-field term

$$m_a = \mathbb{P}\left(b_{i,a} - p_a - \frac{\alpha_r}{f_{a,r}} > \max_{a' \neq a}[b_{i,a'} - p_{a'} - \frac{\alpha_r}{f_{a',r}}] \right).$$

If $a^* \in \arg\max_{a' \neq a}\{p_{a'} + \frac{\alpha_r}{f_{a',r}}\}$, then

$$m_a = \mathbb{P}\left(b_{i,a} - b_{i,a*} \in (p_a - p_{a*} + \frac{\alpha_r}{f_{a,r}} - \frac{\alpha_r}{f_{a*,r}}, +\infty) \right).$$

If X and Y are independent and distributed on $[\underline{b}, \bar{b}]$, then the PDF of $X - Y$ is given by the convolution between the distribution of X with the distribution of $(-Y)$. By inverting the mean-field m_a as a function of p_a, one gets the inverse demand curve, given an expression of the price. In the case of two airlines, the mean-field can be explicitly computed as $m_1 = 1 - F_{b_{i,1} - b_{i,2}}\left(p_1 - p_2 + \frac{\alpha_r}{f_{1,r}} - \frac{\alpha_r}{f_{2,r}}\right)$. In particular, if $p_1 - p_2 + \frac{\alpha_r}{f_{1,r}} - \frac{\alpha_r}{f_{2,r}} \in (\underline{b}, \bar{b})$ and $b_{i,1} - b_{i,2}$ follows the uniform distribution $U[\underline{b}, \bar{b}]$, then

$$m_1 = \frac{\bar{b} - [p_1 - p_2 + \frac{\alpha_r}{f_{1,r}} - \frac{\alpha_r}{f_{2,r}}]}{\bar{b} - \underline{b}}. \tag{17.1}$$

17.2.2 Airline's Payoff

We quantify the cost to the airlines. Let sf denote the number of seats per flight, which is a choice variable for the airline. sf determines the size the aircraft to choose. Then, the airline cost per fight is

$$c_{a,r} + \beta_a \frac{m_a I}{f_{a,r}} + 4\omega_r \sum_{a'} f_{a',r},$$

where I is the total number of passengers. Thus, the total cost for airline a on route r is

$$f_{a,r}[c_{a,r} + \beta_a \frac{m_a I}{f_{a,r}} + 4\omega_r \sum_{a'} f_{a',r}].$$

Flights would have empty seats when demand realizations are low. The payoff of airline a in a fully connected network is determined by

$$\pi_a = 3[m_a I p_a - f_{a,r} c_{a,r} - \beta_a m_a I - 4\omega_r f_{a,r} \sum_{a'} f_{a',r}],$$

where we have a multiplication by 3 because of the three different nonstop flights: AH, BH, and AB. Thus,

$$\pi_a = 3[m_a I (p_a - \beta_a) - f_{a,r}(c_{a,r} + 4\omega_r \sum_{a'} f_{a',r})].$$

We now consider the case of the hub network. In that case, we have only two flights: AH and BH. The passengers of the AB will have a connecting flight at the hub. Airlines congestion cost on a route is given by $3\omega_r f_{a,r} \sum_{a'} f_{a',r}$ because congestion on a route is caused by aircraft movements both at the hub $2\sum_{a'} f_{a',r}$ and at the spoke airport $\sum_a f_{a,r}$. The payoff of airline a in the hub network is determined by

$$\pi_a^h = 2[m_a I (p_a - \beta_a) - f_{a,r}(c_{a,r} + 3\omega_r \sum_{a'} f_{a',r})] + \text{connecting margin},$$

where the connecting margin is $(p_{a,AHB} - 2\beta_a)m_{a,AHB}I$, with $p_{a,AHB}$ the fare for the people connecting at the hub and $m_{a,AHB}$ the mass of those people. Note that, in the hub network, the prices $p_{a,AH} = p_{a,BH}$ but the price $p_{a,AHB}$ is different because there is no direct flight between A and B.

17.3 Demand–Supply Interaction Model

$$\begin{array}{ccc} Airlines & \xleftarrow{\quad Aircrafts \quad} & Demand \\ {\scriptstyle Fares}\downarrow & & \uparrow{\scriptstyle mass\ action} \\ Users & \xrightarrow[QoE]{cheap/time} & MF \end{array}$$

Bi-level Mean-Field-Type Game: Duopoly
Here, we consider two competing airlines over the networks. The payoff of passenger i is

$$P_{i,a} = W_i + s_i + b_{i,a} - \alpha_r \frac{1}{f_{a,r}} - \omega_r \sum_a f_{a,r} - p_a.$$

The mean-field of airline 1 is the demand

$$m_1 I = [\frac{\bar{b} - [p_1 - p_2 + \frac{\alpha_r}{f_{1,r}} - \frac{\alpha_r}{f_{2,r}}]}{\bar{b} - \underline{b}}]I.$$

Fully Connected Network
The payoff of airline 1 is

$$\pi_1 = 3[m_1 I(p_1 - \beta_1) - f_{1,r}(c_{1,r} + 4\omega_r \sum_{a'} f_{a',r})]$$

in the FC network. Airline 1 maximizes π_1 over $(p_1, f_1) \in \mathbb{R}^2_+$ (roughly speaking, f_i should be a nonnegative integer). The first-order optimality condition yields

$$\partial_{p_1}\pi_1(f, p) = -\frac{3I}{\bar{b} - \underline{b}}(p_1 - \beta_1) + 3m_1 I = 3I[m_1 - \frac{(p_1 - \beta_1)}{\bar{b} - \underline{b}}] = 0, \tag{17.2}$$

$$\partial_{f_1}\pi_1(f, p) = 3I(p_1 - \beta_1)\frac{[\frac{\alpha_r}{f_{1,r}^2}]}{\bar{b} - \underline{b}} - 3(c_{1,r} + 4\omega_r \sum_{a'} f_{a',r}) - 3f_{1,r}4w_r = 0. \tag{17.3}$$

The concavity of the payoff with respect to p_1 and with respect to f_1 suggests that the first-order condition is sufficient. However, a projection to positive orthant is required to get the solution. The interior solution yields

$$\bar{b} + \beta_1 - [2p_1 - p_2 + \frac{\alpha_r}{f_{1,r}} - \frac{\alpha_r}{f_{2,r}}] = 0,$$

$$\bar{b} + \beta_2 - [2p_2 - p_1 + \frac{\alpha_r}{f_{1,r}} - \frac{\alpha_r}{f_{2,r}}] = 0.$$

Taking the sum, we obtain $2\bar{b} + \beta_1 + \beta_2 - (p_1 + p_2) - 2[\frac{\alpha_r}{f_{1,r}} - \frac{\alpha_r}{f_{2,r}}] = 0$, and by taking the difference, one gets $\beta_1 - \beta_2 = (p_1 - p_2)$. In particular, in the symmetric case $\beta_1 = \beta_2 = \beta$ and

$$p_1 = p_2 = \bar{b} + \beta,$$

and the equilibrium fare is the positive root of

$$I\bar{b}\frac{\alpha_r}{\bar{b} - \underline{b}} - f_r^2(c_{1,r} + 8f_r\omega_r) - 4f_r^3\omega_r = 0$$

which can be written as

$$-I\bar{b}\frac{\alpha_r}{\bar{b} - \underline{b}} + f_r^2 c_{1,r} + 12 f_r^3\omega_r = 0, \quad f_r > 0. \tag{17.4}$$

There is a unique positive solution to the system (17.4).

Hub Network

In the hub network, the payoff is $\pi_a^h = 2[m_a I(p_a - \beta_a) - f_{a,r}(c_{a,r} + 3\omega_r \sum_a f_{a,r})] + (p_{a,AHB} - 2\beta_a)m_{a,AHB}I_{AHB}$. The maximization yields

$$\partial_{p_1^h}\pi_a^h = 2I[-\frac{(p_a - \beta_a)}{\bar{b} - \underline{b}} + m_a]$$

$$\partial_{p_{1,AHB}}\pi_a^h = I_{AHB}[-\frac{(p_a - 2\beta_a)}{\bar{b} - \underline{b}} + m_a],$$

which implies that, in the symmetric case, $\beta_1 = \beta_2 = \beta$ and

$$p_1^h = p_2^h = \bar{b} + \beta,\ p_{1,AHB} = \bar{b} + 2\beta.$$

We deduce that

$$p_{a,AHB} < p_{a,AH} + p_{a,HB} = 2(\bar{b} + \beta),$$

which is consistent with the airline strategy at the hub. The hub should be designed in a way such that if a passenger takes a flight from A to B with one extra stop at the hub of that airline or partner airlines, the flight ticket price of that passenger is less than the direct flight ticket.

We now look at the traffic at the hub,

$$\partial_{f^h_{a,r}} \pi^h_a = 2I(p_a - \beta_a)\frac{[\frac{\alpha_r}{f^2_{a,r}}]}{\bar{b} - \underline{b}} - 2(c_{a,r} + 3\omega_r \sum_{a'} f_{a',r}) - 2f_{a,r}3w_r + y = 0,$$

which becomes $I\frac{\bar{b}}{\bar{b}-\underline{b}}\alpha_r - c_r(f^h)^2 - 6\omega_r(f^h)^3 + \frac{y}{2}(f^h)^2 = 0$ where $y = I_{AHB}\frac{\bar{b}}{(\bar{b}-\underline{b})(f^h)^2}\alpha_r$. Thus,

$$-\tilde{I}\frac{\bar{b}}{\bar{b}+\underline{b}}\alpha_r + c_r(f^h)^2 + 6\omega_r(f^h)^3 = 0. \tag{17.5}$$

FC Network vs. Hub Network

We now compare the two traffic f^h and f. The two positive roots can be compared without computing explicitly the solutions. In the hub case, the coefficients of f^3 is much smaller in (17.5); it means that one needs more traffic, i.e., $f^h_{a,r} > f_{a,r}$:

$$\begin{aligned} &-I\bar{b}\frac{\alpha_r}{\bar{b}-\underline{b}} + f_r^2 c_{1,r} + 12 f_r^3 \omega_r = 0, \quad f_r > 0, \\ &-\tilde{I}\frac{\bar{b}}{\bar{b}-\underline{b}}\alpha_r + c_r(f^h)^2 + 6\omega_r(f^h)^3 = 0, \quad f^h > 0. \end{aligned} \tag{17.6}$$

However, if we look at the total number of flights per airline in hub case, we have $2 * (2f^h) = (f_{AH} + f_{BH}) + (f_{HA} + f_{HB})$ flights. But the total of flights in FC network case is $(f_{AH} + f_{HA}) + (f_{BH} + f_{HB}) + (f_{AB} + f_{BA}) = 2 * (3f)$. The hub network has fewer number of flights in total if and only if $2f^h < 3f$.

We now look at the symmetric equilibrium payoff of the airlines in both types of networks. In the fully connected network, the equilibrium payoff of the airline is

$$\pi_a = 3[\frac{\bar{b}^2 I}{\bar{b} - \underline{b}} - c_r f - 8w_r f^2].$$

In the hub network, the equilibrium payoff is

$$\pi^h_a = 2[\frac{\bar{b}^2 I}{\bar{b} - \underline{b}} - c_r f^h - 6w_r(f^h)2] + \bar{b} m_{a,AHB} I_{AHB}.$$

Thus, the difference between the two payoffs is

$$\pi^h_a - \pi_a = c_r[3f - 2f^h] + 12w_r[2f^2 - (f^h)^2]. \tag{17.7}$$

From (17.7), if $f > \frac{1}{\sqrt{2}} f^h$,, the total number of flights in hub network is much less than the one in FC network. In this case, $f > \frac{2}{3} f^h$. Hence, both $[3f - 2f^h]$ and $[2f^2 - (f^h)^2]$ are positive. The total number of flights in hub network is much less than the one in FC network. Thus, the capacity of the hub network is less in the sense of Braess as the number of flights is smaller. However, the revenues of both airlines at the hub network have increased assuming that $w_r > 0$. Now, if the weight $w_r = 0$,, then by (17.7), the relation $f > \frac{2}{3} f^h$ is sufficient to give the hub network advantage.

This is a negation of the Braess paradox and it is due savings via the hub. However, the Braess paradox may disappear depending on the congestion and the intensity of the flights. The hub network is not always superior in terms of airlines revenues.

If $f \in [\frac{2}{3} f^h, \frac{1}{\sqrt{2}} f^h]$ or equivalently if $f^h \in [\sqrt{2} f, \frac{3}{2} f]$, then the congestion created at the hub is too high and may affect negatively both passengers and airlines, for example, for $f = 20$ and $f^h = 29$, the payoff difference $\pi_a^h - \pi_a = 6c_r - 12 \times 41 w_r$ is negative for $c_r < 6 \times 41 w_r$ and positive or equal otherwise.

The hub network has been widely advertised claiming that it reduces the number of flights. What we have shown above is that the congestion created at the hub can be high and in that case the revenue of both airlines goes down when adopting the hub network. This is a crowding effect (or mean-field effect) at the hub.

Enhanced Hub Network

As the hub network alone may create huge congestion at the hub, we suggest to examine the "enhanced hub network" in which few direct flights have been added between A and B in addition to hub network. However, this new network should be designed carefully. Clearly, each of the airlines receives an extra revenue from the direct flights AB and BA. The congestion is a bit less than the one at the hub. The enhanced network is closer to the FC network. The main difference now is that enhanced hub network has few number of direct flight than the FC network. If the congestion is low at the hub, no direct flight between A and B is required for the airline. In the presence of congestion terms and when the traffic belongs to $[\frac{2}{3} f^h, \frac{1}{\sqrt{2}} f^h]$, the benefit from the direct flights (with higher prices) may compensate the loss at congestion and also lower a bit the traffic at hub. Note that the passengers and the airlines can use distributed strategic learning algorithms to estimate best-response strategies. The passengers have to choose their airline to flight with. Their decision will based on their own payoff. However, the current payoff here is a random variable.

17.4 Enhanced Communication Networks

What has been presented above is not limited to transportation networks. It can be used for other types of networks including communication networks. The enhanced

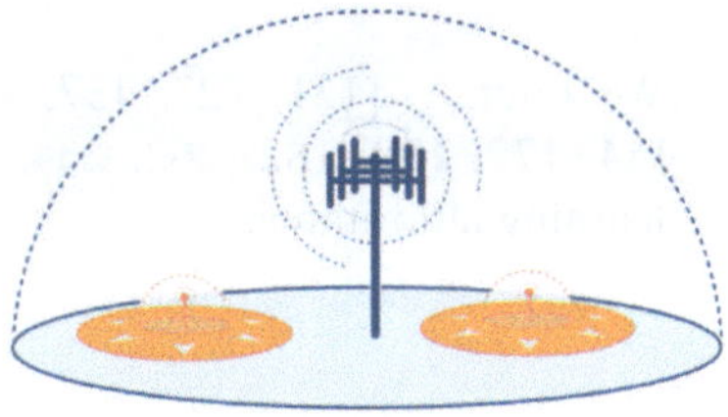

Fig. 17.5 Diagram from Qualcomm: enhanced network. Mixture of small cells (hotspots) and macro base stations

network approach is an idea already adopted in LTE-Advanced Heterogeneous Networks (HetNets) using a mix of macro, pico, femto, and relay base stations (Fig. 17.5). If a user is closer to a small base station (hotspot), one may be attempted to connect to femto station (if open) and save some power. However, if several users are connected to femto, the interference becomes high, and it may be better to connect to macro base stations which are far away (higher-power consumption is required). It is interesting to see that the hub network which corresponds to the femtocell networks may not be dominant, and hence, a combination macro-cell and femtocell network has been proposed.

17.5 Notes

Hub networks are present in many transportation and (relay) communication networks. In these networks, hubs usually act as sorting, transshipment, and consolidation (relay) points. Instead of sending flows directly between all origin–destination pairs, hubs consolidate flow in order to take advantage from the underlying economics. However, this may create a huge congestion between airlines if not well-designed [127, 401]. In standard transportation problems when the locations of the facilities are determined, each demand node receives service from its nearest facility. For hub location problems, when it comes to allocation decisions, the nearest allocation strategy—assigning each demand node to its nearest hub—does not necessarily yield optimal solutions. Similarly, routing data/packets to the nearest relay base stations may lead to suboptimal solutions in the presence of interference and congestion. Thus, the optimal allocations of demand centers to the located hubs must also be determined. A better solution is between hub network and FC network leading to an "enhanced hub network." In the context of communication network, it means that if the interference at the relay is too high, the SINR is very low, and the delay is infinite. Hence, it may be better for the mobile users to connect direct to the macro BS at the expense of higher price (high-power consumption to reach to macro BS).

(continued)

We refer to [121, 223, 437, 441] for evolutionary routing games and to [54, 179, 180, 182, 262, 264, 422, 423, 454, 461, 462, 478] for strategic learning algorithms.

Part VI
Wrap-up

Chapter 18
Conclusion and Future Directions

This chapter concludes the second volume and presents some future directions.

18.1 Conclusions

As we wrap up our journey through this second volume on mean-field-type game theory: applications, we have explored how powerful the MFTG is in providing us with the tools that help us solve various real-world problems in many different areas. In engineering, we have seen how MFTGs make wireless networks work better, improve security, and manage device power efficiently. They even help protect against cyberattacks and malware and make sure devices stay powered up longer.

In computer science, we have looked at how MFTGs can help make cloud networks run more smoothly and improve the way auctions work. We have also seen how they are useful in understanding how different computer systems interact, especially when it comes to machine intelligence. We have also learned how MFTGs can help us make our environment cleaner and smarter, especially in cities. They are useful for managing pollution and fishing in a sustainable way. In biochemistry, these games help us understand and fight diseases, including the COVID-19 pandemic. In social sciences, we have explored how MFTGs can help us understand the dynamics of relationships, like in marriage.

Lastly, in the world of network economics, finance, and insurance, we have seen how MFTGs can help make networks work more efficiently, manage electricity prices, and even change the way we use blockchain technology. They can also make airline networks operate better.

T. Başar et al., *Mean-Field-Type Game Theory II*, Static & Dynamic Game Theory: Foundations & Applications, https://doi.org/10.1007/978-3-032-07031-9_18

18.2 Future Directions

As we conclude our exploration of the wide-ranging applications of MFTGs, we look ahead and consider the exciting future directions in this field. The potential for further advancements and discoveries is immense, and below is a listing of some promising avenues for future research and application. The future of MFTGs is marked by their ability to bring data-driven, adaptive, and strategic risk-aware decision-making to real-world scenarios. We mention applications of mean-field-type games in underserved areas facing unique challenges that require innovative solutions. MFTGs can play a transformative role in addressing these challenges, offering tailored applications in various domains:

- Agriculture with Manure Production: MFTG can be applied to optimize sustainable farming practices in underserved agricultural regions. By modeling manure production and distribution, these risk-aware games can help maximize nutrient utilization, reduce environmental impact, and enhance crop yields, ensuring food security.
- Blockchain in Low-Connectivity and Low-Energy Areas: In regions with limited connectivity and energy resources, blockchain-enabled microgrids can provide reliable access to electricity. MFTG can optimize energy allocation, load balancing, and payment systems within these microgrids, ensuring equitable energy distribution and efficient utilization.
- Supply Chain Traceability: Blockchain technology, coupled with MFTG, can be used to establish transparent and tamper-proof supply chains for agricultural products in underserved areas. This ensures fair pricing, reduces fraud, and promotes economic empowerment for local farmers.
- Machine Intelligence in Internet-Deprived and Energy-Scarce Areas:
  - Offline Machine Learning: MFTG can facilitate offline machine learning models that do not rely on continuous internet connectivity. These models can be used for tasks such as healthcare diagnostics, agriculture optimization, and disaster prediction, even in remote areas with limited access to the internet.
  - Energy-Efficient Edge Computing: In regions with energy constraints, MFTG can optimize the deployment of energy-efficient edge computing solutions. This enables data processing and machine intelligence at the edge of the network, reducing the need for continuous electricity supply.
  - Disaster Response and Early Warning Systems: MFTG can aid in the development of localized, energy-efficient early warning systems for natural disasters, such as floods and wildfires. These systems can operate autonomously, providing critical information to underserved communities.
  - Financial Inclusion: MFTG can optimize microfinance initiatives in underserved areas, helping individuals and small businesses access capital and financial services. These games can tailor lending and repayment strategies to suit the unique economic conditions of these regions.

  - Healthcare Access: In areas with limited healthcare infrastructure, MFTG can optimize the deployment of machine intelligence-assisted telemedicine services and mobile clinics. These systems can provide remote healthcare consultations and diagnostics, improving access to medical care.
  - Offline e-Learning Platforms: MFTG can support the development of offline e-learning platforms that do not rely on continuous internet connectivity. These platforms can deliver educational content to underserved areas in local languages and in audio, promoting lifelong learning and skill development.

- Water Resource Management: MFTG can be applied to optimize the distribution of clean water in regions facing water scarcity. These risk-aware games can help manage water resources efficiently, ensuring access to safe drinking water for underserved populations.
- Psychological MFTGs: Future research can employ psychological MFTGs [45, 365, 375, 456, 478] to design personalized mental health interventions. These models can adapt strategies based on individual emotional states, enhancing their effectiveness in promoting mental well-being.
- Data-Driven MFTGs: The future holds the potential for data-driven MFTGs to play a pivotal role in the autonomy of systems, such as self-driving cars and drones. These systems can learn from real-time data to make decisions in complex, dynamic environments.
- Distributionally Robust MFTGs: Distributionally robust MFTG can be applied to optimize supply chain resilience and machine intelligence cybersecurity. These models can accommodate uncertain disruptions [184, 425, 440, 451, 478, 480] and adapt supply chain strategies to minimize risks while maintaining efficiency.
- Data-Driven Risk-Aware MFTGs: Future developments can focus on enhancing risk-aware MFTGs in the financial sector. These models can continuously adapt investment strategies based on real-time market data to mitigate risks and maximize returns.
- Strategic Learning and Human Feedback: In various domains, including healthcare and customer service, the integration of strategic learning and human feedback in MFTGs can enable more effective collaboration between humans and machine intelligence. These systems can learn from human feedback to make improved decisions.
- Explainable Machine Intelligence (xMI): In critical applications such as healthcare and finance, there is a growing need for interpretable machine intelligence models. Future research can focus on developing MFTGs such as mean-field-type transformers, mean-field-type diffusion transformers, and mean-field-type filters over networks that not only make decisions but also provide clear explanations for those decisions, enhancing transparency and trust.
- Climate Change Mitigation: MFTGs can be used to model and optimize policies for mitigating the negative impacts of climate change. These models can accommodate the behaviors of various stakeholders, including governments, industries, and individuals, to design effective and equitable climate strategies.

- Human-Centric Urban Planning: Applying MFTGs to urban planning can lead to human-centric solutions for transportation and city development. These models can optimize public transit, reduce congestion, and enhance overall quality of life in urban areas.
- Personalized Treatments: In healthcare, MFTGs can contribute to the advancement of precision medicine. These models can tailor treatment plans for individuals based on their genetic and clinical data, optimizing healthcare outcomes.
- Cross-Disciplinary Collaborations: One future direction lies in fostering collaborations between experts from various fields. Combining insights from engineering, computer science, economics, ecology, and more can lead to innovative solutions that address complex, real-world challenges comprehensively.
- Strategic Learning Integration: The integration of machine learning techniques with MFTGs holds great promise. Combining the predictive power of machine learning with the strategic optimization of MFTGs can lead to more adaptive and intelligent systems across domains.
- Sustainable Technologies: In the context of environmental and energy-related applications, future research can focus on developing mean-field game models for optimizing renewable energy integration, reducing carbon emissions, and advancing sustainable technologies in smart cities and beyond.
- Healthcare and Epidemiology: Given the success of MFTGs in modeling infectious diseases, there is a potential for further research in epidemiology and global healthcare planning. Models can be refined and expanded to better predict and manage disease outbreaks and vaccine distribution.
- Ethical Considerations: As MFTGs are applied in areas like economics and blockchain technology, it becomes important to explore the ethical implications. Future research can investigate the data integrity, information integrity, machine integrity, and ethical aspects of decision-making and resource allocation within these systems.
- Real-Time Applications: Advancements in real-time data analytics and communication networks can enable the implementation of mean-field-type game strategies in dynamic, rapidly changing environments such as autonomous vehicles, internet-of-things and internet-of-people networks, and smart manufacturing.
- Education and Outreach: Promoting the understanding of MFTGs among a broader audience is crucial. Educational initiatives, workshops, and outreach programs can help bridge the gap between academic research and practical applications, fostering greater adoption.
- Personalized Education: MFTGs can be applied in the field of education to create personalized learning experiences. These models can adapt teaching strategies to the unique needs and learning styles of students, improving educational outcomes.
- Human–Machine Collaboration: Exploring the synergy between human decision-makers and intelligent systems based on MFTGs can lead to more efficient and effective decision-making processes, particularly in critical domains like healthcare and finance.

- Regulatory Frameworks: In industries like finance and blockchain, establishing regulatory frameworks that accommodate the application of MFTGs is essential. Research can focus on designing policies and regulations that promote fairness and security.
- Quantum MFTGs: As quantum computing technology advances, there is a potential for exploring quantum MFTGs. These could open up entirely new possibilities for optimization in complex systems.

The future of MFTGs is bright and full of opportunities for innovation and positive impact. Researchers, practitioners, and policymakers will shape this field's trajectory, addressing some of the most pressing challenges of our time and improving decision-making processes across a wide spectrum of industries and domains.

References

1. Mahyar Abbasian, Iman Azimi, Amir M Rahmani, and Ramesh Jain. Conversational health agents: A personalized LLM-powered agent framework. *arXiv preprint arXiv:2310.02374*, 2023.
2. Hernan Abeledo and Uriel G Rothblum. Paths to marriage stability. *Discrete applied mathematics*, 63(1):1–12, 1995.
3. S. Adlakha, R. Johari, G. Weintraub, and A. Goldsmith. Oblivious equilibrium for large-scale stochastic games with unbounded costs. *Proceedings of the IEEE Conference on Decision and Control*, 2008.
4. Orna Agmon Ben-Yehuda, Muli Ben-Yehuda, Assaf Schuster, and Dan Tsafrir. The rise of RaaS: the resource-as-a-service cloud. *Communications of the ACM*, 57(7):76–84, 2014.
5. A. Ahmad, N. Ul Hassan, M. Assaad, and H. Tembine. Joint power control and rate adaptation for video streaming in wireless networks with time varying interference. *Preprint*, 2016.
6. Ayaz Ahmad, Naveed Ul Hassan, Mohamad Assaad, and Hamidou Tembine. Joint power control and rate adaptation for video streaming in wireless networks with time-varying interference. *IEEE Transactions on Vehicular Technology*, 65(8):6315–6329, 2015.
7. Ayaz Ahmad, Mubashir Husain Rehmani, Hamidou Tembine, Osama A Mohammed, and Abbas Jamalipour. Ieee access special section editorial: Optimization for emerging wireless networks: IoT, 5g, and smart grid communication networks. *IEEE Access*, 5:2096–2100, 2017.
8. Marianne Akian, Luca Ganassali, Stéphane Gaubert, and Laurent Massoulié. Probabilistic and mean-field model of covid-19 epidemics with user mobility and contact tracing. *arXiv preprint arXiv:2009.05304*, 2020.
9. Anna Aksamit, Kaustav Das, Ivan Guo, Kihun Nam, and Zhou Zhou. Switching to a green and sustainable finance setting: a mean field game approach. *arXiv preprint arXiv:2503.06967*, 2025.
10. Negin Alisoltani, Mostafa Ameli, Megan M. Khoshyaran, and Jean-Patrick Lebacque. Mass evacuation planning based on mean field games theory. In *International Conference on Traffic and Granular Flow*, pages 69–76. Springer Nature Singapore, 2022.
11. Carlos Alos-Ferrer and Ana B Ania. The evolutionary stability of perfectly competitive behavior. *Economic Theory*, 26:497–516, 2005.
12. E. Altman, T. Başar, I. Menache, and H. Tembine. A dynamic random access game with energy constraints. *IEEE/ACM Proc. 7th International Symposium on Modeling and Optimization in Mobile, Ad Hoc, and Wireless Networks (WiOpt'2009).*, 2009.

T. Başar et al., *Mean-Field-Type Game Theory II*, Static & Dynamic Game Theory: Foundations & Applications, https://doi.org/10.1007/978-3-032-07031-9

13. Eitan Altman, Rachid El-Azouzi, Yezekael Hayel, and H. Tembine. Evolutionary power control games in wireless networks. *Lecture Notes in Computer Science, 2008, Volume 4982/2008, 930–942*, 2008.
14. Eitan Altman, Rachid El-Azouzi, Yezekael Hayel, and Hamidou Tembine. Evolutionary power control games in wireless networks. In *NETWORKING 2008 Ad Hoc and Sensor Networks, Wireless Networks, Next Generation Internet: 7th International IFIP-TC6 Networking Conference Singapore, May 5–9, 2008 Proceedings 7*, pages 930–942. Springer, 2008.
15. Eitan Altman, Rachid El-Azouzi, Yezekael Hayel, and Hamidou Tembine. The evolution of transport protocols: An evolutionary game perspective. *Computer Networks*, 53(10):1751–1759, 2009.
16. Eitan Altman, Yezekael Hayel, Hamidou Tembine, and Rachid El-Azouzi. Markov decision evolutionary games with expected average fitness. *Evolutionary Ecology Research*, 11(4):677–689, 2009.
17. Eitan Altman, Anurag Kumar, Chandramani Kishore Singh, and Rajesh Sundaresan. Spatial SINR games combining base station placement and mobile association. In *INFOCOM*, pages 1629–1637. IEEE, 2009.
18. Paul R Amato. Studying marriage and commitment with survey data. In *Handbook of measurement issues in family research*, pages 53–65. Routledge, 2013.
19. David M Ambrose, Megan Griffin-Pickering, and Alpár R Mészáros. Kinetic-type mean field games with non-separable local Hamiltonians. *Journal of the London Mathematical Society*, 111(6):e70202, 2025.
20. Saurabh Amin, Galina A Schwartz, and Hamidou Tembine. Incentives and security in electricity distribution networks. In *International Conference on Decision and Game Theory for Security*, pages 264–280. Springer, 2012.
21. Neverov Andrei and Olga Krivorotko. Neural nets for modelling of scenarios and control of epidemics. In *First Conference of Mathematics of AI*, 2025.
22. Abdul Fatir Ansari, Lorenzo Stella, Caner Turkmen, Xiyuan Zhang, Pedro Mercado, Huibin Shen, Oleksandr Shchur, Syama Sundar Rangapuram, Sebastian Pineda Arango, Shubham Kapoor, et al. Chronos: Learning the language of time series. *arXiv preprint arXiv:2403.07815*, 2024.
23. Danilo Ardagna, Barbara Panicucci, and Mauro Passacantando. A game theoretic formulation of the service provisioning problem in cloud systems. In *Proceedings of the 20th international conference on World wide web*, pages 177–186, 2011.
24. ARISTA. Cloud networking: A network approach that meets the requirements of cloud computing. *CQ2*, 2011.
25. Yuri Ashrafyan and Diogo Gomes. A fully-discrete semi-lagrangian scheme for a price formation MFG model. *Dynamic Games and Applications*, pages 1–31, 2025.
26. Mohamad Assaad, Ayaz Ahmad, and Hamidou Tembine. Risk-sensitive resource control approach for delay limited traffic in wireless networks. In *2011 IEEE Global Telecommunications Conference-GLOBECOM 2011*, pages 1–5. IEEE, 2011.
27. Alexander Aurell, Rene Carmona, Gokce Dayanikli, and Mathieu Lauriere. Optimal incentives to mitigate epidemics: a stackelberg mean field game approach. *SIAM Journal on Control and Optimization*, 60(2):S294–S322, 2022.
28. Ugur Aydin and Naci Saldi. Robustness and approximation of discrete-time mean-field games under discounted cost criterion. *Mathematics of Operations Research*, 2025.
29. Mohamad Aziz and Peter E Caines. Computational investigations of decentralized cellular network optimization via mean field control. In *53rd IEEE Conference on Decision and Control*, pages 5560–5567. IEEE, 2014.
30. Mohamad Aziz and Peter E Caines. A mean field game computational methodology for decentralized cellular network optimization. *IEEE transactions on control systems technology*, 25(2):563–576, 2016.
31. G.V.R. Babu and K.N.V. Vara Prashad. Mann iteration converges faster than ishikawa iteration for the class of zamfirescu operators. *Fixed Point Theory and Applications, Article ID 49615*, 2006.

32. François Baccelli and Bartek Blaszczyszyn. *Stochastic Geometry and Wireless Networks, Volume I - Theory*, volume 1 of *Foundations and Trends in Networking, Vol. 3: No 3–4, pp 249–449*. NoW Publishers, 2009. Stochastic Geometry and Wireless Networks, Volume II - Applications; see http://hal.inria.fr/inria-00403040.
33. T Başar. Time consistency and robustness of equilibria in noncooperative dynamic games. *In F. Van der Ploeg and A. de Zeeuw, editors, Dynamic Policy Games in Economics, pp. 9–54. North Holland*, 1989.
34. T Başar and G J Olsder. Dynamic noncooperative game theory. *Society for Industrial and Applied Mathematics*, 1998.
35. Yu Bai, Di Zhou, and Zhen He. Optimal pursuit strategies in missile interception: Mean field game approach. *Aerospace*, 12(4):302, 2025.
36. Megan Bailey, U Rashid Sumaila, and Marko Lindroos. Application of game theory to fisheries over three decades. *Fisheries Research*, 102(1–2):1–8, 2010.
37. J. Baillon and R.E. Bruck. The rate of asymptotic regularity is $o(1/\sqrt{n})$. *in: Theory and applications of nonlinear operators of accretive and monotone type, Lecture Notes in Pure and Appl. Math. 178, Dekker, New York*, pages 51–81, 1996.
38. J.-B. Baillon, P.L. Combettes, and R. Cominetti. There is no variational characterization of the cycles in the method of periodic projections. *Journal of Functional Analysis, Vol. 262(1), pp. 400–408*, 2012.
39. Reginald A Banez, Hamidou Tembine, Lixin Li, Chungang Yang, Lingyang Song, Zhu Han, and H Vincent Poor. Mean-field-type game-based computation offloading in multi-access edge computing networks. *IEEE Transactions on Wireless Communications*, 19(12):8366–8381, 2020.
40. Mohit Bansil, Alpár R Mészáros, and Chenchen Mou. Global well-posedness of displacement monotone degenerate mean field games master equations. *SIAM Journal on Control and Optimization*, 63(2):993–1021, 2025.
41. J. Barreiro-Gomez, S. E. Choutri, and H. Tembine. Risk-awareness in multi-level building evacuation with smoke: Burj Khalifa case study. *Automatica*, 129(109625), 2021.
42. Julian Barreiro-Gomez. Inter-and intra-population linear quadratic differential games. *IEEE Transactions on Control of Network Systems*, 2025.
43. Julian Barreiro-Gomez, Salah Eddine Choutri, and Hamidou Tembine. Risk-awareness in multi-level building evacuation with smoke: Burj Khalifa case study. *Automatica*, 129:109625, 2021.
44. Julian Barreiro-Gomez, Florian Dörfler, and Hamidou Tembine. Distributed robust population games with applications to optimal frequency control in power systems. In *2018 Annual American Control Conference (ACC)*, pages 5762–5767. IEEE, 2018.
45. Julian Barreiro-Gomez, Tyrone E Duncan, and Hamidou Tembine. Co-opetitive linear-quadratic mean-field-type games. *IEEE Transactions on Cybernetics*, 50(12):5089–5098, 2019.
46. Julian Barreiro-Gomez and Hamidou Tembine. Constrained evolutionary games by using a mixture of imitation dynamics. *Automatica*, 97:254–262, 2018.
47. Julian Barreiro-Gomez and Hamidou Tembine. Blockchain token economics: A mean-field-type game perspective. *IEEE Access*, 7:64603–64613, 2019.
48. Julian Barreiro-Gomez and Hamidou Tembine. Constrained mean-field-type games: Stationary case. In *2019 IEEE 58th Conference on Decision and Control (CDC)*, pages 2208–2213. IEEE, 2019.
49. Julian Barreiro-Gomez and Hamidou Tembine. *Mean-Field-Type Games for Engineers*. CRC Press, 2021.
50. Julian Barreiro-Gomez, Hamidou Tembine, Leonardo Stella, Dario Bauso, and Patrizio Colaneri. Risk-aware control and games in engineering. In *2020 59th IEEE Conference on Decision and Control (CDC)*, pages 3860–3870. IEEE, 2020.
51. Guilherme A Barreto. Time series prediction with the self-organizing map: A review. *Perspectives of neural-symbolic integration*, pages 135–158, 2007.

52. Tamer Başar. Nash equilibria of networked games with delayed coupling in the high population regime. *Central European Journal of Operations Research*, 33(2):415–427, 2025.
53. Dario Bauso, Ben Mansour Dia, Boualem Djehiche, Hamidou Tembine, and Raul Tempone. Mean-field games for marriage. *PloS one*, 9(5):e94933, 2014.
54. Dario Bauso, Jian Gao, and Hamidou Tembine. Distributionally robust games: f-divergence and learning. In *Proceedings of the 11th EAI International Conference on Performance Evaluation Methodologies and Tools*, pages 148–155, 2017.
55. Spencer Becker-Kahn. Notes on the mathematics of large transformer language model architecture. *Preprint*, 2023.
56. Michel Benaim and Jean-Yves Le Boudec. A class of mean field interaction models for computer and communication systems. *Perform. Eval.*, 65(11–12):823–838, 2008.
57. A. Bensoussan, J. Frehse, and Yam S.C.P. Mean field games and mean field type control theory. *SpringerBriefs in Mathematics, Springer*, 2013.
58. Alain Bensoussan. New mathematical problems in management science. *Annual Reviews in Control*, 53:1–5, 2022.
59. Alain Bensoussan, SingRu Celine, and Zhongfeng Yan. A mean-variance approach to capital investment optimization. *SIAM Journal on Financial Mathematics*, 10(1):156–180, 2019.
60. Alain Bensoussan, Joohyun Kim, Bohan Li, and Sheung Chi Phillip Yam. Linear quadratic extended mean field games and control problems. *Journal of Optimization Theory and Applications*, 206(2):1–47, 2025.
61. Alain Bensoussan, Guiyuan Ma, Chi Chung Siu, and Sheung Chi Phillip Yam. Dynamic mean–variance problem with frictions. *Finance and Stochastics*, 26(2):267–300, 2022.
62. Alain Bensoussan, Ho Man Tai, and Sheung Chi Phillip Yam. Mean field type control problems, some Hilbert-space-valued FBSDEs, and related equations. *ESAIM: Control, Optimisation and Calculus of Variations*, 31:33, 2025.
63. H. Berestycki and H. Brezis. Recent contributions to nonlinear partial equations. *Pitman*, 1981.
64. J. Bergin and D. Bernhardt. Anonymous sequential games with aggregate uncertainty. *Journal of Mathematical Economics*, 21:543–562, 1992.
65. Chiara Bernardini. Mass concentration for ergodic choquard mean-field games. *ESAIM: Control, Optimisation and Calculus of Variations*, 31:34, 2025.
66. Jules Berry and Fabio Camilli. Stationary mean field games on networks with sticky transition conditions. *ESAIM: Control, Optimisation and Calculus of Variations*, 31:18, 2025.
67. Yuxuan Bian, Xuan Ju, Jiangtong Li, Zhijian Xu, Dawei Cheng, and Qiang Xu. Multi-patch prediction: Adapting LLMs for time series representation learning. *arXiv preprint arXiv:2402.04852*, 2024.
68. Ofelia Bonesini, Luciano Campi, and Markus Fischer. Correlated equilibria for mean field games with progressive strategies. *Mathematics of Operations Research*, 50(2):1072–1111, 2025.
69. Modibo Bouare, Sidy Danioko, Mariam Dembele, Abdoulaye Diallo, Boubacar Diallo, Abdoulaye Diarra, Bourama Doumbia, Ndeye Molinier, Astou Sidibe, Allahsera Tapo, and Hamidou Tembine. *Machine Intelligence in Africa in 20 Questions*. Sawa Editions, National Library of Mali, 06 2023.
70. Dietrich Braess. Über ein paradoxon aus der verkehrsplanung. *Unternehmensforschung*, 12:258–268, 1968.
71. Marcello Braglia, Davide Castellano, and Roberto Gabbrielli. A novel game theory based exit selection model in emergency conditions. *Advances in Complex Systems*, 16(07):1350018, 2013.
72. Yann Brenier. Décomposition polaire et réarrangement monotone des champs de vecteurs. *C. R. Acad. Sci. Paris Sér. I Math.*, 305(19):805–808, 1987.
73. H. Brezis. Analyse fonctionnelle: théorie et applications. *Masson*, 1983.
74. H. Brezis. Functional analysis, Sobolev spaces and PDEs. *Springer, xiii+599p.*, 2010.
75. Gustaf Bruze, Michael Svarer, and Yoram Weiss. The dynamics of marriage and divorce. *Journal of Labor Economics*, 33(1):123–170, 2015.

76. Michael A Buice and Carson C Chow. Beyond mean field theory: statistical field theory for neural networks. *Journal of Statistical Mechanics: Theory and Experiment*, 2013(03):P03003, 2013.
77. Giuseppe Buttazzo and Filippo Santambrogio. A model for the optimal planning of an urban area. *SIAM J. Math. Anal.*, 37(2):514–530 (electronic), 2005.
78. Peter E Caines and Arman C Kizilkale. Recursive estimation of common partially observed disturbances in MFG systems with application to large scale power markets. In *52nd IEEE Conference on Decision and Control*, pages 2505–2512. IEEE, 2013.
79. Luciano Campi, Federico Cannerozzi, and Fanny Cartellier. Coarse correlated equilibria in linear quadratic mean field games and application to an emission abatement game. *Applied Mathematics & Optimization*, 91(1):1–40, 2025.
80. Shuchao Cao, Weiguo Song, and Wei Lv. Modeling pedestrian evacuation with guiders based on a multi-grid model. *Physics Letters A*, 380(4):540–547, 2016.
81. Shuchao Cao, Weiguo Song, Wei Lv, and Zhiming Fang. A multi-grid model for pedestrian evacuation in a room without visibility. *Physica A: Statistical Mechanics and its Applications*, 436:45–61, 2015.
82. G. Carlier and I. Ekeland. The structure of cities. *J. Global Optim.*, 29(4):371–376, 2004.
83. G. Carlier and I. Ekeland. Equilibrium structure of a bidimensional asymmetric city. *Nonlinear Anal. Real World Appl.*, 8(3):725–748, 2007.
84. G. Carlier, C. Jimenez, and F. Santambrogio. Optimal transportation with traffic congestion and Wardrop equilibria. *SIAM J. Control Optim.*, 47(3):1330–1350, 2008.
85. R. Carmona and F. Delarue. Probabilistic theory of mean field games with applications i-ii. *Springer, ISBN 978-3-319-59820-8*, 2018.
86. René Carmona, Quentin Cormier, and H Mete Soner. Synchronization in a Kuramoto mean field game. *Communications in Partial Differential Equations*, 48(9):1214–1244, 2023.
87. Rene Carmona, Gokce Dayanikli, and Mathieu Lauriere. Mean field models to regulate carbon emissions in electricity production. *Dynamic Games and Applications*, 12(3):897–928, 2022.
88. René Carmona, François Delarue, and Daniel Lacker. Mean field games of timing and models for bank runs. *Applied Mathematics & Optimization*, 76:217–260, 2017.
89. PG Caryl. The mathematics of marriage: Dynamic nonlinear models. *British Journal of Mathematical & Statistical Psychology*, 58(1):186, 2005.
90. John P Caughlin and Ted L Huston. The affective structure of marriage. *The Cambridge handbook of personal relationships*, pages 131–155, 2006.
91. Vladimir Ceperic and Tomislav Markovic. Transforming time-series data for improved LLM-based forecasting through adaptive encoding. *International Journal of Simulation–Systems, Science & Technology*, 25(1), 2024.
92. Neha Chacko and Viju Chacko. Paradigm shift presented by large language models (LLM) in deep learning. *Advances in emerging computing technologies*, 40, 2023.
93. Patrick Chan and Ronnie Sircar. Fracking, renewables, and mean field games. *SIAM Review*, 59(3):588–615, 2017.
94. Yuanyuan Chang, Dena Firoozi, and David Benatia. Large banks and systemic risk: insights from a mean-field game model. *Journal of Systems Science and Complexity*, 38(1):460–494, 2025.
95. Pranav Kumar Chaudhary. Ai, ML, and large language models in cybersecurity. *Preprint*, 2024.
96. Samia Ben Chekroun, Essaid Sabir, Abdellatif Kobbane, Hamidou Tembine, El-Houssine Bouyakhf, and Khalil Ibrahimi. A distributed open-close access for small-cell networks: A random matrix game analysis. In *2015 International Wireless Communications and Mobile Computing Conference (IWCMC)*, pages 320–325. IEEE, 2015.
97. Ali Chelli, Hamidou Tembine, and Mohamed-Slim Alouini. A coalition formation game for transmitter cooperation in OFDMA uplink communications. In *2014 IEEE Global Communications Conference*, pages 4197–4202. IEEE, 2014.

98. Dezhi Chen, Qi Qi, Zirui Zhuang, Jingyu Wang, Jianxin Liao, and Zhu Han. Mean field deep reinforcement learning for fair and efficient UAV control. *IEEE Internet of Things Journal*, 8(2):813–828, 2020.
99. Jieyu Chen, Tianxing Shi, and Nan Li. Pedestrian evacuation simulation in indoor emergency situations: Approaches, models and tools. *Safety science*, 142:105378, 2021.
100. Tian Chen, Xinwei Feng, and Yunxiao Jia. Linear-quadratic mean-field backward stackelberg game with mixed terminal perturbation. *Applied Mathematics & Optimization*, 91(2):28, 2025.
101. Yi-Chang Chiu, Hong Zheng, Jorge Villalobos, and Bikash Gautam. Modeling no-notice mass evacuation using a dynamic traffic flow optimization model. *Iie Transactions*, 39(1):83–94, 2007.
102. Wonjoon Choi, Horst W Hamacher, and Suleyman Tufekci. Modeling of building evacuation problems by network flows with side constraints. *European Journal of Operational Research*, 35(1):98–110, 1988.
103. Salah Eddine Choutri and Tembine Hamidou. A stochastic maximum principle for Markov chains of mean-field type. *Games*, 9(4):84, 2018.
104. Dantong Chu, Kenneth Tsz Hin Ng, Sheung Chi Phillip Yam, and Harry Zheng. Mean field analysis of two-party governance: Competition versus cooperation among leaders. *Automatica*, 173:112028, 2025.
105. Marco Cirant, Fanze Kong, Juncheng Wei, and Xiaoyu Zeng. Critical mass phenomena and blow-up behaviors of ground states in stationary second order mean-field games systems with decreasing cost. *Journal de Mathématiques Pures et Appliquées*, 198:103687, 2025.
106. Abdoul Karim Cissé and Hamidou Tembine. Cooperative mean-field type games. *IFAC Proceedings Volumes*, 47(3):8995–9000, 2014.
107. Amazon.com's cloud-computing platform. Amazon web services. *(AWS) http://aws.amazon.com/documentation/ec2/*, 2012.
108. Asaf Cohen and Ethan Zell. Asymptotic Nash equilibria of finite-state ergodic Markovian mean field games. *Mathematics of Operations Research*, 2025.
109. Beau Coker, Wessel P Bruinsma, David R Burt, Weiwei Pan, and Finale Doshi-Velez. Wide mean-field Bayesian neural networks ignore the data. In *International Conference on Artificial Intelligence and Statistics*, pages 5276–5333. PMLR, 2022.
110. P. L. Combettes and J-C. Pesquet. Deep network network structures solving variational inequalities. *Set-Valued and Variational Analysis*, 28:pp 491–518, 2020.
111. P. L. Combettes and J-C. Pesquet. Lispchitz certificates for layered network structures driven by averaged activation operators. *SIAM Journal on Mathematics of Data Science, vol. 2, Issue 2,*, 2020.
112. J Cook. The mean-field theory of a q-state neural network model. *Journal of Physics A: Mathematical and General*, 22(12):2057, 1989.
113. Leon N Cooper and Christopher L Scofield. Mean-field theory of a neural network. *Proceedings of the National Academy of Sciences*, 85(6):1973–1977, 1988.
114. Thomas J Cova and Justin P Johnson. A network flow model for lane-based evacuation routing. *Transportation research part A: Policy and Practice*, 37(7):579–604, 2003.
115. Carolina Crespi, Georgia Fargetta, Mario Pavone, Rocco A Scollo, and Laura Scrimali. A game theory approach for crowd evacuation modelling. In *International Conference on Bioinspired Methods and Their Applications*, pages 228–239. Springer, 2020.
116. Ross Cressman and Vlastimil Křivan. Using chemical reaction network theory to show stability of distributional dynamics in game theory. *Journal of Dynamics and Games*, 9(4):351–371, 2022.
117. Gianluca Crippa, Chloé Jimenez, and Aldo Pratelli. Optimum and equilibrium in a transport problem with queue penalization effect. *Adv. Calc. Var.*, 2(3):207–246, 2009.
118. Antonio De Paola, Vincenzo Trovato, David Angeli, and Goran Strbac. A mean field game approach for distributed control of thermostatic loads acting in simultaneous energy-frequency response markets. *IEEE Transactions on Smart Grid*, 10(6):5987–5999, 2019.

119. I de Zarza, J de Curto, Gemma Roig, and Carlos T Calafate. LLM multimodal traffic accident forecasting. *Sensors*, 23(22):9225, 2023.
120. M. Debbah and H. Tembine. Reacting to the interference field, mechanisms and games for dynamic spectrum allocation. *Cambridge University Press*, 2013.
121. Ahmad H. Dehwah, H. Tembine, and Christian G. Claudel. A decentralized routing scheme based on a zero-sum game to optimize energy in solar powered sensor networks. *IPSN 2014: 285–286*, 2014.
122. Jodi Dianetti, Salvatore Federico, Giorgio Ferrari, and Giuseppe Floccari. Multiple equilibria in mean-field game models of firm competition with strategic complementarities. *Quantitative Finance*, pages 1–15, 2025.
123. Ming-Hui Ding, Hongyu Liu, and Guang-Hui Zheng. Determining a stationary mean field game system from full/partial boundary measurement. *SIAM Journal on Mathematical Analysis*, 57(1):661–681, 2025.
124. Vinayak Dixit and Brian Wolshon. Evacuation traffic dynamics. *Transportation research part C: emerging technologies*, 49:114–125, 2014.
125. B Djehiche, T Başar, and H Tembine. Mean-field-type game theory: foundations and new directions. *Springer*, 2024.
126. B Djehiche, A Tcheukam, and H Tembine. A mean-field game of evacuation in a multi-level building, special session 118: Mean field games and applications. In *The 11th AIMS Conference on Dynamical Systems, Differential Equations and Applications, Orlando, Florida, USA*, 2016.
127. B. Djehiche, A. Tcheukam, and H. Tembine. Mean-field-type games in engineering. *AIMS Electronics and Electrical Engineering*, 1(1):18–73, 2017.
128. B. Djehiche and H. Tembine. Covid-19: A data-driven mean-field-type game theory perspective. *SIAM News*, 2022.
129. Boualem Djehiche, Julian Barreiro-Gomez, and Hamidou Tembine. Mean-field-type games for blockchain-based distributed power networks. *Book chapter in Beyond Traditional Probabilistic Methods in Economics, Editors: Vladik Kreinovich, Nguyen Ngoc Thach,Nguyen Duc Trung, Dang Van Thanh, Pages 45–64, Part of the Studies in Computational Intelligence book series (SCI, volume 809), Springer*, 2019.
130. Boualem Djehiche, Julian Barreiro-Gomez, and Hamidou Tembine. Mean-field-type games for blockchain-based distributed power networks. In *Beyond Traditional Probabilistic Methods in Economics 2*, pages 45–64. Springer, 2019.
131. Boualem Djehiche, Julian Barreiro-Gomez, and Hamidou Tembine. Price dynamics for electricity in smart grid via mean-field-type games. *Dynamic Games and Applications*, 10:798–818, 2020.
132. Boualem Djehiche and Roxana Dumitrescu. Zero-sum mean-field Dynkin games: characterization and convergence. *Mathematics of Operations Research*, 2025.
133. Boualem Djehiche, Alain Tcheukam, and Hamidou Tembine. A mean-field game of evacuation in a multi-level building. *Special Session 118: Mean field games and applications, The 11th AIMS Conference on Dynamical Systems, Differential Equations and Applications, July 1 - July 5, Orlando, Florida, USA*, 2016.
134. Boualem Djehiche, Alain Tcheukam, and Hamidou Tembine. A mean-field game of evacuation in a multi-level building. *17th International Symposium on Dynamic Games and Applications, Urbino, Italy, July 12–15*, 2016.
135. Boualem Djehiche, Alain Tcheukam, and Hamidou Tembine. Mean-field games for evacuation: Simulation results. *17th International Symposium on Dynamic Games and Applications, July 12–15, Urbino, Italy*, 2016.
136. Boualem Djehiche, Alain Tcheukam, and Hamidou Tembine. A mean-field game of evacuation in multilevel building. *IEEE Trans. Automat. Contr.*, 62(10):5154–5169, 2017.
137. Boualem Djehiche, Alain Tcheukam, and Hamidou Tembine. A mean-field game of evacuation in multilevel building. *IEEE Transactions on Automatic Control*, 62(10):5154–5169, 2017.

138. Boualem Djehiche, Alain Tcheukam, and Hamidou Tembine. Mean-field-type games in engineering. *AIMS Electronics and Electrical Engineering*, doi: https://doi.org/10.3934/ElectrEng.2017.1.18, 1:18–73, 2017.
139. Boualem Djehiche and Hamidou Tembine. The outcomes of generative ai are exactly the Nash equilibria of a non-potential game. *Partial Identification in Economics and related topics, Springer*, pages 57–80, 2024.
140. Jing Dong, Baoxiang Wang, and Yaoliang Yu. Last-iterate convergence in regularized Graphon mean field game. In *Proceedings of the AAAI Conference on Artificial Intelligence, 39,13*, pages 13779–13787, 2025.
141. Zihan Dong, Xinyu Fan, and Zhiyuan Peng. Fnspid: A comprehensive financial news dataset in time series. *arXiv preprint arXiv:2402.06698*, 2024.
142. Safaa Driouech, Essaid Sabir, and Hamidou Tembine. Self-organized device-to-device communications as a non-cooperative quitting game. In *2017 International Conference on Wireless Networks and Mobile Communications (WINCOM)*, pages 1–8. IEEE, 2017.
143. P. Dubey, A. Mas-Colell, and M. Shubik. Efficiency properties of strategic market games: An axiomatic approach. *Journal of Economic Theory*, 22:339–362, 1980.
144. Annette Dumas and Filippo Santambrogio. Deterministic mean field games with jumps and mixed variational structure. *ESAIM: Control, Optimisation and Calculus of Variations*, 31:38, 2025.
145. Tyrone E Duncan, Bozenna Pasik-Duncan, and Hamidou Tembine. Mean-field-type games driven by Rosenblatt process: non-quadratic cost. *Preprint*, 2023.
146. Tyrone E Duncan, Bozenna Pasik-Duncan, and Hamidou Tembine. Multi-layer mean-field-type games driven by Rosenblatt process: non-quadratic cost. *Preprint*, 2023.
147. Tyrone E Duncan, Bozenna Pasik-Duncan, and Hamidou Tembine. Mean-field-type games driven by Rosenblatt processes. *CODIT'2024*, 2024.
148. Yannick Eich, Christian Fabian, Kai Cui, and Heinz Koeppl. Bounded rationality equilibrium learning in mean field games. In *Proceedings of the AAAI Conference on Artificial Intelligence, 39, 13*, pages 13796–13804, 2025.
149. Vijay Ekambaram, Arindam Jati, Nam H Nguyen, Pankaj Dayama, Chandra Reddy, Wesley M Gifford, and Jayant Kalagnanam. TTMs: Fast multi-level tiny time mixers for improved zero-shot and few-shot forecasting of multivariate time series. *arXiv preprint arXiv:2401.03955*, 2024.
150. Hajar El Hammouti, Essaid Sabir, and Hamidou Tembine. A fully distributed satisfactory power control for QOS self-provisioning in 5g networks. In *2017 14th IEEE Annual Consumer Communications & Networking Conference (CCNC)*, pages 688–691. IEEE, 2017.
151. Hajar El Hammouti, Essaid Sabir, and Hamidou Tembine. QOS-aware tactical power control for 5g networks. In *Ubiquitous Networking: Third International Symposium, UNet 2017, Casablanca, Morocco, May 9–12, 2017, Revised Selected Papers 3*, pages 25–37. Springer, 2017.
152. Karthik Elamvazhuthi and Spring Berman. Mean-field models in swarm robotics: A survey. *Bioinspiration & Biomimetics*, 15(1):015001, 2019.
153. Johnson Elane. The math of marriage. *Creative Nonfiction*, 59:16–19, 2016.
154. Hajar Elhammouti, Essaid Sabir, Mustapha Benjillali, Loubna Echabbi, and Hamidou Tembine. Self-organized connected objects: Rethinking QOS provisioning for IoT services. *IEEE Communications Magazine*, 55(9):41–47, 2017.
155. Romuald Elie, Emma Hubert, and Gabriel Turinici. Contact rate epidemic control of covid-19: an equilibrium view. *Mathematical Modelling of Natural Phenomena*, 15:35, 2020.
156. J. Epstein. The intelligence of a machine (C. Wall-Romana, trans.). *Univocal Publishing*, 1946/2014.
157. Jean Epstein. l'intelligence d'une machine, les classiques. *Publisher: Le Editions Jacques Melot*, 1946.
158. Lars Ericson, Xuejun Zhu, Xusi Han, Rao Fu, Shuang Li, Steve Guo, and Ping Hu. Deep generative modeling for financial time series with application in var: A comparative review. *arXiv preprint arXiv:2401.10370*, 2024.

159. Cong Fang, Jason Lee, Pengkun Yang, and Tong Zhang. Modeling from features: a mean-field framework for over-parameterized deep neural networks. In *Conference on learning theory*, pages 1887–1936. PMLR, 2021.
160. Xi Fang, Weijie Xu, Fiona Anting Tan, Jiani Zhang, Ziqing Hu, Yanjun Qi, Scott Nickleach, Diego Socolinsky, Srinivasan Sengamedu, and Christos Faloutsos. Large language models on tabular data–a survey. *arXiv preprint arXiv:2402.17944*, 2024.
161. Emmanuel Farhi and Jean Tirole. Too domestic to fail liquidity provision and national champions. *Review of Economic Studies*, 2024.
162. Cheng Feng, Long Huang, and Denis Krompass. Only the curve shape matters: Training foundation models for zero-shot multivariate time series forecasting through next curve shape prediction. *arXiv preprint arXiv:2402.07570*, 2024.
163. Siqi Feng, Guangchen Wang, Hua Xiao, Zhuangzhuang Xing, and Huanjun Zhang. Linear quadratic optimal control problems for conditional mean-field stochastic differential equations under partial information. *International Journal of Robust and Nonlinear Control*, 2025.
164. Giorgio Ferrari and Ioannis Tzouanas. Stationary mean-field games of singular control under Knightian uncertainty. *arXiv preprint arXiv:2505.08317*, 2025.
165. Rita Ferreira, Diogo Gomes, and Vardan Voskanyan. Weak-strong uniqueness for solutions to mean-field games. *Proceedings of the American Mathematical Society*, 2025.
166. Gadi Fibich and Nir Gavish. Numerical simulations of asymmetric first-price auctions. *Games and Economic Behavior*, 73(2):479–495, 2011.
167. Gadi Fibich and Nir Gavish. Asymmetric first-price auctions: A dynamical-systems approach. *Mathematics of Operations Research*, 37(2):219–243, 2012.
168. Ian Foster, Yong Zhao, Ioan Raicu, and Shiyong Lu. Cloud computing and grid computing 360-degree compared. In *2008 grid computing environments workshop*, pages 1–10. Ieee, 2008.
169. David M Frank and Sahotra Sarkar. Group decisions in biodiversity conservation: implications from game theory. *PLoS One*, 5(5):e10688, 2010.
170. Z. El Oula Frihi, J. Barreiro-Gomez, S. E. Choutri, and H. Tembine. Toolbox to simulate and mitigate covid-19 propagation. *SoftwareX, 100673*, 2021.
171. Zahrate El Oula Frihi, Salah Eddine Choutri, Julian Barreiro-Gomez, and Hamidou Tembine. Hierarchical mean-field type control of price dynamics for electricity in smart grid. *Journal of Systems Science and Complexity*, 35(1):1–17, 2022.
172. Fanzhe Fu, Junru Chen, Jing Zhang, Carl Yang, Lvbin Ma, and Yang Yang. Are synthetic time-series data really not as good as real data? *arXiv preprint arXiv:2402.00607*, 2024.
173. Yichao Fu, Peter Bailis, Ion Stoica, and Hao Zhang. Break the sequential dependency of LLM inference using lookahead decoding. *arXiv preprint arXiv:2402.02057*, 2024.
174. Masaaki Fujii and Masashi Sekine. Mean field equilibrium asset pricing model with habit formation. *Asia-Pacific Financial Markets*, pages 1–52, 2025.
175. Jean J. Gabszewicz and Jacques-Francois Thisse. Location. In R.J. Aumann and S. Hart, editors, *Handbook of Game Theory with Economic Applications*, volume 1 of *Handbook of Game Theory with Economic Applications*, chapter 9, pages 281–304. Elsevier, 1992.
176. David Gale and Lloyd S Shapley. College admissions and the stability of marriage. *The American Mathematical Monthly*, 120(5):386–391, 2013.
177. Hao Gao, Wuchen Li, Miao Pan, Zhu Han, and H Vincent Poor. Analyzing social distancing and seasonality of covid-19 with mean field evolutionary dynamics. In *2020 IEEE Globecom Workshops (GC Wkshps*, pages 1–6. IEEE, 2020.
178. Hao Gao, Wuchen Li, Miao Pan, Zhu Han, and H Vincent Poor. Modeling covid-19 with mean field evolutionary dynamics: social distancing and seasonality. *Journal of Communications and Networks*, 23(5):314–325, 2021.
179. J. Gao and H. Tembine. Distributionally robust games for deep generative learning. *IEEE World Congress on Computational Intelligence Windsor Convention Centre, Rio de Janeiro, Brazil 08–13 July*, 2018.
180. J. Gao and H. Tembine. Distributionally robust games: Wasserstein metric. *International Joint Conference on Neural Networks (IJCNN), Rio de Janeiro, Brazil, July*, 2018.

181. Jian Gao and Hamidou Tembine. Distributed mean-field-type filters for big data assimilation. In *2016 IEEE 18th International Conference on High Performance Computing and Communications; IEEE 14th International Conference on Smart City; IEEE 2nd International Conference on Data Science and Systems (HPCC/SmartCity/DSS)*, pages 1446–1453. IEEE, 2016.
182. Jian Gao and Hamidou Tembine. Bregman learning for generative adversarial networks. In *2018 Chinese Control And Decision Conference (CCDC)*, pages 82–89. IEEE, 2018.
183. Jian Gao and Hamidou Tembine. Distributed mean-field-type filters for traffic networks. *IEEE Transactions on Intelligent Transportation Systems*, 20(2):507–521, 2018.
184. Jian Gao, Yida Xu, Julian Barreiro-Gomez, Massa Ndong, Michalis Smyrnakis, and Hamidou Tembine. Distributionally robust optimization. *Book Optimization Algorithms-Examples*, page 1, 2018.
185. Shanghua Gao, Teddy Koker, Owen Queen, Thomas Hartvigsen, Theodoros Tsiligkaridis, and Marinka Zitnik. Units: Building a unified time series model. *arXiv preprint arXiv:2403.00131*, 2024.
186. Xiaohu Ge, Haoming Jia, Yi Zhong, Yong Xiao, Yonghui Li, and Branka Vucetic. Energy efficient optimization of wireless-powered 5g full duplex cellular networks: A mean field game approach. *IEEE Transactions on Green Communications and Networking*, 3(2):455–467, 2019.
187. Yingqiang Ge, Wenyue Hua, Kai Mei, Juntao Tan, Shuyuan Xu, Zelong Li, Yongfeng Zhang, et al. Openagi: When LLM meets domain experts. *Advances in Neural Information Processing Systems*, 36, 2024.
188. Felix A Gers, Douglas Eck, and Jürgen Schmidhuber. Applying LSTM to time series predictable through time-window approaches. In *International conference on artificial neural networks*, pages 669–676. Springer, 2001.
189. H. Ghazzai, H. Tembine, and M.-S. Alouini. Mobile user association for heterogeneous networks using optimal transport theory. *In Proceedings of The Sixth International Conference on Communications and Networking, ComNet'2017, March 29th - April 1st*, 2017.
190. Hakim Ghazzai, Hamidou Tembine, and Mohamed-Slim Alouini. Mobile user association for heterogeneous networks using optimal transport theory. In *2017 Sixth International Conference on Communications and Networking (ComNet)*, pages 1–6. IEEE, 2017.
191. Daria Ghilli, Fausto Gozzi, Cristiano Ricci, and Giovanni Zanco. A mean field game model with non-local spatial interactions and resources accumulation. *arXiv preprint arXiv:2506.01200*, 2025.
192. Daria Ghilli, Cristiano Ricci, and Giovanni Zanco. A mean field game model for covid-19 with human capital accumulation. *Economic Theory*, pages 1–28, 2023.
193. Ali Gholipour, Babak N Araabi, and Caro Lucas. Predicting chaotic time series using neural and neurofuzzy models: a comparative study. *neural processing letters*, 24:217–239, 2006.
194. Soumyadip Ghosh, Yingdong Lu, Tomasz Nowicki, and Edith Zhang. On representations of mean-field variational inference. *Journal of Applied Analysis*, 2025.
195. Mononito Goswami, Konrad Szafer, Arjun Choudhry, Yifu Cai, Shuo Li, and Artur Dubrawski. Moment: A family of open time-series foundation models. *arXiv preprint arXiv:2402.03885*, 2024.
196. John M Gottman, James D Murray, Catherine C Swanson, Rebecca Tyson, and Kristin R Swanson. *The mathematics of marriage: Dynamic nonlinear models*. mit press, 2005.
197. P Jameson Graber and Alain Bensoussan. Existence and uniqueness of solutions for Bertrand and Cournot mean field games. *Applied Mathematics & Optimization*, 77(1):47–71, 2018.
198. C. Graham. Chaoticity for multiclass systems and exchangeability within classes. *J. Appl. Probab.*, 45(4):1196–1203, 2008.
199. Carlos Groba, Antonio Sartal, and Gustavo Bergantiño. Optimization of tuna fishing logistic routes through information sharing policies: A game theory-based approach. *Marine Policy*, 113:103795, 2020.

200. Nate Gruver, Marc Finzi, Shikai Qiu, and Andrew G Wilson. Large language models are zero-shot time series forecasters. *Advances in Neural Information Processing Systems*, 36, 2024.
201. Guohui Guan, Jiaqi Hu, and Zongxia Liang. N-player and mean field games among fund managers considering excess logarithmic returns. *Annals of Operations Research*, pages 1–29, 2025.
202. Guohui Guan, Zongxia Liang, and Yi Xia. Robust mean-variance stochastic differential reinsurance and investment games under volatility risk and model uncertainty. *Scandinavian Actuarial Journal*, pages 1–41, 2025.
203. Junbiao Guan, Kaihua Wang, and Fangyue Chen. A cellular automaton model for evacuation flow using game theory. *Physica A: Statistical Mechanics and its Applications*, 461:655–661, 2016.
204. M. Gul, I. A. Khan, G. Zaman, A. Rahman, J. Mir, S. A. Ali Biabani, M. Saleem, and A. AlTurkey. A game-theoretic approach to safe crowd evacuation in emergencies. *Computers, Materials & Continua*, 79(1), 2024.
205. Ramakrishna Gummadi, Peter Key, and Alexandre Proutiere. Repeated auctions under budget constraints: Optimal bidding strategies and equilibria. In *the Eighth Ad Auction Workshop*, volume 4, 2012.
206. Xin Guo and Renyuan Xu. Stochastic games for fuel follower problem: N versus mean field game. *SIAM Journal on Control and Optimization*, 57(1):659–692, 2019.
207. Phyllis N Hallenbeck. An analysis of power dynamics in marriage. *Journal of Marriage and the Family*, pages 200–203, 1966.
208. Safa Hamdoun, Abderrezak Rachedi, Hamidou Tembine, and Yacine Ghamri-Doudane. Efficient transmission strategy selection algorithm for M2M communications: An evolutionary game approach. In *2016 IEEE 15th International Symposium on Network Computing and Applications (NCA)*, pages 286–293. IEEE, 2016.
209. Hajar El Hammouti, Essaid Sabir, and Hamidou Tembine. A satisfactory power control for 5g self-organizing networks. *arXiv preprint arXiv:1606.07904*, 2016.
210. Hajar El Hammouti, Essaid Sabir, and Hamidou Tembine. A fully distributed satisfactory power control for QOS self-provisioning in 5g networks. *International Symposium on Ubiquitous Networking, May 9–12*, 2017.
211. Hajar El Hammouti, Essaid Sabir, and Hamidou Tembine. QOS-aware tactical power control for 5g networks. *Book chapter in Ubiquitous Networking, pp 25–37, Lecture Notes in Computer Science book series (LNCS, volume 10542), Revised papers from UNET*, 2017.
212. Ahmed Farhan Hanif, H. Tembine, Mohamad Assaad, and Djamal Zeghlache. Mean-field games for resource sharing in cloud based networks. *IEEE/ACM Transactions on Networking, 24(1): 624–637 (2016)* DOI: https://doi.org/10.1109/TNET.2014.2387100, 2016.
213. Ahmed Farhan Hanif, Hamidou Tembine, Mohamad Assaad, and Djamal Zeghlache. Cloud networking mean-field games. In *2012 IEEE 1st International Conference on Cloud Networking (CLOUDNET)*, pages 46–50. IEEE, 2012.
214. Ahmed Farhan Hanif, Hamidou Tembine, Mohamad Assaad, and Djamal Zeghlache. Cloud networking mean field games. *IEEE CloudNet'12, International Conference on Cloud Networking, November 28–30, Paris, France*, 2012.
215. Ahmed Farhan Hanif, Hamidou Tembine, Mohamad Assaad, and Djamal Zeghlache. Distributed stochastic learning for continuous power control in wireless networks. In *2012 IEEE 13th International Workshop on Signal Processing Advances in Wireless Communications (SPAWC)*, pages 199–203. IEEE, 2012.
216. Ahmed Farhan Hanif, Hamidou Tembine, Mohamad Assaad, and Djamal Zeghlache. Distributed power control in femto cells using Bayesian density tracking. In *2013 51st Annual Allerton Conference on Communication, Control, and Computing (Allerton)*, pages 1388–1393. IEEE, 2013.
217. Ahmed Farhan Hanif, Hamidou Tembine, Mohamad Assaad, and Djamal Zeghlache. Mean-field games for resource sharing in cloud-based networks. *IEEE/ACM Transactions on Networking*, 24(1):624–637, 2015.

218. Rögnvaldur Hannesson. Game theory and fisheries. *Annu. Rev. Resour. Econ.*, 3(1):181–202, 2011.
219. Tao Hao and Jie Xiong. Stochastic differential games of mean-field dynamics and second-order Bellman–Isaacs equations on the Wasserstein space. *Acta Mathematica Sinica, English Series*, pages 1–35, 2025.
220. Mohammad Mehedi Hassan, Biao Song, and Eui-Nam Huh. Distributed resource allocation games in horizontal dynamic cloud federation platform. In *2011 IEEE International Conference on High Performance Computing and Communications*, pages 822–827. IEEE, 2011.
221. Naveed Ul Hassan, Mohamad Assaad, and Hamidou Tembine. Distributed h-infinty-based power control in a dynamic wireless network environment. *IEEE Communications Letters*, 17(6):1124–1127, 2013.
222. Mohammad Hayajneh, Massa Ndong, Najah Abu Ali, and Hamidou Tembine. Non-asymptotic linear growth of energy efficiency in distributed autonomous d2d mimo wireless communications. *IEEE Access*, 8:105914–105921, 2020.
223. Y. Hayel and H. Tembine. Information dissemination using epidemic routing with delayed feedback. *4th IEEE International Conference on Mobile Ad-Hoc and Sensor Systems, MASS'07, Oct 9–11, Pisa, Italy.*, 2007.
224. Y. Hayel, H. Tembine, E. Altman, and R. El-Azouzi. Markov decision evolutionary games with individual energy management. *Annals the International Society of Dynamic Games 2010*, 2010.
225. Xihao He. On the limit theory of mean field optimal stopping with non-Markov dynamics and common noise. *Stochastic Processes and their Applications*, page 104681, 2025.
226. Jehad Hedel and Nga Nguyen. Optimal charging control for electric vehicle fleet using mean field game theory. In *2025 IEEE Texas Power and Energy Conference (TPEC)*, pages 1–6. IEEE, 2025.
227. Jeffry L Hirst and Noah A Hughes. Reverse mathematics and marriage problems with unique solutions. *Archive for Mathematical Logic*, 54:49–57, 2015.
228. Jeffry L Hirst and Noah A Hughes. Reverse mathematics and marriage problems with finitely many solutions. *Archive for Mathematical Logic*, 55(7–8):1015–1024, 2016.
229. Felix Höfer and H Mete Soner. Synchronization games. *Mathematics of Operations Research*, 2025.
230. Harold Hotelling. Stability in competition. *The Economic Journal*, 39(153):41–57, 1929.
231. Minyi Huang, Peter E Caines, and Roland P Malhamé. Individual and mass behaviour in large population stochastic wireless power control problems: centralized and Nash equilibrium solutions. In *42nd IEEE international conference on decision and control (IEEE cat. No. 03CH37475)*, volume 1, pages 98–103. IEEE, 2003.
232. Minyi Huang, Peter E Caines, and Roland P Malhamé. Uplink power adjustment in wireless communication systems: A stochastic control analysis. *IEEE Transactions on Automatic Control*, 49(10):1693–1708, 2004.
233. Minyi Huang, Roland P Malhame, and Peter E Caines. Stochastic power control in wireless communication systems with an infinite horizon discounted cost. In *Proceedings of the 2003 American Control Conference, 2003.*, volume 2, pages 963–968. IEEE, 2003.
234. Pengyan Huang, Guangchen Wang, and Shujun Wang. A large-population stochastic differential game with terminal state constraint and common noise. *IEEE Transactions on Automatic Control*, 2025.
235. Ziyu Huang and Shanjian Tang. Mean field games with major and minor agents: the limiting problem and Nash equilibrium. *Stochastics*, 97(1):81–114, 2025.
236. Roger L Hughes. A continuum theory for the flow of pedestrians. *Transportation Research, Part B: Methodological*, 36(6):507–535, 2002.
237. Cang Hui. Unlocking patterns of nature-the marriage of mathematics and ecology. *Citeseer*, 2015.
238. Vincenzo Ignazio and Michele Ricciardi. A second-order mean field games model with controlled diffusion. *Partial Differential Equations and Applications*, 6(2):1–34, 2025.

239. Oleg Imanuvilov and Masahiro Yamamoto. Global Lipschitz stability for an inverse coefficient problem for a mean field game system. *Mathematical Methods in the Applied Sciences*, 48(3):3858–3870, 2025.
240. S. Ishikawa. Fixed points by a new iteration method. *Proc. Amer. Math. Soc.*, 44(1):147–150, 1974.
241. S. Ishikawa. Fixed points and iteration of a nonexpansive mapping in a Banach space. *Proc. American Mathematical Society*, 59(1), 1976.
242. Javier Izquierdo, I Montalvo, R Pérez, and VS Fuertes. Forecasting pedestrian evacuation times by using swarm intelligence. *Physica A: Statistical Mechanics and its Applications*, 388(7):1213–1220, 2009.
243. JVirajith Jalaparti, Giang Nguyen, Indranil Gupta, and Matthew Caesar. Cloud resource allocation games. *Springer Science Business Media, LLC*, 2009.
244. Boutaina Jebari, Khalil Ibrahimi, Mounir Ghogho, and Hamidou Tembine. Bivo: A decentralized oracle solution for authenticity in blockchain-based IoT networks. *accepted and to appear at IEEE Internet of Things*, 2024.
245. Rajarama Mohan Jena, Snehashish Chakraverty, and Dumitru Baleanu. On the solution of an imprecisely defined nonlinear time-fractional dynamical model of marriage. *Mathematics*, 7(8):689, 2019.
246. Furong Jia, Kevin Wang, Yixiang Zheng, Defu Cao, and Yan Liu. Gpt4mts: Prompt-based large language model for multimodal time-series forecasting. *Proceedings of the AAAI Conference on Artificial Intelligence*, 38(21):23343–23351, 2024.
247. Jincheng Jiang, Wei Tu, Hui Kong, Wei Zeng, Rui Zhang, and Milan Konecny. Large-scale urban multiple-modal transport evacuation model for mass gathering events considering pedestrian and public transit system. *IEEE Transactions on Intelligent Transportation Systems*, 23(12):23059–23069, 2022.
248. Yushan Jiang, Zijie Pan, Xikun Zhang, Sahil Garg, Anderson Schneider, Yuriy Nevmyvaka, and Dongjin Song. Empowering time series analysis with large language models: A survey. *arXiv preprint arXiv:2402.03182*, 2024.
249. Ming Jin, Shiyu Wang, Lintao Ma, Zhixuan Chu, James Y Zhang, Xiaoming Shi, Pin-Yu Chen, Yuxuan Liang, Yuan-Fang Li, Shirui Pan, et al. Time-LLM: Time series forecasting by reprogramming large language models. *arXiv preprint arXiv:2310.01728*, 2023.
250. Ming Jin, Yifan Zhang, Wei Chen, Kexin Zhang, Yuxuan Liang, Bin Yang, Jindong Wang, Shirui Pan, and Qingsong Wen. Position paper: What can large language models tell us about time series analysis. *arXiv preprint arXiv:2402.02713*, 2024.
251. Mingyu Jin, Hua Tang, Chong Zhang, Qinkai Yu, Chengzhi Liu, Suiyuan Zhu, Yongfeng Zhang, and Mengnan Du. Time series forecasting with LLMs: Understanding and enhancing model capabilities. *arXiv preprint arXiv:2402.10835*, 2024.
252. Mohammed Jouhari, Khalil Ibrahimi, Hamidou Tembine, and Jalel Ben-Othman. Underwater wireless sensor networks: A survey on enabling technologies, localization protocols, and internet of underwater things. *IEEE Access*, 7:96879–96899, 2019.
253. Mohammed Jouhari, Khalil Ibrahimi, Hamidou Tembine, Mohammed Benattou, and Jalel Ben Othman. Signaling game approach to improve the MAC protocol in the underwater wireless sensor networks. *International Journal of Communication Systems*, 32(13):e3971, 2019.
254. B. Jovanovic and R. W. Rosenthal. Anonymous sequential games. *Journal of Mathematical Economics*, 17:77–87, 1988.
255. Maryam Kamgarpour and Hamidou Tembine. A Bayesian mean field game approach to supply demand analysis of the smart grid. In *2013 First International Black Sea Conference on Communications and Networking (BlackSeaCom)*, pages 211–215. IEEE, 2013.
256. Verena Kantere, Debabrata Dash, Gregory Francois, Sofia Kyriakopoulou, and Anastasia Ailamaki. Optimal service pricing for a cloud cache. *IEEE Transactions on Knowledge and Data Engineering*, 23(9):1345–1358, 2011.
257. L. V. Kantorovich. On the transfer of masses. *Dokl. Akad. Nauk.*, 37(2):227–229, 1942. Translated in Management Science, Vol. 5, pp. 1–4, 1959.

258. Ryo Karakida, Shotaro Akaho, and Shun-Ichi Amari. Universal statistics of fisher information in deep neural networks: Mean field approach. In *The 22nd International Conference on Artificial Intelligence and Statistics*, pages 1032–1041. PMLR, 2019.
259. Gaurav S. Kasbekar, Eitan Altman, and Saswati Sarkar. A hierarchical spatial game over licenced resources. In *GameNets'09: Proceedings of the First ICST international conference on Game Theory for Networks*, pages 70–79, Piscataway, NJ, USA, 2009. IEEE Press.
260. Kalliopi Kastampolidou and Theodore Andronikos. Game theory and other unconventional approaches to biological systems. In *Handbook of Computational Neurodegeneration*, pages 1–18. Springer, 2021.
261. M. Khan and H. Tembine. Energy-efficiency networking games. *Green It: Technologies and Applications, Part 1, 211–226, Springer Verlag*, 2011.
262. M. Khan, H. Tembine, and A. Vasilakos. Game dynamics and cost of learning in heterogeneous 4g networks. *IEEE Journal on Selected Areas in Communications, vol 30, 1, pp 198–213, January*, 2012.
263. M. A. Khan and H. Tembine. Random matrix games in wireless networks. *IEEE Global High Tech Congress on Electronics (GHTCE 2012), November 18–20, 2012, Shenzhen, China*, 2012.
264. Manzoor Khan and Hamidou Tembine. Meta-learning for realizing self-x management of future networks. *IEEE Access Journal, vol.5, pp. 19072–19083, August*, 2017.
265. Manzoor Ahmed Khan and Hamidou Tembine. Energy-efficiency networking games. In *Green IT: Technologies and Applications*, pages 211–226. Springer, 2011.
266. Manzoor Ahmed Khan, Hamidou Tembine, and Athanasios V Vasilakos. Game dynamics and cost of learning in heterogeneous 4g networks. *IEEE Journal on Selected Areas in Communications*, 30(1):198–213, 2011.
267. Manzoor Ahmed Khan, Hamidou Tembine, and Athanasios V Vasilakos. Evolutionary coalitional games: design and challenges in wireless networks. *IEEE Wireless Communications*, 19(2):50–56, 2012.
268. Samee Ullah Khan and Ishfaq Ahmad. Non-cooperative, semi-cooperative, and cooperative games-based grid resource allocation. In *Proceedings 20th IEEE International Parallel & Distributed Processing Symposium*, pages 10–pp. IEEE, 2006.
269. Yubin Kim, Xuhai Xu, Daniel McDuff, Cynthia Breazeal, and Hae Won Park. Health-LLM: Large language models for health prediction via wearable sensor data. *arXiv preprint arXiv:2401.06866*, 2024.
270. Abdellatif Kobbane, M-A Koulali, Hamidou Tembine, Mohammed El Koutbi, and Jalel Ben-Othman. Dynamic power control with energy constraint for multimedia wireless sensor networks. In *2012 IEEE International Conference on Communications (ICC)*, pages 518–522. IEEE, 2012.
271. Vassili Kolokoltsov. The evolutionary game of pressure (or interference), resistance and collaboration. *Mathematics of Operations Research*, 42(4):915–944, 2017.
272. Vassili N Kolokoltsov. Dynamic quantum games. *Dynamic Games and Applications*, 12(2):552–573, 2022.
273. Vassili N Kolokoltsov. Quantum mean-field games. *The Annals of Applied Probability*, 32(3):2254–2288, 2022.
274. Vassili N Kolokoltsov. On quantum stochastic master equations. *Electronic Journal of Probability*, 30:1–21, 2025.
275. Vassili N Kolokoltsov and Alain Bensoussan. Mean-field-game model for Botnet defense in cyber-security. *Applied Mathematics & Optimization*, 74:669–692, 2016.
276. Vassili N Kolokoltsov and Oleg A Malafeyev. Corruption and Botnet defense: a mean field game approach. *International Journal of Game Theory*, 47:977–999, 2018.
277. Peter M Kotelenez and Thomas G Kurtz. Macroscopic limits for stochastic partial differential equations of Mckean–Vlasov type. *Probability theory and related fields*, 146(1–2):189, 2010.

278. Mohammed-Amine Koulali, Abdellatif Kobbane, Mohammed El Koutbi, Hamidou Tembine, and Jalel Ben-Othman. Dynamic power control for energy harvesting wireless multimedia sensor networks. *EURASIP Journal on Wireless Communications and Networking*, 2012(1):1–8, 2012.
279. M.A. Krasnoselskij. Two remarks on the method of successive approximations (Russian). *Uspehi Mat. Nauk. 10*, 1(63):123–127, 1955.
280. Maurice Kraus, Felix Divo, David Steinmann, Devendra Singh Dhami, and Kristian Kersting. United we pretrain, divided we fail! representation learning for time series by pretraining on 75 datasets at once. *arXiv preprint arXiv:2402.15404*, 2024.
281. Jorn Kunsemoller and Holger Karl. A game-theoretical approach to the benefits of cloud computing. In *Economics of Grids, Clouds, Systems, and Services: 8th International Workshop, GECON 2011, Paphos, Cyprus, December 5, 2011, Revised Selected Papers 8*, pages 148–160. Springer, 2012.
282. Aldis Run Larusdottir and AS Dederichs. Evacuation dynamics of children–walking speeds, flows through doors in daycare centers. In *Pedestrian and evacuation dynamics*, pages 139–147. Springer, 2011.
283. Samson Lasaulce and Hamidou Tembine. *Game theory and learning for wireless networks: fundamentals and applications*. Academic Press, 2011.
284. Samson Lasaulce and Hamidou Tembine. Power allocation games and learning in mimo multiuser channels. *Chapter 8, Pages 211–244, In: PART III: Case Studies, Game Theory and Learning for Wireless Networks: Fundamentals and Applications, Elsevier, Academic Press*, 2011.
285. Razvan-Andrei Lascu, Mateusz B Majka, and Lukasz Szpruch. Entropic mean-field min-max problems via best response flow. *Applied Mathematics & Optimization*, 91(2):48, 2025.
286. Mael Le Treust, Hamidou Tembine, Samson Lasaulce, and Mérouane Debbah. Coverage games in small cells networks. In *2010 Future Network & Mobile Summit*, pages 1–8. IEEE, 2010.
287. Bernard Lebrun. First price auctions in the asymmetric n bidder case. *International Economic Review*, 40(1):125–142, 1999.
288. Bernard Lebrun. Uniqueness of the equilibrium in first-price auctions. *Games and Economic Behavior*, 55(1):131–151, 2006.
289. Jean Lee, Nicholas Stevens, Soyeon Caren Han, and Minseok Song. A survey of large language models in finance (FinLLMs). *arXiv preprint arXiv:2402.02315*, 2024.
290. Wonjun Lee, Siting Liu, Hamidou Tembine, Wuchen Li, and Stanley J. Osher. Controlling propagation of epidemics via mean-field control. *SIAM J. Appl. Math. 81(1): 190–207*, 2021.
291. M. LeTreust, H. Tembine, S. Lasaulce, and M. Debbah. Coverage games in small-cells networks. *Proc. of the Future Network and Mobile Summit (FUNEMS'10), Italy.*, 2010.
292. Chun Li, Yunyun Yang, Hui Liang, and Boying Wu. Robust PCL discovery of data-driven mean-field game systems and control problems. *IEEE Transactions on Circuits and Systems I: Regular Papers*, 68(6):2651–2664, 2021.
293. Lixin Li, Qianqian Cheng, Kaiyuan Xue, Chungang Yang, and Zhu Han. Downlink transmit power control in ultra-dense UAV network based on mean field game and deep reinforcement learning. *IEEE Transactions on Vehicular Technology*, 69(12):15594–15605, 2020.
294. Min Li, Na Li, and Zhen Wu. Linear–quadratic mean-field game for stochastic systems with partial observation. *Automatica*, 171:111821, 2025.
295. Na Li, Yilin Wei, and Qingfeng Zhu. Stochastic linear-quadratic mean-field games of controls for delayed systems with jump diffusion. *Journal of Optimization Theory and Applications*, 206(3):1–34, 2025.
296. Yuxuan Liang, Haomin Wen, Yuqi Nie, Yushan Jiang, Ming Jin, Dongjin Song, Shirui Pan, and Qingsong Wen. Foundation models for time series analysis: A tutorial and survey. *arXiv preprint arXiv:2403.14735*, 2024.
297. Weichen Liao, Xiaoping Zheng, Lisheng Cheng, Ying Zhao, Yuan Cheng, and Yafei Wang. Layout effects of multi-exit ticket-inspectors on pedestrian evacuation. *Safety science*, 70:1–8, 2014.

298. Zhonghua Liao and Qi Lü. On inverse problems for mean field games with common noise via Carleman estimate. *Inverse Problems*, 41(4):045009, 2025.
299. Alex Tong Lin, Samy Wu Fung, Wuchen Li, Levon Nurbekyan, and Stanley J Osher. Alternating the population and control neural networks to solve high-dimensional stochastic mean-field games. *Proceedings of the National Academy of Sciences*, 118(31):e2024713118, 2021.
300. Runzi Lin, Zhenhui Xu, Xiaoliang Huang, Jinwu Gao, Hong Chen, and Tielong Shen. Optimal scheduling management of the parking lot and decentralized charging of electric vehicles based on mean field game. *Applied Energy*, 328:120198, 2022.
301. P. L. Lions. Cours jeux à champ moyens et applications. *College de France*, 2010.
302. Chenxi Liu, Sun Yang, Qianxiong Xu, Zhishuai Li, Cheng Long, Ziyue Li, and Rui Zhao. Spatial-temporal large language model for traffic prediction. *arXiv preprint arXiv:2401.10134*, 2024.
303. Fengao Liu, Mei Yu, Bo Wei, and Feng Xiao. Mean field games of energy storage devices: A finite difference analysis. *Asian Journal of Control*, 2023.
304. Haoxin Liu, Zhiyuan Zhao, Jindong Wang, Harshavardhan Kamarthi, and B Aditya Prakash. Lstprompt: Large language models as zero-shot time series forecasters by long-short-term prompting. *arXiv preprint arXiv:2402.16132*, 2024.
305. Heng Liu, Dianjie Lu, Guijuan Zhang, Xiao Hong, and Hong Liu. Recurrent emotional contagion for the crowd evacuation of a cyber-physical society. *Information Sciences*, 575:155–172, 2021.
306. Hongyu Liu and Catharine WK Lo. Determining state space anomalies in mean field games. *Nonlinearity*, 38(2):025010, 2025.
307. Hongyu Liu, Catharine WK Lo, and Shen Zhang. Decoding a mean field game by the cauchy data around its unknown stationary states. *Journal of the London Mathematical Society*, 111(5):e70173, 2025.
308. Kang Liu and Laurent Pfeiffer. A mesh-independent method for second-order potential mean field games. *IMA Journal of Numerical Analysis*, 45(2):1226–1266, 2025.
309. Lei Liu, Shuo Yu, Runze Wang, Zhenxun Ma, and Yanming Shen. How can large language models understand spatial-temporal data? *arXiv preprint arXiv:2401.14192*, 2024.
310. Qiujia Liu, Linjun Lu, Yijing Zhang, and Miaoqing Hu. Modeling the dynamics of pedestrian evacuation in a complex environment. *Physica A: Statistical Mechanics and its Applications*, 585:126426, 2022.
311. Yong Liu, Guo Qin, Xiangdong Huang, Jianmin Wang, and Mingsheng Long. Autotimes: Autoregressive time series forecasters via large language models. *arXiv preprint arXiv:2402.02370*, 2024.
312. Yong Liu, Haoran Zhang, Chenyu Li, Xiangdong Huang, Jianmin Wang, and Mingsheng Long. Timer: Transformers for time series analysis at scale. *arXiv preprint arXiv:2402.02368*, 2024.
313. Zichang Liu, Jue Wang, Tri Dao, Tianyi Zhou, Binhang Yuan, Zhao Song, Anshumali Shrivastava, Ce Zhang, Yuandong Tian, Christopher Re, et al. Deja vu: Contextual sparsity for efficient LLMs at inference time. In *International Conference on Machine Learning*, pages 22137–22176. PMLR, 2023.
314. Siu Ming Lo, Hui-Chi Huang, Peng Wang, and KK Yuen. A game theory based exit selection model for evacuation. *Fire Safety Journal*, 41(5):364–369, 2006.
315. Ali Akbar Rezaei Lori and Piyush Grover. Topological bifurcations in a mean-field game. In *Proc. R. Soc. A., 481, 2310*, page 20240348. The Royal Society, 2025.
316. Jun Lu, Jinbiao Wu, and Bixuan Yang. Optimal strategies of regular-singular mean-field delayed stochastic differential games. *Annals of Operations Research*, 344(1):175–216, 2025.
317. X. Luo and H. Tembine. Evolutionary coalitional games in random access control. *32nd IEEE International Conference on Computer Communications INFOCOM miniconference, April 14–19, Turin, Italy*, 2013.
318. Xin Luo and Hamidou Tembine. Evolutionary coalitional games for random access control. In *2013 Proceedings IEEE INFOCOM*, pages 535–539. IEEE, 2013.

319. Xin Luo and Hamidou Tembine. Evolutionary coalitional games for random access control. *Annals of Operations Research, ANOR 2017, 34 pages*, DOI: https://doi.org/10.1007/s10479-016-2198-0, 2017.
320. RT Ma, Dah Ming Chiu, JC Lui, Vishal Misra, and Dan Rubenstein. On resource management for cloud users: A generalized Kelly mechanism approach. *Electrical Engineering, Tech. Rep*, 2010.
321. W. R. Mann. Mean value methods in iteration. *Proc. Amer. Math Soc.*, 4:506–510, 1953.
322. Yishay Mansour, S Muthukrishnan, and Noam Nisan. Doubleclick ad exchange auction. *arXiv preprint arXiv:1204.0535*, 2012.
323. Robert C Marshall, Michael J Meurer, Jean-Francois Richard, and Walter Stromquist. Numerical analysis of asymmetric first price auctions. *Games and Economic behavior*, 7(2):193–220, 1994.
324. Diego Marti Mason, Leonardo Stella, and Dario Bauso. Evolutionary game dynamics for crowd behavior in emergency evacuations. In *2020 59th IEEE Conference on Decision and Control (CDC)*, pages 1672–1677. IEEE, 2020.
325. Thibaut Mastrolia and Jiacheng Zhang. Agency problem and mean field system of agents with moral hazard, synergistic effects and accidents. *Journal of Optimization Theory and Applications*, 205(3):1–32, 2025.
326. D.R. McDonald. Lecture notes on mean field convergence. *Dept. of Mathematics, University of Toronto*, 2007.
327. Song Mei, Theodor Misiakiewicz, and Andrea Montanari. Mean-field theory of two-layers neural networks: dimension-free bounds and kernel limit. In *Conference on Learning Theory*, pages 2388–2464. PMLR, 2019.
328. Song Mei, Andrea Montanari, and Phan-Minh Nguyen. A mean field view of the landscape of two-layer neural networks. *Proceedings of the National Academy of Sciences*, 115(33):E7665–E7671, 2018.
329. Qingfeng Meng, Shaohong Tan, Zhen Li, Bingyao Chen, and Weixiang Shi. A review of game theory application research in safety management. *IEEE access*, 8:107301–107313, 2020.
330. François Mériaux, Samson Lasaulce, and Hamidou Tembine. Stochastic differential games and energy-efficient power control. *Dynamic Games and Applications*, 3:3–23, 2013.
331. Bryan L Mesmer and Christina L Bloebaum. Incorporation of decision, game, and Bayesian game theory in an emergency evacuation exit decision model. *Fire Safety Journal*, 67:121–134, 2014.
332. John A Miller, Mohammed Aldosari, Farah Saeed, Nasid Habib Barna, Subas Rana, I Budak Arpinar, and Ninghao Liu. A survey of deep learning and foundation models for time series forecasting. *arXiv preprint arXiv:2401.13912*, 2024.
333. Martha Mitsopoulou, Nikolaos I Dourvas, Georgios Ch Sirakoulis, and Katsuhiro Nishinari. Spatial games and memory effects on crowd evacuation behavior with cellular automata. *Journal of Computational Science*, 32:87–98, 2019.
334. Azhar Mohd Ibrahim, Ibrahim Venkat, Philippe De Wilde, Muhammad Rabani Mohd Romlay, and Alala Bahamid. The role of crowd behavior and cooperation strategies during evacuation. *Simulation*, 98(9):737–751, 2022.
335. G. Monge. Mémoire sur la théorie des déblais et des remblais. *In Histoire de l'Académie Royale des Sciences de Paris*, pages 666–704, 1781.
336. Gaspard Monge. Mémoire sur la théorie des déblais et des remblais. *Histoire de l'Académie Royale des Sciences*, 1871.
337. James D Montgomery. The mathematics of marriage: Dynamic nonlinear models, 2004.
338. Goudiaby MS, Dia BM, Diagne ML, and Tembine H. Cooperative game for fish harvesting and pollution control. *Games, Aug 19;12(3):65*, 2021.
339. Shanmugavelayutham Muthukrishnan. Ad exchanges: Research issues. In *International workshop on internet and network economics*, pages 1–12. Springer, 2009.
340. Nidal Nasser, Ahmed El Ouadrhiri, Mohamed El Kamili, Asmaa Ali, and Muhammad Anan. Crowd management services in Hajj: a mean-field game theory approach. In *2019 IEEE Wireless Communications and Networking Conference (WCNC)*, pages 1–7. IEEE, 2019.

341. Subigya Nepal, Arvind Pillai, William Campbell, Talie Massachi, Eunsol Soul Choi, Xuhai Xu, Joanna Kuc, Jeremy F Huckins, Jason Holden, Colin Depp, et al. Contextual ai journaling: Integrating LLM and time series behavioral sensing technology to promote self-reflection and well-being using the mindscape app. In *Extended Abstracts of the CHI Conference on Human Factors in Computing Systems*, pages 1–8, 2024.
342. Tianyang Nie and Ke Yan. Mean-field partial information non-zero sum stochastic differential games. *Applied Mathematics & Optimization*, 91(2):37, 2025.
343. N Nisan, T Roughgarden, E Tardos, and V Vazirani. *Algorithmic Game Theory, 776 p.* Cambridge University Press, New York, ISBN:0521872820, 2007.
344. Charles N Noussair, Daan van Soest, and Jan Stoop. Cooperation in a dynamic fishing game: A framed field experiment. *American Economic Review*, 105(5):408–413, 2015.
345. Tomohiro Ogawa, Kango Yoshioka, Ken Fukuda, and Takeshi Morita. Prediction of actions and places by the time series recognition from images with multimodal LLM. In *2024 IEEE 18th International Conference on Semantic Computing (ICSC)*, pages 294–300. IEEE, 2024.
346. Yuna Oh and Jun Moon. A feedback-type optimal solution for partially-observed linear-quadratic risk-sensitive optimal control problem of mean-field type stochastic systems. *IEEE Transactions on Automatic Control*, 2025.
347. Kenniy Olorunnimbe and Herna Viktor. Ensemble of temporal transformers for financial time series. *Journal of Intelligent Information Systems*, pages 1–25, 2024.
348. Yohance AP Osborne and Iain Smears. Finite element approximation of time-dependent mean field games with nondifferentiable Hamiltonians. *Numerische Mathematik*, 157(1):165–211, 2025.
349. Wynn P. On a device for computing the em(sn) transformation. *MTAC*, 10:91–96, 1956.
350. Zijie Pan, Yushan Jiang, Sahil Garg, Anderson Schneider, Yuriy Nevmyvaka, and Dongjin Song. S2ip-LLM: Semantic space informed prompt learning with LLM for time series forecasting. *arXiv preprint arXiv:2403.05798*, 2024.
351. S. Papatheodorou, M. Smyrnakis, H. Tembine, and A. Tzes. Path planning and task assignment for data retrieval from wireless sensor nodes relying on game-theoretic learning. *5th international conference on Control Decision and Information Technologies (CoDIT)*, 2018.
352. Sotiris Papatheodorou, Michalis Smyrnakis, Hamidou Tembine, and Anthony Tzes. Game theoretic optimal position computation of collaborating agents for visual area coverage. *SETN 2018: 32:1–32:7*, 2018.
353. Rohit Parasnis, Amol Sakhale, Ryosuke Kato, Massimo Franceschetti, and Behrouz Touri. A case for the age-structured sir dynamics for modelling covid-19. In *2021 60th IEEE Conference on Decision and Control (CDC)*, pages 5508–5513. IEEE, 2021.
354. Neal Parikh and Stephen Boyd. Block splitting for large-scale distributed learning. In *Neural Information Processing Systems (NIPS), Workshop on Big Learning*, 2011.
355. Daniel Ricardo Parisi and Claudio Oscar Dorso. Microscopic dynamics of pedestrian evacuation. *Physica A: Statistical Mechanics and its Applications*, 354:606–618, 2005.
356. Abhay Dutt Paroha and Aakash Chotrani. A comparative analysis of timeGPT and time-LLM in predicting esp maintenance needs in the oil and gas sector. *International Journal of Computer Applications*, 975:8887, 2024.
357. Samir M Perlaza, Hamidou Tembine, and Samson Lasaulce. How can ignorant but patient cognitive terminals learn their strategy and utility? In *2010 IEEE 11th International Workshop on Signal Processing Advances in Wireless Communications (SPAWC)*, pages 1–5. IEEE, 2010.
358. Samir M Perlaza, Hamidou Tembine, Samson Lasaulce, and Mérouane Debbah. Satisfaction equilibrium: A general framework for QOS provisioning in self-configuring networks. In *2010 IEEE Global Telecommunications Conference GLOBECOM 2010*, pages 1–5. IEEE, 2010.
359. Viktoriya Petrakova and Olga Krivorotko. Mean field game for modeling of covid-19 spread. *Journal of Mathematical Analysis and Applications*, 514(1):126271, 2022.

360. Mary Phuong and Marcus Hutter. Formal algorithms for transformers. *CoRR abs/2207.09238*, 2022.
361. Frank Plastria. Static competitive facility location: An overview of optimisation approaches. *European Journal of Operational Research*, 129(3):461 – 470, 2001.
362. Ho Fai Po and Chi Ho Yeung. Re-routing game: The inadequacy of mean-field approach in modeling the herd behavior in path switching. *Physica A: Statistical Mechanics and its Applications*, 658:130266, 2025.
363. Alex Possajennikov. Evolutionary foundations of aggregate-taking behavior. *Economic Theory*, 21:921–928, 2003.
364. Eugene B Postnikov. Long tails of mean-field covid-19 epidemic curves: Implications of a hidden metapopulational dynamics. *Modeling, Control and Drug Development for COVID-19 Outbreak Prevention*, pages 119–151, 2022.
365. Brian Powers, Michalis Smyrnakis, and Hamidou Tembine. Empathy in bimatrix games. *arXiv preprint arXiv:1708.01910*, 2017.
366. Buckdahn R., J. Li, and Peng S. Mean-field backward stochastic differential equations and related partial differential equations. *Preprint*, 2007.
367. Arthur Ransome. On epidemics studied by means of statistics of disease. *British Medical Journal*, 2:306(406), 1868.
368. Nageswara SV Rao, Stephen W Poole, Fei He, Jun Zhuang, Chris YT Ma, and David KY Yau. Cloud computing infrastructure robustness: A game theory approach. In *2012 International Conference on Computing, Networking and Communications (ICNC)*, pages 34–38. IEEE, 2012.
369. Kashif Rasul, Arjun Ashok, Andrew Robert Williams, Hena Ghonia, Rishika Bhagwatkar, Arian Khorasani, Mohammad Javad Darvishi Bayazi, George Adamopoulos, Roland Riachi, Nadhir Hassen, et al. Lag-llama: Towards foundation models for probabilistic time series forecasting. *Preprint*, 2024.
370. Kui Ren, Nathan Soedjak, and Kewei Wang. Unique determination of cost functions in a multipopulation mean field game model. *Journal of Differential Equations*, 427:843–867, 2025.
371. Lu Ren, Yuxin Jin, Wang Yao, Xiao Zhang, and Guanghui Jiao. Large-scale group opinion evolution with coexistence of influential individuals and strongly organized groups based on mean field games. *IEEE Transactions on Network Science and Engineering*, 2025.
372. Alexandros Rigos, Erik Mohlin, and Enrico Ronchi. The cry wolf effect in evacuation: A game-theoretic approach. *Physica A: Statistical Mechanics and its Applications*, 526:120890, 2019.
373. Charles F Roos. A mathematical theory of competition. *American Journal of Mathematics*, 47(3):163–175, 1925.
374. Charles F Roos. A dynamical theory of economics. *Journal of Political Economy*, 35(5):632–656, 1927.
375. Giulia Rossi, Alain Tcheukam, and Hamidou Tembine. How much does users' psychology matter in engineering mean-field-type games. *arXiv preprint arXiv:1702.05355*, 2017.
376. Amal Roy, Chandramani Singh, and Y Narahari. Recent advances in modeling and control of epidemics using a mean field approach. *arXiv preprint arXiv:2208.14765*, 2022.
377. Yusra Bibi Ruhomally, Maheshsingh Mungur, Abdel Anwar Hossen Khoodaruth, Vishwamitra Oree, and Muhammad Zaid Dauhoo. Assessing the impact of contact tracing, quarantine and red zone on the dynamical evolution of the covid-19 pandemic using the cellular automata approach and the resulting mean field system: A case study in mauritius. *Applied Mathematical Modelling*, 111:567–589, 2022.
378. Maskin E. S. and J. G. Riley. Asymmetric auctions. *Rev. Econom. Stud.*, 67:413–438, 2000.
379. Essaid Sabir, Majed Haddad, and Hamidou Tembine. Joint strategic spectrum sensing and opportunistic access for cognitive radio networks. In *2012 IEEE Global Communications Conference (GLOBECOM)*, pages 1368–1373. IEEE, 2012.
380. Naci Saldi. Linear mean-field games with discounted cost. *Mathematics of Operations Research*, 2025.

381. Sina Sanjari, Subhonmesh Bose, and Tamer Başar. Incentive designs for stackelberg games with a large number of followers and their mean-field limits. *Dynamic Games and Applications*, 15(1):238–278, 2025.
382. K. P. R. Sastry and G. V. R. Babu. Approximation of fixed-points of strictly pseudocontractive mappings on arbitrary closed, convex sets in a BANACH space. *Proceedings of the American Mathematical Society*, 128(10):2907–2909, 2000.
383. Mark E Schaffer. Evolutionarily stable strategies for a finite population and a variable contest size. *Journal of theoretical biology*, 132(4):469–478, 1988.
384. Galina A Schwartz, Hamidou Tembine, Saurabh Amin, and S Shankar Sastry. Electricity demand shaping via randomized rewards: A mean field game approach. In *Allerton Conference on Communication, Control, and Computing*, 2012.
385. Shigen Shen, Chenpeng Cai, Yizhou Shen, Xiaoping Wu, Wenlong Ke, and Shui Yu. Joint mean-field game and multiagent asynchronous advantage actor-critic for edge intelligence-based IoT malware propagation defense. *IEEE Transactions on Dependable and Secure Computing*, 2025.
386. Dongmei Shi, Wenyao Zhang, and Binghong Wang. Modeling pedestrian evacuation by means of game theory. *Journal of Statistical Mechanics: Theory and Experiment*, 2017(4):043407, 2017.
387. Hamid Shiri, Jihong Park, and Mehdi Bennis. Massive autonomous UAV path planning: A neural network based mean-field game theoretic approach. In *2019 IEEE Global Communications Conference (GLOBECOM)*, pages 1–6. IEEE, 2019.
388. Arvind V Shrivats, Dena Firoozi, and Sebastian Jaimungal. A mean-field game approach to equilibrium pricing in solar renewable energy certificate markets. *Mathematical Finance*, 32(3):779–824, 2022.
389. Haotian Si, Changhua Pei, Hang Cui, Jingwen Yang, Yongqian Sun, Shenglin Zhang, Jingjing Li, Haiming Zhang, Jing Han, Dan Pei, et al. Timeseriesbench: An industrial-grade benchmark for time series anomaly detection models. *arXiv preprint arXiv:2402.10802*, 2024.
390. A. Silva, H. Tembine, E. Altman, and M. Debbah. Optimal mobile association on hybrid networks: the downlink case. *In Proceedings of 49th IEEE Conference on Decision and Control, CDC, December*, 2010.
391. A. Silva, H. Tembine, E. Altman, and M. Debbah. Optimum and equilibrium in assignment problems with congestion: mobile terminals association to base stations. *IEEE Transactions on Automatic Control, Volume 58 (2013),Issue 8 (August)*, 2013.
392. A. Silva, H. Tembine, M. Debbah, and E. Altman. Uplink spatial games on cellular networks. *Proc. of 48th Annual Allerton Conference on Communication, Control, and Computing, Illinois, IL, USA, September 29-October 1*, 2010.
393. Alonso Silva, Hamidou Tembine, Eitan Altman, and Merouane Debbah. Spatial games and global optimization for the mobile association problem: the downlink case. In *49th IEEE Conference on Decision and Control (CDC)*, pages 966–972. IEEE, 2010.
394. Alonso Silva, Hamidou Tembine, Eitan Altman, and Mérouane Debbah. Optimum and equilibrium in assignment problems with congestion: Mobile terminals association to base stations. *IEEE Transactions on Automatic Control*, 58(8):2018–2031, 2013.
395. H. A. Simon. Rational choice and the structure of the environment. *Psychological Review*, 63(2):129–138, 1956.
396. Maurice Sion. On general minimax theorems. *Pacific Journal of Mathematics*, 8(1):171–176, 1958.
397. Justin Sirignano and Konstantinos Spiliopoulos. Mean field analysis of neural networks: A central limit theorem. *Stochastic Processes and their Applications*, 130(3):1820–1852, 2020.
398. Justin Sirignano and Konstantinos Spiliopoulos. Mean field analysis of deep neural networks. *Mathematics of Operations Research*, 47(1):120–152, 2022.
399. Niranjan Sitapure and Joseph Sang-Il Kwon. Exploring the potential of time-series transformers for process modeling and control in chemical systems: an inevitable paradigm shift? *Chemical Engineering Research and Design*, 194:461–477, 2023.

400. Alain Tcheukam Siwe and Hamidou Tembine. Energy cost saving tips in distributed power networks. In *Smart Grid as a Solution for Renewable and Efficient Energy*, pages 26–48. IGI Global, 2016.
401. Alain Tcheukam Siwe and Hamidou Tembine. Mean-field-type games on airline networks and airport queues: Braess paradox, its negation, and crowd effect. In *2016 13th International Multi-Conference on Systems, Signals & Devices (SSD)*, pages 595–600. IEEE, 2016.
402. Alain Tcheukam Siwe and Hamidou Tembine. Network security as public good: A mean-field-type game theory approach. *13th International Multi-Conference on Systems, Signals & Devices. Conference on Systems, Automation & Control. March 21–24- Leipzig, Germany, pp. 598–603*, 2016.
403. J Maynard Smith and George R Price. The logic of animal conflict. *Nature*, 246(5427):15–18, 1973.
404. Fei Song, Zhe Wang, Jun Li, Long Shi, Wen Chen, and Shi Jin. Dynamic trajectory and power control in ultra-dense UAV networks: A mean-field reinforcement learning approach. *IEEE Transactions on Wireless Communications*, 2025.
405. Xiao Song, Liang Ma, Yaofei Ma, Chen Yang, and Hang Ji. Selfishness-and selflessness-based models of pedestrian room evacuation. *Physica A: Statistical Mechanics and its Applications*, 447:455–466, 2016.
406. Xiao Song, Zenghui Zhang, Gongzhuang Peng, and Guoqiang Shi. Effect of authority figures for pedestrian evacuation at metro stations. *Physica A: Statistical Mechanics and its Applications*, 465:599–612, 2017.
407. Wynn C Stirling. *Satisficing Games and Decision Making: with applications to engineering and computer science*. Cambridge University Press, 2003.
408. Wynn C Stirling and Michael A Goodrich. Satisficing games. *Information Sciences*, 114(1–4):255–280, 1999.
409. Nicholas Stroh. Trackgpt–a generative pre-trained transformer for cross-domain entity trajectory forecasting. *arXiv preprint arXiv:2402.00066*, 2024.
410. Ussif Rashid Sumaila. A review of game-theoretic models of fishing. *Marine policy*, 23(1):1–10, 1999.
411. Wei Sun and Theodore B Trafalis. Risk-aware controller implementation for risk-sensitive mean field games through a game-theoretic differential dynamic programming approach. *International Journal of Control*, pages 1–9, 2025.
412. Manaswini Swamy, Arunima Shukla, and James Purtilo. LLM-based stock market trend prediction. *Preprint*, 2023.
413. A. S. Sznitman. Topics in propagation of chaos. *In P.L. Hennequin, editor, Springer Verlag Lecture Notes in Mathematics 1464, Ecole d'Eté de Probabilités de Saint-Flour XI (1989)*, pages 165–251, 1991.
414. Amirheckmat Taghizadeh, Hamed Kebriaei, and Dusit Niyato. Mean field game for equilibrium analysis of mining computational power in blockchains. *IEEE Internet of Things Journal*, 7(8):7625–7635, 2020.
415. Ludovic Tangpi and Shichun Wang. Optimal bubble riding: A mean field game with varying entry times. *Finance and Stochastics*, 29(2):343–398, 2025.
416. Alain Tcheukam, Boualem Djehiche, and Hamidou Tembine. Evacuation of multi-level building: Design, control and strategic flow. *35th Chinese Control Conference (CCC2016), July 26–28 Southwest China, Chengdu*, 2016.
417. Alain Tcheukam, Boualem Djehiche, and Hamidou Tembine. Evacuation of multi-level building: Design, control and strategic flow. In *2016 35th Chinese Control Conference (CCC)*, pages 9218–9223. IEEE, 2016.
418. Alain Tcheukam and Hamidou Tembine. Energy cost saving strategies in distributed power networks. In *MATEC Web of Conferences*, volume 50, page 02002. EDP Sciences, 2016.
419. Alain Tcheukam and Hamidou Tembine. Mean-field-type games for distributed power networks in presence of prosumers. In *2016 Chinese Control and Decision Conference (CCDC)*, pages 446–451. IEEE, 2016.

420. Alain Tcheukam and Hamidou Tembine. Spatial mean-field games for combatting corruption propagation. *28th Chinese Control and Decision Conference (CCDC), Yinchuan, China, May 28 -30*, 2016.
421. H. Tembine. Dynamic bargaining solutions for opportunistic spectrum access. *IFIP (International Federation For Information Processing) WIRELESS DAYS, December*, 2009.
422. H. Tembine. Distributed strategic learning for wireless engineers. *Lecture notes, 450 pages*, April 2010.
423. H. Tembine. Mean field stochastic games: Simulation, dynamics and applications. *Lecture notes, Supelec*, December 2010.
424. H. Tembine. Dynamic robust games in mimo systems. *IEEE Transactions on Systems, Man, Cybernetics, Part B, 99,Volume: 41, Issue: 4, pp. 990–1002, August*, 2011.
425. H. Tembine. Hybrid mean field game dynamics in large populations. *American Control Conference, ACC 2011, San Francisco, California, US.*, 2011.
426. H. Tembine. A mean-field stochastic game approach to power management. *18th World Congress of the International Federation of Automatic Control (IFAC)*, 2011.
427. H. Tembine. Non-asymptotic mean-field games. *International Conference on Network Games, Control and Optimization (NETGCOOP), November 28–30, Avignon, France*, 2012.
428. H. Tembine. Fast distributed strategic learning for global optima in queueing access games. *19th World Congress of the International Federation of Automatic Control (IFAC), Cape Town, South Africa, 24–29 August*, 2014.
429. H. Tembine. Nonasymptotic mean-field games. *IEEE Transactions on Systems, Man, and Cybernetics, Part B: Cybernetics, vol. 44, number 12, December*, 2014.
430. H. Tembine. Quantile-based mean-field games with common noise. *Preprint*, 2017.
431. H. Tembine. Covid-19: Data-driven mean-field-type game perspective. *Games, 11, 51*; https://doi.org/doi:10.3390/g11040051, 2020.
432. H. Tembine, E. Altman, R. ElAzouzi, and Y. Hayel. Battery state-dependent access control in solar-powered broadband wireless networks. *in Proc. of Second International EuroFGI Conference on Network Control and Optimization (NETCOOP), Proceedings of Springer LNCS, September*, 2008.
433. H. Tembine, E. Altman, R. ElAzouzi, and Y. Hayel. Stable networking games. *In Proc. Forty-Sixth Annual Allerton Conference on Communication, Control, and Computing, Allerton Retreat Center, Monticello, Illinois, September*, 2008.
434. H. Tembine, E. Altman, R. ElAzouzi, and Y. Hayel. Stochastic population games with individual independent states and coupled constraints. *Proceedings of the 3rd International Conference on Performance Evaluation Methodologies and Tools*, 2008.
435. H. Tembine, E. Altman, R. ElAzouzi, and Y. Hayel. Correlated evolutionarily stable strategies in random medium access control. *International Conference on Game Theory for Networks, Gamenets*, 2009.
436. H Tembine, E Altman, Y Hayel, and R El-Azouzi. Evolutionary games. *WP 2.1–Paradigms collection and foundations*, page 45, 2008.
437. H. Tembine and A. Azad. Dynamic routing games: An evolutionary game theoretic approach. *in IEEE CDC-ECC, 50th IEEE Conference on Decision and Control and European Control Conference, December 12–15, Orlando,Florida*, 2011.
438. H. Tembine and Minyi Huang. Mean field difference games. *CDC-ECC, 50th IEEE Conference on Decision and Control and European Control Conference, Orlando, Florida.*, December 12–15 2011.
439. H. Tembine, Minyi Huang, Roland P. Malhamé, and Peter E. Caines. Mean field convergence of stochastic dynamic games. *Unpublished manuscript*, August 2010.
440. H. Tembine, A. Kobbane, and M. El Koutbi. Dynamic robust power allocation games under channel uncertainty and time delays. *Elsevier Computer Communications*, 2011.
441. H. Tembine, A. Kobbane, and K. Ibrahimi. Minority game applied to selfish routing. *2007, JOSTIC'07, Feb.24, Rabat, Morocco.*, 2007.

442. H. Tembine, A. Kobbane, and M. El Koutbi. Robust power allocation games under channel uncertainty and time delays. *IFIP (International Federation For Information Processing) Wireless Days, Venice, Italy*, 2010.
443. H. Tembine, R. Tempone, and Pedro Vilanova. Mean-field interaction in chemical reaction networks. *Allerton Conference on Communication, Control and Computing, 28–30 Sept. pp. 991–997, Print ISBN: 978-1-4577-1817-5*, 2011.
444. H. Tembine, P. Vilanova, and M. Debbah. Noisy mean field stochastic games with network applications. Report, Ecole superieure d'electricite, Paris, France, 2010.
445. H. Tembine, P. Vilanova, and M. Debbah. Noisy mean field game model for malware propagation in opportunistic networks. *2nd International ICST Conference on Game Theory for Networks, Gamenets, April 16–18 - Shanghai, China*, 2011.
446. H. Tembine, P. Vilanova, and M. Debbah. Noisy mean field model for malware propagation in opportunistic networks. *Gamenets*, 2011.
447. H. Tembine, P. Vilanova, and M. Debbah. Noisy mean field game model for malware propagation in opportunistic networks. *Lecture Notes of the Institute for Computer Sciences, Social Informatics and Telecommunications Engineering, Vol. 75, pp 459–474*, 2012.
448. Hamidou Tembine. Dynamic bargaining solutions for opportunistic spectrum access. In *2009 2nd IFIP Wireless Days (WD)*, pages 1–6. IEEE, 2009.
449. Hamidou Tembine. Population games with networking applications. *PhD Thesis, University of Avignon*, 2009.
450. Hamidou Tembine. Evolutionary networking games. *Game Theory for Wireless Communications and Networking, Auerbach Publications*, pages 133–158, 2010.
451. Hamidou Tembine. Dynamic robust games in mimo systems. *IEEE Transactions on Systems, Man, and Cybernetics, Part B (Cybernetics)*, 41(4):990–1002, 2011.
452. Hamidou Tembine. Large-scale games in large-scale systems. *arXiv preprint arXiv:1111.2285*, 2011.
453. Hamidou Tembine. A mean-field stochastic game for battery state-dependent power management. *IFAC Proceedings Volumes*, 44(1):4839–4844, 2011.
454. Hamidou Tembine. Fast distributed strategic learning for global optima in queueing access games. *IFAC Proceedings Volumes*, 47(3):7055–7060, 2014.
455. Hamidou Tembine. Mean-field-type game theory. *Special issue on Nonasymptotic Mean-Field-Type Game Theory*, 2014.
456. Hamidou Tembine. Distributed massive mimo network games: Risk and altruism. In *2015 54th IEEE Conference on Decision and Control (CDC)*, pages 3481–3486. IEEE, 2015.
457. Hamidou Tembine. Risk-sensitive mean-field-type games with lp-norm drifts. *Automatica*, 59:224–237, 2015.
458. Hamidou Tembine. Uncertainty quantification in mean-field-type teams and games. In *2015 54th IEEE Conference on Decision and Control (CDC)*, pages 4418–4423. IEEE, 2015.
459. Hamidou Tembine. Mean-field-type optimization for demand–supply management under operational constraints in smart grid. *Energy Systems*, 7(2):333–356, 2016.
460. Hamidou Tembine. Mean-field-type games. *AIMS Math*, 2(4):706–735, 2017.
461. Hamidou Tembine. *Distributed strategic learning for wireless engineers*. CRC Press, 2018.
462. Hamidou Tembine. Deep learning meets game theory: Bregman-based algorithms for interactive deep generative adversarial networks. *IEEE Transactions on Cybernetics, 14 pages, volume 50 , Issue: 3 , March, pp. 1132–1145*, 2020.
463. Hamidou Tembine. Master adjoint systems in mean-field-type games. *Preprint*, 2020.
464. Hamidou Tembine. Master adjoint systems in mean-field-type games. *in Communications in Information and Systems, vol.21, 4, pp. 623–650*, 2021.
465. Hamidou Tembine. Master adjoint systems in mean-field-type games. *CoScience: Collective Science Platform, Self-Science Journal sjscience.org*, 2022.
466. Hamidou Tembine. Gpt meets game theory. *CRC Press, Taylor and Francis*, 2025.
467. Hamidou Tembine, Eitan Altma, and Rachid El-Azouzi. Asymmetric delay in evolutionary games. In *2nd International ICST Conference on Performance Evaluation Methodologies and Tools*, 2010.

468. Hamidou Tembine, Eitan Altman, and Rachid El-Azouzi. Delayed evolutionary game dynamics applied to medium access control. In *2007 IEEE International Conference on Mobile Adhoc and Sensor Systems*, pages 1–6. IEEE, 2007.
469. Hamidou Tembine, Eitan Altman, Rachid El-Azouzi, and Yezekael Hayel. Battery state-dependent access control in solar-powered broadband wireless networks. In *Network Control and Optimization: Second Euro-NF Workshop, NET-COOP 2008 Paris, France, September 8–10, 2008.*, pages 121–129. Springer, 2008.
470. Hamidou Tembine, Eitan Altman, Rachid El-Azouzi, and Yezekael Hayel. Evolutionary games with random number of interacting players applied to access control. In *2008 6th International Symposium on Modeling and Optimization in Mobile, Ad Hoc, and Wireless Networks and Workshops*, pages 344–351. IEEE, 2008.
471. Hamidou Tembine, Eitan Altman, Rachid El-Azouzi, and Yezekael Hayel. Battery state-dependent access control in solar-powered broadband wireless networks. In *Network Control and Optimization: Second Euro-NF Workshop, NET-COOP 2008 Paris, France, September 8–10, 2008. Revised Selected Papers 2*, pages 121–129. Springer, 2009.
472. Hamidou Tembine, Eitan Altman, Rachid El-Azouzi, and Yezekael Hayel. Evolutionary games in wireless networks. *IEEE Transactions on Systems, Man, and Cybernetics, Part B (Cybernetics)*, 40(3):634–646, 2009.
473. Hamidou Tembine, Eitan Altman, Rachid El-Azouzi, and Yezekael Hayel. Multiple access game in ad-hoc network. In *1st International ICST Workshop on Game theory for Communication networks*, 2010.
474. Hamidou Tembine, Eitan Altman, Rachid El-Azouzi, and Yezekael Hayel. Bio-inspired delayed evolutionary game dynamics with networking applications. *Telecommunication Systems*, 47:137–152, 2011.
475. Hamidou Tembine, Eitan Altman, Rachid ElAzouzi, and Yezekael Hayel. Stable networking games. In *2008 46th Annual Allerton Conference on Communication, Control, and Computing*, pages 1124–1131. IEEE, 2008.
476. Hamidou Tembine, Eitan Altman, Rachid ElAzouzi, and Yezekael Hayel. Stochastic population games with individual independent states and coupled constraints. In *Proceedings of the 3rd International Conference on Performance Evaluation Methodologies and Tools*, pages 1–10, 2008.
477. Hamidou Tembine, Eitan Altman, Rachid ElAzouzi, and Yezekael Hayel. Correlated evolutionarily stable strategies in random medium access control. In *2009 International Conference on Game Theory for Networks*, pages 212–221. IEEE, 2009.
478. Hamidou Tembine, Massa NDong Issa Bamia, Bakary Coulibaly Oumar Issiaka Traore, Moussa Traore, Moussa Sanogo, Mamadou Eric Sangare, Salif Kante, Daryl Noupa Yongueng, Hafiz Tiomoko Ali, Malik Tiomoko, Frejus Laleye, Boualem Djehiche, Wesmanegda Elisee Dipama, Idris Baba Saje, Hammid Mohammed Ibrahim, Moumini Sanogo, Marie Coursel Nininahazwe, Abdul-Latif Siita, Haine Mhlongo, Teddy Nelvy Dieu Merci Kouka, Mariam Serine Jeridi, Mutiyamuogo Parfait Mupenge, Lekoueiry Dehah, Abdoul Aziz Bio Sidi D Bouko, Wilfried Franceslas Zokoue, Odette Richette Sambila, Alina RS Mbango, Mady Diagouraga, Oumarou Moussa Sanoussi, Gizachew Dessalegn, Mohamed Lamine Samoura, and Bintou Laetitia Audrey Coulibaly. Breaking the barriers of text-hungry and audio-deficient ai. *AI MALI*, 2025.
479. Hamidou Tembine, Manzoor Khan, and Issa Bamia. Mean-field-type transformer. *Mathematics, 12, 22, 3506, 52 pages*, 2024.
480. Hamidou Tembine, Abdellatif Kobbane, and Mohammed El Koutbi. Robust power allocation games under channel uncertainty and time delays. In *2010 IFIP Wireless Days*, pages 1–5. IEEE, 2010.
481. Hamidou Tembine, Abdellatif Kobbane, et al. Dynamic robust power allocation games under channel uncertainty and time delays. *Computer communications*, 34(12):1529–1537, 2011.
482. Hamidou Tembine and Ousmane Kodio. Reverse ishikawa-nesterov learning scheme for fractional mean-field games. *IFAC-PapersOnLine*, 50(1):8090–8096, 2017.

483. Hamidou Tembine, Samson Lasaulce, and Marc Jungers. Joint power control-allocation for green cognitive wireless networks using mean field theory. In *2010 Proceedings of the Fifth International Conference on Cognitive Radio Oriented Wireless Networks and Communications*, pages 1–5. IEEE, 2010.
484. Hamidou Tembine, Jean Yves Le Boudec, Rachid ElAouzi, and Eitan Altman. From mean field interaction to evolutionary game dynamics. In *2009 7th International Symposium on Modeling and Optimization in Mobile, Ad Hoc, and Wireless Networks*, pages 1–5. IEEE, 2009.
485. Hamidou Tembine, Raul Tempone, and Pedro Vilanova. Mean field games for cognitive radio networks. In *2012 American Control Conference (ACC)*, pages 6388–6393. IEEE, 2012.
486. Hamidou Tembine, Raul Tempone, and Pedro Vilanova. Mean-field learning: a survey. *arXiv preprint arXiv:1210.4657*, 2012.
487. Hamidou Tembine, Raul Tempone, and Pedro Vilanova. Mean-field learning for satisfactory solutions. In *52nd IEEE Conference on Decision and Control*, pages 4871–4876. IEEE, 2013.
488. Hamidou Tembine, Pedro Vilanova, Mohamad Assaad, and Mérouane Debbah. A mean field stochastic game approach for dynamic access cognitive networks. *Supelec, France and Alcatel-Lucent en radio-flexible, Chaire*, 2010.
489. Hamidou Tembine, Pedro Vilanova, Mohamad Assaad, and Merouane Debbah. Mean field stochastic games for sinr-based medium access control. In *Gamecomm2011*, pages 10–pages, 2011.
490. Hamidou Tembine, Pedro Vilanova, and Mérouane Debbah. Noisy mean field stochastic games with network applications. *Report, Ecole superieure delectricite, Paris*, 2010.
491. Fei Teng and Frédéric Magoulès. A new game theoretical resource allocation algorithm for cloud computing. In *Advances in Grid and Pervasive Computing: 5th International Conference, GPC 2010, Hualien, Taiwan, May 10–13, 2010. Proceedings 5*, pages 321–330. Springer, 2010.
492. Patara Trirat, Yooju Shin, Junhyeok Kang, Youngeun Nam, Jihye Na, Minyoung Bae, Joeun Kim, Byunghyun Kim, and Jae-Gil Lee. Universal time-series representation learning: A survey. *arXiv preprint arXiv:2401.03717*, 2024.
493. NV Trusov. The group behavioral characteristics in a mean field game model with a turnpike effect. *Lobachevskii Journal of Mathematics*, 46(1):341–350, 2025.
494. Gordon Tullock, James M Buchanan, and Robert D Tollison. Efficient rent seeking. *Toward a Theory of the Rent-seeking Society*, 97:112, 1980.
495. Ashish Vaswani, Noam Shazeer, Jakob Uszkoreit Niki Parmar, Llion Jones, Aidan N Gomez, Lukasz Kaiser, and Illia Polosukhin. Attention is all you need. *In Advances in neural information processing systems*, 2017.
496. Tomas Veloz, Pablo Razeto-Barry, Peter Dittrich, and Alejandro Fajardo. Reaction networks and evolutionary game theory. *Journal of mathematical biology*, 68:181–206, 2014.
497. Tina Verma and Arvind Kumar Gupta. Network synchronization, stability and rhythmic processes in a diffusive mean-field coupled seir model. *Communications in Nonlinear Science and Numerical Simulation*, 102:105927, 2021.
498. Hendrik Vermuyten, Jeroen Beliën, Liesje De Boeck, Genserik Reniers, and Tony Wauters. A review of optimisation models for pedestrian evacuation and design problems. *Safety science*, 87:167–178, 2016.
499. P. Vilanova, H. Tembine, and M. Debbah. Noisy mean field stochastic games with network applications. *ICIAM - 7th International Congress on Industrial and Applied Mathematics, ICIAM Vancouver, MS403 Multiscale Approximations Of Kinetic Monte Carlo Simulations.*, 2011.
500. Cédric Villani. *Optimal transport*, volume 338 of *Grundlehren der Mathematischen Wissenschaften [Fundamental Principles of Mathematical Sciences]*. Springer-Verlag, Berlin, 2009. Old and new.
501. Cédric Villani et al. *Optimal transport: old and new*, volume 338. Springer, 2009.

502. Cornelia Von Krüchten and Andreas Schadschneider. Empirical study on social groups in pedestrian evacuation dynamics. *Physica A: Statistical Mechanics and its Applications*, 475:129–141, 2017.
503. Jinhuan Wang, Lei Zhang, Qiongyu Shi, Peng Yang, and Xiaoming Hu. Modeling and simulating for congestion pedestrian evacuation with panic. *Physica A: Statistical Mechanics and its Applications*, 428:396–409, 2015.
504. Jun Wang, Wenjie Du, Wei Cao, Keli Zhang, Wenjia Wang, Yuxuan Liang, and Qingsong Wen. Deep learning for multivariate time series imputation: A survey. *arXiv preprint arXiv:2402.04059*, 2024.
505. J. G. Wardrop. Some theoretical aspects of road traffc research. *In Proc. of the Institute of Civil Engineers, Pt. II:*, pages 325–378, 1952.
506. Guiyi Wei, Athanasios V Vasilakos, Yao Zheng, and Naixue Xiong. A game-theoretic method of fair resource allocation for cloud computing services. *The journal of supercomputing*, 54:252–269, 2010.
507. Xiaoli Wei and Xiang Yu. Continuous time q-learning for mean-field control problems. *Applied Mathematics & Optimization*, 91(1):10, 2025.
508. G. Y. Weintraub, L. Benkard, and B. Van Roy. Oblivious equilibrium: A mean field approximation for large-scale dynamic games. *Advances in Neural Information Processing Systems*, 18, 2005.
509. Piotr Wiecek. Multiple-population discrete-time mean field games with discounted and total payoffs: Approximation of games with finite populations. *Dynamic Games and Applications*, 15(1):329–357, 2025.
510. Gerald Woo, Chenghao Liu, Akshat Kumar, Caiming Xiong, Silvio Savarese, and Doyen Sahoo. Unified training of universal time series forecasting transformers. *arXiv preprint arXiv:2402.02592*, 2024.
511. Shengqiong Wu, Hao Fei, Leigang Qu, Wei Ji, and Tat-Seng Chua. Next-gpt: Any-to-any multimodal LLM. *arXiv preprint arXiv:2309.05519*, 2023.
512. Biteng Xu, Jinbiao Wu, and Zongkui Fu. Optimal regulators for mean-field stochastic system with state-dependent weight costs. *Optimal Control Applications and Methods*, 2025.
513. Mengwei Xu, Wangsong Yin, Dongqi Cai, Rongjie Yi, Daliang Xu, Qipeng Wang, Bingyang Wu, Yihao Zhao, Chen Yang, Shihe Wang, et al. A survey of resource-efficient LLM and multimodal foundation models. *arXiv preprint arXiv:2401.08092*, 2024.
514. Ruimin Xu, Kaiyue Dong, Jingyu Zhang, and Ying Zhou. Linear-quadratic-gaussian mean-field games driven by poisson jumps with major and minor agents. *AIMS Mathematics*, 10(5):11086–11110, 2025.
515. Y. Xu and H. Tembine. Distributed representations of bids in lowest unique bid auctions. *30th Chinese Control and Decision Conference (2018 CCDC), June 9–11, Shenyang, China*, 2018.
516. Y. Xu and H. Tembine. Generative model of bid sequences in lowest unique bid auctions. *30th Chinese Control and Decision Conference (CCDC), June 9–11, Shenyang, China*, 2018.
517. Y. Xu and H. Tembine. Lowest unique bid auctions with resubmission opportunities. *10th International Conference on Agents and Artificial Intelligence (ICAART)*, 2018.
518. Yida Xu and Hamidou Tembine. Risk-sensitive lowest unique bid auctions: Empirical and analytical investigations. *20th World Congress of the International Federation of Automatic Control (IFAC) , Toulouse, France, 9–14 July*, 2017.
519. Hongyi Yang, Jingzhi Liu, Geng Li, Jianming Zhang, Ling Jiang, and Shoulian Yang. Distributed intelligent power distribution optimization method based on mean field game theory. In *2025 IEEE 5th International Conference on Power, Electronics and Computer Applications (ICPECA)*, pages 818–822. IEEE, 2025.
520. Ryosuke Yano and Hisayasu Kuroda. Mean-field game analysis of crowd evacuation using the Cristiani-Santo-Menci method. *Physical Review E*, 108(1):014119, 2023.
521. Batuhan Yardim, Semih Cayci, and Niao He. A variational inequality approach to independent learning in static mean-field games. *ACM/IMS Journal of Data Science*, 2025.

522. Jiexia Ye, Weiqi Zhang, Ke Yi, Yongzi Yu, Ziyue Li, Jia Li, and Fugee Tsung. A survey of time series foundation models: Generalizing time series representation with large language mode. *arXiv preprint arXiv:2405.02358*, 2024.
523. Jessica AM Yeates, Christian Hilbe, Martin Zwick, Martin A Nowak, and Niles Lehman. Dynamics of prebiotic rna reproduction illuminated by chemical game theory. *Proceedings of the National Academy of Sciences*, 113(18):5030–5035, 2016.
524. H. Yin, P. G. Mehta, S. P. Meyn, and U. V. Shanbhag. Synchronization of coupled oscillators is a game. *in Proc. of American Control Conference*, 2010.
525. Hidekazu Yoshioka. A mean field Jacobi process for modeling sustainable tourism. *Dynamic Games and Applications*, pages 1–33, 2025.
526. Xinli Yu, Zheng Chen, Yuan Ling, Shujing Dong, Zongyi Liu, and Yanbin Lu. Temporal data meets LLM–explainable financial time series forecasting. *arXiv preprint arXiv:2306.11025*, 2023.
527. Yong Yu, Bogdan Danila, John A. Marsh, and Kevin E. Bassler. Optimal transport on wireless networks. *at http://arxiv.org/abs/physics/0703261*, 2007.
528. Mohamed Ridha Zenaidi, Zouheir Rezki, Hamidou Tembine, and Mohamed-Slim Alouini. Performance limits of energy harvesting communications under imperfect channel state information. In *2016 IEEE International Conference on Communications (ICC)*, pages 1–6. IEEE, 2016.
529. Jun Zhan, Junqi Dai, Jiasheng Ye, Yunhua Zhou, Dong Zhang, Zhigeng Liu, Xin Zhang, Ruibin Yuan, Ge Zhang, Linyang Li, et al. Anygpt: Unified multimodal LLM with discrete sequence modeling. *arXiv preprint arXiv:2402.12226*, 2024.
530. Cheng Zhang, Nilam Nur Amir Sjarif, and Roslina Ibrahim. Deep learning models for price forecasting of financial time series: A review of recent advancements: 2020–2022. *Wiley Interdisciplinary Reviews: Data Mining and Knowledge Discovery*, 14(1):e1519, 2024.
531. Dezhen Zhang, Gaoyue Huang, Chengtao Ji, Huiying Liu, and Ying Tang. Pedestrian evacuation modeling and simulation in multi-exit scenarios. *Physica A: Statistical Mechanics and its Applications*, 582:126272, 2021.
532. Enze Zhang, Huaze Tang, Xiao-Ping Zhang, and Wenbo Ding. Mean-field aided qmix: A scalable and flexible q-learning approach for large-scale agent groups. In *ICASSP 2025–2025 IEEE International Conference on Acoustics, Speech and Signal Processing (ICASSP)*, pages 1–5. IEEE, 2025.
533. Huanshui Zhang, Weiyan Ge, Wing Shing Wong, and Peter E Caines. A stochastic approximation approach to the robust power control problem. In *42nd IEEE International Conference on Decision and Control (IEEE Cat. No. 03CH37475)*, volume 3, pages 2601–2605. IEEE, 2003.
534. Huanshui Zhang, Wing Shing Wong, Weiyan Ge, and Peter E Caines. A stochastic approximation approach to the power-control problem. *IEEE transactions on communications*, 55(5):878–886, 2007.
535. Jingguo Zhang, Xianjin Yang, Chenchen Mou, and Chao Zhou. Learning surrogate potential mean field games via gaussian processes: a data-driven approach to ill-posed inverse problems. *arXiv preprint arXiv:2502.11506*, 2025.
536. Kuan Zhang, Yawen Xie, Nian Liu, and Siqi Chen. Customized mean field game method of virtual power plant for real-time peak regulation. *IEEE Transactions on Sustainable Energy*, 2025.
537. Limao Zhang, Mengjie Liu, Xianguo Wu, and Simaan M AbouRizk. Simulation-based route planning for pedestrian evacuation in metro stations: A case study. *Automation in Construction*, 71:430–442, 2016.
538. Qilei Zhang and John H Mott. An exploratory assessment of LLM's potential toward flight trajectory reconstruction analysis. *arXiv preprint arXiv:2401.06204*, 2024.
539. Xiao Zhang, Ruoyu Xiang, Chenhan Yuan, Duanyu Feng, Weiguang Han, Alejandro Lopez-Lira, Xiao-Yang Liu, Sophia Ananiadou, Min Peng, Jimin Huang, et al. Dólares or dollars? Unraveling the bilingual prowess of financial LLMs between Spanish and English. *arXiv preprint arXiv:2402.07405*, 2024.

540. Xiyuan Zhang, Ranak Roy Chowdhury, Rajesh K Gupta, and Jingbo Shang. Large language models for time series: A survey. *arXiv preprint arXiv:2402.01801*, 2024.
541. Yaomin Zhang, Haijun Zhang, and Keping Long. Energy efficient resource allocation in cache based terahertz vehicular networks: A mean-field game approach. *IEEE Transactions on Vehicular Technology*, 70(6):5275–5285, 2021.
542. Yaozhong Zhang, Meiyan Ding, Yu Du, Fulun Peng, Jing Wang, and Qiming Yang. UAV swarm intelligent interception driven by mission-empowered mean-field gate recurrent reinforcement learning. *IEEE Transactions on Aerospace and Electronic Systems*, 2025.
543. Yu Zhang, Zhuo Jin, Jiaqin Wei, and George Yin. A hybrid deep learning method for finite-horizon mean-field game problems. *Automatica*, 179:112384, 2025.
544. Huaqin Zhao, Zhengliang Liu, Zihao Wu, Yiwei Li, Tianze Yang, Peng Shu, Shaochen Xu, Haixing Dai, Lin Zhao, Gengchen Mai, et al. Revolutionizing finance with LLMs: An overview of applications and insights. *arXiv preprint arXiv:2401.11641*, 2024.
545. Weicheng Zhao and Jingguo Zhang. A mean-field conventional asset penalizing game: A study of green premium. In *2025 IEEE Symposium on Computational Intelligence for Financial Engineering and Economics Companion (CiFer Companion)*, pages 1–5. IEEE, 2025.
546. Linjiang Zheng, Xiaoli Peng, Linglin Wang, and Dihua Sun. Simulation of pedestrian evacuation considering emergency spread and pedestrian panic. *Physica A: Statistical Mechanics and its Applications*, 522:167–181, 2019.
547. Xiaoping Zheng and Yuan Cheng. Conflict game in evacuation process: A study combining cellular automata model. *Physica A: Statistical Mechanics and its Applications*, 390(6):1042–1050, 2011.
548. Xiaoping Zheng, Tingkuan Zhong, and Mengting Liu. Modeling crowd evacuation of a building based on seven methodological approaches. *Building and environment*, 44(3):437–445, 2009.
549. Tian Zhou, Peisong Niu, Liang Sun, Rong Jin, et al. One fits all: Power general time series analysis by pretrained LM. *Advances in neural information processing systems*, 36:43322–43355, 2023.
550. Q. Zhu, H. Tembine, and T. Başar. Evolutionary games for hybrid additive white gaussian noise multiple access. *in Proc. Global Telecommunications Conference IEEE GLOBECOM, November 30–4,*, 2009.
551. Q. Zhu, H. Tembine, and T. Başar. Evolutionary games for multiple access control. *Advances in Dynamic Game Theory and Applications, Annals of Dynamic Games, vol. 12, Birkhauser*, 2012.
552. Mohamed Ridha Znaidi, Zouheir Rezki, Hamidou Tembine, and Mohamed-Slim Alouini. Performance analysis of communications under energy harvesting constraints with noisy CSI. *ICC*, 2016.
553. Vickrey, William. Counterspeculation, auctions, and competitive sealed tenders. *The Journal of finance*, 16(1):8–37, 1961.

Index

T. Başar et al., *Mean-Field-Type Game Theory II*, Static & Dynamic Game Theory: Foundations & Applications, https://doi.org/10.1007/978-3-032-07031-9

MIX
Papier aus verantwortungsvollen Quellen
Paper from responsible sources
FSC® C105338

If you have any concerns about our products,
you can contact us on
ProductSafety@springernature.com

In case Publisher is established outside the EU,
the EU authorized representative is:
Springer Nature Customer Service Center GmbH
Europaplatz 3, 69115 Heidelberg, Germany

Printed by Libri Plureos GmbH
in Hamburg, Germany